T0205275

Communications in Computer and Information Science 1870

Rationale

The CCIS series is devoted to the publication of proceedings of computer science conferences. Its aim is to efficiently disseminate original research results in informatics in printed and electronic form. While the focus is on publication of peer-reviewed full papers presenting mature work, inclusion of reviewed short papers reporting on work in progress is welcome, too. Besides globally relevant meetings with internationally representative program committees guaranteeing a strict peer-reviewing and paper selection process, conferences run by societies or of high regional or national relevance are also considered for publication.

Topics

The topical scope of CCIS spans the entire spectrum of informatics ranging from foundational topics in the theory of computing to information and communications science and technology and a broad variety of interdisciplinary application fields.

Information for Volume Editors and Authors

Publication in CCIS is free of charge. No royalties are paid, however, we offer registered conference participants temporary free access to the online version of the conference proceedings on SpringerLink (http://link.springer.com) by means of an http referrer from the conference website and/or a number of complimentary printed copies, as specified in the official acceptance email of the event.

CCIS proceedings can be published in time for distribution at conferences or as post-proceedings, and delivered in the form of printed books and/or electronically as USBs and/or e-content licenses for accessing proceedings at SpringerLink. Furthermore, CCIS proceedings are included in the CCIS electronic book series hosted in the SpringerLink digital library at http://link.springer.com/bookseries/7899. Conferences publishing in CCIS are allowed to use Online Conference Service (OCS) for managing the whole proceedings lifecycle (from submission and reviewing to preparing for publication) free of charge.

Publication process

The language of publication is exclusively English. Authors publishing in CCIS have to sign the Springer CCIS copyright transfer form, however, they are free to use their material published in CCIS for substantially changed, more elaborate subsequent publications elsewhere. For the preparation of the camera-ready papers/files, authors have to strictly adhere to the Springer CCIS Authors' Instructions and are strongly encouraged to use the CCIS LaTeX style files or templates.

Abstracting/Indexing

CCIS is abstracted/indexed in DBLP, Google Scholar, EI-Compendex, Mathematical Reviews, SCImago, Scopus. CCIS volumes are also submitted for the inclusion in ISI Proceedings.

How to start

To start the evaluation of your proposal for inclusion in the CCIS series, please send an e-mail to ccis@springer.com.

Haijun Zhang · Yinggen Ke · Zhou Wu ·
Tianyong Hao · Zhao Zhang · Weizhi Meng ·
Yuanyuan Mu
Editors

International Conference on Neural Computing for Advanced Applications

4th International Conference, NCAA 2023 Hefei, China, July 7–9, 2023 Proceedings, Part II

Editors

Haijun Zhang
Harbin Institute of Technology
Shenzhen, China

Yinggen Ke
Chaohu University
Hefei, China

Zhou Wu
Chongqing University
Chongqing, China

Tianyong Hao
South China Normal University
Guangzhou, China

Zhao Zhang
Hefei University of Technology
Hefei, China

Weizhi Meng
Technical University of Denmark
Kongens Lyngby, Denmark

Yuanyuan Mu
Chaohu University
Hefei, China

ISSN 1865-0929 ISSN 1865-0937 (electronic)
Communications in Computer and Information Science
ISBN 978-981-99-5846-7 ISBN 978-981-99-5847-4 (eBook)
https://doi.org/10.1007/978-981-99-5847-4

This Springer imprint is published by the registered company Springer Nature Singapore Pte Ltd.
The registered company address is: 152 Beach Road, #21-01/04 Gateway East, Singapore 189721, Singapore

Paper in this product is recyclable.

Preface

Neural computing and Artificial Intelligence (AI) have become hot topics in recent years. To promote multi-disciplinary development and application of neural computing, a series of NCAA conferences was initiated on the theme of "*make the academic more practical*", providing an open platform for academic discussions, industrial showcases, and basic training tutorials. This volume contains the papers accepted for this year's International Conference on Neural Computing for Advanced Applications (NCAA 2023). NCAA 2023 was organized by Chaohu University and co-organized by Shandong Jianzhu University and Wuhan Textile University, and it was supported by Springer and Cancer Innovation. After the effects of COVID-19, the mainstream part of NCAA 2023 was turned into a hybrid event, with mainly offline participants, in which people could freely connect to keynote speeches and presentations face to face.

NCAA 2023 received 211 submissions, of which 84 high-quality papers were selected for publication in this volume after double-blind peer review, leading to an acceptance rate of just under 40%. These papers were categorized into 13 technical tracks: *Neural network (NN) theory, NN-based control systems, neuro-system integration and engineering applications; Machine learning and deep learning for data mining and data-driven applications; Computational intelligence, nature-inspired optimizers, and their engineering applications; Neuro/fuzzy systems, multi-agent control, decision making, and their applications in smart construction and manufacturing; Deep learning-driven pattern recognition, computer vision and its industrial applications; Natural language processing, knowledge graphs, recommender systems, and their applications; Neural computing-based fault diagnosis and forecasting, prognostic management, and cyber-physical system security; Sequence learning for spreading dynamics, forecasting, and intelligent techniques against epidemic spreading; Multimodal deep learning for representation, fusion, and applications; Neural computing-driven edge intelligence, machine learning for mobile systems, pervasive computing, and intelligent transportation systems; Applications of data mining, machine learning and neural computing in language studies; Computational intelligent fault diagnosis and fault-tolerant control, and their engineering applications; Other Neural computing-related applications.*

The authors of each paper in this volume have reported their novel results of computing theory or applications. The volume cannot cover all aspects of neural computing and advanced applications, but may still inspire insightful thoughts for readers. We hope that more secrets of AI will be unveiled and academics will drive more practical developments and solutions to real-world applications.

July 2023

Haijun Zhang
Yinggen Ke
Zhou Wu
Tianyong Hao
Zhao Zhang
Weizhi Meng
Yuanyuan Mu

Organization

Honorary Chairs

John MacIntyre	University of Sunderland, UK
Tommy W. S. Chow	City University of Hong Kong, China

General Co-chairs

Haijun Zhang	Harbin Institute of Technology, China
Yinggen Ke	Chaohu University, China
Zhou Wu	Chongqing University, China

Program Co-chairs

Tianyong Hao	South China Normal University, China
Zhao Zhang	Hefei University of Technology, China
Weizhi Meng	Technical University of Denmark, Denmark

Organizing Committee Co-chairs

Rongqi Yu	Chaohu University, China
Yanxia Sun	University of Johannesburg, South Africa
Mingbo Zhao	Donghua University, China

Local Arrangement Co-chairs

Yuanyuan Mu	Chaohu University, China
Peng Yu	Chaohu University, China
Yongfeng Zhang	Jinan University, China

Registration Co-chairs

Yaqing Hou	Dalian University of Technology, China
Jing Zhu	Macau University of Science and Technology, China
Shuqiang Wang	Chinese Academy of Sciences, China
Weiwei Wu	Southeast University, China
Zhili Zhou	Nanjing University of Information Science and Technology, China

Publication Co-chairs

Kai Liu	Chongqing University, China
Yu Wang	Xi'an Jiaotong University, China
Yi Zhang	Fuzhou University, China
Bo Wang	Huazhong University of Science and Technology, China
Xianghua Chu	Shenzhen University, China

Publicity Co-chairs

Fei He	Coventry University, UK
Xiao-Zhi Gao	University of Eastern Finland, Finland
Choujun Zhan	South China Normal University, China
Zenghui Wang	University of South Africa, South Africa
Yimin Yang	University of Western Ontario, Canada
Zili Chen	Hong Kong Polytechnic University, China
Reza Maleklan	Malmö University, Sweden
Sinan Li	University of Sydney, Australia

Sponsor Co-chairs

Wangpeng He	Xidian University, China
Bingyi Liu	Wuhan University of Technology, China
Cuili Yang	Beijing University of Technology, China
Guo Luo	Nanfang College of Sun Yat-sen University, China
Shi Cheng	Shaanxi Normal University, China

Forum Chair

Jingjing Cao	Wuhan University of Technology, China

Tutorial Chair

Jicong Fan	Chinese University of Hong Kong, Shenzhen, China

Competition Chair

Chengdong Li	Shandong Jianzhu University, China

NCAA Steering Committee Liaison

Jianghong Ma	Harbin Institute of Technology, China

Web Chair

Xinrui Yu	Harbin Institute of Technology, China

Program Committee Members

Dong Yang	City University of Hong Kong, China
Sheng Li	University of Georgia, USA
Jie Qin	Swiss Federal Institute of Technology (ETH), Switzerland
Xiaojie Jin	Bytedance AI Lab, USA
Zhao Kang	University of Electronic Science and Technology, China
Xiangyuan Lan	Hong Kong Baptist University, China
Peng Zhou	Anhui University, China
Chang Tang	China University of Geosciences, China
Dan Guo	Hefei University of Technology, China
Li Zhang	Soochow University, China
Xiaohang Jin	Zhejiang University of Technology, China
Wei Huang	Zhejiang University of Technology, China

Chao Chen	Chongqing University, China
Jing Zhu	Macau University of Science and Technology, China
Weizhi Meng	Technical University of Denmark, Denmark
Wei Wang	Dalian Ocean University, China
Jian Tang	Beijing University of Technology, China
Heng Yue	Northeastern University, China
Yimin Yang	University of Western Ontario, Canada
Jianghong Ma	Harbin Institute of Technology, China
Jicong Fan	Chinese University of Hong Kong (Shenzhen), China
Xin Zhang	Tianjin Normal University, China
Xiaolei Lu	City University of Hong Kong, China
Penglin Dai	Southwest Jiaotong University, China
Liang Feng	Chongqing University, China
Xiao Zhang	South-Central University for Nationalities, China
Bingyi Liu	Wuhan University of Technology, China
Cheng Zhan	Southwest University, China
Qiaolin Pu	Chongqing University of Posts and Telecommunications, China
Hao Li	Hong Kong Baptist University, China
Junhua Wang	Nanjing University of Aeronautics and Astronautics, China
Yu Wang	Xi'an Jiaotong University, China
BinQiang Chen	Xiamen University, China
Wangpeng He	Xidian University, China
Jing Yuan	University of Shanghai for Science and Technology, China
Huiming Jiang	University of Shanghai for Science and Technology, China
Yizhen Peng	Chongqing University, China
Jiayi Ma	Wuhan University, China
Yuan Gao	Tencent AI Lab, China
Xuesong Tang	Donghua University, China
Weijian Kong	Donghua University, China
Zhili Zhou	Nanjing University of Information Science and Technology, China
Yang Lou	City University of Hong Kong, China
Chao Zhang	Shanxi University, China
Yanhui Zhai	Shanxi University, China
Wenxi Liu	Fuzhou University, China
Kan Yang	University of Memphis, USA
Fei Guo	Tianjin University, China

Wenjuan Cui	Chinese Academy of Sciences, China
Wenjun Shen	Shantou University, China
Mengying Zhao	Shandong University, China
Shuqiang Wang	Chinese Academy of Sciences, China
Yanyan Shen	Chinese Academy of Sciences, China
Haitao Wang	China National Institute of Standardization, China
Yuheng Jia	City University of Hong Kong, China
Chengrun Yang	Cornell University, USA
Lijun Ding	Cornell University, USA
Zenghui Wang	University of South Africa, South Africa
Xianming Ye	University of Pretoria, South Africa
Yanxia Sun	University of Johannesburg, South Africa
Reza Maleklan	Malmö University, Sweden
Xiaozhi Gao	University of Eastern Finland, Finland
Jerry Lin	Western Norway University of Applied Sciences, Norway
Xin Huang	Hong Kong Baptist University, China
Xiaowen Chu	Hong Kong Baptist University, China
Hongtian Chen	University of Alberta, Canada
Gautam Srivastava	Brandon University, Canada
Bay Vo	Ho Chi Minh City University of Technology, Vietnam
Xiuli Zhu	University of Alberta, Canada
Rage Uday Kiran	University of Aizu, Japan
Matin Pirouz Nia	California State University Fresno, USA
Vicente Garcia Diaz	University of Oviedo, Spain
Youcef Djenouri	University of South-Eastern Norway, Norway
Jonathan Wu	University of Windsor, Canada
Yihua Hu	University of York, UK
Saptarshi Sengupta	Murray State University, USA
Wenxiu Xie	City University of Hong Kong, China
Christine Ji	University of Sydney, Australia
Jun Yan	Yidu Cloud, China
Jian Hu	Yidu Cloud, China
Alessandro Bile	Sapienza University of Rome, Italy
Jingjing Cao	Wuhan University of Technology, China
Shi Cheng	Shaanxi Normal University, China
Xianghua Chu	Shenzhen University, China
Valentina Colla	Scuola Superiore S. Anna, Italy
Mohammad Hosein Fazaeli	Amirkabir University of Technology, Iran
Vikas Gupta	LNM Institute of Information Technology, Jaipur
Tianyong Hao	South China Normal University, China

Hongdou He	Yanshan University, China
Wangpeng He	Xidian University, China
Yaqing Hou	Dalian University of Technology, China
Essam Halim Houssein	Minia University, Egypt
Wenkai Hu	China University of Geosciences, China
Lei Huang	Ocean University of China, China
Weijie Huang	University of Hefei, China
Zhong Ji	Tianjin University, China
Qiang Jia	Jiangsu University, China
Yang Kai	Yunnan Minzu University, China
Andreas Kanavos	Ionian University, Greece
Zhao Kang	Southern Illinois University Carbondale, USA
Zouaidia Khouloud	Badji Mokhtar Annaba University, Algeria
Chunshan Li	Harbin Institute of Technology, China
Dongyu Li	Beihang University, China
Kai Liu	Chongqing University, China
Xiaofan Liu	City University of Hong Kong, China
Javier Parra Arnau	Karlsruhe Institute of Technology, Germany
Santwana Sagnika	Kalinga Institute of Industrial Technology, India
Atriya Sen	Rensselaer Polytechnic Institute, USA
Ning Sun	Nankai University, China
Shaoxin Sun	Chongqing University, China
Ankit Thakkar	Nirma University, India
Ye Wang	Chongqing University of Posts and Telecommunications, China
Yong Wang	Sun Yat-sen University, China
Zhanshan Wang	Northeastern University, China
Quanwang Wu	Chongqing University, China
Xiangjun Wu	Henan University, China
Xingtang Wu	Beihang University, China
Zhou Wu	Chongqing University, China
Wun-She Yap	Universiti Tunku Abdul Rahman, Malaysia
Rocco Zaccagnino	University of Salerno, Italy
Kamal Z. Zamli	Universiti Malaysia Pahang, Malaysia
Choujun Zhan	South China Normal University, China
Haijun Zhang	Harbin Institute of Technology, Shenzhen, China
Menghua Zhang	Jinan University, China
Zhao Zhang	Hefei University of Technology, China
Mingbo Zhao	Donghua University, China
Dongliang Zhou	Harbin Institute of Technology, Shenzhen, China
Guo Luo	Nanfang College of Sun Yat-sen University, China
Chengdong Li	Shandong Jianzhu University, China

Yongfeng Zhang	Jinan University, China
Kai Yang	Wuhan Textile University, China
Li Dong	Shenzhen Technology University, China

Contents – Part II

Neural Computing-Based Fault Diagnosis and Forecasting, Prognostic Management, and Cyber-Physical System Security

Sequence Learning for Spreading Dynamics, Forecasting, and Intelligent Techniques Against Epidemic Spreading (2)

Other Neural Computing-Related Topics

Other Neural Computing-Related Topics

Contents – Part I

Deep Learning-Driven Pattern Recognition, Computer Vision and Its Industrial Applications

Improved YOLOv5s Based Steel Leaf Spring Identification

Conglin Gao(✉), Shouyin Lu(✉), Xinjian Gu, and Xinyun Hu

Department of Information and Electrical Engineering, Shandong Jianzhu University, Jinan 250101, China

2394167399@qq.com, lusy@sdjzu.edu.cn

Abstract. When the steel leaf spring grasping robot grasps the steel leaf spring, it needs to identify the steel leaf spring first and then obtain the spatial position information of the target steel leaf spring, so that the robot can obtain the best motion trajectory and improve the efficiency of steel leaf spring grasping. In this paper, the identification method of steel leaf springs is studied. The images of the steel leaf springs are first acquired using the camera, and then the steel leaf springs are recognized using a modified YOLOv5s network. The experimental results demonstrate the effectiveness of the improved YOLOv5s network, with a 60.4% reduction in model size, has an average detection speed of 78.7 f/s and an average detection accuracy of 90.7%.The steel leaf spring identification method studied in this paper meets the requirements of the steel leaf spring gripping robot in terms of speed and accuracy, and provides a reference for the identification of other industrial parts.

Keywords: Steel Leaf Spring Gripping Robot · Steel Leaf Springs · YOLO-v5s · Target Identification

1 Introduction

An approximately equal-strength elastic beam, formed by combining several unequal-length but equal-width alloy spring leafs, is the steel leaf spring. It is the most widely used elastic element in automotive suspension. It has simple structure, reliable work, low cost and easy maintenance. It is both the elastic element of the suspension and the guiding device of the suspension. It transmits various forces and moments, determines the trajectory of the wheel and has a certain frictional damping effect. Therefore, it is widely used on non-independent suspensions.

In the current industrial production process of steel leaf springs, most factories are still manually sorting, handling, assembling and detecting defects of steel leaf springs and other operations, there are many drawbacks in this form, such as high labor intensity of workers, poor operating environment, low production efficiency, manual sorting is subjective and can not work in some high-speed assembly lines or production sites with harsh environments. In recent years, with the increasing cost of labor and the increasing market demand for product quality, many companies have gradually adopted robots for

H. Zhang et al. (Eds.): NCAA 2023, CCIS 1870, pp. 3–17, 2023.
https://doi.org/10.1007/978-981-99-5847-4_1

sorting, handling and other activities of steel leaf springs, and the prerequisite for these tasks is the accurate identification of steel leaf springs.

After the breakthrough of deep learning technology, the technical bottleneck faced by traditional target detection algorithms in feature extraction has been effectively solved [1–3]. Deep learning techniques have been widely used in the field of computer vision, among which the YOLO series is one of the most classical deep learning algorithms [4, 5]. It has achieved good results on various industrial problems [6–8]. At the same time, people continue to improve the YOLO series to make it meet the various needs of people. The literature [9] improved the mosaic data enhancement method of YOLOv5, which resulted in a large degree of accuracy improvement. The literature [10] used a combination of MobileNetV2 backbone network replacement and attention mechanism to achieve a lightweight study of the YOLO series, which led to a greater degree of optimization of both transmission rate and model size. The literature [11] uses MobileNet as the backbone feature extraction network, embeds CBAM and proposes a pruning-then-distillation model compression scheme, which is effective.

In this paper, we propose an improved YOLOv5s-based recognition method for steel leaf springs. Firstly, a camera is used to acquire images of steel leaf springs, and then the steel leaf springs are recognized by the improved YOLOv5s network.In this paper, we use migration learning to train and optimize the YOLOv5s model on the steel leaf spring training set. To improve the generalization ability of the network model, we introduce the attention module in the Neck structure. This effectively strengthens the focus on the target and reduces the influence brought by the environment. To ensure the real-time detection of the steel leaf spring target, we use the combination of ShuffleNetV2 and the Focus module in the backbone network. It replaces the original YOLOv5s backbone network together.

2 YOLOv5 Structure and Method Flow

2.1 Steel Leaf Spring Visual Identification Process

Since the image background of steel leaf springs in factories is very complex, there may be many interferences in the shooting background, and the steel leaf springs taken by cameras at different distances and angles are also diverse, so it is necessary to adopt a stable and effective network model. Firstly, as a target detection algorithm, YOLOv5 belongs to the category of single-stage models, which is the product of continuous integration and improvement on the basis of the original YOLO algorithm, which is characterized by high flexibility, fast detection speed [12–14] and short training time. Secondly, YOLOv5 has good test results on COCO [15] with PASCAL VOC dataset. Therefore, in this paper, YOLOv5s is used as the main body for the identification of steel leaf springs.

This paper proposes a rapid identification method for steel leaf springs based on improved YOLOv5 algorithm, in view of the difficulty of visual recognition of steel leaf springs caused by the differences in front and side views, and inconsistent length and appearance due to certain curvature of steel leaf springs in laboratory and factory simulated environments. The basic flow chart is shown in Fig. 1.

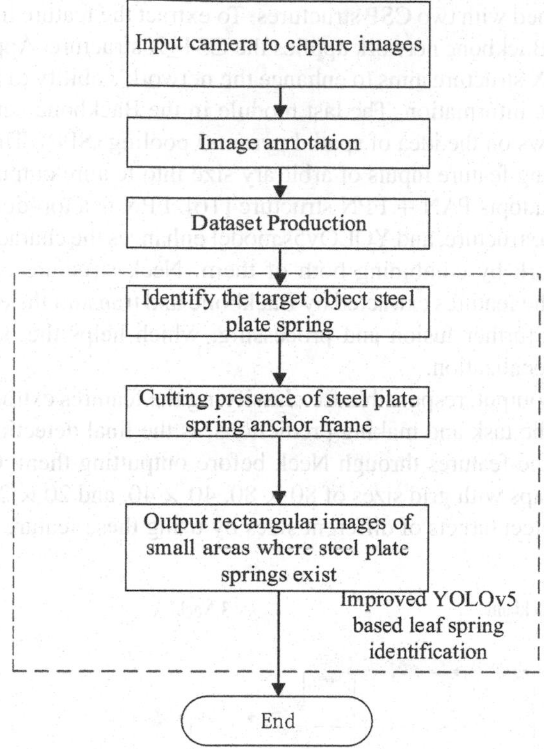

Fig. 1. Flow chart of the proposed method.

2.2 YOLOv5s Network Structure

Consisting of four main components, the network structure of YOLOv5s includes input, Backbone network, Neck network, and Prediction output [14]. Figure 2 shows the network structure diagram of YOLOv5s. The YOLOv5s network model usually preprocesses the input images with Mosaic data augmentation, adaptive anchor frame calculation, and adaptive image scaling at the input side. One of the Mosaic data enhancements aims to enrich the dataset while improving the generalization capability of the model. In the YOLO algorithm, the adaptive anchor frame calculation will give an initial set of anchor frames of length and width for different data sets. Adaptive image scaling, i.e., shrinking or enlarging the image to fit the network structure and performing operations such as normalization, finally transforms the input image into a $640 \times 640 \times 3$ tensor.

The backbone network mainly comprises the Focus module, the CBL module, and the C3 and SPP modules. The main function of the Focus module is to perform slicing operations on the input image, which enables the original image to be sampled without losing any information during the sampling at twice the rate. Performing convolution, BN, and activation function operations on the input feature map, the CBL module is responsible for this task. It uses LeakyRelu as the activation function, which is a modification of the ReLU function of the "modified linear unit". The C3 module has basically the same structural role as the CSP architecture. The difference with other YOLO versions is that

YOLOv5s is designed with two CSP structures. To extract the feature information of the input samples, the Backbone network applies the C3T_X structure. Applied in the Neck network, the C3F_X structure aims to enhance the network's ability to fuse features and retain richer feature information. The last module in the Backbone network is the SPP module, which draws on the idea of spatial pyramid pooling (SPP). The SPP module is capable of converting feature inputs of arbitrary size into feature outputs of fixed size.

Neck network adopts PAN + FPN structure [16], FPN is a top-down structure and PAN is a bottom-up structure, and YOLOv5s model enhances the characterization ability of backbone network by combining both of them. Neck structure can compress the dimensionality of the features extracted by Backbone and transmit the extracted features to the output after further fusion and processing, which helps the output to perform classification and localization.

The Prediction output, responsible for classifying the features extracted by the backbone network for the task and making predictions, is the final detection part that compresses and fuses the features through Neck before outputting them. Generating three scales of feature maps with grid sizes of 80 × 80, 40 × 40, and 20 × 20, the Prediction output is able to detect targets of different sizes by using these feature maps.

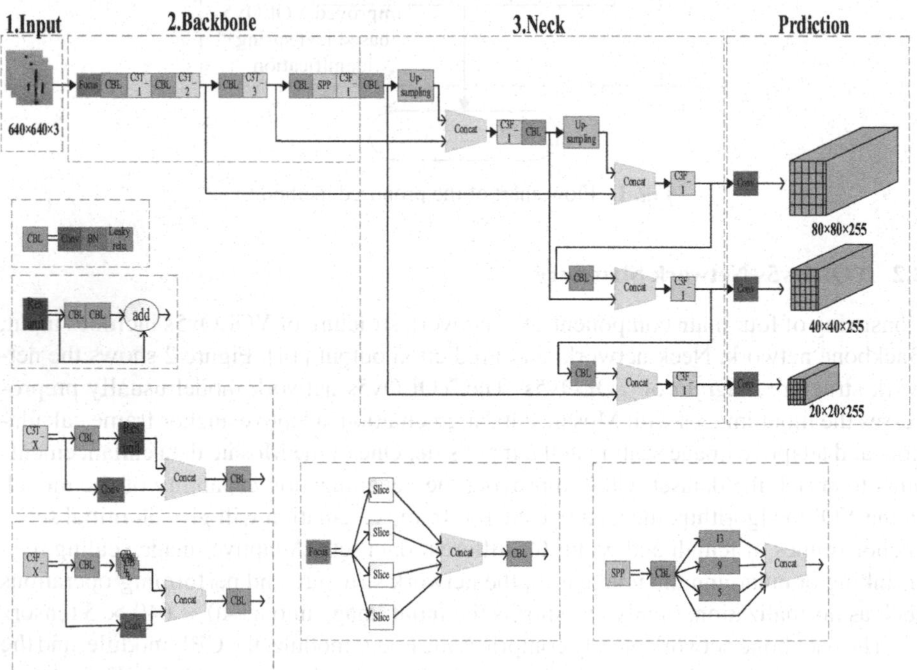

Fig. 2. YOLOv5s network structure diagram.

3 YOLOv5 Recognition Algorithm Improvement

3.1 YOLOv5 Steel Leaf Spring Recognition Based On Migration Learning

Transfer learning is a machine learning method that uses existing knowledge to solve a problem in a different but related domain from the original domain [17].The principle of transfer learning is the process of transferring knowledge from a known domain (source domain) to an unknown domain (target domain), allowing for better learning in the target domain. In the training of convolutional neural networks, migration learning can accelerate the training of the target domain model by the model parameters of the already trained source domain, which solves the problem of excessive training volume, difficulty and slow speed of the target domain due to learning from scratch. It also solves the problem of lack of data sets in the target domain.

The migration algorithm based on the steel leaf spring model uses a pre-training mode, and in this paper, the migration training is performed using the parameters of the YOLOv5 model pre-trained based on CCPD, and the specific migration logic is shown in Fig. 3 [18]. The shallow network in deep learning is mostly used to learn similar appearance and other low-level features, such as the car license plate and the steel leaf spring in Fig. 3, both of them are rectangular in the picture taken by the 2-dimensional camera, so the rectangle is the low-level feature of the car license plate and the steel leaf spring; both the car license plate and the steel leaf spring have color, which can be regarded as their medium-level features; the high-level feature of the car license plate has texture (the number can also be regarded as a texture), while the high-level feature of the steel leaf spring can be regarded as the color black or earth gray. Along with the gradual deepening of the neural network hierarchy, neural networks can learn specific features of special objects to achieve target detection. By using model-based deep migration learning can save the time cost of identifying steel leaf springs, and also improve the robustness and generalization ability of the model.

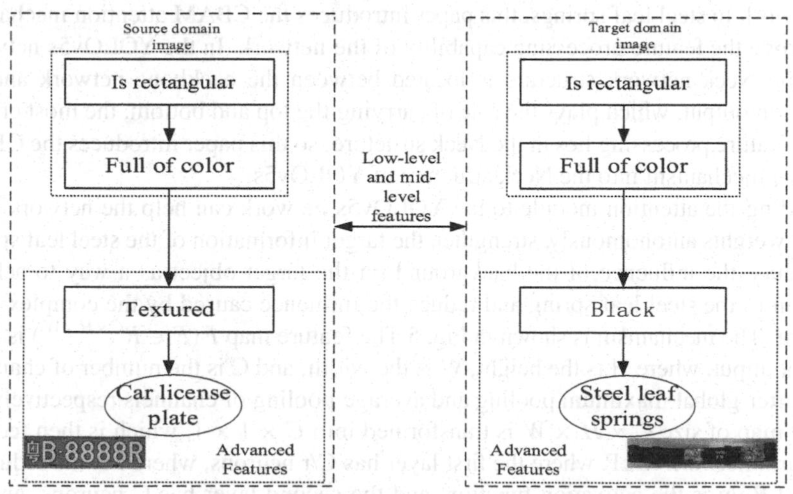

Fig. 3. Learning logic diagrams based on model transfer.

3.2 CBAM Convolutional Attention Mechanism

CBAM Attention Mechanism Module [19] is a lightweight convolutional attention mechanism module proposed by Sanghyun Woo et al. The module consists of two main parts: the first being the Channel Attention Module (CAM) and the second being the Spatial Attention Module (SAM). CBAM not only enhances the characterization of the network by combining the first and second parts, but also suppresses irrelevant noise information, which in turn can improve the detection accuracy of the network. And the CBAM module is able to infer the attention map along the channel attention module and spatial attention module sequentially and multiply the attention map with the input feature map for adaptive feature optimization [20]. The two attention modules, channel attention module and spatial attention module, can be combined either in parallel or sequentially. However, placing the channel attention module in front and the spatial attention module in the back typically yields better results. Figure 4 illustrates the structure of the CBAM attention module.

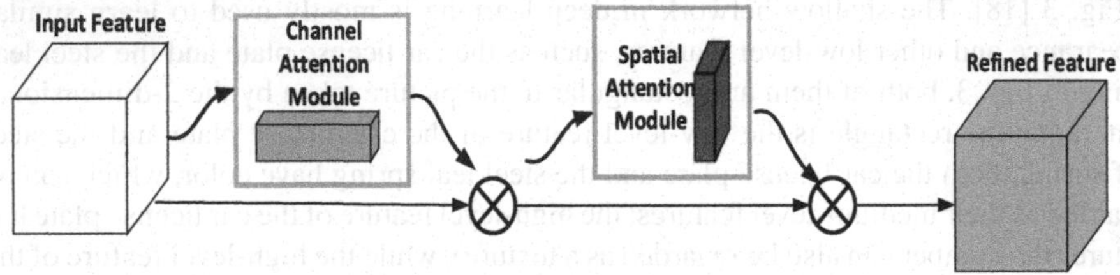

Fig. 4. CBAM network structure.

Since the environment of the steel leaf spring manufacturing plant is complex, which inevitably causes interference to the recognition of steel leaf springs, in order to improve the salience of steel leaf springs in the complex environment and enhance the attention of the network to steel leaf springs, this paper introduces the CBAM attention mechanism to improve the feature processing capability of the network. In the YOLOv5s network, since the Neck network structure is located between the backbone network and the Prediction output, which plays the role of carrying the top and bottom, the most critical part of feature processing lies in the Neck structure, so this paper introduces the CBAM attention mechanism into the Neck structure of YOLOv5s.

Adding the attention module to the YOLOv5s network can help the network learn feature weights autonomously, strengthen the target information of the steel leaf spring, and reduce the influence of the background on the target object as a way to achieve attention to the steel leaf spring and reduce the influence caused by the complex environment. The mechanism is shown in Fig. 5.The feature map $F(F \in R^{C \times H \times W})$ is input from the input, where H is the height, W is the width, and C is the number of channels. Then, after global maximum pooling and average pooling of channels respectively, the feature map of size $C \times H \times W$ is transformed into $C \times 1 \times 1$, which is then fed into the neural network MLP, where the first layer has C/r neurons, where r is the reduction rate and Relu is the activation function, and the second layer has C neurons, and the results obtained are summed up after completion and fed into the Sigmoid function to obtain the weight coefficient M_c, where The weight coefficients are calculated as shown in Eq. (1), after multiplying the initial input with the weight coefficients M_c, the new features after scaling are obtained.

$$M_cF = \sigma(W_1(W_0(F_{avg}^C)) + W_1(W_0(F_{\max}^C))) \tag{1}$$

where σ denotes the Sigmoid function; avg denotes global average pooling; max denotes maximum pooling; $W_0 = R^{\frac{C}{r}\times C}$; $W_1 = R^{C\times\frac{C}{r}}$; F_{avg}^C denotes the average pooling characteristic of size $C \times 1 \times 1$; and $F_{\max}^C$ denotes the maximum pooling characteristic of size G.

The spatial attention module divides the result obtained from the channel attention module into two channel descriptions of $1 \times H \times W$ by maximum pooling and average pooling, then the tensor is stacked together by Concat connection calculation, followed by a convolution operation and a Sigmoid on the stacked features to obtain the weight coefficients M_s. The weight coefficients are calculated as shown in Eq. (2). The scaled new features are then obtained by multiplying the input from the previous step with the weight coefficients.

$$M_sF) = \sigma(f^{7\times7}([F_{avg}^s, F_{\max}^s])) \tag{2}$$

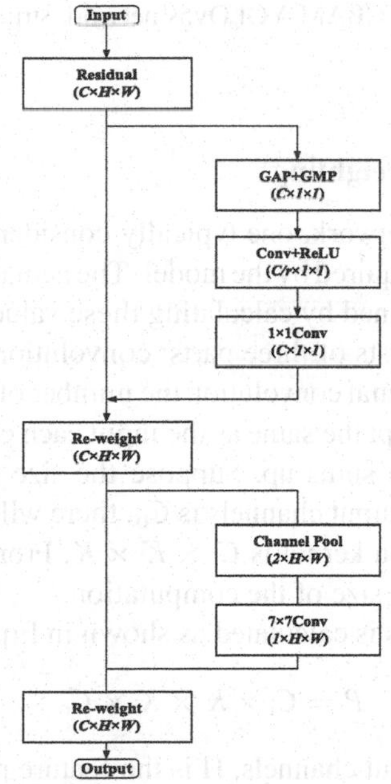

Fig. 5. CBAM attention mechanism.

The network structure diagram of the Neck part after adding the CBAM module is shown in Fig. 6. The CBAM-YOLOv5s model adds a CBAM module before the CBL module on the Neck side and after the C3 module. The attention mechanism is added between the two feature fusions to enhance the network's attention to the steel leaf spring, reduce the environmental influence, obtain more effective information, and improve the target detection accuracy.

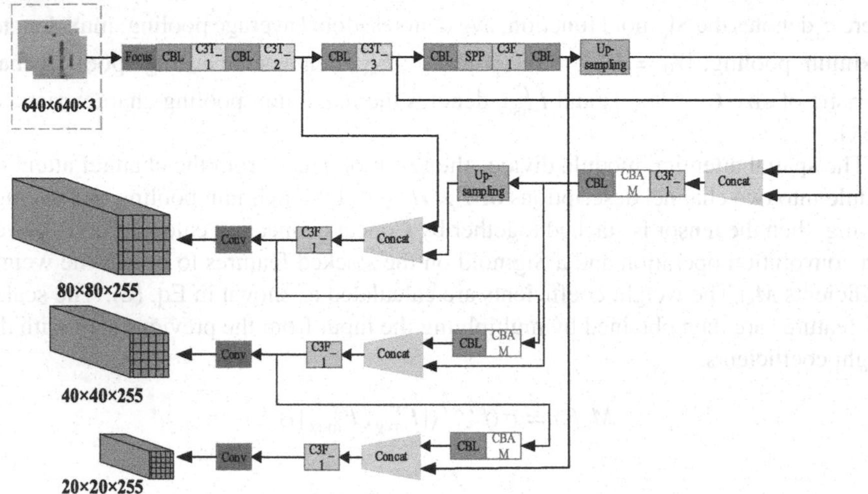

Fig. 6. CBAM-YOLOv5s network structure.

3.3 Network Model Lightweighting

To gauge the lightness of a network, one typically considers the number of parameters and amount of computation required by the model. The actual parameters and complexity of the network can be determined by calculating these values.

Normal convolution consists of three parts: convolution layer, batch normalization and activation function. In normal convolution the number of channels of the convolution kernel of the feature map is kept the same as the input each channel does the convolution operation separately and then sums up. Suppose the size of the input feature map is $C_i \times H \times W$, the number of output channels is C_o, there will be C_o convolution kernels, and the size of the convolution kernel is $C_i \times K \times K$. From this, we can calculate the number of parameters and the size of the computation.

The number of parameters is calculated as shown in Eq. (3):

$$P = C_i \times K \times K \times C_o \tag{3}$$

where C_i is the number of input channels, H is the feature map height, W is the feature map width, K is the width and height dimension of the convolution kernel, and C_o is the number of output channels.

The calculated quantities are shown in Eq. (4):

$$F = C_i \times H \times W \times K \times K \times C_o \tag{4}$$

Deeply separable convolution can be divided into two primary processes: channel-by-channel convolution and point-by-point convolution. Channel-by-channel convolution obtains multiple feature maps by layer-by-layer operation of the input channel number and multiple single-channel convolution kernels, and then performs weighted fusion in the depth direction by point-by-point convolution, and finally achieves the information fusion between different channels. Similarly, we can also get the number of parameters and computation of DW convolution.

The number of parameters is shown in Eq. (5):

$$P_{DW} = C_i \times K \times K + C_i \times C_o \tag{5}$$

The calculated quantities are shown in Eq. (6):

$$F_{DW} = C_i \times H \times W \times K \times K + C_i \times C_o \times H \times W \tag{6}$$

From the above equation, we have:

$$\frac{P_{DW}}{P} = \frac{1}{C_o} + \frac{1}{K^2} = \frac{F_{DW}}{F} < 1 \tag{7}$$

From Eq. (7), it can be seen that the number of parameters and computation of profoundly separable convolution are smaller than those of ordinary convolution, and compared with ordinary convolution, the use of profoundly separable convolution can reduce the computation as well as the model size more significantly to achieve light weight.

Grouped convolution involves dividing the input feature maps into G groups of equal size by channel, and then applying regular convolution to each group. Since the number of channels of each input feature map is $\frac{C_i}{G}$, the number of channels of each convolution kernel is also reduced to $\frac{C_i}{G}$ after grouping. After group convolution, C_o is the number of channels obtained by splicing the output of group G. The number of parameters is shown in Eq. (8):

$$P_{GC} = C_o \times \frac{C_i}{G} \times K \times K \tag{8}$$

The calculated quantities are shown in Eq. (9):

$$F_{GC} = \frac{C_i}{G} \times H \times W \times K \times K \times C_o \tag{9}$$

The above equation shows that parameter count and computation of the normal convolution is the largest among these convolutions, while the depth-separable convolution and grouped convolution both have different degrees of reduction compared to the normal convolution. The number of parameters and the computation of grouped convolution are 1/G of the normal convolution. ShuffleUnitV2 uses the group convolution and deep convolution, which can greatly reduce parameter count and the computational effort of the model.

The backbone network of YOLOv5s contains multiple deep convolutional modules, resulting in excessive computation. In order to design a recognition algorithm for steel leaf springs, it is important to ensure that it can accurately recognize steel leaf springs in a factory environment, but also to compress the size of the algorithm model as much as possible, making the monitoring model lightweight and easy to deploy in hardware devices later. In this paper, we use ShuffleNetV2 to downscale the YOLOv5s model and reduce the model parameters and computational effort. ShuffleUnit is the basic component unit of ShuffleNetV2 [21], and its structure is shown in Fig. 7. Where stride = 1 and stride = 2 correspond to the cases of (a) and (b) in the figure, respectively. The depth convolution layer is DWConv, the point convolution layer is PWConv, the batch normalization is BN, and the ordinary convolution is Conv. Channel Split represents the slicing of the feature map in the channel dimension, Channel Shufle represents the random disruption of the feature map order in the channel dimension and then integration, and Concat represents the stitching of the feature map in the channel direction. ShuffleUnit relies on convolutional steps for downsampling operations. When stride = 1, the feature map is first channel split into left and right branches; the left branch is left unprocessed and the right branch is connected to Concat by triple convolution. The input feature map size is kept the same as the output feature map size. When stride = 2, the feature map does not require channel splitting, and its input feature map is twice as long and wide as the output feature map, respectively.

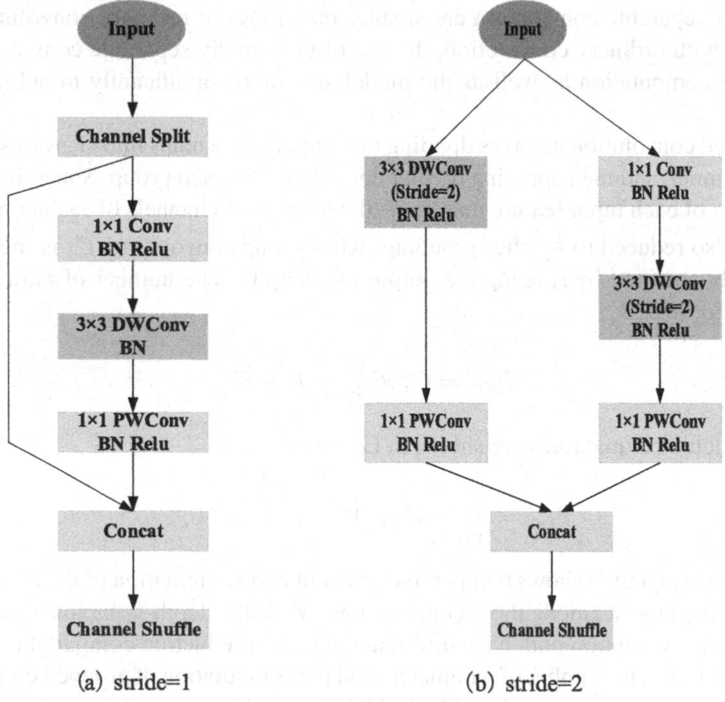

Fig. 7. ShuffeUnit structure.

The Focus module can slice and convolve the input image. This module is capable of slicing the 640 × 640 × 3 input image to generate a 320 × 320 × 12 feature map. This feature map can then undergo a 3 × 3 convolution operation, resulting in an output channel of 32, and finally become a 320 × 320 × 32 feature map. From the above operation, it can be seen that the most intuitive effect of the image after passing through the Focus module is that it is downsampled, but unlike the ordinary convolutional downsampling, the information is not lost after the Focus module is downsampled, so the improved backbone network still selects the Focus module in the YOLOv5s backbone network for downsampling.

Before the improvement, the backbone network primarily consisted of the Focus module, CBL module, C3 module, and SPP module; the improved backbone network mainly consists of Focus module, ShuffleUnit-1, ShuffleUnit-2 and SPP module. Although the improved network structure still maintains the connections of the original model, it not only makes the computational and parametric quantities of the backbone network significantly reduced, but also reduces the demand for hardware devices in the network model and improves the real-time performance of detection. Figure 8 illustrates the overall structure of the improved Backbone. The feature map is reduced to 1/2 of the network input feature map after Focus, 1/8 of the input network after two ShuffleUnit-2s, 1/16 after one ShuffleUnit-2, and finally 1/32 after another ShuffleUnit-2.

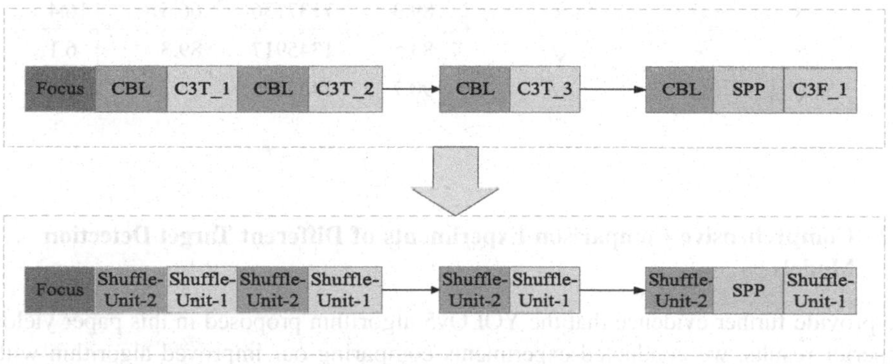

Fig. 8. Improved network.

4 Experimental Results and Analysis.

4.1 Ablation Experiments

Ablation experiments were conducted to assess the effectiveness of each improved part of the algorithm presented in this paper, based on the original algorithm. The experimental results are shown in Table 1, where the symbol "√" indicates that the improved strategy was employed in the network model, whereas the symbol " × " indicates that the improved strategy was not utilized. The improved strategy, as can be seen from Table 1, the steel leaf spring detection accuracy of the original algorithm is 87.2%, and the number of model parameters is larger, 7053910; after adopting migration learning, the steel leaf spring detection accuracy is improved by 5.5%, and the model parameters and detection rate as well as GFLOPs are not affected; after introducing the attention mechanism in the original network, the steel leaf spring detection accuracy is improved by 2.1%. As the number of model parameters increases, the detection rate tends to decrease. After adding the lightweight module to the original network, the accuracy of steel leaf spring detection decreases significantly, but the number of model parameters decreases by 5707993 and the speed of steel leaf spring detection also increases. Finally, by combining the three improvement techniques mentioned above, the steel leaf spring accuracy reaches 90.7%, which is a 3.5% improvement compared to the original network model, which shows that the improved network not only increases the detection accuracy and detection speed, but also reduces the number of parameters very effectively.

Table 1. Ablation experiments.

Serial number	Transfer learning	Attentional Mechanisms	Lightweight model	mAP/%	Number of participants	Detection rate/f.s^{-1}	GFLOPs
1	×	×	×	87.2	7053910	73.8	16.3
2	√	×	×	92.7	7053910	73.8	16.3
3	×	√	×	89.3	7137136	66.5	16.4
4	×	×	√	83.5	1345917	89.3	6.1
5	√	√	√	90.7	1457256	78.7	6.5

4.2 Comprehensive Comparison Experiments of Different Target Detection Models

To provide further evidence that the YOLOv5 algorithm proposed in this paper yields superior results, we conducted experiments comparing our improved algorithm with other mainstream algorithms. The experimental results are presented in Table 2. According to Table 2, the improved YOLOv5s algorithm proposed in this paper achieves a detection accuracy of 90.7% for steel leaf springs, which is 3.5% higher than that of the original YOLOv5s algorithm and significantly outperforms other detection models. Moreover, the model size of this method is only 5.7 MB, which has a significant advantage over other mainstream algorithms in terms of volume. The improved model has a detection speed of 78.7 f/s, and although the introduction of the CBAM module into the improved model does not result in the optimal detection speed, the model can still meet the real-time requirements for the recognition of steel leaf springs in industry. Furthermore, further optimization of computational complexity can reduce hardware requirements for network model training and detection. Therefore, the excellent accuracy and real-time performance of the proposed algorithm in identifying steel leaf springs satisfy the requirements for industrial applications.

Table 2. Comparison experiments.

Detection Model	Backbone	Model Size (MB)	Detection Rate/f.s^{-1}	mAP/%
SSD	VGGnet	185.7	43.2	75.3
YOLOv3	Darknet53	120.3	81.1	79.7
YOLOv4	CSPDarknet53	235.3	32.0	82.1
YOLOv4-Tiny	CSPDarknet53-Tiny	24.2	85.3	74.8
YOLOv5s	CSPDarknet	14.4	73.8	87.2
YOLOv5s-MobileNet	MobileNet	7.3	80.6	81.5
Improvements to YOLOv5s	Shufflenetv2	5.7	78.7	90.7

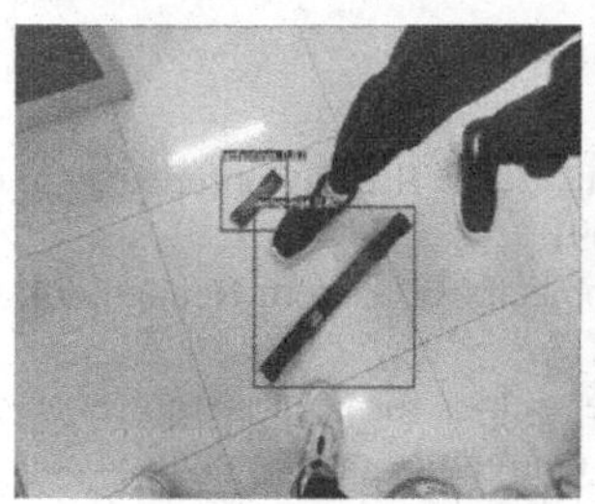
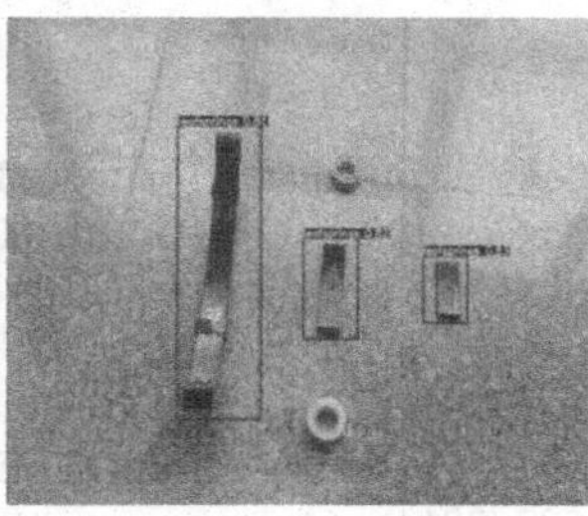
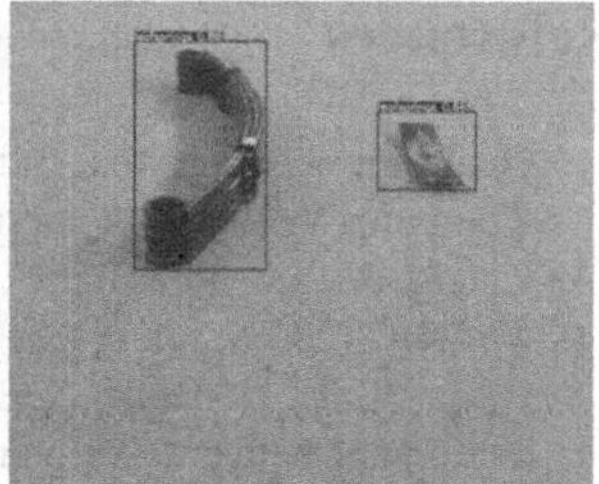

(a) YOLOv5s model detection results

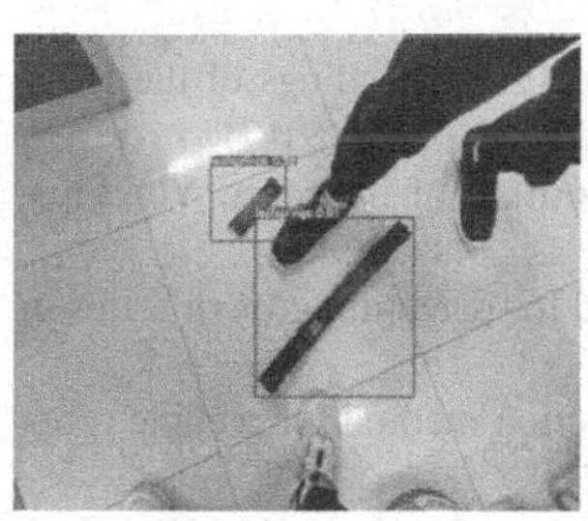
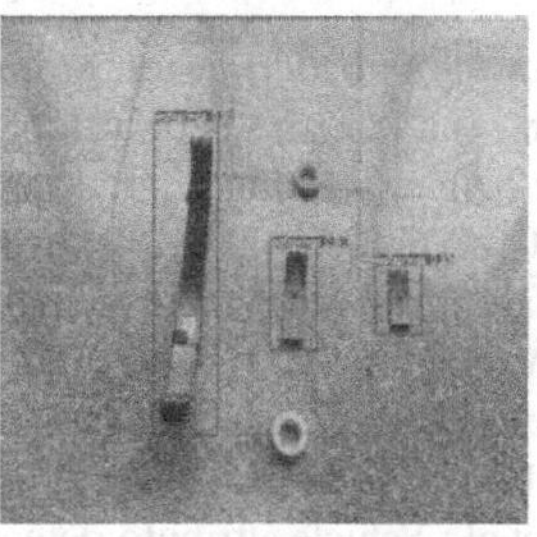
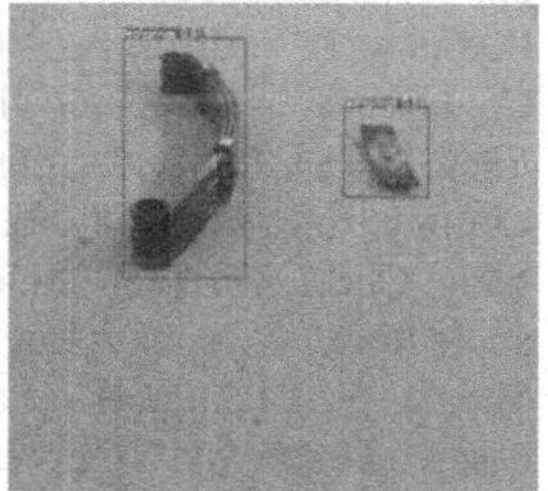

(b) Improved YOLOv5s model detection results

Fig. 9. Leaf spring target detection results.

In order to verify the effect of steel leaf spring detection under different environments, the steel leaf spring images were selected for testing with a confidence threshold of 0.7 and an IOU threshold of 0.45. The results of steel leaf spring detection are shown in Fig. 9. It can be seen that the improved YOLOv5s model can accurately detect the corresponding steel leaf springs, and the detection accuracy is significantly improved compared to the YOLOv5s model.

5 Summary

The recognition of steel leaf springs is the first step to realize the industrial application of steel leaf spring grasping technology. For the current research on the recognition of steel leaf springs, there are few studies, this study proposes an improved YOLOv5s target detection algorithm for steel leaf springs, which uses migration learning to enhance the robustness and generalization ability of the model on the basis of YOLOv5s, and adds an attention mechanism module to enhance the focus of the network model on the target, reduce the influence of invalid information, and lighten the network. The experimental results show that the improved YOLOv5s network, with a 60.4% reduction in model size, achieves an average detection speed of 78.7 f/s and an average detection accuracy of 90.7%. The identification method of steel leaf spring studied in this paper meets the industrial requirements in terms of speed and accuracy, and provides a reference for the identification of other industrial parts.

References

1. Quattoni, A., Torralba, A.: Recognizing indoor scenes. In: 2009 IEEE Conference on Computer Vision and Pattern Recognition, pp. 413–420 (2009)
2. Cheng, X., Lu, J., Feng, J., et al.: Scene recognition with objectness. Pattern Recogn. **74**, 474–487 (2018)
3. Girshick, R., Donahue, J., Darrell, T., et al.: Rich feature hierarchies for accurate objectdetection and semantic segmentation. Proceedings of the IEEE conference on computer vision and pattern recognition, 580–587 (2014)
4. Redmon, J., Divvala, S., Girshick, R., et al.: You only look once: unified, real-time object detection. In: Proceedings of the IEEE Conference on Computer Vision and Pattern Recognition, pp. 779–788 (2016)
5. Redmon, J., Farhadi, A.: YOLO9000: better, faster, stronger. In: Proceedings of the IEEE Conference on Computer Vision and Pattern Recognition, pp. 7263–7271 (2017)
6. Yang, Y., Li, D.: Lightweight helmet wearing detection algorithm of improved YOLOv5. Comput. Eng. Appl. **58**(09), 201–207 (2022)
7. Yang, Q., Li, W., Yang, X., et al.: Improved YOLOv5 method for detecting growth status of apple flowers. Comput. Eng. Appl. **58**(04), 237–246 (2022)
8. Yan, B., Fan, P., Lei, X., et al.: A real-time apple targets detection method for picking robot based on improved YOLOv5. Remote Sens. **13**(9), 1619 (2021)
9. Liu, J., Zhong, G., Huang, S., et al.: Vehicle attribute detection based on improved YOLOv5. Appl. Electron. Tech. **48**(7), 19–24 (2022)
10. Li, R., Qian, H., Guo, J., et al.: Lightweight target detection algorithmbased on M-YOLOv4 model. Foreign Electron. Meas. Technol. **41**(04), 15–21 (2022)
11. Li, Y., Zhang, C., Zhao, Y., et al.: Research on lightweight obstacle detection model based on model compression. Laser J. **43**(09), 38–43 (2022)
12. Avazov, K., Mukhiddinov, M., Makhmudov, F., et al.: Fire detection method in smart city environments using a deep-learning-based approach. Electronics **11**(1), 73 (2021)
13. Liu, S., Zhang, N., Yu, G.: Lightweight security wear detection method based on YOLOv5. Wirel. Commun. Mob. Comput. **2022** (2022)
14. Krizhevsky, A., Sutskever, I., Hinton, G., E.: Imagenet classification with deep convolutional neural networks. Commun. ACM **60**(6), 84–90 (2017)

15. Lin, T.Y., Maire, M., Belongie, S., et al.: Microsoft coco: common objects in context. In: Computer Vision–ECCV 2014: 13th European Conference, Zurich, Switzerland, 6–12 September 2014, Proceedings, Part V 13, pp. 740–755 (2014)
16. Liu, J., Zhong, G., Huang, S., et al.: Vehicle attribute detection based on improved YOLOv5. Appl. Electron. Techn. **48**(7), 19–24, 29 (2022)
17. Weiss, K., Khoshgoftaar, T.M., Wang, D.D.: A survey of transfer learning. J. Big data **3**(1), 1–40 (2016)
18. Jiang, P., Ergu, D., Liu, F., et al.: A review of yolo algorithm developments. Procedia Comput. Sci. **199**, 1066–1073 (2022)
19. Woo, S., Park, J., Lee, J., Y., et al.: CBAM: convolutional block attention module. In: Proceedings of the European Conference on Computer Vision (ECCV), pp. 3–19 (2018)
20. Lin, S., Liu, M., Tao, Z.: Detection of underwater treasures using attention mechanism and improved YOLOv5. Trans. Chin. Soc. Agr. Eng. (Trans. CSAE) **37**(18), 307–314 (2021)
21. Quattoni, A., Torralba, A.: Recognizing indoor scenes. In: 2009 IEEE Conference on Computer Vision and Pattern Recognition, pp. 413–420 (2009)

A Bughole Detection Approach for Fair-Faced Concrete Based on Improved YOLOv5

Bazhou Li[1,2], Feichao Di[1,2(✉)], Yang Li[1,2], Quan Luo[3], Jingjing Cao[3], and Qiangwei Zhao[3]

[1] CCCC Wuhan Harbor Engineering Design and Research Institute Co., Ltd., Wuhan 430040, China
1065953878@qq.com

[2] Hubei Provincial Key Laboratory of New Materials and Maintenance and Reinforcement Technology for Offshore Structures, Wuhan, China

[3] School of Transportation and Logistics Engineering, Wuhan University of Technology, Wuhan 430063, China

Abstract. Surface bughole is a major quality defect of the concrete surface, which has nonnegligible impact on the evaluation of surface quality of fair-faced concrete. However, most existing deep learning methods use image segmentation to detect bugholes or other defects on the surface of formed concrete, there is a lack of a method that can detect bughole on the concrete surface instantly during the pouring process. In addition, the inability to effectively detect small-scale bughole is also an issue that cannot be ignored. This application scenario requires a method with high detection accuracy, fast inference speed and ease of deployment. Aiming at these problems, this paper proposes an improved YOLOv5 network, we propose a detector scale (DS) that is added in the model to detect small size bughole on the fair-faced concrete surface with high accuracy. The concatenations are introduced for the feature transmission during the backbone and head part of the model. This allows for the fusing of low-level and high-level features and improves the perception of the detection model on minor flaws. We also construct a dataset of fair-faced concrete surface bugholes and compare our modified YOLOv5 with the baseline version. Our proposed model has improved mAP@0.5 and mAP@.5:.95 to 89.9% and 65.7%, its performance is superior to YOLOv5, while also retaining good inference speed with approximately 13.4 ms to detect one 1280 × 1024 images on single GPU.

Keywords: Fair-faced concrete · Deep learning · Bughole · YOLOv5 · Object detection

1 Introduction

The concept of green environmental protection is increasingly valued by the international community, thus green building materials came into being. Among

H. Zhang et al. (Eds.): NCAA 2023, CCIS 1870, pp. 18–31, 2023.
https://doi.org/10.1007/978-981-99-5847-4_2

them, fair-faced concrete is widely used as a kind of green concrete in construction projects which is characterized by one-time molding, no external decoration, and direct use of the natural texture of concrete after molding as the finishing effect.

There are strict requirements for the quality of concrete surfaces in the building industry, primarily concerning flatness, tint, and the absence of bughole (surface bubbles). Bughole, as a major quality defect of the concrete surface, is a small pit on the concrete surface after pouring. It is characterized by regular or irregular pits with near circular shapes and dimensions ranging from a few millimeters to 15 mm, which is usually distributed around the concrete surface. The bughole are mainly a common surface defect caused by careless construction details such as the air-entraining effect of the external permeability agent, excessive laitance, short vibration time, and unclean formwork cleaning. The generation of air bughole is almost inevitable and is difficult to be removed completely. Therefore, bughole are considered to be a major defect that affects the appearance quality of concrete and worsens the appearance of concrete structures. In the early stage, manual evaluation is the main method. Bughole on the concrete surface are observed directly by the human eye, or the inspector manually calculates the number and measures the diameter of bughole to obtain the percentage of bughole areas on the surface to evaluate the concrete surface quality [1]. Besides, it can also be measured and evaluated relative to the bughole control scale recommended by the International Commission on Concrete (CIB). Above methods are time-consuming and easily subjectively affected. Therefore, the development of objective automatic detection and evaluation methods based on computer vision has important research value and practical significance for promoting the development of fair-faced concrete technology.

To address these issues, many computer image processing-based methods have been introduced as pioneer improvements. Lemaire et al. proposed a computer-assisted image processing method to detect the size and distribution of bughole on the concrete surface, and used the reference scale of bughole recommended by CIB to classify the appearance quality of concrete with bughole on the surface [2]. Ozkul et al. proposed to use of the principle of differential pressure method to detect bughole on the concrete surface. They developed a bughole detection device whose edge contacts the concrete surface with bughole. The existence of bughole causes the pressurized gas in the device to escape, so as to detect bughole on the concrete surface [3]. Silva et al. developed an expert system with image analysis to classify the concrete surface condition according to the existence of bughole [4]. Peterson et al. discussed a thresholding method based on concrete surface image analysis. They used a flat panel scanner connected to a personal computer to scan the concrete surface and analyzed the scanned image to detect surface bughole [5]. Zhu et al. developed an automatic detection method that can detect the color difference between bughole and concrete surface, automatically calculate the number of bughole and the area of color difference area, and evaluate the surface quality of concrete by using a threshold [6]. Liu et al. established a method using the image processing toolbox based on

MATLAB, OTSU image threshold segmentation technology was used to extract the characteristics of bughole on the concrete surface, and proposed an empirical equation based on the relationship between the bughole reference scale and the bughole area ratio recommended by CIB [7]. Yoshitake et al. developed an image processing method using color images to detect bughole distributed on the concrete surface [8]. However, the direct use of image processing technology for bughole detection has several shortcomings. First of all, insufficient generalization. The algorithm is customized for certain dataset and the performance on the new dataset may be unsatisfactory. Secondly, insufficient robustness. Noise such as illumination, shadows, and other different surface imperfections, the detection of the image processing algorithms may be inaccurate. These factors all constrain the development of detection methods using image processing technology.

One promising solution is deep-learning algorithm. With the development of deep learning technology in recent years, detection methods based on deep convolution neural network (DCNN) have gradually emerged. Wei et al. introduced a modified Mask R-CNN, which achieved instance-level recognition for concrete surface bughole, then the ratio of pixels to the actual size was determined by accurately comparing the size of the image with the actual area, which ultimately achieving quantitative output of bubble area and maximum diameter [9,10]. Yao et al. constructed a DCNN classifier with several layers and trained network with their self-built bughole dataset composed of 4K small size images (28×28 pixel resolutions), their final comparative experiment results indicated that the proposed DCNN method had much better performance than the Otsu method and the LoG method that can avoid the interference of cracks, color-differences, and nonuniform illumination on the concrete surface [11]. Sun et al. implemented a modified DeepLabv3+ to detect cracks and bugholes on concrete surfaces, which has adopted measures such as using separable convolution to replace low-level normal convolution, reducing the expansion ratio in ASPP module, and adding weight value to channel dimensions to improve the detection accuracy [12]. Wei et al. adopted an AlexNet-like network inserted with an inception module as feature extraction layer to detect concrete specimens surface images that have undergone grayscale processing, contrast enhancement, and OSTU threshold segmentation, which combined the respective advantages of image processing and DCNN [13].

This paper is based on a number of research and development status and deep learning development history and research progress of existing concrete appearance quality defect detection, we found that the previously proposed methods are all based on the use scenario of bughole detection on the formed concrete surface, and there is no method for real-time bughole detection during the pouring process. In the actual construction process, if the surface bughole defects of the formed pouring body do not meet the requirements, manual repair of the surface is required, which will increase the cost. For large poured bodies such as bridge piers, the cost of secondary repair is much higher. So there is a need for a method that can detect bughole during the concrete pouring process. This method can effectively and immediately monitor the surface state of the pouring body. When the bughole data exceeds the specified value, an early warning message can be issued to prompt the increase of vibration to eliminate bughole. Therefore, this demand puts forward new requirements for bughole detection methods with high detection accuracy, fast inference speed and ease of deployment. Most existing deep learning based methods use semantic segmentation technology for bughole detection, which is time-consuming and not suitable for real-time detection situations. For example, method based on the Mask R-CNN takes 3 s to detect an image with a resolution of 3024 × 3024 pixels on a single GPU, which is obviously too slow to meet the requirement of real-time detection [10]. Based on these considerations, the YOLOv5 network is introduced for the detection of bughole. We have also modified the network to have better detection performance.

The main contributions of this paper are as follows:

- We establish the data sampling system and construct a fair-faced concrete bughole dataset including 980 images with more than 2750 bounding boxes.
- We propose a modified YOLOv5 network. A new detection scale is proposed to detect small-size bughole and concatenations are built between the backbone and head of YOLOv5 to improve information transmission.
- We conduct experiments on the proposed model using the constructed dataset, the experimental results indicate that our model can achieve higher detection accuracy while maintaining good inference speed.

2 Model Design

2.1 The Network Structure of YOLOv5

As a one-stage anchor-based object detection model [14,15], YOLOv5 is modified from the previous works [16,17]. It comes in five versions: YOLOv5n, YOLOv5s, YOLOv5m, YOLOv5l, YOLOv5x, which deepen the complexity of the model in turn. The model version we used for modifying and comparing is YOLOv5s, no further distinction will be made in the following text and all denoted by YOLOv5. Generally, the structure of YOLOv5 is described to consist of 3 parts, backbone, neck, and head. As the YOLOv5 network structure shown in Fig. 1, CBS module, CSPDarknet53 (CSP) module [16] and Spatial Pyramid Pooling (SPP) module formed its backbone for feature extraction. The CBS module, composed of convolution, normalization, and silu activation operations, is used for feature maps extraction. The CSP module divides the input feature map into two channels with different convolutional structures and then fuses the features. This operation reduces the repeated gradient information, thereby accelerating the inference speed of the entire feature extraction part while maintaining the best possible detection accuracy. PANet [18] as its neck to accelerate the transmission of low-level features and momentous feature information to subsequent network layers via a top-down path, and the YOLO head which is of default feature map sizes 20 × 20 × c (channel), 40 × 40 × c, and 80 × 80 × c for multi-scale prediction of big, medium and small objects of potential targets.

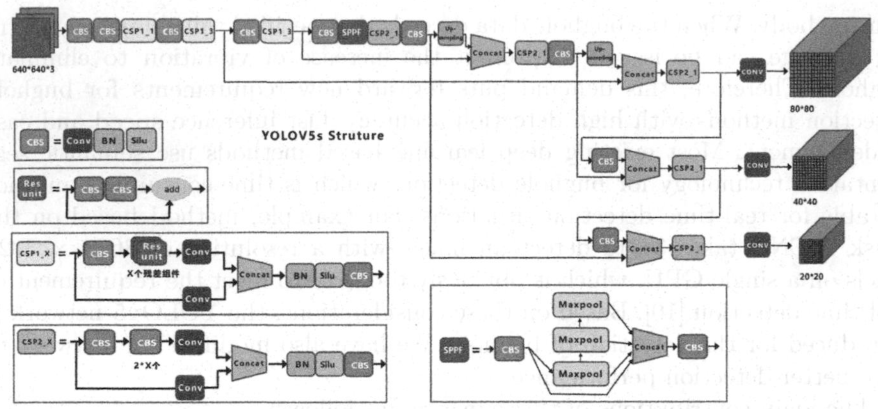

Fig. 1. The network structure of YOLOv5

2.2 Network Structure Improvement

In this section, we will go through our proposed network in depth, as well as provide an outline of our proposed technique. In Fig. 1, we show the network structure of YOLOv5 for comparison with our improvements. The modified YOLOv5 network structure is shown in Fig. 2.

As shown in Fig. 2, there are two improvements for YOLOv5, including a detection scale (DS) and concatenation, which are indicated in the figure by a blue dashed box and red arrow lines respectively. Specifically, the DS is introduced to enable the model to have better detection ability for small-sized targets. When input is transmitted layer by layer down during the feature extraction process of the network, fine-grained information in the input will gradually be lost after passing through multiple convolutional layers, and some small instances may even disappear, while semantic information such as categories will become more prominent, so DS composed of low-level backbone layers and detector head can help obtain more fine-grained information. Futhermore, the concatenation between backbone and head part are proposed to fully exploit the generic or texture features of the image. These concatenation increase the connectivity between backbone and head, which can improve the flow of graphics information across the network.

As shown in Fig. 1, unmodified YOLOv5 model used PANet [18] as the method of parameter aggregation from different backbone levels for different detector levels. It enhances localization information flow in lower layers via a top-down path and shortens the information path between low-level and high-level features. While there is a problem that it predicts boxes at only three different scales, the size of the feature map of the three scales are 80×80, 40×40 and 20×20. But the size of the input image from our dataset is 1280×1024, which will even be padded to 1280×1280 and then resized to 640×640 before being sent into the model.

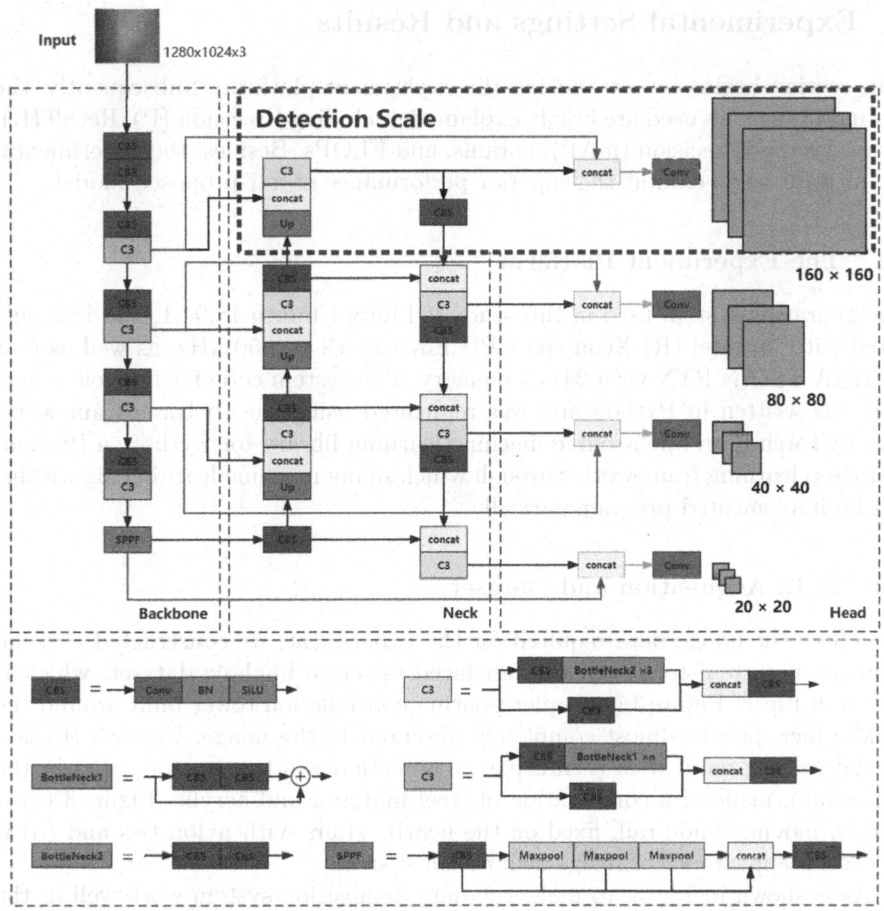

Fig. 2. The modified YOLOv5 network structure. (Color figure online)

As a big feature map size corresponds to a small receptive field according to the theory of receptive field, it will be hard for unmodified YOLOv5 to detect instances accurately whose size is smaller than the receptive field. This means the scale (80 × 80) used for small-size object detection will have its limitation of detecting capability of 8 × 8, which is 16 × 16 being mapped to the original input image. While many small bugholes in our constructed dataset small bughole whose size is smaller than 16 × 16, which is shown in Fig. 4, so we added a detection scale of 160 × 160 feature map to get better detection accuracy. As shown in Fig. 2, a group of upsampling, concatenation and C3 blocks is added in the neck part, and a group of concatenation, C3 block, and CBS blocks are added in the head part, which is shown in the blue dashed box in the figure. Besides, more concatenations are inserted into the model between the backbone and head part, marked with red arrow lines in Fig. 2.

3 Experimental Settings and Results

This section begins by introducing the experiment platform. Subsequently, the evaluation metrics used are briefly explained, including Precision (P), Recall (R), mean Average Precision (mAP), params, and FLOPs. Besides, the experimental are introduced to exhibit the superior performance of our proposed model.

3.1 The Experiment Platform

The operating system used in this study is Linux Ubuntu 18.04.1, and it is outfitted with an Intel (R) Xeon (R) CPU E5-2678 v3 @ 2.50 GHz, as well as four NVIDIA TITAN RTX with 24 GB memory. The system code for bughole detection was written in Python and was replicated using the PyTorch framework. The PyTorch is an open-source machine learning library for Python, a Python-first deep learning framework, through which many machine learning algorithms can be implemented programmatically.

3.2 Data Acquisition and Dataset

To collect the image data required for the experiment, we constructed a visual acquisition system to obtain the fair-faced concrete bughole dataset, which is shown in Fig. 3. Figure 3 (a) is pier pouring construction tower built around the middle pier, pier is almost completely obscured in the image. Figure 3 (b) is a special casting mold with a transparent detection window (as indicated in the red frame), made of a combination of steel material and acrylic. Figure 3 (c) is camera moving guide rail, fixed on the nearby stairs with nylon ties and (d) is an overall appearance of image acquisition system.

As is shown in Fig. 3, to make the data acquisition system work well in the construction site which is affected by dust, liquid splash, vibration, and other negative factors like inconsistent illumination, shadow, and other noises on the concrete surface, as well as the image data required for the experiment does not require high pixel resolution, we finally adopted an industrial camera to collect images and the brand of the camera is iRAYPLE. During the data collection process, the frame rate for image collection is 30FPS and the size of image is 1280×1024. In addition, to enable the camera to rise on the detection plane of 5 m high, we also built a 5.2-m track, then we used a computer, PLC, and wireless transceiver, as shown in Fig. 3 (d), to control the wireless electric motor to move on the track through RS485 communication protocol. The camera is fixed on the motor to achieve up and down movement, which enabled us to complete the data acquisition. In this process, the distance between the camera lens and the detection plane (the transparent detection window plane) is about 25 cm.

Fig. 3. Illustrations on the actual construction site and image acquisition system settings. (Color figure online)

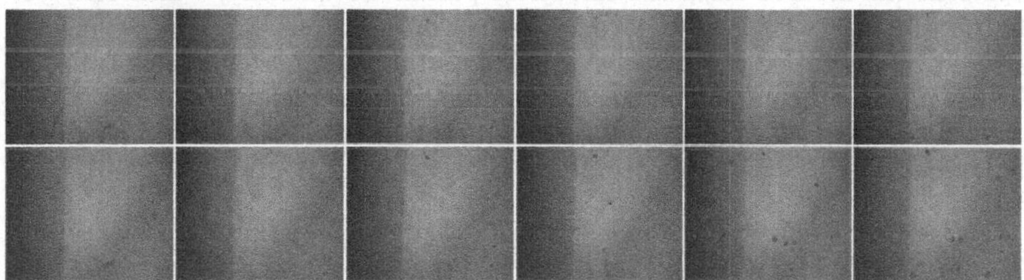

Fig. 4. Part data in dataset (Color figure online)

The concrete building for image collection is a square pier, and its pouring section is a square ring. In the process of cement pouring, the construction sequence is to sprinkle concrete at several points of the circular section in turn and then insert the vibration device for vibration operation in turn. As our target sampling data is the bughole image appearing on the detection plane during the vibration process, We need to control the camera to move up and down to collect as much bughole image data as possible during the vibration process. Finally, we collected video data from a total of 12 vibration processes collected by industrial cameras and mobile phones, with a total duration of 22 min.

Then we proceed with the data processing process, first, we sampled one image per seven frames from these video data and got a total of 980 images. Then Labelme was used to label the masks of the cement area and bughole area, masks in all the images are carefully and accurately labeled manually, and we got the data set for the experiment. Our dataset includes 980 images with more than 2750 bounding boxes, part of the images in dataset are shown in Fig. 4. The green square area is the cement area, and the bughole area is marked with red pixels. It can be observed that the distribution of bubbles is relatively random, with most bubbles having a shape that is approximately circular or elliptical.

3.3 Evaluation Metrics

Positive samples are generally considered in the object detection task if the Intersection over Union (IoU) between the predictions and the labels is greater than the threshold. Otherwise, it is considered a negative sample. These analyses classify detection results as True Positive (TP), False Positive (FP), True Negative (TN), and False Negative (FN). All metrics used in this paper are introduced, including Precision (P), Recall (R), mean Average Precision (mAP), params, and FLOPs.

Precision is the proportion of correctly recognized photos in the validation set that are true positives, whereas recall denotes the percentage of images in the validation set that are correctly detected as positive samples. mAP is the Average Precision (AP) for several categories, and mAP@.5:.95 is the average mAP over various IoU thresholds (from 0.5 to 0.95, step 0.05). Params directly represent the model's size, also known as parameter quantity, which refers to the total number of parameters to be trained during model training. Typically, be used to determine the model's size (computational space complexity). The FLOPs indicate the number of floating point operations (amount of computation in the model), which can be used to measure the calculation time complexity of the model.

3.4 Experimental Results and Analysis

Considering factors such as training duration, optimal convergence state of the model, and existing equipment conditions, we set batch size 48 after several attempts to get better experimental results and the image enhancement settings is mosaic enhancement in the YOLOv5 model. The Stochastic Gradient Descent (SGD) optimizer was used to train network parameters in the training process of the model, lr0 is 0.01, lrf is 0.01, momentum is 0.937 and weight-decay is 5e−4. The weight parameters of the convolutional layer and the deconvolution layer were transferred from YOLOv5 model weight pretrained on the ImageNet dataset. The loss function in network training adopts the BCEWithLogitsLoss and CIoU loss, which are the default loss function in YOLOv5. Besides, details about labels in our self built dataset and anchor settings are shown in Fig. 5.

Figure 5 (a) is the histogram on each class of training set instances, “shuini” is cement area and “qipao” is bughole area. Figure 5 (b) is a schematic diagram of the size of the anchor frame, and the results of many small size boxes in the figure also provide us with an additional basis for proposing a detection scale (DS). Figure 5 (c) is the density distribution diagram of anchor frame center point position and (d) is the density distribution diagram for the ratio of anchor frame width to height to image size. From (c) and (d) it can be observed that the size of the anchor is mainly distributed at the maximum and minimum ends, and the anchor center point locate more in the center of the image, with a relatively uniform distribution around.

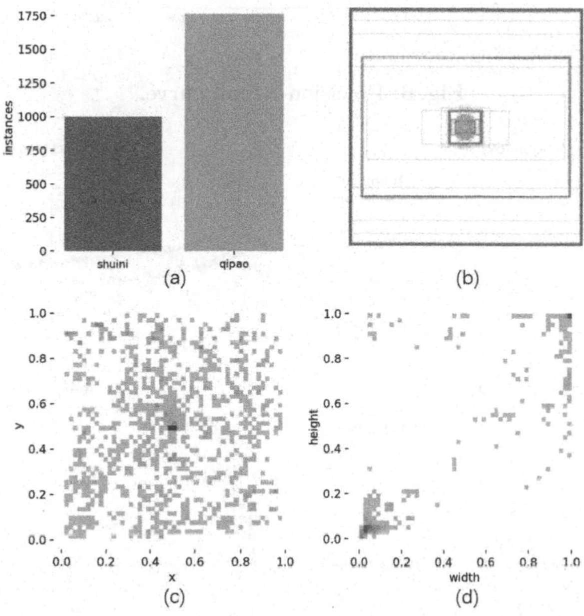

Fig. 5. Labels correlogram.

When training the model, we chose YOLOv5 as our experiments baseline, first, we added the detection scale (DS) into the model, and then the concatenations between the backbone and head part were built. The PR (Precision-Recall) curve and loss curve of the training and validation processes are shown in Fig. 6 and Fig. 7 respectively. As shown in Fig. 7, (a) is location loss, (b) is objectness loss and (c) is classification loss. When the number of iterations is about 500, 3 kind of loss value changes stably. About 560th epoch, loss values have their best convergence effect.

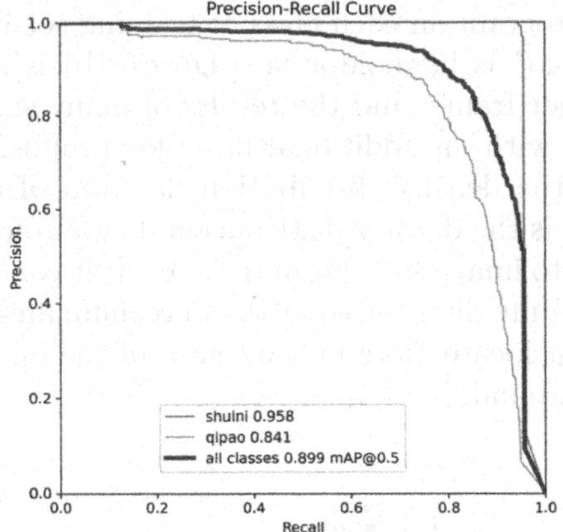

Fig. 6. Precision-Recall curve.

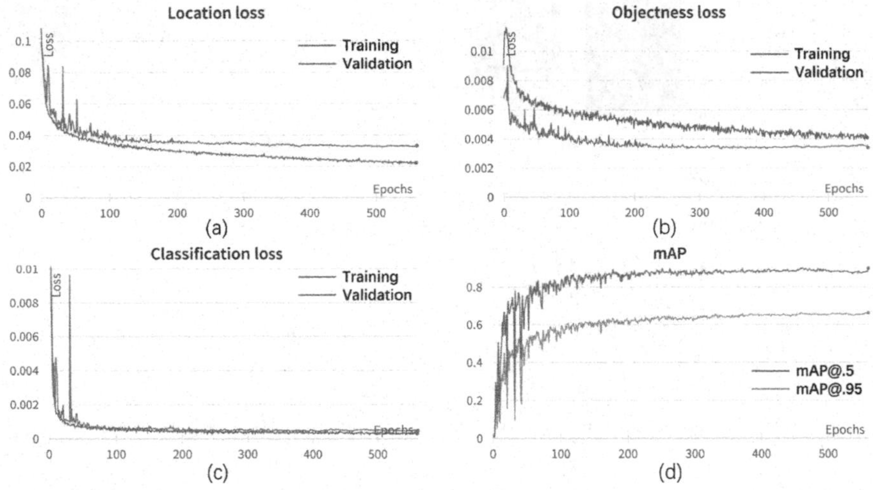

Fig. 7. Loss curve and mAP curve.

Experiment results are shown in Table 1. It can be observed that after inserting DS and concatenation, mAP@0.5 and mAP@.5:.95 achieve considerable improvement by 1.24% and 1.46% respectively. Meanwhile, params increased slightly from 7.03M to 7.19M and GFLOPs also increased a little from 16 GLFOPs to 18.9 GFLOPs. However, considering the small size of the initial model YOLOv5, this increase can be ignored. As shown in table, inference time per image on our proposed model is 13.4 ms (single GPU), this is crucial for us to deploy the trained model in the actual detection scene, which can improve the frame rate in the real-time detection process. As a comparison, method based on Mask R-CNN [10] can achieve pixel level segmentation on one image of 3024 ×

3024 pixels with 3 s time consumption (single GPU), which is obviously too slow to meet the requirement of real-time detection. Comparison of bughole detection result on validation part of self-built dataset is shown in Fig. 8, (a) is the original input concrete surface image, (b) is the groundtruth of bughole defect, it is necessary to note that the manually annotated masks are not entirely accurate that there may be errors or missing annotations about bughole defect, (c) is the result of YOLOv5, (d) is the result of our proposed model. Experiment results show that our model can identify the small defect information with higher accuracy. It is worth mentioning that more details are marked with red dashed box in Fig. 8, we found that our model has the potential to detect bughole that have not been manually labeled. It means that some FP (False-Positive) samples might be TP (True-Positive) samples, but were not manually labeled correctly. This also provided us with ideas for improving the model in the future.

Table 1. Comparison of indicators between YOLOv5 and the network in this paper; DS means detection scale; Concat means concatenation

	DS	Concat	Epoch	Precision	Recall	mAP@0.5	mAP@.5:.95	Params	GFLOPs	ms/img
YOLOv5	×	×	212	0.9292	0.8293	0.8875	0.6426	7.03M	16	10.6
	√	×	521	0.9075	0.8624	0.8963	0.6542	7.19M	18.9	13.4
Ours	√	√	561	0.8956	0.8651	0.8999	0.6572	7.19M	18.9	13.4

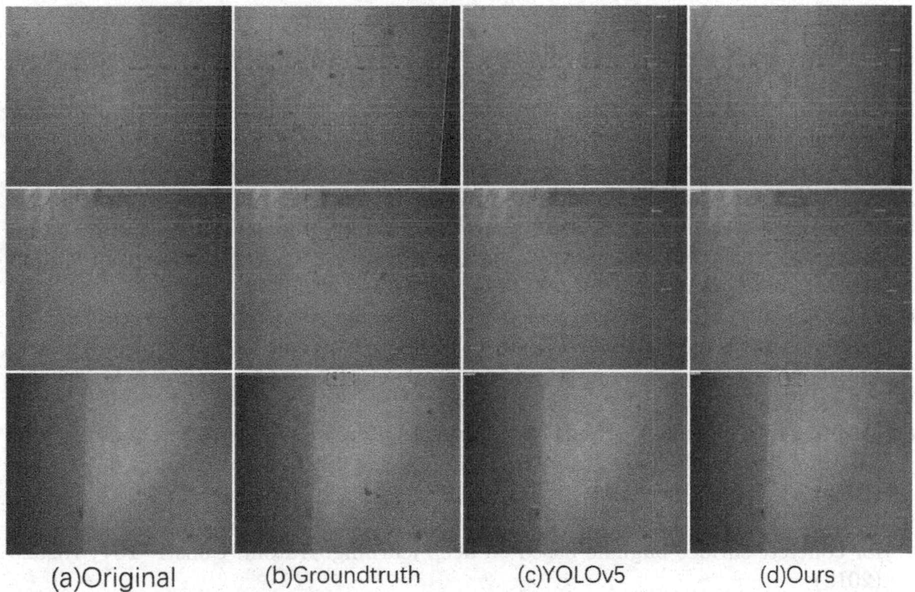

Fig. 8. Comparison of bughole detection result on validation set

4 Conclusion

However, the existing approaches are all used to detect the formed concrete surface, there is a lack of a method that can detect bughole on the concrete surface instantly during the pouring process. This application scenario requires a method with high detection accuracy, fast inference speed and ease of deployment. In this paper, an improved YOLOv5 network model is proposed to solve the problems of the difficult detection of small bughole defects on concrete surface and meet the requirements being used instantly during the pouring process. A detector scale is added to the YOLOv5 model to improve the acquisition of low-level fine-grained features and concatenations are added between the backbone and head parts to achieve an effective fusion of low-level features and high-level features.

The model was trained and tested on the dataset we constructed. The experimental results indicate that our improved network has better defect detection performance in small defects on the concrete surface compared to YOLOv5. At the same time, it has good deployability with approximately 13.4 ms to detect a 1280 × 1024 images on a GPU. Further improving the prediction accuracy of the model and optimizing the time complexity of the algorithm will be the focus of future research.

References

1. Samuelsson, P.: Voids in concrete surfaces. J. Proc. **67**(11), 868–874 (1970)
2. Lemaire, G., Escadeillas, G., Ringot, E.: Evaluating concrete surfaces using an image analysis process. Constr. Build. Mater. **19**(8), 604–611 (2005)
3. Ozkul, T., Kucuk, I.: Design and optimization of an instrument for measuring bughole rating of concrete surfaces. J. Franklin Inst. **348**(7), 1377–1392 (2011)
4. Da Silva, W., Štemberk, P.: Expert system applied for classifying self-compacting concrete surface finish. Adv. Eng. Softw. **64**, 47–61 (2013)
5. Peterson, K., Carlson, J., Sutter, L., et al.: Methods for threshold optimization for images collected from contrast enhanced concrete surfaces for air-void system characterization. Mater. Charact. **60**(7), 710–715 (2009)
6. Zhu, Z., Brilakis, I.: Machine vision-based concrete surface quality assessment. J. Constr. Eng. Manag. **136**(2), 210–218 (2010)
7. Liu, B., Yang, T.: Image analysis for detection of bugholes on concrete surface. Constr. Build. Mater. **137**, 432–440 (2017)
8. Yoshitake, I., Maeda, T., Hieda, M.: Image analysis for the detection and quantification of concrete bugholes in a tunnel lining. Case Stud. Const. Mater. **8**, 116–130 (2018)
9. Wei, F., Yao, G., Yang, Y., et al.: Instance-level recognition and quantification for concrete surface bughole based on deep learning. Autom. Constr. **107**, 102920 (2019)
10. Wei, F., Shen, L., Xiang, Y., et al.: Deep learning-based automatic detection and evaluation on concrete surface bugholes. CMES-Comput. Model. Eng. Sci. **131**(2), 619–637 (2022)
11. Yao, G., Wei, F., Yang, Y., et al.: Deep-learning-based bughole detection for concrete surface image. In: Advances in Civil Engineering 2019 (2019)

12. Sun, Y., Yang, Y., Yao, G., et al.: Autonomous crack and bughole detection for concrete surface image based on deep learning. IEEE Access **9**, 85709–85720 (2021)
13. Wei, W., Ding, L., Luo, H., et al.: Automated bughole detection and quality performance assessment of concrete using image processing and deep convolutional neural networks. Constr. Build. Mater. **281**, 122576 (2021)
14. Fan, J., Huo, T., Li, X.: A review of one-stage detection algorithms in autonomous driving. In: 2020 4th CAA International Conference on Vehicular Control and Intelligence, Hangzhou, China, pp. 210–214. IEEE (2020)
15. Ren, S., He, K., Girshick, R., et al.: Faster R-CNN: towards real-time object detection with region proposal networks. IEEE Trans. Pattern Anal. Mach. Intell. **39**(6), 1137–1149 (2017)
16. Redmon, J., Farhadi, A.: Yolov3: an incremental improvement. arXiv preprint arXiv:1804.02767 (2018)
17. Author, F.: Yolov4: optimal speed and accuracy of object detection. arXiv preprint arXiv:2004.10934 (2020)
18. Liu, S., Qi, L., Qin, H., et al.: Path aggregation network for instance segmentation. In: Proceedings of the IEEE Conference on Computer Vision and Pattern Recognition, Salt Lake City, UT, USA, pp. 8759–8768. IEEE (2018)

UWYOLOX: An Underwater Object Detection Framework Based on Image Enhancement and Semi-supervised Learning

Yue Zhou, Deshun Hu, Cheng Li, and Wangpeng He(✉)

Xidian University, Xi'an, China
{yz,dshu,licheng812}@stu.xidian.edu.cn, hewp@xidian.edu.cn

Abstract. Deep learning-based underwater optical object detection is of great significance to underwater environment exploration and underwater scientific research. However, the quality of underwater images obtained by optical imaging devices is poor and the available labeled data is lacking lead to lower mean Average Precision of underwater object detection. Therefore, in this paper, an novel underwater object detection framework, UWYOLOX, is proposed to solve the problems mentioned above. The key is to present a joint learning-based underwater image enhancement module (JLUIE) and an improved semi-supervised learning method USTAC (Underwater STAC) for underwater object detection. Firstly, JLUIE and YOLOX-Nano share the detection loss for training, so that JLUIE can adaptively enhance each image for better detection performance. Then, USTAC, which can make full use of unlabeled data for training, is introduced to further improve the mean Average Precision of object detection. Experimental results show that the mAP of UWYOLOX is 1.77% higher than YOLOX-nano, which demonstrates the effectiveness of UWYOLOX. Moreover, USTAC can improve the mAP of the model obtained by supervised learning using 800 labeled data by 7.46%. Thus the effectiveness of USTAC is further demonstrated.

Keywords: Underwater object detection · Image enhancement · Joint learning · Semi-supervised learning · YOLOX-Nano

1 Introduction

With the development of economy and the increasing lack of resources, underwater environment exploration and underwater scientific research have attracted much attention in recent years. Deep learning-based underwater optical object detection is of great significance to underwater exploration and research. Due to the complexity and diversity of underwater environment, underwater object detection faces more difficulties and challenges than general object detection. In recent years, many works have been carried out to improve the performance of underwater object detection. For example, Chia-Hung Yeh et al. [1] proposed a depth model for jointly learning color conversion of underwater images and object detection, where the purpose of the color conversion module is to convert a color image into a corresponding grayscale image, thereby reducing the impact of

H. Zhang et al. (Eds.): NCAA 2023, CCIS 1870, pp. 32–45, 2023.
https://doi.org/10.1007/978-981-99-5847-4_3

color absorption on object detection. Fenglei Han et al. [2] proposed a method combining max-RGB and Gray World to enhance underwater images, and then proposed a method based on convolutional neural network to solve the problem of weak illumination, so as to improve the performance of object detection. Tien-Szu Pan et al. [3] proposed the Multi-scale ResNet (M-ResNet) in order to solve the problem of large-scale variation in underwater object detection, which uses multi-scale operation to accurately detect objects of various sizes. Long Chen et al. [4] proposed the Sample-WeIghted hyPEr Network (SwipeNet) to solve the problem that some objects in the underwater scene are small and fuzzy. SwipeNet is composed of high-resolution and semantic-rich super feature maps, which can significantly improve the detection accuracy of small objects. In addition, Wei-Hong Lin et al. [5] proposed a data augmentation strategy called ROIMIX to solve the problem that underwater organisms are close to each other or even overlap. ROIMIX simulates the characteristics of underwater organisms such as overlap, occlusion and blur, which makes the model more robust.

Moreover, the performance of deep learning-based object detectors depends on the amount of the labeled data [6]. However, the cost of manually labeling underwater images is very high. Therefore, studying how to use unlabeled data to improve the performance of underwater object detectors is of great importance. In recent years, semi-supervised learning (SSL) has received more and more attention. The advantage of SSL is that it can use unlabeled data for training, which can improve the performance of the model in the case of lacking labeled data, and greatly reduce the dependence of deep learning-based model on labeled data. At present, most of the research on semi-supervised learning focuses on image classification [7], such as the two popular SSL methods based on pseudo label and consistency regularization [8–11], and the SSL method combining pseudo label and consistency regularization [7]. Learning from the experience of classification, the research on semi-supervised object detection is also developing. Jisoo Jeong et al. [12] proposed a consistency-based semi-supervised object detection method (CSD), similar to consistency regularization (CR) in semi-supervised image classification, which uses consistency constraints as a tool to improve the detection performance by making full use of the available unlabeled data. Kihyuk Sohn et al. [6] proposed a semi-supervised object detection framework (STAC) combining self-training and consistency regularization. STAC has two stages of training inspired by Noisy Student [11], and is combined with Faster R-CNN, which achieves good results on MSCOCO dataset. Qiang Zhou et al. [13] proposed a fully end-to-end semi-supervised object detection framework (Instant-Teaching), which generate pseudo labels in real time in training.

In the existing underwater object detection methods, image enhancement and semi-supervised learning are not fully used to improve the performance of underwater object detection. Therefore, an underwater object detection framework UWYOLOX is proposed in this paper, in which a joint learning-based underwater image enhancement module (JLUIE) and an improved semi-supervised learning method for underwater object detection (USTAC), which is based on a general semi-supervised learning method STAC, are introduced into YOLOX-Nano [14]. Specifically, JLUIE consists of a CNN-based image enhancement module (Pe-PP) and a parameter prediction module (IE), where Pe-PP can adaptively predict the hyperparameters of IE so that IE can adaptively enhance each image. Then, JLUIE is trained jointly with YOLOX-Nano to guarantee the enhanced information beneficial to object detection. In addition, in USTAC, Mosaic [15] data augmentation is introduced into the semi-supervised learning method STAC, and the semi-supervised learning method is combined with the single-stage anchor-free object detector YOLOX, which achieves encouraging results in underwater object detection. Moreover, with regard to the deployment requirements, YOLOX-Nano is choosed as the base model, which is equipped with the most advanced technologies, such as decoupled head, Anchor-Free, and SimOTA, and performs well in small-size models. To sum up, the contributions of this paper can be summarized as the following three points:

1. A joint learning-based image enhancement module (JLUIE) is proposed, which can adaptively enhance each image for better detection performance.
2. A novel parameter prediction module based on PeleeNet [16] is proposed, which can predict the hyperparameters of the image enhancement module (IE).
3. Based on STAC, an improved semi-supervised learning method for underwater object detection (USTAC) is proposed, which can make full use of unlabeled data for training.

2 UWYOLOX

This section details the proposed underwater object detection framework UWYOLOX. This framework introduces a joint learning-based image enhancement module (JLUIE) and an improved semi-supervised learning method for underwater object detection (USTAC) on the basis of YOLOX-Nano. The structure of UWYOLOX is shown in Fig. 1. First, according to USTAC, the labeled and unlabeled images are put into the same batch as the input, which is divided into two paths. The images in one path is resized to 416 × 416 as the input of the image enhancement module (IE), and the images in the other path is resized to 304 × 304 as the input of the parameter prediction module (Pe-PP). IE and Pe-PP form the JLUIE. Second, the hyperparameters required by IE are predicted by Pe-PP according to the input images, and then IE enhances the images in a batch. Next, the enhanced images are input into YOLOX-Nano for object detection. The supervised and unsupervised loss are calculated respectively by using the detection results of the labeled and unlabeled images, and the total detection loss is the weighted sum of the supervised loss and the unsupervised loss. Finally, backpropagation is performed according to the total detection loss, and the parameters of Pe-PP and YOLOX-Nano are updated.

2.1 Joint Learning-Based Image Enhancement Module (JLUIE)

In underwater scenes, due to the scattering and absorption of light by water, the images obtained by optical imaging devices are often of poor quality, which makes object detection more difficult. To solve this problem, a joint learning-based image enhancement module (JLUIE) is proposed. The module includes a parameter prediction module (Pe-PP) and an image enhancement module (IE) as shown in Fig. 1. Pe-PP adaptively predicts the hyperparameters required by IE for each image. Then, IE enhances the underwater images, and the processed images are used as the input of YOLOX-Nano. By jointly training, that is, take the detection loss as the total loss, the Pe-PP which can predict the

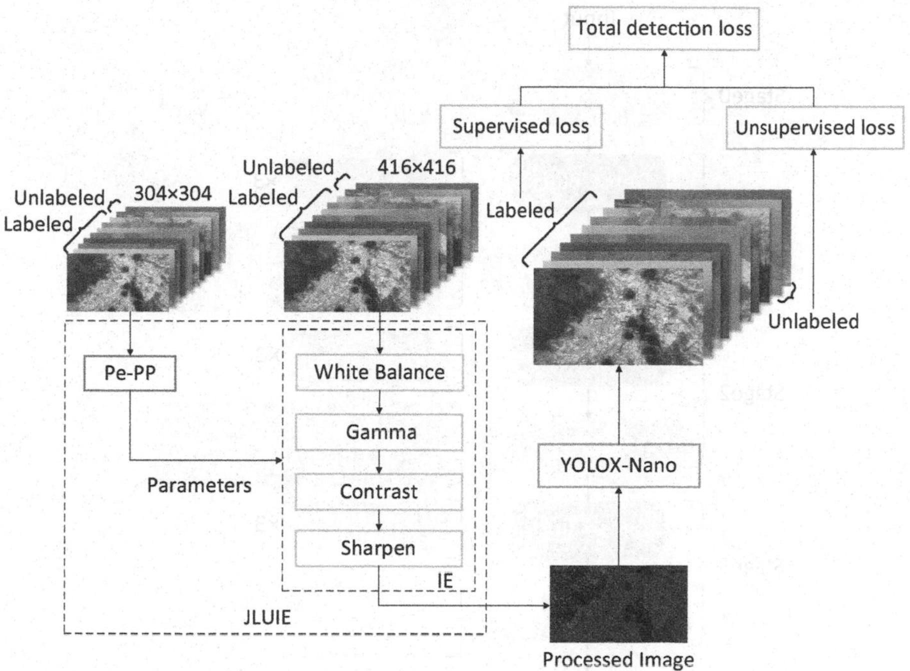

Fig. 1. The structure of JLUIE

appropriate image enhancement hyperparameters is finally obtained, so that JLUIE can enhance the potential information in the image which is beneficial to object detection.

Parameter Prediction Module (Pe-PP). In general, it is necessary to manually optimize the hyperparameters according to the characteristics of different images based on rich experience when performing image enhancement such as White Balance, Gamma Correction, Contrast Adjustment and Sharpening. This method is not only time-consuming and labor-intensive, but also not suitable for the image enhancement module embedded in the object detection system. Therefore, a convolutional neural network, Peleenet, is introduced to predict the hyperparameters required by IE for different underwater images.

Peleenet is an efficient network built by traditional convolutions. This network is a variant of DenseNet [17] proposed by Robert J. Wang et al. The structure of PeleeNet is shown in Fig. 2. The whole network is composed of five stages. Stage 0 consists of a Steam module, whose structure is shown in Fig. 3(a). This structure can effectively improve the feature expression ability of the network without increasing the amount of calculation. Stages 1, 2, 3 and 4 are composed of Dense module and Transition layer, which are responsible for feature extraction. The Dense module is composed of multiple Dense layers. The structure of one Dense layer is shown in Fig. 3(b). A two-way structure is used to obtain receptive fields of different scales: one way uses a 3 × 3 convolution, and the other way uses two stacked 3 × 3 convolutions, which is helpful to learn the visual features of large objects.

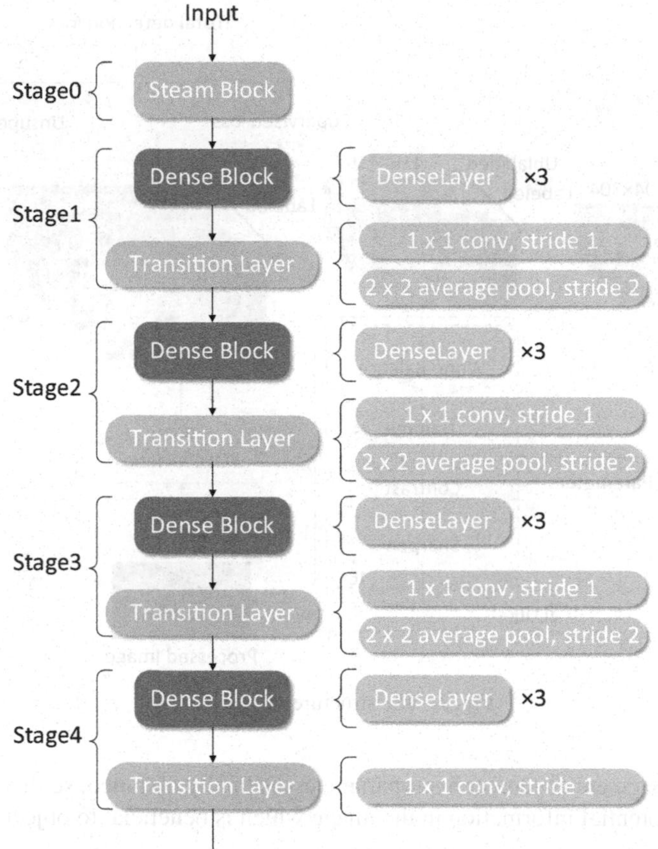

Fig. 2. The structure of Pe-PP

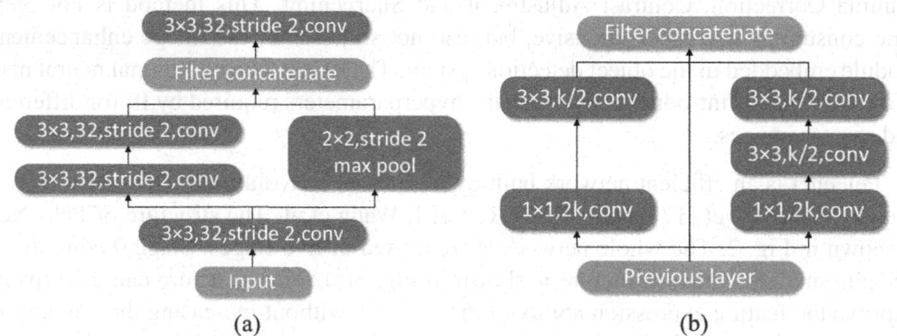

Fig. 3. (a) The structure of Steam module. (b) The structure of Dense layer.

Image Enhancement Module (IE). Four image filters, White Balance, Gamma, Contrast and Sharpen, in [18] are selected to enhance the underwater images, and their effectiveness for underwater object detection has been demonstrated in this paper. The above four filters are all differentiable to ensure that Pe-PP can be trained by backpropagation. In addition, the hyperparameters in the above four image filters are independent of the resolution of the image to be processed, so that the resolution of the input images of Pe-PP can be resized to 304 × 304 to save computing resources, and the actual resolution of the images processed by IE is 416 × 416, which is consistent with that of the input images of YOLOX-Nano.

The mapping function for the White Balance image filter is

$$\mathrm{P_o} = (\mathrm{W_r r_i}, \mathrm{W_g g_i}, \mathrm{W_b b_i}) \tag{1}$$

where $P_i = (r_i, g_i, b_i)$ is the value of input pixel and $P_o = (r_o, g_o, b_o)$ is the value of output pixel. *(r,g,b)* represent the values of the red, green, and blue color channels, respectively. W_r, W_g, W_b are the coefficients of the three color channels of red, green and blue, and are the corresponding multiplication relationship with r_i, g_i, b_i. The mapping is a multiplicative transformation.

The mapping function of the Gamma image filter is

$$\mathrm{P_o} = \mathrm{P_i^G} \tag{2}$$

where P_i is the value of input pixel, P_o is the value of output pixel, and G is the value of Gamma. The mapping is a power transformation.

The mapping function of the Contrast image filter is

$$\mathrm{P_o} = \alpha \cdot \mathrm{En(P_i)} + (1 - \alpha) \cdot \mathrm{P_i} \tag{3}$$

where P_i is the value of input pixel, P_o is the value of output pixel, and α is a linear interpolation between the original image and the fully enhanced image. In the formula (3), $En(P_i)$ is defined as follows:

$$\mathrm{En(P_i)} = \mathrm{P_i} \times \frac{\mathrm{EnLum(P_i)}}{\mathrm{Lum(P_i)}} \tag{4}$$

where

$$\mathrm{Lum(P_i)} = 0.27\mathrm{r}_i + 0.67\mathrm{g_i} + 0.06\mathrm{b_i} \tag{5}$$

$$\mathrm{EnLum(P_i)} = \frac{1}{2}(1 - \cos(\pi \times (\mathrm{Lum(P_i)}))) \tag{6}$$

According to the unsharpen mask technique [19], the mapping function of the Sharpen image filter is

$$\mathrm{F}(\mathrm{x}, \lambda) = \mathrm{I}(\mathrm{x}) + \lambda(\mathrm{I}(\mathrm{x}) - \mathrm{Gau}(\mathrm{I}(\mathrm{x}))) \tag{7}$$

where *I(x)* is the input image, *Gau(I(x))* is the Gaussian filter, and λ is a positive scale factor. The degree of sharpening can be adjusted by changing the value of λ.

The above four image filters are connected in series in the order of White Balance, Gamma, Contrast and Sharpen to form an image enhancement module (IE), as shown in Fig. 1. The hyperparameters corresponding to each image filter are shown in Table 1. White Balance image filter has 3 hyperparameters, Gamma image filter has 1 hyperparameter, Contrast image filter has 1 hyperparameter, and Sharpen image filter has 1 hyperparameter, totaling 6 hyperparameters. The specific value of the hyperparameters corresponding to different input images are predicted by Pe-PP.

Table 1. The hyperparameters corresponding to different image filters

Filter	Parameters
White Balance	W_r, W_g, W_b
Gamma	G
Contrast	α
Sharpen	λ

Joint Learning. In this paper, a new solution, joint learning, is proposed to the problem, "how image enhancement contributes to deep learning-based underwater object detection". The joint learning-based image enhancement module (JLUIE) and the object detection network (YOLOX-Nano) are trained as a whole, and the detection loss is used as the total loss. The purpose of joint learning is to iteratively optimize Pe-PP with the goal of minimizing the detection loss, so that the trained Pe-PP can predict the appropriate hyperparameters for IE, which makes the IE module be able to enhance the potential information beneficial to object detection. Experimental results show that the joint learning-based image enhancement module (JLUIE) can effectively improve the mean Average Precision of the object detector (YOLOX-Nano).

2.2 Improved Semi-supervised Learning Method for Underwater Object Detection (USTAC)

In this paper, an improved semi-supervised learning method based on STAC for underwater object detection is proposed, which combines pseudo label-based self-training and data augmentation-driven consistency regularization. The structure of USTAC is shown in Fig. 4, which includes three stages. In Stage 1, all available labeled data are used for training to obtain a teacher model, and supervised loss is used. In Stage 2, the teacher model obtained in Stage 1 is used for predicting the pseudo labels of unlabeled data. The pseudo label includes a bounding box and a class label, and the format of the pseudo label is consistent with that of manual label. In Stage 3, both labeled data and unlabeled data are used to train. In the training process, two data augmentation strategies, Mosaic and random affine transformation, are used for labeled data. Global color transformation, CoarseDropout, Mosaic and random affine transformation are used to augment unlabeled data (images and pseudo label). Finally, the network is optimized by minimizing the weighted sum of the supervised loss and the unsupervised loss. USTAC has two hyperparameters, *conf* (confidence threshold) and λ_u (unsupervised loss weight).

The confidence of the final retained pseudo label can control by adjusting *conf*, and the weight of the unsupervised loss in the total loss can be changed by adjusting λ_u.

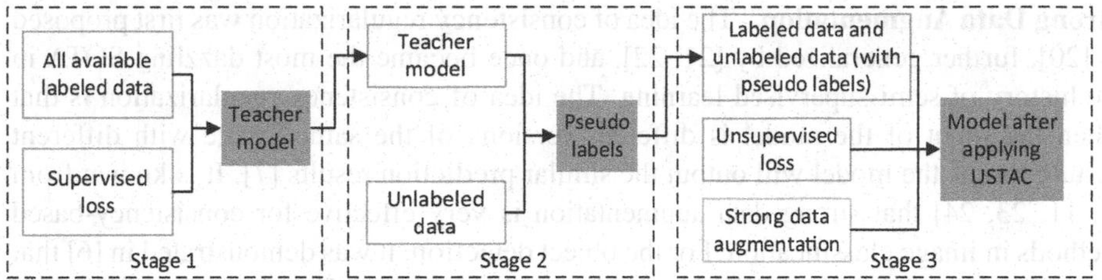

Fig. 4. The structure of USTAC

Training a Teacher Model. As shown in Fig. 4, in Stage 1, all available labeled data is used to train, and the trained model is used as the teacher model. The loss used in this phase of training is the supervised loss, that is, the detection loss of YOLOX-Nano, which is defined as follows:

$$\mathrm{Loss_s} = \mathrm{l_s(x_l, c^*, r^*, o^*)} = \frac{\mathrm{L_{cls}(c, c^*)} + \lambda \mathrm{L_{reg}(r_i, r_i^*)} + \mathrm{L_{obj}(o_i, o_i^*)}}{\mathrm{N_{pos}}} \tag{8}$$

where x_l is the labeled image, c^* is the ground-truth class probability, r^* is the ground-truth box coordinates, and o^* is the ground-truth confidence. L_{cls} represents the classification loss, L_{reg} represents the localization loss, L_{obj} represents the obj loss, λ represents the balance coefficient of the localization loss, and N_{pos} represents the number of Anchor Points assigned as positive samples.

Generating Pseudo Labels. The unlabeled data is input into the teacher model for prediction, and the corresponding pseudo label is generated. The prediction process here includes backbone, FPN, the forward propagation of head, and the post-processing process. The post-processing process of YOLOX-Nano is mainly divided into two steps: the first step is to perform the first round of screening on the bounding boxes by using the confidence, that is, filter out the bounding boxs with confidence less than the confidence threshold. The second step is to perform Non-Maximum Suppression (NMS) on the output of the first step to delete the repeated bounding boxs for the same object. The confidence in the first step is defined as follows:

$$\mathrm{conf}_i = \mathrm{obj\_conf_i} \times \mathrm{class\_conf_i} \tag{9}$$

where i is an index of a bounding box. $Conf_i$ represents the confidence of a bounding box, obj_conf_i represents the probability that there is an object in the bounding box, and $class\_conf_i$ represents the probability that the object in the bounding box is a certain class. The bounding boxs are filtered by setting a confidence threshold. For example, if the $conf_i$ of a bounding box is greater than the confidence threshold, the bounding box is retained, otherwise it is discarded. The *conf* in USTAC is the confidence threshold here. Therefore, in this paper, the retention of pseudo labels is adjusted by changing the value of *conf*. The larger the value of *conf* is, the higher the standard for the prediction result to be retained as a pseudo label is, and the higher the confidence of the reserved pseudo label is.

Strong Data Augmentation. The idea of consistency regularization was first proposed in [20], further generalized by [21, 22], and once became the most dazzling SOTA in the history of semi-supervised learning. The idea of consistency regularization is that when the input of the model is different versions of the same image with different perturbations, the model will output the similar prediction results [7]. It is known from [7, 11, 23, 24] that strong data augmentation is very effective for consistency-based methods in image classification. For the object detection, it was demonstrated in [6] that strong data augmentation is still effective for consistency-based methods. In this paper, a group of effective data augmentation strategies are determined for strong augmentation in underwater scenes: global color transformation, CoarseDropout, Mosaic, and affine transformation.

1. Global color transformation: Equalize, Solarize, RandomBrightness, Contrast, Sharpen, Posterize.
2. CoarseDropout: Randomly set the pixels of the rectangular area in the image to zero.
3. Mosaic: Four images are spliced into one image as training data.
4. Affine transformation: Rotation, Scale, Shear, Translation.

Each image and corresponding pseudo labels are transformed in the following order: firstly, random global color transformation is carried out according to a certain probability. Then, CoarseDropout is applied. Next, Mosaic is carried out, that is, three images are randomly selected and spliced with the current traversal image into the same image. Finally, a random affine transformation is performed on the Mosaic processed image according to a certain probability.

Unsupervised Loss. In Stage 3, the total loss which is the weighted sum of the supervised loss and the unsupervised loss is used for training, where the unsupervised loss is defined as follows:

$$\mathrm{Loss_u} = \mathrm{l_u}(\mathrm{A}(\mathrm{x_u}, \mathrm{r_u^*}), \mathrm{c_u^*}, \mathrm{o_u^*}) = \mathrm{l_s}(\mathrm{x_A}, \mathrm{r_A^*}, \mathrm{c_u^*}, \mathrm{o_u^*}) \tag{10}$$

where x_u represents the unlabeled image, A represents the strong data augmentation, and ${c_u}^*$, ${r_u}^*$, ${o_u}^*$ respectively represents the ground-truth class probability, the ground-truth box coordinates, and the ground-truth confidence, which obtained by the pseudo label. x_A, ${r_A}^*$ respectively represents the unlabeled image and the pseudo bounding box after applying strong data augmentation. $l_u(A(x_u, r_u^*), c_u^*, o_u^*) = l_s(x_A, r_A^*, c_u^*, o_u^*)$ gives the relationship between the supervised loss and the unsupervised loss.

Therefore, the total loss of USTAC is:

$$\mathrm{Loss} = \mathrm{Loss_s} + \mathrm{Loss_u} = \mathrm{l_s}(\mathrm{x_l}, \mathrm{c}^*, \mathrm{r}^*, \mathrm{o}^*) + \lambda_\mathrm{u}\mathrm{l_u}(\mathrm{A}(\mathrm{x_u}, \mathrm{r}^*), \mathrm{c}^*, \mathrm{o}^*) \tag{11}$$

where $Loss_s$ and $Loss_u$ represents the supervised loss and the unsupervised loss respectively. λ_u represents the weight of the unsupervised loss.

3 Experiments

All experiments in this paper are based on the URPC2022 dataset, which is the official dataset of the Underwater Perception Competition of the 2022 National Underwater Robot Competition. The URPC2022 dataset consists of a training set and a testing set. There are 9000 images in the training set (labeled data) and 1500 images in the testing set (unlabeled data). And five categories of objects are labeled in the training set, which are "holothurian", "echinus", "scallop", "starfish" and "waterweeds". The "waterweeds" is ignored in the experiment of this paper. In addition, 9000 labeled data in the training set are divided into three parts of "train", "val" and "test" according to 7:2:1, which are used to train, validate and test respectively.

3.1 Implementation Details

In this paper, Pytorch is used for experiments and a 3080Ti GPU is used for training. The default experimental protocol of YOLOX-Nano is used. In the training process, the resolution of the images inputed to YOLOX-Nano was adjusted to 416 × 416, and the default multi-scale training strategy and data augmentation strategy of YOLOX-Nano are used. CSPDarknet, the backbone of YOLOX-Nano, is initialized with the pre-training weights on the MSCOCO dataset. Train the model for 100 epochs with batchsize set to 8. Except for special instructions, all experimental settings in this paper are consistent with those in this section.

3.2 Experiment Results

UWYOLOX. In this section, the proposed methods is combined with the YOLOX-Nano and the effectiveness of JLUIE and USTAC is demonstrated. Figure 5 shows the experimental results before and after introducing the proposed methods into YOLOX-Nano. It can be found from Fig. 5 that after introducing JLUIE, the mAP@0.5 is increased by 1.58%, and the mAP@0.5:0.95 is increased by 1.02%, which demonstrates the effectiveness of JLUIE for improving the performance of underwater object detection. Then, introduce USTAC on the basis of introducing JLUIE, which increases the mAP@0.5 by 0.19% and increases the mAP@0.5:0.95 by 0.77%. It can be found that USTAC can improve the mAP, but the effect is not obvious. This is because the performance of USTAC is limited here for the amount of unlabeled data is only 1500 in URPC2022 dataset. In the latter experiments, the amount of unlabeled data and labeled data is adjusted, which further demonstrates the effectiveness of USTAC.

JLUIE. In this section, ablation experiments are conducted to demonstrate the effectiveness of Pe-PP and the four image filters in JLUIE. Table 2 records the results of the ablation experiments. A total of 7 experiments were conducted, which are represented by numbers 1–7. In particular, the parameter prediction module used in Experiments 1–6 is CNN-PP, a simple convolutional neural network proposed in [18]. Experiment 1 was performed with the original YOLOX-Nano, and the results were used as the baseline for the ablation experiments. In Experiments 2, 3, 4 and 5, the White Balance, Gamma, Contrast and Sharpen image filters in IE are retained respectively. By

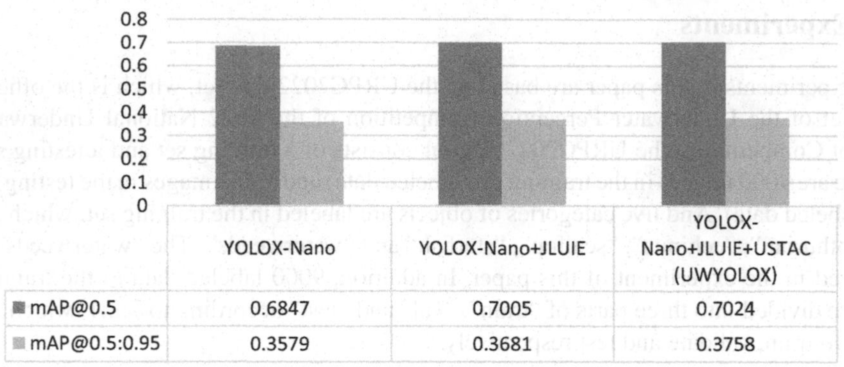

Fig. 5. The mAP of YOLOX-Nano before and after introducing the proposed methods

observing the experimental results, it can be found that each image filter can improve the mAP when used alone, which demonstrates the effectiveness of each image filter in JLUIE. However, the increase of mAP@0.5 and mAP@0.5:0.95 is not obvious. In Experiment 6, the mAP using four image filters in series is significantly better than using only one image filter. Compared with the baseline, the mAP@0.5 is improved by 1.34%, and the mAP@0.5:0.95 is improved by 0.74%, which demonstrates that the performance of underwater object detection can be significantly improved by performing four image enhancement operations in turn. Experiment 7 used Pe-PP as the parameter prediction module, compared with CNN-PP (Experiment 6), the mAP@0.5 increased by 0.24%, the mAP@0.5:0.95 increased by 0.28%, which proved that Pe-PP module is more suitable for parameter prediction. In summary, ablation experiments demonstrate the effectiveness of each component in JLUIE.

Table 2. The results of ablation experiments related to Pe-PP and the four image filters in JLUIE

	Pe-PP	WB	G	C	S	mAP@0.5	mAP@0.5:0.95
1						0.6847	0.3579
2		√				0.6895	0.3621
3			√			0.6889	0.3624
4				√		0.6877	0.3581
5					√	0.6909	0.3593
6		√	√	√	√	0.6981	0.3653
7	√	√	√	√	√	**0.7005**	**0.3681**

USTAC. In this section, the effectiveness of USTAC is further demonstrated by comparing the results of supervised learning and semi-supervised learning (using USTAC) under three different data dividing schemes. Then, the effectiveness of Mosaic data augmentation strategy in USTAC is demonstrated by ablation experiments.

Dataset Dividing. The UCPR2022 dataset includes 9000 labeled data and 1500 unlabeled data. In this paper, the 9000 labeled data are further divided into three parts: "train", "val" and "test", which contain 6300, 800 and 1800 images respectively. In this section, 6300 labeled data in "train" and 1500 unlabeled data in UCPR2022 dataset are merged (7800 in total), and they are divided into labeled data and unlabeled data according to different schemes, of which labeled data only comes from "train".

The three dividing schemes are as follows:

1. 6300 (labeled) + 1500 (unlabeled)
2. 3900 (labeled) + 3900 (unlabeled)
3. 800 (labeled) + 7000 (unlabeled)

Hyperparameter Setting. USTAC has two hyperparameters: confidence threshold (*conf*) and the weight of unsupervised loss (λ_u). Considering that the accuracy of the teacher model is low due to limited training data (up to 6300, down to 800), *conf* was set to 0.3 in order to retain more bounding boxes predicted by the teacher model. In the case of a certain *conf*, the optimal value of λ_u under different data dividing schemes are found.

Analysis of Results. Table 3 records the experimental results under the three data dividing schemes, and the experimental results are obtained with the optimal value of λ_u under each data dividing scheme. When the dividing scheme is 1, 2 and 3, the USTAC improves the mAP@0.5 by 0.19%, 2.52% and 7.46% respectively compared with the supervised learning. By comparing the experimental results of the three data dividing schemes, it is found that the performance of USTAC is significantly improved with the increase of the ratio of the amount of unlabeled data to the amount of labeled data. It can be concluded that the performance of USTAC is related to the ratio of the amount of unlabeled data to the amount of labeled data, and the larger the ratio, the better the performance of USTAC.

Table 3. Experimental results of supervised learning and semi-supervised learning (using USTAC) under three data dividing schemes

Dividing sheme	Unlabeled data/labeled data	Supervised (mAP@0.5)	USTAC (mAP@0.5)	λ_u
1	0.24	0.7005	0.7024	0.03
2	1	0.6583	0.6835	0.5
3	8.75	0.5351	0.6097	2

Table 4 records the experimental results of USTAC with and without Mosaic data augmentation under the data dividing scheme 2. Compared with supervised learning, USTAC can improve the mAP@0.5 by 0.74% and the mAP@0.5:0.95 by 0.83% without Mosaic data augmentation. When using Mosaic data augmentation, USTAC can improve the mAP@0.5 by 2.52% and the mAP@0.5:0.95 by 1.64%. It can be concluded that introducing Mosaic data augmentation to the process of strong augmentation for pseudo labels can effectively improve the performance of USTAC.

Table 4. Experimental results of USTAC with and without Mosaic data augmentation

Method	mAP@0.5	mAP@0.5:0.95
Supervised	0.6583	0.3401
USTAC (without Mosaic)	0.6657	0.3484
USTAC (with Mosaic)	**0.6835**	**0.3565**

4 Discussion and Conclusion

In this paper, a novel underwater object detection framework, UWYOLOX, is proposed. This framework introduces a joint learning-based image enhancement module (JLUIE) and an improved semi-supervised learning method for underwater object detection (USTAC), and the experimental results demonstrate the effectiveness of the proposed methods. Moreover, JLUIE and USTAC proposed in this paper can also be combined with other object detectors besides YOLOX-Nano.

To further improve the performance of underwater object detection, future work will focus on how to make full use of image enhancement, semi-supervised learning and unsupervised learning.

Acknowledgements. This research is supported financially by the Natural Science Basic Research Program of Shaanxi (Grant No. 2023JCYB289, 2022JQ412), the Fundamental Research Funds for the Central Universities (Grant No. ZYTS23102), the Shaanxi Key Laboratory Open Project (Grant No. 300102253508) and the National Natural Science Foundation of China (Grant No. 52175112).

References

1. Yeh, C.H., Lin, C.H., Kang, L.W., et al.: Lightweight deep neural network for joint learning of underwater object detection and color conversion. IEEE Trans. Neural Networks Learn. Syst. **33**(11), 6129–6143 (2021)
2. Han, F., Yao, J., Zhu, H., et al.: Underwater image processing and object detection based on deep CNN method. J. Sens. (2020)
3. Pan, T.S., Huang, H.C., Lee, J.C., et al.: Multi-scale ResNet for real-time underwater object detection. SIViP **15**, 941–949 (2021)
4. Chen, L., Liu, Z., Tong, L., et al.: Underwater object detection using Invert Multi-Class Adaboost with deep learning. In: 2020 International Joint Conference on Neural Networks (IJCNN), pp. 1–8. IEEE (2020)
5. Lin, W.H., Zhong, J.X., Liu, S., et al.: RoIMix: proposal-fusion among multiple images for underwater object detection. In: ICASSP 2020–2020 IEEE International Conference on Acoustics, Speech and Signal Processing (ICASSP), pp. 2588–2592. IEEE (2020)

6. Sohn, K., Zhang, Z., Li, C.L., et al.: A simple semi-supervised learning framework for object detection. arXiv preprint arXiv:2005.04757 (2020)
7. Sohn, K., Berthelot, D., Carlini, N., et al.: Fixmatch: Simplifying semi-supervised learning with consistency and confidence. In: Advances in Neural Information Processing Systems, vol. 33 (2020)
8. Reddy, Y., Viswanath, P., Reddy, B.E.: Semi-supervised learning: a brief review. Int. J. Eng. Technol. **7**(1.8), 81 (2018)
9. Van, Engelen, J.E., Hoos, H.H.: A survey on semi-supervised learning. Mach. learn. **109**(2), 373–440 (2020)
10. Lee, D.H.: Pseudo-label: The simple and efficient semi-supervised learning method for deep neural networks. Workshop Challenges Represent. Learn. ICML **3**(2), 896 (2013)
11. Xie, Q., Luong, M.T., Hovy, E., et al.: Self-training with noisy student improves imagenet classification. In: Proceedings of the IEEE/CVF Conference on Computer Vision and Pattern Recognition, pp. 10687–10698 (2020)
12. Jeong, J., Lee, S., Kim, J., et al.: Consistency-based semi-supervised learning for object detection. In: Advances in Neural Information Processing Systems, vol. 32 (2019)
13. Zhou, Q., Yu, C., Wang, Z., et al.: Instant-teaching: an end-to-end semi-supervised object detection framework. In: Proceedings of the IEEE/CVF Conference on Computer Vision and Pattern Recognition, pp. 4081–4090 (2021)
14. Ge, Z., Liu, S., Wang, F., et al.: Yolox: exceeding yolo series in 2021. arXiv preprint arXiv: 2107.08430 (2021)
15. Bochkovskiy, A., Wang, C.Y., Liao, H.Y.M.: Yolov4: optimal speed and accuracy of object detection. arXiv preprint arXiv:2004.10934 (2020)
16. Wang, R.J., Li, X., Ling, C.X.: Pelee: a real-time object detection system on mobile devices. In: Advances in Neural Information Processing Systems, vol. 31 (2018)
17. Huang, G., Liu, Z., Van, Der, Maaten, L., et al.: Densely connected convolutional networks. In: Proceedings of the IEEE Conference on Computer Vision and Pattern Recognition, pp. 4700–4708 (2017)
18. Liu, W., Ren, G., Yu, R., et al.: Image-adaptive YOLO for object detection in adverse weather conditions. In: Proceedings of the AAAI Conference on Artificial Intelligence, pp. 1792–1800 (2022)
19. Polesel, A., Ramponi, G., Mathews, V.J.: Image enhancement via adaptive unsharp masking. IEEE Trans. Image Process. **9**(3), 505–510 (2000)
20. Bachman, P., Alsharif, O., Precup, D.: Learning with pseudo-ensembles. In: Advances in Neural Information Processing Systems, vol. 27 (2014)
21. Sajjadi, M., Javanmardi, M., Tasdizen, T.: Regularization with stochastic transformations and perturbations for deep semi-supervised learning. In: Advances in Neural Information Processing Systems, vol. 29 (2016)
22. Laine, S., Aila, T.: Temporal ensembling for semi-supervised learning. arXiv preprint arXiv: 1610.02242 (2016)
23. Berthelot, D., Carlini, N., Cubuk, E.D., et al.: Remixmatch: semi-supervised learning with distribution alignment and augmentation anchoring. arXiv preprint arXiv:1911.09785 (2019)
24. Xie, Q., Dai, Z., Hovy, E., et al.: Unsupervised data augmentation for consistency training. In: Advances in Neural Information Processing Systems, vol. 33 (2020)

A Lightweight Sensor Fusion for Neural Visual Inertial Odometry

Yao Lu[1], Xiaoxu Yin[2], Feng Qin[3], Ke Huang[4], Menghua Zhang[1(✉)], and Weijie Huang[1(✉)]

[1] School of Electrical Engineering, University of Jinan, Jinan 250000, China
zhangmenghua@mail.sdu.edu.cn, cse_huangwj@ujn.edu.cn
[2] Qilu Aerospace Information Research Institute, Jinan, China
[3] Zaozhuang Vocational College of science and technology, Zaozhuang, China
[4] School of Information Science and Engineering, Shandong Normal University, Jinan 250000, China

Abstract. In recent years, the performance of visual inertial odometry (VIO) based on deep learning has shown significant advantages over traditional geometric methods. However, all existing methods estimate each pose through visual and inertial measurements, which involves a large amount of computational redundancy, resulting in huge time costs and hardware damage when training and deploying on devices. In order to maintain accuracy while reducing the number of training parameters, an improved algorithm based on Visual-Selective-VIO is proposed. To reduce the number of network parameters and maintain the training accuracy, a unique attention mechanism is designed for the visual branch and a lightweight pose estimation module. By improving the visual branch, we serialize the information of attention feature maps, covering both channel and spatial dimensions. Then, we multiply these two feature maps with the original input feature maps for adaptive feature correction. This method improves the sensitivity of the model to channel features and enables more accurate image localization. Experimental results show that our algorithm maintains accuracy with a 10% reduction in network parameters compared to advanced VIO algorithm, making it more suitable for training large-scale datasets and deployment in practical applications.

Keywords: Visual inertial odometry · Gate recurrent unit · Adaptive learning

1 Introduction

Humans can perceive their own motion in space through a variety of multimodal fusion methods. Optic flow (vision) and proprioception (inertial sensors) are the two most important sensory information for humans to perceive their self-motion [1].

H. Zhang et al. (Eds.): NCAA 2023, CCIS 1870, pp. 46–59, 2023.
https://doi.org/10.1007/978-981-99-5847-4_4

Estimating six degrees of freedom (6-DOF) motion is one of the important technical challenges faced by robots and autonomous driving. The advantages of visual cameras, such as low cost and ease of operation, have made them widely used in these fields. Over the past decade, with the development of visual odometers and visual synchronous localization and mapping (VSLAM) [7], this challenge has gradually gained attention and exploration. These technologies provide an important background and foundation for visual based 6-DOF motion estimation. 6-DOF motion estimation has shown impressive results. However, while methods such as DSO [8] and ORB-SLAM [9] have achieved high precision and real-time positioning in large-scale environments, there is still much room for improvement in positioning accuracy under non-textured environments, image blurring, and extreme lighting conditions. In the fields of computer vision, robotics, and autonomous driving, visual-inertial odometry based on the fusion of visual information and inertial sensor information is currently a topic of strong research interest [2–6]. Compared with traditional visual odometry, the visual-inertial odometry system includes additional IMU information, which can improve the motion tracking performance of mobile agents in non-textured environments or under extreme lighting conditions, and provide more accurate and robust attitude information. At the same time, the low cost, high performance, and all-time domain advantages of camera and inertial sensor fusion are widely used in the fields of robotics, drones, and smart phones. However, traditional visual-inertial odometry methods (not based on deep learning) heavily rely on manual intervention for fault case analysis and system initialization selection, and require careful parameter tuning for various specific environments. Deploying such a system with rapid calibration in fast-moving scenarios still faces significant challenges.

In recent years, with the continuous development and successful application of deep learning methods in various computer vision tasks [15–17], deep learning and data-driven VIO methods [7,10–14] have attracted widespread attention and demonstrated strong competitiveness in some complex and specific scenarios.

Compared to traditional geometric based methods, deep learning based VIO solutions utilize deep neural networks (DNNs) to extract higher quality features. These solutions are trained on large-scale datasets to learn better fusion of visual and inertial features, and to filter out abnormal sensor data. However, training large-scale datasets requires a significant amount of time and resource costs. In order to reduce the number of network parameters and maintain the accuracy of training, we propose an architecture that combines GRU and CBAM.

By using this combination architecture, we can reduce the number of network parameters while maintaining training accuracy. Experiments have shown that our designed method can effectively improve the accuracy of training. The advantage of this method is that it can better capture the correlation between images and inertial data, and can automatically filter out interference from abnormal sensors.

Our research results indicate that VIO solutions based on deep learning have better performance and adaptability compared to traditional methods. By training on large-scale datasets, we can enable the model to learn richer feature representations and reduce training costs by optimizing the network structure. This will provide higher accuracy and efficiency for the application of VIO technology, and provide more reliable solutions for future visual navigation and positioning tasks.

In this paper, our main contributions are summarized as follows:

- A novel framework to reduce the parameter size of the training network is proposed, which can improve the efficiency of deployment on devices, and reduce computational costs. The method has been fully compared with other advanced algorithms and provides a new solution for training VIO large-scale datasets.
- The complementary advantages between Gate Recurrent Unit and Convolutional Block Attention Module are discovered in VIO field in this article.
- Our method is extensively tested on the KITTI Odometry dataset, and achieves good performance in terms of adaptability.

2 Relate Work

2.1 VO

The VO algorithm estimates the incremental self motion of the camera. A traditional VO algorithm, involves extracting features from an image, matching features between the current image and subsequent images, and then calculating optical flow. Motion can be calculated using optical flow. The fast semi direct monocular visual odometer (SVO) algorithm (Forster, Pizzoli, and Scaramuzza 2014) is an example of the most advanced VO algorithm. Its design is to directly operate on image patches without relying on slow feature extraction, thus achieving fast and robust performance. On the contrary, it uses a probability depth filter on the patch of the image itself. Then update the depth filter by aligning the entire image. This algorithm runs in real-time on embedded platforms and has high computational efficiency. However, its probability formula makes it difficult to tune, and it also requires a bootstrap process to initiate this process. As expected, its performance largely depends on the hardware used to prevent tracking failures - typically requiring the use of a global shutter camera above 50 fps to ensure accurate mileage estimation [24,25].

2.2 Traditional VIO Methods

In recent years, VIO has become a highly focused method that integrates camera and IMU data into a pose estimator, with the ability to provide higher robustness and accuracy in complex and dynamic environments. In the past few decades, tightly coupled VIO systems can be mainly divided into two categories: filter based methods and optimization based methods. Among the filter based methods, representative ones include MSCKF [19] and ROVIO [20]. MSCKF combines geometric constraints and IMU measurements in a multi-state constrained extended Kalman filter (EKF), which has low computational complexity and provides accurate attitude estimation in large scale real world environments. ROVIO uses EKF to fuse IMU data and photometric errors, which is another popular filter based VIO method. Among the optimization based methods, representative ones include OKVIS [21] and VINS [5]. OKVIS is a keyframe based VIO system, while VINS is a tightly coupled method based on nonlinear optimization that achieves high precision mileage measurement by integrating pre integrated IMU measurements and feature observations. These VIO methods have made significant progress in combining camera and IMU data, and have been widely applied in the field of attitude estimation. They exhibit excellent performance and robustness in different environments and application scenarios.

2.3 Deep Learning-Based VIO

With the development of hardware, deep learning based methods have achieved significant success in computer vision applications, including VIO. VINet [7] is the first end-to-end trainable depth learning VIO method, which learns attitude regression from image sequences and IMU measurements through supervised learning. In this method, the long short memory (LSTM) network is introduced to model the correlation of temporal motion. Subsequently, Chen et al. [10] proposed two different masking techniques to selectively fuse visual and inertial features. ATVIO [11] adopts an attention based fusion function and applies adaptive loss for attitude regression. Recent research has also proposed a self supervised learning framework to learn 6-DoF self motion without ground annotation during training. Shamwell et al. [12] proposed VIOLearner, which estimates attitude through a multi-level error correction view synthesis method. DeepVIO [13] improves VIO's attitude estimation through additional optical flow self supervision. In addition, Almalioglu et al. [14] demonstrated a self supervised VIO method based on depth estimation. Mingyu Yang and Yu Chen [18] proposed an adaptive method for disabling dynamic visual modality for visual selective VIO. This method can effectively fuse visual and IMU information in specific environments, thereby improving the accuracy and efficiency of localization. This study provides valuable ideas and methods for the further development of the VIO field. These deep learning based methods have made significant progress in the field of VIO and provide new ideas and technical means for achieving more accurate and efficient attitude estimation. They are of great significance for solving complex visual navigation and positioning problems, and are expected to provide more reliable solutions for future robots and autonomous navigation systems.

3 Method

The main network structure proposed in this article mainly consists of Inertial Encoder, Visual Encoder, fusion module, pose estimation module, and Decision Module. The Inertial Encoder consists of Conv1d, BatchNorm1d, and LeakyReLU, while the Visual Encoder consists of Conv2d, CBAM, BatchNorm2d, and LeakyReLU, with CBAM connected after downsampling at each

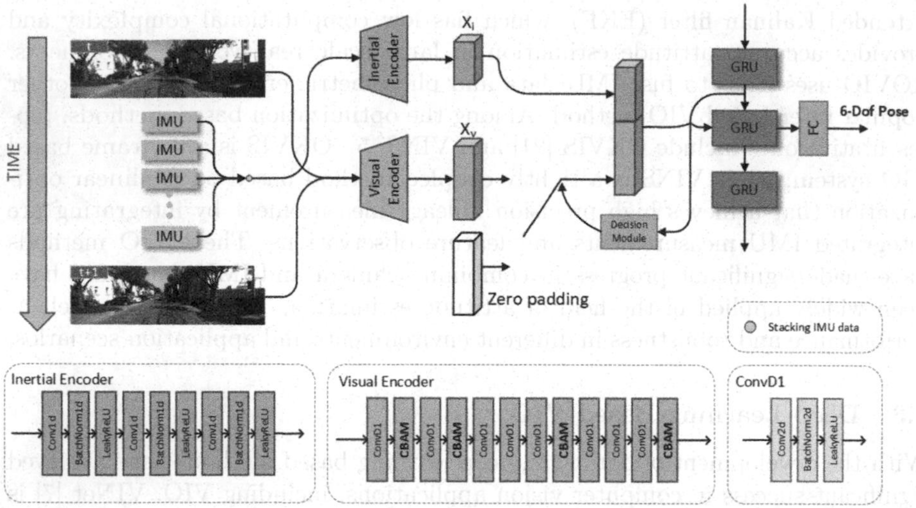

Fig. 1. Proposed architecture for pose estimation

layer. The pose estimation module consists of GRU and FC. The imu features and visual features output by the Encoder are input into the fusion network for fusion output, and then input into the pose estimation module to output 6-DoF. As shown in Fig. 1. Here, this article draws inspiration from the decision network approach proposed by Mingyu Yang, Yu Chen et al. [18] in Visual Selective VIO and designs the Decision Module. During the training process, we use Gumbel Softmax distribution to sample decisions from the decision module to ensure that the entire system is end-to-end differentiable. In the reasoning process, the decision is sampled through the Bernoulli distribution controlled by the policy network.Once the decision module determines the use of visual modality, the current image will be processed by a visual encoder and the obtained visual features, along with inertial features, will be provided to the attitude estimation module for regression GRU attitude estimation. However, if the decision module decides to disable the visual encoder, the input of zero padding will be passed to the GRU to ensure the continuity of the calculation process. This design enables the system to perform flexible input processing according to the instructions of the decision module, while maintaining the trainability and effectiveness of the algorithm.

3.1 Attention Mechanism for the Visual Branch

In order to improve the representation ability and performance of the CNN network, we introduce the Convolutional Block Attention Module(CBAM) module in our algorithm, as shown in Fig. 2. Traditional convolutional networks only focus on local information and often ignore global information, which leads to poor performance. Therefore, the CBAM module can better focus on the global information of the monocular camera image. The channel attention module compresses the spatial dimensions while keeping the channel dimension unchanged, focusing on meaningful information in the image. The spatial attention module compresses the channel dimensions while keeping the spatial dimensions unchanged, focusing on the position information of the target. By using these two modules, the computational performance of the model is significantly improved with a small increase in computational and parameter complexity. The formula of Channelx Attention Module as (1):

$$\begin{aligned} \mathrm{M_c(F)} &= \sigma(MLP(AvgPool(\mathrm{F}))+MLP(MaxPool(\mathrm{F}))) \\ &= \sigma(\mathrm{W_1}(\mathrm{W_0}(\mathrm{F^c_{avg}}))+\mathrm{W_1}(\mathrm{W_0}(\mathrm{F^c_{max}}))) \end{aligned} \tag{1}$$

The formula of Spatial Attention Module as (2):

$$\begin{aligned} \mathrm{M_s(F)} &= \sigma(f^{7*7}[AvgPool(\mathrm{F});MaxPool(\mathrm{F})]) \\ &= \sigma(f^{7*7}[\mathrm{F^s_{avg}};\mathrm{F^s_{max}}]) \end{aligned} \tag{2}$$

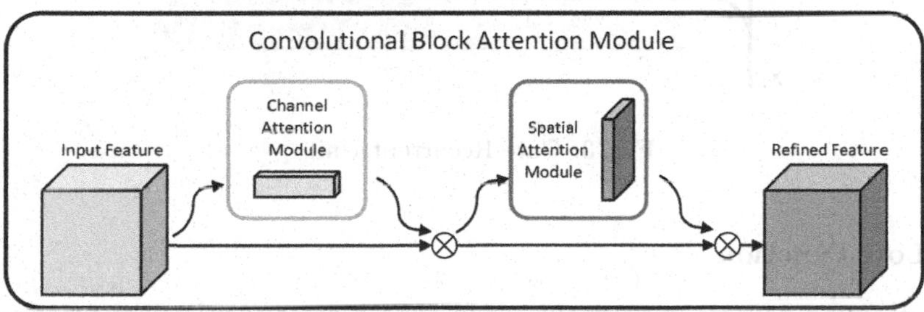

Fig. 2. Convolution Block Attention Module

3.2 Lightweight Pose Estimation Module

To address the challenge of reducing the number of parameters in a model while maintaining its computational performance, we have introduced the Gate Recurrent Unit (GRU) module, as illustrated in Fig. 3. This module has fewer parameters, which greatly improves hardware computation and time costs, providing a significant advantage for engineering practical applications. By incorporating a two-layer GRU-based pose estimation Recurrent Neural Network (RNN) in our study, we aim to enhance the accuracy and efficiency of our model.

The GRU module is a type of RNN that utilizes gating mechanisms to control the flow of information. Specifically, it employs two gates, namely the update gate and reset gate, which work together to regulate the memory content of the cell. The update gate determines the proportion of new information that should be retained in the cell state, while the reset gate controls the amount of old information that should be discarded.

By utilizing the GRU module, we can effectively reduce the number of parameters in our model without sacrificing its performance. This is particularly useful in practical applications where computational resources are limited, and efficiency is crucial. Our two-layer GRU-based pose estimation RNN leverages the advantages of the GRU module to improve the accuracy and robustness of our model, making it suitable for a wide range of applications, including autonomous driving, robotics, and augmented reality.

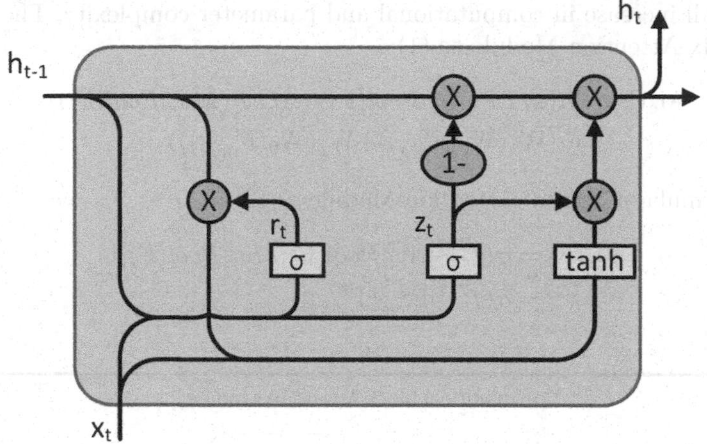

Fig. 3. Gate Recurrent Unit

3.3 Loss Function

In our training process, we use the mean square error (MSE) loss function to minimize the attitude estimation error, which is defined by formula (3). This loss function plays a key role in the training process, helping us measure the accuracy of attitude estimation. By applying MSE loss to attitude estimation error, our goal is to reduce the model's error in this task and improve its performance:

$$\mathcal{L}_{\mathrm{pose}} = \frac{1}{3(\mathrm{T}-1)} \sum_{\mathrm{t}=1}^{\mathrm{t}-1} (\|\hat{\mathrm{v}}_{\mathrm{t}} - \mathrm{v}_{\mathrm{t}}\|_2^2 + \alpha \|\hat{\varphi}_{\mathrm{t}} - \varphi_{\mathrm{t}}\|_2^2) \tag{3}$$

In the formula, T is the length of the training sequence. $\mathrm{v}_t$and φ_t represent the ground truth translation vector and rotation vector, respectively. α is the weight that balances the translation loss and rotation loss, and it is set to 100 according to the previous supervised VO/VIO methods. Additionally, we apply an extra penalty factor C to the use of each visual encoder to encourage disabling of visual feature computation. During the training process, we compute the average penalty and define it as the efficiency loss:

$$\mathcal{L}_{eff} = \frac{1}{T-1} \sum_{t}^{t-1} \mathrm{Cd}_t \tag{4}$$

Finally, we train the end-to-end system to comprehensively consider the sum of attitude estimation loss and efficiency loss (Eq. 5), in order to achieve a balance of computational efficiency while maintaining good accuracy.

$$\mathcal{L} = \mathcal{L}_{pose} + \mathcal{L}_{eff} \tag{5}$$

During the training process, we not only focus on the accuracy of attitude estimation, but also consider the computational efficiency of the system. We combine attitude estimation loss with efficiency loss to find a balance point. This balance point enables our system to perform efficiently while providing accurate attitude estimation.

4 Experiment

4.1 Dataset

We tested our method on the KITTI Odometry dataset [23], which is a highly influential VO/VIO dataset in the field. The dataset includes 22 stereo video sequences, out of which sequences 00–10 provide ground truth trajectories, while sequences 11–22 are used for evaluation without ground truth. To follow the procedure described in [22], sequence 03 was excluded as it lacked raw data. We trained our model on sequences 00, 01, 02, 04, 06, 08, and 09, and tested it on sequences 05, 07, and 10. The dataset's left monocular images were used for this purpose, with the frequency of image and ground truth poses is 10 Hz and the frequency of IMU data is 100 Hz.

4.2 Experimental Setup and Details

This architecture was implemented using PyTorch and trained on NVIDIA 3090Ti GPU. During the training process, we adjusted all images to a size of 512 × 256. The sequence length of training was set as 11. We inserted 11 frames of IMU data between every two images, resulting in an input dimension of 6 × 11 for IMU data. The visual encoder used a pre trained FlowNet-S network for optical flow estimation, as detailed in reference [22]. CBAM was added after each downsampling layer, and a fully connected (FC) layer was added at the end to generate 512 dimensional visual features. We used three one-dimensional convolutions for the IMU data branch and one FC layer, generating 256 dimensional inertial features. The attitude estimation network consisted of two GRU layers, which has 1024 gate units. The hidden state of the last GRU layer was used to estimate a 6-degree of freedom attitude through two layers of MLP at every time step. The training process was divided into two major stages which included warm-up and joint-training stage. When it was at the warm-up stage, a random strategy was used to train the visual encoder, inertial encoder, and attitude estimation network for 40 epoches, and the output of the visual encoder was used with a 50% probability. The learning rate was set to 7×10^{-5} during this stage. When it was at the warm-up stage the joint-training stage, all end-to-end components (including the policy network) were trained for 40 epoches with a learning rate of 7×10^{-6}, followed by another 20 epoches with a learning rate set at 1×10^{-7}. The batch size was set as 32. During the training, the visual information of the first frame was used to ensure effective initial pose estimation. The implementation and training settings of this architecture ensured effective processing and feature extraction of images and IMU data, and optimized the performance of each component through joint training to achieve accurate attitude estimation.

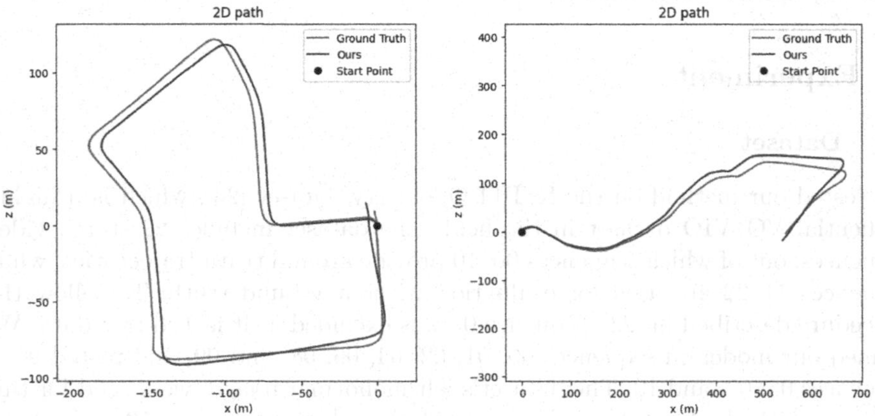

Fig. 4. Ground truth trajectories and motion trajectories on KITTI sequences 07 and 10.

In order to thoroughly evaluate the accuracy of our odometry estimates, we have computed the root mean square error (RMSE) of the estimated translation and rotation vectors for the entire trajectory. This is a widely used metric for evaluating the overall performance of odometry systems. In addition, we have also assessed the relative translation and rotation errors, denoted by t_{rel} and r_{rel}, respectively. These metrics are used to evaluate the accuracy of odometry estimates for various subsequence path lengths, as described in [23].

The RMSE metric provides a comprehensive assessment of the performance of our odometry system by considering the error of both the translation and rotation vectors. By computing the RMSE for the entire trajectory, we can obtain an overall measure of the system's accuracy. This allows us to compare the performance of our system against other state-of-the-art methods.

In addition to the RMSE, we have also evaluated the relative translation and rotation errors. These metrics are particularly useful when assessing the performance of odometry systems for specific subsequence path lengths. By analyzing the t_{rel} and r_{rel} metrics for various subsequence path lengths, we can gain a better understanding of the performance of our system in different scenarios. This helps us to identify any potential weaknesses in our system and devise strategies to improve its accuracy and robustness.

Overall, the combination of the RMSE and relative translation and rotation errors provides a comprehensive and detailed evaluation of the accuracy of our odometry system. By thoroughly analyzing these metrics, we can identify areas for improvement and further optimize our system to meet the requirements of various practical applications.

4.3 Main Result

We evaluated our method on the KITTI dataset and compared it with the full modal baseline, GRU-only, and CBAM-only approaches. To ensure a fair comparison, we trained our proposed model and the other three models using the same optimizer and common hyperparameters, including the number of epochs and learning rate. We tested the models on the KITTI dataset and calculated the average usage of the visual encoder and the average root mean square error (RMSE) of translation and rotation. Table 1 summarizes the results.

Table 1. The relative translational t_{rel} & rotationalr_{rel} error, and visual encoder usage of the baseline model and the overall parameter quantity of the network

	Seq.05			Seq.07			Seq.10			The amount of parameters
	t_{rel}	r_{rel}	usage	t_{rel}	r_{rel}	usage	t_{rel}	r_{rel}	usage	
Baseline	2.8431	1.0804	27.1475	2.7688	2.2087	29.4813	3.6016	1.7061	31.7765	48.454376M
Only GRU	8.2991	4.1528	**22.037**	11.2387	7.7288	23.5669	14.1928	6.8264	**22.6856**	44.518120M
Only CBAM	3.7812	1.6506	31.1707	**3.280**	3.2786	33.6670	**3.5754**	1.8408	37.1143	48.664988M
Ours	**3.6371**	**1.6365**	27.8724	3.3949	**2.8249**	**21.929**	5.6794	**1.7744**	29.8582	**44.728732M**

We conducted multiple experiments and found that with the addition of GRU, the usage of visual encoder and total parameter count of the network both decreased as expected. When CBAM was introduced, The incorporation of CBAM has resulted in more comprehensive features covering the object to be recognized, leading to improved object recognition probability. This suggests that the attention mechanism has effectively trained the network to prioritize key information for improved recognition. But in this paper, adding a separate attention mechanism to the visual side can increase the dominance of visual information, resulting in a decrease in accuracy. Therefore, the GRU was added to suppress overly strong visual features to achieve a balanced effect. Moreover, through comparison, we found that our method increased by 0.6–1% in terms of relative translation/rotation error, but the overall parameter count of the network decreased by 7%.

In Fig. 5, we present a visual explanation of the frequency and vehicle speed used by the visual encoder on sequence 07. The description in the upper left corner uses color coding, with darker colors indicating lower usage and lighter colors indicating higher usage, to demonstrate the use of visual encoders in

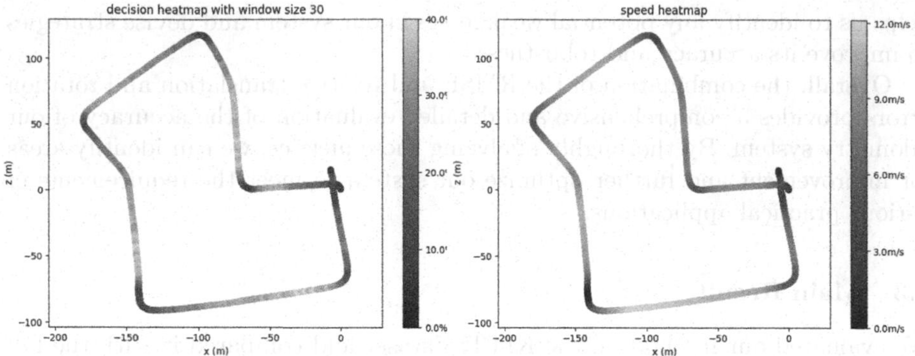

Fig. 5. The visualization of the learning strategy for sequence 07. The image in the upper left corner shows the mapping of visual encoder usage, showing the local usage of each time step. The top right corner displays a proxy vehicle speed chart. Strategic networks tend to activate visual encoders more frequently during fast linear movements, while reducing their use during slow movements and turns.

local areas. At the same time, in the upper right corner, we show the speed changes of the agent at each time step, with darker colors indicating lower speeds. Through this strategy, the system can dynamically adjust the utilization of visual encoders based on different motion states to more effectively handle different scenes and actions. The visualization of this learning strategy helps us understand the decision-making process of the system under different conditions and its adaptability to visual encoders.

Observing the diagram, it can be observed that there is a clear correlation between the use of visual modality and vehicle speed and turning angle. When the vehicle is moving slowly or making turns, the strategy network uses less visual mode. This may be because the perception of environmental details during slow driving and turning is not as critical, resulting in a relatively low level of activation of the visual encoder. However, we observed that the visual encoder was activated more frequently when the proxy vehicle was traveling rapidly in a straight line.

This behavior can be explained by the inherent property of direct measurement of angular velocity in IMU. Compared to angle estimation based on visual features, using IMU to estimate turning angles is relatively easy. This is because the angular velocity can be calculated through a simple first-order integration. However, for the estimation of the translation process, additional IMU measurement is required. Because IMU can only measure the acceleration, which is the second derivative of the translation, the velocity constraint needs to be initialized. Therefore, relying solely on IMU for estimation usually results in significant errors when the vehicle is moving rapidly. To reduce this error, the strategy network frequently uses visual modalities to provide additional information.

In summary, the visual explanation in Fig. 5 reveals the relationship between the frequency of visual encoder usage and vehicle speed. It displays the changes in the activation level of visual modes under different driving states, as well as the role of visual encoders in vehicle control. This is very valuable for a deep understanding of the decision-making process and perception ability of autonomous driving systems.

5 Conclusion

In this paper, a novel method is proposed to address the challenge of integrating VIO algorithms into devices more easily. Our method reduces model parameters by introducing GRU and improves accuracy by incorporating CBAM. Additionally, the visual modality can be opportunistically disabled when visual information is not critical, reducing computational cost and power consumption. Our experiments demonstrate that our method provides approximately 10% reduction in parameter computation with no significant performance degradation. Moreover, the learned policy is interpretable and exhibits scene-dependent adaptive behavior. Our adaptive learning strategy is model independent, so it can be easily applied in other deep VIO systems. The universality of this strategy enables it to quickly migrate to different systems and frameworks without requiring significant modifications and adaptation. This provides researchers and developers with a flexible and efficient method to utilize adaptive learning strategies in their own deep VIO systems, thereby improving the performance and robustness of attitude estimation. This portability and ease of use make our learning strategy a valuable tool that can promote further research and application in the field of deep VIO.

Acknowledgement. This work was supported by the Youth Foundations of Shandong Province under Grant Nos. ZR202102230323 and ZR2021QF130, the National Natural Science Foundation of China under Grant No. 62273163, and the Key R & D Project of Shandong Province under Grant No. 2022CXGC010503.

References

1. Fetsch, C.R., Turner, A.H., DeAngelis, G.C., Angelaki, D.E.: Dynamic reweighting of visual and vestibular cues during self-motion perception. J. Neurosci. **29**(49), 15601–15612 (2009)
2. Forster, C., Carlone, L., Dellaert, F., Scaramuzza, D.: Onmanifold preintegration for real-time visual Cinertial odometry. IEEE Trans. Rob. **33**(1), 1–21 (2017)
3. Leutenegger, S., Lynen, S., Bosse, M., Siegwart, R., Furgale, P.: Keyframe-based visual Cinertial odometry using nonlinear optimization. Int. J. Robot. Res. **34**(3), 314–334 (2015)
4. Li, M., Mourikis, A.I.: High-precision, consistent EKF based visual-inertial odometry. Int. J. Robot. Res. **32**(6), 690–711 (2013)

5. Qin, T., Li, P., Shen, S.: VINS-MONO: a robust and versatile monocular visual-inertial state estimator. IEEE Trans. Rob. **34**(4), 1004–1020 (2018)
6. Clark, R., Wang, S., Wen, H., Markham, A., Trigoni, N.: ViNet: visual-inertial odometry as a sequence-to-sequence learning problem. In: Proceedings of the AAAI Conference on Artificial Intelligence, vol. 31 (2017)
7. Cadena, C., et al.: Past, present, and future of simultaneous localization and mapping: toward the robust-perception age. IEEE Trans. Rob. **32**(6), 1309–1332 (2016)
8. Engel, J., Koltun, V., Cremers, D.: Direct sparse odometry. IEEE Trans. Pattern Anal. Mach. Intell. **40**(3), 611–625 (2017)
9. Mur-Artal, R., Tard®s, J.D.: Orb-slam2: an open-source slam system for monocular, stereo, and RGB-D cameras. IEEE Trans. Robot. **33**(5), 1255–1262 (2017)
10. Chen, C., Rosa, S., Miao, Y., et al.: Selective sensor fusion for neural visual-inertial odometry. In: Proceedings of the IEEE/CVF Conference on Computer Vision and Pattern Recognition, pp. 10542–10551 (2019)
11. Liu, L., Li, G., Li, T.H.: AtVio: attention guided visual-inertial odometry. In ICASSP 2021–2021 IEEE International Conference on Acoustics, Speech and Signal Processing (ICASSP), pp. 4125–4129. IEEE (2021)
12. Shamwell, E.J., Leung, S., Nothwang, W.D.: Vision-aided absolute trajectory estimation using an unsupervised deep network with online error correction. In: 2018 IEEE/RSJ International Conference on Intelligent Robots and Systems (IROS), pp. 2524–2531. IEEE (2018)
13. Han, L., Lin, Y., Du, G., Lian, S.: Deepvio: self-supervised deep learning of monocular visual inertial odometry using 3D geometric constraints. In: 2019 IEEE/RSJ International Conference on Intelligent Robots and Systems (IROS), pp. 6906–6913. IEEE (2019)
14. Almalioglu, Yasin, et al.: SelfVIO: self-supervised deep monocular Visual CInertial Odometry and depth estimation, pp. 119–136. Neural Networks, 150 (2022)
15. Krizhevsky, A., Sutskever, I., Hinton, G.E.: ImageNet classification with deep convolutional neural networks. Commun. ACM **60**(6), 84–90 (2017)
16. Simonyan K, Zisserman A.: Very deep convolutional networks for large-scale image recognition. arXiv preprint arXiv:1409.1556 (2014)
17. Ren, S., et al.: Faster R-CNN: towards real-time object detection with region proposal networks. In: Advances in Neural Information Processing Systems, vol. 28 (2015)
18. Yang, M., Chen, Y., Kim, H.S.: Efficient deep visual and inertial odometry with adaptive visual modality selection. In: Computer Vision CECCV 2022: 17th European Conference, Tel Aviv, Israel, October 23–27, pp. 233–250. Proceedings, Part XXXVIII (2022)
19. Mourikis, A.I., Roumeliotis, S.I.: A multi-state constraint Kalman filter for vision-aided inertial navigation. In: Proceedings 2007 IEEE International Conference on Robotics and Automation, pp. 3565–3572. IEEE (2007)
20. Bloesch, M., Omari, S., Hutter, M., Siegwart, R.: Robust visual inertial odometry using a direct EKF-based approach. In: 2015 IEEE/RSJ International Conference on Intelligent Robots and Systems (IROS), pp. 298–304. IEEE (2015)
21. Leutenegger, S., Furgale, P., Rabaud, V., et al.: Keyframe-based visual-inertial slam using nonlinear optimization. In: Proceedings of Robotis Science and Systems (RSS) 2013 (2013)
22. Chen, C., et al.: Selective sensor fusion for neural visual-inertial odometry. In: Proceedings of the IEEE/CVF Conference on Computer Vision and Pattern Recognition, pp. 10542–10551 (2019)

23. Geiger, A., Lenz, P., Urtasun, R.: Are we ready for autonomous driving? The kitti vision benchmark suite. In: 2012 IEEE Conference on Computer Vision and Pattern Recognition, pp. 3354–3361. IEEE (2012)
24. Forster, C., Carlone, L., Dellaert, F., Scaramuzza, D.: IMU preintegration on manifold for efficient visual-inertial maximum-a-posteriori estimation. In: Robotics: Science and Systems XI (2015)
25. Forster, C., Pizzoli, M., Scaramuzza, D.: SVO: fast semi-direct monocular visual odometry. In: 2014 IEEE International Conference on Robotics and Automation (ICRA), pp. 15–22. IEEE (2014)

A Two-Stage Framework for Kidney Segmentation in Ultrasound Images

Zhengxuan Song[1], Xun Liu[2], Yongyi Gong[3], Tianyong Hao[4], and Kun Zeng[1(✉)]

[1] School of Computer Science, Sun Yat-sen University, Guangzhou, China
songzhx6@mail2.sysu.edu.cn, zengkun2@mail.sysu.edu.cn

[2] Department of Nephrology, The Third Affiliated Hospital of Sun Yat-sen University, Guangzhou, China

[3] School of Information Science and Technology, Guangdong University of Foreign Studies, Guangzhou, China
gongyongyi@gdufs.edu.cn

[4] School of Computer Science, South China Normal University, Guangzhou, China
haoty@m.scnu.edu.cn

Abstract. In this paper, we propose an algorithm framework for kidney segmentation in ultrasound images. Due to the characteristics of kidney ultrasound images, such as high noise, heterogeneous structure, low contrast, multiple artifacts, and relatively fixed shape, accurately segmenting clear and complete kidney structures from the images is still a challenging task. Our framework consists of two parts: shape aware dual-task multi-scale fusion network and self-correction. The first part uses a U-shape structure with multi-scale feature cross fusion skip connections to perform segmentation and can simultaneously predict the segmentation map and the shape-aware prediction level set function, we use a dual-task consistency module to constrain the shape of the segmentation map, enabling the network to learn the target area more accurately, through dual-task consistency supervised learning, we obtain a pre-trained model. The second part uses an iterative aggregation strategy for the pre-trained model to optimize it and reduce the noise and other issues in the prediction results. Experimental results show that our algorithm framework outperforms several state-of-the-art methods on kidney ultrasound datasets.

Keywords: Kidney ultrasound images · Shape aware · Self-correction · Dual-task consistency · Multi-scale Fusion

1 Introdution

Ultrasound is a commonly used diagnostic method in kidney disease [1], which can obtain information such as kidney size and morphology. By segmenting the target area of kidney ultrasound images, information can be obtained to assist in evaluating kidney function. However, due to the characteristics of ultrasound imaging, there are still many challenges in kidney ultrasound image segmentation [2]. These characteristics mainly include noise, relatively fixed shape,

H. Zhang et al. (Eds.): NCAA 2023, CCIS 1870, pp. 60–74, 2023.
https://doi.org/10.1007/978-981-99-5847-4_5

heterogeneous structure, low contrast, and artifacts. In the past decade, deep learning technology has achieved remarkable results in assisting medical image processing. However, there are still some problems in the existing deep learning segmentation methods under the condition of limited sample size and annotation accuracy of datasets. For example, as shown in Fig. 1, Unet [3]and Swin-Unet [4] are currently widely used classic methods based on convolution and transformers. However, due to the lack of a global receptive field in the convolution method, it is prone to incorrectly segmenting targets and edge external noise when processing ultrasound images with heterogeneous structures. Transformer-based methods are susceptible to insufficient feature extraction of local areas due to their partition strategy, resulting in a large amount of edge noise and internal hole problems in segmentation results.

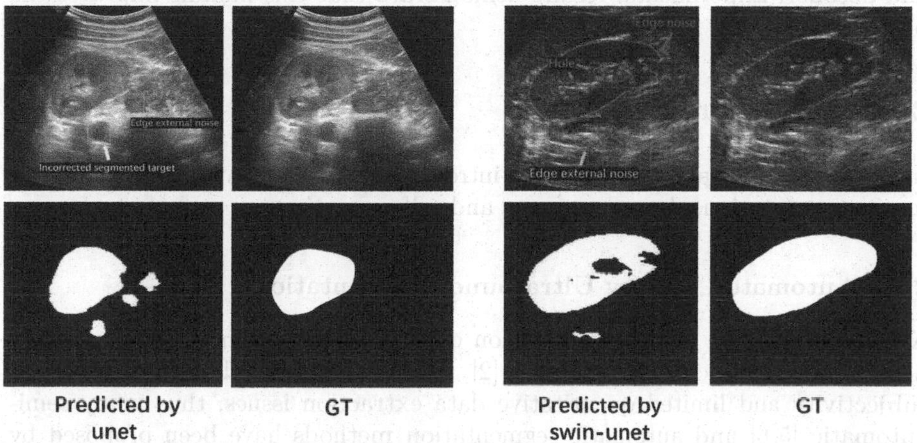

Fig. 1. The traditional Unet model has issues with incorrectly segmented targets and edge external noise in kidney ultrasound images. The Swin-Unet model based on Transformer also has problems with edge noise, internal holes, and edge external noise. The correct segmentation results are shown by the ground truth.

In the paper, we propose a shape-aware self-correcting learning framework to address segmentation problems. The core of this framework is to introduce the self-correction iterative learning strategy for pre-trained models into the shape-aware segmentation network. The framework is divided into two parts: the first part is a shape-aware dual-task multi-scale fusion segmentation network, which applies geometric shape constraints to improve the performance of supervised learning; the second part is the self-correction iterative learning method for pre-trained models, which optimizes the model's robustness and accuracy, eliminates segmentation problems such as edge noise, internal holes, and external noise, and obtains a corrected segmentation map.

Our contributions can be summarized as follows: (1) We propose a shape-aware self-correcting learning framework that applies geometric shape constraints to pixel-level kidney segmentation maps using a shape-aware pre-trained model and solves the problem of edge noise through self-correction iterative learning strategies; (2) We constructed a shape-aware dual-task multi-scale fusion segmentation network. By jointly predicting the segmentation map and the level set function representing the kidney contour, and maximizing the consistency of the dual tasks to improve the performance of supervised learning. Experimental results demonstrate that our method outperforms various state-of-the-art methods; (3) We design a channel-spatial attention module(CSA) module in the segmentation network, which uses channel and spatial attention to enhance the network's attention to the features of interest in the skip connection layer, and also strengthens the semantic consistency between the deep network encoder and decoder. Experimental results demonstrate that this module improves the performance of the network.

2 Relate Works

In this section, we provide a detailed introduction to the related works on automatic segmentation, shape awareness and self-correction.

2.1 Automated Kidney Ultrasound Segmentation

Kidney ultrasound image segmentation can be performed in three ways: manual, semi-automatic, and automatic [2]. Traditional manual segmentation has subjectivity and limited quantitative data extraction issues, thus many semi-automatic [5,6] and automatic segmentation methods have been proposed by researchers, including the use of deformable models [2,7], active contour models [8], and shape priors [9]. However, these methods still require manual preprocessing and have low segmentation accuracy, making automatic segmentation methods based on deep learning [3,10,11] a research hotspot. Among them, methods based on improvements to the UNet model [12–14] have shown good segmentation performance for kidney ultrasound images. In recent years, state-of-the-art algorithms such as TransUnet [15], SETR [16], and SwinUnet [4] have demonstrated the potential of Transformers in medical image segmentation. For example, in breast ultrasound image segmentation, Zhuang *et al.* [17] used residual swin-transformer [18] blocks as feature extraction modules for the encoder and decoder. Some methods consider refining segmentation accuracy from skip-connection layers, such as UNet++ proposed by Zhou *et al.* [19], which introduced convolution-based nested connections to narrow the semantic gap between the encoder and decoder networks. Wang *et al.* [20] proposed a multi-scale channel-crossing attention network, which effectively fits the low-level information of the encoder and the high-level semantic and resolution information of the decoder.

2.2 Level-Set Function

Due to the relatively fixed morphology and heterogeneous structures of ultrasound images, some research focuses on shape and appearance. For example, Yin *et al.* [21] proposed a subsequent boundary distance regression and pixel classification network for automatic segmentation of the kidney. Sun *et al.* [22] proposed a method that introduced shape flow and texture flow to simultaneously capture shape information, and Ma *et al.* [23] discussed the method of incorporating distance transform maps into CNN to improve segmentation. While Li *et al.* [24], Xue *et al.* [25], Luo *et al.* [26] constructed a shape-aware multi-task network semi-supervised learning framework using labeled and unlabeled data. However, these segmentation networks still lack accuracy.

2.3 Self-correction

Some works consider noise annotation in the segmentation dataset as a limiting factor for segmentation performance. Zou *et al.* [27] proposed a segmentation method that is robust to noisy annotation, which uses noisy annotations and noise detection methods to dynamically correct model predictions. Li *et al.* [28] proposed a label correction method that can repair noisy labels, making the model and labels more robust and accurate during self-correction learning cycles.

3 Method

3.1 Overview

Our framework consists of two parts: shape aware dual-task multi-scale fusion network and self-correction. In the first part, a pre-trained model with shape constraints and noisy prediction is obtained through dual-task consistency supervision learning. In the second part, we use a self-correction multi-cycle iterative aggregation strategy to optimize the pre-trained model and noisy prediction, eliminating various noise issues in the prediction results. In the following two sections, we will introduce these two parts separately.

3.2 Shape Aware Dual-Task Multi-scale Fusion Network

In this section, we provide a detailed description of each module component of the shape-aware dual-task multi-scale fusion segmentation network, as shown in Fig. 2, our segmentation network includes the multi-scale feature embedding, channel-wise cross fusion transformer block (CFTB), channel-spatial attention module (CSA), and shape aware dual-task consistency module. We then introduce the loss functions used to train the dual-task network.

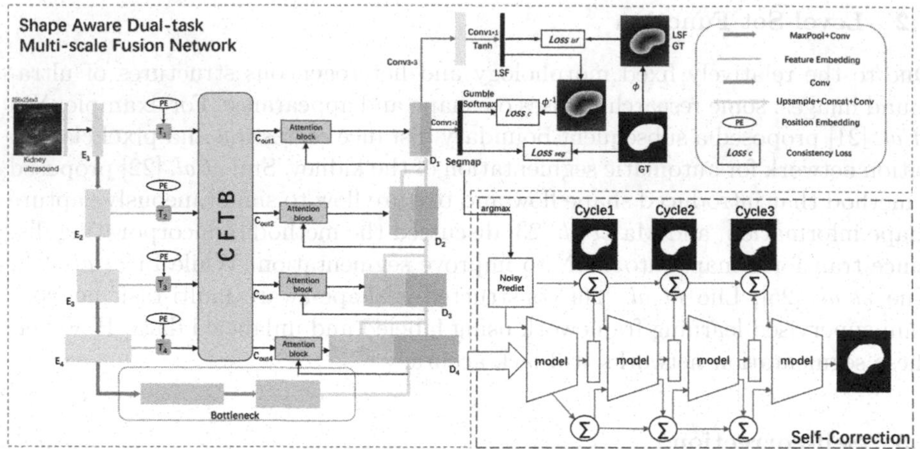

Fig. 2. The workflow of our framework, which is divided into two parts: (1) Shape Aware Dual-task Multi-scale Fusion Network: adopts a U-shape structure with multi-scale feature cross fusion skip connections. In the figure, E1 to E4 represent encoders for extracting image features, and D1 to D4 represent decoders. (2) Self-Correction: optimizes the pre-trained model and noisy predictions through multiple cycles of iterative aggregation strategies.

Multi-scale Feature Embedding. As shown in Fig. 2, The encoder features $E_i \in \mathbb{R}^{\frac{H}{i} \times \frac{W}{i} \times C_i}$ are fed into the skip connections layer, we divide the features E_i at each scale into patches with size $P_i \in \mathbb{R}^{\frac{P}{i} \times \frac{P}{i}}$ and keeping the channel number consistent. We flatten the patches into a sequence to obtain $Sequence_i \in \mathbb{R}^{\frac{H \times W}{P^2} \times C_i}$. To encode the spatial information of each patch, we use a learnable position embedding $PE_i \in \mathbb{R}^{N \times C_i}$, where $N = \frac{H \times W}{P^2}$, and embed it into the sequence $Sequence_i$ to obtain the final sequence input, as follows:

$$T_i = Sequence_i + PE_i \tag{1}$$

Channel-Wise Cross Fusion Transformer Block. We obtain four input tokens $T_i \in \mathbb{R}^{\frac{H \times W}{P^2} \times C_i}, (i = 1, 2, 3, 4)$ through multi-scale feature embedding. Then, a channel-wise cross fusion transformer block consists of multi-headed channel-cross self-attention (MCCA) and multi-layer perceptron (MLP). First, we perform layer normalization, and then concatenate the four tokens to obtain $T_\Sigma = Concat(T_1, T_2, T_3, T_4)$, where $T_\Sigma \in \mathbb{R}^{\frac{H \times W}{P^2} \times C_\Sigma}, C_\Sigma = \sum_{i=1}^{4} C_i$, we use a parameter matrix to transform T_i and T_Σ into queries, keys, and values:

$$Q_i = T_i W_{Q_i}, K = T_\Sigma W_K, V = T_\Sigma W_V, \tag{2}$$

where $W_{Q_i} \in \mathbb{R}^{C_i \times C_i}$, $W_K \in \mathbb{R}^{C_\Sigma \times C_\Sigma}$, $W_Q \in \mathbb{R}^{C_\Sigma \times C_\Sigma}$, the channel-wise cross self-attention can be represented as:

$$CCA_i = Softmax[\Psi(\frac{Q_i^T \times K}{\sqrt{C_\Sigma}})] \times V^T, \tag{3}$$

where Ψ represents the instance normalization operation. Thus, the multi-head channel-wise cross self-attention can be expressed as:

$$MCCA_i = T_i + (\frac{\sum_{n=0}^{N-1} CCA_i^n}{N})^T, \tag{4}$$

where N represents the number of heads. The output of $MCCA$ is obtained by layer normalization, multi-layer perceptron, and residual operation, and the output of the entire block is represented by C_{outi}:

$$C_{outi} = MCCA_i + MLP(LN(MCCA_i)), \tag{5}$$

where LN represents layer normalization operation, and MLP represents multi-layer perceptron. We pass the obtained feature C_{outi} to the next module, the CSA module.

Channel-Spatial Attention Module. Unlike the work of Wang *et al.* [20], as shown in Fig. 3, we use both channel and spatial attention [29] and introduce a residual structure.

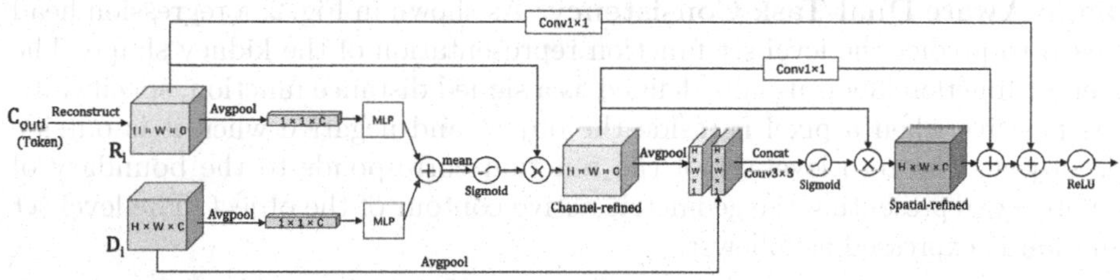

Fig. 3. Channel-spatial Attention Module.

First, the features obtained through reconstruction $R_i \in \mathbb{R}^{H\times W\times C}$ and the output of the decoder upsampled in the $(i+1)^{th}$ layer $D_i \in \mathbb{R}^{H\times W\times C}$ are used as input to the module. The first part is the channel attention, which uses average pooling to aggregate the spatial information of the two input features, obtaining the channel-wise average pooling features $F_{avg}^{R_i} \in \mathbb{R}^{1\times 1\times C}$ and $F_{avg}^{D_i} \in \mathbb{R}^{1\times 1\times C}$. These average pooling features are then passed through an MLP consisting of two linear layers, $L_0^{R_i}$, $L_0^{D_i} \in \mathbb{R}^{C,C/r}$ and $L_1^{R_i}$, $L_1^{D_i} \in \mathbb{R}^{C/r,C}$, respectively. The output results are then averaged, passed through a sigmoid function, and multiplied by the feature R_i, resulting in the channel-refined feature A_c^i. The process is represented as follows:

$$A_c^i = \sigma(\frac{L_0^{R_i}(L_1^{R_i}(F_{avg}^{R_i})) + L_0^{D_i}(L_1^{D_i}(F_{avg}^{D_i}))}{2}) \times R_i, \tag{6}$$

where σ represents the sigmoid function. The second part is spatial attention. We first aggregate the channel-refined feature A_c^i and D_i by average pooling to obtain the spatial-wise average pooling features $S_{avg}^{R_i} \in \mathbb{R}^{H\times W\times 1}$, $S_{avg}^{D_i} \in \mathbb{R}^{H\times W\times 1}$ respectively. Then, we concatenate these two features and perform a standard convolutional layer followed by a sigmoid function. Finally, we multiply the output by the channel-refined feature A_c^i to obtain the spatial-refined featureA_s^i. The equation is as follows:

$$A_s^i = \sigma(Conv_{3\times 3}(Concat(S_{avg}^{D_i}, S_{avg}^{D_i}))) \times A_c^i \tag{7}$$

To reduce the loss of global feature information, we use a convolutional layer to connect the channel-refined feature A_c^i and the reconstructed feature R_i from the transformer block to the output of the CSA module with a residual link, which is represented as follows:

$$CSA_i = A_s^i + Conv_{1\times 1}(A_c^i) + Conv_{1\times 1}(R_i) \tag{8}$$

The output of the CSA module is concatenated with the decoder feature of the ith layer and then processed by convolution and upsampling to generate the decoder feature of the $(i-1)^{th}$ layer.

Shape Aware Dual-Task Consistency. As shown in Fig. 2, a regression head is used to predict the level set function representation of the kidney shape. The level set function [30] is usually defined as a signed distance function, specifically, it is positive when a pixel is inside the object and negative when it is outside the object. The zero level set of the function corresponds to the boundary of the object, representing the geometric active contour of the object. The level set function is expressed as follows:

$$\phi(n) = \begin{cases} -\inf\limits_{y\in\partial\Omega} \|x-y\|_2, \; y \in \Omega_{in} \\ 0, \; y \in \partial\Omega \\ \inf\limits_{y\in\partial\Omega} \|x-y\|_2, \; y \in \Omega_{out} \end{cases}, \tag{9}$$

where $\|x-y\|_2$ represents the Euclidean distance between image pixels x and y, inf represents the minimum lower bound of the set, $\partial\Omega$ represents the contour of the object, and Ω_{in} and Ω_{out} represent the interior and exterior pixel sets of the contour, respectively. We can use Eq. 9 to map labeled segmentation map data to level set function labels for predicting the regression head. To achieve consistency between the two tasks, we need to define a method for converting the level set function back to segmentation maps. Due to the non-differentiability of the level set function $\phi(n)$, we cannot directly use its inverse operation for conversion. Instead, we use a smoothed approximation of the Heaviside function sigmoid function [25,26] as a substitute for the inverse operation of the level set function. The function is defined as follows:

$$\phi^{-1}(n) = \frac{1}{1 + e^{-k\Delta z}}, \tag{10}$$

where z represents the value of the level set function corresponding to a certain pixel and k is set to a very small negative number (set as -1500 in our experiments) in the formula. Negative pixels within the contour are approximately mapped to 1, while positive pixels outside the contour are approximately mapped to 0. As shown in Fig. 4 illustrates how the shape in the segmentation map is constrained by the consistency between the two tasks, and the loss function for the consistency will be introduced in the next section on loss function.

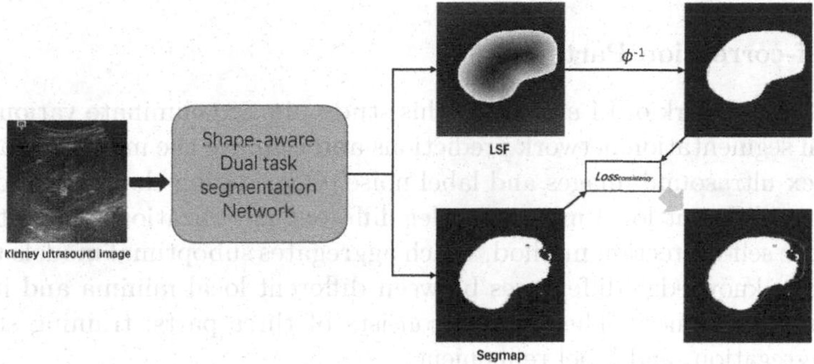

Fig. 4. Dual-task Consistency: illustrates the process of using shape-aware information to constrain the segmentation map.

Loss Function. Firstly, we define D_{seg} and D_{lsf} as the datasets for the segmentation task and the level set function prediction task, respectively. There exists a relationship $(X, Y_{seg}) \in D_{seg}$, $Y_{lsf} \in D_{lsf}$, where there is a mapping $Y_{lsf} = \phi(Y_{seg})$. Our model's loss function is divided into three parts: $\mathcal{L}_{seg}$, $\mathcal{L}_{lsf}$, and $\mathcal{L}_{consistency}$, which represent the loss functions for the segmentation map task, the level set function prediction task, and the consistency between the two tasks, respectively.

The loss function for the segmentation map task is defined as follows:

$$\mathcal{L}_{seg} = \sum \mathcal{L}_{ce}(f_1(X), Y_{seg}) + \sum \mathcal{L}_{dice}(f_1(X), Y_{seg}), \tag{11}$$

where $\mathcal{L}_{ce}$ and $\mathcal{L}_{dice}$ represent the cross-entropy loss and dice loss [31], respectively, $f_1(X)$ is the function form for predicting the segmentation map task. Next, we define the loss function for the level set function prediction task as:

$$\mathcal{L}_{lsf} = \sum \mathcal{L}_{mse}(f_2(X), Y_{lsf}) = \sum \frac{1}{n} \sum_{j=1}^{n} (f_2(X)_j - (Y_{lsf})_j)^2, \tag{12}$$

where $f_2(X)$ is the function form for predicting the level set function task, j represents the jth pixel in the level set function map, and n represents the number of pixels in the map. Using the level set function's approximate inverse operation in Eq. 10, we define a consistency loss function between the two tasks as:

$$\mathcal{L}_{consistency} = \sum \mathcal{L}_{mse}(gumbel\_softmax(f_1(X)), \phi^{-1}(f_2(X))), \tag{13}$$

where $\mathcal{L}_{mse}$ represents the mean square error loss. We use the gumbel softmax function [32] to replace the non-differentiable argmax function to convert the segmentation task's prediction results to a single-channel vector, making the loss function differentiable. Finally, our loss function for the consistency supervision learning is expressed as follows:

$$\mathcal{L} = \lambda_1 \mathcal{L}_{seg} + \lambda_2 \mathcal{L}_{lsf} + \lambda_3 \mathcal{L}_{consistency} \tag{14}$$

3.3 Self-correction Part

Building on the work of Li *et al.* [28], this study aims to eliminate various types of noise in segmentation network predictions and improve the model's robustness to complex ultrasound images and label noise. Considering that the model may converge to different local minima under different initialization parameters, we propose the self-correction method, which aggregates suboptimal models to complement the knowledge differences between different local minima and improve the model's robustness. The method consists of three parts: training strategy, model aggregation, and label refinement.

Training Strategy. We use the pre-trained model and noisy predictions as inputs to self-correction. For each cycle, in order to allow the model to escape from the original local minimum and reach another local minimum, we need a larger learning rate at the beginning and a smaller learning rate at the end. Therefore, we use a cyclic restart annealing learning rate scheduler, which can be expressed as:

$$v = v_{min} + \frac{1}{2}(v_{max} - v_{min}) \times (1 + cos(\frac{T_e}{T}\pi)), \tag{15}$$

where v_{min} and v_{max} represent the lower and upper bounds of the learning rate, T represents the number of epochs in a cycle, and T_e represents the number of epochs relative to the start of the cycle.

Model Aggregation. Firstly, we define the sub-optimal model obtained at the end of the i-th cycle as O. We adjust the parameter weights to aggregate the sub-optimal model O_{i-1} from the previous cycle to obtain the initial model O_i for the next cycle, as follows:

$$O_i = \frac{i}{i+1} O_{i-1} + \frac{1}{i+1} O \tag{16}$$

Since the aggregated model obtained by stacking model parameters may lead to inaccurate estimation of the mean and variance of the refined label by the parameters of Batch Normalization layers, it is necessary to re-estimate the parameters of Batch Normalization layers with new samples after model aggregation [28].

Label Refinement. We actually use the noisy predicted segmentation maps generated by the pre-trained model as the initial training label data for self-correction. With the knowledge complementarity brought by the model aggregation, the masks generated by new suboptimal models contain more refined information. We aggregate the suboptimal masks and the updated mask of the current suboptimal model to generate a new mask that is likely to alleviate or eliminate various noise problems that existed previously. The operation is as follows:

$$\gamma_i = \frac{i}{i+1}\gamma_{i-1} + \frac{1}{i+1}\gamma, \tag{17}$$

where γ_{i-1} is the training label for the current cycle, and γ is the updated mask generated by the suboptimal model O obtained at the end of the current cycle. The refined map γ_i obtained will be used as the training label for the next cycle $(i+1)$.

4 Experiments

4.1 Dataset and Implementation Details

The experiments conducted and the data used in this study have been approved by the ethics committee of a hospital (Affiliated with Sun Yat-sen University Third Hospital). All images were annotated by more than ten students with annotation experience under the guidance and review of a professor (co-first author of this paper) who is experienced in the field of kidney. Each kidney ultrasound image was double-blind annotated and reviewed at least three times.

We performed 4-fold cross-validation on the kidney ultrasound image dataset. We trained our model on an NVIDIA GeForce RTX 4090 server with 24 GB of memory using Pytorch 1.13.1 to implement our algorithm. We used the Rmsprop optimizer to minimize the loss function during the training of this method, with the learning rate initialized to 1e−5, 200 epochs, a batch size of 16, and a uniform input resolution of 256 × 256. We used Dice coefficient (Dice), Jaccard index (Jaccard), Average Symmetric Surface Distance (ASSD), and 95% Hausdorff Distance (95HD) as evaluation metrics.

4.2 Experiment Results

We compared our proposed method with other state-of-the-art methods, which we classified into three types: transformer-based methods such as SwinUNet [4], TransUnet [15], SETR [16], and UCtransNet [20]; shape-aware methods such as SASSNet [24], DTC [26], and SAUNet [22]; and UNet-based methods such as

Table 1. Comparison with state-of-the-art segmentation methods, where SETR uses PUP as decoder and the scale of Swin-Unet is tiny.

Methods	Dice↑(%)	Jaccard↑(%)	ASSD↓(voxel)	95HD↓(voxel)
Swin-Unet-tiny [4]	97.12	94.44	0.050	0.131
TransUNet [15]	97.167	94.51	0.054	0.158
UNet [3]	97.21	94.61	0.042	0.244
SETR-PUP [16]	97.242	94.68	0.0474	0.105
SASSNET [24]	97.245	94.67	0.046	0.114
DTC [26]	97.25	94.65	0.0471	0.115
UNet++ [19]	97.31	94.82	0.0449	0.090
SAUNet [22]	97.33	94.82	0.0446	0.087
UCTransNet [20]	97.36	94.88	0.0441	0.086
Ours	**97.40**	**94.96**	**0.043**	**0.067**
Ours+Self-Correction	**97.43**	**95.03**	**0.046**	**0.066**

UNet [3] and UNet++ [19]. The codes for these methods were obtained from their respective papers, and the network parameters were set in accordance with the corresponding papers. For SETR, we selected the SETR-PUP decoder, and after experimental comparison, we selected the Swin-Unet tiny scale. All these methods maintained consistent training parameters during the training process. The experimental results are shown in Table 1. Due to the use of periodic restart annealing learning rate in the self-correction part, we distinguished the results of the segmentation part from those of the self-correction part in the experimental results. Due to the accuracy of the annotations, the numerical differences in our experimental results are relatively small in order of magnitude, which actually reflects the effectiveness of the network.

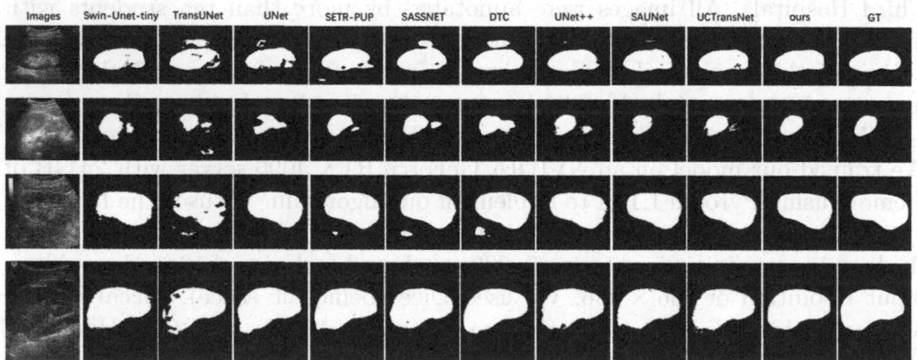

Fig. 5. We compared the segmentation results of 10 methods, including our proposed method, on a kidney ultrasound dataset. The results showed that our method outperformed other methods in terms of edge and shape accuracy.

From the experimental results, our method improved the best results of the previous methods in Dice coefficient and Jaccard from 97.36% to 97.40% and from 94.88% to 94.96%, respectively, in the segmentation part. After adding the self-correction part, the performance of the network was significantly improved, and the Dice coefficient was further improved from 97.40% to 97.43%. Finally, our method achieved the best performance in terms of Dice, Jaccard, ASSD, and 95HD.

There is a serious noise problem in kidney ultrasound images, as shown in Fig. 5. We show the segmentation results of four kidney ultrasound images using 10 different methods. It can be seen that compared to other state-of-the-art methods, our method performs better in preserving the shape, smoothing the edges, and resisting noise interference.

In order to eliminate the noise in the predicted segmentation maps, we show the self-correction process in Fig. 6. It can be seen that the input image in the first cycle has serious noise problems, and each cycle afterwards corrects the predicted results. As can be seen, after 6 cycles, the output of our model is very close to the ground truth, achieving the goal of eliminating multiple types of noise and improving the accuracy and robustness of the model.

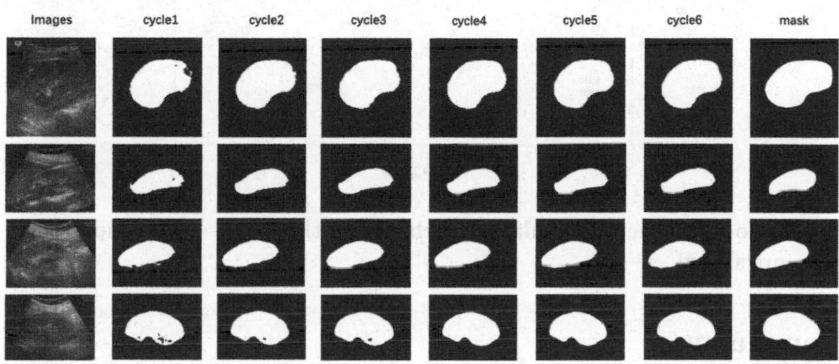

Fig. 6. Displayed are the results of the self-correction process after six cycles, showing that after six cycles, the output of our model gradually approaches the ground truth in terms of accuracy and smoothness.

4.3 Ablation Studies

Our ablation experiments were conducted on a kidney ultrasound dataset, consisting of two parts. The first part compared the effectiveness of each network component, including the channel-spatial attention module(CSA), the dual-task network, and the self-correction structure. As our segmentation network is an improvement on the UCTransNet, we used it as a baseline, and Table 2 shows the results of the network component ablation.

In the second part, we tested the performance of Self-Correction for 10 cycles, where each cycle consists of 5 epochs. The prediction performance of the model after each iteration is shown in Fig. 7. From the experimental results, three cycles can be the optimal number of iterations for self-correction.

Table 2. Ablation experiments on our kidney ultrasound datasets.

Methods	Dice↑(%)	Jaccard↑(%)	ASSD↓(voxel)	95%HD↓(voxel)
baseline(UCTransNet)	97.36	94.88	0.0441	0.086
baseline+CSA	97.38	94.93	0.0439	0.0673
baseline+CSA+Dual-task	97.40	94.96	0.043	0.0670
baseline+CSA+Dual-task+Self-Correction	97.43	95.03	0.046	0.066

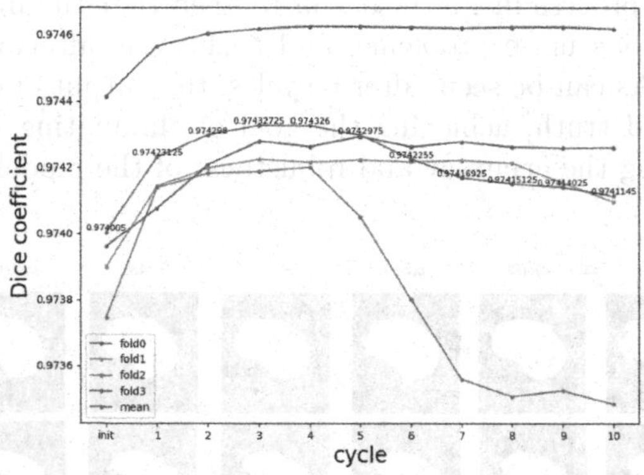

Fig. 7. The model prediction results for each fold in the 4-fold cross-validation of self-correction are shown.

5 Conclusion

In this paper, we propose a two-stage learning framework that integrates the self-correction method into the shape aware dual-task multi-scale fusion network to address the noise problems such as edge noise, holes, edge external noise, and incorrectly segmented targets in kidney segmentation maps. This framework first applies geometric shape constraints to pixel-level kidney segmentation prediction maps to improve the accuracy of the segmentation network's kidney region predictions, and then eliminates noise problems through self-correction learning. Experimental results demonstrate that our framework outperforms various current state-of-the-art methods.

References

1. Levin, A., Stevens, P.E.: Early detection of CKD: the benefits, limitations and effects on prognosis. Journal **7**(8), 446–457 (2011)
2. Torres, H.R., Queiros, S., Morais, P., Oliveira, B., Fonseca, J.C., Vilaca, J.L.: Kidney segmentation in ultrasound, magnetic resonance and computed tomography images: a systematic review. Comput. Methods Programs Biomed. **157**, 49–67 (2018)

3. Ronneberger, O., Fischer, P., Brox, T.: U-Net: convolutional networks for biomedical image segmentation. In: Navab, N., Hornegger, J., Wells, W.M., Frangi, A.F. (eds.) MICCAI 2015. LNCS, vol. 9351, pp. 234–241. Springer, Cham (2015). https://doi.org/10.1007/978-3-319-24574-4_28
4. Cao, H., et al.: Swin-unet: Unet-like pure transformer for medical image segmentation. In: Karlinsky, L., Michaeli, T., Nishino, K. (eds.) ECCV 2022. LNCS, vol. 13803, pp. 205–218. Springer, Cham (2022). https://doi.org/10.1007/978-3-031-25066-8_9
5. Xie, J., Jiang, Y., Tsui, H.: Segmentation of kidney from ultrasound images based on texture and shape priors. IEEE Trans. Med. Imaging **24**(1), 45–57 (2005). https://doi.org/10.1109/TMI.2004.837792
6. Sandmair, M., Hammon, M., Seuss, H., Theis, R., Uder, M., Janka, R.: Semiautomatic segmentation of the kidney in magnetic resonance images using unimodal thresholding. BMC. Res. Notes **9**(1), 1–10 (2016)
7. Marsousi, M., Plataniotis, K.N., Stergiopoulos, S.: An automated approach for kidney segmentation in three-dimensional ultrasound images. IEEE J. Biomed. Health Inform. **21**(4), 1079–1094 (2017). https://doi.org/10.1109/JBHI.2016.2580040
8. Mendoza, C.S., Kang, X., Safdar, N., Myers, E., Peters, C.A., Linguraru, M.G.: Kidney segmentation in ultrasound via genetic initialization and Active Shape Models with rotation correction. In: 2013 IEEE 10th International Symposium on Biomedical Imaging, San Francisco, CA, USA, pp. 69–72 (2013). https://doi.org/10.1109/ISBI.2013.6556414
9. Jokar, E., Pourghassem, H., Linguraru: Kidney segmentation in ultrasound images using curvelet transform and shape prior. In: 2013 International Conference on Communication Systems and Network Technologies, Gwalior, India, pp. 180–185 (2013). https://doi.org/10.1109/CSNT.2013.47
10. Zhao, H., Shi, J., Qi, X., Wang, X., Jia, J.: Pyramid scene parsing network. In: Proceedings of the IEEE Conference on Computer Vision and Pattern Recognition (2017)
11. Chen, L.-C., Zhu, Y., Papandreou, G., Schroff, F., Adam, H.: Encoder-decoder with atrous separable convolution for semantic image segmentation. In: Ferrari, V., Hebert, M., Sminchisescu, C., Weiss, Y. (eds.) ECCV 2018. LNCS, vol. 11211, pp. 833–851. Springer, Cham (2018). https://doi.org/10.1007/978-3-030-01234-2_49
12. Ravishankar, H., Venkataramani, R., Thiruvenkadam, S., Sudhakar, P., Vaidya, V.: Learning and incorporating shape models for semantic segmentation. In: Descoteaux, M., Maier-Hein, L., Franz, A., Jannin, P., Collins, D.L., Duchesne, S. (eds.) MICCAI 2017. LNCS, vol. 10433, pp. 203–211. Springer, Cham (2017). https://doi.org/10.1007/978-3-319-66182-7_24
13. Jackson, P., Hardcastle, N., Dawe, N., Kron, T., Hofman, M.S., Hicks, R.J.: Deep learning kidney segmentation for fully automated radiation dose estimation in unsealed source therapy. Front. Oncol. **8**, 215 (2018)
14. Weerasinghe, N.H., Lovell, N.H., Welsh, A.W., Stevenson, G.N.: Multi-parametric fusion of 3D power Doppler ultrasound for fetal kidney segmentation using fully convolutional neural networks. IEEE J. Biomed. Health Inform. **25**(6), 2050–2057 (2020)
15. Chen, J., et al.: TransuNet: transformers make strong encoders for medical image segmentation. arXiv preprint. arXiv:2102.04306 (2021)
16. Zheng, S., et al.: Rethinking semantic segmentation from a sequence-to-sequence perspective with transformers. In: Proceedings of the IEEE/CVF Conference on Computer Vision and Pattern Recognition, pp. 6881–6890 (2021)

17. Zhuang, X., et al.: Residual Swin transformer Unet with consistency regularization for automatic breast ultrasound tumor segmentation. In: 2022 IEEE International Conference on Image Processing (ICIP), pp. 3071–3075(2022)
18. Liu, Z., et al.: Swin transformer: hierarchical vision transformer using shifted windows. In: Proceedings of the IEEE/CVF International Conference on Computer Vision, pp. 10012–10022 (2021)
19. Zhou, Z., Rahman Siddiquee, M.M., Tajbakhsh, N., Liang, J.: UNet++: a nested U-net architecture for medical image segmentation. In: Stoyanov, D., et al. (eds.) DLMIA/ML-CDS -2018. LNCS, vol. 11045, pp. 3–11. Springer, Cham (2018). https://doi.org/10.1007/978-3-030-00889-5_1
20. Wang, H., Cao, P., Wang, J., Zaiane, O.R.: UctransNet: rethinking the skip connections in u-net from a channel-wise perspective with transformer. In: Proceedings of the AAAI Conference on Artificial Intelligence, pp. 2441–2449 (2022)
21. Yin, S., et al.: Automatic kidney segmentation in ultrasound images using subsequent boundary distance regression and pixelwise classification networks. Med. Image Anal. **60**, 101602 (2020)
22. Sun, J., Darbehani, F., Zaidi, M., Wang, B.: SAUNet: shape attentive U-net for interpretable medical image segmentation. In: Martel, A.L., et al. (eds.) MICCAI 2020. LNCS, vol. 12264, pp. 797–806. Springer, Cham (2020). https://doi.org/10.1007/978-3-030-59719-1_77
23. Ma, J., et al.: How distance transform maps boost segmentation CNNs: an empirical study. In: Medical Imaging with Deep Learning, pp. 479–492. PMLR (2020)
24. Li, S., Zhang, C., He, X.: Shape-aware semi-supervised 3D semantic segmentation for medical images. In: Martel, A.L., et al. (eds.) MICCAI 2020. LNCS, vol. 12261, pp. 552–561. Springer, Cham (2020). https://doi.org/10.1007/978-3-030-59710-8_54
25. Xue, Y., et al.: Shape-aware organ segmentation by predicting signed distance maps. In: Proceedings of the AAAI Conference on Artificial Intelligence, vol. 34, no. 07, pp. 12565–12572 (2020)
26. Luo, X., Chen, J., Song, T., Wang, G., Huang, X.: Semi-supervised medical image segmentation through dual-task consistency. In: Proceedings of the AAAI Conference on Artificial Intelligence, vol. 35, No. 10, pp. 8801–8809 (2021)
27. Zou, H., Gong, X., Luo, J., Li, T.: A robust breast ultrasound segmentation method under noisy annotations. Comput. Methods Programs Biomed. **209**, 106327 (2021)
28. Li, P., Xu, Y., Wei, Y., Yang, Y.: Self-correction for human parsing. IEEE Trans. Pattern Anal. Mach. Intell. **44**(6), 3260–3271 (2020)
29. Woo, S., Park, J., Lee, J.-Y., Kweon, I.S.: CBAM: convolutional block attention module. In: Ferrari, V., Hebert, M., Sminchisescu, C., Weiss, Y. (eds.) ECCV 2018. LNCS, vol. 11211, pp. 3–19. Springer, Cham (2018). https://doi.org/10.1007/978-3-030-01234-2_1
30. Li, C., Xu, C., Gui, C., Fox, M.D.: Level set evolution without re-initialization: a new variational formulation. In: 2005 IEEE Computer Society Conference on Computer Vision and Pattern Recognition (CVPR 2005), vol. 1, pp. 430–436 (2005)
31. Milletari, F., Navab, N., Ahmadi, S.A.: V-net: fully convolutional neural networks for volumetric medical image segmentation. In: 2016 Fourth International Conference on 3D Vision (3DV), pp. 565–571. IEEE (2016)
32. Jang, E., Gu, S., Poole, B.: Categorical reparameterization with gumbel-softmax. arXiv preprint. arXiv:1611.01144 (2016)

Applicability Method for Identification of Power Inspection Evidence in Multiple Business Scenarios

Libing Chen(✉), Wenjing Zhang, and Yiling Tan

Marketing Service Center, State Grid Chongqing Electric Power Company, Chongqing, China
{chenlibing,zhangwj0226}@cq.sgcc.com.cn

Abstract. To address the issues of low accuracy and poor applicability of text recognition in power acquisition images, data enhancement is used to expand samples, YOLOv3 retraining optimization and network compression are used to improve the scope of application, to meet the needs of power inspection evidence identification in multiple business scenarios. Firstly, to avoid insufficient model training due to small sample size, the Transformer model based on text migration is used to generate simulation training samples. Secondly, based on the expanded sample set, the applicability of the YOLOv3 network recognition model is retrained to improve the accuracy of power inspection image recognition. Finally, by combining structural design with knowledge distillation, the retrained YOLOv3 network recognition model is compressed to meet the on-site recognition needs of mobile micro applications. Experiments have proved that the retrained deep YOLOv3 network recognition model has significantly improved the accuracy of recognizing power marketing inspection images. The shallow recognition model effectively improves the timeliness of recognition. Specifically, four types of businesses were selected to carry out pilot applications based on the above methods. The audit time for a single business was less than 1% of the manual. This method effectively improves the efficiency of power inspection.

Keywords: Power material identification · Data augmentation · Yolov3 retraining · Network compression

1 Introduction

At present, with the development of digital transformation of enterprises [1], the power marketing audit has carried out information construction [2]. In the application scenario of marketing audit business, it is necessary to compare and audit unstructured data such as explanatory materials, on-site photos, system screenshots, etc. Traditionally, it relies on manual audit, which has problems such as low efficiency, high work intensity, inconsistent standards. In response to the above issues, in order to accurately identify key information in unstructured power data, this article conducts research on character recognition methods for power audit evidence materials, providing a data foundation for subsequent digital and intelligent audit work.

H. Zhang et al. (Eds.): NCAA 2023, CCIS 1870, pp. 75–89, 2023.
https://doi.org/10.1007/978-981-99-5847-4_6

Currently, commonly used text detection and recognition methods include DBNet text detection algorithm [3], CRNN text recognition algorithm [4], YOLOv3 network [5], etc. For example, reference [6] realizes the recognition of multi-angle and irregular characters by improving the convolutional recurrent neural network (CRNN).Reference [7] improves the DBNet network based on the structural characteristics of Uyghur language to achieve accurate detection. The above methods are all based on specific text recognition requirements, and the text recognition method has been improved to improve the recognition performance. The above methods are based on specific text recognition requirements and have been improved [8]. Therefore, this article adaptively improves the current mainstream YOLOv3 network recognition method based on different types of images obtained from marketing inspection business scene [9], which used to learn the characteristics of power text [10]. However, on-site inspections need to embed the recognition module in mobile devices. The current YOLOv3 network recognition model has problems such as large scale and low real-time performance. Therefore, some institutions and scholars have carried out research on network model compression. The above methods use methods such as network pruning, knowledge distillation [11], parameter quantization, and structural design to compress the deep network individually or in combination [12]. However, in actual marketing inspection application scenarios, the actual amount of image data collected in different business scenarios is relatively small, which often makes it difficult to support the model's retraining and compression parameter adjustment needs [13]. Therefore, the insufficient number of basic training samples leads to poor model recognition performance [14]. The main contributions are as follows:

(1) In view of the current text detection and recognition method is directly applied to character recognition of the power inspection supporting materials, its recognition rate often cannot reach the industrial application standard. Therefore, this paper retrains the YOLOv3 network based on different types of power inspection evidence image samples. By enhancing the recognition and retraining of specific training samples [15], the internal network can learn information that conforms to the characteristics of power collection samples, thereby improving recognition accuracy [16].

(2) The number of power audit evidence samples obtained for different business scenarios is small, and it is difficult to fully train the deep network model. This paper constructs a Transformer model based on text-to-text migration. Perform the same mode simulation generation on the coverage part of the real training sample data input by the Transformer (i.e. the coverage part is data, and the generation is also data) [17]. To ensure that the generated samples can fit the key feature quantities in normal samples (i.e. steel stamp features) [18], this paper uses data augmentation to generate simulation training samples, thereby expanding the training sample set.

(3) In response to the needs of on-site inspection mobile terminal micro-applications, in order to ensure high text detection accuracy and detection speed. This paper adopts the method of combining structure design and knowledge distillation to compress the deep YOLOv3 network recognition mode.

This article expands the training sample set by using the Transformer model of text to text transfer. Then the YOLOv3 network recognition model was optimized by the expanded sample set. Finally, the YOLOv3 network recognition model is compressed using structural design combined with knowledge distillation to improve recognition timeliness.

2 Constructing a Sample Library for Identifying Power Inspection Supporting Materials

In order to solve the problem of data enhancement diversity, this paper adopts the "mask-filling" method, which generates more diverse data based on the Text Transfer Transformer model [19] while maintaining the original training sample structure unchanged. That is to improve on the basis of Transformer's Encoder-Decoder architecture [20]. First, obtain the power description text of a certain type of power text recognition type library, input the encoder module (Encoder), and the Encoder is used as a document word segmentation extraction module, including position encoding and multi-head attention mechanisms, residuals and regularization and feedforward neural networks section, as the word segmentation after document extraction, it intersects with the 183796 commonly used words in the power field preset by the augmentation module (Augment) to obtain text fragments of a certain type of power text recognition type library. Then, perform partial mask processing on the prediction frame area identified by the YOLOv3 network recognition model in the real training sample set, that is, covering a portion of the text fragments. Next, the acquired text fragments of a certain type of power text recognition type library are input into the decoder module (Decoder) of Transformer, and the Decoder training is carried out through feedback adjustment parameters to make it cover the original coverage area filling in the area format, including image pixels and text word segmentation, so as to obtain the simulation training sample set.

It is specifically expressed as related documents based on a certain type of power text recognition type library. The input is a sentence that is contextually related. The Encoder performs sample word segmentation and extraction [10], and pays attention to different information of the input data and the feature enhancement module (Augment) through the multi-head attention mechanism. Through the intersection of commonly used words in the field of electric power, the output of text fragments of different types of libraries is obtained, that is, the enhanced sample pool $Z^{Ns} = \{Z_1^{Ns}, Z_2^{Ns}, Z_3^{Ns}, \cdots\}$

The enhanced sample pool Z^{Ns} is used as the input of the second multi-head attention mechanism in the Transformer module. The Key and Query matrix values in the second multi head attention mechanism in the module correspond to the enhanced feature matrix transformation values of the Encoder and the normal sample extraction feature matrix transformation values, respectively. The Value matrix comes from the output of the previous layer in the Transformer. By enhancing the features, the influence on the distribution of the normal sample feature quantity is realized, and it is adjusted by point multiplication after the Value matrix. Finally, the cross entropy loss function is used:

$$Loss = -[X \ln \hat{X} + (1 - X) \ln(1 - \hat{X})] \tag{1}$$

Calculate the coverage area of the real training samples input and the pattern and pixel error of the expanded training samples by the Transformer's decoder. The parameters of Encoder and Transformer networks are optimized and adjusted by feedback iteration until the loss function converges below the threshold. After the parameters are fixed, the simulation training sample generation module is obtained.

3 Text Recognition Based on YOLOv3 Network

This article is based on the detection method of YOLOv3 network to achieve target positioning and character recognition of images collected in power marketing audit scenes. Taking transformer sign recognition as an example, the specific process is as follows:

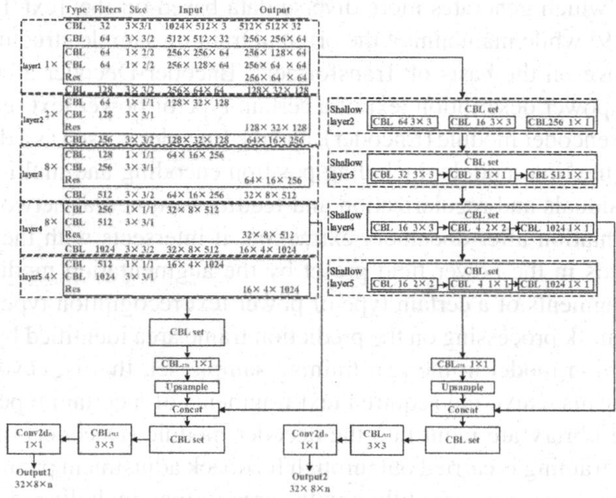

Fig. 1. YOLOv3 network structure diagram

This paper uses the YOLOv3 network to build a recognition model. The feature extraction stage network includes 52 CBL layers, 23 residual layer structures, and 3 feature map outputs of different scales are designed. As shown in Fig. 1, the basic structural unit structure of each network is as follows:

The CBL layer is the network layer of the activation operation represented by the set convolution operation (Convolution), the batch regularization operation (BN, Batch Normal) and the activation function Leaky ReLU. The image sample data is first convolved with a convolution kernels constructed with a weight of $b \times b$. The convolution kernel moves on the image with a defined step size, the gray value of the image pixel is multiplied by the weight in the convolution kernel, and the multiplication value of the same pixel obtained by the RGB three channels of the same convolution kernel is added output eigenvalues as a convolution operation.

The BN operation regularizes the eigenvalues obtained after the convolution operation. The eigenvalues obtained after the batch regularization operation are used to obtain the eigenvalue output of the CBL layer using the activation function LeakyReLU.

A CBLset layer refers to a set of multiple CBL layers. The residual layer (Res) is to add a jump connection on the basis of the original network layer to solve the problem of gradient disappearance in the deep network training process. The upsampling layer (Upsample) refers to the use of interpolation to enlarge the original image, so that it can be displayed on a display device with a higher resolution. The concat layer refers to a multi-scale feature fusion operation, which usually superimposes shallow network features on deep network features to avoid the degradation of contour recognition caused by the deep network training process.

Next, the YOLOv3 network structure diagram outputs $\hat{x}, \hat{y}, \hat{w}, \hat{h}, \hat{p}_1, \cdots, \hat{p}_N$ of each grid unit and each prediction frame, and performs target positioning and character recognition based on the output value. Among them, the target positioning is specifically expressed as a prediction box, and the target localization confidence c is calculated from the predicted position $(\hat{x}, \hat{y})$ and predicted size $(\hat{w}, \hat{h})$ of the prediction box:

$$c = p(Object) * IOU_{pred}^{truth} \tag{2}$$

$$IOU_{pred}^{truth} = \frac{area(t) \cap area(p)}{area(t) \cup area(p)} \tag{3}$$

Among them, $p(Object)$ indicates whether there is a center point of the target $(Object)$ in the grid unit, if it exists, then $p(Object) = 1$, if not, then $p(Object) = 0$. IOU_{pred}^{truth} indicates the intersection ratio of the predicted frame and the real frame area. $area(t)$ indicates the area of the real frame. $area(p)$ indicates The area of the predicted frame is obtained from the predicted position $(\hat{x}, \hat{y})$ and the predicted size $(\hat{w}, \hat{h})$.

Set the target location reliability threshold M, if the target location reliability is $c > M$, add it to the prediction frame set D to be recognized, and complete the character recognition through the prediction category $\hat{p}_1, \cdots, \hat{p}_N$. Specifically, calculate the confidence probability value of the i th type character based on the obtained prediction box, which is $s_i = p_i \times c$.

Among them, p_i represents the conditional probability that the characteristic value in the current prediction frame meets the i nd type of characters, and c represents the confidence level of the current prediction frame.

Take the character class i corresponding to the maximum confidence probability value s_i, which is the character recognition output $C = \arg\max(s_i)$.

Based on the above model structure, the transformer signboard samples taken on site and the simulation training samples are taken together as the sample set, which is divided into the training set and the test set at a ratio of 7:3. Based on the training set, the YOOv3 network recognition model parameters are retrained and optimized, and the parameters are feedback adjusted through the loss function. The process is as follows:

According to the output $\hat{x}, \hat{y}, \hat{w}, \hat{h}, \hat{p}_1, \cdots, \hat{p}_N$ of the YOLOv3 network recognition model, the model loss function is calculated and the parameters are adjusted. The loss function contains three parts, namely, the coordinate size error between the predicted frame and the real frame, the confidence error between the predicted frame and the real frame, and the character recognition classification error.

The calculation formula for the coordinate size error between the predicted box and the real box is:

$$loss_1 = \lambda_{coord} \sum_{i=0}^{S^2} \sum_{j=0}^{B} l_{ij}^{obj} (2 - w_i \times h_i)[(x_i - \hat{x}_i)^2 + (y_i - \hat{y}_i)^2 + (w_i - \hat{w}_i)^2 + (h_i - \hat{h}_i)^2] \tag{4}$$

Among them, λ_{coord} represents the setting position coefficient of the prediction frame, and its value is set to a constant of 5.l_{ij}^{obj} is used to judge whether the j th prediction frame in the i rd grid unit is responsible for the target obj, and the maximum prediction with the real frame IOU_{pred}^{truth} of the target obj The frame is responsible for the target obj, if it is responsible, then $l_{ij}^{obj}=1$, otherwise it is 0. $(\hat{x}_i, \hat{y}_i)$, $(\hat{w}_i, \hat{h}_i)$ represent the position and predicted size of the predicted frame.(x_i, y_i), (w_i, h_i) represent the position and size of the real frame.

The formula for calculating the confidence error between the predicted box and the real box is:

$$loss_2 = \sum_{i=0}^{S^2}\sum_{j=0}^{B} l_{ij}^{obj}(c_i - \hat{c}_i)^2 + \lambda_{noord}\sum_{i=0}^{S^2}\sum_{j=0}^{B} l_{ij}^{noobj}(c_i - \hat{c}_i)^2 \tag{5}$$

where, λ_{noord} represents the coefficient when there is no target obj in the grid unit, and its value is set to a constant 0.5.l_{ij}^{noobj} is used to judge whether the j th prediction box in the i th grid unit is not responsible for the target obj, if not responsible, $l_{ij}^{noobj} = 1$, otherwise 0.c_i indicates the confidence of the real box.$\hat{c}_i$ indicates the confidence of the predicted box.

The formula for calculating category conditional probability error is:

$$loss_3 = \sum_{i=0}^{S^2} l_{ij}^{obj} \sum_{C\in classes} (p_i(C) - \hat{p}_i(C))^2 \tag{6}$$

where, $p_i(C)$ represents the true probability that the i grid unit is class C.$\hat{p}_i(C)$ represents the conditional probability that the i grid unit is predicted to be class C.

$$loss = loss_1 + loss_2 + loss_3 \tag{7}$$

According to the loss function error loss feedback gradient, adjust each weight value in the convolution kernel and update the weight until the error is less than the set threshold. This article uses a test set to test the accuracy of the retrained model. If the accuracy exceeds the set threshold, the YOLOv3 network recognition model is output. If the accuracy is lower than the set threshold, it will be fed back to the training stage for model retraining.

4 Network Compression with Structure Design and Knowledge Distillation

This article describes the model obtained in the previous section as a deep YOLOv3 network recognition model.By using mobile micro applications to capture images on-site, there is a demand for relatively low image resolution, complex background scenes, limited mobile device resources, and high timeliness requirements. Therefore, it is necessary to compress the deep YOLOv3 network recognition model while ensuring recognition accuracy.

This paper uses structural design combined with knowledge distillation to compress the deep YOLOv3 network recognition model. As shown in Fig. 3, the image recognition result of the mobile terminal is obtained by outputting Output2. In the process of extracting the output hidden features of layer 2-layer 5 as shallow network learning parameters, the network structure of layer 2-layer 5 is mainly compressed by structural design decomposition. Taking layer 2 as an example, its output data is $64 \times 16 \times 256$, which can be expressed as a three-dimensional matrix $\kappa_{64,16,256}$, and the structural design decomposes three CBL shallow network representations:

$$\kappa_{64,16,256} \approx A_{64,3\times3} B_{16,3\times3} C_{258,1\times1} \tag{8}$$

Among them, $A_{64,3\times3}$, $B_{16,3\times3}$, and $C_{258,1\times1}$ represent the compressed CBL shallow network.

Based on the hidden data output by layer 1 is the input data of layer 2 shallow network, the output hidden data $64 \times 16 \times 256$ of layer 2 is used as the learning target, and the goal of minimizing the loss function L is the goal, and the gradient feedback adjusts the layer 2 shallow network parameters, as Three concatenated CBL shallow networks with 64 3×3 convolution kernels, 16 3×3 convolution kernels and 256 1×1 convolution kernels, after merging the output values of the 3 shallow networks, as the input hidden data of depth layer 3, that is, replace the original deep layer 2 network layer with the layer 2 shallow network layer, construct a soft parameter tuning loss function and a hard parameter tuning loss function, and jointly feedback and adjust the 3 shallow network parameters. The loss function is:

$$L_{soft}(\hat{p}_i, \hat{p}_i') = -\sum_i \hat{p}_i \log \hat{p}_i' \tag{9}$$

$$L_{hard}\left(\hat{p'}_i, C_i\right) = -\sum_i C_i \log \hat{p'}_i \tag{10}$$

$$L_{cls} = \mu L_{hard} L_{soft}(\hat{p}_i, \hat{p}_i') + (1-\mu) L_{hard}\left(\hat{p'}_i, C_i\right) \tag{11}$$

where, $\hat{p}'$ represents the character recognition probability output after the layer 2 shallow network layer replaces the original deep layer 2 network layer, L_{hard} is the loss value obtained based on the actual character recognition result, C_i is the actual category of the i-th character, and the value is 0 and 1, L_{soft} is the loss value calculated based on the output probability of the deep YOLOv3 network recognition model and the model output probability after layer 2 shallow replacement, using similar cross entropy, the recognition training loss is L_{cls}, and μ is the weight adjustment parameter. The loss function constructed based on the learning objective of the intermediate hidden layer is:

$$L_{reg} = \sum_s \left(h_s - h_s'\right) \log\left(h_s - h_s'\right) \tag{12}$$

where, the learning target loss of the middle hidden layer is L_{reg}, which represents the loss function generated during the learning process of the shallow network on the output of the middle layer, h_s and h_s' respectively represent the output value of the sth hidden layer output by the deep layer and the shallow layer 2, Therefore, the loss function of layer 2 compression parameter tuning is:

$$L = \lambda L_{cls} + (1-\lambda) L_{reg} \tag{13}$$

where, λ is the adjustment ratio. Based on the target loss function L, the feedback parameters are adjusted until the loss function is lower than the set threshold, that is, the compression of the layer 2 shallow network recognition model is completed, and the layer 3-layer 5 network compression is consistent with the principle of layer 2.

After the deep YOLOv3 network recognition model is compressed based on knowledge distillation, Output2 represents the lightweight recognition model of the output result, and the network capacity is compressed from 95.48 MB to 31.53 MB.

5 Experiment and Analysis

This article obtains 2642 electricity marketing inspection images as the sample set by a certain region from August to December 2022. The image categories include system screenshots, energy meters, and transformer identification plates. The specific business application scenarios include user information verification for unassigned meter reading segments, reverse active power anomaly verification, default electricity consumption inspection, and abnormal verification of dismantled energy representation (Fig. 2). Examples of images to be identified are as follows:

Fig. 2. Examples of identification images for different businesses

To demonstrate the effectiveness of the method of this article, the image recognition of transformer identification plates is taken as the specific case in Sects. 5.1 and 5.2. The identified factory number and nameplate capacity data are compared with the user basic file and operational capacity data of the marketing business application system, which is used to identify the super-capacity power users. In Sect. 5.3, we will select four types of business scenarios to demonstrate the effectiveness of method applications.

5.1 Training Sample Augmentation Quality Assessment

Before the training of the deep YOLOv3 network recognition model, the captured pictures were first normalized to a size of 1024×512, 450 pictures were randomly extracted as the training sample set, and the remaining 192 pictures were used as the test sample set. Due to the limited number of training samples obtained based on actual data, it is difficult to meet the needs of sufficient training for deep neural networks. Therefore, this article uses data augmentation to generate training samples based on 450 training sample set images and historical archived texts. The quality of the generated samples is evaluated using two metrics: similarity and diversity. The specific indicators are defined as follows:

(1) Similarity SSIM indicator

$$SSIM\,(X,\widehat{X}) = \frac{(2\mu_X \mu_{\widehat{X}} + \text{cov}_1)(2\sigma_{X\widehat{X}} + \text{cov}_2)}{(\mu_X^2 + \mu_{\widehat{X}}^2 + \text{cov}_1)(\sigma_X^2 + \sigma_{\widehat{X}}^2 + \text{cov}_2)} \tag{14}$$

Among them, the similarity index defines structural information from the perspective of image composition as independent of brightness and contrast, reflecting the properties of object structure in the scene, and models distortion as a combination of three different factors: brightness, contrast, and structure. The mean μ is used as an estimate of brightness, the standard deviation σ as an estimate of contrast, and the covariance cov as a measure of structural similarity. The closer the SSIM index value is to 1, the more similar the training sample image is to the real sample image.

(2) Diversity indicator Distinct (abbreviated as Dist):

$$Dist = -p(C^n|\widehat{C}^n)\log p(C^n|\widehat{C}^n) \tag{15}$$

Among them, there are n types of characters in total in the power transformer sign sample library, C^n represents the actual number category of characters, and $\hat{C}^n$ represents the number of character categories in the generated training samples. The more balanced the character distribution of the generated training samples, the higher the entropy value, and the closer the Dist value is to 0, which means the higher the diversity of the generated samples.

Based on the current research status of generated samples, three methods are selected for comparison, as follows:

Method 1: Generate extended training sample data based on randomly scrambled characters [11].

Method 2: Generate expanded training sample data based on Encoder + GAN, that is, extract training sample text data through Encoder, use it as the data input of GAN.

Method 3: Generate extended training sample data based on Transformer not including enhanced feature modules.

Method 4: The method in this paper generates expanded training sample data.

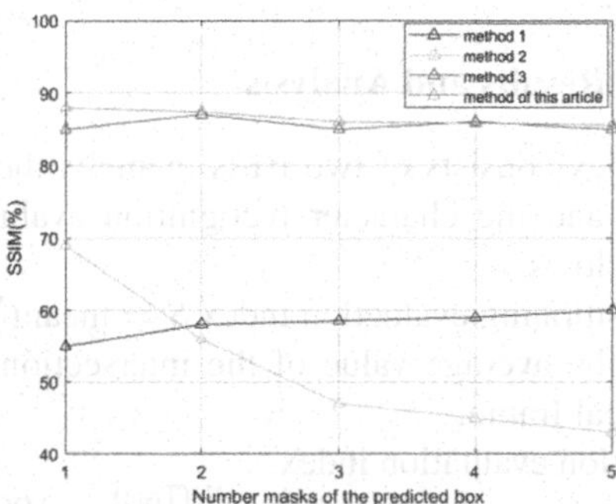

Fig. 3. Generated sample similarity index comparison chart

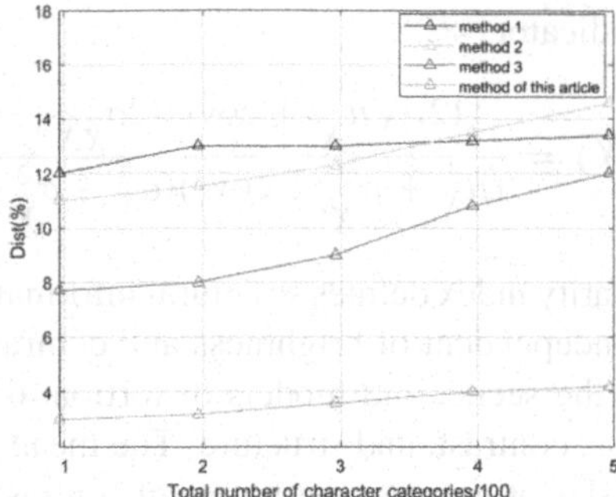

Fig. 4. Comparison of generated sample diversity evaluation indicators

Figure 3 and Fig. 4 show the evaluation results of similarity index SSIM and diversity index Dist under the requirements of text recognition in different scenarios. The similarity SSIM of method 1 is about 55%, and the diversity index Dist is 13%. With the increase of the number of coverage prediction boxes and character categories, its values are relatively stable. This is because the synthetic data is difficult to fit the state of the captured image, and the random extraction also shows a certain proportion of fat characters, so the effect is stable but the simulation effect is poor. In method 2 the generated data is unstable, since the generated data is not sample generation based on actual real sample data. Especially with the increase of the coverage prediction frame and the number of characters, the decrease ratio increases significantly. In method 3, the similarity index SSIM of the generated samples is similar to the method in this paper, but the diversity index Dist is worse than the method in this paper, indicating that the data enhancement extraction module can effectively extract power word segmentation features. Method 4 is the method of this paper, which uses the data enhancement extraction module to directional strengthen the power word segmentation of relevant scenes to maximize the available information, and retains the influence of image features of real training samples, so that in the similarity index SSIM and diversity index Dist, both show its superiority. Therefore, the training samples expanded by the method in this paper have better diversity and rationality.

5.2 Model Recognition Results and Analysis

The model evaluation index consists of two parts, namely, the prediction frame positioning evaluation index and the character recognition evaluation index, which are respectively defined as follows.

(1) Prediction box positioning evaluation index $S = \text{mean}(\sum IOU_{\text{Pr ed}}^{\text{Truth}})$.

Where, S represents the average value of the intersection and union ratio of the predicted frame and the real frame.

(2) Character recognition evaluation index.

Accuracy of text character recognition $F_1 = \frac{N_1(\text{True})}{N_1(\text{Total})} \times 100\%$.

Where, $N_1(\text{True})$ represents the number of correctly recognized text characters, $N_1(\text{Total})$ represents the total number of text characters, and $N_1(\text{Total}) = 13231$.

Text character line recognition accuracy rate$F_2 = \frac{N_2(\text{True})}{N_2(\text{Total})} \times 100\%$

Where, $N_2(\text{True})$ represents the number of correctly recognized text character, $N_2(\text{Total})$ represents the actual total number of text character. $N_2(\text{Total}) = 2313$.

This paper takes the transformer signboard text recognition as an example. As shown in the Fig. 5, it is a sample library for non-transformer signboard text recognition. The current general YOLOv3 method is used to build a recognition model, and use it to recognize the text on the transformer signboard, as shown in the Fig. 6. Since the characters on the transformer signboard contain electric text and embossed features, the current general recognition method is not effective. The specific on-site inspection results are as follows:

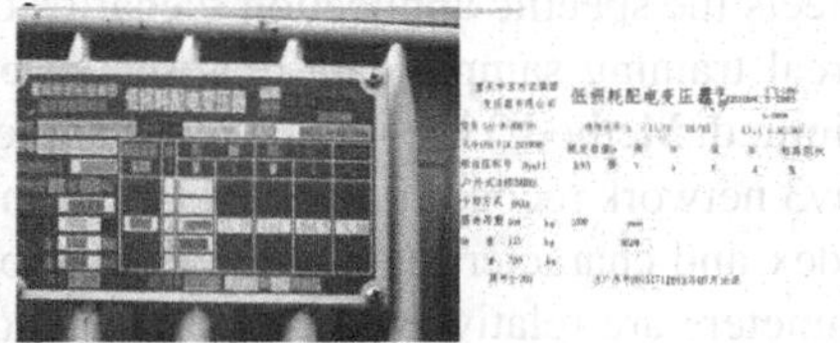

Fig. 5. Without retraining optimization

Fig. 6. Retraining optimization

In this paper, based on the transformer signboard image taken at the current site, the data enhancement method is used to generate training samples to realize the expansion of the sample set, and based on the expanded training sample set, the YOLOv3 network recognition model is trained. The sample of the test recognition result is shown in the figure. The recognition effect of the prediction frame, stamped text and overall character is better than that of the picture, and the specific recognition comparison effect is shown in Table 1:

Table 1. Training parameters of YOLOv3 network recognition model

Recognition result	Prediction box positioning evaluation index S/%	Accuracy of text character recognition F_1/%	Text character line recognition accuracy F_2/%	average recognition time /ms
Method 1	48.73	56.20	70.31	25.92
Method 2	65.31	78.47	88.26	22.88
Method 3	80.57	93.32	98.52	23.90
78.65	91.56	97.84	11.86	78.65

Method 1: The model training optimization is not performed based on the transformer signboard text recognition sample library.

Method 2: Use the YOLOv3 method to build a recognition model, and perform model training and optimization based on the transformer signage text recognition sample library.

Method 3: Use the YOLOv3 method to build a recognition model, expand the transformer signboard text recognition sample library, and then optimize the model training based on real samples and simulation samples.

Method 4: The shallow YOLOv3 network recognition model is constructed by using the structural design described in this article.

It can be seen from Table 1 that method 1 directly uses the general YOLOv3 method to detect transformer signboard characters, and its prediction frame location evaluation index, text character recognition accuracy rate, and text character line recognition accuracy rate cannot meet the application standards of specific power scenarios. For the self-trained text character recognition model represented by methods 2–4, the above three indicators have been significantly improved, indicating that the self-trained network can learn sample text information that meets the specific application scenarios of electric power. However, due to the lack of real training sample data represented by method 2, the effect of model optimization is limited. Method 3 is based on real samples and simulation samples to optimize the YOLOv3 network recognition model twice, and its prediction frame positioning evaluation index and character recognition evaluation index are the best. However, its network parameters are relatively large, and it is not suitable for on-site photography of micro-app mobile terminals. Therefore, method 4 of this paper uses structural design combined with knowledge distillation to compress the model trained by method 3. Compared with the deep YOLOv3 network recognition model, the prediction frame positioning evaluation index, text character recognition accuracy and text character line of the shallow YOLOv3 network recognition model The recognition accuracy rate is reduced to less than 2%. However, the size of the model parameters is only one-third of the original one, and the average recognition time efficiency is improved, which is only half of the time-consuming of the original model. Therefore, method 3 to build a model is suitable for the system side, and method 4 to build a model is suitable for the mobile terminal.

5.3 Application Effect of Intelligent Verification in Power Inspection

To demonstrate the feasibility and applicability of the method proposed in this article, a pilot application was carried out. We select four types of intelligent verification business scenarios, which include user information verification for unassigned meter reading segments, reverse active power anomaly verification, default electricity consumption inspection, and abnormal verification of dismantled energy representation. The specific verification process is described as follows:

Business 1: user information verification for unassigned meter reading segments.

Business 2: Reverse active power anomaly verification.

Business 3: Default electricity consumption inspection.

Business 4: Abnormal verification of dismantled energy representation.

In different business scenarios, according to the processing flow, we select deep or shallow YOLOv3 network recognition models that are retrained based on power image samples. Taking the default electricity consumption inspection as an example, a shallow model is embedded in the mobile and a deep model is used on the system side, both of which recognize the transformer identification plate. The system screenshot identifies the user's electricity consumption data. The above data is input together into

Table 2. Application effect display of power intelligent verification

Business Scenario Category	Manual verification	Intelligent verification application		
	Verification time consumption/s	System screenshot recognition/s	Mobile image recognition/s	Verification time consumption/s
Business 1	74.29	2.39×10^{-2}	/	0.645
Business 2	150.00	2.39×10^{-2}	/	0.7
Business 3	197.20	2.39×10^{-2}	1.186×10^{-2}	1.03
Business 4	188.46	2.39×10^{-2}	1.051×10^{-2}	0.84

the default electricity consumption inspection customization process to automatically identify abnormal users.

As shown in Table 2, this article sets the task volume for four types of businesses, all of which are 50 cases. Compare the time efficiency between the average manual time consumption and the average intelligent verification time consumption of single instance business processing. The intelligent verification time consumption for the four types of business has been reduced to 0.87%, 0.47%, 0.52%, and 0.45% of the manual time consumption, respectively. The efficiency of verification has been increased by 100 times, and the accuracy rate is not less than 80%. Especially in abnormal verification of disturbed energy representation, image recognition is digital recognition, so the accuracy is the highest, which only one verification error. Transformer identification plates are mostly engraved, so the recognition accuracy is low. But if the captured images are clear and standardized, this method can accurately identify and accuracy rate of 82%. Therefore, there are 8 cases of work orders that cannot be automatically verified and entered the manual review process in the default electricity consumption inspection. In summary, the application of this method can effectively reduce labor costs and improve electricity.

6 Conclusion

Aiming at the problems of low accuracy of power marketing inspection image recognition and single applicability of business scenarios, a multi-business scenario-oriented power inspection evidence recognition applicability method is constructed. First, the Transformer model based on text-to-text migration is used to generate simulation training samples, so as to realize the expansion of the training sample set. Subsequently, based on the power real sample set and the simulation sample set, the applicability training of the YOLOv3 network recognition model is carried out to ensure that its internal network can learn information that conforms to the characteristics of the power corpus. Finally, in response to the needs of on-site shooting and recognition of mobile micro-applications, the structure design combined with knowledge distillation is used to compress the YOLOv3 network recognition model. On the premise of ensuring recognition accuracy, the network scale is reduced and the timeliness of recognition is improved.

Experiments have proved the effectiveness and superiority of the method described in this paper. During the on-site inspection of electric power marketing, it accurately identifies and extracts key information from unstructured data, greatly reducing the workload of inspection personnel, and providing digital and intelligible technical support for subsequent audit work.

References

1. Niu, R.K., Zhang, X.L., Wang, Y.J., et al.: Electric power marketing inspection business supervision system based on data mining. J. Jilin Univ. (Inf. Sci.e Ed.) **40**(01), 103–110 (2022)
2. Wang, Y.F., Wang, J.Y., Zhong, L.L., et al.: Text recognition of power equipment nameplates based on deep learning. Electr. Power Eng. Technol. **41**(05), 210–218 (2022)
3. Wang D.L., Kang B., Zhu R.: Text detection method for electrical equipment nameplates based on deep learning [J/OL].J. Graph. 1–9 (2023)
4. Dong, W.Z., Chen, Y., Liang, H.L.: Improved area mark recognition method of DBNet and CRNN. Comput. Eng. Des. **44**(01), 116–124 (2023)
5. Gong, A., Zhang, Y., Tang, Y.H.: Automatic reading method of electric energy meter based on YOLOv3. Comput. Syst. & Appl. **29**(01), 196–202 (2020)
6. Yan Y.J.: Research on natural scene text recognition based on CRNN algorithm. Xidian University (2020)
7. Wang D.Q.: Scene Uyghur character detection system based on improved DBNe. Xinjiang University (2021)
8. Zhou, Y., Zhang, Y., Wang, C., et al.: Detection and identification of digital display meter of distribution cabinet based on YOLOv5 Algorithm. Neural Comput. Adv. Appl. NCAA 2022. Commun. Comput. Inf. Sci. **1637**, 301–315(2022)
9. Qu, C.R., Chen, L.W., Wang, J.S., Wang, S.G.: Research on industrial digital meter recognition algorithm based on deep learning. Appl. Sci. Technol. 1–7 (2022)
10. Long, S., He, X., Yao, C.: Scene text detection and recognition: the deep learning era. Int. J. Comput. Vis. **129**, 1–24 (2020)
11. Zhao X.J., Li H.L.: Deep neural network compression algorithm based on hybrid mechanism [J/OL].J. Comput. Appl. 1–8 (2023)
12. Liu Z., Sun J.D., Wen J.T.: Bearing fault diagnosis method based on multi-dimension compressed deep neural network. J. Electron. Meas. Instr. **36**(07), 189–198 (2022)
13. Liao, M., Wan, Z., Yao, C., et al.: Real-time scene text detection with differentiable binarization. In: AAAI Conference on Artificial Intelligence, pp. 11474–11481 (2020)
14. Gao, X., Deng, F., Yue, X.H.: Data augmentation in fault diagnosis based on the Wasserstein generative adversarial network with gradient penalty. Neurocomputing **396**, 487–494 (2020)
15. Zhang, S., Liu, Y., Jin, L., et al.: Feature enhancement network: a refined scene text detector (2017). arXiv: 171104249
16. Zhu, S., Han, F.: A data enhancement method for gene expression profile based on improved WGAN-GP. Neural Comput. Adv. Appl. NCAA 2021. Commun. Comput. Inf. Sci. **1449**, 242–254 (2021)
17. Ponnoprat, D.: Short-term daily precipitation forecasting with seasonally-integrated autoencoder. Appl. Soft Comput. **102**, 107083 (2021)
18. Chu, J., Cao, J., Chen, Y.: An ensemble deep learning model based on transformers for long sequence time-series forecasting. Neural Comput. Advanced Appl. NCAA 2022. Commun. Comput. Inf. Sci. **1638**, 273–286 (2022)

19. Hwang M.H., Shin J.K., Seo H.J.,et al.: Ensemble-NQG-T5: ensemble neural question generation model based on text-to-text transfer transforme. Appl. Sci. **13**(2) (2023)
20. Qiu, X., Ren, Y., Suganthan, P.N., Amaratunga, G.A.J.: Empirical mode decomposition based ensemble deep learning for load demand time series forecasting. Appl. Soft Comput. **54**, 246–255 (2017)

A Deep Learning Algorithm for Synthesizing Magnetic Resonance Images from Spine Computed Tomography Images using Mixed Loss Functions

Rizhong Huang[1], Menghua Zhang[1](✉), Ke Huang[2], and Weijie Huang[1](✉)

[1] School of Electrical Engineering, University of Jinan, Jinan 250000, China
zhangmenghua@mail.sdu.edu.cn, cse_huangwj@ujn.edu.cn

[2] School of Information Science and Engineering, Shandong Normal University, Jinan 250000, China
huangke@sdnu.edu.cn

Abstract. A new supervised approach based on Generative Adversarial Networks (GAN) is proposed to address the challenging task of image translation between CT and MR images. The proposed algorithm utilizes a mixed loss function with three components to synthesize high-quality MR images from human spine CT images. The loss of structural consistency aims to enhance the structural perception of images and maintain consistency between the converted and original images. The adversarial loss trains the generator to produce realistic images that cannot be distinguished from real MR images by the discriminator. The pixel translation loss preserves all the details of the input CT image while reducing contour loss. By combining these three loss components with their respective weights, the synthesized MR image can retain more details and reduce image distortion. The experimental results demonstrate that our method outperforms current mainstream algorithms in MAE and PSNR evaluation metrics. This approach provides a promising solution for generating high-quality MR images from CT images, which can benefit many applications in the field of medical imaging.

Keywords: Deep learning · Computed tomography · Magnetic resonance imaging · Mixed loss function

1 Introduction

Spinal diseases are becoming increasingly common in young people due to changes in lifestyle. The diagnosis and treatment of spinal diseases have greatly improved with the introduction of computed tomography (CT) and magnetic resonance (MR). CT scans has good imaging effects on bones, providing clear images of bone structures. On the other hand, MR images have powerful soft tissue imaging capabilities, providing clear images of soft tissues, such as nerves and disks. But MR images are difficult to obtain and expensive. Many patients opt for CT scans as their first choice. With the development of deep learning, many scholars have applied it to the field of medical imaging, which

H. Zhang et al. (Eds.): NCAA 2023, CCIS 1870, pp. 90–102, 2023.
https://doi.org/10.1007/978-981-99-5847-4_7

has made image translation work possible. The MR spine images from CT spine images can provide doctors with a large amount of medical diagnostic information in medical diagnosis while reducing the economic pressure on patients.

In the field of image translation, Generative adversarial network (GAN) [1] is particularly favored by many scholars due to its powerful mechanism, as it can perfectly complete the task of synthesizing MR images from spinal CT images. GAN consists of a generator (G) and a discriminator (D), where the generator tries to generate values that deceive the discriminator as real, while the discriminator attempts to differentiate between real and generated values. Through continuous adversarial training, G and D improve their learning abilities until they reach a Nash equilibrium, allowing G to output results that are extremely close to reality. Using the unique adversarial mechanism of GAN, CT images can be synthesized into MR images with better performance. However, there are many challenges in the image translation work of synthesizing MR images from CT images. Firstly, the quality of synthesized MR images highly depends on the training dataset. A dataset with paired CT and MR images can facilitate learning more accurate mapping relationships from CT images to MR images. However, obtaining large number of paired data with different patterns is challenging in the field of spinal medicine, and the available data often cannot successfully match between CT and MR images. Meanwhile, the synthesized MR images may have issues such as low accuracy, lack of detailed information, and translation errors. Further research is needed to address this limitation and improve the application of CT images in spinal medical MR image synthesis.

In order to solve many of the problems mentioned above, we propose a set of mixed loss function to solve the problem of low quality of output images in the process of image translation. The main contributions of this article are as follows:

1. A mixed loss function is proposed to ensure the structural consistency between the translated image and the original image, reduce the loss of image contour translation, and enrich the details of the output image.
2. A successfully matched CT image and MR image dataset are established. This dataset includes 552 pairs of successfully matched CT images and MR image data.

2 Related Work

Generative Adversarial Networks have become a popular image processing method with the potential to generate high-quality images. However, obtaining matching pairs of datasets can be difficult, leading researchers to focus on unsupervised learning with GANs. Zhu et al. proposed an unsupervised image translation model based on the generation countermeasure network (CycleGAN) [2] to solve the problem of the lack of paired data sets between different domains. By introducing a cyclic consistency loss function, the model can conduct unsupervised learning without paired data, effectively improving the consistency of image translation, and greatly enhancing the practicality of the model. Radford et al. [3] proposed a method for image generation using deep convolutional generative adversarial networks (DCGAN) to solve the problems of slow training speed and difficulty in generating realistic images. DCGAN uses convolutional neural networks to construct the generator and discriminator, allowing the network to learn high-quality feature representations from images and creating clearer and more coherent generated images. Moreover, DCGAN eliminates fully connected layers, effectively avoiding overfitting and speeding up model training. Kaneko et al. [4] proposed an improved method for CycleGAN, which solved the problem of difficulty in learning from the source target to the real target. Through step-by-step adversarial loss and optimization of the generator and discriminator, the mapping from the source target to the real target can be completed without relying on parallel data. Karras et al. [5] designed a new GAN training method that addresses the difficulty of generating high-quality images, starting with lower-resolution image and gradually increasing the resolution by adding layers, making training more stable and achieving significant success in generating high-quality and more complex images.

Although unsupervised learning can address the problem of unmatched training data, generating image quality is still challenging in medical imaging. Thus, in this field, supervised learning with GANs using paired datasets is preferred. Mirza et al. [6] proposed an extended model for generating adversarial networks (CGAN), which solves the problem of difficulty in controlling specific attributes of generated images. By introducing additional information y as a condition in the generator and discriminator, images with specific attributes and features can be generated as needed, greatly improving the controllability of the image translation process. Yoon et al. [7] proposed a new GAN training method (OUT-GAN) based on mixed regularization aimed at addressing the collapse phenomenon and sample instability in image translation. Its core idea is to introduce multiple regularization terms during generator training to enhance the model's robustness and stability. TTUR [8–10] is a new GAN training method that dynamically adjusts the learning rates of the generator and discriminator to ensure a balance between them during training, increasing training stability. Also, TTUR introduces gradient penalty techniques to address difficulty in generating high-quality samples, improving the performance of the generator. Wang et al. [11] proposed a method for generating high-resolution images from low-resolution inputs, addressing the difficulty of generating high-resolution images from low-resolution images. This method uses convolutional neural networks to learn the mapping function from low-resolution images to high-resolution images, allowing semantic operations to be performed on generated images by changing input attributes and generating higher-quality high-resolution images. This method has excellent versatility.

3 Method

Our proposed framework consists of a generator G and a discriminator D, as illustrated in Fig. 1. The generator takes a CT image (x) as input and generates an MR image (y), aiming to achieve $y = G(x)$. The discriminator evaluates the realism of the generated image y by receiving both y and Ground-truth MR image as input, and tries to classify them as real or synthetic. The generator and discriminator compete with each other in a game-like process, which we will describe in detail in Sects. 3.1 and 3.2. To train the model, we use a mixed loss function that comprises three parts: adversarial loss L_{ad}, pixel translation loss L_{pi}, and structure consistency loss L_{st}, which are explained in Sects. 3.3, 3.4, and 3.5.

Our framework aims to achieve high-quality image translation from CT to MR, which has important applications in medical imaging. By integrating adversarial training and structural constraints, our model can generate realistic MR images while preserving the underlying anatomical structures of the original CT images.

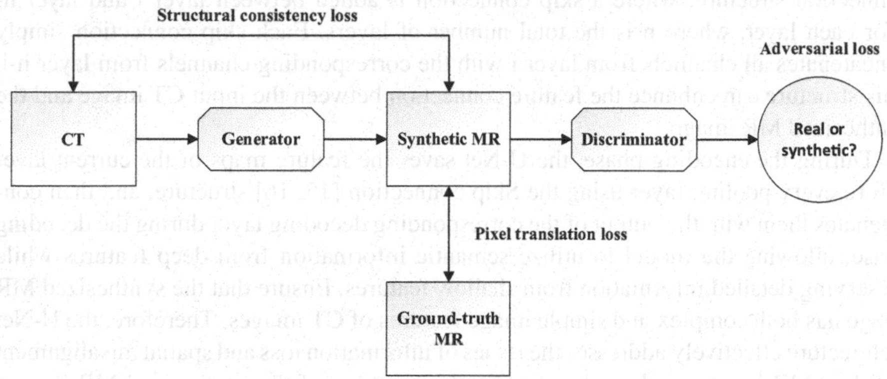

Fig. 1. Framework structure overview

3.1 Generator

The general convolution structure will limit the learning ability of the model due to the weight sharing property. At the same time, the convolution neural network is prone to overfitting when there are many parameters. Therefore, U-Net [12–14] is used in generator structure, as shown in Fig. 2.

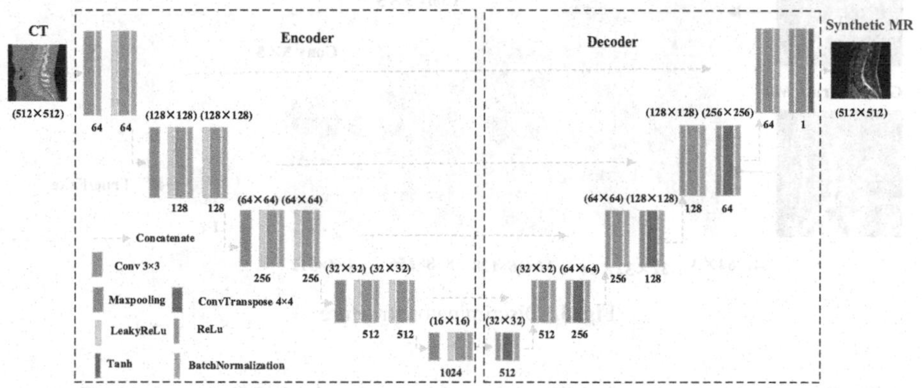

Fig. 2. Generator structure

The U-Net architecture is a symmetric Encoder-Decoder model that consists of repeated 3 × 3 convolutions, each with a BatchNormalization layer and LeakyReLu layer, and maximum pooling layers are used for downsampling with a stride of 2. The feature channels are multiplied by 2 at each downsampling step. The decoder consists of repeated 3 × 3 convolution layers and 4 × 4 ConvTranspose layers, each with ReLu and BatchNormalization layers. In the last layer, a Tanh activation operation is used to map the synthesized MR image. Additionally, the U-Net architecture includes a Skip connection structure, where a skip connection is added between layer i and layer n-i for each layer, where n is the total number of layers. Each skip connection simply concatenates all channels from layer i with the corresponding channels from layer n-i. This structure can enhance the feature connection between the input CT image and the synthesized MR image.

During the encoding phase, the U-Net saves the feature maps of the current layer before every pooling layer using the Skip connection [15, 16] structure, and then concatenates them with the output of the corresponding decoding layer during the decoding phase, allowing the model to utilize semantic information from deep features while preserving detailed information from shallow features. Ensure that the synthesized MR image has both complex and simple image features of CT images. Therefore, the U-Net architecture effectively addresses the issues of information loss and spatial misalignment in CT-to-MR image synthesis, improving the accuracy of the synthesized MR images and the stability of the generator G structure.

3.2 Discriminator

We introduce the PatchGAN [18, 19, 20] discriminator structure on the discriminator D, as shown in Fig. 3.

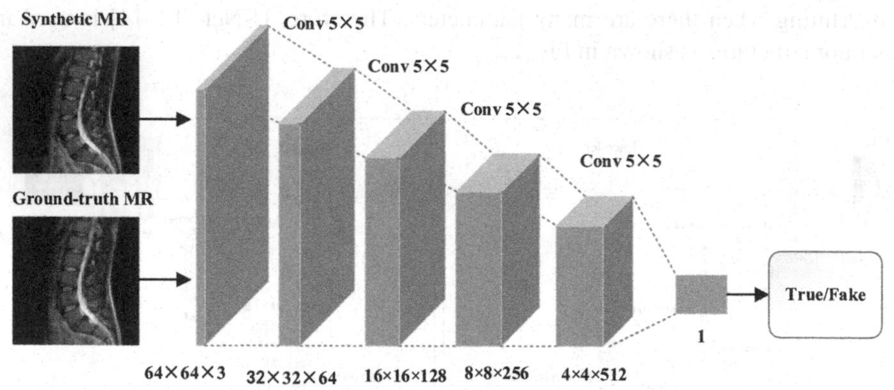

Fig. 3. Discriminator structure

The PatchGAN structure is utilized to divide the Synthetic MR image and Ground-truth image into multiple patches of fixed size (N × N). These patches are then inputted to the discriminator D for assessment. Slicing the image into patches accelerates training, improves global information capture, and allows for a finer understanding of local areas. The discriminator provides a true or false judgment for each patch and averages them to produce the final output. A probability value between 0 and 1 is outputted by the discriminator, denoting the authenticity of the image. If the probability value is close to 1, it indicates that the image is realistic and possibly real. Conversely, a probability close to 0 suggests a fake generated image. Feedback from the discriminator is used to adjust generator parameters, improving the realism and detail clarity of the generated images.

3.3 Mixed loss function

In order to improve the accuracy of MR images after conversion and preserve spine detail information, we propose a new generator mixed loss function $L_G = \lambda_{ad}L_{ad} + \lambda_{pi}L_{pi} + \lambda_{st}L_{st}$. The mixed loss function is divided into three parts: the adversarial loss L_{ad} and its corresponding weight λ_{ad} , the pixel translation loss L_{pi} and its corresponding weight λ_{pi}, the structural consistency loss L_{st} and its corresponding weight λ_{st} , the discriminator's loss function .

Adversarial Loss
The Adversarial Loss minimizes the difference between the generator G and the discriminator D. The generator G is pushed to generate more representative target images, deceiving the discriminator D, which, in turn, is forced to learn how to distinguish generated images from real ones. Consequently, the generator G can progressively improve its ability to generate better target images. The discriminator is defined as:

$$L_{dis} = \frac{\mathbb{E}_{\hat{y}}\left[\left(1 - D\left(\hat{y}\right)\right)^2\right] + \mathbb{E}_x\left[(D(G(x)))^2\right]}{2} \tag{1}$$

where, G is the generator, D is the discriminator, x is the Synthetic MR image, The former term of formula (1) represents the probability that the generated image is judged as fake, and the latter term represents the probability that the generated image is judged as real. Similarly, the optimization objective of the generator is as follows:

$$L_{ad} = \mathbb{E}_x\left[(1 - D(G(x)))^2\right] \tag{2}$$

Pixel Translation Loss
In the task of image translation from CT images to MR images, we aim to generate synthetic MR images from CT images using GAN while retaining the pixel information from the original CT images. Hence, a loss function is required during network training to ensure that the synthesized MR image closely resembles the ground-truth MR image. Pixel translation loss serves as such a loss function by computing the average Euclidean distance between the generated and real images. At each Patch level, the discriminator D determines whether an image is "real" or "synthetic," and this judgment information can be used for fine prediction by interpolating it back to the size of the input synthetic MR image.

$$Q_{pi}(y) = f_{\text{interp.}}(\text{ReLU}(D(y))) \tag{3}$$

where, $f_{\text{interp.}}()$ is a function interpolated back to the synthetic MR image size and min-max normalized, applying ReLU activation function to preserve positive values. On the basis of formula (3), we obtain the L_{pi}:

$$L_{pi} = \mathbb{E}_{x,\hat{y}}\big[\big\|\big(\hat{y} - G(x)\big) \odot Q_{pi}(G(x))\big\|_1\big] \tag{4}$$

where, $\| \ \|_1$ represents the L_1 norm of the Euclidean distance [20–22], L_{pi} focuses on the difference at the pixel level to ensure that the generated image and the real image are as close as possible every pixel.

Structural Consistentcy Loss

In our approach, we aim to ensure that the output Synthetic MR image y is structurally consistent with the input CT image x. However, during our specific training, we observed that there were occasional structural differences between the output Synthetic MR images and CT images. To address this issue, we propose a structural consistency loss. To implement this loss, we use spatially-correlative maps (SCM) to attenuate the structural differences that exist after CT images are replaced by MR images. The SCM map M^d is used to capture the correlation between a pixel and other pixels γ in the patch level. We obtain the C channel feature map N_x of the CT image from the structure encoder. In a two-dimensional patch with an estimated size of $H \times W$, the correlation map mapping between CT image x and query pixel d is defined as follows as:

$$M_x^d = \left(N_x^d\right)^{\top}\left(N_x^{\gamma}\right) \tag{5}$$

where, $N_x^d \in \mathbb{R}^{1\times C}$ is the feature of the query pixel d with C channels, $N_x^d \in \mathbb{R}^{C\times H\times W}$ is the corresponding feature of all the query pixels of each patch in the $H \times W$ area. $M_x^d \in \mathbb{R}^{1\times H\times W}$ captures the spatial correlation of features between the query pixel and other pixels. Randomly select I sampling point on N_x. According to formula (1), get $M_x = \left[M_x^1, M_x^2, M_x^3, \ldots, M_x^I\right] \in \mathbb{R}^{I\times H\times W}$, which is a semi-sparse expression with higher computational efficiency, Similarly, we can get the expression $M_y = \left[M_y^1, M_y^2, M_y^3, \ldots, M_y^I\right] \in \mathbb{R}^{I\times H\times W}$ of M_y. We define the structural consistency loss as the cosine of M_x and M_y :

$$L_{st} = \left\|1 - \cos\left(M_x, M_y\right)\right\|_1 \tag{6}$$

4 Experiment and Analysis

4.1 Paired Dataset and Pretreatment

In our experiment, we used a total of 1104 medical image data from real patients with spinal diseases. These data were divided into 552 groups and each dataset contained successfully paired CT pictures and MR pictures. The size of each picture was 512×512 pixels, including 441 training sets (80%) and 110 test sets (20%). Before training, we performed preprocessing steps to ensure the quality of the data. Firstly, we used 3D Slicer software to pair the pictures that were not successfully paired. Secondly, we corrected

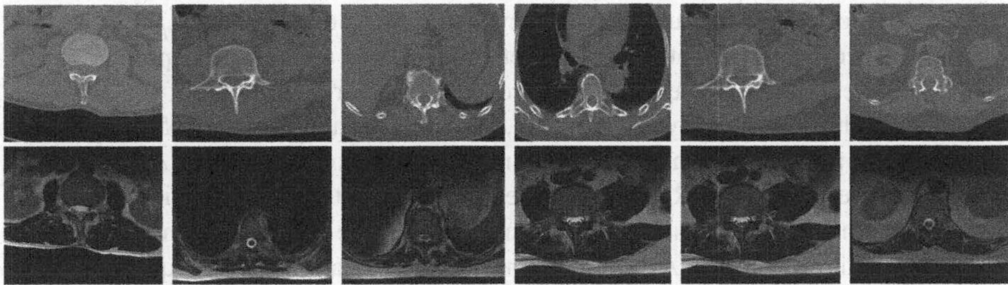

Fig. 4. Transverse sections paired pictures of spine

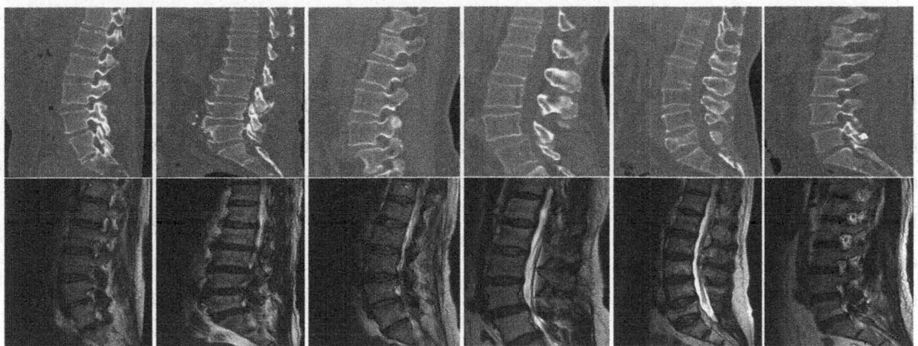

Fig. 5. Sagittal sections paired pictures of spine

the deviation of the spine CT and MR images processed by 3DSlicer [23, 24] software using the N4 ITK [12, 25, 26] algorithm. Finally, spine medicine experts conducted a final verification to ensure the accuracy of the images. To ensure efficient training, we augmented the images used for training and resized all images to uniform grayscale. For images with non-uniform sizes, we cropped all real CT and MR images using a standardized method. Horizontally, the images were sliced at the most ventral part of the vertebral body. Next, in the dorsal direction, they were cut from the center of the dural sac to the ventral endplate of the vertebral body. Finally, a vertical cut was made downward at the same height as the center of the intervertebral disc to ensure that the images required for training had a uniform format. We present several successfully paired CT and MR images, comparing transverse sections (Fig. 4) and sagittal sections (Fig. 5).

4.2 Experimental Setup and Evaluation Methods

During the experimentation phase, our model was trained on the PyTorch framework using 882 images (80%) for training and 220 images (20%) for validation from a total of 1104 images. We utilized two GeForce GTX3090 GPUs in an Ubuntu20.04 environment, and the training process lasted for 200 epochs with a batch size of 2. Both the CT input images and synthesized MR images were kept at a size of 512×512 pixels. To optimize the performance of the model, we determined that the weights for adversarial loss, pixel displacement loss, and structural consistency loss were set to $\lambda_{ad} = 1, \lambda_{pi} = 100$, and $\lambda_{st} = 10$, respectively. The patch size was set to be 64×64 and the query point $I = 512$.

To compare the effectiveness of our improved framework with state-of-the-art experimental algorithms such as CycleGAN [27], cGAN [28], and Pix2pix [25, 26], we utilized common performance metrics including Mean Absolute Error (MAE) and Peak Signal to Noise Ratio (PSNR) to assess the similarity between Synthetic MR images and Ground-truth MR images. The maximum and average values of both metrics were used as evaluation criteria. MAE showed a negative correlation with the quality of generated MR images, while PSNR exhibited a positive correlation. In Table 1, we present a detailed comparison of the performance metrics among different experimental methods.

Table 1. Quantitative evaluation of MAE and PSNR between different methods

Index	MAE↓	MAE↓	PSNR↑	PSNR↑
Methods	Optimal	Average	Optimal	Average
CycleGAN	7.47	69.59	17.67	10.82
Pix2pix	1.84	48.72	39.75	27.67
cGAN	6.16	61.40	18.79	11.06
Ours	1.57	47.33	40.39	29.51

Based on the evaluation of existing methods and our improved framework, as shown in Table 1, we can conclude that our improved framework outperforms mainstream methods. Among them, the best-performing method is Pix2pix. Therefore, in Sect. 4.3, we compare our results with the MR images synthesized by Pix2pix.

4.3 Experimental Result

We utilized images of the lumbar spine segment of the human body to generate MR images of the spine with a transverse section (Fig. 6) and sagittal section (Fig. 7) that effectively demonstrate the comparison with the input CT images.

A comparison of the synthesized magnetic resonance (MR) images of spinal transverse sections shown in Fig. 6 reveals that there are marked differences in the quality of transformation in the intervertebral disc region between Example 1 and Example 2 when using the Pix2pix method. Specifically, the contours of the joint ligaments that encircle the vertebral bones appear somewhat blurry. In contrast, the new method presented in this paper offers more detailed information and clearer contour ranges in this region, resulting in a more complete structure representation.

We compared the synthesized MR images of the spinal sagittal sections after modal conversion in Fig. 7. We found that in Example 1, the Pix2pix method produced relatively blurry images with severe loss of edge information at the junction of vertebrae and intervertebral discs. In contrast, our proposed new method was able to clearly distinguish the boundary between vertebrae and intervertebral discs. In Example 2, the Pix2pix method lost a considerable amount of key details in the area containing vertebrae, spinal

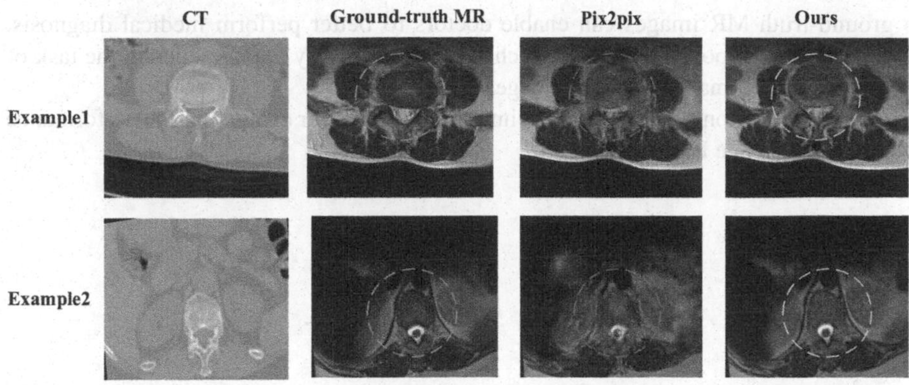

Fig. 6. Comparison of transverse sections of spine

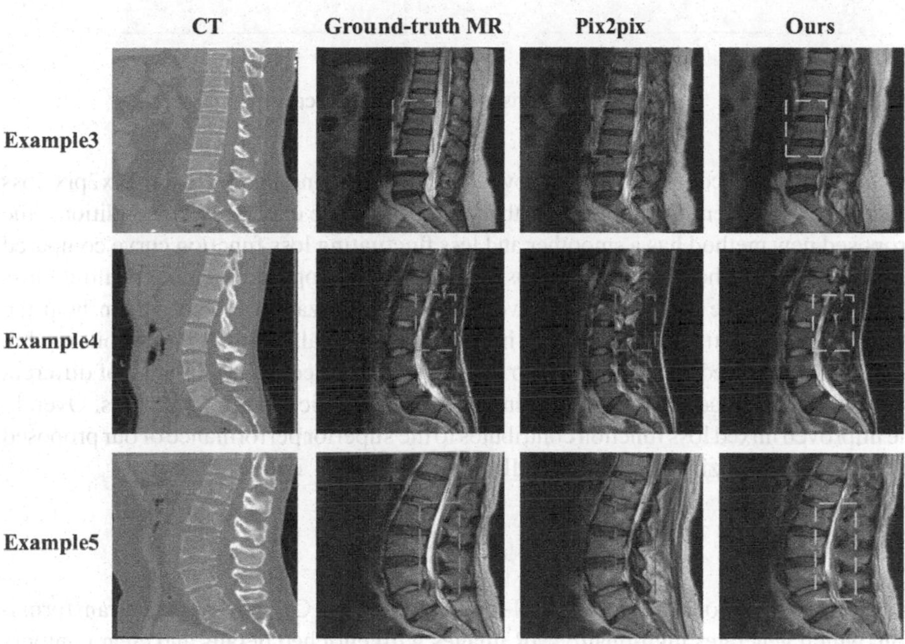

Fig. 7. Comparison of sagittal sections of spine

nerves, and ligaments, making it appear chaotic. Our proposed new method, on the other hand, preserved more detailed information and produced higher quality results in this structurally complex area. In Example 3, the Pix2pix method lost edge information in the area where the spinal cord was included in the vertebral body transformation, resulting in a significant difference from the ground-truth MR image. Our proposed new method demonstrated more accurate information on the spinal cord and vertebral structure, and was more similar to the ground-truth image. In the field of spinal medicine, synthesized MR images that contain richer detailed and edge information and are more similar to ground-truth MR images can enable doctors to better perform medical diagnosis. Therefore, our proposed method can achieve higher quality performance in the task of synthesizing MR images from CT images.

The loss function is also one of the important factors for evaluating the performance of experiments. See Fig. 8 below.

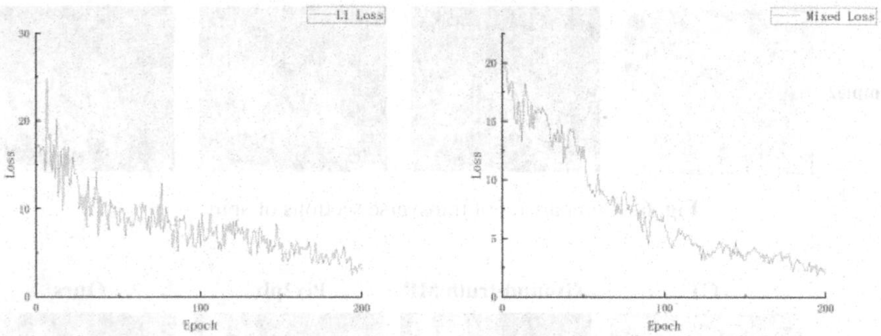

Fig. 8. Comparison of loss function curves

In Fig. 8, we compare our improved mixed loss function with the Pix2pix loss function. It is evident from the figure that under the same experimental conditions, the proposed new method has a smoother and less fluctuating loss function curve compared to the Pix2pix method. As the total loss function. We proposed includes multiple loss functions and these loss functions have certain regularization effects, it can help the model avoid overfitting problems and improve the generalization ability of our model. The improved mixed loss function also allows us to balance the importance of different loss functions and adjust them according to specific application requirements. Overall, the improved mixed loss function contributes to the superior performance of our proposed method in synthesizing high-quality MR images from CT images.

5 Conclusion

In this study, we propose a novel GAN-based method for CT to MR image transformation, which generates high-quality MR images with enriched details and clear contours using real spine images from patients. Our mixed loss function incorporates structural consistency loss, adversarial loss, and pixel-wise translation loss to comprehensively evaluate the model's performance in image transformation tasks. We use different types of losses with distinct penalty mechanisms for samples with varying levels of difficulty, such as edge loss to enhance the accuracy in bounding box regression and reduce detail loss and misrepresentation. The weighted combination of these loss functions enables the model to learn feature representations more effectively and enhances robustness. This method can be applied to other human body parts, such as the brain, lungs, and legs, and has potential applications in spinal medical diagnosis. To improve the generalizability of this method, we recommend introducing larger datasets with diverse pathological cases in future research.

Acknowledgements. This work was supported in part by the National Natural Science Foundation of China under Grant No. 62273163, the Key R&D Project of Shandong Province under Grant No. 2022CXGC010503, and the Youth Foundation of Shandong Province under Grant Nos. ZR202102230323 and ZR2021QF130.

References

1. Goodfellow, I., Pouget-Abadie, J., Mirza, M., et al.: Generative adversarial networks. Commun. ACM. **63**(11), 139–144 (2020)
2. Zhu, J.Y., Park, T., Isola, P., et al.: Unpaired image-to-image translation using cy-cle-consistent adversarial networks. In: Proceedings of the IEEE International Conference on Computer Vsion (ICCV), pp. 2223–2232 (2017)
3. Radford, A., Metz, L., Chintala, S.: Unsupervised representation learning with deep convolutional generative adversarial networks. arXiv preprint arXiv: 1511.06434 (2015)
4. Kaneko, T., Kameoka, H., Tanaka, K., et al.: Cyclegan-vc2: improved cyclegan-based non-parallel voice conversion. In: ICASSP 2019-2019 IEEE International Conference on Acoustics, Speech and Signal Processing (ICASSP), pp. 6820–6824 (2019)
5. Karras, T., Aila, T., Laine, S., et al.: Progressive growing of gans for improved quality, stability, and variation. arXiv preprint 1710.10196 (2018)
6. Mirza, M., Osindero, S.: Conditional generative adversarial nets. arXiv preprint arXiv: 1411.1784 (2014)
7. Yoon, D., Oh, J., Choi, H., et al.: OUR-GAN: One-shot Ultra-high-Resolution Generative Adversarial Networks. arXiv preprint arXiv: 2202.13799 (2022)
8. Heusel, M., Ramsauer, H., Unterthiner, T., et al.: Gans trained by a two time-scale update rule converge to a local nash equilibrium. In: Advances in Neural Information Processing Systems, pp.1–12 (2017)
9. Bynagari, N.B.: GANs trained by a two time-scale update rule converge to a local Nash equilibrium. In: Asian Journal of Applied Science and Engineering (AJASE). 8, 25-34 (2019)
10. Sato, N., Iiduka, H.: Using constant learning rate of two time-scale update rule for training generative adversarial networks. arXiv preprint arXiv: 2201.11989 (2022)
11. Wang, T.C., Liu, M.Y., Zhu, J.Y., et al.: High-resolution image synthesis and semantic manipulation with conditional gans. In: Proceedings of the IEEE Conference on Com-puter Vision and Pattern Recognition, pp. 8798–8807. IEEE (2018)
12. Ronneberger, O., Fischer, P., Brox, T.: U-net: convolutional networks for biomedical image segmentation. In: Medical Image Computing and Computer-Assisted Intervention–MICCAI 2015, pp. 234–241 (2015)
13. Zunair, H., Hamza, A.B.: Sharp U-Net: depthwise convolutional network for biomedical image segmentation. Comput. Biol. Med. (CIBM) **136**, 104699 (2021)
14. Shaziya, H., Shyamala, K., Zaheer, R.: Automatic lung segmentation on thoracic CT scans using U-net convolutional network. In: 2018 International Conference on Com-munication and Signal Processing (ICCSP), pp. 0643–0647 (2018)
15. Wu, D., Wang, Y., Xia, S.T., et al.: Skip connections matter: on the transferability of adversarial examples generated with resnets. arXiv preprint arXiv: 2002.05990 (2020)
16. Drozdzal, M., Vorontsov, E., Chartrand, G., et al.: The importance of skip connections in biomedical image segmentation. In: International Workshop on Deep Learning in Med-ical Image Analysis, International Workshop on Large-Scale Annotation of Biomedical Data and Expert Label Synthesis, pp. 179–187 (2016)
17. Tustison, N., Gee, J.: N4ITK: Nick's N3 ITK implementation for MRI bias field correc-tion. Insight J. 1–8 (2009)

18. Tustison, N.J., Avants, B.B., Cook, P.A., et al.: N4ITK: improved N3 bias correction. IEEE Trans. Med. Imag. **29**(6), 1310–1320 (2010)
19. Mengqiao ,W., Jie, Y., Yilei, C., et al.: The multimodal brain tumor image segmentation based on convolutional neural networks. In: 2017 2nd IEEE International Conference on Computational Intelligence and Applications (ICCIA), pp. 336–339. IEEE (2017)
20. Danielsson, P.E.: Euclidean distance mapping. Comput. Graph. Image Process. **14**(3), 227–248 (1980)
21. Wang, L., Zhang, Y., Feng, J.: On the Euclidean distance of images. IEEE Trans. Pattern Anal. Mach. Intell. **27**(8), 1334–1339 (2005)
22. Liberti, L., Lavor, C., Maculan, N., et al.: Euclidean distance geometry and applications. SIAM Rev. **56**(1), 3–69 (2014)
23. Pieper, S., Halle, M., Kikinis, R.: 3D slicer. In: 2004 2nd IEEE International Symposium on Biomedical Imaging: Nano to Macro (IEEE Cat No. 04EX821), pp. 632–635. IEEE (2004)
24. Kikinis, R., Pieper, S.D., Vosburgh, K.G.: 3D slicer: a platform for subject-specific image analysis, visualization, and clinical support. In: Intraoperative Imaging and Im-age-Guided Therapy, pp. 277–289 (2013)
25. Qu, Y., Chen, Y., Huang, J., et al.: Enhanced pix2pix dehazing network. In: Proceedings of the IEEE/CVF Conference on Computer Vision and Pattern Recognition, pp. 8160–8168 (2019)
26. Wang, X., Yan, H., Huo, C., et al.: Enhancing Pix2Pix for remote sensing image classi-fication. In: 2018 24th International Conference on Pattern Recognition (ICPR), pp. 2332–2336. IEEE (2018)
27. Yang, H., Sun, J., Carass, A., et al.: Unpaired brain MR-to-CT synthesis using a struc-ture-constrained CycleGAN. In: Deep Learning in Medical Image Analysis and Multi-modal Learning for Clinical Decision Support (DLMIA), pp. 174–182 (2018)
28. Loey, M., Manogaran, G., Khalifa, N.: A deep transfer learning model with classical data augmentation and CGAN to detect COVID-19 from chest CT radiography digital images. Neural. Comput. Appl. **32**, 1–13 (2020)

Investigating the Transferability of YOLOv5-Based Water Surface Object Detection Model in Maritime Applications

Yu Guo[1,2](✉), Zhuo Chen[1,2], Qi Wang[1,2], Tao Bao[1,2], and Zexing Zhou[1,2]

[1] China Ship Scientific Research Center, Wuxi 214000, People's Republic of China
guoyu@cssrc.com.cn
[2] Taihu Lake Laboratory of Deep Sea Technological Science, Wuxi 214000, People's Republic of China

Abstract. Object detection on the water surface is crucial for unmanned surface vehicles in maritime environments. Despite the challenges posed by variable lighting and ocean conditions, advancements in this field are necessary. In this paper, we investigate the transferability of YOLOv5-based water surface object detection models in cross-domain scenarios. The evaluation is based on publicly available datasets and two newly proposed datasets, Taihu Trial Dataset(TTD) and Fuxian Trial Dataset(FTD), which contain similar target classes but distinct scene and features. Results from extensive experiments indicate that zero-shot transfer is challenging, but a limited number of samples from the target domain can greatly enhance model performance.

Keywords: Object detection · Model Transferability · Intelligent Perception

1 Introduction

The proliferation of unmanned surface vehicles (USVs) has brought forth numerous applications in marine science, search and rescue missions, monitoring tasks, and beyond. Despite the extensive capabilities of these USVs, their deployment in the complex and dynamic marine environment presents substantial challenges, particularly in terms of perception and navigation. Consequently, there is a pressing need to create an efficient, reliable, and adaptable object detection model to augment the safety and performance of USVs.

In the realm of computer vision, deep learning technology has demonstrated its ubiquity and prowess, with the YOLOv5 algorithm emerging as a particularly efficient tool for rapid object detection. However, it is pertinent to note that datasets can exhibit considerable variability in their image distributions due to a myriad of factors including the source of the images, the timing and location of image collection, scene dynamics, and individual preferences. These

H. Zhang et al. (Eds.): NCAA 2023, CCIS 1870, pp. 103–115, 2023.
https://doi.org/10.1007/978-981-99-5847-4_8

disparities invariably lead to discrepancies across data domains, which can consequently impact the generalization capacity of the models. To gauge the model's adaptability, this study utilizes multiple datasets, each representing a distinct domain.

The central focus of this paper is to scrutinize the transferability of the YOLOv5-based object detection model specifically designed for USVs. We endeavor to evaluate the model's ability to leverage knowledge from disparate dataset domains and its adaptability to a variety of marine environments.

The YOLOv5 algorithm, a one-stage target detection method, is designed to identify multiple targets within a single image by harnessing a solitary convolutional neural network (CNN) for forward propagation. This design translates into faster detection times and enhanced accuracy. To determine the model's transferability, we administer a series of experiments utilizing several datasets that portray various marine conditions, encompassing diverse lighting conditions and scenes. The model's performance is gauged using the mean average precision (mAP) and recall as standard evaluation metrics.

In order to contribute to the understanding of the transferability of the YOLOv5-based USV object detection model, we conduct investigations using publicly accessible datasets in conjunction with two novel datasets, namely the Taihu Trial Dataset (TTD) and Fuxian Trial Dataset (FTD). Each dataset encapsulates distinct scenes, lighting conditions, and target attributes, thereby facilitating a comprehensive evaluation of the model's transferability.

The salient contributions of this study include:

- The introduction of two novel datasets, TTD and FTD, tailored for water object detection.
- Comprehensive experiments and analytics to appraise the transferability of the YOLOv5-based USV object detection model. Our findings reveal that zero-shot transfer remains a formidable challenge, yet a marginal sample size from the target domain can considerably augment model performance.

In summation, this paper contributes valuable insights into the transferability of the YOLOv5-based USV object detection model, thereby establishing a robust foundation for future pursuits in engineering applications.

2 Related Work

2.1 Object Detection

Target detection and recognition is a crucial problem in computer vision, serving as the foundation for more complex visual tasks, such as target tracking and scene understanding. It has a wide range of applications in areas such as video surveillance and intelligent transportation [19].

Current target detection technologies can be broadly categorized into two groups: (1) traditional methods based on manual feature design, such as Haar features with Adaboost [17], HOG features with SVM [11], and DPM [18], and (2) deep learning-based methods. The advent of deep learning has greatly advanced the field of object detection. Compared to traditional methods based on manual feature design, deep learning-based methods offer flexible structures, automatic feature extraction, high detection accuracy, and fast detection speed, making them increasingly popular. Based on the training method of the model, they can be further divided into two categories: single-stage detection algorithms and two-stage detection algorithms.

Single-stage detectors, such as SSD [10], RetinaNet [7], and the YOLO series [1,4,12–14], have become increasingly popular in recent years due to their high inference speed and real-time performance. They perform object detection and classification in a single forward pass, making them suitable for real-time applications. Single-stage algorithms directly extract features from the original image to predict object classification and location, transforming the target frame positioning problem into a regression problem. While single-stage algorithms have a simple structure and fast detection speed, they can suffer from low accuracy for scenes with excessive target density or high target overlap, resulting in frequent missed detections.

Two-stage detection algorithms divide the detection process into two steps. Firstly, they generate regions of interest (ROIs) and extract features, and then classify all the feature maps generated by the ROIs through a series of classifiers and refine the bounding boxes through linear regression. Unlike single-stage algorithms, two-stage algorithms can accurately detect dense targets and small targets, but with a slower detection speed. Representative methods in this category include RCNN [3], Faster RCNN [15], FPN [6], etc.

2.2 Yolov5

The YOLOv5 framework is a cutting-edge solution for target detection, developed by Ultralytics LLC [4]. This open-source framework integrates state-of-the-art advancements in target detection and image segmentation, including the FPN [6] and PAN [9] algorithms. With a focus on both accuracy and real-time performance, the YOLOv5 framework has undergone continuous updates and improvements. This section provides an in-depth overview of the YOLOv5 framework's target detection model structure.

The YOLOv5 algorithm consists of five versions, namely YOLOv5n, YOLOv5s, YOLOv5m, YOLOv5l, and YOLOv5x. These versions differ in the depth and width of the model. A larger number of model parameters results in higher computational complexity and increased accuracy. As shown in Table 1, the width and depth coefficients of the different model versions vary. A larger depth coefficient indicates a model with more convolutional layers, while a larger width coefficient means a model with a larger number of large-scale convolution kernels.

In this section, we take YOLOv5s as an example to illustrate the structure of the YOLOv5 model. Other versions are based on this version but differ in terms of network depth and width. The YOLOv5s model is comprised of four parts: input, backbone network, neck network, and prediction.

Table 1. Comparing Model Parameters and Performance in Various YOLOv5 Versions

YOLOv5 algorithm version	Depth factor	Width factor	COCO mAP@0.5	Params (M)	FLPOS (B)
YOLOv5n	0.33	0.25	45.7	1.9	4.5
YOLOv5s	0.33	0.50	56.8	7.2	16.5
YOLOv5m	0.67	0.75	64.1	21.2	49
YOLOv5l	1.00	1.00	67.3	46.5	109.1
YOLOv5x	1.33	1.25	68.9	86.7	205.7

The input stage is where images are fed into the network. The input image size for YOLOv5s is 608×608. This stage typically includes an image preprocessing stage, where the input image is scaled to the network's input size and undergoes operations such as normalization. During the network training phase, YOLOv5 utilizes mosaic data augmentation to improve both model training speed and network accuracy. Additionally, YOLOv5 introduces an adaptive anchor frame calculation method and adaptive image scaling method.

The backbone network is a high-performance classifier network that extracts general feature representations. YOLOv5 uses the CSP Darknet53 network with a focus structure as the backbone network.

The Neck network, situated between the backbone network and the head network, is designed to further improve feature diversity and robustness. YOLOv5 employs the SPP module and the FPN + PAN module as the Neck network.

The prediction stage generates the target detection result. The number of branches at the output end varies depending on the detection algorithm, typically consisting of both a classification branch and a regression branch. YOLOv5 adopts the processing method of YOLOv4 and replaces the Smooth L1 Loss function with the GIOU Loss [16] to further enhance the algorithm's detection accuracy.

2.3 Public Dataset

The field of target detection has several commonly used datasets, including PASCAL VOC [2], MS COCO [8], and OpenImages [5].

The MS COCO dataset is a comprehensive and widely used benchmark for evaluating the performance of object detection algorithms. Launched in 2014, the dataset contains over 330,000 images and more than 200,000 annotated images that cover 80 object categories, including people, animals, vehicles, and indoor and outdoor scenes. Each image is annotated with object instances, including their location, size, class label, and a descriptive caption of the scene. The MS COCO dataset provides a challenging testbed for evaluating object detection algorithms, as it has become a standard benchmark in the computer vision community. In this paper, we focus on water targets and extract a subset of images from the COCO dataset containing water targets, resulting in 11,189 boat samples without buoy classes.

The OpenImages dataset, released by Google, is currently the largest dataset in the computer vision field, containing 9 million images and 15.8 million detection boxes. This paper selects a subset of the OpenImages dataset, with a total of 11,196 samples, including 10,999 samples of ships and 197 samples of buoys.

PASCAL VOC is another popular benchmark dataset used to evaluate the performance of object detection, classification, and segmentation algorithms in computer vision. First introduced in 2005, the dataset consists of 20 object categories, including people, animals, vehicles, and indoor and outdoor scenes. There are 9963 images in the dataset, containing 24640 annotated objects. This paper focuses on ships, and thus, selects a subset of the VOC dataset with a total of 1403 samples, all of which are ship samples, as there is no buoy class in VOC.

3 Method

3.1 Dataset

In this paper, the focus is on the detection of surface targets, which encompasses three categories: boats, buoys, and digital boards. These categories have been carefully selected to represent the diverse range of objects that are typically found in real-world navigation scenarios. In particular, the digital board category is a novel addition and is specifically designed to represent the unique challenges that arise when navigating unmanned surface vehicles in the ocean.

This paper aims to address the issue of limited types and quantities of water targets in existing public datasets. While the MS COCO and VOC datasets only contain the category of boats, and the OpenImages dataset only includes two categories of boats and buoys, there is a lack of digital board categories, which are crucial for navigation purposes. To overcome this, this paper presents the TTD and FTD datasets, designed specifically for the task of surface target detection.

The images in the TTD dataset were collected from Taihu Lake, Jiangsu Province, China, using an unmanned vehicle equipped with a photoelectric dome camera. The camera is installed on the top of the ship's mast and can rotate in two degrees of freedom, yaw and pitch, with the ability to adjust focus. The images were annotated by the authors, comprising of three categories: boats, buoys, and digital boards. The TTD dataset consists of 13,252 samples in total, with 6,762 boats, 3,460 buoys, and 3,030 digital boards. The dataset includes images captured under various weather and lighting conditions, providing diversity in terms of target shape, color, position, and size.

Similarly, the FTD dataset was collected from Fuxian Lake, Yunnan Province, China, using the same equipment as the TTD dataset. The images were annotated by the authors, with a total of 3,681 samples, including 1,957 boats, 930 buoys, and 794 digital boards. In addition to the difference in illumination and weather, the FTD dataset offers differences in the background, lighting, and target shape compared to the TTD dataset, due to the varying

Table 2. Comparison of Sample Counts across Different Datasets and Classes

From	Dataset	Total samples	Class		
			Boat	Buoy	DigitalBoard
Public	COCO	11189	11189	0	0
	OpenImages	11196	10999	197	0
	VOC	1403	1403	0	0
Ours	TTD	13252	6762	3460	3030
	FTD	3681	1957	930	794

latitude and longitude of the data scene. The ratio of the training set to the validation set for both the TTD and FTD datasets is 5:1.

Overall, the TTD and FTD datasets provide a valuable resource for researchers and practitioners, who can use these datasets to obtain accurate and reliable results for surface target detection. The authors hope that these datasets will advance the field of navigation safety for unmanned surface vehicles, and inform future research and data collection practices (Table 2).

3.2 Metric

The Mean Average Precision (mAP) is a widely used metric for evaluating the performance of object detection algorithms. It combines both localization accuracy and the confidence of the classifier by averaging the precision across all classes.

First, for each ground truth object in the test set, the Intersection over Union (IoU) is calculated between the ground truth bounding box (*gt*) and the predicted bounding box (*d*) for each class:

$$\mathrm{IoU(gt, d)} = \frac{\mathrm{gt} \cap \mathrm{d}}{\mathrm{gt} \cup \mathrm{d}} \tag{1}$$

Then, each predicted object is then assigned to the ground truth object with the highest IoU. Precision (P) and recall (R) are calculated for each class using the following equations:

$$\mathrm{P} = \frac{\mathrm{TP}}{\mathrm{TP} + \mathrm{FP}} \tag{2}$$

$$\mathrm{R} = \frac{\mathrm{TP}}{\mathrm{TP} + \mathrm{FN}} \tag{3}$$

Then the precision and recall values are plotted on a precision-recall curve for each class, and the Average Precision (AP) is calculated for each class by finding the area under the precision-recall curve:

$$\mathrm{AP_i} = \int_0^1 \mathrm{P(R_i \geq r)dr} \tag{4}$$

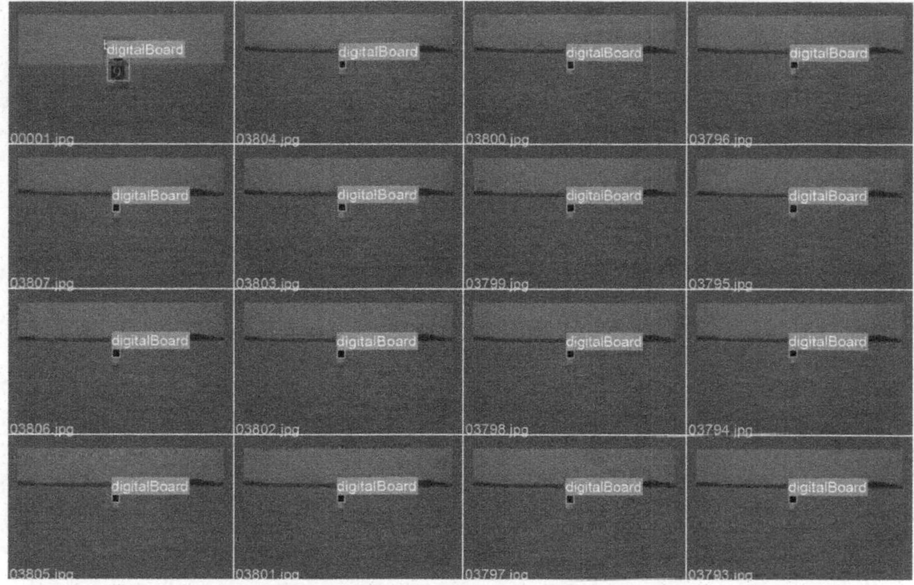

Fig. 1. Illustrative samples from the TTD dataset

Finally, the mAP is calculated as the average of the APs for all classes:

$$\mathrm{mAP} = \frac{1}{\mathrm{N}} \sum_{\mathrm{i}=1}^{\mathrm{N}} \mathrm{AP_i} \tag{5}$$

where N is the number of classes and AP_i is the average precision for class i.

4 Experiments

In this paper we aim to evaluate the transferability of the YOLOv5 model by conducting experiments on various datasets. The datasets used in this paper include three categories of surface targets: boats, buoys, and digital boards. The digital board category is unique to this paper and is not present in any public dataset.

The experiments are divided into two cases: fine-tuning and non-fine-tuning. In the fine-tuning case, a small amount of data from the target domain is added to the training dataset and the model is fine-tuned before evaluation. The results of the experiments are presented in Table 4, where the source domain refers to the training dataset and the target domain refers to the evaluation dataset.

Table 4 show the difficulty in cross-domain prediction for water surface object detection models. For example, the YOLOv5m model, trained on the public dataset and TTD dataset, performed well with mAP@0.5 of 0.99 and mAP@0.5:0.95 of 0.77 without fine-tuning. However, when tested on the FTD

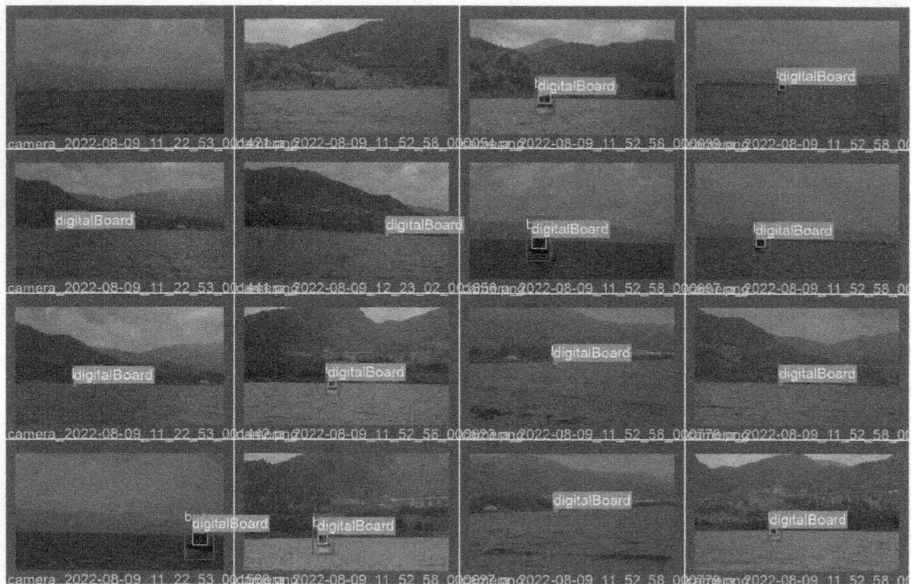

Fig. 2. Illustrative Samples from the FTD Dataset

Table 3. Comparison of Model Performances on Different Source and Target Domains

No.	Source domain	Target domain	Model performance		Model
			mAP@0.5	mAP@0.5:0.95	
1	Public +TTD trainset	TTD val	0.91	0.57	YOLOv5s
2	Public +TTD trainset	FTD val	0.33	0.11	
3	Public +TTD trainset	TTD val	**0.99**	**0.77**	YOLOv5m
4	Public +TTD trainset	FTD val	0.32	0.12	
5	Public +TTD trainset + part of FTD trainset(349)	FTD val	0.94	0.62	YOLOv5m
6	Public +TTD trainset +part of FTD trainset(1959)	FTD val	0.96	0.70	
7	Public +TTD trainset + FTD trainset(3084)	FTD val	**0.97**	**0.73**	

dataset, the mAP@0.5 index dropped to 0.32, and the mAP@0.5:0.95 was only 0.12, indicating a lack of zero-sample transferability.

On the other hand, the results show that fine-tuning with a small amount of data from the target domain can significantly improve the model's performance. The mAP@0.5 index improved to 0.93, and the mAP@0.5:0.95 was 0.62. This suggests that the model has learned the general and common features needed for target detection in the source domain, and fine-tuning with a small amount of data from the target domain can achieve significant improvements.

Furthermore, the results show that the YOLOv5m model, with larger parameters, has better cross-domain transferability than the YOLOv5s model. The rest of the paper will elaborate on the details of each test to further analyze the factors affecting the results (Table 3).

Table 4. Assessment of the Transferability of Different Categories of YOLOv5m Model in Experiment 5

Class	Number of samples	P	R	mAP@0.5	mAP@0.5:0.95
all	4059	0.973	0.896	0.939	0.623
buoy	1038	0.990	0.965	0.984	**0.747**
digitalBoard	889	0.968	0.755	0.848	0.426
boat	2132	0.962	0.969	**0.985**	0.696

Table 5. Assessment of the Transferability of Different Categories of YOLOv5m Model in Experiment 6

Class	Samples	P	R	mAP@0.5	mAP@0.5:0.95
all	4060	0.989	0.927	0.962	0.698
buoy	1039	0.991	0.976	0.988	**0.789**
digitalBoard	889	0.992	0.832	0.907	0.591
boat	2132	0.984	0.974	**0.992**	0.714

Table 6. Assessment of the Transferability of Different Categories of YOLOv5m Model in Experiment 7

Class	Samples	P	R	mAP@0.5	mAP@0.5:0.95
all	4059	0.99	0.933	0.967	0.729
buoy	1038	0.993	0.989	**0.994**	**0.818**
digitalBoard	889	0.992	0.831	0.915	0.622
boat	2132	0.986	0.979	0.992	0.748

The results of Experiment 5, Experiment 6, and Experiment 7 using the YOLOv5m model are presented in Tables 4, 5, and 6 respectively. The source domains for each experiment vary, and the target domains are always the FTD dataset.

Table 4 shows the detection results for three objects (buoy, digitalBoard, and boat), and includes the number of images and labels, as well as the precision rate (P), recall rate (R), average precision (mAP), and other relevant indicators. In Experiment 5, the overall target domain (all) has 4059 labels, with a precision of 0.973, a recall of 0.896, and a mean precision (mAP) of 0.939. The precision rates, recall rates, and mAP values for each object category (buoy, digitalBoard, and boat) are also reported.

Table 5 summarizes the results of Experiment 6, which uses a larger sample size from the FTD dataset, compared to Experiment 5. The results for the overall target domain (all) show a precision rate of 0.989, recall rate of 0.927, and mean precision (mAP) of 0.962. The results for each object category (buoy, digitalBoard, and boat) are also reported.

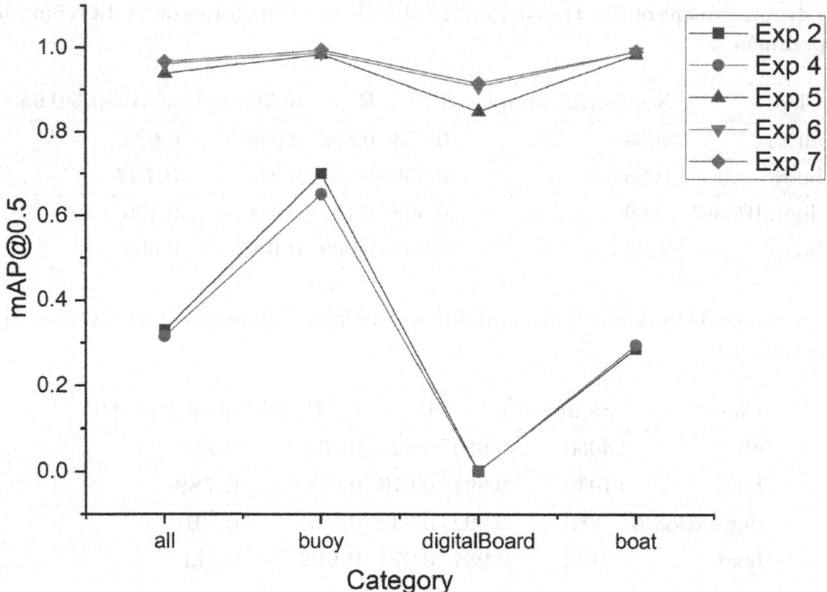

Fig. 3. Comparison of Mean Average Precision Across Different Categories in Multiple Trials

Table 6 displays the results of Experiment 7, which has the largest sample size from the FTD dataset and includes a synthetic dataset. The results for the overall target domain (all) show a precision rate of 0.99, recall rate of 0.933, and mean precision (mAP) of 0.967. The precision rates, recall rates, and mAP values for each object category (buoy, digitalBoard, and boat) are also reported.

Figure 3 illustrates the map@0.5 indicators of the model for all classes, buoy, digital board, and boat, on the same target domain (FTD validation set) in experiments 2, 4, 5, 6, and 7. Among these experiments, the source domains of Experiment 2 and Experiment 4 are the same, both of which consist of public datasets and the TTD training set. The source domains of Experiments 5, 6, and 7 are also mixed with data from the FTD training set, with the least amount of FTD data being added in Experiment 5 and the largest amount being added in Experiment 7. Experiment 2 utilized the YOLOv5s model, while Experiments 4, 5, 6, and 7 used the YOLOv5m model, which has a larger parameter scale.

From the results of these experiments, it can be concluded that the closer the source and target domains are, the better the performance of the model. Conversely, the further the domains are from each other, the worse the performance. When the datasets in the source and target domains do not overlap, the overall performance of the model decreases. However, the buoy class and boat class are less affected by this domain shift, as they have data from public datasets and therefore have a higher level of diversity in scenes and features, strengthening the model's transferability. The digital board class, on the other hand, is the most affected by the shift as it was newly defined for the surface target detection task in the TTD and FTD datasets, and thus has a more limited range of scenes and features in the source domain.

When the source and target domains partially overlap, the performance of the model significantly improves. Comparing Experiment 4 and Experiment 5, the target domains are the same, but Experiment 5 has a small number of FTD training set samples (349) in its source domain. As a result, the overall map@0.5 increased by 0.62, with a 0.33 increase for the buoy category, a 0.84 increase for the digital board category, and a 0.69 increase for the boat category. This demonstrates that even a small amount of target domain data can greatly improve the performance of the model. However, as the amount of doped data increases, the improvement of the model decreases. This is evident in the comparison of Experiment 6 and Experiment 7, which show a small difference in performance.

The results of these experiments have significant implications for practical applications. In real-world scenarios, it is not always necessary to invest a lot of resources into labeling a massive dataset when the scene or task changes. While large datasets can improve the accuracy of the model, the marginal effect of this improvement decreases as the size of the dataset increases. This means that in many applications, only a small amount of user labeling is necessary and existing target detection models can perform well. This is a valuable insight for engineering applications.

5 Conclusion

We explores the transferability of the YOLOv5-based water surface object detection model. The research examines the model's performance when transferring from source domains such as COCO, OpenImages, VOC and self-calibration datasets such as TTD to a target domain. The experiments were conducted using both the YOLOv5s and YOLOv5m models in YOLOv5, with a total of seven controlled experiments.

The results indicate that the model's performance decreases when the difference between the source and target domains is significant. However, as the number of target domain samples in the source domain increases, the performance improves. Additionally, the YOLOv5m model with more parameters performed better than the YOLOv5s model. This suggests that increasing the model's parameters can enhance its generalization ability.

In practical applications, the YOLOv5-based water surface object detection model is likely to encounter difficulties in completely unfamiliar target domains. To address this challenge, this paper recommends the following solutions for future water surface detection models:

- Constructing an adaptive detection model that adjusts and updates parameters in real-time based on different scenarios.
- Adding a limited number of target domain samples to the source domain, which, as the results indicate, can significantly enhance the model's performance on the target domain.

References

1. Bochkovskiy, A., Wang, C.Y., Liao, H.Y.M.: YOLOv4: optimal speed and accuracy of object detection. arXiv preprint arXiv:2004.10934 (2020)
2. Everingham, M., Van Gool, L., Williams, C.K., Winn, J., Zisserman, A.: The pascal visual object classes (voc) challenge. Int. J. Comput. Vision **88**, 303–308 (2009)
3. Girshick, R., Donahue, J., Darrell, T., Malik, J.: Rich feature hierarchies for accurate object detection and semantic segmentation. In: Proceedings of the IEEE Conference on Computer Vision and Pattern Recognition, pp. 580–587 (2014)
4. Jocher, G.: YOLOv5 by Ultralytics. https://doi.org/10.5281/zenodo.3908559. https://github.com/ultralytics/yolov5
5. Kuznetsova, A., et al.: The open images dataset v4: unified image classification, object detection, and visual relationship detection at scale. Int. J. Comput. Vision **128**(7), 1956–1981 (2020)
6. Lin, T.Y., Dollár, P., Girshick, R., He, K., Hariharan, B., Belongie, S.: Feature pyramid networks for object detection. In: Proceedings of the IEEE Conference on Computer Vision and Pattern Recognition, pp. 2117–2125 (2017)
7. Lin, T.Y., Goyal, P., Girshick, R., He, K., Dollár, P.: Focal loss for dense object detection. In: Proceedings of the IEEE International Conference on Computer Vision, pp. 2980–2988 (2017)
8. Lin, T.-Y., et al.: Microsoft COCO: common objects in context. In: Fleet, D., Pajdla, T., Schiele, B., Tuytelaars, T. (eds.) ECCV 2014. LNCS, vol. 8693, pp. 740–755. Springer, Cham (2014). https://doi.org/10.1007/978-3-319-10602-1_48
9. Liu, S., Qi, L., Qin, H., Shi, J., Jia, J.: Path aggregation network for instance segmentation. In: Proceedings of the IEEE Conference on Computer Vision and Pattern Recognition, pp. 8759–8768 (2018)
10. Liu, W., et al.: SSD: single shot multibox detector. In: Leibe, B., Matas, J., Sebe, N., Welling, M. (eds.) ECCV 2016. LNCS, vol. 9905, pp. 21–37. Springer, Cham (2016). https://doi.org/10.1007/978-3-319-46448-0_2
11. Pang, Y., Yuan, Y., Li, X., Pan, J.: Efficient hog human detection. Sig. Process. **91**(4), 773–781 (2011)
12. Redmon, J., Divvala, S., Girshick, R., Farhadi, A.: You only look once: unified, real-time object detection. In: Proceedings of the IEEE Conference on Computer Vision and Pattern Recognition, pp. 779–788 (2016)
13. Redmon, J., Farhadi, A.: YOLO9000: better, faster, stronger. In: Proceedings of the IEEE Conference on Computer Vision and Pattern Recognition, pp. 7263–7271 (2017)
14. Redmon, J., Farhadi, A.: YOLOv3: an incremental improvement. arXiv preprint arXiv:1804.02767 (2018)
15. Ren, S., He, K., Girshick, R., Sun, J.: Faster R-CNN: towards real-time object detection with region proposal networks. Advances in Neural Information Processing Systems 28 (2015)
16. Rezatofighi, H., Tsoi, N., Gwak, J., Sadeghian, A., Reid, I., Savarese, S.: Generalized intersection over union: a metric and a loss for bounding box regression. In: Proceedings of the IEEE/CVF Conference on Computer Vision and Pattern Recognition, pp. 658–666 (2019)
17. Whitehill, J., Omlin, C.W.: Haar features for FACS AU recognition. In: 7th International Conference on Automatic Face and Gesture Recognition (FGR06), pp. 5–101. IEEE (2006)

18. Yan, J., Lei, Z., Wen, L., Li, S.Z.: The fastest deformable part model for object detection. In: Proceedings of the IEEE Conference on Computer Vision and Pattern Recognition, pp. 2497–2504 (2014)
19. Zhao, Z.Q., Zheng, P., Xu, S.T., Wu, X.: Object detection with deep learning: a review. IEEE Trans. Neural Netw. Learn. Syst. **30**(11), 3212–3232 (2019)

Physical-Property Guided End-to-End Interactive Image Dehazing Network

Junhu Wang[1], Suiyi Zhao[1], Zhao Zhang[1](✉), Yang Zhao[1], and Haijun Zhang[2]

[1] School of Computer Science and Information Engineering, Hefei University of Technology, Hefei 230601, China
cszzhang@gmail.com, zhaoyang@pkusz.edu.cn

[2] Department of Computer Science, Harbin Institute of Technology, Shenzhen, People's Republic of China
hjzhang@hit.edu.cn

Abstract. Single image dehazing task predicts the latent haze-free images from hazy images corrupted by the dust or particles existed in atmosphere. Notwithstanding the great progress has been made by the end-to-end deep dehazing methods to recover the texture details, they usually cannot effectively preserve the real color of images, due to lack of constraint on color preservation. In contrast, atmospheric scattering model based dehazing methods obtain the restored images with relatively rich real color information due to unique physical property. In this paper, we propose to seamlessly integrate the properties of physics-based and end-to-end dehazing methods into a unified powerful model with sufficient interactions, and a novel Physical-property Guided End-to-End Interactive Image Dehazing Network (PID-Net) is presented. To make full use of the physical properties to extract the density information of haze maps for deep dehazing, we design a transmission map guided interactive attention (TMGIA) module to teach an end-to-end information interaction network via dual channel-wise and pixel-wise attention. This way can refine the intermediate features of end-to-end information interaction network, and do it a favor to obtain better detail recovery by sufficient interaction. A color-detail refinement sub-network further refines the dehazed images with abundant color and image details to obtain better visual effects. On several synthetic and real-world datasets, our method consistently outperforms other state-of-the-arts for detail recovery and color preservation.

Keywords: Single image dehazing · Texture detail · Color preservation

1 Introduction

Haze is a common natural phenomenon that is formed by the atmospheric dust and particles, and the images captured in haze condition suffer from the content degradation that results in low contrast and color distortion. Single image

H. Zhang et al. (Eds.): NCAA 2023, CCIS 1870, pp. 116–131, 2023.
https://doi.org/10.1007/978-981-99-5847-4_9

dehazing, as a classical low-level vision task [29,33,35], has been attracting much attention [2,5,17]. After dehazing, the restored haze-free images are beneficial to both the visual perception and subsequent high-level vision tasks, such as object detection [3,10], semantic segmentation [12,34] and scene understanding [25]. Current image dehazeing methods can be roughly divided into two types, i.e., atmosphere scattering model (ASM)-based methods and deep learning-based end-to-end methods. ASM-based methods take advantage of the physical properties of image itself to restore haze-free images, while end-to-end methods directly dehaze in a black-box manner.

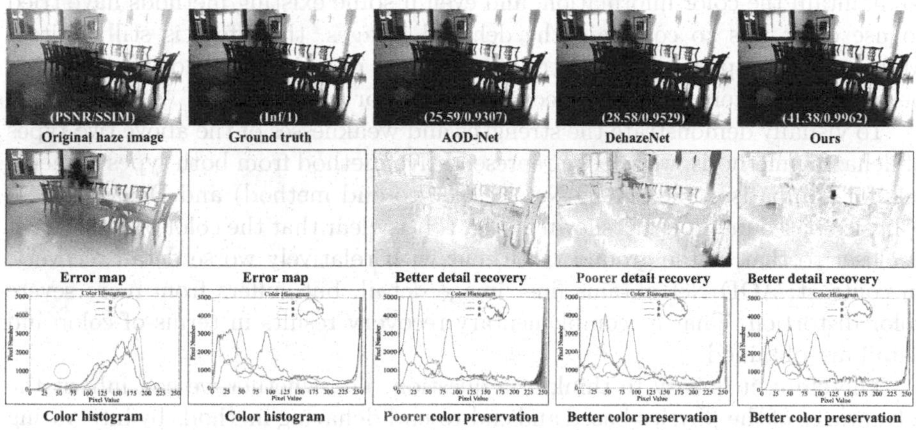

Fig. 1. Comparison of different image dehazing methods, in terms of dehazed images, texture detail characterized by the error maps [37] and the color histogram in RGB color space. For the error maps, the whiter the better. Clearly, for detail recovery, AOD-Net (end-to-end method) is more similar to the ground truth than DehazeNet. For the color preservation, DehazeNet's (physics-based method) color is closer to the ground truth than that of AOD-Net. Furthermore, our method performs the best in terms of both detail recovery and color preservation. (Color figure online)

Earlier works are mainly based on ASM for dehazing and can be considered as physics-based methods, since they make full use of the physical properties, e.g., atmospheric light and transmission maps. ASM can be formulated as

$$I(x) = J(x)t(x) + A(1 - t(x)), \tag{1}$$

where $I(x)$, $J(x)$ respectively denote the hazy image and the haze-free one, A denotes the global atmosphere light, and $t(x)$ denotes the transmission map indicating the intensity of haze. ASM can simulate the imaging principle of haze very realistically, since it could be reasonably decomposed into the incident light attenuation model and the atmosphere light imaging model. Physics-based methods leverage ASM to obtain the haze-free images by estimating the global atmospheric light and transmission maps [1,9,22,26]. As a result, the color of dehazed results for these physics-based methods look more real and natural, since global atmosphere light controls the color of haze images [13]. Yet, ASM has an inherent limitation to obtain accurate detail recovery, since it is inevitable to accumulate estimation errors on physical properties that greatly damage the details of dehazed images.

The recently proposed end-to-end deep methods directly dehaze image in an end-to-end manner [2,15,18,21], without using the physical properties, thus avoiding the accumulation of intermediate errors. Besides, due to the strong learning ability of deep neural networks, high-quality dehazed images with finer detail recovery are obtained. However, this black-box learning manner is hard to maintain the color information, and even if some existing methods have tried to use color loss to constrain the dehazed images, the effect is still limited. Moreover, an inappropriate constraint may cause negative effects on the dehazing performance in terms of detail recovery and color preservation.

To visually demonstrate the strengths and weaknesses of the above two types of dehazing methods, we select a representative method from both types of methods for comparison, i.e., AOD-Net [9] (end-to-end method) and DehazeNet [1] (physics-based method), as shown in Fig. 1. It is clear that the color of DehazeNet is closer to that of the ground truth, but with relatively worse detail recovery. In contrast, AOD-Net obtains finer image detail, but suffers from more severe color distortion. That is, complementary recovery results in terms of color and detail are obtained.

Therefore, it is easy to think of a question, i.e., whether we can inherit the advantages of the physics-based and end-to-end dehazing methods by integrating them, so that the restoration results can be accurate in terms of both detail recovery and color preservation? We provide a positive answer. But how to integrate and interact? Because simple "one plus one" combination not only degrades the dehazing result, but also loses the original property of the method itself, which has been proved by experiments. Therefore, we present an effective interaction strategy that can take full advantages of both, and propose a novel method that delivers stronger abilities for performing detail recovery and color preservation via sufficient information. Overall, the main contributions of this paper are summarized as follows:

- We propose a novel Physical-property Guided End-to-End Interactive Image Dehazing Network (PID-Net for short). PID-Net perfectly integrates the advantages of the end-to-end and physics-based dehazing models, so that it can restore as many details of images as possible and retain color information as much as possible. For dehazing, PID-Net uses a multi-source "coarse-to-fine" pipeline, i.e., it firstly uses an end-to-end information interaction network for coarse dehazing, which further includes an end-to-end dehazing sub-network (E-Net) and a physics-based dehazing sub-network (P-Net), which obtains the initial coarse dehazed images with different properties. Then, a new color-detail refinement sub-network (R-Net) is designed to obtain the final finer recovery result with richer content and color information. To the best of our knowledge, no prior study has been done to investigate the physical-property seamlessly integrated end-to-end interactive network for deep image dehazing.

- One critical problem is how to perform interaction between E-Net and P-Net. We show that equipping E-Net with abundant physical properties is a good solution. Specifically, we design a Transmission Map Guided Interactive Attention (TMGIA) module that performs channel-wise and pixel-wise attention with the transmission map of P-Net for interaction. Thus, we can refine the features of E-Net with abundant haze density information contained in the transmission map, which enforces E-Net to recover better details.
- Merely relying on the TMGIA module to interact E-Net with P-Net is still not enough, because the transmission map is gray-scale feature without color information. That is, the interaction in TMGIA module is still too weak for color preservation. As such, we propose R-Net to make up for this deficiency. R-Net follows the encoder-decoder structure and concatenates features from E-Net and P-Net in different scales during decoding. Based on fully mining and fusing feature information, we obtain better results with abundant color and detail information.

2 Related Work

2.1 Physics-Based Dehazing Methods

Image dehazing using the atmospheric scattering model obtained pleasing result in color preservation. For example, DCP [6] assumes that there is at least one channel for each pixel whose value is close to zero and then estimates transmission maps. DehazeNet [1] utilizes a neural network to learn more accurate transmission map and makes dehazing results better. DCPDN [31] uses a well-designed network to estimate the transmission map and global atmosphere light respectively to obtain haze-free images. LAP-Net [14] progressively learns the transmission map and designs new loss functions to constrain the maps, thus restoring clean images. These methods all aim at estimating the transmission map and global atmosphere light more accurately, thereby recovering haze-free images by Eq. 1. However, such operation is an error amplification process, as there are inevitable errors in the estimated transmission map and atmospheric light, which undoubtedly deprives restored images lose the detailed information.

2.2 End-to-End Deep Dehazing Methods

In recent years, lots of end-to-end image dehazing methods [2,23] have been proposed, which directly obtain restored haze-free images from hazy input. These methods abandoned physics properties and turned to design various modules and networks to get better dehazing results. For example, GriddehazeNet [15] designs three modules and fuses multi-scale features to get haze-free images. FD-GAN [4] proposes an end-to-end network with a fusion-discriminator, and uses high-low frequency to remove haze. KDDN [7] proposes a teacher-student network for

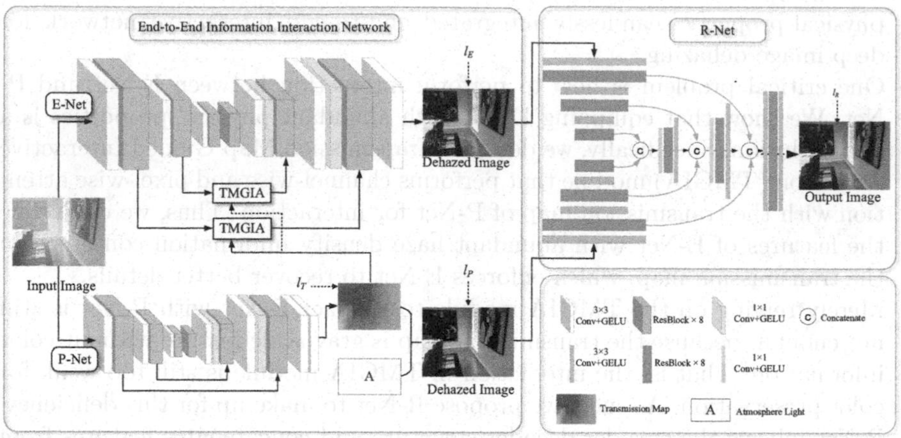

Fig. 2. The overall architecture of our proposed PID-Net, which contains two main components, i.e., an end-to-end information interaction network for coarse dehazing and a color-detail refinement sub-network (R-Net) for obtaining finer enhancement result. The end-to-end information interaction network is consisted of an end-to-end dehazing sub-network (E-Net), a physics-based dehazing sub-network (P-Net) and a transmission map guided interactive attention (TMGIA) module in order to obtain two coarse dehazing results, while R-Net further refines the features to obtain the final visual pleasing results with abundant color information and image details.

dehazing; the teacher network uses clean images to obtain hidden features, and then uses these features to guide the student network to learn a mapping of haze-free images. FFA-Net [20] employs a feature attention module by dual channel and pixel attention, and cascades them for dehazing. These end-to-end methods aim to design a fully end-to-end network and avoid the estimation error of the physical properties, greatly recovering the detail of the images. However, simple end-to-end dehazing is a black-box process with no guarantee of distortion-free color, resulting in severe color deviations in the dehazed results.

3 Proposed Method

We introduce the proposed PID-Net framework in detail. The architecture of PID-Net is illustrated in Fig. 2. As can be seen, PID-Net includes two main components, i.e., end-to-end information interaction network for coarse dehazing and a color-detail refinement sub-network R-Net. Next, we will introduce them in detail, respectively.

3.1 End-to-End Information Interaction Network

This information interaction network exchanges the information between the physics-based sub-network P-Net and end-to-end sub-network E-Net via the TMGIA module.

E-Net. This network mainly takes the responsibility of preserving the image details and produces a coarse version of haze-free images for further processing. We construct E-Net using a UNet-like architecture [24], which includes an encoder and a decoder in the end-to-end network structure. Given a hazy image I_{hazy}, we first get the initial result F of the encoder and the intermediate features F_1, F_2. Then, F_1, F_2 are sent into the TMGIA module to get the fusion features F_1', F_2', which retains abundant physical property. Finally, the features F_1', F_2', and the result F are fed into decoder to obtain the final dehazing results of E-Net. This processes can be formulated as follows:

$$F, F_1, F_2 = Enc(I_{hazy}), \tag{2}$$

$$J_E = Dec(F, F_1', F_2'), \tag{3}$$

where Enc and Dec denote the encoder and decoder of E-Net respectively, J_E denotes the restored haze-free images with E-Net. We use L_1 loss to make J_E close to the ground-truth, which can be formulated as follows:

$$\mathcal{L}_E = \|J_E - J\|_1, \tag{4}$$

where J is the ground-truth and $\|\cdot\|_1$ denotes the L_1 loss.

P-Net. P-Net obtains the coarse haze-free images from another perspective. The goal of P-Net is to make full use of the physical properties of the image itself for dehazing. P-Net contains a trainable module P_{train} and the ASM. The structure design of the module P_{train} is similar to E-Net, i.e., we generally adopt a UNet-like architecture with skip connection. The main difference is that we employ two standard convolutions followed by a GELU activation at the end of P_{train}, aiming to change the channels and estimate the transmission map. Besides, we adopt DCP [6] to compute the global atmosphere light. As a result, given a hazy image I_{hazy}, the whole pipeline of P-Net is that: we first estimate the transmission map t and the global atmosphere light A from hazy input I_{hazy}, and then restore haze-free images by ASM. The processes are defined as

$$t = P_{train}(I_{hazy}), \tag{5}$$

$$J_P = \frac{I_{hazy} - A \cdot (1 - t)}{t}, \tag{6}$$

where J_P is the restored clean images with P-Net and $\cdot$ indicates element-wise multiplication. We still use the L_1 loss on the estimated transmission map t from P-Net to obtain reliable haze density information, which is formulated as

$$\mathcal{L}_P = \|J_P - J\|_1, \tag{7}$$

$$\mathcal{L}_T = \|t - t_{real}\|_1, \tag{8}$$

where t_{real} denotes the ground-truth of transmission map.

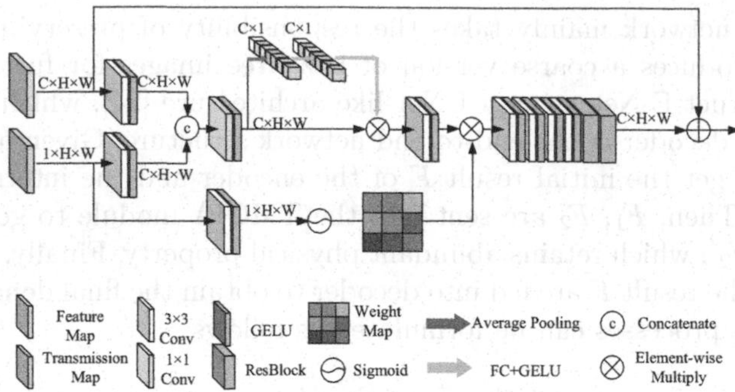

Fig. 3. The overall structure of TMGIA with dual channel-wise and pixel-wise attention.

TMGIA. This elaborate module aims at enhancing the detail and structure in the coarse restored images, and making full use of the physical property for end-to-end learning by sufficient interaction. We use the TMGIA to embed the transmission map information obtained by P-Net to E-Net, and the specific structure is illustrated in Fig. 3. Given the transmission map t yielded by P-Net, we first perform 2× downsampling operation to obtain t'. Together with the intermediate features F_1 and F_2 yielded by E-Net, we then use TMGIA to obtain the fused features ${F_1}'$ and ${F_2}'$. Note that ${F_1}'$ and ${F_2}'$ here denote the same features as Sect. 3.1. The whole process can be formulated as follows:

$$ {F_1}' = TMGIA(F_1, t), \tag{9} $$

$$ {F_2}' = TMGIA(F_2, t'). \tag{10} $$

3.2 Color-Detail Refinement Sub-Network (R-Net)

After obtaining two coarse image dehazing results with sufficient interaction, R-Net uses a multi-source "coarse-to-fine" pipeline that further performs feature refinement. This refinement stage is significant to yield a more visual pleasing result. We design a new structure for R-Net, as shown in Fig. 2. Specifically, the results J_E and J_P obtained by E-Net and P-Net are respectively sent into two encoders (i.e., E_{enc} and P_{enc}) whose structures are the same as encoder of E-Net. For J_E, we obtain three different scale intermediate features form E_{enc}, denoted as F_{E1}, F_{E2} and F_{E3}, where F_{E1} is the original scale and F_{E3} is 4× scale. So does P_{enc} and we denote the features as F_{P1}, F_{P2} and F_{P3} respectively. To extract accurate detail and color information from J_E and J_P, we concatenate the same scale features and then up-sample them for further fusion, which are denoted as D_i (i=1,2). After completely fusing different scale features, R-Net ultimately obtains the haze-free images. The specific process can be formulated as follows:

$$ F_{E1}, F_{E2}, F_{E3} = E_{enc}(J_E), \tag{11} $$

Table 1. Performance comparison on two benchmark datasets: SOTS [11] and Haze4K [16].

Datasets		**SOTS** [11]				**Haze4K** [16]			
Metrics		**PSNR/SSIM↑**	**LPIPS↓**	**CSE↓**	**MSE↓**	**PSNR/SSIM↑**	**LPIPS↓**	**CSE↓**	**MSE↓**
Physics-based methods	DCP [6]	19.63/0.860	0.0692	2.6014	0.8722	19.96/0.860	0.0844	4.2869	1.9221
	DehazeNet [1]	19.82/0.821	0.0420	1.3269	0.6705	19.04/0.846	0.0501	5.2987	1.1228
End-to-end methods	AOD-Net [9]	19.39/0.841	0.0525	2.9175	1.0619	18.99/0.880	0.0591	2.9534	1.1294
	GriddehazeNet [15]	32.14/0.984	0.0074	0.5066	0.0469	23.29/0.930	0.0103	0.0473	0.0160
	FFA-Net [20]	36.36/0.989	0.0048	*0.0656*	0.0169	*26.96*/0.950	**0.0079**	*0.0402*	**0.0093**
	AECR-Net [30]	37.17/0.990	–	–		26.12/*0.956*	0.0160	1.2376	0.2362
	MAXIM [27]	*38.11*/0.991	*0.0027*	0.0872	**0.0124**	–/–	–	–	–
	Ours	**38.19/0.991**	**0.0026**	**0.0560**	*0.0141*	**28.86/0.977**	*0.0099*	**0.0346**	*0.0147*

$$F_{P1}, F_{P2}, F_{P3} = P_{enc}(J_P), \tag{12}$$

$$D_2 = Upsample(F_{E3} \odot F_{P3}), \tag{13}$$

$$D_1 = Upsample(F_{E2} \odot F_{P2} \odot D_2), \tag{14}$$

$$J_R = R(F_{E1} \odot F_{P1} \odot D_1), \tag{15}$$

where $\odot$ is the concatenation operation and $Upsample$ contains two convolutions and a resblcok to upsample the features. R has the same structure as $Upsample$, but it doesn't change the feature scale. We use the L_1 loss and the frequency domain reconstruction loss to make the recovered image close to the ground-truth:

$$\mathcal{L}_R = \|J_R - J\|_1 + 0.1 \times \|FFT(J_R) - FFT(J)\|_1, \tag{16}$$

where J_R denotes the recovery result of R-Net and $FFT(\cdot)$ denotes the fast Fourier transform.

3.3 Loss Function

We train our proposed PID-Net using the weighted average of the above-introduced several losses:

$$\mathcal{L} = \mathcal{L}_R + \lambda_E \mathcal{L}_E + \lambda_P \mathcal{L}_P + \lambda_T \mathcal{L}_T, \tag{17}$$

where λ_E, λ_P and λ_T are adjustable parameters, which are empirically set to 0.6, 0.3 and 0.1 in this paper respectively. By joint optimization, PID-Net can well constrain the color and detail information in the recovery results.

4 Experiments

4.1 Experimental Settings

Datasets. We evaluate each method by using both synthetic and real-word datasets. Synthetic datasets include RESIDE [11] and Haze4K [16]. RESIDE is the most widely-used synthetic dataset for image dehazing, which has an Indoor Training Set (ITS) and a Synthetic Objective Testing Set (SOTS). In this study,

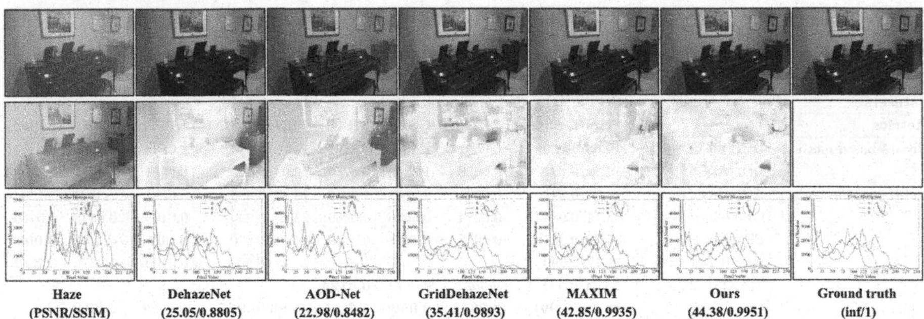

Fig. 4. Visual comparison of different dehazing methods on RESIDE-Indoor dataset. For each method, each column shows the dehazed image, error maps representing texture detail, and the color histogram. The error maps and color histogram of the oringal haze image and ground-truth are also reported for comparison. Clearly, our method performs better than other methods for detail and color recovery.

we use ITS for training and SOTS for testing. Haze4K consists of 3,000 training images and 1,000 testing images. We follow the original partition to train and test our model. Furthermore, to evaluate the generalization ability, we also conduct visual experiments on several real-world images presented in [5].

Evaluation Metrics. For the synthetic datasets, we use five reference metrics to evaluate the performance of different methods, including *Peak Signal Noise Ratio* (PSNR), *Structural Similarity* (SSIM) [28], *Mean Squared Error* (MSE), *Learned Perceptual Image Patch Similarity* (LPIPS) [32] and *Color-Sensitive Error* (CSE) [36]. In this study, upper arrow indicates higher is better, lower arrow is the opposite.

Compared Methods. Seven related state-of-the-art (SOTA) dehazing methods are compared, including two physics-based method (i.e., DCP [6], DehazeNet [1]) and five end-to-end methods (i.e., AOD-Net [9], GriddehazeNet [15], FFA-Net [20], AECR-Net [30] and MAXIM [27]). If the pretrained model of compared methods is available, we will directly use it for evaluation, otherwise, we will retrain it based on the provided source code.

Implementation Details. We implement our method with PyTorch version 1.10 on a Nvidia RTX 3090 with 24G memory. We set the batch size to 4 and train the model with 100 epochs. The images are randomly cropped into 384×384 pixels. The Adam method [8] is used as the optimizer with $\beta_1 = 0.5$ and $\beta_2 = 0.999$. The initial learning rate is set to 0.0001 and it is decreased by half every 10 epochs.

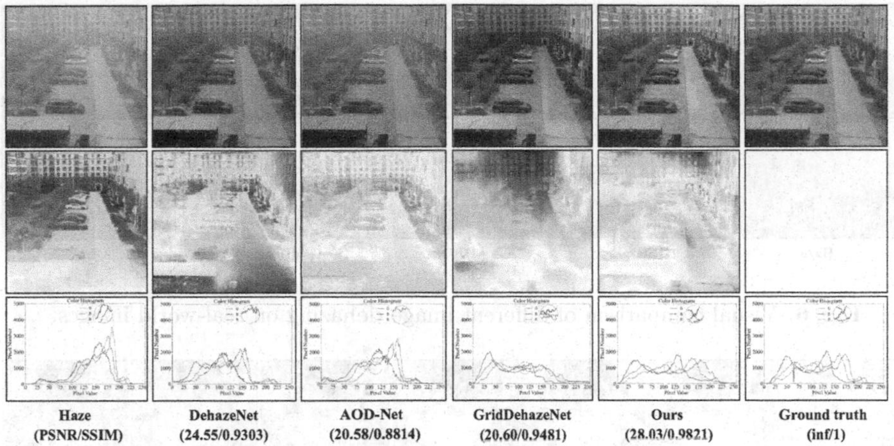

Fig. 5. Visual comparison of image dehazing on Haze4K dataset, in terms of dehazed images, error maps and color histograms.

4.2 Quantitative Dehazing Results

Quantitative Results on SOTS [11]. We first evaluate each method on the SOTS dataset, and the numerical results are described in Table 1. We can see that: 1) End-to-end methods perform better than the physics-based methods in most cases, due to the strong learning ability and the avoidance of error amplification; 2) comparing to other methods, our PID-Net obtains the highest PSNR/SSIM value, which means that our method restores the highest quality images. For the comparison on LPIPS and CSE metrics, our PID-Net still outperforms the other competitors, i.e., our restored haze-free image is more close to the ground-truth in terms of detail and color. Since the LPIPS measures the perception distance between two images while the CSE measures the color difference, it proves the superiority of our method in maintaining abundant texture and color information. Besides, our PID-Net is comparable to MAXIM on the MSE metric, and outperforms all other compared methods.

Quantitative Results on Haze4K [16]. We then evaluate the performance of each method on Haze4K dataset. MAXIM [27] was not involved in the comparison because the authors did not provide the training code. From the results in Table 1, we can see that our method obtains the competitive and even better results than other methods, in which for the PSNR/SSIM metrics, our PID-Net outperforms all competitors. Besides, for the CSE metric, our PID-Net still clearly superior to other competitors, which further proves the excellent color preservation capability of our PID-Net.

Fig. 6. Visual comparison of different image dehazing on real-world images.

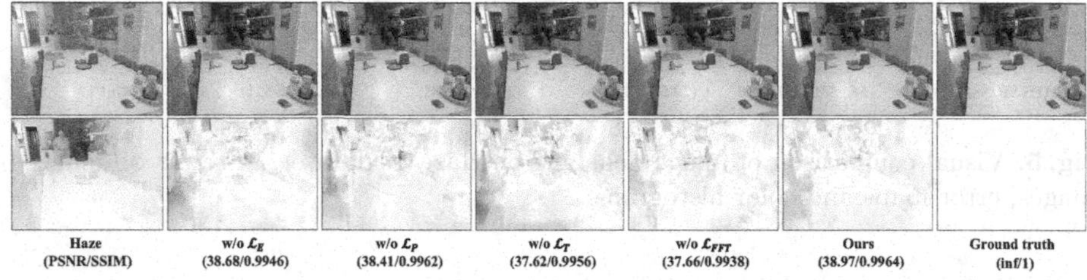

Fig. 7. Effects on the enhanced images and error maps vs. removing different losses on RESIDE-Indoor dataset.

Fig. 8. Effects on the enhanced images and error maps vs. removing different modules on RESIDE-Indoor dataset.

4.3 Visual Image Analysis

Visual Results on Synthetic Datasets. We demonstrate the visual comparison of the dehazing results on RESIDE-Indoor and Haze4K in Figs. 4 and 5 respectively, where we visualize the restored images, the corresponding error maps and the color histograms. On the one hand, the error maps between the restored images and the ground-truth reflect the method's ability of detail recovery in some extent. In practice, error maps can be obtained by calculating the Euclidean distance between two images and the brighter error maps mean the better restored results, which can be referred to [37]. On the other hand, color histograms measured in RGB color space reflect the method's ability of color preservation. As a result, we can see from the figure that our PID-Net obtains

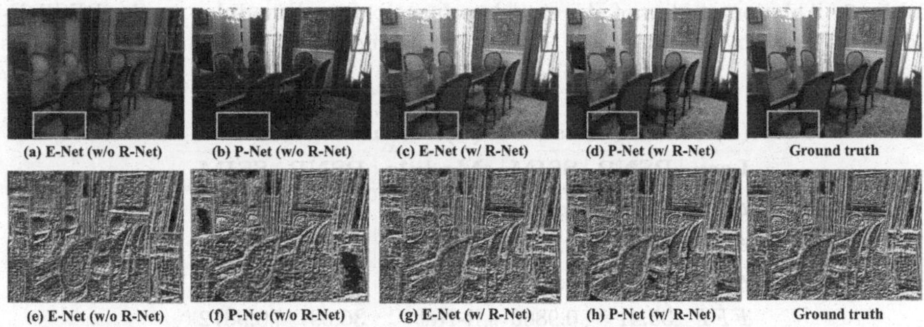

Fig. 9. Dehazed images (top row) and texture details (bottom row). (a) and (b) are the coarse results from E-Net and P-Net respectively, obtained by replacing R-Net with simply add operation. (c) and (d) are the coarse results of E-Net and P-Net respectively based on R-Net. Clearly, without R-Net greatly decreased the recovery performance in terms of both color and detail.

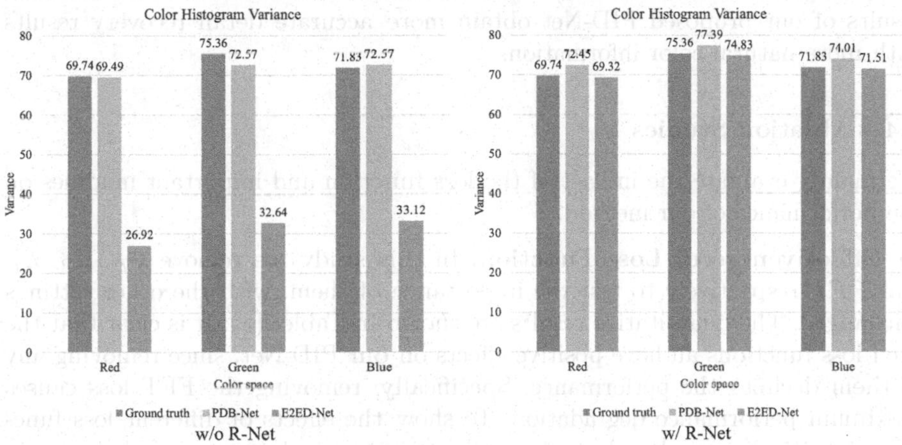

Fig. 10. Color similarity display using color histogram variance calculated in R, G and B channels respectively. (Color figure online)

more accurate restored results than other competitor in terms of detail recovery and color preservation. Specifically, from the error maps and the partial zoom of the color histograms in Fig. 4, we can see that our PID-Net recovers better detail and color, compared to MAXIM [27] whose quantitative performance is closest to ours. The comparison results shown in Fig. 5 also demonstrate the visual superiority of our PID-Net.

Visual Results on Real-World Hazy Images. To evaluate the generalization ability of each method, we compare the visual dehazing effects on several real-world hazy images [5]. We use the model pretrained on ITS dataset [11]

Table 2. Ablation studies of the used loss functions and modules on the RESIDE-Indoor dataset, by removing the component individually with the other settings unchanged.

(a) Remove loss			**(b) Remove module**		
Loss	**PSNR**	**SSIM**	**Module**	**PSNR**	**SSIM**
$\mathcal{L}_E$	37.66	0.9907	E-Net	35.98	0.9744
$\mathcal{L}_P$	37.91	0.9910	P-Net	35.39	0.9889
$\mathcal{L}_T$	37.90	0.9910	TMGIA	37.14	0.9902
FFT	35.41	0.9866	R-Net	36.09	0.9872
Ours	**38.19**	**0.9912**	**Ours**	**38.19**	**0.9912**

to directly test these real-world hazy images, the visual comparison results are shown in Fig. 6. We can find that existing methods either produce accurate restoration results or lose some color information. In contrast, the dehazing results of our proposed PID-Net obtain more accurate detail recovery results with more natural color information.

4.4 Ablation Studies

We mainly evaluate the impact of the loss function and important modules on the performance of our method.

1) Effectiveness of Loss Function. In this study, we remove $\mathcal{L}_T$, $\mathcal{L}_P$, $\mathcal{L}_E$ and FFT respectively to test the importance of them, with the other settings unchanged. The quantitative results are shown in Table 2(a). It is clear that the used loss functions all have positive effects on our PID-Net, since removing any of them declines the performance. Specifically, removing the FFT loss causes maximum performance degradation. To show the effects of different loss functions intuitively, we also show the restored image and error maps between in Fig. 7. Obviously, our PID-Net obtains brighter error maps compared to other models, which means that the used loss function is effective.

2) Effectiveness of E-Net, P-Net and TMGIA. In this study, we remove the E-Net, P-Net and TMGIA module from our method and retrain the model, respectively. The analysis results are shown in Table 2(b) and Fig. 8. We can find that removing E-Net loses the advantage of end-to-end methods in recovering details and has severe negative impact. Compared to remove E-Net, our PID-Net without P-Net obtains the lowest PSNR value and this firmly demonstrates the significance of physical properties in dehazing, as shown in Table 2(b). Removing TMGIA also decreases the performance. That is, each component is important.

3) Effectiveness of R-Net. To prove the effectiveness of R-Net, we replace it by a simple addition operation under the same settings. As shown in Table 2(b), without R-Net significantly decreases the performance. In Fig. 9 and 10, we also visualize the color histogram variance and texture features [18,19], which reflect color and detail information respectively. Although removing R-Net caused severe damages on the results of E-Net and P-Net, they still maintain their advantages for recovering detail and natural color. In contrast, with R-Net, the results of E-Net and P-Net all have obvious improvements in both detail and color, which again demonstrates the effectiveness of R-Net. From Fig. 10, changes of color information can be observed from the color histogram variance. Specifically, with R-Net clearly has a positive effect on color preservation.

5 Conclusion

We improved the image dehazing task by sufficiently exploring the interaction between end-to-end and physics-based methods, so that both detail and color information can be discovered and preserved. Technically, we proposed a novel physical-property guided end-to-end interactive image dehazing network called PID-Net with a multi-source "coarse-to-fine" pipeline. Specifically, our method uses an end-to-end information interaction network to learn initial coarse result, and designs a color-detail refinement network to refine and obtain the final visual pleasing haze-free output with accurate detail recovery and color preservation. Extensive experiments show that our PID-Net obtains more pleasing and realistic dehazing images, with significant improvement compared to other closely-related methods. In the future, we will explore more lightweight dehazing models and trade-off the dehazing performance and model size. Besides, extending the dehazing task to the joint high-level vision task is also an interesting future work.

Acknowledgments. The work described in this paper is partially supported by the National Natural Science Foundation of China (62072151, 61932009), the Anhui Provincial Natural Science Fund for the Distinguished Young Scholars (2008085J30), and the CAAI-Huawei MindSpore Open Fund.

References

1. Cai, B., Xu, X., Jia, K., Qing, C., Tao, D.: DehazeNet: an end-to-end system for single image haze removal. IEEE Trans. Image Process. **25**(11), 5187–5198 (2016)
2. Chen, D., et al.: Gated context aggregation network for image dehazing and deraining. In: Proceedings of the IEEE Winter Conference on Applications of Computer Vision, pp. 1375–1383 (2019)
3. Chen, Y., Li, W., Sakaridis, C., Dai, D., Van Gool, L.: Domain adaptive faster R-CNN for object detection in the wild. In: Proceedings of the IEEE Conference on Computer Vision and Pattern Recognition, pp. 3339–3348 (2018)
4. Dong, Y., Liu, Y., Zhang, H., Chen, S., Qiao, Y.: FD-GAN: generative adversarial networks with fusion-discriminator for single image dehazing. In: Proceedings of the AAAI Conference on Artificial Intelligence, vol. 34, pp. 10729–10736 (2020)
5. Fattal, R.: Dehazing using color-lines. ACM Trans. Graph. **34**(1), 1–14 (2014)
6. He, K., Sun, J., Tang, X.: Single image haze removal using dark channel prior. IEEE Trans. Pattern Anal. Mach. Intell. **33**(12), 2341–2353 (2010)

7. Hong, M., Xie, Y., Li, C., Qu, Y.: Distilling image dehazing with heterogeneous task imitation. In: Proceedings of the IEEE/CVF Conference on Computer Vision and Pattern Recognition, pp. 3462–3471 (2020)
8. Kingma, D.P., Ba, J.: Adam: a method for stochastic optimization. arXiv preprint arXiv:1412.6980 (2014)
9. Li, B., Peng, X., Wang, Z., Xu, J., Feng, D.: AOD-Net: all-in-one dehazing network. In: Proceedings of the IEEE International Conference on Computer Vision, pp. 4770–4778 (2017)
10. Li, B., Peng, X., Wang, Z., Xu, J., Feng, D.: End-to-end united video dehazing and detection. In: Proceedings of the AAAI Conference on Artificial Intelligence, vol. 32 (2018)
11. Li, B., et al.: Reside: a benchmark for single image dehazing. arXiv preprint arXiv:1712.04143 1 (2017)
12. Li, G., Xie, Y., Lin, L., Yu, Y.: Instance-level salient object segmentation. In: Proceedings of the IEEE Conference on Computer Vision and Pattern Recognition, pp. 2386–2395 (2017)
13. Li, Y., Chang, Y., Gao, Y., Yu, C., Yan, L.: Physically disentangled intra-and inter-domain adaptation for varicolored haze removal. In: Proceedings of the IEEE/CVF Conference on Computer Vision and Pattern Recognition, pp. 5841–5850 (2022)
14. Li, Y., et al.: LAP-Net: level-aware progressive network for image dehazing. In: Proceedings of the IEEE/CVF International Conference on Computer Vision, pp. 3276–3285 (2019)
15. Liu, X., Ma, Y., Shi, Z., Chen, J.: GriddehazeNet: attention-based multi-scale network for image dehazing. In: Proceedings of the IEEE/CVF International Conference on Computer Vision, pp. 7314–7323 (2019)
16. Liu, Y., et al.: From synthetic to real: image dehazing collaborating with unlabeled real data. In: Proceedings of the ACM International Conference on Multimedia, pp. 50–58 (2021)
17. Meng, G., Wang, Y., Duan, J., Xiang, S., Pan, C.: Efficient image dehazing with boundary constraint and contextual regularization. In: Proceedings of the IEEE International Conference on Computer Vision, pp. 617–624 (2013)
18. Ojala, T., Pietikainen, M., Harwood, D.: Performance evaluation of texture measures with classification based on Kullback discrimination of distributions. In: Proceedings of 12th International Conference on Pattern Recognition, vol. 1, pp. 582–585 (1994)
19. Ojala, T., Pietikäinen, M., Harwood, D.: A comparative study of texture measures with classification based on featured distributions. Pattern Recogn. **29**(1), 51–59 (1996)
20. Qin, X., Wang, Z., Bai, Y., Xie, X., Jia, H.: FFA-Net: feature fusion attention network for single image dehazing. In: Proceedings of the AAAI Conference on Artificial Intelligence, vol. 34, pp. 11908–11915 (2020)
21. Ren, W., Cao, X.: Deep video dehazing. In: Zeng, B., Huang, Q., El Saddik, A., Li, H., Jiang, S., Fan, X. (eds.) PCM 2017. LNCS, vol. 10735, pp. 14–24. Springer, Cham (2018). https://doi.org/10.1007/978-3-319-77380-3_2
22. Ren, W., Liu, S., Zhang, H., Pan, J., Cao, X., Yang, M.-H.: Single image dehazing via multi-scale convolutional neural networks. In: Leibe, B., Matas, J., Sebe, N., Welling, M. (eds.) ECCV 2016. LNCS, vol. 9906, pp. 154–169. Springer, Cham (2016). https://doi.org/10.1007/978-3-319-46475-6_10
23. Ren, W., et al.: Gated fusion network for single image dehazing. In: Proceedings of the IEEE Conference On Computer Vision and Pattern Recognition, pp. 3253–3261 (2018)

24. Ronneberger, O., Fischer, P., Brox, T.: U-Net: convolutional networks for biomedical image segmentation. In: Proceedings of the International Conference on Medical Image Computing and Computer-Assisted Intervention, pp. 234–241 (2015)
25. Sakaridis, C., Dai, D., Van Gool, L.: Semantic foggy scene understanding with synthetic data. Int. J. Comput. Vision **126**(9), 973–992 (2018)
26. Tang, K., Yang, J., Wang, J.: Investigating haze-relevant features in a learning framework for image dehazing. In: Proceedings of the IEEE Conference on Computer Vision and Pattern Recognition, pp. 2995–3000 (2014)
27. Tu, Z., et al.: Maxim: multi-axis MLP for image processing. In: Proceedings of the IEEE/CVF Conference on Computer Vision and Pattern Recognition, pp. 5769–5780 (2022)
28. Wang, Z., Bovik, A.C., Sheikh, H.R., Simoncelli, E.P.: Image quality assessment: from error visibility to structural similarity. IEEE Trans. Image Process. **13**(4), 600–612 (2004)
29. Wei, Y., et al.: DerainCycleGAN: Rain attentive CycleGAN for single image deraining and rainmaking. IEEE Trans. Image Process. **30**, 4788–4801 (2021)
30. Wu, H., et al.: Contrastive learning for compact single image dehazing. In: Proceedings of the IEEE/CVF Conference on Computer Vision and Pattern Recognition, pp. 10551–10560 (2021)
31. Zhang, H., Patel, V.M.: Densely connected pyramid dehazing network. In: Proceedings of the IEEE Conference on Computer Vision and Pattern Recognition, pp. 3194–3203 (2018)
32. Zhang, R., Isola, P., Efros, A.A., Shechtman, E., Wang, O.: The unreasonable effectiveness of deep features as a perceptual metric. In: Proceedings of the IEEE Conference on Computer Vision and Pattern Recognition, pp. 586–595 (2018)
33. Zhang, Z., Zheng, H., Hong, R., Xu, M., Yan, S., Wang, M.: Deep color consistent network for low-light image enhancement. In: Proceedings of the IEEE/CVF Conference on Computer Vision and Pattern Recognition, pp. 1889–1898 (2022)
34. Zhao, H., Shi, J., Qi, X., Wang, X., Jia, J.: Pyramid scene parsing network. In: Proceedings of the IEEE Conference on Computer Vision and Pattern Recognition, pp. 2881–2890 (2017)
35. Zhao, S., Zhang, Z., Hong, R., Xu, M., Yang, Y., Wang, M.: FCL-GAN: a lightweight and real-time baseline for unsupervised blind image deblurring. In: Proceedings of the ACM International Conference on Multimedia, pp. 6220–6229 (2022)
36. Zhao, S., et al.: CRNet: unsupervised color retention network for blind motion deblurring. In: Proceedings of the 30th ACM International Conference on Multimedia, pp. 6193–6201 (2022)
37. Zheng, C., Shi, D., Liu, Y.: Windowing decomposition convolutional neural network for image enhancement. In: Proceedings of the 29th ACM International Conference on Multimedia, pp. 424–432 (2021)

A Clothing Classification Network with Manifold Structure Based on Second-Order Convolution

Ruhan He[1] and Cheng Quan[2(✉)]

[1] School of Computer Science and Artificial Intelligence, Wuhan Textile University, Wuhan, China

[2] Hubei Provincial Engineering Research Center for Intelligent Textile and Fashion, Wuhan, China

lunaticdiary@foxmail.com

Abstract. Currently, AI-based clothing image classification techniques mostly use traditional deep learning methods, which are based on monocular clothing images for classification. However, the diversity of perspectives of realistic clothing images bring great difficulties and challenges to clothing classification. Moreover, deep convolutional networks have limitations of their own. They treat data as vectors in Euclidean space and fail to make full use of the potential low dimensional non-linear geometric structure information within high-dimensional clothing image data. Therefore, this paper explores and exploits the geometric structure information inherent of clothing image data from the perspective of a non-Euclidean manifold learning method, and designs and implements a clothing classification network with manifold structure based on second-order convolution to classify images using the second-order statistics of clothing features for image classification. Firstly, the input clothing image features extracted by the convolution neural network are pooled with the covariance pooling module to obtain the second-order statistical covariance, which is converted into SPD manifold to characterize the feature information of the clothing image set, and then a complete manifold structure neural network is constructed to enhance the feature representation ability of the model on the geometric intrinsic structure of the clothing image set. The experimental results of this method on the multi view clothing image dataset MVC show that it has good effectiveness, robustness, and accuracy.

Keywords: Clothing Classification · Deep Learning · Manifold Learning · SPD Manifold · Covariance Pooling

1 Introduction

With the rapid development of computer vision technology and the availability of a large number of clothing datasets, the research on smart fashion has made great progress. Smart fashion technology based on deep learning has become one of the most important implementations to help us better understand and grasp fashion trends and reduce the time cost and economic cost in modern online shopping consumption patterns.

H. Zhang et al. (Eds.): NCAA 2023, CCIS 1870, pp. 132–146, 2023.
https://doi.org/10.1007/978-981-99-5847-4_10

Smart Fashion has an extremely wide range of applications. It can detect and analyze fashion elements in images, classify and synthesize clothing images. In this way, we can provide users with personalized recommendations to meet their needs. For example, on e-commerce platforms, we can use images to classify products into different categories so that consumers can search more easily. In addition, we can look at clothing images to find items that are similar to them and provide clothing recommendations. There is even the possibility of online virtual fitting through a clothing image and a user image. In summary, by introducing smart fashion technology, the shopping experience of consumers can be effectively improved. It adds more opportunities for distributors' sales activities and thus achieves greater financial gains. Therefore, enhancing the application of smart fashion technology will bring more benefits to companies.

Clothing classification is one of the fundamental tasks in the field of smart fashion, which can improve our garment detection, garment key point detection, garment segmentation, retrieval and clothing recommendation effectiveness.

However, due to the wide variety of garments, their styles and forms also vary greatly (color, texture, cuff length, pleats, etc.). Therefore, the effect of extracting these feature values by hand alone is often unsatisfactory. And in real life, as clothing is flexible and subject to some objective factors, problems such as deformation and shelter of clothing cannot be avoided. Moreover, different clothing pictures possess diverse perspectives, and the styles of commercially available clothing images vary widely. In addition, some photos taken by cell phones are often unable to fully reflect the true effect of clothing due to the influence of objective conditions and environment. During the continuous deformation of the garments, the garment images will also be shaded to different degrees, and the angles and lighting, etc. will also change. Optimizing the model to extract the features in the input garment images in order to classify them more accurately is an important challenge at present.

Deep learning-based garment classification techniques usually use convolutional neural networks to extract the representation features of garment images and then classify the garments. This classification method is often difficult to extract the geo-metric information embedded inside the image features, because deep neural net-works assume that the data lie in Euclidean space. As a generalization of deep learning in non-Euclidean spaces. Geometric deep learning is a more generalized type of neural network. It expects to follow the geometric agency of the data from the beginning of the network design. In contrast it is able to extract more information about the geometric structure of the data than traditional deep learning models and tends to be more effective in practical applications.

In order to make the algorithm more accurate in classification when applied in real scenarios. In this paper, we focus on optimizing the ability of traditional deep learning-based clothing classification algorithms to represent the internal structure of clothing image sets by converting the clothing image features extracted by CNN to data located on the stream shape. Then, in order to maintain the intrinsic geometric structure of the extracted feature data, the paper feeds the converted popular value data into SPDNET, and performs operations on the SPD matrix to ensure that the dimensionality-reduced data conforms to the stream shape structure. Finally, this paper converts the popular value data output from SPDNET back to Euclidean space and outputs the final clothing classification results through a fully connected network, which effectively improves the classification accuracy.

2 Related Work

2.1 Clothing Classification

In 2016, Liu et al. [1] proposed a large-scale clothing dataset with comprehensive annotations –DeepFashion, and the first comprehensive set of clothing annotation criteria. This innovative solution was widely recognized and also made an important contribution to the development of the garment inspection field. Luo et al. [2] achieved more efficient garment retrieval by fine-tuning the network on the ImageNet dataset to accommodate the features of garments. Liu et al. [3] proposed a new clothing dataset, MVC, which has images of no less than four different views of each garment in the dataset. These data come from various platforms on the web and contain 264 attribute annotations. It can solve the problem of poor generalization of most models when the image viewpoint changes.

In 2019, Ge et al. [4] proposed DeepFashion2, a dataset with comprehensive task definition and rich annotation support for fashion images. Sidnev et al. [5] address the drawback that key point estimation in DeepFashion2 is difficult to use for low-power devices, a single-stage fast garment detection method based on Mask R-CNN [6] and CenterNet [7], a multi-target network. It can be used on low-power devices with relatively high accuracy without any post-processing, and it can solve both garment detection and keypoint detection tasks simultaneously.

In 2021, Lin et al. [8] proposed dynamic coding filter fusion (DCFF). Compact CNNs were derived in a computationally economical and regularization-free manner to achieve efficient image classification. Each filter in DCFF is first given a similarity distribution with the temperature parameter as the filter proxy, on top of which a new Kullback-Leibler scatter-based dynamic coding criterion is proposed to evaluate the importance of the filters. Using filter fusion, i.e., using the weighted average of the specified proxies as the retained filters, replaces the simple retention of the high score filter method. As the temperature parameter approaches infinity, a single thermal similarity distribution is obtained. As a result, the relative importance of each filter can vary with the training of the compact CNN, leading to dynamically variable fusion filters without relying on pre-trained models and introducing sparse constraints. The model classification accuracy is improved while greatly reducing the model parameters.

Liu et al. [9] proposed a multi-label classification method based on cross-attention based transformer decoder. Mallavarapu et al. [10] proposed a 3-stage fine-grained clothing classification model for in-depth classification of fashion attributes in the same category. Previously Eshwar et al. [11] used CNN only to classify clothing in a simple way with a large difference between each class. Iliukovich-Strak et al. [12] proposed a fine-grained clothing classification system with two levels of classes with 10 super-classes such as high heels and slip-on shoes, each containing 14 classes. Xu et al. [13] proposed a fashion image classification method by applying the hierarchical classification structure of hierarchical CNN (H-CNN) on clothing images. By defining three levels of hierarchical classes to reduce the classification errors of labels. For each fashion item, the model outputs three hierarchical labels, e.g., "dress" - "top" - "T-shirt". These results provide novelty and possibilities for fashion apparel classification. In the same year, Qu et al. [14] proposed the multilayer semantic representation network (MSRN). By modeling label relevance and thus exploring local and global semantics of labels, and using label semantics to guide multilayer semantic representation learning through an attention mechanism for multi-label classification of images. Cheng et al. [15] proposed a multi-label Transformer architecture based on window division, intra-window pixel attention, and cross-window attention, which particularly improves the performance of multi-label image classification task. Lin [16] proposed a new training scheme for garment landmark detection: aggregation and fine - tuning. The homogeneity between the landmarks of different classes of garments was exploited. It was used to design a training program that achieved the best results on a clothing landmark detection competition. Zheng et al. [17] mapped the clothing style time recognition problem to a clothing style classification problem under the assumption that clothing styles change from year to year. A new deep neural network was proposed. The network achieves accurate body segmentation by fusing multi-scale convolutional features into a full convolutional network. The segmented part is then subjected to feature learning and fashion classification, avoiding the influence of image background.

In the real situation, the clothing image data actually conforms to the flow structure more. And the above deep learning-based clothing classification methods all consider the clothing image set (video) as data in Euclidean space, which cannot keep the flow structure feature of clothing images. This limits the application of deep learning-based clothing classification methods in practical scenarios. In this paper, we investigate the use of geometric structure information of clothing images to improve the accuracy of clothing classification algorithms by converting clothing image data to stream shapes.

2.2 Manifold Structured Neural Network

Traditional convolutional neural network (CNN) is limited to processing data residing in vector space, while data residing in smooth non-Euclidean space (e.g., Riemannian manifolds) appear in many problem domains. In the past few years, there has been a surge of research on extending CNNs to manifolds using geometric deep learning methods. Masci et al. [18] extracted local regions on shaped manifolds based on local geodesic polar coordinates and then implemented geodesic convolution neural networks (GCNNs) through a cascade of filters and linear and nonlinear operators. Poulenard et al. [19] proposed directional convolution, allowing propagation and association of directional information across layers so that different regions on the shape are associated with each other, and directional convolution is also per-formed in local geodesic polar coordinates.

Ionescu et al. [20] implemented matrix-based backpropagation methods in a deep learning framework. They also proposed a region classification network DeepO2P and a logarithmic backpropagation algorithm based on SPD matrices. Based on [20], several methods were proposed to perform SPD matrix learning. Huang et al. [21] proposed a flow structured network cascaded by three kinds of convolutional layers for nonlinear learning of symmetric positive definite matrices (SPD Matrix). Later Huang et al. [22]

proposed Grnet and successfully extended convolutional neural networks to Grassmannian manifolds. Nguyen et al. [23] modeled human skeletal data as manifolds successfully for symmetric positive definite matrix-based learning neural networks. Qiao et al. [24] proposed an end-to-end deep heterogeneous hashing (DHH) algorithm for learning face images and videos with a unified binary code. The method represents face images as vectors in Euclidean space and models face videos as covariance matrices on Riemannian manifolds. Li et al. [25] proposed to use the manifold structure of the data to normalize the learning of deep action features. Chakraborty et al. [26] proposed a popular value convolutional neural network (ManifoldNet) based on weighted Frechet means that implemented a generalization of CNN to Riemannian manifolds. Brooks et al. [27] proposed a batch normalization technique based on SPD matrices that can effectively avoid overfitting.

3 Method

In order to utilize the second-order statistics of the extracted image features, this paper proposes a clothing classification network with manifold structured based on second-order convolution, which extracts the second-order statistics of image features by covariance pooling and adds a batch normalization (BN) layer to the flow network to speed up the convergence (Fig. 1).

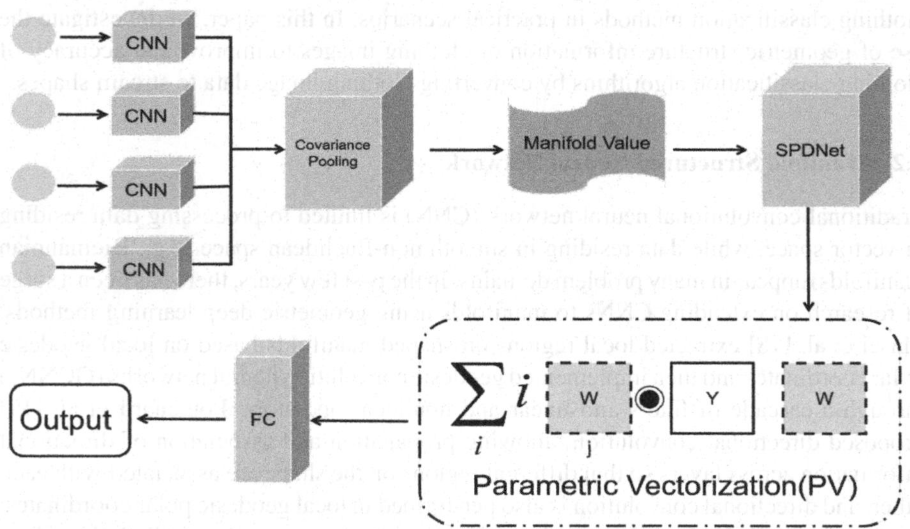

Fig. 1. Network architecture. The network consists of three parts. The first part is a conventional convolution neural network. The second part is the covariance pooling module. The third part is a manifold structured neural network that processes the SPD matrix.

3.1 Covariance Pooling

Deep convolutional neural networks can extract first-order statistics from input images, but cannot extract second-order statistics, such as covariance. In fact, the covariance matrix is a symmetric positive definite matrix, located exactly on the manifold. Psychological research has shown that second-order statistics play an important role in human visual recognition processes. The continuous development and utilization of covariance matrices, i.e. covariance descriptors, in the past have also proven that second-order statistics typically outperform first-order features in visual recognition tasks. However, the internal structure of actual clothing image data is more in line with the manifold structure. This article uses the method of pooling co-variance differences to obtain the required manifold data in order to extract the geometric structure features of clothing data. The following introduces the method of covariance pooling.

Firstly, in the first part of the entire model, a pre trained VGG-16 model is used to extract feature while the training clothing image is fed into this model for convolution. After removing the fully connected layer behind it, the clothing image features output from the convolution are fed into the covariance pooling module to obtain the second-order statistics of the clothing features [28]. Assuming the dimension of the input image data in the network is F × 3 × H × W. The process of covariance pooling is to first extract clothing features from each image (frame) through a convolutional neural network, and then perform covariance pooling on European clothing features to obtain a (F + 1) × (F + 1) tensor, which is located on the Riemannian manifold. Figure 2 shows a schematic diagram of the process of covariance pooling. The following de-scribes the specific calculation method for pooling of collaborative differences.

For the case where the input is a single image of size 3 × H × W, this image is fed into the whole model and again, after applying a VGG-16 model with the final fully connected layer removed, the output of the last standard convolutional layer is assumed to be a (C × H′ × H′) tensor. Next spreading this tensor yields a (C × N) shape matrix $X = [x_1, x_2, \ldots, x_N]$, where N = W × H. The covariance is then calculated according to the following Eq. 1 to obtain a (C × C) shape feature covariance matrix:

$$\Sigma = \frac{1}{N}\sum_{k=1}^{N}(x_k - \mu)(x_k - \mu)^T \tag{1}$$

where $\mu = \frac{1}{N}\sum_{k=1}^{N} x_k$ is the mean vectors of the feature vectors. The resulting covariance matrix Σ encodes the second-order statistics of the image features, but does not encode the first-order statistics as well. However, in practice the first order statistics may also contain valuable information. Finally, to obtain a better representation of the features, the covariance matrix is calculated to incorporate the first-order information of the input image features using the following Eq. 2:

$$C = \begin{bmatrix} \Sigma + \beta^2 \mu\mu^T & \beta\mu \\ \beta\mu^T & 1 \end{bmatrix} \tag{2}$$

The above equation combines the mean values of the features by means of the parameter β, which was set to β = 0.3 in the experiments.

The covariance calculation above is for a single image, but the input data in the experiments are multiple images (which can be thought of as videos). By simply considering the number of images (video frames) as channels, and assuming that the features extracted from each image after passing through the convolutional neural network are $f_i \in \mathbb{R}^c$, i ∈ [1, *n*], the covariance can be calculated as in Eq. 2 above.

The covariance calculation above is for a single image, but the input data in the experiments are multiple images (which can be thought of as videos). The covariance pooling can be performed in the same way as above by transforming the features extracted from each image after the convolutional neural network into $f_i \in \mathbb{R}^d$, i ∈ [1, *n*], where n is the number of images.

The covariance pooling module sits between the convolution neural network and the manifold network, replacing the fully connected layer behind the convolutional neural network. The covariance pooled features are located on Riemann manifolds and will be fed directly into the manifold network as manifold data.

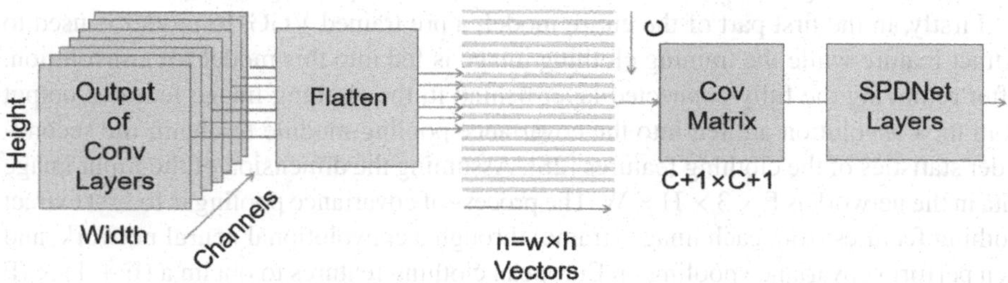

Fig. 2. Schematic diagram of covariance pooling

3.2 Second Order Convolution on SPD Manifolds

As shown in Fig. 3, the Riemann manifold network architecture is consistent with that of a classical neural network, with the first stage being used to learn the relevant representation of the input data points, and the second stage performing the final classification based on the extracted features. The manifold network takes into account the special structure of SPD manifolds and is able to learn information about the geometric structure of the input SPD manifold-valued data. The overall structure of the Streaming Network model is a generalization of the Convolutional Neural Net-work (CNN) to SPD streams, with the three main components acting as convolution, correction of the ReLU (activation function), and Batch Normalization respectively.

The main function of the second-order convolution-based manifold network is to generate a more compact and discriminative SPD matrix. As a generalization of convolution neural networks to SPD manifolds, the second-order convolution-based manifold network maps the input SPD manifold to another SPD matrix of smaller dimensionality, while keeping the output results still residing on the SPD manifold. This mapping, which we call second-order convolution, is implemented by a bilinear mapping layer (BiMap), a batch normalization layer and a linear correction unit. The BiMap layer induces matrix dimensionality reduction by converting the input SPD matrix (usually a covariance matrix derived from the data) into a new SPD matrix with a bilinear mapping using a bilinear mapping. The computational burden of the algorithm for learning the geometric structural features of the SPD matrix is alleviated and the computational process can be represented as follows.

$$X_k = f_b^{(k)}(X_{k-1}; W_k) = W_k X_{k-1} W_k^T \tag{3}$$

where $X_{k-1} \in Sym^+_{d_{k-1}}$ is the input SPD matrix for the kth layer, $W_k \in R_*^{d_k \times d_{k-1}} d_k < d_{k-1}$ is the transformation matrix (weight matrix) and $X_k \in R^{d_k \times d_k}$ is the result matrix calculated. The transformation matrix is a row full rank matrix to ensure that the output is an SPD matrix. The output dimensionality is reduced after the bilinear mapping, but still resides on an SPD manifold.

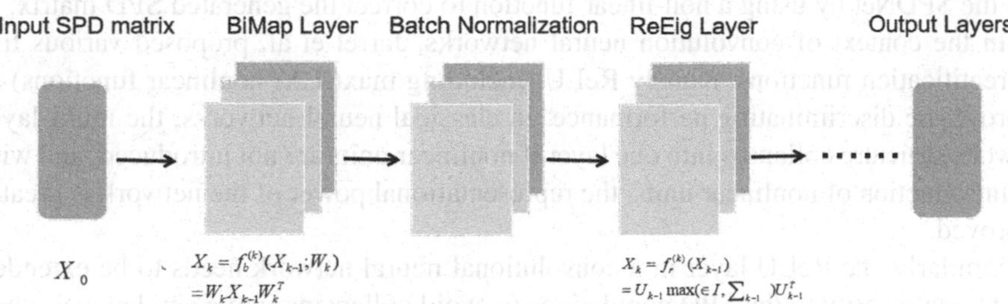

Fig. 3. Schematic diagram of Riemannian manifold network for SPD manifolds

3.3 Riemann Batch Normalization of SPD Matrices

Another important part of a convolution neural network is the batch normalization unit, or BN layer. Extending the classical batch normalization layer in convolution neural networks from Euclidean space to SPD manifolds can accelerate model convergence and avoid model over-fitting. In this paper, the output SPD matrix is normalized by introducing a Riemann mean-based batch normalization method. In this section, the implementation of the Riemann Batch Normalization (RBN) algorithm for the SPD matrix is presented.

In a convolution neural network, batch normalization starts by averaging the data within a batch and then normalizing it, followed by multiplication and addition using a parametric scale transformation and offset. The Riemann batch normalization (RBN) method for a batch of SPD matrices is described below.

The steps of the RBN are divided into four main steps:

1. Finding the batch Riemann mean for each training batch of data.
2. Update the runtime mean of the training set from the calculated batch Riemann mean.
3. Use the updated runtime mean to centralize the training data of the batch.
4. Batch shifting the centralized results to the parametric SPD matrix.

Where in the first step the batch Riemann mean is calculated for that batch of da-ta. The second step calculates the runtime mean, which is the weighted Riemann mean of the current batch Riemann mean and the runtime mean weighted by η. This is equivalent to shifting the runtime mean along the geodesic towards the current batch mean (1 − η). The method extends the batch normalization method in Euclidean space to the space of popular values. In the experiments, the Riemann BN is appended to each layer of the BiMap in the network.

3.4 Rectified Linear Units

As shown in Fig. 3, in addition to the bilinear mapping layer another important part of the Riemann manifold network is the linear correction unit, the ReEig layer, which is a generalization of the ReLU activation function from Euclidean space to the SPD manifold in the classical convolution neural network. The non-linearity is introduced into the SPDNet by using a non-linear function to correct the generated SPD matrix.

In the context of convolution neural networks, Jarret et al. proposed various linear rectification functions, namely ReLU (including max(0, x) nonlinear functions) to improve the discriminating performance. In classical neural networks, the multi-layer network structure collapses into one layer if nonlinear units are not introduced, and with the introduction of nonlinear units, the representational power of the network is greatly improved.

Similarly, the ReLU layer in a convolutional neural network needs to be extended to add nonlinearity to the SPD populations to avoid collapsing into a single layer when stacking multi-layer SPD flow networks. As in Eq. 4, the ReEig layer (first) layer corrects the SPD matrix by a non-linear function f_r that adjusts small positive eigenvalues according to a given a threshold ϵ:

$$X_k = f_r^{(k)}(X_{k-1}) = U_{k-1} max(I, \Sigma_{k-1}) U_{k-1}^T \quad (4)$$

where $X_{k-1} = U_{k-1} \sum_{k-1} U_{k-1}^T$, U_{k-1} and $\sum_{k-1}$ denote the eigenvalues and eigenvectors obtained by eigenvalue decomposition (EIG) of the output matrix X_{k-1} of the previous layer. ϵ is the correction threshold, I is the identity matrix, and $\max\left(\epsilon I, \sum_{k-1}\right)$ is a diagonal matrix A. The diagonal elements in A are defined as

$$A(i,i) = \begin{cases} \sum_{k-1}(i,i), & \sum_{k-1}(i,i) > \epsilon \\ \epsilon, & \sum_{k-1}(i,i) \le \epsilon \end{cases} \quad (5)$$

The ReEig layer is applied to the input of each second-order convolution, regularizing the convolved features and adding non-linearity to the network, effectively preventing multiple layers of second-order convolutions from collapsing into one layer after stacking.

3.5 Parameter Vectorization Layer

Generally, in a convolution neural network, the classification results are ultimately output through a fully connection layer. In contrast, the final output of a second-order convolution-based manifold network is a manifold value data, which requires that the manifold-valued feature of the output of the manifold network must first be mapped to a vector space before it can be sent to a fully-connected layer to output a class probability estimate. In some studies, the use of a Log-Euclidean Riemannian metric has enabled the manifold to be reduced to a flat space via a matrix logarithmic operation log () on the SPD matrix in order to apply Euclidean calculations directly in the flat space, and the Log-Euclidean Riemannian metric of the SPD manifold can effectively reduce computational costs [29].

Therefor the feature value vectorization in a general second-order convolution-based manifold structured neural network is achieved by simply applying a logarithmic mapping to the matrix and then flattening the matrix. The approach is simple and easy to implement, but the disadvantage is that the dimension of the output vector is too large when dealing with matrices of large dimension, leading to an excessive number of parameters in the fully connected layer. In this paper, instead of a straightforward flattening, a second-order representation-based parameter vectorization is used, with the parameter vectorization process following Eq. 6:

$$v_j = \left([W]_{:,j}\right)^T Y[W]_{:,j} = \sum_{i=1}^{d} [W \odot YW]_{i,j} \tag{6}$$

where $W \in \mathbb{R}^{d \times n}$ is the training parameter, $Y \in \mathbb{R}^{d \times d}$ is the output of the previous layer, $\mathrm{v} \in \mathbb{R}^{\mathrm{n}}$ is the feature vector obtained by second order vectorization, and the j th component of the feature vector v is $\mathrm{v_j}$, equal to the transpose of the j th column of the parameter matrix multiplied by Y then multiplied by the j th column of the parameter matrix.

This approach to feature vectorization is more flexible than the previous approach, while preventing too many parameters. The vectorized feature values lie in vector space and are fed directly into the fully connected layer for classification.

4 Experiments

The experimental platform is Ubuntu16 64-bit operating system, Intel CPU i9, NVIDIA GTX 2080Ti GPU, and Pytorch deep learning framework is also used for the experiments. For the experimental parameters, the batch size is set to 128. The optimizer uses Adam.

4.1 Dataset

The MVC dataset was used for the experimental dataset. The MVC dataset [3] can be used for research to solve the problem of poor generalization performance of the clothing classification algorithm when the clothing perspective is transformed. As shown in Fig. 4, there are at least 4 different views for each of the same garment.

In this paper, 17 categories in MVC were selected as classification labels, and 2,000 images were selected for each category, and another 100 image sets of 6,800 images were selected for each category in groups of 4 images with different perspectives. The training dataset is 60%, the validation dataset is 20%, and the last 20% is used as the test set of the model.

4.2 Details

In this experiment, a pre-trained VGG-16 model is used as the backbone network in the first part of the whole model. Four different viewpoint images of the same garment are fed into the backbone network as a set of inputs simultaneously for feature

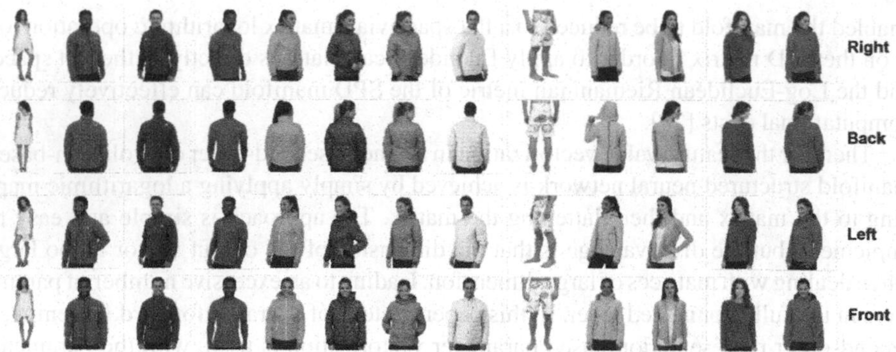

Fig. 4. Different angles of the garment view. There are at least 4 different views for each of the same garment.

extraction. The extracted image features are then fed into the covariance pooling layer to extract the second-order statistics. Our proposed model consists of a 3-layer stack of BiMap+RBN+ReEig. The covariance pooled data is fed directly into the second-order convolution layer. The output of the last layer is fed to a parametric characterization (PV) layer, which maps the manifold value features of the input to R^n. Next, two fully connected layers are applied to this value input R^n, and finally a softmax function is used to output the class probabilities. Experimental results show that this architecture provides the best performance among similar architectures.

The network parameters were tuned using cross-entropy loss and stochastic gradient descent. The momentum was set to 0.9 and the weight decay was adjusted to 0.0001. Before performing the training, we adjusted the learning rate of the model to 0.001 and the batch size to 128, finally the number of iterations was set to 100. When testing was performed, the lowest loss value obtained by the algorithm when selected as the final model result.

4.3 Results

The evaluation metrics used in this experiment are Precision, Recall and F1 metric as evaluation metrics. To test the effectiveness of the proposed algorithm on classification of clothing image sets (videos). We compared it with other algorithms. The compared methods include the commonly used C3D [30], LRCN [31], and SPD-Net [21], and the experimental results are shown in Table 1. Both the proposed method and LRCN, SPD-Net use CNN for image feature extraction. The same VGG-16 network is selected in this experiment. From Table 1, we can see that the proposed algorithm is higher than the existing algorithms in three evaluation metrics: Precision, Recall, and F-1 value. Its Precision, Re-call, and F-1 are 89.87%, 89.52%, and 89.44%, which are 2.45%, 3.05%, and 3.45% higher than LRCN, respectively. Besides, the proposed algorithm has less number of parameters than that of LRCN (about 0.94M). Its number of parameters is more than that of C3D, which is mainly because the VGG-16 network contains about 138M number of parameters. Among the three algorithms, C3D has lower metrics, which may be due to the fact that four images were selected as an image set (video) in the experiment, and the amount of training data is not able to meet the demand of C3D.

Table 1. Classification results on the MVC dataset

Method	Precision	Recall	F-1	Parameters
C3D	67.32 ± 0.09	65	66.44	73.24
LRCN	87.42 ± 0.07	86.47	85.99	140.25
SPD-Net	89.46 ± 0.13	89.12	88.69	138.57
Ours	**89.87 ± 0.10**	**89.52**	**89.44**	**139.31**

The accuracy of the proposed method and LRCN, C3D and SPDNet in the training process is shown in Fig. 5. The C3D algorithm starts to approach the best results after 80 rounds, the LRCN algorithm starts to converge after 40 rounds, and the SPD-Net algorithm starts to converge after 20 rounds. The accuracy of the proposed algorithm is significantly higher than that of the SPD algorithm after several rounds of training, and is more stable and achieves the best results faster than the SPD algorithm.

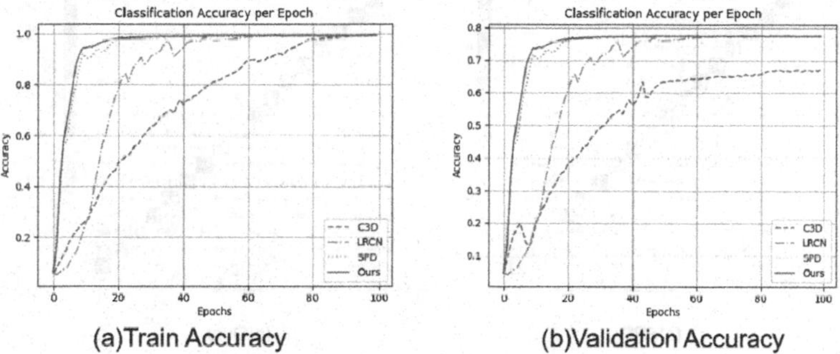

Fig. 5. Accuracy of train and validation

The accuracy of the validation process is shown in Fig. 5. The C3D algorithm starts to approach the best result after 40 rounds of training, the LRCN algorithm starts to approach the best result after 40 rounds of training, and the SPD-Net algorithm starts to converge after 20 rounds of training. In terms of convergence speed, the algorithm proposed in this chapter also achieves the best results faster than SPD.

The confusion matrix of the proposed method and LRCN, C3D and SPDNet on the test set is shown in Fig. 6, where 1–17 denote clothing categories. Among them, 1–8 are men's clothing, in order of coats and outerwear, jeans, pants, shirts and tops, sleepwear, sweaters, swimwear and underwear, while 9–17 are women's clothing, in order of coats and outerwear, dress, jeans, pants, shirts and top, sleepwear, sweaters, swimwear and underwear. Overall, women's clothing is more difficult to classify than men's clothing, probably because women's clothing has more similar geometric features, such as underwear and swimwear. Among the 17 clothing categories, sweaters, coats and outerwear and swimwear are the most difficult to classify. Among these methods, the C3D has the worst classification performance for men's swimwear, the LRCN has the worst classification performance for women's jackets, and the SPD can significantly improve the classification performance for these clothing categories. Compared to the SPD algorithm, our method further improves the classification accuracy for men's coats and outerwear, men's pants and women's sweaters, and our method achieves the highest stability.

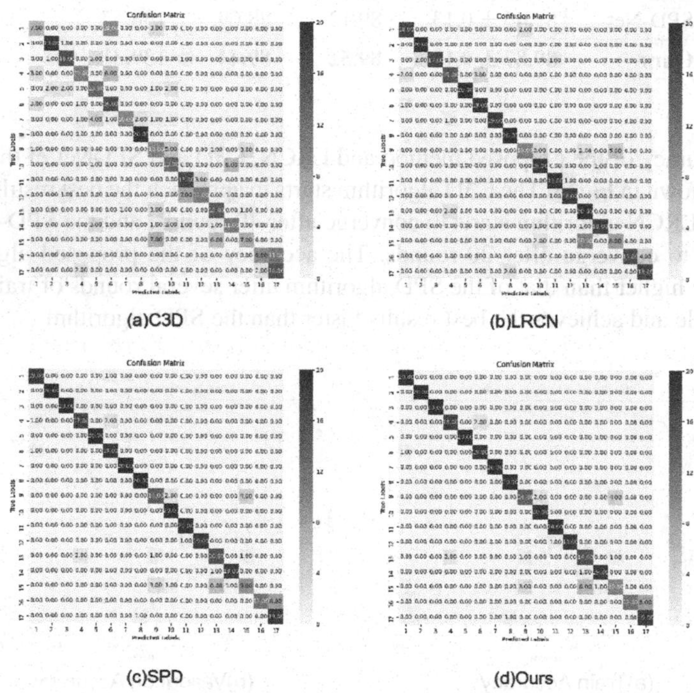

Fig. 6. Confusion matrix of test set

5 Conclusion

In this paper, a clothing classification method based on second-order convolution and covariance pooling is proposed. The covariance pooling method is used to fuse first-order and second-order features of clothing images, transform clothing image features into manifold data and introduce Riemann batch normalization, while a second-order feature vectoring method is used to transform manifold value features into vectors in Euclidean space, the manifold structured neural network based on second-order convolution learns geometric structure information of the images. In general, the method proposed in this paper accelerates model convergence, avoids parameter explosion in the feature vectoring process, effectively improves the accuracy and robustness of clothing classification, and achieves better classification results with an accuracy of 89.87% compared to state-of-the-art work. At present, the acquisition of manifold value clothing data relies on covariance pooling methods. In the future, it is necessary to further obtain clothing datasets that are directly manifold value data. Furthermore, the structure of this network can be further optimized, and more neural network structures can be attempted to be integrated.

References

1. Liu, Z., Luo, P., Qiu, S., Wang, X., Tang, X.: DeepFashion: powering robust clothes recognition and retrieval with rich annotations. In: Computer Vision & Pattern Recognition. IEEE (2016). https://doi.org/10.1109/CVPR.2016.124
2. Xiao, L., Yichao, X.: Exact clothing retrieval approach based on deep neural network. In: 2016 IEEE Information Technology, Networking, Electronic and Automation Control Conference (ITNEC 2016) (2016)
3. Liu, K.H., Chen, T.Y., Chen, C.S.: MVC: a dataset for view-invariant clothing retrieval and attribute prediction. In: Proceedings of the 2016 ACM on International Conference on Multimedia Retrieval, pp. 313–316 (2016)
4. Ge, Y., Zhang, R., Wang, X., Tang, X., Luo, P.: Deepfashion2: a versatile benchmark for detection, pose estimation, segmentation and re-identification of clothing images. In: Proceedings of the IEEE/CVF Conference on Computer Vision and Pattern Recognition, pp. 5337–5345 (2019)
5. Sidnev, A., Krapivin, A., Trushkov, A., Krasikova, E., Kazakov, M.: DeepMark++: CenterNet-based clothing detection (2020). 10.48550
6. He, K., Gkioxari, G., Dollár, P., Girshick, R.: Mask R-CNN. IEEE Trans. Pattern Anal. Mach. Intell. (2017). https://doi.org/10.1109/TPAMI.2018.2844175
7. Duan, K., Bai, S., Xie, L., Qi, H., Huang, Q., Tian, Q.: CenterNet: keypoint triplets for object detection. In: Proceedings of the IEEE/CVF International Conference on Computer Vision, pp. 6569–6578 (2019)
8. Lin, M., Ji, R., Chen, B., Chao, F., Ji, R.: Training compact CNNs for image classification using dynamic-coded filter fusion. IEEE Trans. Pattern Anal. Mach. Intell. (2023)
9. Liu, S., Zhang, L., Yang, X., Su, H., Zhu, J.: Query2label: a simple transformer way to multi-label classification. arXiv preprint arXiv:2107.10834 (2021)
10. Mallavarapu, T., Cranfill, L., Kim, E.H., Parizi, R.M., Morris, J., Son, J.: A federated approach for fine-grained classification of fashion apparel. Mach. Learn. Appl. **6**, 100118 (2021)
11. Eshwar, S.G., Rishikesh, A.V., Charan, N.A., Umadevi, V.: Apparel classification using convolutional neural networks. In: 2016 International Conference on ICT in Business Industry & Government (ICTBIG), pp. 1–5. IEEE (2016)
12. Iliukovich-Strakovskaia, A., Dral, A., Dral, E.: Using pre-trained models for fine-grained image classification in fashion field. In: Proceedings of the First International Workshop on Fashion and KDD, KDD, pp. 31–40 (2016)
13. Seo, Y., Shin, K.S.: Hierarchical convolutional neural networks for fashion image classification. Expert Syst. Appl. **116**, 328–339 (2019)
14. Qu, X., Che, H., Huang, J., Xu, L., Zheng, X.: Multi-layered semantic representation network for multi-label image classification. Int. J. Mach. Learn. Cybern., 1–9 (2023)
15. Cheng, X., et al.: MLTR: multi-label classification with transformer. In: 2022 IEEE International Conference on Multimedia and Expo (ICME), pp. 1–6. IEEE (2022)

16. Lin T H.: Aggregation and finetuning for clothes landmark detection. arXiv preprint arXiv: 2005.00419 (2020)
17. Zhang, Z., Song, C., Zou, Q.: Fusing hierarchical convolutional features for human body segmentation and clothing fashion classification. arXiv preprint arXiv:1803.03415 (2018)
18. Masci, J., Boscaini, D., Bronstein, M., Vandergheynst, P.: Geodesic convolutional neural networks on Riemannian manifolds. In: Proceedings of the IEEE International Conference on Computer Vision Workshops, pp. 37–45 (2015)
19. Poulenard, A., Ovsjanikov, M.: Multi-directional geodesic neural networks via equivariant convolution. ACM Trans. Graph. (TOG) **37**(6), 1–14 (2018)
20. Ionescu, C., Vantzos, O., Sminchisescu, C.: Matrix backpropagation for deep networks with structured layers. In: Proceedings of the IEEE International Conference on Computer Vision, pp. 2965–2973 (2015)
21. Huang, Z., Van Gool, L.: A riemannian network for SPD matrix learning. In: Proceedings of the AAAI Conference on Artificial Intelligence, vol. 31, no. 1 (2017)
22. Huang, Z., Wu, J., Van Gool, L.: Building deep networks on Grassmann manifolds. In: Proceedings of the AAAI Conference on Artificial Intelligence, vol. 32, no. 1 (2018)
23. Huang, Z., Wan, C., Probst, T., Van Gool, L.: Deep learning on lie groups for skeleton-based action recognition. In: Proceedings of the IEEE Conference on Computer Vision and Pattern Recognition, pp. 6099–6108 (2017)
24. Qiao, S., Wang, R., Shan, S., Chen, X.: Deep heterogeneous hashing for face video retrieval. IEEE Trans. Image Process. **29**, 1299–1312 (2019)
25. Li, C., et al.: Deep manifold structure transfer for action recognition. IEEE Trans. Image Process. **28**(9), 4646–4658 (2019)
26. Chakraborty, R., Bouza, J., Manton, J., Vemuri, B.C.: ManifoldNet: a deep network framework for manifold-valued data. arXiv preprint arXiv:1809.06211 (2018)
27. Brooks, D., Schwander, O., Barbaresco, F., Schneider, J.Y., Cord, M.: Riemannian batch normalization for SPD neural networks. In: Advances in Neural Information Processing Systems, 32 (2019)
28. Yu, K., Salzmann, M.: Second-order convolutional neural networks. Clin. Immunol. Immunopathol. (2017). https://doi.org/10.1006/clin.1993.1030
29. Sra, S.: Positive definite matrices and the S-divergence. Proc. Am. Math. Soc. **144**(7), 2787–2797 (2016)
30. Tran, D., Bourdev, L., Fergus, R., Torresani, L., Paluri, M.: Learning spatiotemporal features with 3D convolutional networks. In: Proceedings of the IEEE International Conference on Computer Vision, pp. 4489–4497 (2015)
31. Donahue, J., et al.: Long-term recurrent convolutional networks for visual recognition and description. In: Proceedings of the IEEE Conference on computer Vision and Pattern Recognition, pp. 2625–2634 (2015)

Multi-size Scaled CAM for More Accurate Visual Interpretation of CNNs

Fuyuan Zhang(✉), Xiaohong Xiang(✉), Xin Deng, and Xiaoyu Ding

Chongqing University of Posts and Telecommunications, Nan'an District, Chongqing 400065, China
s210231249@stu.cqupt.edu.cn, {xiangxh,dengxin,dingxy}@cqupt.edu.cn

Abstract. The search for decision bases for image classification neural networks is one of the popular research directions in deep neural network interpretability. These studies highlight regions of interest to image classification models by generating a saliency map that assign contribution values to each pixel in the input image. However, these current methods cannot accurately locate the key features of the target object and tend to include other irrelevant objects in the salient region, resulting in unreliable saliency maps. In addition, the problems of noise and low resolution have also plagued the researchers. To address the above issues, we propose a saliency map generation method based on multi-size scaling of the input image for the CNN-based model. The method scales the input images in multiple sizes and extracts different resolution feature maps and their corresponding gradients from specific convolutional layers, and fuses them as masks of the input images. Then the masked input images are input to the model separately to obtain the weights of each mask, and finally the masks are combined with linear weighting to obtain the saliency map. Experiments show that our method can produce more detailed saliency maps and accurately target regions of interest to the model. It has significant advantages and higher application value than the current CAM-based (class activation mapping) methods.

Keywords: Interpretability · CAM · CNN

1 Introduction

At present, we can see the application of deep neural networks in numerous fields. In the field of vision, deep neural networks have shown superior performance in tasks such as object detection, semantic segmentation, and image classification. However, the fact that deep neural networks contain many hidden layers makes it difficult to explain their decision mechanisms, which limits their application in many critical areas, such as medical, aviation, and autonomous driving. Therefore, many current researches are devoted to finding a way that can explain the decision of deep neural network and intuitively show the decision reasons of deep neural network to users.

H. Zhang et al. (Eds.): NCAA 2023, CCIS 1870, pp. 147–161, 2023.
https://doi.org/10.1007/978-981-99-5847-4_11

The saliency map is an interpretative tool applied to image classification neural networks, which assigns a contribution value to each pixel in the image to indicate the importance of these pixels to the model decision. A popular class of saliency methods are CAM-based methods [4,8,9,15,20,23,24], which rely on extracting activation maps with rich semantic information from convolutional layers and using multiple weighting algorithms to combine these feature maps in a linearly weighted manner to obtain saliency maps. This class of methods has good interpretability, but they are limited by the resolution of the activation maps, which give more noisy and less focused saliency maps. There is another class of methods, which are based on perturbation, they treat the model as a black box and find the saliency regions in the image by perturbing the input image and observing the probability change of the output, RISE [12] is a typical one, Score-CAM [20] and Group-CAM [23] also reflect this idea. These methods are intuitive, but they are computationally expensive.

We propose our method to solve the problem that the saliency regions of existing methods are not focused and cannot accurately lock important features of the object. Our method enlarges the input image in multiple sizes and uses the idea of Grad-CAM to fuse the masks with higher resolution to perturb the input image to get the weights and combine these masks to get the saliency map. Figure 1 illustrates the pipeline of our method. Our contribution is twofold:

- A new visual interpretation method for CNNs is proposed, which gives finer, more focused saliency maps and more accurate localization of object features. It finds salient regions that are more consistent with the decision basis of the model.
- We improve the evaluation metrics and through a more comprehensive assessment, our method demonstrates excellent performance.

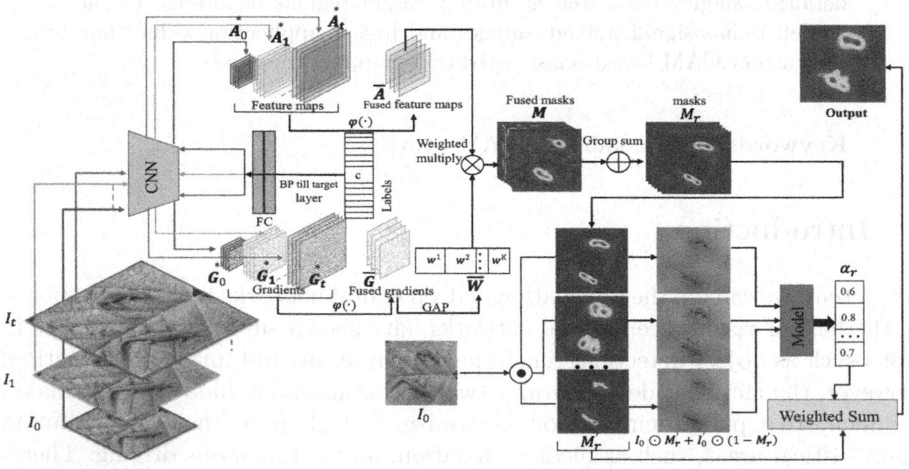

Fig. 1. Pipeline of our method

2 Related Work

In this section, we introduce the current saliency map methods, which can be broadly classified into the following two categories.

2.1 Backpropagation Based Saliency Methods

Zeiler et al. [21] proposed a visualization method based on deconvolution, which inverse maps the values in the feature map back to the pixel space of the input image, thereby indicating which pixels in the image are involved in the decision. Building on this work, the Guided BP [19] proposes to highlight important features of the visualized target by suppressing inputs and values with gradients less than 0 during backpropagation. DeepLIFT [16] and LRP [3] distribute the contribution of the output down to the input by modifying the rules of backpropagation to obtain the correlation score of each pixel in the input image to the output. Simonyan et al. [17] propose using the input image's gradients as a visual interpretation. This method believes that some input pixels play a major role in the prediction results of the network. It directly computes gradients of specified probability score to the input, but the input gradients contain obvious noise and the visualization is fuzzy. SmoothGrad [18] and VarGrad [1] propose an effective method, which adds noise to the input image several times to generate a group of images containing noise. By averaging the results, the final saliency map is smoother. Although these studies have a solid theoretical basis, their visualization results are not easy to understand and noisy for humans. In addition, many of these methods are class-agnostic, i.e., they do not visually interpret the results for a given class. Some studies [2] point out that the reliability of these methods is questionable, they are not sensitive to network parameters, even without a network trained to get similar results.

There exist several methods that exploit the activation information within a model to produce salient maps. For instance, class activation mapping based approaches, including the first proposed Class Activation Mapping (CAM) [24] and its variants such as Grad-CAM [15], Grad-CAM++ [4], XGrad-CAM [8], etc., aim to generate category-distinct saliency maps by weighting different channels of the feature maps with category-specific gradient information obtained through back-propagation. On the other hand, the Relevance-CAM [10] utilizes the relevance scores obtained from Layer-wise Relevance Propagation (LRP) to weight the feature maps, demonstrating that shallow convolutional layers still retain class-relevant information. The CAMERAS [9] framework, which fuses the feature maps and gradients through multi-stage scaling of the input image, generates a higher resolution saliency map. However, this method is only effective in Convolutional Neural Network (CNN) models with residual structures. In general, the CAM-based methods leverage the feature maps, a crucial data component within a model, resulting in good performance and widespread research attention.

2.2 Perturbation-Based Saliency Methods

Perturbation-based saliency methods generate saliency maps without relying on internal model data and instead focus solely on the inputs and outputs of the model. Early approaches, such as the use of a black square [21] to scramble the input image to identify the regions of interest for the model, have been replaced by more sophisticated techniques. For instance, Local Interpretable Model-agnostic Explanations (LIME) [13] employs a proxy model to fit the local decision-making behavior of the target model and thereby assess the sensitivity of various features. RISE [12] utilizes a large number of masks to mask the input image and calculates the probability of the target category as the weight for each mask, which is then combined linearly to produce the saliency map. Another approach, Mask [7] uses gradient descent to identify the mask with the lowest confidence in the target category. Score-CAM [20] and Group-CAM [23] perturb the input image using the feature map as the mask to compute the weight of the feature map and combine it linearly to generate the saliency map. While these methods can provide a direct representation of significant regions in the input image, they often have high computational costs and require optimization.

3 Proposed Approach

Our method address the problems that current saliency methods are ambiguous in locating target objects, weak in highlighting important features, ineffective in recognizing the same class of objects when they appear multiple times in the input image. Our method can be roughly divided into two stages: in the first stage, we extract and fuse feature maps and gradients as masks; In the second stage, we optimize the masks and use them to perturb the input image to obtain the weight of each mask, and combine all the masks and their corresponding weights linearly to obtain the saliency map. In the first stage, the backpropagation method of Grad-CAM [15] is applied our method, and we will give the details of its procedure.

3.1 Grad-CAM

Let $I \in \mathbb{R}^{3\times H\times W}$ be an input image, a CNN based deep neural model $\mathcal{F}$ which has been pre-trained. $\mathcal{F}_c(I)$ denotes a score (before the softmax) on class c with input image I. Given a layer l of model $\mathcal{F}$, feature maps $\overset{*}{\boldsymbol{A}}$ can be extracted from the layer l, when I as an input image enter into model $\mathcal{F}$ and pass a full forward propagation. To get the gradients of $\mathcal{F}_c(I)$ with respect to the kth feature map $A^k \in \overset{*}{\boldsymbol{A}}$, we can pass a backward propagation with respect to $\mathcal{F}_c(I)$. To get the each feature map's importance weight, gradients are global average pooled across dimensions of height and width.

$$w^k = \frac{1}{m\times n}\sum_{i=i}^{m}\sum_{j=1}^{n}\frac{\partial \mathcal{F}_c(I)}{\partial A_{ij}^k(I)} \tag{1}$$

where A_{ij}^k denotes the value of the ith row and j^{th} column in the kth feature map, m and n are respectively the width and height of the gradients matrix. $w^k \in \overset{*}{\boldsymbol{W}}$ is the kth feature map's weight. $\overset{*}{\boldsymbol{W}}$ is a one-dimensional vector of size K, K is the number of channels of $\overset{*}{\boldsymbol{A}}$.

Formally, the Grad-CAM saliency map with respect to l^{th} layer can be calculated by

$$L_{Grad-CAM} = ReLU(\sum_K (w^k A^k)) \tag{2}$$

ReLU means Grad-CAM is only interested in the positive influence on features of class c.

3.2 Masks Generation

Let $I_0 \in \mathbb{R}^{3\times H_0 \times W_0}$ be an original input image. We denote the original input resolution by $\zeta_0 = (H_0, W_0)$. Like Grad-CAM, given an original input image I_0 and an interested class c, $\overset{*}{\boldsymbol{A}}_{\boldsymbol{0}}$ are feature maps extracted from the layer l, $\overset{*}{\boldsymbol{G}}_{\boldsymbol{0}}$ means the gradients of $\mathcal{F}_c(I_0)$ with respect to the feature maps $\overset{*}{\boldsymbol{A}}_{\boldsymbol{0}}$,

$$\overset{*}{\boldsymbol{G}}_{\boldsymbol{0}} = \frac{\partial \mathcal{F}_c(I_0)}{\partial \overset{*}{\boldsymbol{A}}_{\boldsymbol{0}}} \tag{3}$$

where $\overset{*}{\boldsymbol{G}}_{\boldsymbol{0}}$ has K channels, which is the same as the number of channels in $\overset{*}{\boldsymbol{A}}_{\boldsymbol{0}}$.

Then we interpolate the original image I_0 to I_t, t is the tth iteration. The resolution of I_t is $\zeta_t = (H_t, W_t)$. The interpolation function is $\varphi(I_0, \zeta_t)$ which use bilinear interpolation to resize the resolution of I_0 from ζ_0 to ζ_t. Consider $\zeta_{max} = (H_{max}, W_{max})$ as the specified maximum resolution threshold, N is the specified maximum number of iterative steps, ζ_t can be calculated by

$$\zeta_t = \zeta_0 + \lfloor \frac{\zeta_{max}}{N} \rfloor (t-1) \tag{4}$$

The input image resolution is different in each iteration, the resolution of feature maps extracted from the same layer is also different, and the class information contained in feature maps of different resolutions in each iteration is also different, the fusion of activation maps in different resolutions can be included more class-sensitivity information. The fusion can be computed as

$$\overline{\boldsymbol{A}} = \frac{1}{t_{max}} \sum_{t=0}^{t_{max}} \varphi(\overset{*}{\boldsymbol{A}}_t, \zeta_0) \tag{5}$$

where t_{max} is the maximum number of valid iterations. "Valid" means if the most interested class the model outputs in an iteration is c, this iteration is valid, and feature maps in this iteration will be fused, and not vice versa, so $t_{max} \leq N$.

Every iteration, we compute the gradients of $\mathcal{F}_c(I_t)$ with respect to the l layer's feature maps $\overset{*}{A}_t$. The fusion of gradients can be computed as

$$\overline{G} = \frac{1}{t_{max}} \sum_{t=0}^{t_{max}} \varphi(\overset{*}{G}_t, \zeta_0) \tag{6}$$

$\overline{G}$ are global average pooled to get the weights of $\overline{A}$

$$\overline{W} = \frac{1}{m \times n} \sum_{i=i}^{m} \sum_{j=1}^{n} \overline{G} \tag{7}$$

Then we have $\overline{A}$ of shape $[1 \times K \times H_0 \times W_0]$ and its weights $\overline{W}$ of shape $[1 \times K \times 1 \times 1]$, K is the number of channels of lth layer.

Formally, preliminary masks are calculated by

$$M = \overline{A} \cdot \overline{W} \tag{8}$$

3.3 Masks Optimization

Compared with the masks obtained from a single original input image, M obtained after multi-scale amplification of the original image contains richer class information. However, M is preliminary. The shape of M is $[1 \times K \times H_0 \times W_0]$, where K is the number of channels on the last layer of CNN and generally in the hundreds like Score-CAM. Thus, it will significantly increase the calculation cost if directly using M for masks.

We divide M which has K masks into B groups according to their adjacency. Then we sum up all the masks in each group into one mask. The "group sum" can be calculated by

$$M_r = ReLU\left(\sum_{k=r\times g}^{(r+1)\times g-1} M^k\right) \tag{9}$$

where $r \in \{0, 1, 2, \ldots, B-1\}$, $g = K/B$ is the number of feature maps in each group, M^k is the kth feature map in M.

So far, we still have the last step which is scaling each pixel value m_{ij} of M_r into $[0, 1]$ by utilizing Min-Max normalization function,

$$M_r' = \frac{M_r - min(M_r)}{max(M_r) - min(M_r)} \tag{10}$$

Then M_r' is a qualified mask with the same resolution of I_0 to mask the input.

3.4 Saliency Map Generation

Saliency map is a heat map with the same resolution as the original input image I_0. For a given class c of interest, it can show the contribution degree of each pixel in the original image to $\mathcal{F}_c(I_0)$. We use M'_r to mask the input image I_0, the intuition behind it is to preserve the information including the important features of c in the original picture. However, if the mask is directly element-wise multiply with the input image, the resulting image will have too sharp boundaries between the masked regions and the salient regions, which will cause adversarial effects [5].

To avoid the above situation, gaussian blur function is used to blur the input image and replace the masked region to make the boundary smoother. For a single mask M'_r, the masking mode is calculated as follows:

$$I_r = I_0 \odot M'_r + I'_0 \odot (1 - M'_r) \tag{11}$$

where I'_0 is a gaussian blur of I_0, it serves as a baseline image, and is feed into the model, we will get a very low $\mathcal{F}_c(I'_0)$ approaching zero. $\odot$ means hadamard product. Specific gaussian blur function is $guassian\_blur2d(input, kernel\_size, sigma)$, in this paper, following [23], $kernel\_size = 51$ and $sigma = 50$.

The contribution α_r of each M_r can be quantified by feeding I_r into the model to obtain $\mathcal{F}_c(I_r)$ for salient regions,

$$\alpha_r = \mathcal{F}_c(I_r) - \mathcal{F}_c(I'_0) \tag{12}$$

Formally, the final saliency map can be compute as

$$L = ReLU(\sum_r \alpha_r M_r) \tag{13}$$

4 Experiments

In this section, we evaluate our method qualitatively and quantitatively and compare its performance with the current popular CAM-based methods. We first compare the masks generated by our method and Group-CAM [23] to explore the advantages of using the fused masks by our method. Moreover, we subjectively compare and analyze the saliency maps generated by our method with that generated by other methods. Then, we quantitatively evaluate our method and other methods. Finally, we use a sanity check to show that our method is sensitive to model parameters.

4.1 Experiments Setup

In the insertion and deletion experiments, we choose dataset ILSVRC2012 val [14], which contains 50k images, each of which we resize to 224 pixels in length and width and normalize using STD vector [0.229, 0.224, 0.225] and mean vector [0.485, 0.456, 0.406]. All the subjective comparison images in this paper are also from ILSVRC2012 validation dataset. In the localization experiment, we use the PASCAL VOC 2007 test set [6] containing 4952 images from 20

Table 1. Experimental results of insertion (**higher AUC is better**), deletion (**lower AUC is better**) on ILSVRC2012 validation split. The best result in each row is shown as **bold**.

Method	Insertion(AUC)	Deletion(AUC)	Over-all(AUC)
Grad-CAM	53.19	11.52	41.67
Grad-CAM++	51.57	12.16	39.41
XGrad-CAM	52.57	11.53	41.04
Score-CAM	**55.10**	11.43	43.67
Group-CAM	54.61	11.21	43.40
CAMERAS	44.10	**8.01**	36.09
Ours	54.52	9.78	**44.74**

categories and the COCO 2014 validation set containing about 50k images. The pre-trained model used for the insertion and deletion experiments is VGG19 from torchvision. For the localization experiments, we use VGG16 and fine-tune it on both datasets separately. For all the methods that need to extract feature maps, we choose the last convolutional layer of the model. The default parameters of Group-CAM and our method in the experiments are $B = 32$, $\theta = 70$.

4.2 Qualitative Evaluation

We randomly select some representative images for testing. Saliency maps generated by our method are compared with saliency maps generated by current popular CAM-based methods, including Grad-CAM [18], Grad-CAM++ [4], Score-CAM [20], XGrad-CAM [8], Group-CAM [23], CAMERAS [9]. The superiority of our method can be verified by direct comparison.

As shown in Fig. 2, our method produces cleaner visualizations with less highlighting in non-target regions. In the case of multiple occurrences of the same category, saliency maps generated by our method well present the important features of each object, while other methods make different objects lose their independence. In addition, compared with other methods, our method can lock in important features with the least area of highlighting and is more sensitive to the shape of features.

4.3 Insertion and Deletion

Motivation. Insertion and deletion experiments use the pixel ranking information given by the saliency map to determine the ranking of the contribution of each pixel in the input image to a given category. Insertion and deletion experiments are proposed by RISE [12] and can accurately reflect the sensitivity of the saliency map to key features. The principle of Insertion is that when we start to insert the important pixels at the corresponding position of the original image according to the priority of pixel importance given by the saliency map until all

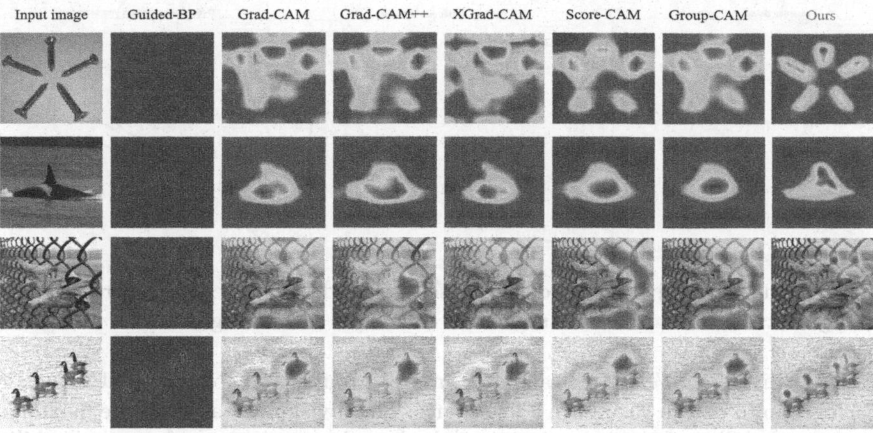

Fig. 2. Saliency maps of CAM-based methods and backpropagation-based method Guided-BP. It is obvious that our method can accurately describe the contour of the target object and highlight the important features of the target.

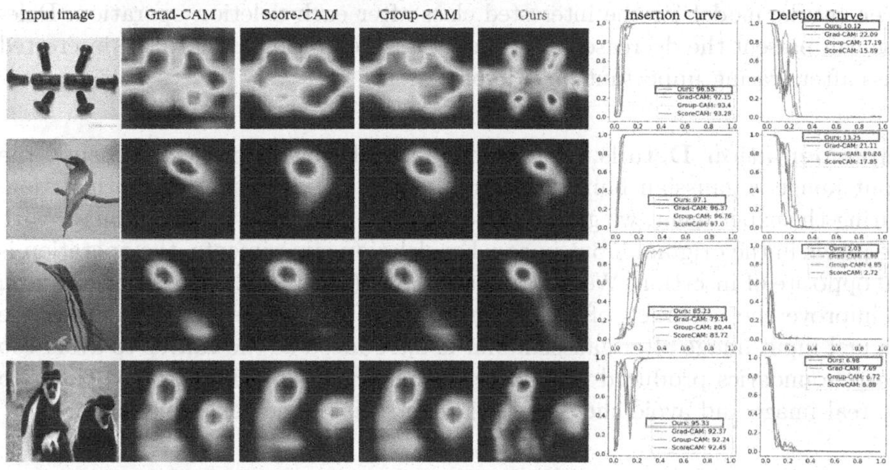

Fig. 3. Grad-CAM, Score-CAM, Group-CAM and MSG-CAM generated saliency maps and corresponding insertion and deletion curves for the 4 representative images selected. For the deletion curve, the saliency map given by a better visualization method should drop as fast as possible, so as to lower the AUC. And the Insertion curve is the exact opposite of the deletion curve.

pixels are inserted completely, the probability score given by the model for the class of interest is recorded in the process of each insertion operation. On the contrary, deletion is to gradually erase the corresponding pixel information from the original input image according to the priority of pixel importance given by the saliency map. Like the insertion, deletion also records the probability score

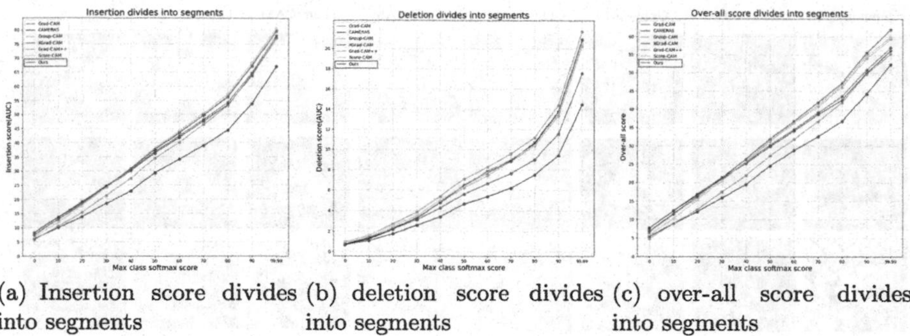

(a) Insertion score divides into segments (b) deletion score divides into segments (c) over-all score divides into segments

Fig. 4. We divide ILSVRC2012 validation into 11 groups, and count the AUC metrics in each group to obtain the above 3 line charts. The division is based on the maximum softmax probability of the input image, and when the maximum softmax probability of the image is at $[0, 10\%)$, the image is classified in the group corresponding to '0' in the horizontal coordinate, and so on for the other groups.

given by the model for the interested class after each deletion operation. Deletion can present the decrease of confidence degree of the model to the interested class after erasing important pixel information.

Implementation Details. It should be noted that in our experiments, the input image is gaussian blurred to serve as the initial image for the insertion. During the experiment, we gradually replace the pixels in the initial image with the pixels in the original input image. For deletion experiments, the operation is the opposite of insertion. We only replace 0.89%(i.e.,224×2) pixels per iteration to improve the accuracy of the experiment. It should be explained that the original input image after gaussian blur is introduced as the canvas to avoid too sharp boundaries produced when pixels are inserted or deleted, to be closer to the real image and avoid the impact of adversarial attack examples [12].

AUC Metrics. We use AUC as a quantitative metric, which calculates the area enclosed by the fold and horizontal coordinates of the change in probability scores for the specified category over the experiment. The higher the AUC(insertion), the better, and the opposite for AUC(deletion). We use AUC(over-all) as an overall indicator, which is obtained by subtracting AUC(insertion) and AUC(deletion), also the higher the better. Figure 3 shows the saliency maps generated by several pictures according to different algorithms, the insertion and deletion curves are drawn according to saliency maps.

Experimental Evaluation. The experimental results of insertion and deletion on 50K images are listed in Table 1. In the insertion experiment, our method only less than 0.6 different from the first place (ScoreCAM). However, our method outperforms ScoreCAM by 1.65 in the deletion experiment. It indicates that prior

Table 2. Pointing game on PASCAL VOC 2007 test split and COCO 2014 validation split. The mean accuracy of our method is higher than other CAM based methods.

Method	PASCAL VOC test Mean Accuracy(%)	COCO Validation Mean Accuracy(%)
Grad-CAM	83.04	55.50
Grad-CAM++	83.21	52.91
XGrad-CAM	86.70	55.93
Score-CAM	73.92	51.20
Group-CAM	82.41	54.14
CAMERAS	87.16	55.40
Ours	**87.20**	**56.32**

methods like ScoreCAM overemphasize unimportant regions while ours can yield more accurate explanations. Overall, our method is the optimum, with its 1 point higher in AUC(over-all) than the second place (ScoreCAM), as it maintains a higher AUC(insertion) while achieving a significantly lower AUC(deletion). In general, our method achieves good results in sparsity, insertion, and deletion.

Split Experiments. To investigate the relationship between the confidence of the model on the target class of the input images and the AUC metric, we split the dataset into 11 groups according to the maximum probability scores and counted the experimental performance of each method in each group. The data results are presented in Fig. 4. It is clear that as the confidence of the model for the target objects increases, the performance of each method begins to show a significant gap, with our method achieving the best performance in the group of images with probability scores greater than 99.99%, and the AUC (over-all) leading the second place by more than 2 points. This means that our method can more accurately find out the basis of the model's decision when the model is very confident about the target object in the input image.

4.4 Localization Evaluation

A saliency map can be used as an important basis for the evaluation of weakly supervised location performance. Pointing game [22] is an experiment to evaluate saliency map positioning ability. In detail, The calculation for the pointing game is $Acc = \frac{hit}{hit+miss}$. We find the point with the highest value in the saliency map generated for a given class and record its coordinates. If the coordinates are inside the annotated bounding box of the class object, a hit is recorded, otherwise, a miss is recorded. To more accurately reflect the localization of the saliency map, we abandon the method of only calculating the coordinates of the maximum value point, and instead calculate the coordinates of the top 100 points in the saliency map, which can eliminate the influence of accidental noise.

The metrics we compare are averaged over *Acc* for each class in the dataset. The saliency methods are tested on in PASCAL VOC test set [6] and COCO 2014 Validation set [11], respectively, and the model used, VGG19, is also fine-tuned on each of these two datasets.

The test results are reported in Table 2. Our method achieves optimality on both datasets, which indicates that our method is more accurate in target localization.

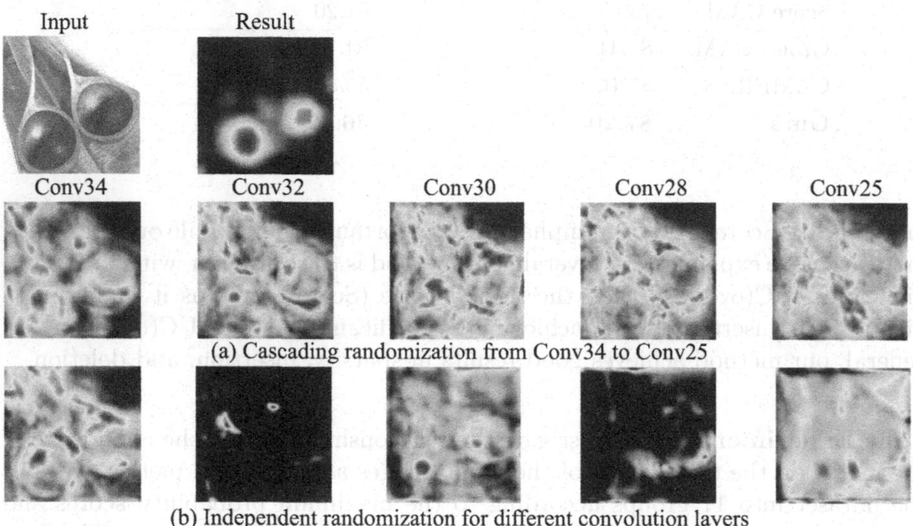

(a) Cascading randomization from Conv34 to Conv25

(b) Independent randomization for different convolution layers

Fig. 5. Sanity check results of our method

4.5 Sanity Check

Finally, we test our method with a sanity check experiment. Some backpropagation-based visualization methods are not sensitive to model parameters [2] which means that models can give similar results regardless of whether they have been trained or not, which defeats the purpose of visualization. Therefore, the ability to pass the sanity check is the criterion to measure the interpretability of the visualization method. Specifically, we apply cascade randomization and independent randomization to the pre-trained VGG19 according to the experimental method in [2]. The cascade randomization starts from the output end of the model and gradually randomizes the parameters of the convolutional layer until the parameters of all the convolutional layers are randomized. The independent randomization also starts near the output of the model and randomizes the parameters of only one individual convolutional layer at a time, leaving the other convolutional layers unchanged. As shown in Fig. 5, The results show that our method is very sensitive to model parameters. When the parameters are randomized by cascade randomization or independent randomization, it will have a significant impact on the saliency map.

5 Conclusion

In this work, we introduce our method to generate a saliency map that is more consistent with the decision-making process of the CNN neural network. It can accurately capture important features of interest to neural networks and give pure and reliable visualization results. Through comprehensive experiments, it is proved that our method has obvious advantages over the existing CAM-based methods. We also find that when the neural network is very confident about the class in a picture, our method can accurately reflect this situation. Finally, our method passes a sanity check, which shows that it is reliable and can reflect the parameters learned by the neural network.

Acknowledgement. This work was supported in part by the Natural Science Foundation of Chongqing under Grant cstc2020jcyj-msxmX0284; in part by the Scientific and Technological Research Program of Chongqing Municipal Education Commission under Grant KJQN202000625.

References

1. Adebayo, J., Gilmer, J., Goodfellow, I., Kim, B.: Local explanation methods for deep neural networks lack sensitivity to parameter values. ArXiv preprint abs/1810.03307 (2018)
2. Adebayo, J., Gilmer, J., Muelly, M., Goodfellow, I.J., Hardt, M., Kim, B.: Sanity checks for saliency maps. In: Bengio, S., Wallach, H.M., Larochelle, H., Grauman, K., Cesa-Bianchi, N., Garnett, R. (eds.) Advances in Neural Information Processing Systems 31: Annual Conference on Neural Information Processing Systems 2018, NeurIPS 2018, 3–8 December 2018, Montréal, Canada, pp. 9525–9536 (2018)
3. Binder, A., Montavon, G., Lapuschkin, S., Müller, K.-R., Samek, W.: Layer-wise relevance propagation for neural networks with local renormalization layers. In: Villa, A.E.P., Masulli, P., Pons Rivero, A.J. (eds.) ICANN 2016. LNCS, vol. 9887, pp. 63–71. Springer, Cham (2016). https://doi.org/10.1007/978-3-319-44781-0_8
4. Chattopadhay, A., Sarkar, A., Howlader, P., Balasubramanian, V.N.: Grad-CAM++: generalized gradient-based visual explanations for deep convolutional networks. In: 2018 IEEE Winter Conference on Applications of Computer Vision (WACV), pp. 839–847. IEEE (2018)
5. Dabkowski, P., Gal, Y.: Real time image saliency for black box classifiers. In: Guyon, I., et al. (eds.) Advances in Neural Information Processing Systems 30: Annual Conference on Neural Information Processing Systems 2017, 4–9 December 2017, Long Beach, CA, USA, pp. 6967–6976 (2017)
6. Everingham, M., Van Gool, L., Williams, C.K., Winn, J., Zisserman, A.: The pascal visual object classes (voc) challenge. Int. J. Comput. Vision **88**(2), 303–338 (2010)
7. Fong, R.C., Vedaldi, A.: Interpretable explanations of black boxes by meaningful perturbation. In: IEEE International Conference on Computer Vision, ICCV 2017, Venice, Italy, 22–29 October 2017, pp. 3449–3457. IEEE Computer Society (2017). https://doi.org/10.1109/ICCV.2017.371

8. Fu, R., Hu, Q., Dong, X., Guo, Y., Gao, Y., Li, B.: Axiom-based grad-CAM: towards accurate visualization and explanation of CNNs. In: 31st British Machine Vision Conference 2020, BMVC 2020, Virtual Event, UK, 7–10 September 2020. BMVA Press (2020)
9. Jalwana, M.A., Akhtar, N., Bennamoun, M., Mian, A.: CAMERAS: enhanced resolution and sanity preserving class activation mapping for image saliency. In: Proceedings of the IEEE/CVF Conference on Computer Vision and Pattern Recognition, pp. 16327–16336 (2021)
10. Lee, J.R., Kim, S., Park, I., Eo, T., Hwang, D.: Relevance-CAM: your model already knows where to look. In: Proceedings of the IEEE/CVF Conference on Computer Vision and Pattern Recognition, pp. 14944–14953 (2021)
11. Lin, T.-Y., et al.: Microsoft COCO: common objects in context. In: Fleet, D., Pajdla, T., Schiele, B., Tuytelaars, T. (eds.) ECCV 2014. LNCS, vol. 8693, pp. 740–755. Springer, Cham (2014). https://doi.org/10.1007/978-3-319-10602-1_48
12. Petsiuk, V., Das, A., Saenko, K.: RISE: randomized input sampling for explanation of black-box models. In: British Machine Vision Conference 2018, BMVC 2018, Newcastle, UK, 3–6 September 2018, pp. 151. BMVA Press (2018)
13. Ribeiro, M.T., Singh, S., Guestrin, C.: "why should I trust you?": explaining the predictions of any classifier. In: Krishnapuram, B., Shah, M., Smola, A.J., Aggarwal, C.C., Shen, D., Rastogi, R. (eds.) Proceedings of the 22nd ACM SIGKDD International Conference on Knowledge Discovery and Data Mining, San Francisco, CA, USA, 13–17 August 2016, pp. 1135–1144. ACM (2016). https://doi.org/10.1145/2939672.2939778
14. Russakovsky, O., et al.: ImageNet large scale visual recognition challenge. Int. J. Comput. Vision **115**(3), 211–252 (2015)
15. Selvaraju, R.R., Cogswell, M., Das, A., Vedantam, R., Parikh, D., Batra, D.: Grad-CAM: visual explanations from deep networks via gradient-based localization. In: IEEE International Conference on Computer Vision, ICCV 2017, Venice, Italy, 22–29 October 2017, pp. 618–626. IEEE Computer Society (2017). https://doi.org/10.1109/ICCV.2017.74
16. Shrikumar, A., Greenside, P., Kundaje, A.: Learning important features through propagating activation differences. In: International Conference on Machine Learning, pp. 3145–3153. PMLR (2017)
17. Simonyan, K., Vedaldi, A., Zisserman, A.: Deep inside convolutional networks: visualising image classification models and saliency maps. In: In Workshop at International Conference on Learning Representations. CiteSeer (2014)
18. Smilkov, D., Thorat, N., Kim, B., Viégas, F., Wattenberg, M.: SmoothGrad: removing noise by adding noise. ArXiv preprint abs/1706.03825 (2017)
19. Springenberg, J.T., Dosovitskiy, A., Brox, T., Riedmiller, M.: Striving for simplicity: the all convolutional net. arXiv preprint arXiv:1412.6806 (2014)
20. Wang, H., et al.: Score-CAM: score-weighted visual explanations for convolutional neural networks. In: Proceedings of the IEEE/CVF Conference on Computer Vision and Pattern Recognition Workshops, pp. 24–25 (2020)
21. Zeiler, M.D., Fergus, R.: Visualizing and understanding convolutional networks. In: Fleet, D., Pajdla, T., Schiele, B., Tuytelaars, T. (eds.) ECCV 2014. LNCS, vol. 8689, pp. 818–833. Springer, Cham (2014). https://doi.org/10.1007/978-3-319-10590-1_53
22. Zhang, J., Bargal, S.A., Lin, Z., Brandt, J., Shen, X., Sclaroff, S.: Top-down neural attention by excitation backprop. Int. J. Comput. Vision **126**(10), 1084–1102 (2018)

23. Zhang, Q., Rao, L., Yang, Y.: Group-CAM: group score-weighted visual explanations for deep convolutional networks. arXiv preprint arXiv:2103.13859 (2021)
24. Zhou, B., Khosla, A., Lapedriza, À., Oliva, A., Torralba, A.: Learning deep features for discriminative localization. In: 2016 IEEE Conference on Computer Vision and Pattern Recognition, CVPR 2016, Las Vegas, NV, USA, 27–30 June 2016, pp. 2921–2929. IEEE Computer Society (2016). https://doi.org/10.1109/CVPR.2016.319

Joint Attention Mechanism of YOLOv5s for Coke Oven Smoke and Fire Recognition Algorithm

Yiming Liu[1], Yunchu Zhang[1,2(✉)], Yanfei Zhou[1], and Xinyi Zhang[1]

[1] School of Information and Electrical Engineering, Shandong Jianzhu University, Jinan 250101, China
yczhang@sdjzu.edu.cn

[2] Shandong Key Laboratory of Intelligent Buildings Technology, Jinan 250101, China

Abstract. Aiming at the requirement of all-weather smoke and fire emission environmental detection in coke plants, this paper proposes the YOLOv5s coke oven smoke and fire recognition algorithm with joint attention mechanism. The algorithm takes YOLOv5s as the base network and adds the attention mechanism module in BackBone to make the network pay more attention to important features and improve the accuracy of target detection; in addition, this paper adds light labels to solve the interference of strong lights on flame recognition based on smoke and fire labels, and solves the smoke and fire detection problem of day and night scenes by triage training and detection. Doing comparison experiments on the self-built dataset, the YOLOv5s model with the joint CBAM module works best. The experimental results show that compared with the original YOLOv5s model, the mAP value of smoke and fire recognition in daytime scenes is improved by 4.4%, and the mAP value of smoke and fire recognition in nighttime scenes is as high as 97.1%.

Keywords: Smoke and fire recognition · YOLOv5s · Attention mechanism · Target detection

1 Introduction

Smoke and fire detection is widely used in fire prevention, security monitoring and other fields. With the development of the economy, environmental pollution has become an urgent global problem. The burning of substances such as white waste, straw and industrial waste can cause serious environmental pollution and reduce the quality of the atmosphere. During the process of coal loading, coal levelling and coke pushing, coke ovens can suffer from smoke running and fires, which not only pollute the atmosphere, but also have a negative impact on the coke oven itself. Therefore, the automatic detection and statistics of smoke and fire during the production of coke ovens is conducive to promoting the safe production of coke ovens and the improvement of process equipment and reducing atmospheric pollution.

H. Zhang et al. (Eds.): NCAA 2023, CCIS 1870, pp. 162–176, 2023.
https://doi.org/10.1007/978-981-99-5847-4_12

Most traditional fire smoke detection methods achieve fire identification based on four characteristics of the fire (i.e. colour, disturbance, localised fire morphology, colour distribution). Shen Shilin [1] a correlation algorithm is proposed using the fire oscillation characteristics and the influence of three components, R, G and B, on the correlation is analysed to provide an effective criterion for visual fire detection systems. Wang Jin [2] a particle swarm optimization algorithm based on two-dimensional maximum full threshold selection method is proposed for fire recognition. The method combines the dynamic and static characteristics of fires with a robust algorithm and high recognition rate and sensitivity, which is applicable to a wide range of fire monitoring. Xiyin Wu [3] a fire detection algorithm that fuses circularity, rectangularity and centre of gravity height coefficients is proposed, and then the fused fire features are fed into a support vector machine for classification. This method greatly improves the fire detection efficiency and robustness, and effectively extracts the suspected fire region.

In recent years, deep learning has been rapidly developing in the field of target detection. Haolin Chen [4] a fire detection method based on the Uo-Net model is proposed, which uses a combined structure of multiple convolutional kernels to reduce the number of channels in the feature extraction network layers. At the same time, the attention graph of the image segmentation network is used to guide the detection model to detect fires, thus improving the performance of the fire detection model. Pei-Hao Chen [5] a hybrid Gaussian modelling approach is used for motion detection, followed by the Adaboost algorithm in integrated learning for suspected fire region extraction from motion images, and finally MobileNetv3, a lightweight neural network, is used to automatically extract suspected fire region features for fire identification. Xinjian Li [6] the network structure of the fire detection model was improved using deep separable convolution, and various data enhancement techniques and edge-based loss functions were used to improve the accuracy; through parameter tuning, a real-time detection of 21 ms on an embedded mobile system was achieved, effectively solving the problem of small-scale fire miss detection. Miao Cunke [7] a multi-feature fusion-based neural network video fire detection method is proposed to analyze the relationship between the fire region and the mean value of YCbCr in the whole video region, and build a convolutional neural network using TensorFlow to achieve accurate recognition of video fires. Sun Weiya [8] a lightweight and efficient video fire detection algorithm based on deep learning convolutional neural network target detection algorithm is proposed. A motion target detection algorithm based on hybrid Gaussian model is used to rationalize the fire target detection results with high efficiency and low resource consumption, and the accuracy rate reaches 98.94% in the video detection experiment of Self-built fire dataset.

Tongjun Liu [9] a straw burning detection method based on smoke and fire detection is proposed, which is based on the YOLOv3 target detection algorithm and is optimized for the straw burning phenomenon with corresponding improvements to achieve the goal of detecting the straw burning phenomenon. To address the problem of low detection performance of smoke and fire detection in complex scenes and environments with high interference, such as farmland, Lin Li [10] the improved SSD model was compared with the classical SSD model, YOLOv3 model and Faster R-CNN model, and the experimental results showed that the improved SSD model improved the mAP by 18.5%, 20.3% and 17.7%, and improved the FPS by 18, 30 and 24 respectively compared with the other three models. The improved SSD model is more suitable for detecting smoke and fire targets in complex scenes and better for detecting small targets. Zhu Xiu [11] the proposed algorithm uses YOLOv4, a one-stage target detection method, as the model framework for fire and smoke detection, which improves the running speed and facilitates the loading of the model on embedded devices such as cameras, enabling real-time detection of fire and smoke, thus promoting the promotion of smoke and fire detection in the environmental protection field.

There is a paucity of research literature on the detection of smoke and fire during coke oven production. At present, YOLOv5 is a relatively mature and stable target detection algorithm. In this paper, only YOLOv5s network model has been improved and optimised, and the next step is to consider the research and improvement of the emerging algorithm of YOLO series, so that it can meet both real-time and accuracy, and better complete the detection of smoke and fire emissions in coke plants around the clock.

In this paper, the YOLOv5s algorithm is improved to meet the requirements of 24/7 environmental protection detection of smoke and fire emissions in the cokemaking process by combining the mainstream attention model SE (Squeeze and Excitation) module and the convolutional block attention CBAM (Convolutional Block Attention Module) module to design A YOLOv5s smoke and fire detection recognition algorithm with a joint attention mechanism. The main contributions are as follows:

(1) In order to solve the problem of nighttime work lighting being mistakenly detected as fires, light category labels are added to the fire and smoke sample category labels when constructing the self-built daytime and nighttime scene datasets respectively.
(2) The web open source dataset and the self-built dataset were trained and detected for classification using the original YOLOv5s model, and the three classification results were compared and analysed. The experimental results show that the best detection results were achieved when the training and test datasets were unified, with an accuracy of 96.8% when both the training and test sets were night scenes.
(3) Comparing the training and testing results of the improved model with the original model, the final results showed that the accuracy of the "YOLOv5s+CBAM" model was 97.1% when the training and testing sets were all night scenes, and the accuracy was 4.4% higher when the training and testing sets were all day scenes compared with the original YOLOv5s model.

2 The Main Problems with Coke Oven Smoke and Fire Detection

In each coke oven the coke side, the machine side and the side of the coal tower are equipped with monitoring points to monitor the coke side, the machine side and the production site at the top of the furnace respectively. An example diagram of a production site is shown in Fig. 1, 2 and 3. Both sides of the coke oven are roughly divided according to the centre line of the coke oven, the side with the coke stopper or coke quenching truck is called the coke oven coke side and the side with the coke pushing or coal loading truck is called the machine side of the coke oven.

The following is a screenshot of the coke side production site monitoring video of the 5# coke oven (see Fig. 1). In Fig. 1, area ① is the furnace door from which the coal cake can enter the inside of the coke oven; area ② is the coke quenching car, which is used to take the coke around 1000 °C exported by the coke stopper car and transport it to the coke quenching chamber for quenching and eventually transporting the quenched coke coal to a designated location; area ③ is the coke stopper car, which can run on rails and is mainly used to open and close the coke side furnace door and guide the hot coke pushed out from the coke oven char chamber to the coke quenching car.

Fig. 1. Production site - coke side of #5 furnace

The following is a screenshot of the machine side production site monitoring video of the #6 coke oven (see Fig. 2). In Fig. 2, area ① is the furnace door, when the door is closed, the charring chamber is isolated from the atmosphere and when it is opened, the coke pushing rod is able to push out the coke; area ② is the coal loading truck, which runs at the top of the coke oven and carries out the coal loading operation at the top of the furnace, loading the empty charring chamber with raw coking coal after the coke has just been pushed out; area ③ is the coke pushing truck, whose main function is to pick and close the machine side furnace door and push the mature red coke from the charring chamber in the coke oven.

Fig. 2. Production site-machine side of 6# furnace

The following is a video screenshot of the production site at the top of the coal tower (see Fig. 3), the area in the red box is the working area at the top of the furnace, i.e. the small furnace door at the top of the coke oven. As the coke oven equipment ages [12], the door seals become less tight and smoke problems often occur. The biggest pollution emission of the coke oven is in the process of coke pushing and coal loading [13]. The coal loading starts until the coal leveling starts, the coal lowering is not yet finished, and the small furnace door is opened early for coal leveling when the coal loading process does not reach the coal leveling volume, which will lead to a large amount of barren gas coming out from the small furnace door.

Fig. 3. Production site - top of the furnace

During the process of coal loading, coke leveling and coke pushing, coke ovens are subject to smoke running and flaming, which not only pollute the atmospheric environment, but also adversely affect the coke oven itself. At present, deep learning is developing rapidly and the target detection methods are becoming more and more mature, but the following problems still need to be solved for the fire and smoke detection in the actual production site of coke ovens.

(1) The environment in the factory production site is relatively complex, surveillance camera for 24 h work. The surveillance images are in colour during the day and in greyscale at night, however one of the more important features of the fires is the colour, so this creates some difficulty in smoke and fire detection and identification. The image below shows a day and night comparison under the surveillance camera (see Fig. 4).
(2) When it is dark in order to improve efficiency, staff set up high brightness lights for illumination and high light levels, which may misidentify the lights as being against the fires. The diagram below shows an example of light disturbance at the coke oven site at night (see Fig. 5).

To address the above issues, this experiment uses the YOLOv5s model with the addition of an attention mechanism to achieve smoke and fire recognition in industrial sites. Attention mechanism [24] is a core technology widely used in natural language processing, statistical learning, image detection, speech recognition and other fields since the rapid development of deep learning. Based on research on human attention, expert scholars have proposed the attention mechanism, which essentially means achieving an efficient allocation of information processing resources.

(a)Day-time coke oven site images (b)Night-time coke oven site images

Fig. 4. Day/night comparison chart

Fig. 5. Light disturbance at the coke oven site at night

3 Design of a Coke Oven Smoke and Fire Detection Model

3.1 YOLOv5s Model

YOLOv5 is a single-stage object detection algorithm that adds some new and improved ideas to YOLOv4, resulting in significant performance gains in both speed and accuracy. On June 25, 2020, Ultralytics [18] released the first official version of YOLOV5, the most advanced object detection technique available today and the strongest in terms of inference speed.

YOLOv5 is divided into four models, of which YOLOv5s [17] as the basis for the other versions, is the smallest file of the four models, and also has the smallest depth and the smallest width of the feature map. YOLOv5s divides the entire network structure into four parts: Input, Backbone, Neck, and Prediction (see Fig. 6).

The input side represents the input image, with a size of 640 * 640 [21]. For different data sets, there is adaptive anchor frame calculation, which adaptively calculates the best anchor frame value for different training sets each time; for different images with different length and width, there is adaptive image scaling, which adaptively adds the least black border to the original image.

The Backbone part contains the Focus structure and the CSP structure, the key part of the Focus structure is the slicing operation, the original $640 \times 640 \times 3$ image is input into the Focus structure, using the slicing operation, first into a $320 \times 320 \times 12$ feature map, then after a convolution operation of 32 convolution kernels, finally into a $320 \times 320 \times 32$ feature map; There are two types of CSP structures in the YOLOv5s

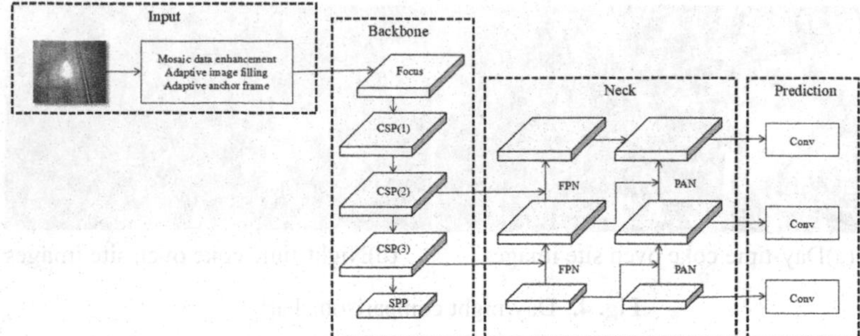

Fig. 6. YOLOv5s network structure

network, one CSP1_X structure is applied to the Backbone backbone network and the other CSP2_X structure is applied in the Neck.

The Neck part is in between the Backbone and Prediction parts, and passes the image features to the Prediction part. The FPN+PAN structure is used for this part, and the CSP2 structure is borrowed from the CSPNet design to enhance the network's feature fusion capability.

The Prediction section performs prediction of image features and generates Bounding boxes and predicted target types. GIOU_Loss (Generalized Intersection over Union) is used as the loss function of the target Bounding box and GIOU_ NMS non-maximum suppression is used for multi-target screening.

3.2 Attention Mechanisms

SE Module. SE Net [22] combining attention with convolutional channels won the Image Net 2017 competition for the classification task. The core of CNN networks is the convolution operator, which uses a convolution kernel to obtain the output feature map from the input feature map, a process that is often accompanied by a reduction in the spatial dimension of the feature map and an increase in the channel dimension. The convolution operation can be thought of as the process of continuously extracting feature information from space and mapping this information to different higher dimensional channels in a differentiated manner. However, the number of channels of the feature map after convolution is large and the model needs to differentiate between different channels. For this reason, SENet proposes the Squeeze and Excitation (SE) module (see Fig. 7).

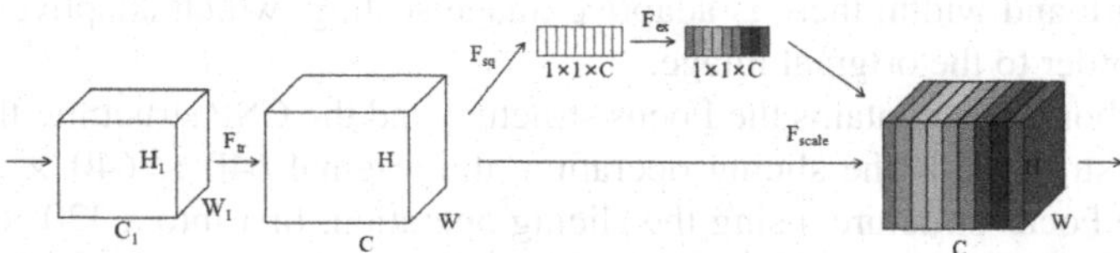

Fig. 7. SE module structure diagram

The SE module first reduces the dimensionality of the input feature map through global average pooling to obtain global features on different channels, then compresses and reorganizes these global features to obtain the weights of different channels, and finally assigns these weights to the respective channel dimensions to obtain the final output features. The SE module essentially reorganises the feature map in the channel dimension by giving higher weights to those channels with important features, so that the model focuses more on the important channel features and ignores the less important ones.

CBAM Module. Sanghyun Woo [23] et al. in 2018 first proposed CBAM, a hybrid attention structure that combines channel attention and spatial attention. The innovation of CBAM is not only that it draws on the idea of SENet to design a new channel attention module through GAP (Global Average Pooling) and GMP(Global Maximum Pooling), but also proposes a hybrid attention module that combines channel attention and spatial attention and can be embedded into other existing network structures. The diagram below shows its exact structure (see Fig. 8).

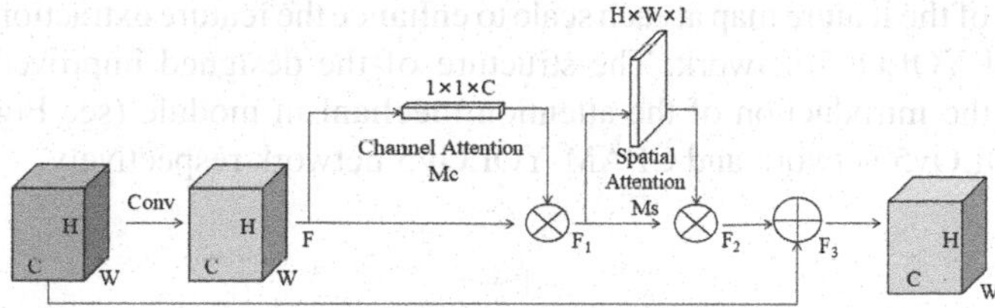

Fig. 8. CBAM module structure diagram

The essence of channel attention is to learn the weight distribution of each feature channel of the feature map and to adaptively strengthen the important channels and constrain the unimportant ones by reorganising the features of different feature channels through channel weighting. The process is to perform GAP and GMP operations on the spatial features on the channels to obtain two one-dimensional tensors with the same number of dimensions as the number of channels. The channel attention weight M_C is obtained by summing the two weight vectors after sigmoid activation. The formula for calculating the channel attention module is shown below.

$$M_C = \sigma(MLP(MaxPool(F)) + MLP(AvgPool(F)))$$

$$= \sigma\left(W_0 F_{avg} + W_1 F_{\max}\right) \quad (1)$$

σ is the sigmoid function, *MLP* is a multi-layer perceptron, and F_{avg} and F_{max} are the outputs after global flat pooling and global maximum pooling respectively.

The essence of spatial attention is to learn the distribution of weights on the feature map space, assigning different weights at different positions on the spatial dimension of the feature map and adaptively enhancing the model's attention to the feature regions. The process is to perform GAP and GMP operations on the input features in the channel dimension to obtain two two-dimensional tensors of the same size as the input features. The two tensors are then stitched together in the channel dimension and reduced in dimensionality by convolution, and finally the attention weights M_S over the spatial domain are obtained after sigmoid activation. The formula for calculating the spatial attention module is shown below.

$$M_S = \sigma\left(f^{7*7}[MaxPool(F); AvgPool(F)]\right)$$
$$= \sigma\left(f^{7*7}[F_{avg}; F_{\max}]\right) \tag{2}$$

The f^{7*7} is a convolutional kernel of size 7 * 7.

3.3 YOLOv5s for Joint Attention Mechanisms

In this paper, the attention mechanism SE network module and CBAM network module are applied at each location of the multi-scale feature output of the backbone feature extraction network of the YOLOv5 network, respectively, for feature rescaling of each channel/space of the feature map at each scale to enhance the feature extraction capability of the original YOLOv5 network. The structure of the designed improved YOLOv5 network with the introduction of the attention mechanism module (see Fig. 9) and is named SE-YOLOv5 network and CBAM-YOLOv5 network respectively.

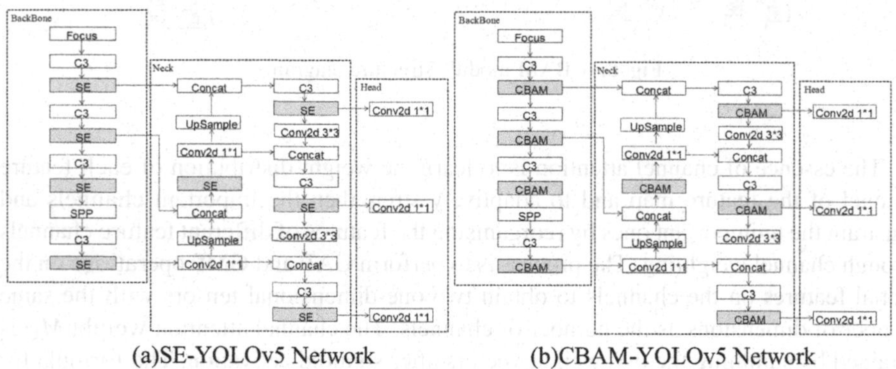

(a)SE-YOLOv5 Network (b)CBAM-YOLOv5 Network

Fig. 9. The model structure after the introduction of the attention mechanism module

4 Experiments and Analysis of Results

4.1 Constructing the Dataset

The dataset used in this paper consists of two parts, one from an open source dataset on the web and the other from a coking plant. The open source dataset from the web is an unlabelled set of 1000 images, all of which are daytime scenes. The dataset from the coke plant is a video surveillance camera shot of the coke oven production site, covering the coke side south, coke side north, machine side south, machine side north and coal tower side of each coke oven. This experiment intercepted 1280 photos from the surveillance videos, including 780 daytime scenes and 500 nighttime scenes. The original data set of 2280 photos covers the fire smoke image data under different lighting conditions.

The original data set was 2280, and after the data enhancement operation the data set was expanded to 10,400, including 2,000 network datasets and 8,400 factory private datasets. Taking into account different lighting conditions, the 10,400 datasets were divided into daytime (high saturation) and nighttime (low saturation) scenes according to their saturation level, with 4,900 daytime scenes and 5,500 nighttime scenes. The data set includes three categories, fire, smog, and light. At the same time, divide the data set into training set, test set and validation set in the ratio of 8:1:1.

4.2 Model Training

The experimental hardware in this paper is a PC with Intel(R) Core(TM) i5-9500F CPU@3.00 GHz, 8 GB RAM, Windows 10 64-bit operating system, Radeon 520 graphics card, python software development environment and Pytorch deep learning framework.

In the training process, the SGD optimization algorithm was used for parameter training, and the specific training parameters were set as follows: the initial learning rate (lr0) was 0.01, the final decay rate (lr0*lrf) was 0.001, the learning rate momentum (momentum) was 0.937, the weight-decay (weight-decay) was 0.0005, the weight file (weights) was yolov5s.pt, the number of iterations (epochs) was 50, the batch-size was 4, the input image resolution (imgsz) was 640px, and the input image resolution (imgsz) was 640px. Weights is yolov5s.pt, the number of iterations (epochs) is 50, the batch-size is 4, the input image resolution (imgsz) is 640px, and GIOU_ Loss is used as the loss function, with loss weights box of 0.05, cls of 0.5, and obj of 1.0.

The Original Model. The problem of surveillance working 24 h a day and the resulting dataset being colour images during the day and greyscale images at night is addressed by a group comparison training approach. This subsection uses the YOLOv5s target detection model, which is trained in three cases:

The first one is mixed training, where the training set is a mixture of daytime and nighttime scenes, and the test set is also a mixture of daytime and nighttime scenes. The second type is split training, where the test set is daytime when the training set is daytime scenes, and the test set is nighttime when the training set is nighttime scenes. The third type is cross-training, in which the test set is nighttime when the training set is daytime scenes and daytime when the training set is nighttime scenes. The details are shown in Table 1.

Attention Mechanisms. In this subsection, the improved YOLOv5s target detection model, i.e., the YOLOv5s model with the addition of the attention mechanism module, is used for training. The YOLOv5s model with the addition of SE module and the YOLOv5s model with the addition of CBAM module are used for training, respectively, and the data set configuration and hyperparameters are kept the same as the original YOLOv5s model, and the training method is chosen as the triage training method.

Table 1. Details of the training and test sets for the three scenarios

	Training set		Test set/Validation set	
	Daytime	Night	Daytime	Night
Mixed training	Factory-built (1600 sheets)	Factory-built (1600 sheets)	Factory-built (200 sheets)	Factory-built (200 sheets)
Split training	Web Open Source (1600 sheets) + Factory Built (1600 sheets)	Factory-built (3200 sheets)	Web Open Source (200 sheets) + Factory Built (200 sheets)	Factory-built (400 sheets)
Cross training	Factory-built (2420 sheets)	Factory-built (2420 sheets)	Factory-built (190 sheets)	Factory-built (190 sheets)

4.3 Comparative Analysis of Experimental Results

Evaluation Metric. In this experiment, mAP (Mean Average Precision) is used as the evaluation index of the target detection model. The precision P represents the proportion of correctly identified meters to all identified meters in the dataset, and the recall R represents the proportion of correctly identified meters to all meters, calculated as follows:

$$P = \frac{TP}{TP + FP} \times 100\% \tag{3}$$

$$R = \frac{TP}{TP + FN} \times 100\% \tag{4}$$

TP indicates the number of correctly identified pyrotechnic areas; FP indicates the number of incorrectly identified pyrotechnic areas; and FN indicates the number of missed pyrotechnic areas. The P-R (Precision-Recall) curve can be plotted according to Eqs. (3) and (4), and the AP (Average Precision) of a single category can be calculated by averaging the Precision values of the P-R curve. mAP is the average of the accuracy AP of each category, and the relevant calculation formula of mAP is as follows:

$$AP_{\mathrm{i}} = \int_0^1 p(r)d(r), i = 1, 2, 3... \tag{5}$$

$$mAP = \frac{\sum AP_i}{n}, i = 1, 2, 3... \tag{6}$$

The p is the precision rate, r is the recall rate, and n is the number of categories.

YOLOv5s Original Model. Since the experimental process was divided into three cases for training, the experimental results were analyzed in three cases accordingly. mAP was selected as the evaluation index, and the combined experimental results of the three cases can be obtained in Table 2.

Table 2. Comparison of results by type of situation

		Number of training sets (sheets)	Number of test sets (sheets)	Number of Validation sets (sheets)	mAP (100%)
Mixed training		3200	400	400	64.3
Split training	Day-Day	3200	400	400	84.3
	Night-Night	3200	400	400	96.8
Cross-training	Day-Night	2420	190	190	20.1
	Night-Day	2420	190	190	37.6

As can be seen from Table 2, the effect of cross-training is poor, with less than 50% mAP value; the mAP is improved by 26.7% when comparing mixed training with cross-training; but overall, the detection effect of triage training is better, among which the effect is especially outstanding when the datasets are all night scenes, and the mAP value of smoke and fire recognition reaches 96.8%, reflecting the good algorithm performance. From the experimental results, it can be seen that the experiments based on YOLOv5s model, and adopt the mode of triage training, and divide the level of saturation of smoke and fire image data into two scenes, daytime and nighttime, and carry out labeling, training as well as testing respectively, can get better detection results.

YOLOv5s Model with Added Attention Mechanism. The YOLOv5s model with the SE module added and the YOLOv5s model with the CBAM module added were trained separately for triage, and the control experiments were done on the same dataset as the original YOLOv5s model, and mAP was selected as the evaluation index. To verify the stability of the YOLOv5 algorithm, the YOLOXs algorithm was added as a control group, and the experimental parameters and data set configuration were kept the same as the original YOLOv5s model. Table 3 shows the comparison of the experimental results.

Table 3. Comparison of experimental results

Models	mAP (100%)		Running time (ms/sheet)
	Daytime scenes	Night scenes	
YOLOXs	85.7	96.9	25.4
YOLOv5s	84.3	96.8	20.1
YOLOv5s+SE	86.6	96.9	24.5
YOLOv5s+CBAM	88.7	97.1	25.2

As shown in Table 3, the YOLOv5s model with the attention module has different degrees of improvement in detection accuracy compared with the original YOLOv5s model, and the attention module CBAM is more effective than SE. When the test set and training set are both night scenes, the mAPs of "YOLOv5s+SE" and "YOLOv5s+CBAM" improve by "0.1%" and "0.3%", respectively, compared with the original model, The mAP of "YOLOv5s+SE" and "YOLOv5s+CBAM" improved by "0.1%" and "0.3%" respectively compared with the original model, and the mAP of the original model for the nighttime scenes was already as high as "97.3%". This fully illustrates the correctness of adding light labels for training, and successfully eliminates the interference of nighttime lights on flames. When the test set and training set are daytime scenes, the mAP of the "YOLOv5s+CBAM" model is improved by 4.4% compared with the original model, which proves that adding the attention mechanism to the algorithm has a significant optimization effect. Also, as can be seen from the table above, the mAP values of YOLOXs are slightly improved compared to the original YOLOv5s model, but the model is too large, resulting in a significant slowdown in running time compared to the original YOLOv5s model, which is not conducive to real-time detection, which fully illustrates the necessity of choosing to use YOLOv5s as the base model for improvement.

Visual Analysis of the Results. From the above section, it can be seen that the experimental results of the YOLOv5s algorithm model with the CBAM module are better, and the mAP value is as high as 97.1% when the data sets are all night scenes. The following figure shows the visualization results of the experiments using the YOLOv5s algorithm model with the CBAM module, from which it can be seen that the detection of "flame", "smoke" and "light" The confidence level for the detection of all three types of labels is considerable(see Fig. 10).

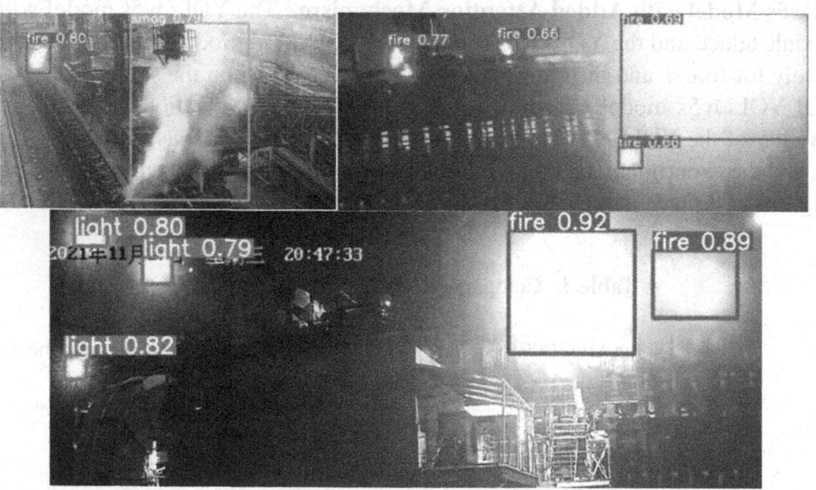

Fig. 10. Visualization results of the "YOLOv5s+CBAM" experiment

5 Conclusion

In order to achieve the target detection of coke oven flame smoke and meet the requirements of environmental protection detection of smoke and fire emission in the coking process around the clock, this paper designs a YOLOv5s coke oven smoke and fire detection recognition algorithm with a joint attention mechanism. Firstly, the YOLOv5s network model is used for experiments, and the results show that the best detection effect is achieved when the scenes of training and test sets are unified, in which the case that both training and test sets are night scenes is particularly outstanding, and the mAP reaches 96.8%. For the problem of light interference, we choose to use add plus light labels for recognition. For the problem that the target is easily disturbed by background information, the attention module is chosen to be introduced to YOLOv5s, and the final result is that the mAP using YOLOv5s+CBAM model training is as high as 97.1% when the training and test sets are all night scenes, and the mAP when the data set is all day scenes is improved by 4.4% compared with the training result using the original YOLOv5s model. The experimental results show that the YOLOv5s algorithm model using the joint attention mechanism has good robustness and reliability for coke oven flame smoke detection and recognition.

As YOLOv5 is currently a relatively mature and stable target detection algorithm, this paper only improves and optimises the YOLOv5s network model. The next step is to consider the research and improvement of the emerging algorithms of the YOLO series so that they can meet both real-time and accuracy and better complete the detection of all-weather smoke and fire emissions from coke plants. At the same time, this experimental dataset scenario only involves the production site picture of a coke oven in a coking plant, which is relatively single and cannot be applied to the smoke and fire detection in various industries. The next step will be to consider increasing the number of scenes in the dataset beyond the production site of this plant, so as to better realise the smoke and fire detection of industrial sites and meet more stringent environmental requirements.

References

1. Shen, S., et al.: A fire identification method based on video image correlation. J. Saf. Environ. **7**(6), 96–99 (2007)
2. Wang, J., Yu, W., Han, T.: Fire detection based on infrared video image. J. SJTU **42**(12), 1979–1987 (2008)
3. Wu, X., et al.: Fire detection algorithm based on multi-feature fusion. J. Int. Syst. **10**(2), 240–247 (2015)
4. Chen, H., et al.: Fire detection method based on UO-net model. J. JOU **29**(4), 8–15 (2020)
5. Chen, P., Xiao, D., Liu, H.: Video fire detection algorithm based on deep learning. J. Combust. Sci. Technol. **27**(6), 695–700 (2021)
6. Li, X., et al.: Lightweight fire detection method based on CNN in complex scenes. J. Pattern Recognit. Artif. Int. **34**(5), 415–421 (2021)
7. Miao, C., Yang, L., Jiang, Y.: Video image fire detection method based on neural network. J. Comput. Inf. Technol. **2021**(04), 71–74 (2021)
8. Sun, W., et al.: Efficient video fire detection algorithm based on motion features. J. Data. Acquis. Process. **36**(6), 1276–1285 (2021)

9. Liu, T.: Research on the application of pyrotechnic detection method based on deep learning in straw burning ban. D. HU (2020)
10. Li, L., Cao, L.: Research on fireworks image detection in farmland based on improved SSD algorithm. J. Road Eng. **52**(05), 783–789 (2022)
11. Zhu, Y., et al.: Fire smoke detection algorithm for lightweight network. J. Appl. Technol. **49**(2), 1–7 (2022)
12. Bai, Y., Zhao, H., Liu, H.: Analysis and control measures of "smoke and fire" in 7.63m coke oven coal. J. Manag Technol. Entertain. **12**, 294 (2013)
13. Tan, S., et al.: Real-time detection of mask wearing based on YOLOv5 network model. J. Laser **42**(2), 147–150 (2021)
14. Lin, S., Chen, J., Huang, S.: Research on student behavior detection based on deep learning. J. Mult. Netw. Teach. **06**, 237–240 (2022)
15. Chen, J., Li, L.: Video detection of illegal mining based on YOLOv5 neural network model. J. Water Conserv. Tech. Support **08**, 61–63+119+124 (2021)
16. Zhou, Y., et al.: Research on target detection algorithm of mobile robot based on YOLOv5. J. Equip. Man. Tech. **08**, 15–18 (2021)
17. Lu, S., Feng, J., Duan, P.: Review of video fire identification methods. J. Telecommun. Technol. **37**(03), 179–184+200 (2013)
18. Cao, J., Qin, Y., Ji, X.: Review of fire detection algorithms based on video. J. Data Acquis. Process. **35**(01), 35–52 (2020)
19. Anshul, G., Abhishek, S., Anuj, K.: Video fire and smoke based fire detection algorithms: a literature review. J. Fire Technol. **56** (2020)
20. Ma, L., et al.: Research on object detection algorithm based on YOLOv5s. J. Commnu. Knowl. Technol. **17**(23), 100–103 (2021)
21. Chen, S., et al.: Research on fire detection based on YOLO neural network and machine vision. J. Technol. Inf. **03**, 107–110 (2022)
22. Hu, J., Shen, L., Sun, G.: Squeeze-and-excitation networks. In: Proceedings of the IEEE Conference on Computer Vision and Pattern Recognition, pp. 7132–7141 (2018)
23. Woo, S., Park, J., Lee, J.Y., Kweon, I.S.: CBAM: convolutional block attention module. In: Ferrari, V., Hebert, M., Sminchisescu, C., Weiss, Y. (eds.) ECCV 2018. LNCS, vol. 11211, pp. 3–19. Springer, Cham (2018). https://doi.org/10.1007/978-3-030-01234-2_1
24. Ren, H., Wang, X.: A review of attentional mechanisms. J. Comput. Appl. **41**(S1), 1–6 (2021)

Natural Language Processing, Knowledge Graphs, Recommender Systems, and Their Applications

An Enhanced Model Based on Recurrent Convolutional Neural Network for Predicting the Stage of Chronic Obstructive Pulmonary Diseases

Zhanjie Mai[1], Pengjiu Yu[2], Chunli Liu[2], Qun Luo[2], Li Wei[2](✉), and Tianyong Hao[1]

[1] School of Computer Science, South China Normal University, Guangzhou, China
{2019022613,haoty}@m.scnu.edu.cn
[2] The First Affiliate Hospital of Guangzhou Medical University, Guangzhou, China
chunli@gird.cn

Abstract. Chronic Obstructive Pulmonary Disease (COPD) is the third leading cause of death in China and has caused serious affect to health and life quality. However, the disease stage of COPD patient is difficult to be accurately assessed because of the dynamic changes of patient conditions and complex risk factors. Therefore, exploring a rapid and accurate method to predict disease stages of COPD patients is of great significance. This study proposes an enhanced recurrent convolutional neural networks model for predicting COPD patient's disease stage for assistant disease prevention and treatment. Data was collected from The First Affiliate Hospital of Guangzhou Medical University, which had standardized disease registration and follow-up management for 5108 patients with COPD. Our enhanced recurrent convolutional neural network consists of a bidirectional LSTM layer to extract the global features, a convolutional layer to extract the local features, and finally choose the semantic vectors of text representations through the attention mechanism and max-pooling layer. The performance of the proposed model is evaluated on a real-world electronic medical record of 5108 patients with COPD to predict their disease stages, which achieved 93.2% in terms of accuracy, outperforming a list of baseline models. This paper proposes an enhanced recurrent convolutional neural network model and evaluated it on a real-world clinical dataset containing around 5,000 patients with COPD. The experiment results demonstrated the model superior performance across all evaluation metrics, achieving the best performance.

Keywords: Chronic obstructive pulmonary disease · electronic medical record · prediction · recurrent convolutional neural networks

1 Introduction

COPD is a significant public health concern in China, drawing increasing attention from healthcare systems. It is projected to become one of the top three leading causes of death globally by 2030. Moreover, it estimates that COPD is the fourth leading cause of

H. Zhang et al. (Eds.): NCAA 2023, CCIS 1870, pp. 179–190, 2023.
https://doi.org/10.1007/978-981-99-5847-4_13

death in the world and the third leading cause of death in China according to the study by Global Burden of Disease (GBD) [1]. In the latest study of Adeloye et al. [2], it explained the prevalence and disease status of COPD in 65 countries and regions, pointing out that respiratory diseases as COPD as an example will make a huge impact on human health in the future, and more than three-quarters of all COPD cases worldwide will occur in developing countries. The prevalence and impact of COPD on the population highlight the urgent need for effective prevention, diagnosis, and management strategies to address this growing health challenge [3]. COPD is a condition that can be prevented and treated, which is characterized by a long-term obstruction of airflow. The disease is primarily caused by the chronic inflammation of the airways and lung tissue in response to harmful gases or particles, notably tobacco smoke. The prevalence rates of COPD according to the data of Global Health Data Exchange Database from 1997 to 2020 is shown in Fig. 1, which shows a clear trend of increasing from 1997 to 2020.

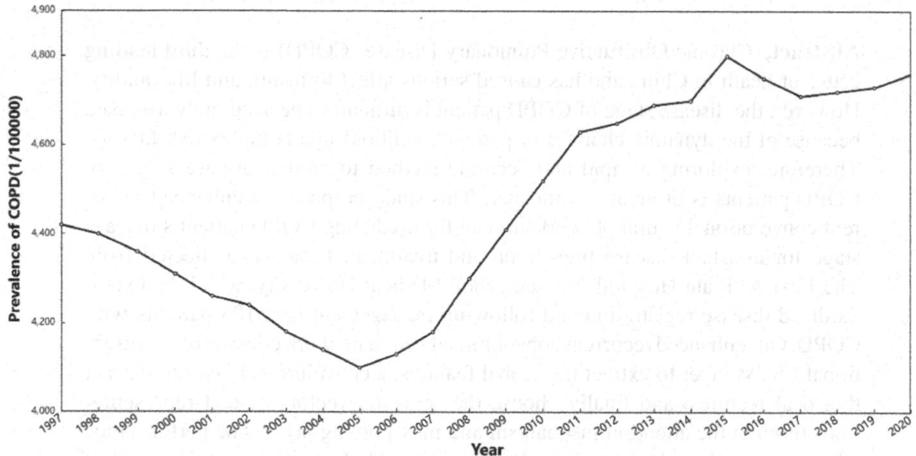

Fig. 1. The prevalence of COPD in China by year.

Electronic Medical Records (EMR) contain valuable information in unstructured text format, encompassing various details like family history, physician description of the clinical symptoms, and personal physical indicators. These pieces of information are usually found within textual descriptions provided by clinicians, such as progress notes and formalized discharge summaries. These records hold hidden insights that can contribute to a comprehensive understanding of a patient's medical history and aid in providing effective information for doctors to estimate patients' disease [4]. Therefore, methods can be developed to extract essential information such as patient characteristics from patients' EMR for analysis. Over the past few years, mathematical models have been designed to predict the incidence of COPD. However, the diagnosis of COPD is mainly based on FEV1 [5] and has significant difficulties according to other symptoms. There are large individual differences between patients, and sometimes there is a gap between symptoms and objective examination. Therefore, patients' subjective statements in EMR may lead to wrong staging, which can lead to delay or even improper treatment. The diagnosis of COPD is commonly determined by comprehensive analysis according to clinical manifestations, risk factor contact history, signs and laboratory examination [6]. The diagnosis is subjective and complex, or even based on individual understanding of the diagnostic criteria by doctors. The difference of practical experience and knowledge level may also lead to the different diagnosis results on the disease stage. On the other hand, the standard of judging the disease stage of COPD patient has its own defects. The staging of a patient's disease is a qualitative evaluation, but the condition of the disease changes dynamically. Therefore, qualitative grading is difficult to make a suitable evaluation for each different patient, and the severity of the patient's condition in the same period cannot be clearly distinguished. In addition, the pathogenic factors of COPD are complex and diverse, and a single index is difficult to accurately assess the condition, which requires patients to carry out multiple examinations and confirm the diagnosis, and lacks a simple and effective comprehensive evaluation index [7]. As a disease with high morbidity, the lung structure has been irreversibly changed, the clinical treatment is restricted, and the therapeutic effect is limited. Therefore, developing a rapid and accurate evaluation method is of great significance to improve the clinical efficacy of COPD.

Nowadays, machine learning models have been widely used in the medical domain. In 2019, Lin et al. [8] used medical insurance data to build a Bayesian network model to predict the disease stage of COPD patients. Wang et al. [9] employed medical pulmonary function tests data of patients to construct a COPD identification model, and the results proved that the efficacy of their proposed machine learning-based approach in accurately identifying patients with acute COPD. In 2022, Pu et al. [10] extracted 72 parameters from patients' medical records to establish a random forest model. To determine the optimal threshold, the model was trained with a 1% interval to identify the best FEV value for distinguishing acute COPD from normal symptoms. The performance of the model achieved 85% accuracy compared with the diagnosis of versus expert. Their research suggested that applying machine learning methods could help better assess the clinical situation of COPD patients. However, the methods mentioned above may cause the issues of low accuracy, the need for manual feature selection, and the inability to represent text semantics evidently. In recent years, deep learning has emerged as a successful approach in the medical field for extracting valuable information from EMR data. Numerous studies have focused on predicting the risk of COPD in patients. These efforts aim to address the high global incidence rate of COPD by leveraging deep learning techniques to improve prediction accuracy and enhance preventive measures. Compared with machine learning, the deep learning networks can learn the semantic information in the text, automatically extract the features in the text, finally achieve better performance.

Therefore, this paper targets at proposing an enhanced model for staging the patient status with COPD. The performances of different models are compared on a COPD EMR dataset. The contribution of this paper lies in the following aspects: 1) A COPD dataset is constructed based on the EMRs of approximately 5,000 patients from one of top hospitals in China. 2) an enhanced recurrent convolutional neural network is proposed for predicting the disease stage of patient suffering from COPD. 3) Evaluation and result demonstrate the effectiveness of our proposed model by comparing with a list of models on the dataset.

2 Methods

To predict the disease stage of patient suffering from COPD, an enhanced recurrent convolutional neural network is proposed by focusing on the clinical description of the patient conditions in EMRs. By investigating the EMRs of COPD patients, patients with COPD were divided into three categories including remission, general and acute exacerbation based on clinical diagnostic criteria. To address the issue that some records may contain missing values, all missing data has been transformed into structured data through the process of data cleaning and data imputation. The specific steps in the process includes collecting clinical data, pre-processing data, dividing patient data into training set and test set, the network model training, and evaluating prediction performance. The overall workflow of the model construction and processing is shown in Fig. 2.

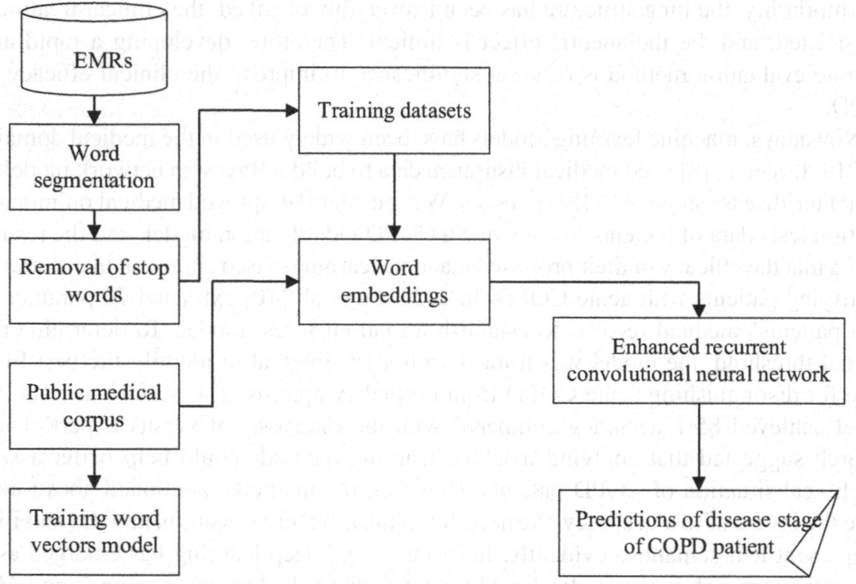

Fig. 2. The overall workflow of the model construction and processing using EMRs data of patients with COPD.

2.1 Enhanced Recurrent Convolutional Neural Network

Although CNN model has advantage in extracting deep information from the text. However, it has difficulties in setting convolutional kernel size. For traditional recurrent neural network model, it achieves bad performance on long-term dependence of sentences. To overcome the limitations of the aforementioned models, we proposed enhanced recurrent convolutional neural network based on RCNN [11] proposed by Lai et al. First, a bi-LSTM structure is applied to capture the contextual information to the greatest extent possible when learning word representations. Moreover, we set three convolution sublayers are used to extract feature of the document through the different kernel sizes to achieve the best performance. Ultimately, the architecture of the model is shown in Fig. 3, consisting of an embedding layer, a bidirectional LSTM layer to extract the global features, a convolutional layer to extract the local features, and finally choose the semantic vectors of text representations through the attention mechanism and max-pooling layer.

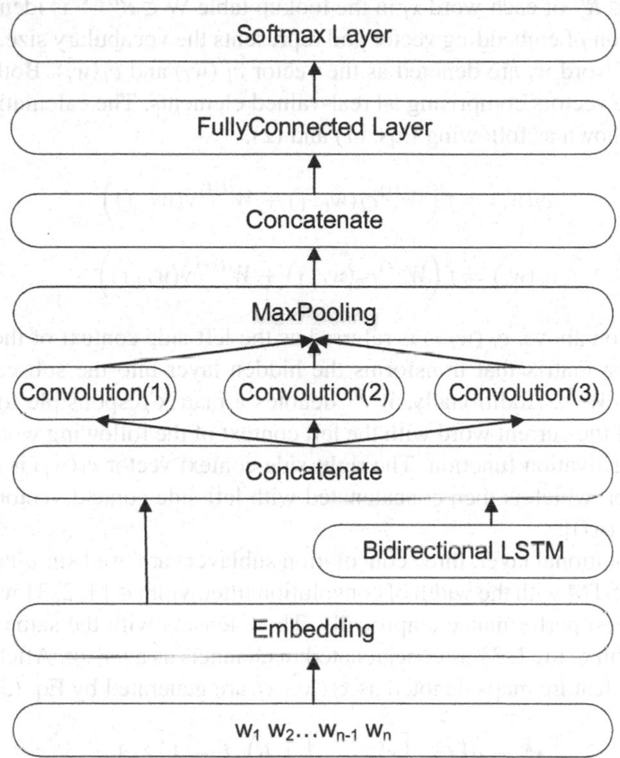

Fig. 3. The architecture of the enhanced recurrent convolutional neural network.

2.2 Pre-training Word Embedding

In this study, textual sentences in EMR describe the basic disease symptoms and patients' specific information which may imply useful information behind text. Due to the nature of Chinese characters, word segmentation is required for Chinese EMR text preprocessing. We use Jieba (https://pypi.org/project/jieba/), a commonly used tool, to segment Chinese text into words with a specified medical dictionary for further processing. After that, for each document, stop words and special symbols are removed in the text.

To effectively utilize the formalization of language or mathematical descriptions to represent words in EMRs is the aim of text representation [12]. One-hot text representation and distributed text representation currently commonly are used mainly included in text representation methods [13]. One-hot text representation is characterized by having a single element set to 1 in each row of the feature matrices, with all other elements set to 0. However, a significant drawback of one-hot representation is its data sparsity, which leads to high data processing consumption [14].

Formally, a document D containing m sentences is denoted as the input of the network, in which every sentence includes a sequence of words $w_1, w_2 \ldots w_n$, where n is the number of the words in sentence. In the embedding layer of the model, word or phrase embeddings $x_i \in R^L$ of each word x_i in the lookup table W $\in R^{L \times V}$ is identified, where L is the dimension of embedding vector and represents the vocabulary size. The left and right context of word w_i are denoted as the vector c_l (w_i) and $c_r(w_i)$. Both c_l (w_i) and $c_r(w_i)$ are dense vectors comprising $|k|$ real-valued elements. The calculation of c_l (w_i) and $c_r(w_i)$ are shown as following Eqs. (1) and (2),

$$c_l(w_i) = f\left(W^{(l)}c_l(w_{i-1}) + W^{(sl)}\mathrm{v}(w_{i-1})\right) \tag{1}$$

$$c_r(w_i) = f\left(W^{(r)}c_r(w_{i+1}) + W^{(sr)}\mathrm{v}(w_{i+1})\right) \tag{2}$$

In formulation above, c_l (w_{i-1}) is referred as the left-side context of the word w_{i-1}. $W^{(l)}$ represents a matrix that transforms the hidden layer into the subsequent hidden layer, similar to $W^{(r)}$. Additionally, $W^{(sl)}$ denotes a matrix responsible for combining the semantics of the current word with the left context of the following word. Function f is a non-linear activation function. The right-side context vector $c_r(w_i)$ is calculated in a similar manner, which is then concatenated with left-side context vector $c_l(w_i)$ to be $\boldsymbol{x}_i = [c_l(wi); cr\ (wi)]$.

In the convolutional layer, three convolution sublayers are used simultaneously produced by a Bi-LSTM with the width of convolution filter width $\in \{1, 2, 3\}$ which is tested to achieve the best performance empirically. Three tensors with the same shape corresponding to width $\in \{1, 2, 3\}$ as concatenated in channels as a tensor. After convolution operation, three feature maps denoted as x_i^1, x_i^2, x_i^3 are generated by Eq. (3),

$$x_{ij}^k = g\left(f^k \cdot \left[x_{i,j:j+k,:}^x\right] + b\right), k = 1, 2, 3, \tag{3}$$

where RELU [15] is used as the activation function g shown in Eq. (4) and b.

$$RELU(x) = \max(0, x) \tag{4}$$

After semantic vectors of the word w_i is obtained, the tanh activation function is incorporated with the model. $y_i^{(2)}$ is a latent semantic vector, as shown in Eq. (5)

$$y_i^{(2)} = \tanh\left(W^{(2)}x_i + b^{(2)}\right) \tag{5}$$

To enhance the learning capability of the mode the Rectified Linear Unit (ReLU) is incorporated as an activation function. ReLU function can improve the efficiency of the model by reducing the number of computations and shortening the learning period. Additionally, ReLU sets the output of certain neurons to zero, effectively preventing overfitting.

After that, our model is capable of converting text with varying lengths into a fixed-length vector representation by incorporating a pooling layer. This pooling layer enables the model to capture information from the entire text, regardless of its length. It helps to summarize the text and extract essential features, facilitating effective analysis and understanding of the text data. Though there are other types of pooling layers, such as average pooling layers proposed by Lin et al. [16], we use max-pooling, as shown in Eq. (6), since it seeks the most important latent semantic factors in text. The output layer, similar to traditional neural networks, is defined as Eq. (7).

$$\mathbf{y}^{(3)} = \max_{i=1}^{n} \mathbf{y}_i^{(2)} \tag{6}$$

$$y^{(4)} = W^{(4)} y^{(3)} + b^{(4)} \tag{7}$$

The softmax classifier is applied to y (4), which converts the output of multiple neurons into probability between 0 and 1 for each category of patients. As shown in Eq. (8), the absolute value of the classification result is associated with that specific category.

$$p_i = \frac{\exp\left(\mathbf{y}_i^{(4)}\right)}{\sum_{k=1}^{n} \exp\left(\mathbf{y}_k^{(4)}\right)} \tag{8}$$

2.3 Evaluation Metrics

The evaluation aims to measure whether the main outcome predicts the disease stage of patients suffering from COPD correctly. Recall, precision and F1 score are normally applied to evaluate the performance of the model in machine learning and statistical analysis, particularly in binary classification tasks. Nevertheless, the measures above are inadequacy when dealing with multiple classes. In this paper, macro-average evaluation metrics are also utilized. Firstly, the metrics of each class are calculated, and then average the metrics. The measures are defined by the Eq. (9–11).

$$Macro\text{-}P = \frac{1}{n} \sum_{j=1}^{n} Precision_i \tag{9}$$

$$Macro\text{-}R = \frac{1}{n} \sum_{i=1}^{n} Recall \tag{10}$$

$$Macro\text{-}F1 = \frac{1}{n} \sum_{=1}^{n} F1_i \tag{11}$$

3 Experiments and Results

3.1 Dataset

A dataset containing EMRs of 5,000 COPD patients is obtained from the First Affiliated Hospital of Guangzhou Medical University. According to diagnostic criteria by clinicians, the patients are divided into three disease stages: stable stage of COPD, acute exacerbation of COPD, and COPD. An EMR of a patient presented as a fixed-format table that describes a patient with various field-value records including basic information, condition of diseases, specialist inspections, body check, etc.

In the dataset, each patient has an average of 4.5 medical records, including first course records and medical records during treatment. The first course records focus on symptoms, past medical history, and examinations at admission. In detail, there are 1,928 cases of acute exacerbation, 2,135 cases of stabilization and 1045 cases of ordinary COPD. In each EMR, text content can be divided into two sections, while the first contains a patient's physiological characteristics and admission symptoms and the latter contains clinician's diagnosis of the patient. The average sentence length of EMR is 18.2 and maximum sentence length is 45 (Table 1).

3.2 Baseline Methods

Our method is compared with a list of deep learning methods as baselines including the following methods.

FastText: Joulin et al. [17] proposed FAST TEXT to finish simple text classification problems. In contrast to unsupervised trained word vectors obtained from word2vec, this baseline approach utilizes averaged word features to create robust sentence representations. After these sentence vectors are summed, a softmax classification layer is applied to generate predictions. This methodology aims to leverage the collective information of words within a sentence to improve the accuracy and effectiveness of the classification process.

TextRNN: Liu et al. [18] proposed three RNN based architectures to model text sequence with multi-task learning. Experimental results showed that models can achieve better performances of text classification than NBOW which summed the word vectors and applies a non-linearity followed by a softmax classification layer.

CNN: Convolutional Neural Network [19] concatenated the word embeddings in a predefined window. Formally, the text matrix undergoes convolution using filters of various lengths. The width of each filter is set to match the length of the word vector. The resulting vector obtained from each filter is then subjected to max-pooling. Consequently, each filter corresponds to a numerical value. These filter outputs are concatenated together to form a vector that represents the sentence. This methodology allows for capturing meaningful features and patterns across different sections of the text, leading to an effective sentence representation.

3.3 Parameters Setting

Our model as the enhanced RCNN is implemented using Scikit-learn, TensorFlow, and Keras. During the training process, the number of training epochs were set to 100 and

Table 1. An example of the electronic medical record of a patient with COPD from the First Affiliated Hospital of Guangzhou Medical University.

Basic demographic information of patients	*Age: XX; Gender: XX; ...*
Cause of admission	*Repeated shortness of breath, chest tightness for 1 year*
Detailed description of the condition	*The patient had no obvious cause of shortness of breath after exercise since August last year, with rapid walking* 500 m *and smelling and hot air, chest tightness, occasional mild wheezing at night, no cough, expectoration, no fever, shivering, no hemoptysis. In the past year, the patient's symptoms were repeated and progressive aggravation. The outpatient department was admitted to our department with shortness of breath and chest tightness. Since the onset of the disease, the patient has no fever, mental ability, poor appetite, general sleep, normal stool, no significant weight loss, labor capacity decline*
Specialist inspections	*Both sides of respiratory rhythm were symmetrical, palpation was normal, double lung percussion was clear, auscultation was clear, dry and wet rale was not reached*
Body check	*Patients has no percussion pain in the liver and spleen, no percussion pain in the liver area, no percussion pain in the double kidney area, negative moving turbidimetric sound, and the sound. There were no deformities in the spine and limbs, normal movement and normal muscle strength. Physiological reflex exists, but pathological reflex is not induced*
Identification of disease	*Acute stage of COPD*

utilized early stopping until the cross entropy starts to converge. To update the model weights and optimize the performance, backpropagation and the Adam update rule are employed, which is a gradient-based optimization technique [20]. The hyper parameters of our model are listed in Table 2.

A number of experiments were conducted to evaluate the performance of our model. First, we compared the performance of the proposed model with four baseline models: Fast text classification, CNN, TextBiRNN (Feed-Forward Networks with Attention). For both the proposed and the baseline models, we employed the five-fold cross-validation strategy on 80% of the data for model training, 10% of the data for validation test, the rest 10% for testing. The experiments were repeated ten times and the final performances were averaged on the ten repetitions.

Table 2. The hyper parameters of our enhanced RCNN model.

Parameter	Description	Value
d_w	Dimension of word embedding	100
l_r	Learning rate	0.001
B	Batch size	64
n	Number of epochs	10
K_l	kernel size	[1, 2, 3]

To test the stability of our model, the performance of our proposed model during training with different epochs is presented. As shown in Fig. 4, the model performs roughly the same on the training and test set during the training process. The train loss reached about 0.14, which converges around on the fifth epoch. The results show that the model has achieved relative stable performance during training after two epochs.

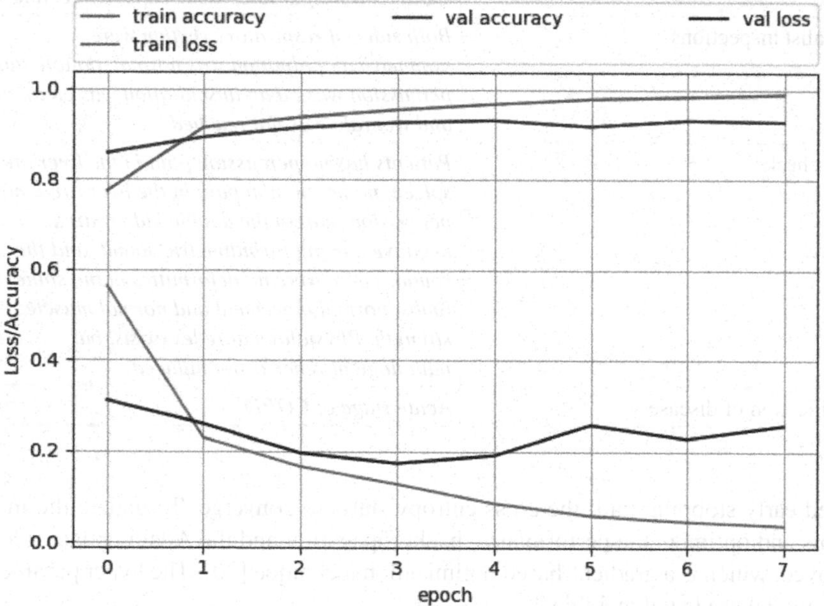

Fig. 4. The performance of our model in the training process on the training set and validation set.

3.4 Descriptive Analysis

The comparison result of all the models is presented in Table 3. TextRNN achieved an accuracy of 0.895, a precision of 0.900, a recall of 0.900, and a F1-score of 0.900.

CNN achieved improvements on every metric with an accuracy of 0.925, a precision of 0.932, a recall of 0.900, and a F1-score 0.904, the highest performance among the baseline methods. A comparison of the RCNN and our Enhanced RCNN model indicated the model improvement. The accuracy value had been improved from 0.893 to 0.932 (4.4%). The precision value had been improved from 0.892 to 0.912 (2.2%). The recall value had been improved from 0.900 to 0.943 (4.8%). The F1-score value, as a balanced metric demonstrated that our Enhanced RCNN model achieved the best performance compared with RCNN and all the other baseline methods.

Table 3. The performance comparison of all the models on the testing dataset

Models	Accuracy	Macro-P	Macro-R	Macro-F1
FastText	0.887	0.878	0.890	0.892
CNN	0.925	0.932	0.922	0.904
RCNN	0.893	0.892	0.900	0.900
TextRNN	0.895	0.900	0.900	0.900
Enhanced RCNN	0.932	0.912	0.943	0.920

4 Conclusions

COPD is a multifaceted respiratory disorder that poses significant economic burdens on both patients and society. In this study, we present an enhanced recurrent convolutional neural network model to enhance the prediction accuracy of disease stages in COPD patients. Our model's performance was evaluated using a real-world clinical dataset comprising approximately 5,108 COPD patients. The experimental results showcased the superiority of our proposed model over the baseline models across all evaluation metrics. These findings underscore the remarkable potential of our model in accurately predicting the disease stages of COPD patients, highlighting its significant contribution to improving clinical outcomes and patient care.

References

1. Nici, L., Donner, C., Wouters, E., et al.: American thoracic society/European respiratory society statement on pulmonary rehabilitation. Am. J. Respir. Crit. Care Med., 1390–1413 (2016)
2. Adeloye, D., Song, P., Zhu, Y., et al.: Global, regional, and national prevalence of, and risk factors for, chronic obstructive pulmonary disease (COPD) in 2019: a systematic review and modelling analysis. Lancet Respir. Med., 447–458 (2022)
3. Vogelmeier, C.F., Criner, G.J., Martinez, F.J., et al.: Global strategy for the diagnosis, management, and prevention of chronic obstructive lung disease 2017 report: GOLD executive summary. Am. J. Respir. Crit. Care Med., 557–582 (2017)

4. Zeng, Q.T., Goryachev, S., Weiss, S., et al.: Extracting principal diagnosis, co-morbidity and smoking status for asthma research: evaluation of a natural language processing system. BMC Med. Inform. Decis. Mak., 1–12 (2010)
5. Liang, Z., Zhang, G., Huang, J.X., Hu, Q.V.: Deep learning for healthcare decision making with EMRs. In: IEEE International Conference on Bioinformatics and Biomedicine, pp. 546–559(2014)
6. Rabe, K.F., Hurd, S., Anzueto, A., et al.: Global strategy for the diagnosis, management, and prevention of chronic obstructive pulmonary disease: GOLD executive summary. Am. J. Respir. Crit. Care Med., 532–555 (2017)
7. Behara, R., Agarwal, A., Fatteh, F., Furht, B.: Predicting hospital readmission risk for COPD using EHR information. In: Handbook of Medical and Healthcare Technologies, pp. 297–308 (2013)
8. Lin, S., Zhang, Q., Chen, F., et al.: Smooth Bayesian network model for the prediction of future high-cost patients with COPD. Int. J. Med. Inform., 147–155 (2019)
9. Wang, C., Chen, X., Du, L., et al.: Comparison of machine learning algorithms for the identification of acute exacerbations in chronic obstructive pulmonary disease. Comput. Methods programs Biomed., 128–156 (2020)
10. Pu, Y., Zhou, X., Zhang, D., et al.: Re-defining high risk COPD with parameter response mapping based on machine learning models. Int. J. Chronic Obstr. Pulm. Dis., 2471–2483 (2022)
11. Lai, S., Xu, L., Liu K., Zhao, J.: Recurrent convolutional neural networks for text classification. In: Proceedings of the AAAI Conference on Artificial Intelligence, pp. 29–36 (2016)
12. Wei, Z., Miao, D., Chauchat J.H., et al.: N-grams based feature selection and text representation for Chinese text classification. Int. J. Comput. Intell. Syst., 365–374 (2012)
13. Chen, X., Xu, L., Liu, Z., et al.: Joint learning of character and word embedding. In: Proceedings of the 24th International Conference on Artificial Intelligence, pp. 1236–1242, July 2015
14. Collobert, R., Weston, J., Bottou, L., et al.: Natural language processing (almost) from scratch. J. Mach. Learn. Res., 2493–2537 (2012)
15. Glorot, X., Bordes A., Bengio, Y.: Deep sparse rectifier neural networks. In: JMLR Workshop and Conference Proceedings, pp. 315–323 (2011)
16. Lin, M., Chen, Q., Yan, S.: Network in network. ArXiv Preprint ArXiv: 1312.4400 (2013)
17. Joulin, A., Grave, E., Bojanowski, P., Mikolov, T.: Bag of tricks for efficient text classification. ArXiv Preprint ArXiv: 1607.01759 (2016)
18. Tang, D., Qin, B., Liu, T.: Document modeling with gated recurrent neural network for sentiment classification. In: Proceedings of the 2015 Conference on Empirical Methods in Natural Language Processing, pp. 1422–1432 (2015)
19. Kalchbrenner, N., Grefenstette, E., Blunsom, P.: A convolutional neural network for modelling sentences. ArXiv Preprint ArXiv: 1404.2188 (2014)
20. Duchi, J., Hazan, E., Singer, Y.: Adaptive subgradient methods for online learning and stochastic optimization. J. Mach. Learn. Res., 13–27 (2014)

Hybrid Recommendation System with Graph Neural Collaborative Filtering and Local Self-attention Mechanism

Ao Zhang[1], Yifei Sun[1](✉), Shi Cheng[2], Jie Yang[1], Xin Sun[1], Zhuo Liu[1], Yifei Cao[1], Jiale Ju[1], and Wenya Shi[1]

[1] School of Physics and Information Technology, Shaanxi Normal University, Xi'an, China
{aozhang,yifeis,jieyang2021,sunxin_,zhuoliu,yifeic,jujiale, wenyas}@snnu.edu.cn

[2] School of Computer Science, Shaanxi Normal University, Xi'an, China
cheng@snnu.edu.cn

Abstract. As the problem of information overload becomes increasingly serious, traditional recommendation algorithms are difficult to efficiently provide assistance to users. In this context, more and more scholars have applied deep learning methods to recommendation systems, such as deep reinforcement learning, self-attention mechanisms, graph neural collaborative filtering and so on. Experimental results have shown that these methods significantly improve the performance of systems. However, most studies directly apply existing models to recommendation algorithms, wasting a lot of potential features. In recent years, the latest studies have optimized model performance through method fusion. In this paper, we propose a hybrid model NGCF-A that combines graph neural collaborative filtering and self-attention mechanisms. In a system based on graph neural collaborative filtering, local self-attention layers are added to embedding propagation layer, so that we can capturing high level associations between users and items. In addition, we add an implicit feedback transformation method and a dataset splitting strategy that conforms to change in interest, aiming to make better use of features. The experimental results show that this model achieves better performance than traditional graph neural collaborative filtering model, and the new implicit feedback transformation method further optimizes the model.

Keywords: Graph neural collaborative filtering · Local Self-Attention mechanism · Implicit feedback transformation method

1 Introduction

Recommendation system is a technology that provides personalized services to users by analyzing their interactive behavior, preferences, interests and other information [1–3]. It is widely used in e-commerce, social networking, short video recommendation and other fields [4–6]. Traditional collaborative filtering algorithms are based on the idea of birds of a feather flock together, and recommend similar users or similar items

H. Zhang et al. (Eds.): NCAA 2023, CCIS 1870, pp. 191–201, 2023.
https://doi.org/10.1007/978-981-99-5847-4_14

by calculating similarity [7, 8]. However, collaborative filtering algorithms are highly dependent on user interaction behavior. Therefore, in the absence of behavioral data for new users or new items, CF based methods often fall into a cold start problem and unable to effectively recommend [9, 10]. Existing users have only interacted with a small number of items, and sparse data has greatly reduced accuracy. In addition, some classic methods, such as content based recommendation and matrix decomposition [11–13], have different limitations and cannot meet actual needs. Therefore, it is necessary to make up for the shortcomings of traditional methods.

In the era of explosive growth in data volume and severe overload of information, more and more scholars have begun to apply deep learning methods to recommendation systems [14, 15], and have achieved good results. Khanduri et al. [16] proposed a hybrid recommendation system based on graph networks and collaborative filtering, using graph networks to analyze the strength of connections between nodes, while using collaborative filtering to find similar user preferences, and combining the two methods to obtain the recommendation list. Seng et al. [17] proposed a hybrid model that combines graph neural collaborative filtering and item temporal sequence relationships, and designed a sliding window strategy that divides the item temporal sequence into subsequence groups, deeply describing the dynamic changes in user interests, enhancing the ability to express interactive behavior. Abeer Aljohani et al. [18] proposed a GAT4Rec model based on self-attention mask learning, which uses auxiliary information to capture user interest change signals, filter out excellent subsequences of user historical interaction sequences, and help the model learn more effective long-term dependencies. EE YEO KEAT et al. [19] proposed a deep reinforcement learning model based on Deep Q-Network, and applied the model to solve multi-objective optimization problems, making up for some of the disadvantages of probabilistic multi-objective algorithms based on evolutionary algorithms, effectively improving the quality of recommendations.

This paper proposes a hybrid model, NGCF-A, which adds local self-attention mechanism to a graph neural collaborative filtering model to better capture the dependency relationships between nodes. In addition, in the data preprocessing stage, in order to better utilize existing data, we improves the implicit feedback transformation method. In traditional method, all rating data is often treated as 1 to convert explicit rating data into implicit interactions. In this case, even if the user gives a poor evaluation of a certain item, the model still considers this data as important information due to the presence of interaction records. Therefore, we have set a threshold for the scoring data and only consider ratings greater than the threshold as valid interaction records. Experimental results have shown that this has improved the quality of recommendations. In order to capture changes in interest, this paper also attempts to improve the dataset splitting strategy by dividing the training and testing sets based on timestamps to simulate real recommendation scenarios. However, this strategy did not achieve the desired results, and the reasons will be explained below.

The rest of this paper is organized as follows. The second section introduces the preliminary work. The third section introduces the specific details of the model. The fourth section introduces the experimental results of this model. The fifth section is the final conclusion.

2 Preliminaries

2.1 NGCF

Graph neural collaborative filtering (NGCF) is a recommendation model based on graph neural network (GNN). NGCF models the user-item interaction as a bipartite graph, and uses GNN to learn the low dimensional embedded representation of users and items. This method can capture the high-level connection between users and items, and improve the performance of recommendation results [20–22]. NGCF consists of three parts: embedding layer, embedding propagation layer, and prediction layer. Next, we will introduce their functions in detail.

Embedding Layer: The embedding layer is responsible for converting users and items into embedding vector inputs, the specific workflow will be explained below.

Embedding Propagation Layer: The embedding propagation layer is the core part of NGCF, used to perform graph convolution operations on user-item interaction graphs. At this layer, the initial embeddings of users and items are updated through multi-layer graph convolutional layers to achieve deep message transmission, thereby capturing the higher-order interrelationships between users and items. The output of the embedding propagation layer is a user and item embedding matrix updated through graph convolution operations.

Prediction Layer: The function of the prediction layer is to generate a prediction recommendation list based on the updated user and item embeddings. It calculates the inner product between the user vector and the item vector, and maps it to the rating space to obtain a predicted rating.

2.2 Local Self-attention

Local self-attention is a method that only focuses on the local context of the input sequence [23–25]. Unlike global self-attention, local self-attention does not calculate the relationships between all elements in the sequence, but only focuses on a small number of neighboring elements around each element. This method can reduce computational complexity and memory requirements, making the self-attention mechanism more suitable for processing long sequence data. In local self-attention, for each element in the sequence, we only focus on the elements within a fixed size window around it.

3 Proposed Model

In this section, we will specifically introduce the proposed graph neural collaborative filtering recommendation model that integrates local self-attention. The flowchart of this model is shown in Fig. 1, which will be introduced one by one.

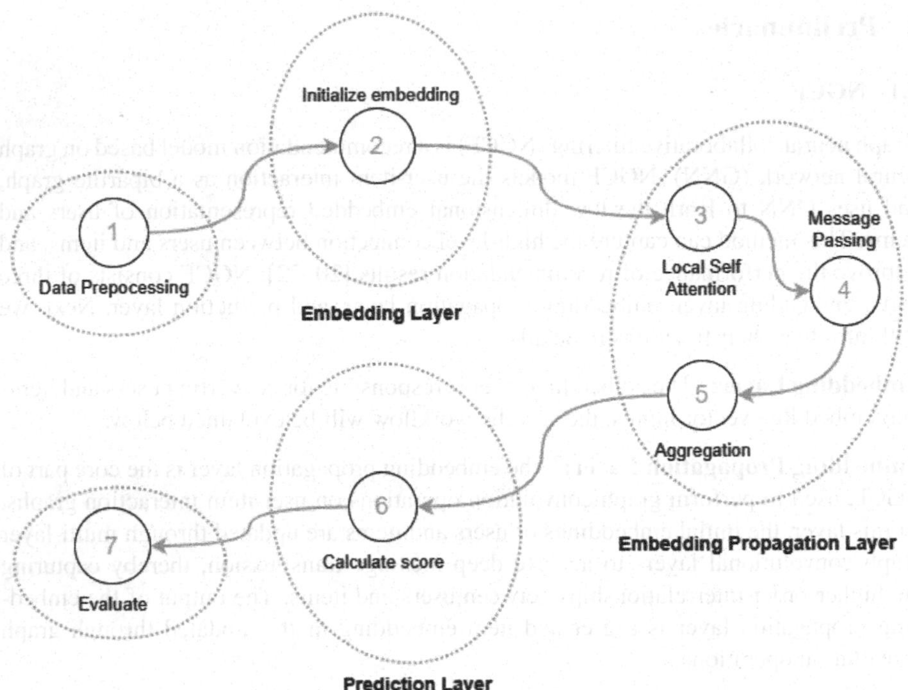

Fig. 1. The flowchart of NGCF-A model.

3.1 Data Preprocessing

In the data preprocessing stage, NGCF often discards timestamp data, which makes it difficult for the model to capture dynamic changes in user interests through timestamps. Therefore, we have added a new dataset splitting strategy to the model, dividing the earlier 80% of timestamp data into training set and the later 20% into test set to better simulate actual recommendation scenarios.

In addition, when processing rating data, NGCF will convert all existing rating data to 1, converting explicit feedback into implicit feedback, and transforming regression problems into binary classification problems, making the model easier to learn and able to capture potential user interests. However, this method ignores the user's preference for the item, and item which be rated 1 will receive the same importance as item which be rated 5 simply because they have interaction records. Therefore, we have added a new implicit feedback transformation method, setting a threshold of 3 for the rating data, treating rating data above the threshold as positive samples and rating data below the threshold as negative samples, enabling the model to better understand users' interests and item characteristics, thereby recommending more suitable items for users.

Next, we use the processed user-time interaction data to build a Laplacian matrix. First, a adjacency matrix A is generated. The rows of the matrix represent users and the columns represent items. If the user's score on an item is greater than the threshold, the corresponding matrix element is 1, otherwise it is 0. Next, a degree matrix D is generated, which is a diagonal matrix. The diagonal elements represent the number of connections between each node and other nodes. If the node is a user, this element represents the number of items interacting with that user. If the node is a item, this element represents the number of users interacting with this item. Finally, Laplacian matrix L is calculated by adjacency matrix A and degree matrix D. There are two methods: undirected graph and normalization. We usually choose normalized Laplacian matrix, as shown in Formula 1:

$$L = D^{-\frac{1}{2}} * A * D^{-\frac{1}{2}} \tag{1}$$

3.2 Embedding Layer

After data preprocessing, the embedding layer randomly initializes the initial embedding vectors of users and items, generating an embedding vector matrix. In order to capture the structural information in the user-item interaction diagram, the embedded vector matrix is multiplied with the normalized Laplacian matrix L to obtain the embedded matrix C, which is the representation of users and items. Before input into the embedded propagation layer, we apply the node dropout operation to matrix C to avoid overfitting.

3.3 Embedding Propagetion Layer

Embedding propagation layer is a core step in NGCF, and the local self-attention mechanism is also applied to this layer. After being processed by the embedding layer, the user-item interaction data is represented as a bipartite graph, in which nodes represent users and items, while edges represent interactions between users and items. Each edge has a weight that represents the user's level of interest in the item. The key steps in embedding propagation layer are message passing and aggregation. The message passing mechanism is shown in Formula 2:

$$m_{u \leftarrow i} = \frac{1}{\sqrt{|N_u||N_i|}}(w_1 e_i + w_2(e_i \cdot e_u)) \tag{2}$$

where $m_{u \leftarrow i}$ represents the message to be transmitted, e_u and e_i denote the embedded user and item features, w denote the weight of the message transmission, N_u and N_i denote normalization coefficients. As the propagation path length increases, the message gradually decays. This step obtains neighborhood information by calculating the similarity of the inner product between e_u and e_i, and then integrates its own node information for transmission back.

The message aggregation mechanism is shown in Formula 3:

$$e_u = LeakyReLU\left(m_{u \leftarrow u} + \sum_{i \in N_u} m_{u \leftarrow i}\right) \tag{3}$$

where $m_{u \leftarrow u}$ is the information of its own node, and $LeakyReLU$ denotes the activation function. After this step, the message propagation of user's neighborhood is completed, the user node is updated, and the first level transmission and aggregation are completed.

In the stage of message transmission and aggregation, each node collects the information of its neighbor nodes, and aggregates them through the nonlinear activation function to get the updated node representation. At this stage, we added a local self-attention mechanism to calculate the attention score of neighboring node information, enabling nodes to aggregate based on the importance of neighboring nodes, thereby better capturing the dependency relationships between nodes. The specific calculation steps are as follows: first, for each node, locate the neighboring nodes within the fixed size window around it. In our work, the window size is set to 5. Next, calculate the attention weight between adjacent nodes and the current node, and implement it by calculating the dot product and applying the Softmax function. Finally, the calculated attention weights are used to weight and sum the messages of neighboring nodes, and the weighted scores are passed back.

We used a three-layer embedding propagation layer to capture higher-order interrelationships between users and items. Both layers performed local self-attention calculations, message passing, and message aggregation operations, outputting an updated user and item embedding matrix.

3.4 Prediction Layer

The function of the prediction layer is to generate recommendation predictions based on updated user and item embeddings. The specific process is: the prediction layer calculates the inner product between user embedding and item embedding, and maps it to the rating space to obtain the predicted score. Predictive ratings can be used to sort items and generate recommendation lists for each user.

3.5 Evaluate

After the model training is completed and the prediction layer completes the scoring prediction, we output the results to the test set for evaluation. The specific evaluation metrics will be provided in the following text.

4 Experiments

This paper selects the Movielens latest small dataset to evaluate recommendation quality, which includes 100837 ratings from 610 users on 9742 movies. Movielens is a well-known film rating dataset in the industry which includes features such as user occupation and film category. The model proposed in this paper can also be used in fields such as music, news, short video and e-commerce recommendation. After considering the sparsity of the dataset and the form of rating data, we chose the classic Movielens dataset.

4.1 Evaluation Metrics

To measure the performance of the recommendation model, we used five evaluation metrics: *Epoch*, *NDCG@K*, *HR@K*, *Recall@K*, *Precision@K*, the details are as follows:

$$NDCG@K = \frac{DCG@K}{IDCG@K} \tag{4}$$

$$HR@K = \frac{\sum_{i=1}^{K} hit(i)}{K} \tag{5}$$

$$\mathrm{Re}call@K = \frac{TP}{TP + FN} \tag{6}$$

$$\Pr ecision@K = \frac{TP}{TP + FP} \tag{7}$$

where *Epoch* means how many generations of operations have passed to stop training, *K* denotes the length of the recommendation list. In our work, when the difference in *BPR Loss* between two epochs is less than 0.00001, the model ends training. *NDCG* denotes normalized discounted cumulative gain, and *HR* denotes hit ratio. These two metrics are widely used to measure the performance of recommendation models. In addition, in order to analyze the impact of recommendation list length *K* on recommendation results, we set *K* values to 5, 10, and 20, respectively. Some other parameter settings are shown in Table 1.

Table 1. Parameter settings for the model.

Parameter	Parameter value
Embedding size	64
Batch size	256
Dropout ratio	0.1
Window size	5
Layer size	3

4.2 Results

We compared the performance of four algorithms on a given dataset. They are NGCF (graph neural collaborative filtering), NGCF-A (NGCF model incorporating local self-attention mechanism), NGCF-A-I (a NGCF-A model with implicit feedback transformation method added), and NGCF-A-T (a NGCF-A model with timestamp segmentation method added). We have drawn a comparison chart of four metrics except for *Epoch*, and the performance of the algorithm on *Epoch* is directly presented in the following text. We also considered the impact of recommendation list length *K* on the results. The experimental results are shown in Fig. 2. Due to the poor performance of NGCF-A-T in most metrics, we did not draw it in the figure. The complete data will be provided below.

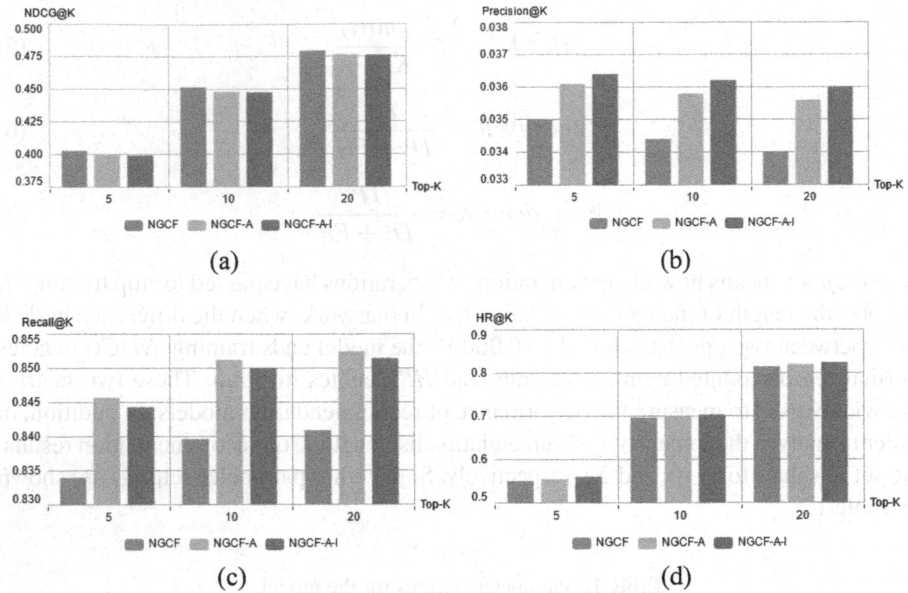

Fig. 2. Performance of different algorithms on different metrics.

4.3 Analyse

Based on the experimental results, we can observe that the convergence speed of the three improved methods we proposed is much higher than that of the traditional NGCF model. As the length of the recommendation list *K* increases, all algorithms have achieved better results. The reason for it may be that the recommendation list length is too short, and the model tends to recommend popular items while ignoring the diversity of user interests. Therefore, as the length of the list increases, the true interests of users are discovered, and all metrics are improved.

For the other four metrics, NGCF only achieved better performance on *NDCG* and lagged behind our proposed model in other metrics. This indicates that the model NGCF-A proposed in this paper, has better performance than NGCF. After replacing the new implicit feedback conversion method, the performance of the NGCF-A-I model has been further improved. However, after applying the method of dividing the dataset by timestamp, NGCF-A-T performs poorly. The possible reason is that dividing the dataset by timestamp makes the dataset more sparse, and it is difficult to capture user interest changes through simple methods, making it difficult for the model to fully capture user long-term and short-term interests. The complete experimental data is shown in Table 2.

Specifically, NGCF outperforms other models by about 0.8% in *NDCG*, NGCF-A-I outperforms NGCF by about 5% in *Precision* and 0.4% in *HR*, and NGCF-A outperforms NGCF by about 1.4% in *Recall*. In summary, NGCF only achieved a slight lead in *NDCG* and has poor overall performance. Although NGCF-A-I performs slightly lower in *Recall* than NGCF-A, they only differ by 0.15%.

Table 2. Complete experimental results.

Top-k	Metrics	NGCF	NGCF-A	NGCF-A-I	NGCF-A-T
5	Epoch	65	53	54	**51**
	NDCG	**0.4028**	0.3996	0.3992	0.2098
	HR	0.5502	0.5523	**0.5527**	0.3231
	Precision	0.0350	0.0361	**0.0364**	0.0220
	Recall	0.8340	**0.8457**	0.8443	0.6889
10	Epoch	63	52	52	**50**
	NDCG	**0.4517**	0.4481	0.4476	0.2587
	HR	0.6963	0.6986	**0.6989**	0.4695
	Precision	0.0344	0.0358	**0.0362**	0.0218
	Recall	0.8396	**0.8512**	0.8501	0.6943
20	Epoch	62	53	53	**51**
	NDCG	**0.4793**	0.4764	0.4760	0.2856
	HR	0.8143	0.8161	**0.8165**	0.5884
	Precision	0.0340	0.0356	**0.0360**	0.0217
	Recall	0.8409	**0.8524**	0.8514	0.6954

In fact, the time required for the model to run is equally important. In real life, slow running time can greatly reduce work efficiency. Due to the addition of timestamp based dataset partitioning in NGCF-A-T, more time is required during the data preprocessing. For other methods, NGCF-A-I filters some negative data and runs faster than NGCF-A. In summary, NGCF-A-I achieved the best performance in our experiment.

5 Conclusion

In this paper, we propose a graph neural collaborative filtering model that integrates the local self-attention mechanism (NGCF-A). In the embedding propagation layer stage, we add a local self-attention mechanism, which allows neighboring nodes of a node to have different importance scores. The hybrid model effective improve recommendation accuracy.

In addition, we propose two improvement strategies based on NGCF-A. Experiments show that the model with new implicit feedback transformation method, NGCF-A-I, achieves the best performance. The new implicit feedback method filters out some items that do not meet user interests, although it increases sparsity, the model can better capture user real preferences.

NGCF-A-T did not achieve the desired results, but time series is also an important feature in recommendation scenarios. So our future work aims to better utilize time data. We consider to use more complex partitioning strategies, such as sliding window methods or user based partitioning methods. So that we can learn more about the dynamic evolution of user interests. In addition, we want to apply the model we proposed to other fields, such as music recommendation and commerce recommendation. In this work, we need to adjust the model to improve its applicability to different scenarios.

Acknowledgement. This work was supported by the Natural Science Basic Research Plan in Shaanxi Province of China (Program No. 2022JM-381,2017JQ6070) National Natural Science Foundation of China (Grant No. 61703256), Foundation of State Key Laboratory of Public Big Data (No. PBD2022-08) and the Fundamental Research Funds for the Central Universities (Program No. GK202201014, GK202202003, GK201803020).

References

1. Shang, M.-S.. Fu, Y., Chen, D.-B.: Personal recommendation using weighted bipartite graph projection. In: 2008 International Conference on Apperceiving Computing and Intelligence Analysis, pp. 198–202 (2008)
2. Tong, L.: Personal recommendation based on community partition of bipartite network. In: 2015 International Conference on Cloud Computing and Big Data (CCBD), pp. 336–341 (2015)
3. Sun, J., Zhu, Z., Wang, Y.: Research on personalized recommendation case organization. In: 2010 International Conference on Innovative Computing and Communication and 2010 Asia-Pacific Conference on Information Technology and Ocean Engineering, pp. 312–315 (2010)
4. Yan, L.: Personalized recommendation method for e-commerce platform based on data mining technology. In: 2017 International Conference on Smart Grid and Electrical Automation (ICSGEA), pp. 514–517 (2017)
5. Li, X.: Research on the application of collaborative filtering algorithm in mobile e-commerce recommendation system. In: 2021 IEEE Asia-Pacific Conference on Image Processing, Electronics and Computers (IPEC), pp. 924–926 (2021)
6. Xiaona, Z.: Personalized recommendation model for mobile e-commerce users. In: 2021 13th International Conference on Measuring Technology and Mechatronics Automation (ICMTMA), pp. 707–710 (2021)
7. Chen, Y.-W., Xia, X., Shi, Y.-G.: A collaborative filtering recommendation algorithm based on contents' genome. In: IET International Conference on Information Science and Control Engineering 2012 (ICISCE 2012), pp. 1–4 (2012)
8. Shrivastava, N., Gupta, S.: Analysis on item-based and user-based collaborative filtering for movie recommendation system. In: 2021 5th International Conference on Electrical, Electronics, Communication, Computer Technologies and Optimization Techniques (ICEECCOT), pp. 654–656 (2021)
9. Embarak, O.H.: A method for solving the cold start problem in recommendation systems. In: 2011 International Conference on Innovations in Information Technology, pp. 238–243 (2011)
10. Gaspar, P., Kompan, M., Koncal, M., Bielikova, M.: Improving the personalized recommendation in the cold-start scenarios. In: 2019 IEEE International Conference on Data Science and Advanced Analytics (DSAA), pp. 606–607 (2019)
11. Paireekreng, W.: Mobile content recommendation system for re-visiting user using content-based filtering and client-side user profile. In: 2013 International Conference on Machine Learning and Cybernetics, pp. 1655–1660 (2013)

12. Pan, M., Yang, Y., Mi, Z.: Research on an extended SVD recommendation algorithm based on user's neighbor model. In: 2016 7th IEEE International Conference on Software Engineering and Service Science (ICSESS), pp. 81–84 (2016)
13. Shah, K., Salunke, A., Dongare, S., Antala, K.: Recommender systems: an overview of different approaches to recommendations. In: 2017 International Conference on Innovations in Information, Embedded and Communication Systems (ICIIECS), pp. 1–4 (2017)
14. Huang, G.: E-commerce intelligent recommendation system based on deep learning. In: 2022 IEEE Asia-Pacific Conference on Image Processing, Electronics and Computers (IPEC), pp. 1154–1157 (2022)
15. Zarzour, H., Alsmirat, M., Jararweh, Y.: Using deep learning for positive reviews prediction in explainable recommendation systems. In: 2022 13th International Conference on Information and Communication Systems (ICICS), pp. 358–362 (2022)
16. Khanduri, S., Prabakeran, S.: Hybrid recommendation system with graph based and collaborative filtering recommendation systems. In: 2022 IEEE 2nd Mysore Sub Section International Conference (MysuruCon), pp. 1–7 (2022)
17. Seng, D., Li, M., Zhang, X., Wang, J.: Research on neural graph collaborative filtering recommendation model fused with item temporal sequence relationships. IEEE Access **10**, 116972–116981 (2022)
18. Aljohani, A., Rakrouki, M.A., Alharbe, N., Alluhaibi, R.: A self-attention mask learning-based recommendation system. IEEE Access **10**, pp. 93017–93028 (2022)
19. Keat, E.Y., et al.: Multiobjective deep reinforcement learning for recommendation systems. IEEE Access **10**, 65011–65027 (2022)
20. Hou, Y.: Application of neural graph collaborative filtering in movie recommendation system. In: 2021 IEEE International Conference on Electronic Technology, Communication and Information (ICETCI), pp. 113–116 (2021)
21. Sangeetha, M., et al.: Predicting personalized recommendations using GNN. In: 2022 6th International Conference on Computing Methodologies and Communication (ICCMC), pp. 228–234 (2022)
22. Liang, Z., Ding, H., Fu, W.: A survey on graph neural networks for recommendation. In: 2021 International Conference on Culture-oriented Science & Technology (ICCST), pp. 383–386 (2021)
23. Zhao, J., Zhao, P., Zhao, L., Liu, Y., Sheng, V.S., Zhou, X.: Variational self-attention network for sequential recommendation. In: 2021 IEEE 37th International Conference on Data Engineering (ICDE), pp. 1559–1570 (2021)
24. Yin, Y., Huang, C., Sun, J., Huang, F.: Multi-head self-attention recommendation model based on feature interaction enhancement. In: ICC 2022 - IEEE International Conference on Communications, pp. 1740–1745 (2022)
25. Shi, X., Xu, M., Hu, J.: Long and short-term neural network news recommendation model based on self-attention mechanism. In: 2021 International Conference on Computer Information Science and Artificial Intelligence (CISAI), pp. 30–34 (2021)

MAMF: A Multi-Level Attention-Based Multimodal Fusion Model for Medical Visual Question Answering

Shaopei Long[1], Zhenguo Yang[2], Yong Li[1], Xiaobo Qian[1], Kun Zeng[3], and Tianyong Hao[1(✉)]

[1] School of Computer Science, South China Normal University, Guangzhou, China
{shaopei-lauv,lycutter,haoty}@m.scnu.edu.cn

[2] School of Computer Science, Guangdong University of Technology, Guangzhou, China
yzg@gdut.edu.cn

[3] School of Computer Science, Sun Yat-Sen University, Guangzhou, China
zengkun2@mail.sysu.edu.cn

Abstract. Medical Visual Question Answering (VQA) targets at accurately answering clinical questions about images. The existing medical VQA models show great potential, but most of them ignore the influence of word-level fine-grained features which benefit filtering out irrelevant regions in medical images more precisely. We present a Multi-level Attention-based Multimodal Fusion model named MAMF, aiming at learning a multi-level multimodal semantic representation for medical VQA. First, we develop a Word-to-Image attention and a Sentence-to-Image attention to obtain the correlations of word embeddings and question feature to image feature. In addition, we propose an attention alignment loss which contributes to adjust the weights of image regions gained from word embeddings and question feature to emphasize relevant regions for improving the quality of predicted answers. Results on VQA-RAD and PathVQA datasets suggest that our MAMF significantly outperforms the related state-of-the-art baselines.

Keywords: Medical Visual Question Answering · Multimodal fusion · Attention mechanism · Deep learning · Medical image

1 Introduction

Visual Question Answering (VQA) has obtained extensive attention from numerous scholars dedicated to research Computer Vision (CV) [1, 2] or Natural Language Processing (NLP) [3, 4] in the past few years. As a specific domain of VQA, the purpose of medical VQA is to answer diagnostically a question asked on a medical image. An outstanding medical VQA model can profit both clinicians and sick person. It can provide subsidiary analysis for clinical diagnoses and therapeutics for doctors. In addition, a Medical-VQA system helps ask for medical consultation whenever patients need. Therefore, developing a medical VQA model helps relieve the burden of healthcare

H. Zhang et al. (Eds.): NCAA 2023, CCIS 1870, pp. 202–214, 2023.
https://doi.org/10.1007/978-981-99-5847-4_15

and make medical diagnoses and treatment more efficient. Although medical VQA has tremendous potential, researches on medical VQA still face many challenges. Compared with general VQA, medical VQA is more challenging. In the foremost, well-annotated medical VQA datasets for training model are extraordinarily rare, since they are time-consuming and strenuous to gain precise annotations by clinicians. For example, the manually annotated dataset VQA-RAD [5] includes varied types of questions but it contains only 315 radioactive pictures. Furthermore, some general VQA models cannot be adopted to develop Medical-VQA systems. The reason is that they always utilize extremely complex visual feature extraction modules such as Faster R-CNN [6] and ResNet-101 [7], which included a great deal of arguments and demanded to be trained with large datasets. The direct employment of these models may result in the overfitting issue. Furthermore, clinical questions are not only harder to be understanded for the VQA system as they are about professional medical knowledge, but also needed to be answered precisely as they are relevant to safety and health.

Some previous works [8, 9] attempted to utilize general VQA models and fine-tuned them on Medical-VQA datasets. Nevertheless, medical images and clinical questions were quite different from those of general VQA. Raghu et al. [10] proposed to transfer knowledge from general VQA, but they gained a subtle improvement. Nguyen et al. [11] employed Model-Agnostic Meta-Learning (MAML) [12] to obtain weights of the visual feature extractor. In addition, they utilized Convolutional Denoising Auto-Encoder (CDAE) [2] to make model more robust. Though these groundbreaking medical VQA works pushed forward the research field, they only focused on making better the feature extractor, while ignored inference module. Zhan et al. [13] concentrated on enhancing the inference ability of models. Specifically, they devised a Question-Conditioned Reasoning (QCR) module to identify the importance of each word. Besides, they proposed a task-conditioned reasoning (TCR) strategy to enlarge the difference of reference abilities for close-ended and open-ended tasks accordingly. Nevertheless, owing to the limitation of medical data, it can only obtain rough fusion features. Li et al. [14] designed two reasoning modules to obtain fine-grained relations between words and image regions. But they ignored the relationships between Word-to-Image attention and Sentence-to-Image attention, which make them unable to gain more fine-grained semantic information.

In order to gain a multi-level multimodal fusion feature, we design a Multi-level Attention-based Multimodal Fusion (MAMF) model by developing a Word-to-Image (W2I) attention and a Sentence-to-Image (S2I) to model the relations of both word embeddings and question feature to the image feature for medical VQA. The W2I attention is adopted to word-level fine-grained reasoning, while the S2I attention is applied to sentence-level coarse-grained reasoning. Besides, we propose an Attention Alignment Loss (AAL) to concentrate on adjusting the weights of the image regions learned from word embeddings and question feature to lay stress on crucial image regions and obtain multi-level multimodal semantic representation to predict the high-quality answer.

To sum up, our contributions are as follows:

1) A novel Multi-level Attention-based Multimodal Fusion (MAMF) model is proposed by developing a Word-to-Image (W2I) attention and a Sentence-to-Image (S2I) attention to capture word-level and sentence-level inter-modality relations of them, as well as to learn a multi-level multimodal semantic representation for medical VQA.

2) An Attention Alignment Loss (AAL) is designed to adjust the importance of the image regions obtained from word embeddings and question feature to identify the relevant and crucial image regions.
3) The evaluations on VQA-Rad and PathVQA datasets show that our proposed MAMF significantly superior to the related state-of-the-art baselines.

2 Related Work

VQA has aroused great research interest among scholars since Antol et al. [15] proposed the first VQA task. VQA models in general domain adopted various methods for extracting image feature and question feature. As for image feature extraction module, researchers commonly utilized object detectors like simple CNNs [16], SSD [17], and Faster-RCNN [6]. As for question feature extractor, they usually adopted models like GTP-3 [12], Bert [3] and RoBerta [18]. After that, the extracted features were aggregated by using bilinear pooling model like Multimodal Compact Bilinear Pooling [19], Multimodal Low-rank Bilinear Pooling [20] or Bilinear Attention Network (BAN) [21] to obtain a fusion feature. The feature was transmitted to the classifier to predict the answer.

However, these models could not be simply adopted to develop a Medical-VQA system, owing to the limitation of medical data. Therefore, Nguyen et al. [11] utilized a meta-learning algorithm MAML [12] and CDAE [2] to obtain weight initialization of visual feature extractor to learn visual features. Do et al. [22] proposed a multiple meta-model quantifying (MMQ) algorithm to learn meta-annotation. Nevertheless, they ignored the reasoning module of the models, which led to limit their performances. Consequently, Zhan et al. proposed a question-conditioned reasoning (QCR) module to adjust the weights of words and task-conditioned reasoning (TCR) method to learn inference abilities for close-ended tasks and open-ended tasks respectively. Gong et al. [23] designed a novel multi-task learning paradigm. However, this needed large-scale medical data. Bo et al. [24] adopted contrastive learning to gain several cumbersome models and train an unsophisticated student model by distilling these models and fine-tuning on VQA-RAD dataset.

Various attention mechanisms were also adopted in the medical VQA field. Vu et al. [25] proposed a multi-glance attention method to obtain the most related image regions. Sharma et al. [26] proposed a MedFuseNet to utilize a co-attention mechanism to improve the quality of fusion feature. However, these previous works neglected to learn multi-level multimodal feature representations which limited their performance. In this paper, we develop a Word-to-Image (W2I) attention and a Sentence-to-Image (S2I) attention to concentrate on learning a multi-level multimodal semantic representation.

3 Methods

3.1 Problem Formulation

Medical VQA is defined as a multiclassification problem. Given an image I and a question q, the output is the predicted answer $\hat{a}$. The both I and q are input into model f to obtain the predicted answer:

$$\hat{a} = \arg\max_{a \in A} f(a|I, q, \theta), \tag{1}$$

where A and a denote candidate answers and one of them, separately, and θ denotes all parameters.

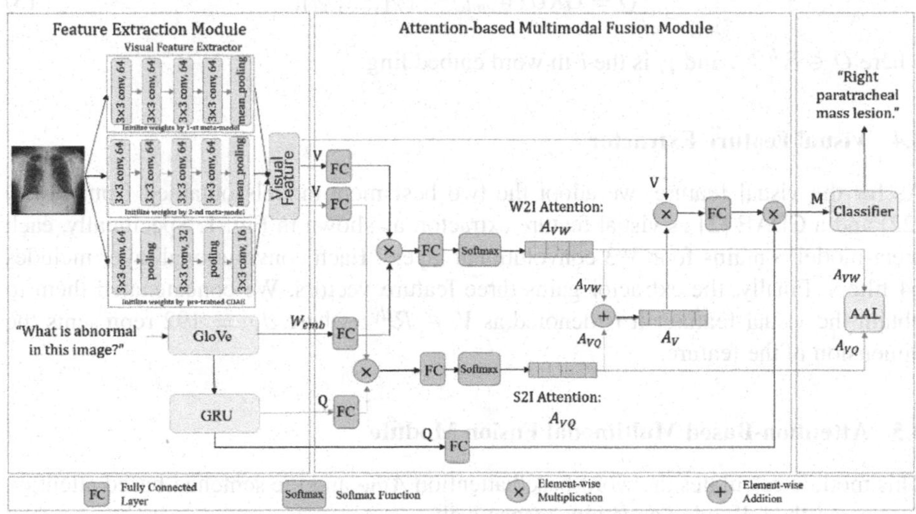

Fig. 1. Overview framework of our proposed MAMF. Each medical image gains three 64-D vector through a CDAE encoder and two meta-model. The vectors are concatenated to generate the visual feature V. GloVe and GRU are adopted to produce the word embedding sequence W_{emb} and the semantic feature Q. A_{VW} and A_{VQ} are W2I attention weight and S2I attention weight respectively, and M is a fusion feature.

3.2 Overview of Our Proposed Model

The structure of MAMF is shown in Fig. 1. Overall, the model includes a visual feature extractor, a word embedding module GloVe [27], a question embedding module GRU [28], an attention-based multimodal fusion module and a classifier. Glove is adopted to convert every word to a 300-dimension word. Then we utilized GRU to generates question feature. The visual feature extractor utilizes the Convolutional Denoising Auto-Encoder [2] and two meta-models obtained from Multiple Meta-model Quantifying (MMQ) [22]. The attention-based multimodal fusion module is adopted to model the relations between visual feature and word embeddings, and between visual feature and question feature, respectively. Finally, the classifier is adopted to classify multimodal semantic representations and then provide predicted answers to the Medical-VQA tasks.

3.3 Word Embedding and Question Representation

In the foremost, given a question q who has l words, GloVe [27] is adopted to generate a word embedding sequence. $w_i \in \mathbb{R}^{d_w}$ express the i-th word vector:

$$W_{\text{emb}} = WordEmbedding(q) = [w_1, ..., w_l]. \tag{2}$$

The word embedding $W_{emb} \in \mathbb{R}^{d_w \times l}$ is then sent to Gated Recurrent Unit (GRU) [28] whose dimension is d_G to gain the semantic feature:

$$Q = GRU(W_{emb}) = [\gamma_1, ..., \gamma_l], \tag{3}$$

where $Q \in \mathbb{R}^{d_G \times l}$, and γ_i is the i-th word embedding.

3.4 Visual Feature Extractor

As for the visual feature, we adopt the two best meta-models obtained from MMQ [22] and a CDAE [2] as visual feature extractor, as shown in Fig. 1. Specifically, each meta-model contains four 3*3 convolutional layers. Each convolutional layer includes 64 filters. Finally, the extractor gains three feature vectors. We concatenated them to obtain the visual feature. It is denoted as $V \in \mathcal{R}^{d_V}$, where $d_V = 192$ represents the dimension of the feature.

3.5 Attention-Based Multimodal Fusion Module

This module calculates the word-based attention A_{VW} and the sentence-based attention A_{VQ} using the following equations respectively.

$$A_{VW} = \text{softmax}(l \times w_1 \times ((w_2 \times V) \circ (w_3 \times W_{\text{emb}})) + b), \tag{4}$$

$$A_{VQ} = \text{softmax}(l\prime \times w_1' \times ((w_2' \times V) \circ (w_3' \times Q)) + b'), \tag{5}$$

where l and w_x represent the weight matrix and a fully connected layer, respectively, and b denotes a scalar. Besides, $\circ$ indicates element-wise multiplication. The softmax functions in Eq. (4) and Eq. (5) are adopted to normalize the attention weights.

The attention weight of image feature is computed as:

$$A_V = A_{VQ} + A_{VW}. \tag{6}$$

The attention weight A_V and visual feature V are then element-wise multiplied to obtain the visual feature,

$$V' = A_V \circ V. \tag{7}$$

The visual feature and the question feature are both sent to fully connected layers. The vectors from the fully connection layers are element-wise multiplied together to obtain the joint embedding M. M is then sent to a classifier. The predicted answer $\hat{a}$ has the highest probability among the candidate answers. The accuracy is computed as follows:

$$Accuracy = \frac{1}{n_{Test}} \sum^{Test} (Onehot(\arg\max(\hat{a})) \cdot a), \tag{8}$$

where a denotes the correct answer of the task.

3.6 Loss Function

The predicted answers are utilized to obtain binary cross entropy loss during training,

$$L_{CE} = -\frac{1}{n_{Train}} \sum_{i=1}^{n_{Train}} (a \log(\hat{a}) - (1-a)\log(1-\hat{a})). \tag{9}$$

In addition, an Attention Alignment Loss (AAL) is proposed to align the word-based attention and the sentence-based attention to emphasize relevant and crucial image regions. The loss function is computed as follows:

$$L_{AAL} = -\frac{1}{n_{\text{Train}}} \sum_{i=1}^{n_{\text{Train}}} \left\| A_{VQ} - A_{VW} \right\|^2. \tag{10}$$

At last, the final loss function is calculated as follows:

$$\text{Loss} = \alpha L_{AAL} + L_{CE}, \tag{11}$$

where α is a weighting parameter.

4 Experiments

4.1 Datasets

The prevalent medical VQA datasets are adopted to evaluate our proposed MAMF: (1) VQA-RAD [5]: It contains 3,515 question-answer pairs and 315 radiology images. Some questions arc related to the same image. The clinicians or patients ask various questions about position, presence, organ and others. (2) PathVQA [29]: It contains 32,799 question-answer pairs, including "how", "what","where" and other types. There are 3,328 medical images obtained from the PEIR digital library and 1,670 pathological images selected from several medical literatures. The answer types of two datasets are classified as close-ended and open-ended. The close-ended answers are "yes/no" or several options, while the open-ended answers are free-form texts. The question-answer pairs of PathVQA dataset are generated by a semi-automated approach using image captions and then manually reviewed and modified by clinicians.

4.2 Experiment Settings

All experiments are performed on the Ubuntu 20.04.4 server with NVIDIA GTX 1080 GPU based on PyTorch library in version 1.8. We adopt Adam optimizer to optimize our model. The learning rate is set to 1e–4 and batch size is set to 128. For semantic textual features, each question contains 12 words. GloVe [27] is utilized to generate the word embeddings. They are input into GRU [28] to gain question feature. As for visual representations, each 128-dimensional image is input into 2 quantified meta-models obtained from the MMQ [22] and a Convolutional Denoising Auto-Encoder, which generates 3 vectors. The enhanced visual feature is produced by concatenating these vectors. We adopt accuracy, precision, recall and F1-score (denoted as Acc, P, R, F1) as evaluation metrics.

4.3 Baseline Models

The medical VQA baselines including MAML, BiAN, MEVF [11], MMQ [22], CR [13] and CMSA [26] are reimplemented by using the open-source codes. The brief descriptions of baselines are in Table 1.

Table 1. The brief descriptions of baseline models.

Models	Descriptions
MAML	It utilized model-agnostic meta-learning method to obtain semantic representations
BiAN	It utilized ImageNet [30] to initialize the weights of the visual feature extractor
MEVF	It adopted MAML [12] and CDAE [2] to extract visual feature, and then used BAN to fuse them with question features
MMQ	It designed a multiple meta-model quantifying module to utilize meta-annotation
CR	It adopted a QCR module to improve fusion feature and proposed a TCR strategy
CMSA	It introduced a Cross-Modal Self-Attention module to effectively obtain the crucial semantic information

4.4 Results

The results of our proposed MAMF and other baseline models on the VQA-RAD test set are shown in the Table 2. The results of baseline models are re-implemented using available codes. From the table, it suggests that MAMF significantly superior to other state-of-the-art baselines. MAMF gains the best overall accuracy 74.94%, precision 82.39%, recall 74.94% and F1-score 78.02%. As for close-ended tasks and open-ended tasks, we also achieve the best performances except precision of the open-ended. Although we utilize the MMQ methods to enhance our image feature extractor, the reason may be that our model reduces the prediction probability of the true positive samples during fusion stage. The tasks corresponding to open-ended questions are harder for medical VQA models to answer correctly, since their answers can be free-form text. However, our proposed model MAMF still outperforms other baselines benefitting from the W2I attention, S2I attention, and AAL.

We also perform experiments on PathVQA dataset. Compared with VQA-RAD dataset, PathVQA have more diversities. It can verify the robustness of our proposed MAMF. The result is shown in Table 3. Our proposed MAMF gains the best performances reaching the best accuracy 54.28%, precision 65.82%, recall 54.28% and F1-score 52.38% on the entire test set. MAMF obtains dramatic improvement on the open-ended questions compared with other baseline models. The reasons of this improvement are as follows: First, MAMF builds word-level correlation representation of word embeddings and image feature, which filters unrelated regions in the image and retains essential ones for predicting answer. Second, our proposed AAL aligns the attention weights of regions in the image learned from the W2I attention and Q2I attention to recognize essential words and image regions for reasoning.

Table 2. Results on the VQA-RAD.

Models	Overall (%)				Open-ended (%)				Closed-ended (%)			
	Acc	P	R	F1	Acc	P	R	F1	Acc	P	R	F1
MEVF	67.18	71.89	63.19	66.09	49.72	65.14	42.46	43.16	78.68	78.55	76.84	77.45
MMQ	71.80	82.17	72.06	75.71	60.90	**84.45**	61.45	61.71	79.01	81.08	79.04	80.39
CR	71.60	77.67	68.96	72.31	60.10	57.69	56.11	56.18	79.01	77.49	80.95	79.15
CMSA	73.17	79.73	73.17	75.35	61.45	73.17	61.45	60.71	80.88	82.38	80.88	81.46
MAMF	**74.94**	**82.39**	**74.94**	**74.94**	**65.36**	78.23	**65.36**	**65.81**	**81.25**	**83.63**	**81.25**	**82.28**

Table 3. Results on the PathVQA.

Models	Overall (%)				Open-ended (%)				Closed-ended (%)			
	Acc	P	R	F1	Acc	P	R	F1	Acc	P	R	F1
BiAN	35.60	37.32	35.60	37.39	2.90	0.40	2.90	0.06	68.20	82.46	68.20	79.12
MAML	42.90	45.87	42.90	46.32	5.90	7.57	5.90	8.17	79.50	84.57	79.50	84.49
MEVF	44.80	40.28	44.80	40.84	8.10	2.01	8.10	2.50	81.40	83.31	81.40	81.99
MMQ	48.80	45.14	48.80	45.36	13.40	7.51	13.40	7.61	84.00	83.76	84.00	83.51
MAMF	**54.28**	**65.82**	**54.28**	**52.38**	**22.49**	**46.12**	**22.49**	**18.93**	**85.75**	**85.87**	**85.75**	**85.78**

4.5 Ablation Study

Several ablation experiments are conducted to verify the effectiveness of each part of MAMF. The experiment results are shown in Table 4 and Table 5. We remove W2I attention, S2I attention and AAL successively. The performances of MAMF without W2I attention and MAMF without S2I attention datasets dramatically decreased compared with the complete form of MAMF. Without the W2I attention, the model cannot establish word-level correlations between the word embeddings and image feature. Thus, it can only use the coarse sentence-level multimodal semantic representations to roughly reason. Without the S2I attention, the model can neither properly understand the meaning of questions nor predict the high-quality answers. These two ablation instances show the effectiveness of the W2I attention and S2I attention. As for the model MAMF without AAL, it also obtains poor performances on the two datasets. As discussed in Sect. 3.5, AAL is used to align the W2I attention and S2I attention, which helps locate crucial image regions to optimize the model. Furthermore, the complete form of MAMF obtains the best performance. Consequently, our proposed MAMF gains a satisfactory performance that utilizes the W2I attention and S2I attention to obtain the multi-level semantic information of image from word-level feature and sentence-level feature in the question, respectively, and employs the AAL to maximize the similarity of the relevant regions obtained from the W2I and S2I attention respectively.

Table 4. Ablation experiments on the VQA-RAD.

Models	Overall (%)				Open-ended (%)				Closed-ended (%)			
	Acc	P	R	F1	Acc	P	R	F1	Acc	P	R	F1
MAMF w/o W2I	72.72	80.38	72.72	75.52	63.12	77.60	63.12	63.81	79.04	82.24	79.04	80.36
MAMF w/o S2I	72.28	80.45	72.28	75.17	60.33	75.05	60.33	61.10	80.14	82.90	80.14	81.34
MAMF w/o AAL	73.39	79.85	73.39	75.73	63.12	72.51	63.12	63.57	80.14	82.11	80.14	81.00
MAMF	**74.94**	**82.39**	**74.94**	**74.94**	**65.36**	**78.23**	**65.36**	**65.81**	**81.25**	**83.63**	**81.25**	**82.28**

Table 5. Ablation experiments on the PathVQA.

Models	Overall (%)				Open-ended (%)				Closed-ended (%)			
	Acc	P	R	F1	Acc	P	R	F1	Acc	P	R	F1
MAMF w/o W2I	52.17	51.14	52.19	50.88	18.67	17.50	18.67	16.53	85.37	85.44	85.37	85.39
MAMF w/o S2I	51.97	49.48	51.97	49.31	18.55	14.59	18.55	14.00	85.13	85.69	85.13	85.17
MAMF w/o AAL	51.82	49.92	51.82	50.25	18.37	15.16	18.37	15.60	85.04	85.25	85.04	85.08
MAMF	**54.28**	**65.82**	**54.28**	**52.38**	**22.49**	**46.12**	**22.49**	**18.93**	**85.75**	**85.87**	**85.75**	**85.78**

Table 6. α changes from 0 to 2.0 in Eq. (11).

Model	Type	α										
		0	0.2	0.4	0.6	0.8	1.0	1.2	1.4	1.6	1.8	2.0
MAMF	Open-ended (%)	63.12	63.13	63.10	60.33	63.12	63.88	**65.36**	63.70	61.50	61.66	60.10
	Closed-ended (%)	80.14	**81.99**	77.90	80.14	79.04	80.37	81.25	80.10	80.90	80.07	79.01
	Overall (%)	73.39	74.50	72.00	72.28	72.72	73.39	**74.94**	73.60	73.20	72.28	71.60

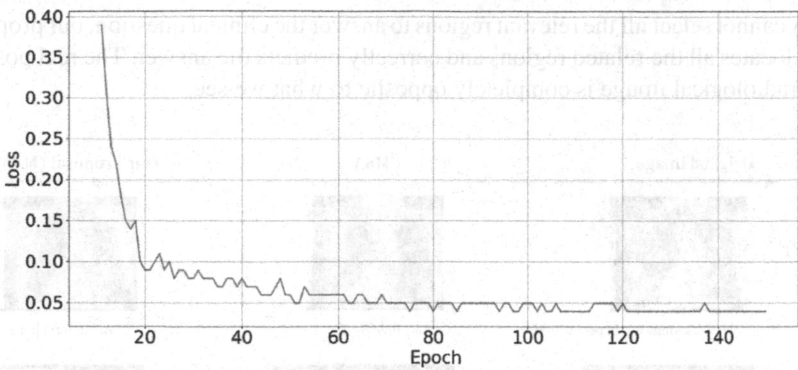

Fig. 2. The loss curve of MAMF.

4.6 Hyperparameter Analysis

We allocate distinct values of the hyperparameter α in the AAL in Eq. (11) and conduct experiments on the VQA-RAD dataset, as shown in Table 6. The overall task and open-ended task can gain the best performances when α is 1.2. Therefore, α is set to 1.2 during training our proposed model.

We train MAMF for 150 epochs. The loss curve and accuracy curve of MAMF are shown in Fig. 2 and Fig. 3, respectively. As shown from the Fig. 2, MAMF gains a relatively stable state after about approximately 150 epochs. From the Fig. 3, we can see that the accuracy curve also slowly becomes stable. Consequently, the value of hyperparameter epochs is set to 150 during training.

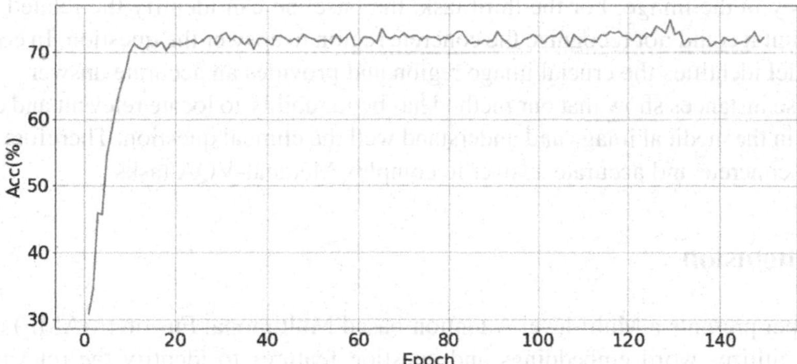

Fig. 3. The accuracy curve of MAMF.

4.7 Qualitative Evaluation

The qualitative evaluation of our proposed MAMF and the best baseline CMSA on the VQA-RAD dataset is shown in Fig. 4. For the first VQA task, while the baseline model

CMSA cannot select all the relevant regions to answer the clinical question, our proposed model locates all the related regions and correctly predicts the answer. The real position of the radiological image is completely opposite to what we see.

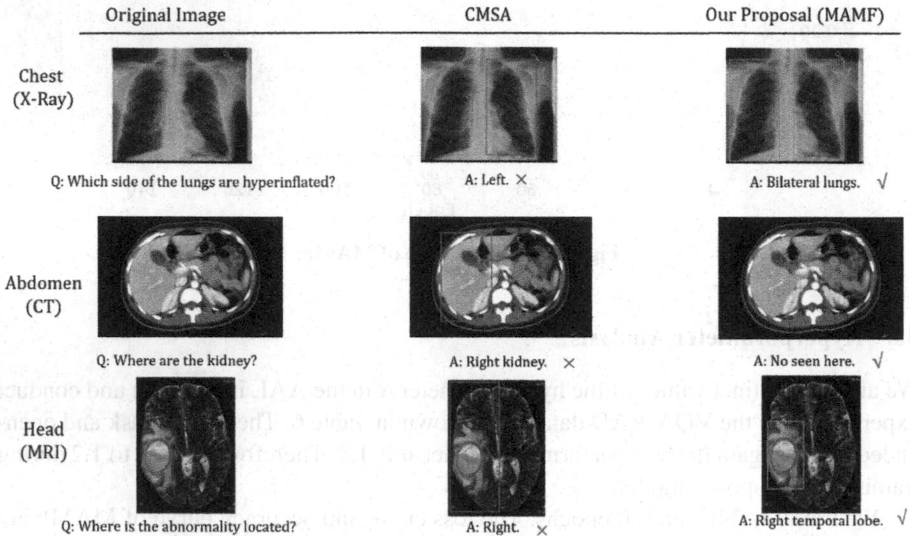

Fig. 4. Visualization of performances of our presented model MAMF and the baseline CMSA.

Therefore, "left" in the answer means the right region of the image. As for the second task, the CMSA identified the liver as the kidney, while our method finds that there is no kidney in the image. For the third task, the baseline can identify the related image region, but it could not recognize the concrete region to answer the question. In contrast, our model identifies the crucial image region and provides an accurate answer.

These instances show that our method has better ability to locate relevant and crucial regions in the medical image and understand well the clinical question. Therefore, it can provide concrete and accurate answer to complex Medical-VQA tasks.

5 Conclusion

This paper presents a Multi-level Attention-based Multimodal Fusion (MAMF) model. MAMF utilizes word embeddings and question features to identify the relevant and key regions of medical image by adopting a W2I attention and a S2I attention. It then contributes to obtain a multi-level multimodal semantic representation. Moreover, we propose an attention alignment loss to align the word-based attention and sentence-based attention to recognize relevant and crucial regions in medical images. This model is beneficial for clinicians in diagnosing different diseases. It also can help patients obtain the answers of health-related questions. Additionally, our model significantly outperforms related state-of-the-art baselines.

Acknowledgements. The work is supported by grants from Humanities and Social Sciences Research Foundation of the Ministry of Education, "Intelligent Analysis and Evaluation of Learning Effection Based on Multi-Modal Data" (No. 21YJAZH072).

References

1. Finn, C., Abbeel, P., Levine, S.: Model-agnostic meta-learning for fast adaptation of deep networks. In: Proceedings of the International Conference on Machine Learning, pp. 1126–1135 (2017)
2. Masci, J., Meier, U., Cireşan, D., Schmidhuber, J.: Stacked convolutional auto-encoders for hierarchical feature extraction. In: Proceedings of the International Conference on Artificial Neural Networks, pp. 52–59 (2011)
3. Devlin, J., Chang, M.W., Lee, K., Toutanova, K.: BERT: pre-training of deep bidirectional transformers for language understanding. arXiv preprint arXiv:1810.04805 (2018)
4. Hao, T., Li, X., He, Y., Wang, F.L., Qu, Y.: Recent progress in leveraging deep learning methods for question answering. Neural Comput. Appl. **34**, 2765–2783 (2022)
5. Lau, J.J., Gayen, S., Ben Abacha, A., Demner-Fushman, D.: A dataset of clinically generated visual questions and answers about radiology images. Sci. Data **5**, 1–10 (2018)
6. Ren, S., He, K., Girshick, R., Sun, J.: Faster R-CNN: towards real-time object detection with region proposal networks. Adv. Neural Inf. Process. Syst. **28** (2015)
7. He, K., Zhang, X., Ren, S., Sun, J.: Deep residual learning for image recognition. In: Proceedings of the IEEE/CVF Conference on Computer Vision and Pattern Recognition, pp. 770–778 (2016)
8. Abacha, A.B., Gayen, S., Lau, J.J., Rajaraman, S., Demner-Fushman, D.: NLM at ImageCLEF 2018 visual question answering in the medical domain. In: Working Notes of CLEF 2018 - Conference and Labs of the Evaluation Forum (CEUR Workshop Proceedings, Vol. 2125). CEUR WS.org, Avignon, France (2018)
9. Abacha, A.B., Hasan, S.A., Datla, V.V., Liu, J., Demner-Fushman, D., Müller, H.: VQA-Med: overview of the medical visual question answering task at ImageCLEF 2019. In: Working Notes of CLEF 2019 - Conference and Labs of the Evaluation Forum (CEUR Workshop Proceedings, vol. 2380). CEUR-WS.org, Lugano, Switzerland (2019)
10. Raghu, M., Zhang, C., Kleinberg, J., Bengio, S.: Trans-fusion: understanding transfer learning for medical imaging. In: Advances in Neural Information Processing Systems 32: Annual Conference on Neural Information Processing Systems, pp. 3342–3352. NeurIPS, Vancouver, BC, Canada (2019)
11. Nguyen, B.D., Do, T.-T., Nguyen, B.X., Do, T., Tjiputra, E., Tran, Q.D.: Overcoming data limitation in medical visual question answering. In: Shen, D., Liu, T., Peters, T.M., Staib, L.H., Essert, C., Zhou, S., Yap, P.-T., Khan, A. (eds.) MICCAI 2019. LNCS, vol. 11767, pp. 522–530. Springer, Cham (2019). https://doi.org/10.1007/978-3-030-32251-9_57
12. Brown, T., et al.: Language Models are Few-Shot Learners. arXiv preprint arXiv:2005.14165 (2020)
13. Zhan, L.M., Liu, B., Fan, L., Chen, J., Wu, X.M.: Medical visual question answering via conditional reasoning. In: Proceedings of the 28th ACM International Conference on Multimedia, pp. 2345–2354 (2020)
14. Li, Y., et al.: A Bi-level representation learning model for medical visual question answering. J. Biomed. Inform. **134**, 104183 (2022)
15. Antol, S., et al.: VQA: visual question answering. In: Proceedings of the IEEE International Conference on Computer Vision, pp. 2425–2433 (2015)

16. Zhou, B., Tian, Y., Sukhbaatar, S., Szlam, A., Fergus, R.: Simple baseline for visual question answering. arXiv preprint arXiv:1512.02167 (2015)
17. Liu, W., et al.: SSD: single shot multibox detector. In: Leibe, B., Matas, J., Sebe, N., Welling, M. (eds.) ECCV 2016. LNCS, vol. 9905, pp. 21–37. Springer, Cham (2016).https://doi.org/10.1007/978-3-319-46448-0_2
18. Yang, Z., Dai, Z., Yang, Y., Carbonell, J., Salakhutdinov, R.R., Le, Q. V.: XLNet: generalized autoregressive pretraining for language understanding. arXiv preprint arXiv:1906.08237 (2019)
19. Fukui, A., Park, D.H., Yang, D., Rohrbach, A., Darrell, T., Rohrbach, M.: Multimodal compact bilinear pooling for visual question answering and visual grounding. arXiv preprint arXiv:1606.01847 (2016)
20. Kim, J.H., On, K.W., Lim, W., Kim, J., Ha, J.W., Zhang, B.T.: Hadamard product for low-rank bilinear pooling. arXiv preprint arXiv:1610.04325 (2016)
21. Kim, J.H., Jun, J., Zhang, B.T.: Bilinear attention networks. arXiv preprint arXiv:1805.07932 (2018)
22. Do, T., Nguyen, B.X., Tjiputra, E., Tran, M., Tran, Q.D., Nguyen, A.: Multiple Meta-Model Quantifying for Medical Visual Question Answering. arXiv preprint arXiv:2105.08913 (2021)
23. Gong, H., Chen, G., Liu, S., Yu, Y., Li, G.: Cross-Modal Self-Attention with Multi-Task Pre-Training for Medical Visual Question Answering. arXiv preprint arXiv:2105.00136 (2021)
24. Liu, Bo., Zhan, L.-M., Wu, X.-M.: Contrastive Pre-training and representation distillation for medical visual question answering based on radiology images. In: de Bruijne, M., Cattin, P.C., Cotin, S., Padoy, N., Speidel, S., Zheng, Y., Essert, C. (eds.) MICCAI 2021. LNCS, vol. 12902, pp. 210–220. Springer, Cham (2021). https://doi.org/10.1007/978-3-030-87196-3_20
25. Vu, M.H., Löfstedt, T., Nyholm, T., Sznitman, R.: A question-centric model for visual question answering in medical imaging. IEEE Trans. Med. Imaging **39**(9), 2856–2868 (2020)
26. Sharma, D., Purushotham, S., Reddy, C.K.: MedFuseNet: an attention-based multimodal deep learning model for visual question answering in the medical domain. Sci. Rep. **11**(1), 1–18 (2021)
27. Pennington, J., Socher, R., Manning, C.D.: GloVe: global vectors for word representation. In: Proceedings of the 2014 Conference on Empirical Methods in Natural Language Processing, pp. 1532–1543 (2014)
28. Cho, K., Van Merriënboer, B., Bahdanau, D., Bengio, Y.: On the properties of neural machine translation: encoder-decoder approaches. In: Proceedings of SSST@EMNLP 2014, Eighth Workshop on Syntax, Semantics and Structure in Statistical Translation, pp. 103–111. Association for Computational Linguistics, Doha, Qatar (2014)
29. He, X., Zhang, Y., Mou, L., Xing, E., Xie, P.: PathVQA: 30000+ questions for medical visual question answering. arXiv preprint arXiv:2003.10286 (2020)
30. Krizhevsky, A., Sutskever, I., Hinton, G.E.: ImageNet classification with deep convolutional neural networks. In: Proceedings of Advances in Neural Information Processing Systems, pp. 1097–1105 (2012)

ASIM: Explicit Slot-Intent Mapping with Attention for Joint Multi-intent Detection and Slot Filling

Jingwen Chen[1], Xingyi Liu[2], Mingzhi Wang[1(✉)], Yongli Xiao[1], Dianhui Chu[1], Chunshan Li[1], and Feng Wang[2]

[1] Harbin Institute of Technology, Weihai, China
{chenjw,wangmz,xiaoyl,chudh,lics}@hit.edu.cn
[2] Weichai Power Co., Ltd., Weifang, China
{liuxingy,wangfeng05}@weichai.com

Abstract. The accurate analysis of a user's natural language statement, including their potential intentions and corresponding slot tags, is crucial for cognitive intelligence services. In real-world applications, a user's statement often contains multiple intentions, and most existing approaches either mainly focus on the single-intent research problems or utilizes an overall encoder directly to capture the relationship between intents and slot tags, which ignore the explicit slot-intent mapping relation. In this paper, we propose a novel Attention-based Slot-Intent Mapping Method (ASIM) for joint multi-intent detection and slot filling task. The ASIM model not only models the correlation among sequence tags while considering the mutual influence between two tasks but also maps specific intents to each semantic slot. The ASIM model can balance multi-intent knowledge to guide slot filling and further increase the interaction between the two tasks. Experimental results on the MixATIS dataset demonstrate that our ASIM model achieves substantial improvement and state-of-the-art performance.

Keywords: intent detection · slot filling · multi-intent · deep learning · attention

1 Introduction

The accurate analysis of the potential intentions in a user's natural language statement, as well as the corresponding slot tags that correspond to these intentions, is very important for cognitive intelligence services. Intent detection and slot filling are two core components of the cognitive service system. Intent detection aims to output the real intent of the user and solve the problem of what the user wants to do. Slot filling is to mark the important words in user input, which is to extract the details needed in service provision. Take the statement "Help me book a flight ticket from Beijing to New York." as an example, intention detection can be regarded as the text classification problem, which needs to output a user's real intention, i.e., "booking air tickets". Slot filling task can be

H. Zhang et al. (Eds.): NCAA 2023, CCIS 1870, pp. 215–228, 2023.
https://doi.org/10.1007/978-981-99-5847-4_16

thought of as a sequence labeling problem, requiring an output of a sequence, e.g., "O, O, O, O, O, B-from_location, O, B-to_location I-to_location".

In past research work, intent detection and slot-filling tasks were normally considered as independent tasks, but later researchers [2] considered that there was a correlation between them since the two tasks always appear together in the same conversation and have a mutual influence. Although many joint models have achieved good performance, these methods are based on the assumption that user utterance contains only simple single intent. However, in a real scenario, this is not the case. According to Gangadharaiah R et al. [2], 52% of user sentences in Amazon's customer service system involve multiple intents. Hence, in the process of real utterance, users will be faced with changing intent halfway or involving multiple intents in one sentence. For example, "Play jay Chou's latest single. No, just play the music video for the new song". In this case, the user suddenly changed his initial request. Or the user may have more than one demand, for example, "How much is the air ticket to Beijing during the National Day holiday, and tell me the recent weather in Beijing." Therefore, it is necessary to accurately identify all the intents in the user's utterance, which is very important work to provide information for the subsequent services.

In the early time, multi-intent detection is regarded as a text multi-label classification problem. However, the most multi-label classifier can work with long text, whereas the multi-intention detection task always works with short user utterances. Compared with single-intent detection, in a short text, there exist three main problems in multi-intent detection: 1. How to find out that users' utterances have multiple intents, and what is the difference between multi-intent utterances and single-intent utterances; 2. How to find out the number of intents hidden in the utterances after confirming that the utterance is multi-intent; 3. How to accurately identify all user intents.

In addition to the above problems, the multi-intent model still presents a unique challenge: how to effectively incorporate multiple intents knowledge to guide the slot prediction, since each word in a sentence has a different relevance for a different intent. Reference [3] proposed a slot gate mechanism, which flows intent information to the slot-filling task. Reference [4] proposed a new self-attention mechanism model, which enhanced the gate mechanism through intent information. The model used the intent information as the gate state information of the slot. Despite the promising performance, most of the previous works directly use multi-intent knowledge to predict the tags appearing of each slot word in a sentence, which would introduce part of noise information. Take the utterance "How much is the air ticket from New York to Beijing during the National Day holiday and tell me the recent weather in Beijing." for example (Fig. 1), if multiple intents information is directly used to guide slot filling of all words in a sentence, irrelevant information will be introduced and lead to poor performance. As shown in Fig. 1 (a), for the word "New" and "York", the intent "GetWeather" is almost irrelevant. Obviously, using the same intent knowledge to predict all slot tags may bring ambiguity. Therefore, for different words in one sentence, how to introduce more detailed intent information is a crucial problem of intent detection and slot filling model.

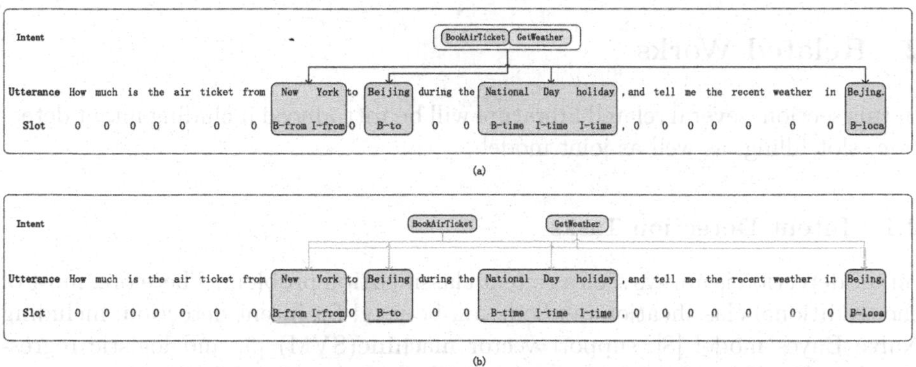

Fig. 1. Gangadharaiah et al. [2]: Utterance-level intent is shared by all the slots(a) vs ASIM: Word-level intent-slot mapping information is used by each slot(b)

In this paper, a novel Attention-based Slot-Intent Mapping Method (ASIM) is proposed, which not only can model the correlation among sequence tags while considering the mutual influence between two tasks, but also can map specific intent to each semantic slot. Unlike most existing models that implicitly share information between intent detection and slot filling through shared encoders, our ASIM model adopts respective encoders for intent detection and slot filling, which achieves the first information sharing by exchanging hidden state information between the two task encoders. More than that, our ASIM model constructs another information interaction in the decoder stage, where the importance coefficient of each word and intent information is calculated through the attention mechanism, which refines the guidance effect of multi-intent information on word slot filling.

We conclude the main contributions of this paper:

- the proposed ASIM model will explore the multi-intent detection task and slot filling task together, which can construct two information-sharing mechanisms in the encoder and decoder module to capture the correlation between intent detection and slot filling as well as analyze users' slot-intent mapping on the word level.
- We use an attention mechanism to balance the degree of closeness between multiple intents and words to guide slot filling in the decoder stage.
- we conducted experiments on MixATIS data-set to validate our hypothesis and the results illustrate that our ASIM model achieved better performance than other intent detection model.

The rest of the paper is organized as follows. In Sect. 2, several related literature will be introduced including intent detection, slot filling, as well as joint model. Section 3 mainly discuss more details about ASIM model. In Sect. 4, we conduct the experiments to verify the performance of ASIM model from different perspectives. Finally, we conclude our work and give the further plan.

2 Related Works

In this section, several related literature will be introduced including intent detection, slot filling, as well as joint model.

2.1 Intent Detection Tasks

Intent detection is always seen as a text classification problem. Therefore, most of the traditional classification methods can be used for intent detection, including Naive Bayes model [8], support vector machine(SVM) [9] and logistic regression [10]. Traditional methods can be divided into rule-based template semantic recognition methods [11] and statistics-based classification algorithms [12]. Even without a lot of training data, the rule-based template methods still achieve good results. However, the template needs to be formulated by experts, and the template reconstruction requires a lot of economic cost and time cost to adopt this method. The classification algorithms based on statistics need to extract the key information of corpus, so these methods need a lot of training data. Therefore, traditional intent detection methods cannot meet higher requirements. With the great success of artificial neural networks in other fields, intent detection methods based on deep neural networks have become popular. With the success of convolutional neural network (CNN) in the field of computer vision, researchers [13] employ CNN network to determinate the 5-gram features of sentence, and maximum pooling was applied to generate the feature embedding vector of words. As in [14], Recurrent Neural Network (RNN) and Long Short Term Memory (LSTM) are applied to intent detection according to the sequential nature of user utterances.

2.2 Slot Filling Tasks

Slot filling task can be formulated as a sequence labeling problem. The previous methods to solve the slot filling problem are mainly divided into three categories.

1) Dictionary approach [15]. This method searches for dictionary keywords mainly through string matching. Since a large number of corpus is needed to construct the dataset, this method consumes manpower and faces the problem of data scarcity.

2) Rule-based approach [16–18]. This method marks keywords in user utterance by rule matching. Because domain experts are required to make rules, so the costs are high. In addition, scalability is poor. With the gradual increase of user's requirements, experts are needed to constantly improve the existing rules, and rule conflicts are easy to occur.

3) Traditional machine learning method [19–22]. This method takes artificially labeled corpus as the training set and optimizes model parameters through multiple training to minimize the target loss function. Not only a large number of labeled training data are required, but also features are manually constructed.

With the high-speed development of deep neural network [23,24]. many AI algorithms have also been applied to slot filling, such as recurrent neural network (RNN), convolutional neural network (CNN) and various combinations of traditional machine learning methods [25].

2.3 Joint Model

Considering intent-slot relation and information-sharing mechanism between intent and slot tasks, the researchers began to train the two tasks together. The joint model is not only take advantage of information interaction between two tasks, but also simplifies the training process by training only one model. The early research literature in this field is the CNN+Tri-CRF method [13], which utilizes CNN networks as a shared encoder to integrate the intent detection and slot filling tasks, and then employs a CRF layer to handle dependencies among slot tags. Guo et al. [27] proposed a joint training methods of the recursive neural network (RecNNs) for intent detection and slot filling tasks. Zhang et al. [28,29] employ a Gated recurrent unit (GRU) to learn the representation of each time step in RNN and predict the label of each slot tags. Liu et al. [1] proposed introducing attention to the alignment-based RNN models which can add additional information to the intent detection and slot filling tasks. Goo ea al. [3] utilize a slot gate structure to learn the relationship between intent and slot attention vectors and achieve better semantic segment results by the global optimization. Wang et al. [5] employ a Bi-model based RNN network structures to handle the cross-impact between the intent detection and slot filling tasks. The key points of Bi-model are two inter-connected bidirectional LSTMs structure and two different cost functions in an asynchronous training. Qin et al. [26,30] propose two attention mechanism-based models that adopt Stack Propagation which can directly employ the intention embedding as input for slot filling, and capture the semantic information of intent. In recent years, pre-trained language models [31–33] have significantly enhanced the performance of many natural language processing (NLP) applications. Chen ta al. [34] investigates BERT pre-trained model to address the poor generalization capability on intent detection and slot filling. Zhang et al. [35] design a effective encoder-decoder framework to improve the performance of intent detection and slot filling tasks.

3 Methodology

In this part, We will begin by defining the joint intent detection and slot filling tasks. Then, we will give the detail of the Attention-based Slot-Intent Mapping model (ASIM), which calculates the correlation between multiple intents and the current word, and then uses the information of multiple intents to guide slot filling. The model we proposed, increases the mutual influence between two tasks.

3.1 Problem Definition

The input sequence is defined as $x = (x_1, x_2, x_3, ...x_n)$. Intent detection is treated as a classification problem, and final output is intent label $y^1 = (y_1^1, y_2^1, y_3^1, ...y_m^1)$, where m is the number of the intents the input sequence contains. Slot filling is treated as a sequence labeling problem, and the final output is $y^2 = (y_1^2, y_2^2, y_3^2, ...y_n^2)$.

3.2 ASIM Model

Figure 2 illustrates the network structure of the ASIM model, in which intent detection and slot filling use different encoders and decoders respectively. The left part of the network is designed for intent detection and the right part is designed for slot filling. In the bottom part, the two encoders read and encode the input sentence. Then the encoded information is passed to the decoder for outputting the predicted intents and slot tags. The subsequent sentences give the details of how the encoders and decoders operate for intent detection and slot filling.

3.3 Encoder

BiLSTM. Considering a specific relationship between intent detection and slot filling, most studies use shared encoders to share information between intent detection and slot filling tasks. However, these approaches are not only poorly interpretable but also not obvious for intent detection and slot filling information flow. Hence, to explicitly describe the interaction between intent detection and slot filling, the proposed ASIM uses two encoders corresponding to intent detection and slot filling, respectively.

BiLSTM consists of two LSTM units. For the input sequence $x = (x_1, x_2, x_3, ...x_n)$, BiLSTM obtains the forward hidden state vector $\overrightarrow{h^i} = (\overrightarrow{h_1^i}, \overrightarrow{h_2^i}, ..., \overrightarrow{h_n^i})$ from x_1 to x_n, and obtains the backward hidden state vector $\overleftarrow{h^i} = (\overleftarrow{h_1^i}, \overleftarrow{h_2^i}, ..., \overleftarrow{h_n^i})$ from x_n to x_1. The final hidden state vector $H^i = (h_1^i, h_2^i, ..., h_n^i)$ is obtained by concatenating forward hidden state vector and backward hidden state vector, where $i = 1$ corresponds to the task of intent detection and $i = 2$ corresponds to slot filling.

Attention Mechanism. Generally, sentences with multiple intentions are longer than those with a single intent. As the length of text increases, although BiLSTM can capture information from both sides of the sentence, it still causes some information loss. In addition, the correlation between the current tag and

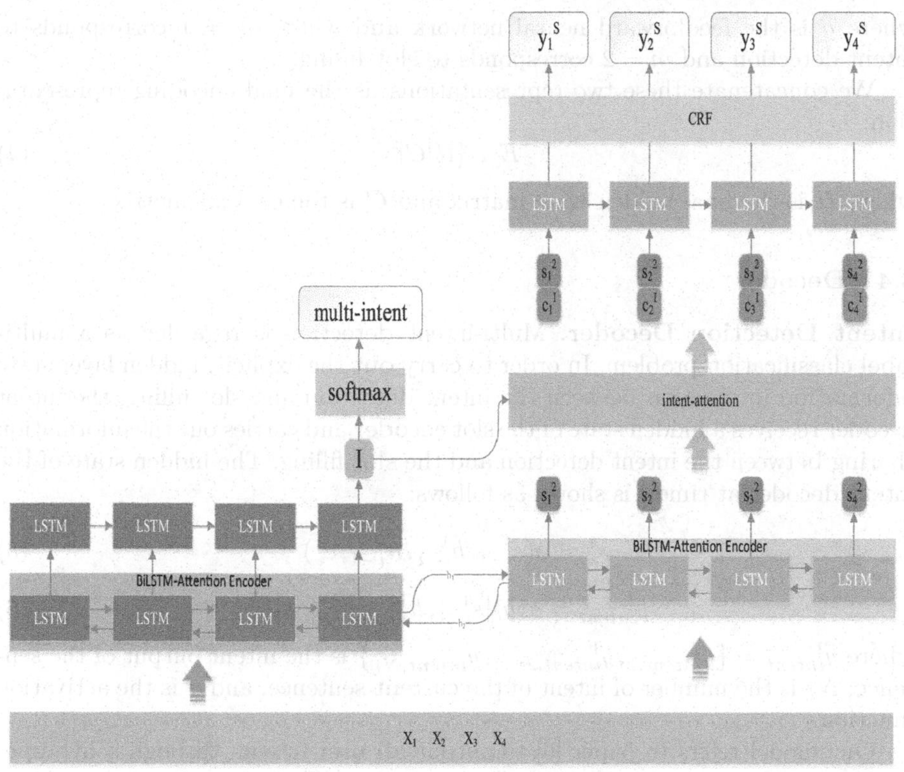

Fig. 2. The structure of the ASIM model.

the other tags in the sentence is not the same. Therefore, a self-attention mechanism is added in the encoder stage of slot filling in this paper, aiming to assign different importance degrees to other words related to the current word when encoding the current word information. The attention mechanism not only makes up for the information loss generated by BiLSTM but also obtains the correlation information between the current tag and the other tags in one sentence. The following is the introduction of the attention mechanism.

For each hidden state h_i^m, the context vector c_i^m is obtained by calculating the weighted sum of the hidden states:

$$c_i^m = \sum_{j=1}^{n} a_{i,j}^m h_j^m \tag{1}$$

The attention score can be obtained from the following formula:

$$a_{i,j}^m = \frac{exp(e_{i,j}^m)}{\sum_{k=1}^{n} exp(e_{i,k}^m)} \tag{2}$$

$$e_{i,j}^m = g(s_{i-1}^m, h_k^m) \tag{3}$$

where g is the feedforward neural network and where $m = 1$ corresponds to intent detection and $m = 2$ corresponds to slot filling.

We concatenate these two representations as the final encoding representation:

$$E = [H|C] \tag{4}$$

where H is the final hidden state matrix and C is the context matrix.

3.4 Decoder

Intent Detection Decoder. Multi-intent detection is regarded as a multi-label classification problem. In order to carry out the explicit, hidden layer state information interaction between the intent detection and slot filling, the intent decoder receives a hidden state of the slot encoder and carries out the information sharing between the intent detection and the slot filling. The hidden state of the intent decoder at time i is shown as follows:

$$s_i^1 = \phi(s_{i-1}^1, h_{i-1}^1, h_{i-1}^2, c_i^1) \tag{5}$$

$$y_{intent}^1 = \sigma(\hat{y}_i^1 | s_{i-1}^1, h_{i-1}^1, h_{i-1}^2, c_i^1) \tag{6}$$

where $y_{intent}^1 = \{y_{intent,1}^1, y_{intent,2}^1, ..., y_{intent,N_I}^1\}$ is the intent output of the sentence, N_I is the number of intent of the current sentence, and σ is the activation function.

Our model refers to paper [6] to output all user intents through a hyperparameter t_u. Suppose the prediction intent is $I = (I_1, I_2, I_3, ...I_n)$, I_i represents $y_{I_i}^I$ greater than t_u, where t_u is the hyperparameter obtained by fine-tuning the validation data set. For example, if $y^I = \{0.9, 0.3, 0.6, 0.7, 0.2\}$ and $t_{0.5}$, then we can get $I = (1, 3, 4)$.

Slot Filling Decoder. The intent encoder hidden state h_{i-1}^1 and the slot encoder hidden state h_{i-1}^2 are utilized for slot filling:

$$s_i^2 = \varphi(h_{i-1}^2, h_{i-1}^1, s_{i-1}^2, c_i^2) \tag{7}$$

$$y_i^2 = \sigma(\hat{y}_n^2 | h_{i-1}^2, h_{i-1}^1, s_{i-1}^2, c_i^2) \tag{8}$$

where σ is the activation function and the s_{i-1}^2 is the hidden state of slot decoder.

Slot-Intent Mapping with Attention. The core of this part is to balance the degree of closeness between multi-intent and the slot and use the balanced multi-intent information to guide the slot filling. The concrete implementation is as follows.

Firstly, according to the current word and the predicted multiple intents, we calculate the degree of closeness between each intent and the current word and get the score of each intent, which is used as the attention score of the currently hidden unit:

$$a_{i,j} = \frac{exp(e_{i,j}^I)}{\sum_{k=1}^{m} exp(exp(e_{i,k}^I))} \tag{9}$$

$$e_{i,j}^{I} = g(I_{i,j}, h_k^2) \tag{10}$$

where I is the predicted multiple intents. h_i^2 is the current slot hidden state. The calculated weight a_i^I is the weight of intent, which represents the degree of closeness of the corresponding intent to the current word.

By summing up all the weighted predicted intents, the context vector of intents c_i^I to the current word is obtained:

$$c_i^I = \sum_{j=1}^{m} a_{i,j}^I I_{i,j} \tag{11}$$

where c_i^I represents the integration of all the information related to the current sentence intent, which is used to guide the slot filling. The output of the slot filling decoder is:

$$s_i^2 = \varphi(e_i^2, h_{i-1}^2, h_{i-1}^1, s_{i-1}^2, c_i^I) \tag{12}$$

$$y^2 = \sigma(\hat{y}_i^2 | e_i^2, h_{i-1}^2, h_{i-1}^1, s_{i-1}^2, c_i^I) \tag{13}$$

where σ is the activation function.

CRF. If the model without CRF is used for slot filling, the slot label with the highest score of each label is selected as the slot label of the word. However, in practical applications, the tag with the highest score may not always be the most suitable one. In order to solve this issue, a CRF layer is added after the slot filling decoder.

The CRF layer will model several dependencies of the slot tags to ensure that the predicted slot is more suitable so as to increase the accuracy of correct slot prediction. These constraints can be learned automatically through the CRF layer during data training.

For sentence 1 "please give me the flight times the morning on united airline for september twentieth from philadelphia to san francisco". The true tag of the phrase "flight times" is "$B - flight\_time\ I - flight\_time$", but the slot tag predicted by the model without CRF is "$O\ I - flight\_time$". For sentence 2 "what type of ground transportation is available at philadelphia airport and then how many first class flights does united have today". The true tag of the phrase "philadelphia airport" is "$B - airport\_name\ I - airport\_name$", but the slot tag predicted by the model without CRF is "$B - city\_name\ I - airport\_name$". The model with CRF can reduce the errors mentioned in these two sentences in most cases.

The CRF layer can learn some constraints for correctly predicting slots. For BIO-tagged data, the possible constraints are:

1) Instead of "$I - X$", an X element should begin with "$B - X$" or "O". For example, tag "$I - flight\_time$" in sentence 1 should not be the beginning of the "flight_time" element. Thus, the model that add CRF layers can correctly predict "$B - flight\_time$" as the beginning of the element "flight_time".
2) For slot label sequence "$B - label_1\ I - label_2\ I - label_3$...", $label_1$, $label_2$ and $label_3$ should be the same entity category. As shown in sentence 2, "$B - city\_name\ I - airport\_name$" is clearly wrong.

3.5 Asynchronous Training

We employ two different cost functions to train the ASIM model with an asynchronous fashion. We define the loss function of intention network is $\mathcal{L}_1$, and the loss function of slot filling networks is $\mathcal{L}_2$. $\mathcal{L}_1$ and $\mathcal{L}_2$ are formulated as:

$$\mathcal{L}_1 \triangleq -\sum_{i=1}^{k} \hat{y}_{intent}^{1,i} log(y_{intent}^{1,i}) \tag{14}$$

and

$$\mathcal{L}_2 \triangleq -\sum_{j=1}^{n}\sum_{i=1}^{m} \hat{y}_j^{2,i} log(y_j^{2,i}) \tag{15}$$

where k denotes the number of intent label types, m represents the number of semantic tag types, n is the length of a word sequence.

4 Experimental Results

4.1 The Data-Set Description

To assess the efficiency of the proposed ASIM model, experiments are carried out on MixATIS with multiple intents. Due to the scarcity of multi-intent data sets, reference [6] constructed multi-intent data set MixATIS data on commonly used single-intent data set ATIS. ATIS has 656 words, 18 intents, and 130 slot labels. By using conjunctions "and" to combine sentences with various intentions, a sentence can have one to three intentions, in which the proportion of each number of intents is [0.3,0.5,0.2]. The number of training sets, verification sets, and test sets of the final MixATIS data set is 18000, 1000 and 1000, respectively.

4.2 Baselines

1) Attention BiRNN. Liu et al. [2] propose introducing attention to RNN model and bring additional information to the intent detection and slot filling tasks.

2) Slot-Gated Atten. Goo et al. [3] use a slot-gated based RecNNS to explicitly consider the information-sharing between the two tasks.

3) Bi-Model. Wang et al. [4] employ two inter-connected bidirectional LSTMs structure and two different cost functions to improve model performance.

4) SF-ID Network. Niu et al. [6] design bi-directional interrelated network to model direct correlation between intent detection and slot filling.

4.3 The Experiment Design

This paper deals with the data set MixATIS as in [7]. The identifier "UNK" represents those that occur in the test data but not in the training data, and the number is represented by a string of varying lengths "DIGIT" based on its digits.

The ASIM model uses deep learning PyTorch framework for training. In the training stage, the word feature dimension d_C is 300, the maximum sentence length is 130, and the BiLSTM unit dimension d is 200. The threshold value $t_u = 2$.

4.4 The Experiment Results

We first employ the MixATIS benchmark datasets to show the performance of the ASIM model. The specific results are shown in Table 1. Compared with the previous benchmark models, the model proposed in this paper improves slot F_1 and Intent acc on MixATIS data set and intent detection has a big improvement. It can be seen from the results that the Intent Acc is 1.6% higher than that of Bi-Model. It can be analyzed that the probable reason is that the attention mechanism we added in the encoder captures the important sentence information so that the content of sentence important information contained in the embedding vector is increased. Besides, the slot and intent mapping module also play a positive role in improving the Intent Acc. Since intent detection and slot filling are related to each other, the improvement of one task can also have a positive influence on the other.

Table 1. Comparison of experimental results

Model	MixATIS	
	Slot (F1)	Intent (Acc)
Attention BiRNN	86.6	71.6
Slot-Gated	**88.1**	65.7
Slot-gated Inten	86.7	66.2
Bi-Model	85.5	72.3
SF-ID	87.7	63.7
the ASIM model	**87.19**	**73.90**

Ablation Experiments. Model modification. Table 2 shows the ablation experiment results. As can be seen from Table 2, both the attention mechanism of intent detection and slot filling added in the encoder and the Slot-Intent mapping module have a positive effect on the experimental results. The attention mechanism of the encoder part improves the two tasks. Compared with the Bi-model, the slot $F1$ score is improved by 2.21, and the Intent Acc is improved by 1.1% in Bi-Model(with encoder Attention). The reason is that the attention mechanism added in the encoder captures the information of the important words in the sentence and reduces the information loss caused by the LSTM model. Therefore, the decoder can receive input vectors containing more information about those important words. The slot-Intent Mapping module does not clearly improve slot filling but improves intent detection by 0.3% in Bi-Model(with slot-intent Mapping). The possible reason for this phenomenon is the Slot-Intent Mapping module significantly increases the interaction between intent detection and slot filling, which not only provides guidance to each other but also potentially introduces a bit of error information. But in general, the model proposed in this paper achieves good results in intent detection and slot filling.

Table 2. Comparison of ablation results

Model	MixATIS	
	Slot (F1)	Intent (Acc)
Bi-Model	85.50	72.30
Bi-Model (with encoder Attention)	**87.71**	73.40
Bi-Model (with slot-intent Mapping)	85.50	72.60
the ASIM model	**87.19**	**73.90**

5 Conclusions

In this paper, A novel attention-based slot-intent mapping (ASIM) model was proposed for joint multi-intent detection and slot filling tasks. The ASIM model not only can model the correlation among sequence tags while considering the mutual influence between two tasks, but also can map specific intention to each semantic tag. In particular, this ASIM uses two encoding structure to achieve more obvious information interaction between the two tasks and uses an attention mechanism to balance the degree of closeness between multiple intents and words to guide slot filling in the decoder stage. Then, the interaction between the two tasks is mutually reinforcing. A CRF layer is added after the slot filling decoder which can model several dependencies of the slot tags to ensure that the predicted labels are more suitable. After experimental verification on a real-world dataset, the ASIM model has achieved the best performance than other state-of-art methods.

References

1. Liu, B., Lane, I.: Attention-Based Recurrent Neural Network Models for Joint Intent Detection and Slot Filling (2016)
2. Gangadharaiah, R., Narayanaswamy, B.: Joint multiple intent detection and slot labeling for goal-oriented dialog. In: Proceedings of the 2019 Conference of the North (2019)
3. Goo, C.-W., et al.: Slot-gated modeling for joint slot filling and intent prediction. In: Proceedings of NAACL (2018)
4. Li, C., Li, L., Qi, J.: A selfattentive model with gate mechanism for spoken language understanding. In: Proceedings of EMNLP (2018)
5. Wang, Y., Shen, Y., Jin, H.: A bi-model based RNN semantic frame parsing model for intent detection and slot filling. In: Proceedings of NAACL (2018)
6. Haihong, E., Niu, P., Chen, Z., Song, M.: A novel bi-directional interrelated model for joint intent detection and slot filling. In: Proceedings of ACL (2019)
7. Qin, L., Xu, X., Che, W., Liu, T.: AGIF: an adaptive graph-interactive framework for joint multiple intent detection and slot filling. In: EMNLP Findings (2020)
8. Mccallum, A., Nigam, K.: A comparison of event models for Naive Bayes text classification. In: AAAI-98 Workshop on Learning for Text Categorization, pp. 41–48 (1998)
9. Haffner, P., Tur, G., Wright, J.H.: Optimizing SVMs for complex call classification. In: IEEE International Conference on Acoustics, pp. 632–635 (2003)
10. Genkin, A., Lewis, D.D., Madigan, D.: Large-scale Bayesian logistic regression for text categorization. Technometrics **49**(3), 291–304 (2007)
11. Dowding, J., et al.: Gemini: a natural language system for spoken-language understanding (1994)
12. Pengju, Y.: Research on Natural Language Understanding in Conversational Systems. Tsinghua University, Beijing (2002)
13. Xu, P., Sarikaya, R.: Convolutional neural network based triangular CRF for joint intent detection and slot filling. In: 2013 IEEE Workshop on Automatic Speech Recognition and Understanding, pp. 78–83 (2013)
14. Ravuri, S., Stolcke, A.: Recurrent neural network and LSTM models for lexical utterance classification. In: Interspeech (2015)
15. Wang, Q.: Biological named entity recognition combining dictionary and machine learning. Dalian University of Technology (2009)
16. Collins, M., Singer, Y.: Unsupervised models for named entity classification. In: Joint SIGDAT Conference on Empirical Methods in Natural Language Processing and Very Large Corpora, pp. 100–110 (1999)
17. Cucerzan, S., Yarowsky, D.: Language independent named entity recognition combining morphological and contextual evidence. In: Joint SIGDAT Conference on Empirical Methods in Natural Language Processing and Very Large Corpora, pp. 90–99 (1999)
18. Mikheev, A., Moens, M., Grover, C.: Named entity recognition without gazetteers. In: Proceedings of the Ninth Conference on European Chapter of the Association for Computational Linguistics. Association for Computational Linguistics, pp. 1–8 (1999)
19. Wei, L., Mccallum, A.: Rapid development of Hindi named entity recognition using conditional random fields and feature induction. ACM Trans. Asian Lang. Inf. Process. **2**(3), 290–294 (2003)

20. Bikel, D.M., Miller, S., Schwartz, R., et al.: Nymble: a high-performance learning name-finder. Anlp, pp. 194–201 (1998)
21. Bikel, D.M., Schwartz, R., Weischedel, R.M.: An algorithm that learns what's in a name. Mach. Learn. **34**(1–3), 211–231 (1999)
22. Borthwick, A., Grishman, R.: A maximum entropy approach to named entity recognition. Graduate School of Arts and Science. New York University (1999)
23. Yao, K., Zweig, G., Hwang, M.-Y., Shi, Y., Yu, D.: Recurrent neural networks for language understanding. In: Interspeech (2013)
24. Mesnil, G., He, X., Deng, L., Bengio, Y.: Investigation of recurrentneural-network architectures and learning methods for spoken language understanding. In: Interspeech (2013)
25. Yao, K., Peng, B., Zhang, Y., Yu, D., Zweig, G., Shi, Y.: Spoken language understanding using long short-term memory neural networks. In: SLT (2014)
26. Qin, L., Che, W., Li, Y., et al.: A stack-propagation framework with token-level intent detection for spoken language understanding (2019)
27. Guo, D., Tur, G., Yih, W.-T., Zweig, G.: Joint semantic utterance classification and slot filling with recursive neural networks. In: 2014 IEEE Spoken Language Technology Workshop (SLT), pp. 554–559 (2014)
28. Zhang, X., Wang, H.: A joint model of intent determination and slot filling for spoken language understanding. In: IJCAI, vol. 16, pp. 2993–2999 (2016)
29. Liu, B., Lane, I.: Joint online spoken language understanding and language modeling with recurrent neural networks. arXiv preprint arXiv:1609.01462 (2016)
30. Qin, L., Ni, M., Zhang, Y., Che, W.: Cosda-ml: multi-lingual code-switching data augmentation for zero-shot cross-lingual NLP. arXiv preprint arXiv:2006.06402 (2020)
31. Devlin, J., Chang, M.-W., Lee, K., Toutanova, K.: BERT: pre-training of deep bidirectional transformers for language understanding. In: NAACL-HLT (1) (2019)
32. Sun, Y., et al.: Ernie 2.0: a continual pre-training framework for language understanding. In: Proceedings of the AAAI Conference on Artificial Intelligence, vol. 34, no. 05, pp. 8968–8975 (2020)
33. Yang, Z., Dai, Z., Yang, Y., Carbonell, J., Salakhutdinov, R.R., Le, Q.V.: Xlnet: generalized autoregressive pretraining for language understanding. In: Advances in Neural Information Processing Systems, vol. 32 (2019)
34. Chen, Q., Zhuo, Z., Wang, W.: BERT for joint intent classification and slot filling. arXiv preprint arXiv:1902.10909 (2019)
35. Zhang, Z., Zhang, Z., Chen, H., Zhang, Z.: A joint learning framework with BERT for spoken language understanding. IEEE Access **7**, 168:849–168:858 (2019)

A Triplet-Contrastive Representation Learning Strategy for Open Intent Detection

Guanhua Chen[1], Qiqi Xu[1], Choujun Zhan[1], Fu Lee Wang[2], Kuanyan Zhu[3], Hai Liu[1(✉)], and Tianyong Hao[1]

[1] School of Computer Science, South China Normal University, Guangzhou, China
{2021023267,xuqiqi,haoty}@m.scnu.edu.cn, namelh@gmail.com

[2] Hong Kong Metropolitan University, Hong Kong, China
pwang@hkmu.edu.hk

[3] Aberdeen Institute of Data Science and Artificial Intelligence, South China Normal University, Foshan, China
kuanyanzhu@qq.com

Abstract. Open intent detection aims to correctly classify known intents and identify unknown intents that never appear in training samples, thus it is of practical importance in dialogue systems. Discriminative intent representation learning is a key challenge of open intent detection. Previous methods usually restrict known intent features to compact regions to learn the representations, which assumes that open intent is outside regions. However, open intent can be distributed among known intents. To address this issue, this paper proposes a triplet-contrastive learning strategy to learn discriminative semantic representations and differentiate between similar open intents and known intents. Further, a method named Triplet-Contrastive Adaptive Boundary (TCAB) is proposed, which leverages the triplet-contrastive learning strategy and an adaptive decision boundary method to detect open intent. Extensive experiments on three benchmark datasets show that our method achieves substantial improvements compared with a list of baseline methods.

Keywords: Open intent detection · Triplet-contrastive learning · Adaptive decision boundary

1 Introduction

Accurately identifying users' intents from questions plays a vital role in dialogue systems. Many deep neural network-based intent detection models have achieved rapid development in recent years [1–5]. They usually work with a closed-world assumption that all intent classes are accessible in training stage. However, due to the diversity and uncertainty of users' needs, covering all intent categories is impossible. In addition, users' utterances may exist with unknown (open) intents, and a system may not recognize them. Therefore, to explore users' potential needs and improve users' satisfaction, it is crucial to effectively detect open intents. To this end, the open intent classification task was proposed and had received a lot of attention in recent years [6–9].

H. Zhang et al. (Eds.): NCAA 2023, CCIS 1870, pp. 229–244, 2023.
https://doi.org/10.1007/978-981-99-5847-4_17

Recent researches have attempted to address this problem by developing efficient intent classification models. An intuitive approach was to use a threshold of predicted probability of a K-class classifier to determine whether a sample belongs to the unknown intent. Hendrycks and Gimpel [10] proposed a baseline that utilized softmax probability as a confidence score to detect out-of-distribution samples. However, due to the overfitting of deep learning methods, the K-class classifier would predict known intent with overconfidence and even a sample of unknown intent would be classified into the K-class known intents with a high probability. Therefore, these threshold-based methods were insufficient to effectively distinguish known and unknown intents. To this end, subsequent work employed outlier detection algorithms to adjust the decision boundary. For example, Lin et al. [6] proposed to first learn discriminative deep features with large margin cosine loss, and then applied a local outlier factor [11] (LOF) to detect open intents. Xu et al. [7] proposed a generative distance-based classifier and introduced a Mahalanobis distance under Gaussian discriminant analysis (GDA) to detect OOD intents. Zhang et al. [12] proposed a post-processing method to learn adaptive decision boundary for open intent classification. Instead of combining a K-class classifier with outlier detection algorithms, some other work considered directly training a (K+*1*)-class classifier for open intent detection. Cheng et al. [13] proposed a (K+*1*)-class classification framework, which generated pseudo outliers via soft labeling and manifold mixup strategies. Nevertheless, this approach did not take into account the data distribution of fine-grained open classes. There are still two main challenges currently in open intent detection. One is how to learn discriminative representations based on known intents to detect these unknown intents. The second is how to construct decision boundaries to effectively distinguish known intents from unknown intents.

This paper focuses on the first challenge and proposes a representation learning strategy named triplet-contrastive learning. Previous methods commonly learn discriminative features by pulling samples belonging to the same class together while pushing apart samples from different classes, which aims to minimize intra-class variance and maximize inter-class variance. They implicitly assume that regions of semantic features are compact in a feature space. However, open intents can be distributed among known intents and exhibit less differentiation among known intents. To solve this problem, we design the triplet-contrastive learning strategy to learn discriminative intent representations and differentiate between similar open and known classes. Specially, we mine farthest instances with same class and nearest instances with different class for each utterance, then optimize distances of them through a triplet-contrastive loss. For the task of open intent detection, we propose a method named Triplet-Contrastive Adaptive Boundary (TCAB). It utilizes the triplet-contrastive strategy to generate intent representations, and an adaptive boundary method to calibrate decision boundaries for each known intent.

In summary, the contributions of our paper are as follows:

- A triplet-contrastive strategy is designed by incorporating farthest-nearest distance relationship into the contrastive learning objective, which is conducive to distinguishing among similar open and known classes.

- A method named Triplet-Contrastive Adaptive Boundary (TCAB) is proposed for the task of open intent detection, learning intent representations by leveraging triplet-contrastive learning and constructs adaptive boundaries.
- Extensive experiments conducted on three real-world datasets show that the proposed method achieves better and more robust performance than state-of-the-art methods.

2 Related Work

Intent detection is an important component of dialogue systems that aims to identify the potential purposes of users. Most of existing classification models work under the closed-world assumption, which often goes against practical systems deployed in an open environment, as practical systems often encounter queries out of domain (OOD) of support intents. Therefore, identifying open intents has received a lot of attention recently [9,12,14].

Current methods could be classified into two major types. The first type of methods directly trained a (*K+1*)-class classifier with one additional class to address open intent detection, where the additional class was the unseen intent. For example, Ryu et al. [15] proposed to use only in-domain (IND) sentences to build a generative adversarial network for OOD detection. Zheng et al. [16] proposed a model to generate pseudo OOD samples, and used unlabeled data to enhance OOD detection performance. Some subsequent work used automatic generation of pseudo-OOD data for open intent detection. For example, Choi et al. [17] proposed a method to generate OOD samples using only IND training datasets automatically. Zhan et al. [8] generated a set of pseudo outliers to train a (*K+1*)-class classifier for open intent detection. Although the (*K+1*)-way method was easy to accept and did not require modification of model structure, collecting suitable OOD samples was time-consuming and labor-intensive.

The second type of methods mainly used outlier detection algorithms to open intent detection and could be further divided into three categories: threshold-based, post-processing, and joint optimization. Threshold-based methods used a threshold of predicted probability to determine whether a sample belongs to open intent. For example, DOC [18] built a multi-class classifier with a output layer of sigmoids to enhance unknown class detection. The post-processing methods focused on designing a specific model architecture for representation learning, and applied outlier detection algorithms to detect the open intent. For example, Yan et al. [19] proposed a semantic-enhanced Gaussian mixture model (SEG) for discriminative feature learning, and then applied LOF to detect unknown intents. Zeng et al. [20] proposed a supervised contrastive learning model to learn discriminative semantic intent representations, and employed GDA for OOD detection. Joint optimization methods jointly learned discriminative features and decision boundaries. Zhang et al. [21] proposed a joint optimization framework DA-ADB for open intent classification. It learned discriminative intent features with distance-aware strategy and appropriate decision boundaries adaptive to feature space for open intent classification.

The above methods mainly focus on limiting feature distribution in the feature learning stage or downstream detection stage, and assume semantic feature regions as compact regions in feature space, which means open intents only exist between different known intents and not within the known intent distributions. However, open intents can appear within and around distributions of different known intents, and they differ less from known intents. To identify those open intents that differ little from known intents, our proposed method just utilizes a farthest neighbor with same class as positive and a nearest neighbor with different class as negative, then uses these samples for triplet-contrastive learning.

3 Method

3.1 Definition

The definition of the open intent detection task and some notations are introduced. Given an intent label set I and a dataset D, the intent label set $I = I^{Known} \cup I_{K+1}$ consists of a known intent label set and an open intent I_{K+1}. $I^{Known} = \{I_i\}_{i=1}^{K}$ is the known intent label set and K is the number of known intent. The remaining labels in the initial label set $I \setminus I^{Known}$ are set uniformly as the open intent I_{K+1}. The dataset $D = \{D^{Tr}, D^{Val}, D^{Te}\}$ consists of a training, validation, and testing set, each subset containing a set of labeled samples (u_i, y_i), where u_i is an utterance sample and y_i is the ground-truth intent label of u_i. The training set and validation set only contain known intent samples, and their label sets are I^{Known}. The label set of the testing set is I, containing known intents and the open intent. The goal [12] of open intent detection is to learn a model based on the training set D^{Tr} and the validation set D^{Val}, and then apply it to the testing set D^{Te} to correctly identify known intents and detect the open intent.

3.2 The Architecture of TCAB

A method TCAB is proposed for open intent detection. The method is composed of two components: triplet-contrastive representation learning and adaptive decision boundary learning. The first component first utilizes users' utterances as input to generate intent representations and then selects farthest instances with same class and nearest instances with difference class for triplet-contrastive learning. The second component employs AB to learn appropriate decision boundaries for known intents and applies them for open intent classification. Figure 1 illustrates the overall architecture of our proposed method.

3.3 Feature Extraction

We use the pretrained language model BERT [22] as the model backbone for extracting features of intents. Given the ith utterance u_i, all token embeddings $[CLS, T_1, T_2, ..., T_M] \in \mathbb{R}^{(M+1)\times H}$ are generated from the last hidden layer of

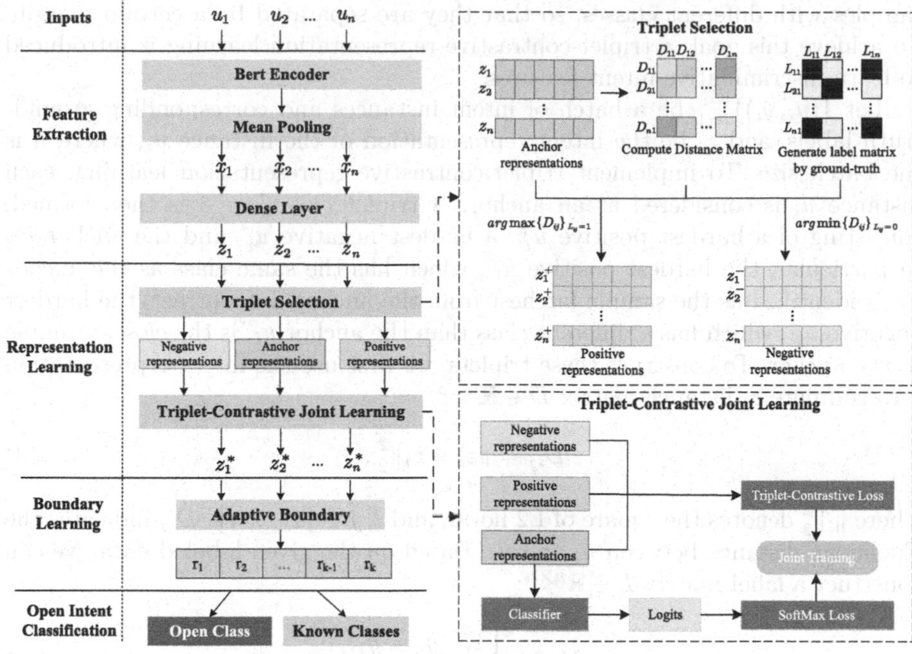

Fig. 1. The overall architecture of our proposed method.

BERT. Same as previous work [12,21,23], the mean pooling layer is applied to generate sentence semantic representation $x_i \in \mathbb{R}^H$:

$$x_i = \text{MeanPooling}\left([CLS, T_1, T_2, ...T_M]\right), \tag{1}$$

where CLS is a special classification token, M is the sequence length and H is the dimension of the hidden layer. To enhance the feature extraction capability, we feed x_i into a dense layer h to acquire the intent representation $z_i \in \mathbb{R}^D$:

$$z_i = h(x_i) = \text{ReLU}\left(W_h x_i + b_h\right), \tag{2}$$

where D is the dimension of intent representation, $W_h \in \mathbb{R}^{H \times D}$ and $b_h \in \mathbb{R}^D$ denote the weight and bias term of the layer h respectively.

3.4 Triplet-Contrastive Representation Learning

Due to lack of open intent samples in training stage, directly training a model for open intent detection is unappliable. The previous methods indirectly reduce the risk of open spaces by pulling samples of known intents with same class closer together and pushing samples with different classes apart. However, these methods may increase the risk of misidentifying open intents that are similar to known intents. To mitigate the risk, we aim to pull the farthest intent samples belonging the same class together, while pushing them away from the nearest intent samples with different classes, so that they are separated by a certain margin. To achieve this goal, a triplet-contrastive representation learning is introduced to learn discriminative intent features.

Let $\{(u_i, y_i)\}_{i=1}^n$ be a batch of intent instances and corresponding ground-truth labels, and z_i be the intent representation of the instance u_i, where n is the batch size. To implement triplet-contrastive representation learning, each instance u_i is considered as an anchor. A triplet (u_i, u_i^+, u_i^-) is then formed, consisting of a hardest positive u_i^+, a hardest negative u_i^- and the anchor u_i. In particular, the hardest positive u_i^+, which has the same class as the anchor u_i, is identified as the sample farthest from the anchor. In contrast, the hardest negative u_i^-, which has a different class than the anchor u_i, is the closest sample to the anchor. To construct these triplets, we first use the intent representation z to compute a distance matrix $D \in \mathbb{R}^{n \times n}$:

$$D_{ij} = \|z_i - z_j\|_2^2, \tag{3}$$

where $\|\cdot\|_2^2$ denotes the square of L2 norm, and $i, j \in \{1, ..., n\}$. D_{ij} indicates the Euclidean distance between u_i and u_j. Based on the given labeled data, we can construct a label matrix $L \in \mathbb{R}^{n \times n}$:

$$L_{ij} := \begin{cases} 1, & y_i = y_j, \\ 0, & y_i \neq y_j \end{cases} \tag{4}$$

where $i, j \in \{1, ..., n\}$. Then, we utilize the distance matrix D and label matrix L with intent representation z to select hardest positive and hardest negative for each anchor. The selection strategy is formulated as:

$$z_i^+ = arg\max_{z_j} \{D_{ij}\}_{L_{ij}=1}, \tag{5}$$

$$z_i^- = arg\min_{z_j} \{D_{ij}\}_{L_{ij}=0}, \tag{6}$$

where z_i^+ and z_i^- respectively denote the representation of the hardest positive u_i^+ and the hardest negative u_i^-. For each triplet, we aim to satisfy following constraint:

$$\|z_i - z_i^-\|_2^2 - \|z_i - z_i^+\|_2^2 \geqslant \alpha, \tag{7}$$

which means the distance between u_i and u_i^- greater than u_i and u_i^+ at least a margin α. Hence, these triplets are fed into a triplet-contrastive loss which is computed by triplet loss [24]:

$$L_{TC} = \frac{1}{N} \sum_{i=1}^{N} \max\left(0, \|z_i - z_i^+\|_2^2 - \|z_i - z_i^-\|_2^2 + \alpha\right), \tag{8}$$

where N is the number of training samples. When $\|z_i - z_i^+\|_2^2 + \alpha > \|z_i - z_i^-\|_2^2$, the loss is positive and network focuses on penalizing these triplets.

To better leverage the label information, we combine the softmax loss L_S with the triplet-contrastive loss L_{TC} to acquire final optimization objective L:

$$L_S = -\frac{1}{N}\sum_{i=1}^{N} \log \frac{\exp\left(\phi\left(z_i\right)^{y_i}\right)}{\sum_{j=1}^{K} \exp\left(\phi\left(z_i\right)^j\right)}, \tag{9}$$

$$L = L_S + \lambda L_{TC}, \tag{10}$$

where $\phi\left(\cdot\right)$ denotes a linear classifier and $\phi\left(z_j\right)^j$ denotes the classification logits of the jth class. λ is a hyperparameter to control the balance between the softmax loss L_S and the triplet-contrastive loss L_{TC} in Eq. (10). Through triplet-contrastive learning, the initial representation z can be tuned to more discriminative representation z^*, and they are further used to learn decision boundaries.

3.5 Open Intent Detection

An adaptive boundary method is proposed to learn decision boundary for known intents. We follow the approach of Zhang et al. [12] to construct spherical decision boundaries for each known class. After representation learning, training sample u_i into the model again to acquire its representation z_i^*. Based on the representation and corresponding label (z_i^*, y_i), central vector $(c_1, ...c_K)$ can be calculated and boundary radius $(r_1, ...r_K)$ can be learned for each known class. Specially, the central vector $c_k \in \mathbb{R}^D$ is the average vector of each class:

$$c_k = \frac{1}{|A_k|} \sum_{\left(z_i^*, y_i\right) \in A_k} z_i^*, \tag{11}$$

where A_k and $|A_k|$ are all samples of the kth class and corresponding quantity. The boundary radius r_k is learned by a neural network with a boundary parameter $\widehat{r_k} \in \mathbb{R}$, and Softplus activation function is used as a mapping between r_k and $\widehat{r_k}$:

$$r_k = \log\left(1 + \exp\left(\widehat{r_k}\right)\right). \tag{12}$$

Then, a boundary loss L_b with two hinge loss terms is introduced to balance empirical and open space risks [25]:

$$\begin{aligned} L_b = \frac{1}{N}\sum_{i=1}^{N} \big[&\max\left(0, \|z_i^* - c_{y_i}\|_2 - r_{y_i}\right) \\ &+ \max\left(0, r_{y_i} - \|z_i^* - c_{y_i}\|_2\right)\big]. \end{aligned} \tag{13}$$

In testing phase, a testing example is treated as open class if it lies outside all decision boundaries.

$$y = \begin{cases} \text{open, if } \|z^* - c_k\| > r_k, \forall k \in \{1, ..., K\}, \\ arg \max\limits_{I_i \in I^{Known}} \phi\left(I_i|x\right), \text{otherwise.} \end{cases} \tag{14}$$

4 Experiments

4.1 Datasets and Evaluation Metrics

To evaluate the effectiveness of our proposed method, we conducted experiments on three benchmark datasets, including BANKING, OOS, and SNIPS. The detailed statistics are shown in Table 1.

BANKING is a dataset in banking domain. It contains 77 fine-grained intent categories and 13,083 customer service queries [26].

OOS is a dataset that covers 150 intent classes over 10 domains [27], containing 22,500 in-scope queries and 1,200 out-of-scope queries.

SNIPS is a dataset about personal voice assistant. It contains 14,484 samples and 7 intent categories in different domains [28].

Table 1. Statistics of the three datasets.

Dataset	#Class	#Training	#Validation	#Testing	Avg of sentence length
BANKING	77	9003	1000	3080	11.91
OOS	150	15000	3000	5700	8.31
SNIPS	7	13084	700	700	9.05

Following previous work [6,12,18,21], we use accuracy score (ACC) and macro F1-score (F1) as metrics to evaluate overall performance. Besides, macro F1-score over known classes (Known) and the open class (Open) are used to evaluate the ability of distinguishing known intents and the open intent.

4.2 Baselines and Settings

We compare our method with following state-of-the-art open intent classification methods:

OpenMax is an open set detection method in computer vision [29]. It replaces the softmax layer with OpenMax and uses the Weibull distribution to calibrate confidence score.

MSP uses maximum softmax probability of samples as confidence score to judge whether they belong to the open intent [10]. We use the same confidence threshold (0.5) as in [6,12].

DOC builds a 1-vs-rest sigmoid classifier for known classes and uses Gaussian fitting to determine a threshold [18].

LMCL uses large margin cosine loss to learn discriminative features and applies LOF to detect the open intent [7].

SEG incorporates class semantic information into a Gaussian mixture distribution to learn discriminative features [19].

ADB is a post-processing method to learn adaptive decision boundaries [12].

DA-ADB is a variant of ADB [21], which utilizes distance-aware strategy to learn discriminative intent features.

Following the same setting as previous work [7,12,18], all datasets are divided into training, validation, and testing sets. The number of known classes varies in the ratio of 25%, 50%, and 75%. The remaining classes are treated as the open class and removed from the training and validation set. Both known classes and the open class are used in testing stage. For each experimental setting, we report the average performance of ten runs with different random seeds.

To make a fair comparison, we employ BERT-Base [22] as model backbone of all baselines. To speed up training process, we freeze all parameters of BERT except for the last transformer layer. The feature dimension D is set as 768, the training batch size and learning rate are 128 and 2e−5 respectively. For margin hyperparameter α, we set the value as 2.0, 3.0, 0.5 for BANKING, OOS, SNIPS respectively. The tradeoff parameter λ is set as 1.0 for all settings. All experiments are trained 100 epochs and the best model is selected based on the performance on the validation set with early stopping.

4.3 Main Results and Ablation Study

We present the results of method comparison in Table 2 and Table 3, where the best results are highlighted in bold. Specially, Table 2 shows the overall performance of macro-F1 score and accuracy score over all classes. Table 3 shows the performance of known classes and the open class in macro-F1 score.

Our method TCAB achieves the best performance on the three datasets and outperforms all the baselines. Compared with the best baseline DA-ADB, our method improves ACC on BANKING by 1.05%, 1.62%, and 1.62%, on OOS by 1.37%, 1.17%, and 1.07%, on SNIPS by 4.42%, 2.57%, and 1.23% in 25%, 50% and 75% settings respectively, which shows the effectiveness of our method.

For the performance on fine-grained classes, our method and outperforms all other baselines on both known classes and open class. Our method learns more discriminative features with the aid of triplet-contrastive learning, which is beneficial for open intent classification. Especially on SNIPS dataset, our method outperforms the best baseline ADB by 5.38%, 4.24%, and 2.18% on the open class and 2.21%, 1.35%, and 0.68% on known classes in 25%, 50% and 75% settings respectively. This shows that our method is not only effective in detecting the open class, but also better in classifying known classes.

To explore the effect of each component in TCAB, we conduct ablation studies on the BANKKING dataset. Results in Table 4 show that our method outperforms all ablation models, and indicate that all components contribute to the final performance. Further, through observing the performance by removing triplet-contrastive learning, the accuracy drops about 4%, 5%, 2% on the three

Table 2. Overall performance of open intent classification with different known class ratio (25%, 50% and 75%) on the three datasets. † means $p < 0.05$ under t-test.

	Methods	BANKING		OOS		SNIPS	
		F1	ACC	F1	ACC	F1	ACC
25%	OpenMax	54.76	50.14	63.73	71.17	49.68	59.57
	MSP	51.27	43.88	53.06	56.27	37.65	28.57
	DOC	66.45	71.94	76.20	86.49	55.10	45.63
	LMCL	65.01	70.00	77.41	87.46	69.46	66.70
	SEG	51.86	50.60	49.38	56.27	70.03	66.02
	ADB	71.62	78.85	77.19	87.59	71.65	66.30
	DA-ADB	73.65	81.09	79.95	89.49	52.75	47.09
	TCAB	**74.54**	**82.14†**	**81.66†**	**90.86†**	**74.92†**	**71.12†**
50%	OpenMax	74.48	65.10	80.32	80.42	65.50	59.82
	MSP	72.45	61.54	72.71	66.90	63.34	57.97
	DOC	78.09	74.42	83.75	85.06	78.15	72.50
	LMCL	75.73	71.22	83.21	84.71	82.12	78.91
	SEG	63.30	55.24	49.38	56.27	71.07	65.71
	ADB	80.90	78.86	85.05	86.64	82.91	79.01
	DA-ADB	82.60	81.64	85.64	87.96	77.35	73.69
	TCAB	**83.73†**	**83.26†**	**86.94†**	**89.13†**	**84.83**	**81.54†**
75%	OpenMax	84.90	78.46	78.13	79.30	73.63	72.20
	MSP	84.54	77.55	83.88	77.04	72.56	71.49
	DOC	83.67	78.93	87.88	85.90	83.11	79.96
	LMCL	80.73	74.26	86.48	84.57	86.51	84.29
	SEG	69.74	64.87	48.11	47.86	76.67	73.53
	ADB	85.96	81.08	88.53	86.32	87.57	84.94
	DA-ADB	85.68	81.18	88.47	87.46	84.29	81.57
	TCAB	**86.92†**	**82.80†**	**89.65†**	**88.53†**	**88.50†**	**86.17†**

known classes settings respectively. This presents that triplet-contrastive learning can learning better intent representation. In addition, through observing the performance by removing adaptive boundary, the accuracy drops by 20.82%, 13.57%, 3.37% with three settings respectively. This reflects that constructing decision boundary is effective to detect the open intent.

Table 3. Performance of known classes and open class with different known class ratio (25%, 50% and 75%) on the three datasets. † means $p < 0.05$ under t-test.

	Methods	BANKING		OOS		SNIPS	
		Known	Open	Known	Open	Known	Open
25%	OpenMax	54.91	51.78	63.34	78.38	43.56	61.92
	MSP	51.75	42.17	52.81	62.83	56.48	0.00
	DOC	65.83	78.23	75.81	91.08	65.32	34.67
	LMCL	64.42	76.36	77.03	91.85	70.64	67.12
	SEG	51.73	54.28	48.99	64.08	70.99	68.10
	ADB	70.94	84.56	76.80	91.84	73.46	68.03
	DA-ADB	72.97	86.49	79.60	93.20	56.93	44.38
	TCAB	**73.87**	**87.31†**	**81.32†**	**94.14†**	**75.67†**	**73.41†**
50%	OpenMax	75.02	53.86	80.29	82.34	78.31	14.23
	MSP	73.14	45.91	72.82	64.04	77.40	7.12
	DOC	78.24	72.44	83.70	87.33	83.97	54.89
	LMCL	75.93	67.97	83.16	87.24	84.97	70.73
	SEG	63.82	43.63	62.43	60.05	78.97	39.46
	ADB	80.96	78.44	85.00	88.65	85.93	70.83
	DA-ADB	82.61	82.10	85.58	90.14	80.05	66.58
	TCAB	**83.72†**	**83.84†**	**86.89†**	**91.09†**	**87.28**	**75.07†**
75%	OpenMax	85.43	54.18	78.14	77.47	86.15	11.03
	MSP	85.16	48.15	84.04	65.81	85.77	6.52
	DOC	84.01	64.00	87.92	83.76	89.40	51.67
	LMCL	81.27	49.56	86.51	82.65	89.88	69.65
	SEG	70.28	38.21	48.13	45.24	85.48	32.59
	ADB	86.20	66.47	88.58	83.92	90.86	71.13
	DA-ADB	85.69	69.51	88.49	86.09	87.42	68.62
	TCAB	**87.19†**	**70.21†**	**89.67†**	**87.14†**	**91.54†**	**73.31†**

Table 4. Ablation study results on the BANKING dataset.

	Setting	Known	Open	F1	Acc
25%	TCAB	**73.87**	**87.31**	**74.54**	**82.14**
	w/o adaptive boundary	60.41	67.29	60.75	61.86
	w/o triplet-contrastive	70.1	84.28	70.81	78.41
50%	TCAB	**83.72**	**83.84**	**83.73**	**83.26**
	w/o adaptive boundary	77.12	62.79	76.75	69.69
	w/o triplet-contrastive	80.86	78.48	80.8	78.78
75%	TCAB	**87.19**	**70.21**	**86.92**	**82.8**
	w/o adaptive boundary	86.22	56.18	85.71	79.43
	w/o triplet-contrastive	86.15	66.51	85.82	80.97

4.4 Analysis

We explore the effect of the tradeoff parameter λ on the performance of our method. We conduct experiments by varying λ from 0.2 to 1.0 step by 0.2 on the three datasets with different known class proportions. As shown in Fig. 2, TCAB achieves a relatively stable and robust performance, which indicates the robustness of our method. As λ increases, the accuracy shows an upward trend on almost all of the settings. When $\lambda = 1.0$, TCAB achieves the best or competitive performance on all settings. Though $\lambda = 0.6$ and 0.8 is slightly better on BANKING and OOS with 75% known class ratio, it is lower on the other two settings, since a larger λ allows an excessive effect of the triplet-contrastive strategy and could be learn more discriminative features. The discriminative features help to learn compact decision boundaries, which is more effective when the sample of known intents is fewer. However, this may be disadvantageous for identifying known intents.

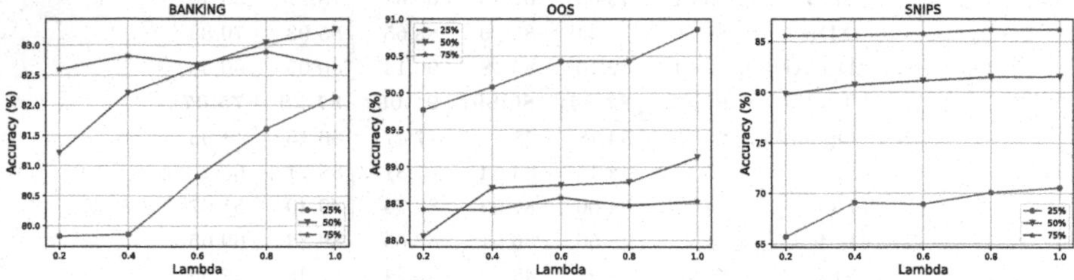

Fig. 2. Effect of λ on the three datasets with different known class proportions.

As a hyperparameter described in Eq. (7), the setting of α affects the distance between positive and negative samples on the representation learning. Figure 3 shows the results of using different α values on the three datasets. In general, a larger α leads to a better performance as it moves positive and negative samples farther apart, allowing the model to better distinguish between similar samples to enhance the discrimination of intent representations. For the BANKING dataset, the open intent detection performance shows an increasing trend and it achieves the highest ACC when $\alpha = 2.0$ at the 25% and 50% settings. A stable performance is achieved at the 75% setting. By gradually increasing the α value from 0 to 3.0, the performance of OOS gradually reaches the highest ACC in all settings. However, when α is greater than 0.5, the model performance shows a decreasing trend on SNIPS with 25% ratio. The reason may be that model with a large α pay more attention to expanding the interval between positive and negative samples, causing the training loss not converge.

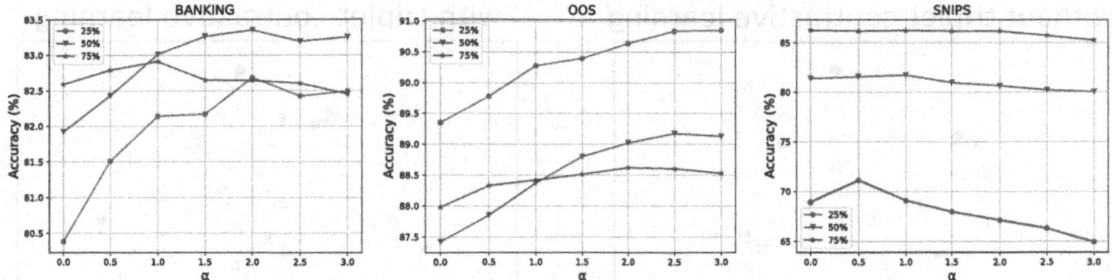

Fig. 3. Effect of α on the three datasets with different known class proportions.

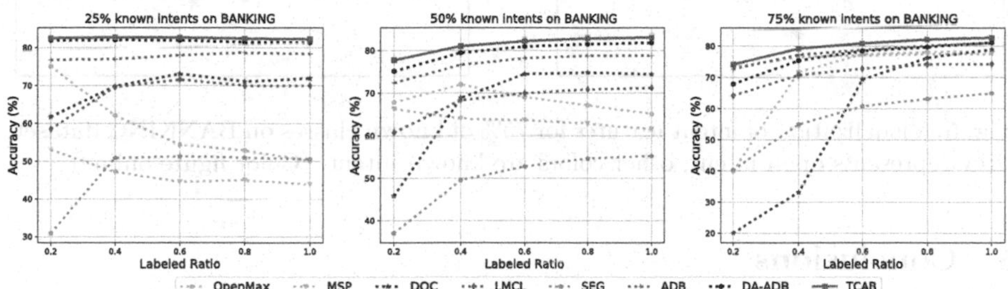

Fig. 4. Effects of labeled ratio on BANKING with different known class proportions.

We explore the effect of labeled data by varying the labeled ratio on the BANKING dataset, where the known class proportions are 25%, 50%, and 75%, and the range of labeled ratios are 0.2, 0.4, 0.6, 0.8, and 1.0. As shown in Fig. 4, our method achieves the best performance on all settings. Besides, Our method, ADB and DA-ADB keep a robust performance under different labeled ratios, while our method achieves a better performance. MSP and DOC perform better with small labeled ratios. Because as the amount of labeled data increases, these methods identify the open class as known classes with a high confidence. Since OpenMax only uses positive training samples to calculate the centroids of each class, it is easily influenced by the number of positive samples. In addition, LMCL and SEG perform better with more labeled data, since they rely heavily on the prior knowledge of labeled data.

We use t-SNE [30] to visualize the learned features on BANKING dataset with 75% ratio in Fig. 5, where open intent is in gray color, and known intents are represented by other colors. From these results, the learned features with triplet-contrastive strategy are more discriminative. The features of known classes are intra-class compact and inter-class separable. The distribution of the open class is also relatively more compact. This shows the effectiveness of the triplet-contrastive learning.

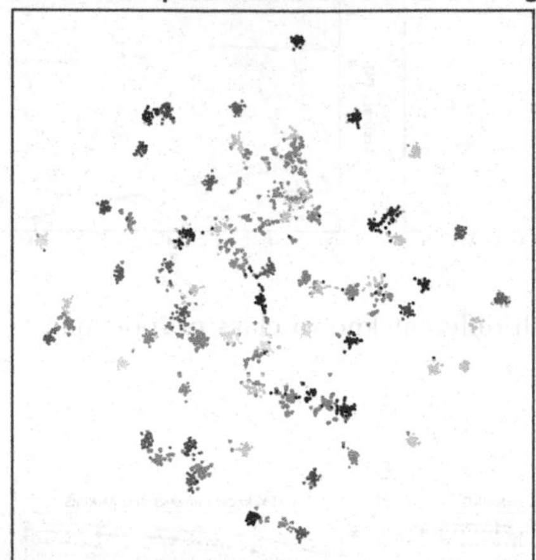

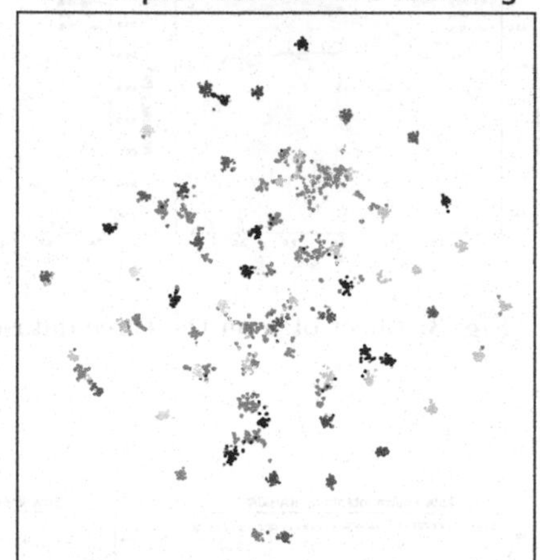

Fig. 5. Visualization of intent features for 75% of known classes on BANKING dataset. Grey represents open intent, other colors are known intent. (Color figure online)

5 Conclusions

This paper proposes a triplet-contrastive representation learning strategy, which constructs a triplet for each anchor through selecting a hard positive and a hard negative, and uses these triplets with triplet-contrastive loss to learn discriminative features by pulling anchor and positive together and pushing anchor and negative apart. In addition, a method named TCAB is proposed for open intent classification task. It utilizes triplet-contrastive strategy to acquire intent representations, and learns adaptive boundary of known intents. Compared with state-of-the-art methods, our method not only achieves consistent improvements on three intent benchmark datasets, but also yields robust performance with different ratios of known classes and labeled data.

Acknowledgements. The work described in this paper was substantially supported by a grant from the Research Grants Council of the Hong Kong Special Administrative Region, China (UGC/FDS16/E09/22).

References

1. Weld, H., Huang, X., Long, S., Poon, J., Han, S.C.: A survey of joint intent detection and slot filling models in natural language understanding. ACM Comput. Surv. (CSUR) (2021)
2. Liu, H., Liu, Y., Wong, L.P., Lee, L.K., Hao, T.: A hybrid neural network BERT-cap based on pre-trained language model and capsule network for user intent classification. In: Complexity 2020 (2020)
3. Luo, Y., Huang, Z., Wong, L.P., Zhan, C., Wang, F.L., Hao, T.: An early prediction and label smoothing alignment strategy for user intent classification of

medical queries. In: International Conference on Neural Computing for Advanced Applications, pp. 115–128 (2022)
4. Liu, Y., Hao, T., Liu, H., Mu, Y., Weng, H., Wang, F.L.: OdeBERT: one-stage deep-supervised early-exiting BERT for fast inference in user intent classification. ACM Trans. Asian Low-Resource Lang. Inf. Process. **22**(5), 1–18 (2023)
5. Hao, T., Li, X., He, Y., Wang, F.L., Qu, Y.: Recent progress in leveraging deep learning methods for question answering. In: Neural Computing and Applications, pp. 1–19 (2022)
6. Lin, T.E., Xu, H.: Deep unknown intent detection with margin loss. In: Proceedings of the 57th Annual Meeting of the Association for Computational Linguistics. pp. 5491–5496 (2019)
7. Xu, H., He, K., Yan, Y., Liu, S., Liu, Z., Xu, W.: A deep generative distance-based classifier for out-of-domain detection with mahalanobis space. In: Proceedings of the 28th International Conference on Computational Linguistics, pp. 1452–1460 (2020)
8. Zhan, L.M., Liang, H., Liu, B., Fan, L., Wu, X.M., Lam, A.Y.: Out-of-scope intent detection with self-supervision and discriminative training. In: Proceedings of the 59th Annual Meeting of the Association for Computational Linguistics and the 11th International Joint Conference on Natural Language Processing (Volume 1: Long Papers), pp. 3521–3532 (2021)
9. Zhou, W., Liu, F., Chen, M.: Contrastive out-of-distribution detection for pre-trained transformers. In: Proceedings of the 2021 Conference on Empirical Methods in Natural Language Processing, pp. 1100–1111 (2021)
10. Hendrycks, D., Gimpel, K.: A baseline for detecting misclassified and out-of-distribution examples in neural networks. arXiv preprint arXiv:1610.02136 (2016)
11. Breunig, M.M., Kriegel, H.P., Ng, R.T., Sander, J.: LoF: identifying density-based local outliers. In: Proceedings of the 2000 ACM SIGMOD International Conference on Management of Data, pp. 93–104 (2000)
12. Zhang, H., Xu, H., Lin, T.E.: Deep open intent classification with adaptive decision boundary. In: Proceedings of the AAAI Conference on Artificial Intelligence, vol. 35, pp. 14374–14382 (2021)
13. Cheng, Z., Jiang, Z., Yin, Y., Wang, C., Gu, Q.: Learning to classify open intent via soft labeling and manifold mixup. IEEE/ACM Trans. Audio Speech Lang. Process. **30**, 635–645 (2022)
14. Shu, L., Benajiba, Y., Mansour, S., Zhang, Y.: Odist: open world classification via distributionally shifted instances. In: Findings of the Association for Computational Linguistics: EMNLP 2021, pp. 3751–3756 (2021)
15. Ryu, S., Koo, S., Yu, H., Lee, G.G.: Out-of-domain detection based on generative adversarial network. In: Proceedings of the 2018 Conference on Empirical Methods in Natural Language Processing, pp. 714–718 (2018)
16. Zheng, Y., Chen, G., Huang, M.: Out-of-domain detection for natural language understanding in dialog systems. IEEE/ACM Trans. Audio Speech Lang. Process. **28**, 1198–1209 (2020)
17. Choi, D., Shin, M.C., Kim, E., Shin, D.R.: Outflip: generating examples for unknown intent detection with natural language attack. In: Findings of the Association for Computational Linguistics: ACL-IJCNLP 2021, pp. 504–512 (2021)
18. Shu, L., Xu, H., Liu, B.: Doc: deep open classification of text documents. In: Proceedings of the 2017 Conference on Empirical Methods in Natural Language Processing, pp. 2911–2916 (2017)

19. Yan, G., et al.: Unknown intent detection using gaussian mixture model with an application to zero-shot intent classification. In: Proceedings of the 58th Annual Meeting of the Association for Computational Linguistics, pp. 1050–1060 (2020)
20. Zeng, Z., et a.: Modeling discriminative representations for out-of-domain detection with supervised contrastive learning. In: Proceedings of the 59th Annual Meeting of the Association for Computational Linguistics and the 11th International Joint Conference on Natural Language Processing (Volume 2: Short Papers), pp. 870–878 (2021)
21. Zhang, H., Xu, H., Zhao, S., Zhou, Q.: Towards open intent detection. arXiv preprint arXiv:2203.05823 (2022)
22. Kenton, J.D.M.W.C., Toutanova, L.K.: BERT: pre-training of deep bidirectional transformers for language understanding. In: Proceedings of NAACL-HLT, pp. 4171–4186 (2019)
23. Lin, T.E., Xu, H., Zhang, H.: Discovering new intents via constrained deep adaptive clustering with cluster refinement. In: Proceedings of the AAAI Conference on Artificial Intelligence, vol. 34, pp. 8360–8367 (2020)
24. Schroff, F., Kalenichenko, D., Philbin, J.: Facenet: a unified embedding for face recognition and clustering. In: Proceedings of the IEEE Conference on Computer Vision and Pattern Recognition, pp. 815–823 (2015)
25. Scheirer, W.J., de Rezende Rocha, A., Sapkota, A., Boult, T.E.: Toward open set recognition. IEEE Trans. Pattern Anal. Mach. Intell. **35**(7), 1757–1772 (2012)
26. Casanueva, I., Temčinas, T., Gerz, D., Henderson, M., Vulić, I.: Efficient intent detection with dual sentence encoders. In: Proceedings of the 2nd Workshop on Natural Language Processing for Conversational AI, pp. 38–45 (2020)
27. Larson, S., et al.: An evaluation dataset for intent classification and out-of-scope prediction. In: Proceedings of the 2019 Conference on Empirical Methods in Natural Language Processing and the 9th International Joint Conference on Natural Language Processing (EMNLP-IJCNLP), pp. 1311–1316 (2019)
28. Coucke, A., et al.: Snips voice platform: an embedded spoken language understanding system for private-by-design voice interfaces. arXiv preprint arXiv:1805.10190 (2018)
29. Bendale, A., Boult, T.E.: Towards open set deep networks. In: Proceedings of the IEEE Conference on Computer Vision and Pattern Recognition, pp. 1563–1572 (2016)
30. van der Maaten, L., Hinton, G.: Visualizing data using t-SNE. J. Mach. Learn. Res. **9**, 2579–2605 (2008)

A User Intent Recognition Model for Medical Queries Based on Attentional Interaction and Focal Loss Boost

Yuyu Luo[1], Yi Xie[1], Enliang Yan[1], Lap-Kei Lee[2], Fu Lee Wang[2], and Tianyong Hao[1](✉)

[1] School of Computer Science, South China Normal University, Guangzhou, China
{2020022977,2020023050,yanenliang,haoty}@m.scnu.edu.cn

[2] School of Science and Technology, Hong Kong Metropolitan University, Hong Kong, China
{lklee,pwang}@hkmu.edu.hk

Abstract. Pre-trained language models such as BERT and RoBERTa have obtained new state-of-the-art results in the user intent recognition task. Nevertheless, in the medical field, the models frequently neglect to make full use of label information and seldom take the difficulty of intent recognition for each query sentence into account. In this paper, a new user intent recognition model based on Text-Label Attention Interaction and Focal Loss Boost named TAI-FLB is proposed to identify user intents from medical query sentences. The model focuses on incorporating a text-to-label attention interaction mechanism based on label embedding to exploit the information from labels. Moreover, during training process, the loss contribution of difficult samples with unclear intention is increased to shift model focus towards difficult samples in medical query statements. Experimental evaluation was performed on two publicly available datasets KUAKE and CMID. The results demonstrated that the proposed TAI-FLB model outperformed other baseline methods and demonstrated its effectiveness.

Keywords: Intent Recognition · Attentional Interaction · Focal Loss Boost · Label Embedding

1 Introduction

The user intent recognition task is one of fundamental tasks in the field of natural language processing, aiming to identify the purpose and needs that users want to express from query sentences [1]. Recently, the application of information technology to healthcare industry has promoted the development of healthcare informatization and brought better healthcare services to public. In particular, online health communities provide a convenient way for users to seek medical and health information service [2]. Users can seek help by describing their health conditions or posting inquiries. For example, a query "How should elderly people with high blood pressure be treated?" contains a user's query about the treatment of high blood pressure, and the predefined user intent category of the query as "treatment options" accordingly. Similarly, a query "Can diabetics take metformin?" corresponds to the predefined intent label "medical advice".

H. Zhang et al. (Eds.): NCAA 2023, CCIS 1870, pp. 245–259, 2023.
https://doi.org/10.1007/978-981-99-5847-4_18

With the development of artificial intelligence technology, pre-trained language models are widely used in the user intention recognition task and have achieved state-of-the-art results. Most of existing intention recognition methods are based on pre-trained language models such as BERT [3] and RoBERTa [4] to obtain contextually relevant text representations in order to extract semantic information embedding from utterances. The training data are usually from information recommendation, consumer services and other domains. However, due to the high complexity of query sentences and the large number of domain-specific terms in the medical domain, it is difficult to extract abundant fine-grained features from medical query sentences. Ineffective use of contextual information to disambiguate polysemous words, and inconsistent difficulty of intent recognition for each query sentence become major challenges. Thus, how to use natural language processing methodology to accurately extract intention from user query sentences has become the main focus [5].

In a supervised medical intent recognition task, both query sentences and predefined intent labels contain abundant semantic information. Most conventional intention recognition models encode labels using one-hot encoding, which enables a vector representation of labels as a purely symbolic representation without any semantic information. The labels are only utilized in the computation of model loss but fails to play a full role in the forward propagation process. Label embedding [6, 7] aims to map predefined label sequences, label description texts or label hierarchy relations to the same semantic space as the word embedding of query sentences. It has been proven to be an efficient way to support model training. Therefore, label information can be utilized for modeling in both text representation phase and prediction phase [8]. Motivated by these, we encode both query sentences and label sequences simultaneously in an intention recognition model. And inspired by the co-attention adopted in the multi-modal learning [9, 10], we propose to incorporate attentional interactions into the intention recognition task during label embedding encoding in order to utilize label embedding information more effectively.

In addition, some medical query sentences contain colloquial expressions, mis-expressions or unclear information when users ask for symptoms that are uncommon or difficult to be expressed by simple descriptions. The user intent of these query sentences is unclear. It results that some samples are more difficult for the model to accurately identify intent than others. Intuitively, some easily recognized samples may obtain incorrectly up-weighted, or difficult samples with unclear intention obtain erroneously down-weighted. Thus, it is inefficient to train a model reasonably on all samples using the widely applied cross-entropy loss or Focus Loss [11] as the objective function. To take the difficulty of intent recognition for each sample into account and increase the loss contribution of difficult samples, we improve the Focus Loss to enhance the accuracy of the intention recognition task.

In summary, the major contribution of the paper lies on three-fold:

1) An attentional interaction model TAI-FBL based on label embedding is proposed for obtaining word-level semantic information between query sentences and label sequences.

2) A Focal Loss Boost is proposed to shift model focus towards difficult samples with unclear intention in medical query sentences and increases the loss contribution of difficult samples during model training.
3) Experiments on two standard datasets demonstrate that the proposed TAI-FLB model outperforms state-of-the-art baseline methods, verifying its effectiveness.

2 Related Work

Traditional methods for user intent recognition use feature engineering techniques such as N-grams [12] and bag-of-words (BoW) [13] as the feature extractor, and then apply machine learning algorithms as the classifier. In recent years, models that were based on neural networks, such as Convolutional Neural Networks (CNN) [14] and Recurrent Neural Networks (RNN) [15], have been applied in the intent recognition task to extract more semantic features from text. Pre-trained language models have also been widely applied to intent recognition. Compared with traditional approaches, large pre-trained models such as BERT [3] and RoBERTa [4] recently have achieved substantial improvement in terms of performance thanks to their powerful encoding ability. Pre-trained models represent text sequences as a standard structure consisting of word vectors, positional embeddings and token embeddings to obtain dynamic word vector representations [16]. Many researches combined pre-trained models and neural networks for the intent recognition task. Guo et al. [17] utilized BERT to extract global features between words and sentences and input the features to a capsule network. Liu et al. [18] combined RoBERTa & CNN and utilized CNN to perform convolutional operations on features extracted by RoBERTa to capture important semantic information. All these methods were applicable at obtaining semantic information in query sentences. However, these supervised intent recognition models only used query sentences as input information and neglected to utilize label sequences in text representation stage.

To alleviate this issue, some researchers motivated by the idea of word embedding [19], encoded labels by mapping predefined sequences of labels, labelled description texts, or labelled hierarchy relations into the same semantic space as the embedding representation of query sentences. Wang et al. [7] and Du et al. [20] adopted label embedding frameworks for more efficient text classification. It had proved that label information had a positive role in the process of extracting semantic features. Liu et al. [9] used labels as part of input information and incorporated it into a Bi-LSTM for aggregation with query sentences. It enabled the semantic information of label sequences to be one of the bases for classification decisions. Wang et al. [21] adopted a hierarchical framework to obtain label information and modelled the relationship between query sentences and label sequences through a dynamic routing mechanism. However, most of label categories contained in user intent datasets did not have caste relationships. It was necessary to further explore how to fully utilize label information in the feature extraction process of the intention recognition model. Co-attention [22, 23] was widely applied in multi-modal learning between source and target to jointly produce attended representations and to focus on relevant parts of both. Therefore, this paper proposes to embed pre-defined label set into the same space as word embedding and then adopt an attentional interaction mechanism to produce a text-attended label representation.

As for user intent recognition tasks in the medical domain, the variability of intent recognition difficulty was another issue that requires continued research. Some researchers applied dynamic early exits in multiple stacks of Transformer encoders and matched the corresponding amount of computation to the recognition difficulty of query sentences. FastBERT [24] and DeeBERT [25] added sub-classifiers after each Transformer encoder, causing samples to exit training earlier if their probability entropy was less than a threshold. However, the threshold was set manually and experiments were required to select applicable parameters. LeeBERT [26] performed distillation learning for each Transformer layer and assigned a set of learnable weights to each sub-classifier. ELF [27] superimposed losses for samples that did not satisfy early exit criterion, resulting in an increase for the average loss contribution of difficult samples. This approach failed to differentiate the confidence inconsistency of the sample's fit results in each Transformer layer.

In dynamic early exit networks above, difficult samples exited later and accumulated a higher overall loss. But as for medical queries, ordinary users lacked medical expertise and might not express their intentions as accurate and concise as clinicians did. They might have misrepresentations and colloquial expressions. These query sentences contained user intent that was not sufficiently clear, and some intent information was even difficult to recognize manually. It was not sufficient to improve the accuracy of intent recognition in the medical field by dynamic early exit networks. In order to directly increase the contribution of difficult samples to the overall model loss, we propose Focal Loss Boost, which enables the model to increasingly focus on difficult samples in medical query sentences and further improves accuracy under inconsistent intent recognition difficulty.

3 The TAI-FLB Model

A new intention recognition model based on a Text-Label Attention Interaction Mechanism and Focal Loss Boost as TAI-FLB is proposed to identify user intents from medical queries. The model encodes query sentences and pre-defined label sequences to obtain query embedding and label embedding. In order to utilize the label sequences more effectively according to the characteristics of downstream tasks, the model further encodes the query sentences based on the label description information to obtain the text-attended label representation for measuring query sentences relevance to labels. In addition, we propose a Focal Loss Boost to the intention recognition task by improving Focal Loss that maintains the initial weights of easily recognized samples during training, as well as to enhance model in focusing on difficult samples to improve the accuracy of intention recognition. Figure 1 presents the overall architecture of the model.

3.1 Query Sentence Embedding and Label Embedding

The model processes query sentences and label sequences simultaneously during intent feature extraction by embedding both into the same semantic space. The model utilizes the pre-trained language model RoBERTa as encoder in order to extract semantic features in query sentences and label sequences. In addition, the model applies multi-layer

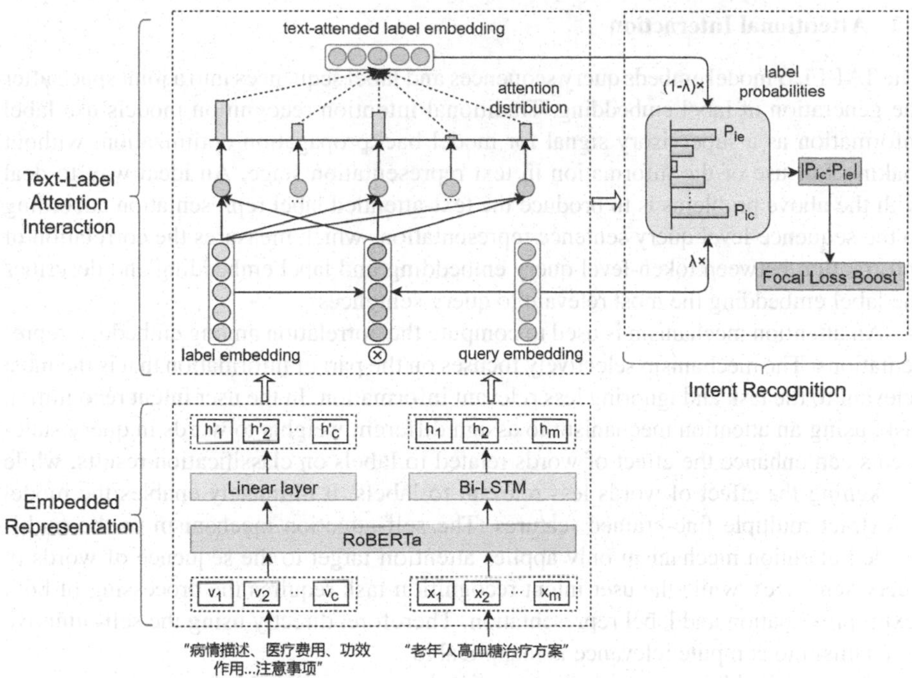

Fig. 1. The overall architecture of the TAI-FLB model

Bi-LSTM to encode the query sentences to capture contextual information. Since the structure of label sequences is relatively simple, in which individual labels are composed of one or two words. Encoding labels using complex neural networks may cause the overfitting issue. Therefore, our model does not apply over complex structures to extract semantic information from label sequences.

For a query sequence $X = \{x_1, x_2, \ldots\ldots, x_m\} \in R^m$ and a label sequence $V = \{v_1, v_2, \ldots\ldots, v_c\} \in R^c$, where m is the maximum sequence length of dataset and c is the number of classes. Firstly, the query embedding $X_{rb} \in R^{m\times d}$ and label embedding $L_{emb} \in R^{c\times d}$ are obtained based on RoBERTa, and d is the dimension of hidden feature space to represent the text. The TAI-FLB model applies multilayer Bi-LSTM neural network to encode the query sentence to obtain the contextual representation $X_{bl} \in R^{m\times d}$ as shown in Eq. (1). The model concatenates the representations obtained from the two stages to obtain final query sentence representation $X_{emb} \in R^{m\times 2d}$, which preserves positional embedding information and contextual sequencing information. X_{emb} is a vector representation that combines syntactic, semantic and text structure information, as shown in Eq. (2).

$$X_{bl} = Bi-LSTM\,(X_{rb}) \tag{1}$$

$$X_{emb} = Concat[X_{rb};\ X_{bl}] \tag{2}$$

3.2 Attentional Interaction

The TAI-FLB model embeds query sequences and label sequences into a joint space after the generation of label embedding. Traditional intention recognition models use label information as a supervisory signal for model backpropagation optimization, without making full use of the information in text representation stage. An ideal way to deal with the above problems is to produce the text-attended label representation according to the sequence-level query sentence representation, which measures the correlation of information between token-level query embedding and label embedding and describes the label embedding the most relevant to query sentences.

An attention mechanism is used to compute the correlation among embedded representations. The mechanism selectively focuses on the part of information that is the most relevant to the task and ignoring less relevant information. In the user intent recognition task, using an attention mechanism to assign different weights to words in query statements can enhance the effect of words related to labels on classification results, while weakening the effect of words less relevant to labels. It ultimately enables the model to extract multiple fine-grained features. The self-attention mechanism or the multi-headed attention mechanism only applies attention target to the sequence of words in query sentences, while the user intent recognition task requires the processing of both text representation and label representation. Therefore, directly using the self-attentive mechanism to compute relevance is inapplicable.

The model adds an attentional interaction layer to use label information in the text representation stage, and improves the self-attention mechanism into a joint attentional interaction mechanism. The attentional interaction mechanism utilizes both the query embedding X_{emb} and the label embedding L_{emb} as input sequences to the attentional interaction layer, which results in a text-attended label embedding representation. Firstly, the similarity matrix $S \in R^{d \times c}$ is obtained by calculating the inner product between words, which contains the relevance information of each query word to each label, and the calculation is shown in Eq. (3). Subsequently, the model obtains the attention weight vector G by normalizing the similarity matrix, which is calculated as shown in Eq. (4). m is the number of words in the text and m_i is a vector representation of the randomly initialized context.

$$S = X_{emb} L_{emb}^T \tag{3}$$

$$G = \frac{\exp(S \times m_i)}{\sum_{m=1}^{M} (\exp(S \times m_i))} \tag{4}$$

Afterwards, the TAI-FLB calculates text-attended label embedding features by combining the attention weight vector with the embedded L_{emb} of the label sequence. In this process, the model establishes direct associations with the most relevant labels by assigning weights to each word in the query statement. These weights are updated during the backpropagation process by referring to the contextual context. As shown in Eq. (5), the model obtains text-attended label embedding features X_{text_label} as supplementary information. X_{text_label} contains both abundant textual multivariate semantic features and semantic information in label sequences. By embedding query sentences and label sequences into the same semantic space for attentional interaction, the model reinforces the role of labels with higher relevance to user intents during training process. Both explicit and implicit correlations between text representations and labels are generated for better intent recognition.

$$X_{text\_label} = (\frac{G}{\sqrt{d}}) \times L_{emb} \quad (5)$$

3.3 Focal Loss Boost

There are easily recognized samples with clear intent and difficult samples with unclear intent from intention query sentences in the medical domain. The text structure of the easily recognized samples is simple, with distinct features and low loss value. It contains intention information that can be accurately recognized by the model after forward training and back propagation optimization. In contrast, difficult samples are defined as query sentences that contain colloquial expressions, misrepresentations or redundant information. The loss value of such query statements is usually high, but the overall loss contribution to model training is small. It results that difficult samples are more challenging for the model to accurately identify intent than others. When the intention recognition model uses traditional cross-entropy loss function as the objective function, the focus of the model is set as whether the model accurately predicts sample class. At this point, it fails to solve the problem of inconsistent difficulty of intention recognition.

Focal Loss focuses model attention during training on the variability of the difficulties of intention recognition. During training process, a model focuses more attention on difficult samples that are more likely to produce false identification results, and increases the loss contribution of difficult samples. Specifically, Focal Loss adds a weighting factor α_c and a modulation factor $(1 - p_{ic})^{\gamma}$ to the cross-entropy loss and the calculation is shown in Eq. (6).

$$L_{fl} = -\alpha_c(1 - P_{ic})^{\gamma} \log P_{ic} \quad (6)$$

$\alpha_c \in [0,1]$ is the weight of the c-th class sample. P_{ic} is the predicted probability that sample i belongs to category c to indicate the probability that the model correctly predicts the sample. When the intention category of the sample is incorrectly predicted and the value of P_{ic} is small, the confidence level of the prediction result is low. Modulation factor converges to 1 and the original loss contribution is not affected. When the intention category of samples is correctly predicted and P_{ic} converges or equals to 1, the confidence of the prediction result is high and the modulation factor converges to 1. Thus, the loss contribution is reduced accordingly. Focal Loss defines γ as a focusing parameter. The Focal Loss is calculated in the same way as the cross-entropy loss function when $\gamma = 0$. With the increase of γ, the modulation factor has different treatment for easily recognized samples and difficult samples. γ adjusts the weights of easily identifiable samples as a way to indirectly bias the focus of the model to difficult samples that are not correctly predicted. However, easily recognized samples and difficult samples are dynamic concepts in Focal Loss. During the iterative process, reducing the model attention to samples with high confidence may enable these samples to become less confident, and transform from easily recognizable samples to difficult ones. The convergence of the model is slower and the optimization of the model thus is affected.

In order to incorporate the notion of inconsistent intent recognition difficulty during training, we propose Focal Loss Boost, named FLB, as the improvement of the Focal Loss to optimize model training process. By setting a prediction probability threshold to define easily recognized samples and difficult samples, it shifts model focus towards difficult samples and reduce the influence of the model on the training process of easily recognized samples. The FLB is calculated using Eq. (7) and Eq. (8).

$$FLB = -\alpha_t(1 - \varphi(P_{ic}, K))^{\gamma} \log P_{ic} \tag{7}$$

$$\varphi(P_{ic}, K) = \begin{cases} P_{ic} - K, |P_{ic} - P_{ie}| < K \\ y_{ic}, |P_{ic} - P_{ie}| \geq K \end{cases} \tag{8}$$

K is the prediction probability threshold. P_{ic} is the maximum prediction probability of the sample. P_{ie} is the sub-maximum prediction probability of the sample. The model determines the sample as an easily identifiable sample when the absolute difference value between the maximum prediction probability and the sub-maximum prediction probability is higher than K. The value of $\varphi(P_{ic}, K)$ is taken as the probability distribution of the true label. The FLB is calculated in the same way as the cross-entropy loss does, and the focus of the model is adjusted to whether it accurately predicts the label category of the sample. In contrast, when the absolute value is lower than K, the model determines the sample as a difficult sample. The value of $\varphi(P_{ic}, K)$ is taken as the difference between the maximum prediction probability and the prediction probability threshold. FLB incorporates the notion of inconsistent intent recognition difficulty and enables easily recognized samples to maintain the original training process, and further enhances the weight of difficult samples during training iterations. As a result, the samples with incorrectly predicted intent categories correspond to increase loss contribution and acquire more model attention during the next training round.

The TAI-FLB model utilizes FLB as the objective function and performs supervised learning of original text representation and text-attended label representation separately during the training process. $Loss_{union}$ is the overall loss after weighted sum of the original text representation *loss_or* and the text-attended label representation *loss_mix*. Equation (9) shows the calculation process of $Loss_{union}$. λ is a weighting factor, which is used to adjust the degree of influence of *loss_or* and *loss_mix* on the overall loss contribution.

$$Loss_{union} = \lambda \times loss\_or + (1 - \lambda) \times loss\_mix \tag{9}$$

4 Experiments and Results

4.1 Datasets

Experiments were conducted based on two publicly available Chinese medical standard datasets KUAKE and CMID [28]. The KUAKE dataset contained 10,800 medical query sentences with predefined labels in 11 user intent categories, such as *diagnosis*, *cause*, *method*, *advice*, *result*, *effect*, and *price*. The CMID dataset had 36 categories of user intent labels, including *definition, prevention, infectivity, price, side effect, recovery time, prevention*, etc. Since missing values existed in original data of the CMID dataset and certain categories contained extreme small size of samples, we supplemented the missing values by referring to their contextual values and removed the categories containing samples less than 20. Eventually, 13,249 sentences with predefined labels in 31 categories of the CMID were utilized. The statistics of the two datasets are shown in Table 1.

Table 1. Statistics of the KUAKE and CMID datasets

Dataset	#Categories	#Training	#Validation	#Test	# Maximum words per sentence
KUAKE	11	6,931	1,955	1,994	60
CMID	31	6,093	4,078	3,078	100

4.2 Evaluation Metrics

Four widely used metrics are applied to evaluate the user intent classification task: Accuracy, Precision, Recall and F1-score. Accuracy is the percentage of correctly predicted samples over total samples. Precision is the proportion of correctly predicted positive samples over all samples that predicted to be positive. Recall is the proportion of correctly predicted positive samples over all relevant samples. F1-score is as a harmonic average of Precision and Recall. The calculations of the metrics are as Eq. (10)–(12).

$$Accuracy = \frac{TP + TN}{TP + FP + TN + FN} \tag{10}$$

$$Precision = \frac{TP}{TP + FP}, Recall = \frac{TP}{TP + FN} \tag{11}$$

$$F1 - score = \frac{2 \times Precision \times Recall}{Precision \times Recall} \tag{12}$$

TP denotes the number of correctly predicted positive samples, TN denotes the number of correctly predicted negative samples, FP denotes the number of incorrectly predicted positive samples, while FN denotes the number of incorrectly predicted negative samples.

4.3 Parameter Settings

The EP-LSA strategy was implemented on the Pytorch 1.7.1 framework including the pre-trained models RoBERTa and Bi-LSTM. The dimension of user intent features was 768, the epoch was 10, and the dropout rate was 0.1. The strategy was optimized by Adam for training with the learning rate of $2e^{-5}$ for the RoBERTa and Bi-LSTM. The batch size was 16. The focusing parameters γ was 0.2, which was selected by set of experiments.

Certain parameters were set differently on the two datasets. The maximum sequence length of the KUAKE dataset was limited to 60, while the length of the CMID dataset was limited to 100. Since the query representation and text-attended label embedding representation had different degrees of contribution to downstream tasks, the hyper-parameter λ was set to 0.4 for the KUAKE dataset and was set to 0.7 for the CMID dataset.

4.4 Baseline Methods

The TAI-FLB model was compared with a set of baseline methods on the KUAKE and CMID datasets to verify its effectiveness on the user intent recognition task.

1) **RBERT-C** [18]: The model utilized a RoBERTa pre-trained model as sentence encoder, combined CNN as feature extractor, as well as used a maximum pooling operation to enhance feature representation capability.
2) **RoBERTa+RCNN** [29]: The joint model combined a Bi-LSTM bidirectional loop structure and a maximum pooling operation to maximize the capture of contextual information, and retained a large range of sequential information when learning word representations.
3) **RoBERTa+RCNN+Focal Loss** [11]: The model reduced the weight of easily classified samples with the help of Focal Loss, allowing to focus more on difficulty classified samples during training.
4) **RoBERTa+Bi-LSTM** [30]: The model utilized RoBERTa pre-trained word vectors as input, extracted contextual features by Bi-LSTM.
5) **FastBERT** [24]: The model connected branch classifiers after each layer of Transformer sub-network to perform dynamic early exits for samples, and trained branch classifiers using self-distillation.

4.5 The Result

The performance of our TAI-FLB model was compared with the baseline methods on the KUAKE and CMID datasets and the results are shown in Table 2. Both the RoERTa+RCNN and RoBERTa+RCNN+Focal Loss utilized a bidirectional loop structure of Bi-LSTM and a maximum pooling of convolutional layer in RCNN to extract features from query sentences and achieved an improvement performance compared with the RBERT-C. The RoBERTa+RCNN achieved an F1-score of 0.819 and 0.414 on the KUAKE and CMID datasets, respectively. The RoBERTa+RCNN+Focal Loss obtained an F1-score of 0.817 on the KUAKE dataset and an F1-score of 0.427 on the CMID dataset, slightly higher than that of the RoBERTa+RCNN. Our TAI-FLB model achieved an accuracy of 0.825 and an F1-score of 0.824 on the KUAKE dataset, while achieved an accuracy of 0.545 and an F1-score of 0.461 on the CMID dataset. The TAI-FLB model had an improvement ratio of 1.2% on F1-score and 0.9% on accuracy compared to the RoBERTa + Bi-LSTM on the KUAKE dataset. In addition, the improvement was more significant on the CMID dataset as 1.8% on F1-score and 1.1% on accuracy compared to the RoBERTa + Bi-LSTM. The results verified that TAI-FLB model enhanced the

Table 2. The performance comparison on the KUAKE and CMID datasets

Methods	Datasets	Precision	Recall	F1-score	Accuracy
RBERT-C	KUAKE	0.778	0.820	0.798	0.798
	CMID	0.420	0.419	0.400	0.484
RoERTa+RCNN	KUAKE	0.799	0.839	0.819	0.817
	CMID	0.442	0.390	0.414	0.530
RoERTa+RCNN +Focal Loss	KUAKE	0.803	0.819	0.817	0.818
	CMID	0.475	0.388	0.427	0.532
FastBERT	KUAKE	0.822	0.793	0.807	0.814
	CMID	0.478	0.402	0.437	0.524
RoERTa +Bi-LSTM	KUAKE	0.799	0.826	0.812	0.816
	CMID	0.522	0.413	0.443	0.534
TAI-FLB	KUAKE	**0.801**	**0.841**	**0.824**	**0.825**
	CMID	**0.489**	**0.406**	**0.461**	**0.545**

contextual representation capability by utilizing label information in the text representation stage. In addition, the TAI-FLB model used FLB as the objective function, which enabled more effective training for difficult samples with unclear intention.

To verify the effectiveness of the FLB, multiple comparison experiments were conducted. As shown in Table 3, the TAI-FLB model improved F1-score by 0.2% compared to the model without FLB on the KUAKE dataset, while improved F1-score by 0.5% on the CMID dataset. An additional experiment using the baseline models of RoBERTa-RCNN and FastBERT verified the similar performance improvement. The RoBERTa-RCNN model improved accuracy by 1.1% compared to the model without FLB and improved F1-score by 1.6% on the CMID dataset. The FastBERT model improved F1-score by 0.9% compared to the model without FLB and by improved F1-score 1.1% on the CMID dataset. In addition, the enhancement of F1-score for the three models on the CMID dataset was more obvious than that on the KUAKE dataset, which might be caused by the relatively large proportion of difficult samples in the CMID dataset. Enhancing model focus on difficult samples was beneficial to improve the accuracy of intention recognition for difficult samples. In summary, FLB had a positive impact on model performance by defining the difficulty of intention recognition of samples, which allowed both easily recognized samples to maintain a natural iterative process and difficult samples with increased weights.

To observe the changes of the model after incorporating the attention interaction mechanism, we analyzed the label weights of query statements. Figure 2 showed the label weights of each word in the query sentence "*How should elderly people with hypertension be treated*" from the KUAKE dataset. It contained user intent about how to deal with hypertension symptoms for the elderly, and the predefined user intent category was "*method*". Deeper colors and larger values in the graph indicated stronger relevance of the words to the corresponding labels.

Table 3. The performance comparison of the effects of the TAI-FLB model and the baseline model using the FLB

Methods	Datasets	Precision	Recall	F1-score	Accuracy
RoBERTa-RCNN	KUAKE	0.799	0.839	0.819	0.817
	CMID	0.442	0.390	0.414	0.530
RoBERTa-RCNN+FLB	KUAKE	0.803	0.819	0.819	0.819
	CMID	0.479	0.389	0.430	0.541
FastBERT	KUAKE	0.822	0.793	0.807	0.814
	CMID	0.478	0.402	0.437	0.524
FastBERT+FLB	KUAKE	0.796	0.841	0.816	0.818
	CMID	0.486	0.414	0.448	0.532
TAI	KUAKE	0.798	0.837	0.822	0.821
	CMID	0.469	0.400	0.456	0.541
TAI-FLB	KUAKE	0.801	0.841	0.824	0.825
	CMID	0.489	0.406	0.461	0.545

The query sentence contained the words "*elderly people*", "*hypertension*", "*how*", "*treated*" and so on, which had different correlation coefficients with the predefined labels. "*elderly people*" had the highest correlation with the label "*method*" and the

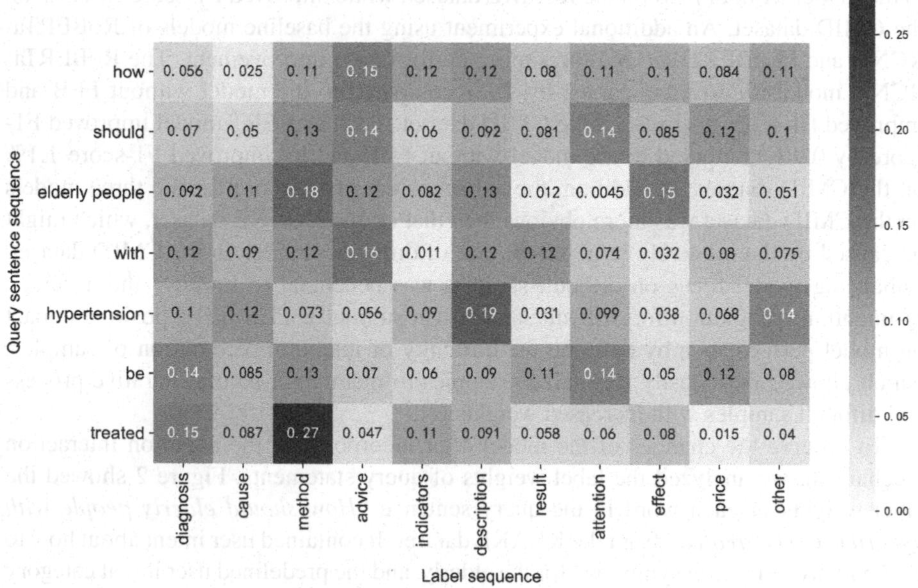

Fig. 2. Visual analysis of attentional interaction structure

weight as 0.18. "*hypertension*" had the highest correlation with the label "*description*" and the weight as 0.19. "*treated*" had the highest correlation with the label "*method*" with the weight as 0.27. Therefore, the words "*elderly people*", "*hypertension*", "*treated*" and the labels most relevant with these words would play a more important role for intent feature extraction. The result indicated that attentional interaction mechanisms made full use of labeling information in the process of feature representation. The model assigned the most relevant label weights to each word by calculating the relevance of the word to all labels. Finally, the text-attended label embedding features were obtained. It was beneficial for the model to obtain word-level semantic information and understand the interpretation of each word based on obtaining fine-grained features of contextual representation.

5 Conclusions

This paper proposed a TAI-FLB model for the user intent recognition from medical queries. The model introduced an attention interaction mechanism with label embedding, where query sentences and label sequences were encoded to obtain text-attended label representation. We incorporated the notion of inconsistent intent recognition difficulty during training. The easily recognized samples maintained natural iterative process, and the weights of the difficult samples were increased during training. Experiments on two standard datasets demonstrated the model was effective for user intent recognition from medical queries compared with a list of state-of-the-art baseline methods.

Acknowledgements. The work described in this paper was substantially supported by a grant from the Research Grants Council of the Hong Kong Special Administrative Region, China (UGC/FDS16/E09/22).

References

1. Xie, W., Gao, D., Hao, T.: A feature extraction and expansion-based approach for question target identification and classification. In: Wen, J., Nie, J., Ruan, T., Liu, Y., Qian, T. (eds.) CCIR 2017. LNCS, vol. 10390, pp. 249–260. Springer, Cham (2017). https://doi.org/10.1007/978-3-319-68699-8_20
2. Cai, R., Zhu, B., Ji, L., Hao, T., Yan, J., Liu, W.: An CNN-LSTM attention approach to understanding user query intent from online health communities. In: 2017 IEEE International Conference on Data Mining Workshops (ICDMW), pp. 430–437 (2017)
3. Devlin, J., Chang, M.W., Lee, K., Toutanova, K.: BERT: pre-training of deep bidirectional transformers for language understanding. Arxiv Preprint Arxiv:1810.04805 (2018)
4. Cui, Y., Che, W., Liu, T., et al.: Pre-training with whole word masking for Chinese BERT. IEEE/ACM Trans. Audio 3504–3514 (2021)
5. Hao, T., Li, X., He, Y., Wang, F.L., Qu, Y.: Recent progress in leveraging deep learning methods for question answering. Neural Comput. Appl. 1–19 (2022)
6. Zhang, H., Xiao, L., Chen, W., Wang, Y., Jin, Y.: Multi-task label embedding for text classification. Arxiv Preprint Arxiv:1710.07210 (2017)
7. Wang, G., et al.: Joint embedding of words and labels for text classification. Arxiv Preprint Arxiv:1805.04174 (2018)

8. Liu, N., Wang, Q., Ren, J.: Label-embedding bi-directional attentive model for multi-label text classification. Neural Process. Lett. 375–389 (2021)
9. Liu, M., Liu, L., Cao, J., Du, Q.: Co-attention network with label embedding for text classification. Neurocomputing 61–69 (2022)
10. Lu, J., Batra, D., Parikh, D., Lee, S.: ViLBERT: pretraining task-agnostic visiolinguistic representations for vision-and-language tasks. In: Advances in Neural Information Processing Systems, pp. 13–23 (2019)
11. Lin, T.Y., Goyal, P., Girshick, R., He, K., Dollár, P.: Focal loss for dense object detection. In: Proceedings of the IEEE International Conference on Computer Vision, pp. 2980–2988 (2017)
12. Brown, P.F., Della Pietra, V.J., Desouza, P.V., Lai, J.C., Mercer, R.L.: Class-based n-gram models of natural language. Comput. Linguist. 467–480 (1992)
13. Harris, Z.S.: Distributional structure. Word, 146–162 (1954)
14. Johnson, R., Zhang, T.: Deep pyramid convolutional neural networks for text categorization. In: Proceedings of the 55th Annual Meeting of the Association for Computational Linguistics, pp. 562–570 (2017)
15. Liu, P., Qiu, X., Huang, X.: Recurrent neural network for text classification with multi-task learning. Arxiv Preprint Arxiv:1605.05101 (2016)
16. Qiu, X., Sun, T., Xu, Y., Shao, Y., Dai, N., Huang, X.: Pre-trained models for natural language processing: a survey. Sci. China Technol. Sci. 1872–1897 (2020)
17. Guo, H., Liu, T., Liu, F., Li, Y., Hu, W.: Chinese text classification model based on BERT and capsule network structure. In: 2021 7th IEEE International Conference on Big Data Security on Cloud, pp. 105–110 (2021)
18. Liu, Y., Liu, H., Wong, L.-P., Lee, L.-K., Zhang, H., Hao, T.: A hybrid neural network RBERT-C based on pre-trained RoBERTa and CNN for user intent classification. In: Zhang, H., Zhang, Z., Wu, Z., Hao, T. (eds.) NCAA 2020. CCIS, vol. 1265, pp. 306–319. Springer, Singapore (2020). https://doi.org/10.1007/978-981-15-7670-6_26
19. Mikolov, T., Chen, K., Corrado, G., Dean, J.: Efficient estimation of word representations in vector space. Arxiv Preprint Arxiv:1301.3781 (2013)
20. Du, C., Chen, Z., Feng, F., Zhu, L., Gan, T., Nie, L.: Explicit interaction model towards text classification. In: Proceedings of the AAAI Conference on Artificial Intelligence, pp. 6359–6366 (2019)
21. Wang, X., Zhao, L., Liu, B., Chen, T., Zhang, F., Wang, D.: Concept-based label embedding via dynamic routing for hierarchical text classification. In: Proceedings of the 59th Annual Meeting of the Association for Computational Linguistics and the 11th International Joint Conference on Natural Language Processing, pp. 5010–5019 (2021)
22. Seo, M., Kembhavi, A., Farhadi, A., Hajishirzi, H.: Bidirectional attention flow for machine comprehension. Arxiv Preprint Arxiv:1611.01603 (2016)
23. McCann, B., Keskar, N.S., Xiong, C., Socher, R.: The natural language decathlon: multitask learning as question answering. Arxiv Preprint Arxiv:1806.08730 (2018)
24. Liu, W., Zhou, P., Zhao, Z., Wang, Z., Deng, H., Ju, Q.: FastBERT: a self-distilling BERT with adaptive inference time. Arxiv Preprint Arxiv:2004.02178 (2020)
25. Xin, J., Tang, R., Lee, J., Yu, Y., Lin, J.: DeeBERT: dynamic early exiting for accelerating BERT inference. Arxiv Preprint Arxiv:2004.12993 (2020)
26. Zhu, W.: LeeBERT: Learned early exit for BERT with cross-level optimization. In: Proceedings of the 59th Annual Meeting of the Association for Computational Linguistics and the 11th International Joint Conference on Natural Language Processing, pp. 2968–2980 (2021)
27. Duggal, R., Freitas, S., Dhamnani, S., Chau, D.H., Sun, J.: ELF: an early-exiting framework for long-tailed classification. Arxiv Preprint Arxiv:2006.11979 (2020)
28. Chen, N., Su, X., Liu, T., Hao, Q., Wei, M.: A benchmark dataset and case study for Chinese medical question intent classification. BMC Med. Inform. Decis. Mak. 1–7 (2020)

29. Aldahdooh, J., Tanoli, Z., Jing, T.: R-BERT-CNN: drug-target interactions extraction from biomedical literature. In: Proceedings of the BioCreative VII Workshop, pp. 102–106 (2021)
30. Lin, D., Cao, D., Lin, S., Qu, Y., Ye, H.: Extraction and automatic classification of TCM medical records based on attention mechanism of BERT and Bi-LSTM. Comput. Sci. 416–420 (2020)

Neural Computing-Based Fault Diagnosis and Forecasting, Prognostic Management, and Cyber-Physical System Security

Multiscale Redundant Second Generation Wavelet Kernel-Driven Convolutional Neural Network for Rolling Bearing Fault Diagnosis

Fengxian Su(✉), Shuwei Cao, Tianheng Hai, and Jing Yuan

School of Mechanical Engineering, University of Shanghai for Science and Technology, 516 JunGong Road, Shanghai 200093, China

sfx202103@163.com

Abstract. Fault diagnosis of rolling bearings is crucial in areas related to rotating machinery and equipment applications. If faults are detected in time at an early stage, it can guarantee the safe and effective operation of the equipment, saving valuable time and high maintenance costs. Traditional fault diagnosis techniques have achieved remarkable results in rolling bearing fault detection, but they rely heavily on expert knowledge to extract fault features. Manual extraction of features in the face of massive industrial data exhibits poor timeliness. In recent years, with the development and wide application of deep learning, data-driven mechanical fault diagnosis methods are becoming a hot topic of discussion among related researchers. Among them, Convolutional Neural Network (CNN) is an effective deep learning method. In this study, a new method of multiscale redundant second generation wavelet kernel-driven convolutional neural network for rolling bearing fault diagnosis is proposed, called RW-Net. By performing two layers of redundant second generation wavelet decomposition on the input time-domain signal in the shallow layer of the network, the network can automatically extract fault features with rich information. The proposed method is validated by Case Western Reserve University (CWRU) bearing test data, and the average fault identification accuracy is 99.4%, which verifies the feasibility and effectiveness of the proposed method.

Keywords: Deep learning · CNN · Redundant second generation wavelet · Fault diagnosis · Rolling bearing

1 Introduction

With the rapid development of modern industry, mechanical equipment has become the cornerstone of promoting productivity and facilitating economic growth [1, 2]. Among them, rolling bearings are widely used in various mechanical equipment, as key components of rotating mechanical equipment, in the mechanical transmission process has the role of load-bearing weight and reducing friction. Rolling bearings are prone to pitting, spalling, cracks, and other local failures under harsh working conditions such as high load, strong impact, and high temperature. The local failure of rolling bearings is one of the main causes of rotating machinery faults. If not found in time, it will have an impact

H. Zhang et al. (Eds.): NCAA 2023, CCIS 1870, pp. 263–278, 2023.
https://doi.org/10.1007/978-981-99-5847-4_19

on the safe operation of mechanical equipment, and even cause serious economic losses and personal casualties [3]. Therefore, the research of accurate and effective rolling bearing fault diagnosis and health monitoring methods is of great significance to reducing downtime, preventing safety accidents, and guaranteeing the safe and effective operation of equipment.

Generally, mechanical fault diagnosis techniques are divided into three parts: signal acquisition, feature extraction, and fault identification. Among them, the latter two of these steps are very important and greatly influence the accuracy of the final diagnosis [4]. Usually, the vibration signals of rolling bearings are complex and non-smooth with high background noise. It is a great challenge to effectively extract the representative fault features in the vibration signal for fault diagnosis [5]. Several researchers have achieved notable results in the field of rotating machinery fault diagnosis using signal processing-related techniques. Li et al. use the variational modal decomposition (EMD) method for feature extraction of fault signals. The authors solved the problem of information loss and over-decomposition and verified the effectiveness of the proposed method using high-speed locomotive wheelset bearings [6]. Chen et al. reveal the essence of wavelet transform inner product matching in rotating machinery fault diagnosis through simulation and field test experiments [7]. Ming et al. propose the spectral auto-correlation analysis method and applied it to the feature extraction of early faint faults of rolling bearings [8]. However, these traditional methods rely heavily on a priori knowledge and expert knowledge. This limits the wide usage of traditional fault diagnosis methods. Compared to traditional signal processing techniques, intelligent fault diagnosis is a new development in mechanical fault detection technology [9].

The development of intelligent machinery fault diagnosis has benefited from the rapid development of sensing technology, computer technology, and data storage technology in recent years. These technologies provide the technology for data acquisition, transmission, and storage in manufacturing systems [10, 11]. For the intelligent diagnosis model based on a neural network, its network structure contains large amount of learnable parameters. In order to fully training the network parameters, it requires a lot of engineering data. Most of the actual engineering data are generated during the fault-free period. Only a small fraction of the fault data is generated when a machine breaks down, and the data used for neural network training is artificially processed, and it contains a dataset corresponding to the real labels. Manual processing of data is time-consuming and laborious. Based on the above, how to obtain higher accuracy with relatively less training data is a hot topic worthy of study.

When a rolling bearing malfunctions during operation, it can cause the dynamic signal to contain non-stationary components. Wavelet transforms with good time-frequency multi-resolution characteristics is a powerful tool for dynamic signal analysis. However, in engineering practice, how to select the appropriate wavelet basis function from the library of wavelet basis functions to match the signal to be analyzed becomes a difficult problem. Hence, the second generation wavelet transform was born, which constructs predictors and updaters to match the signal to be analyzed adaptively by lifting methods. Nevertheless, the second generation wavelet transform has a splitting operation, which makes the number of points of the approximation signal decrease to half for each second generation wavelet decomposition performed on the signal. As the number of decompositions increases, the approximation signal contains less and less information, the signal is also prone to distortion. The redundant second generation wavelet (i.e. RSGW) transform overcomes this shortcoming. There is no sampling operation in the signal decomposition process. The length of the approximation and detail signals are the same as the original signal, and the information in the decomposed signal is redundant [12, 13]. Gao et al. use RSGW for signal noise reduction to improve the SNR [14]. Jiang et al. interpolate the initial prediction operator and updating operator of the RSGW to obtain the redundant prediction operator and updating operator corresponding to the number of decomposition layers, and the experiment shows that the method can accurately extract the signal features [15].

Inspired by this, this paper combines RSGW with convolution layers to form a deep CNN (i.e., RW-Net) driven by multiscale RSGW convolution kernels. The first layer of the network is the Conv1, in which multiscale RSGW transform is performed. The more layers of RSGW decomposition, the clearer the feature extraction and the less noise. The kernel of this convolution layer is the RSGW kernel, which is obtained by interpolating the second generation wavelets to complement the zeros. Therefore, the deep CNN driven by multiscale RSGW convolution kernels has the following advantages: 1) Multiple RSGW transform can be performed within the Conv1, and the decomposition result is always the same length as the original signal, which is suitable for engineering data with small sample lengths. 2) The RSGW transform realizes the translational invariance of the signal and can extract and retain richer dynamic fault features, which can effectively enhance the overall fault diagnosis performance of RW-Net. The contributions of this paper are as follows:

1) A new rolling bearing fault diagnosis method is proposed.
2) RW-Net uses the collected time series as input to achieve end-to-end fault identification without expert knowledge, reducing the complexity and timeliness of fault diagnosis.
3) The RW-Net model is developed and applied to the engineering case to verify the validity of the model in comparison with classical and popular networks. The RSGW layers proposed in this model have universal applicability and can be applied to almost any network. In addition, compared with the traditional convolutional neural network, RW-Net has only two training parameters, which saves computing storage space and improves the convergence speed of the network training.

The rest of the paper is organized as follows. Section 2 introduces the theoretical foundation knowledge of RSGW transform and convolutional neural networks. Section 3 explains the design of wavelet kernels and the construction of RSGW layers. Section 4 introduces the effectiveness and result analysis of RW-Net in the experiment. Section 5 is the conclusion.

2 Theoretical Foundation

2.1 RSGW Transform

In the RSGW transform algorithm, there are only two steps, prediction and updating, and the splitting operation is removed compared to the second generation wavelet transform. The RSGW performs a predict-and-update operation on the input signal $\hat{s}^{(k)}$. The RSGW transform is schematically shown in Fig. 1.

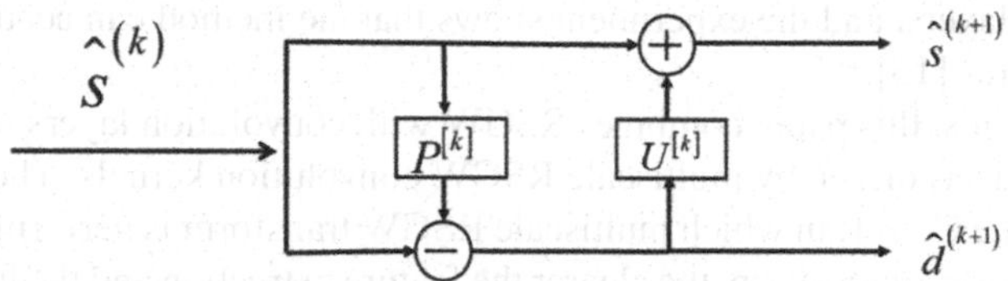

Fig. 1. Schematic diagram of RSGW transform

Prediction: The redundant predictor after interpolation zero padding is used to predict the signal, and the prediction error is defined as the detail signal of the RSGW transforms:

$$\hat{d}_i^{(k+1)} = \hat{s}_i^{(k)} - \sum_i p_r^{[k]} \hat{s}^{(k)}_{r+i-2^{k-1}N}. \tag{1}$$

Updating: Based on the detail signal, the redundant updater $U^{[k]}$ after interpolation zero padding is used to update the detail signal, and the updated signal $\hat{s}_i^{(k+1)}$ is defined as the approximation signal of the RSGW:

$$\hat{s}_i^{(k+1)} = \hat{s}_i^{(k)} + \sum_i u_l^{[k]} \hat{d}^{(k)}_{l+i-2^{k-1}\tilde{N}}. \tag{2}$$

2.2 Fundamental Theory of CNN

Convolutional neural networks contain two parts: feature extraction and feature selection, where the convolution, activation, and pooling layers perform feature extraction on the input signal and the fully connected layer filter the extracted features. The back-propagation algorithm calculates the gradient values of the variable parameters of the network and the adaptive optimization function to update the mode parameters so that the output of the network corresponds to the true labels of the input signals. As the input signal passes through the convolution layer, the filters (also called convolution kernels) in the convolution layer progressively scan the input matrix in specific steps to obtain a matrix of smaller size and containing fault features. The value of the convolution kernels determines the feature type extracted by the convolution layers. During the convolution operation, the convolution kernels are always the same batch, which is what makes the convolution layer different from the normal network layers. Therefore, one of the most important features of the convolution layer is weight sharing, which reduces the trainable parameters. In addition, the output channels of the redundant second generation wavelet convolution layer mentioned in this paper are equal to the number of redundant second generation wavelet convolution kernels. The calculation of the convolution layer is as follows [16]:

$$Y^{l(i,j)} = \sigma_i^l * X^{l(j)} = \sum_{j'=0}^{W-1} \sigma_i^{l(j')} X^{l(j+j')}. \tag{3}$$

where $\sigma_i^{l(j')}$ is the j' th weight on the i th convolution kernel of the l th layer. $X^{l(r^j)}$ denotes the j th convolved region of the l th layer. W indicates the width of the convolution kernel. $Y^{l(i,j)}$ denotes the result of the i th convolution kernel in the layer l with the convolved region on the input signal X.

The activation layer is a nonlinear mapping of the values output from the convolution layer to a specific interval using an activation function. When the excitation input reaches a certain strength, the neuron is activated. Since the convolution operations in the convolution layer are linear, the complexity of the neural network and its ability to fit the target are greatly reduced if there is no activation function for nonlinear operations.

The role of the pooling layer is to extract higher dimensional fault features through pooling operations, thus reducing the computation and making the data representation more obvious. The common pooling operations are maximum pooling and average pooling [17]. Maximum pooling is defined as representing the maximum value of the pooled region in the data as this region, while average pooling is defined as representing the average value of the pooled region in the data as this region.

Fully connected layers are structures in which neurons in this layer are connected two by two with neurons in the upper layer, but not between neurons in this layer. The role of the fully connected layer is to enhance the nonlinear mapping capability of the network, to limit the size of the network, and to classify the features extracted by the convolutional and pooling layers. Converts the pooling layer output data into a one-dimensional vector, which is then used as input to the fully connected layers. The output length of the fully connected layer is the number of labels recognized by the neural network.

3 The Proposed Method

As shown in Fig. 2, the RSGW convolution layer will be explained according to the convolution calculation method and the construction of the convolution kernel. The convolution calculation method is the RSGW transform in signal processing theory. The construction of the convolution kernel will be constrained and designed according to the vanishing wavelet theory. The construction of the convolution kernel will be constrained and designed according to the vanishing moment in wavelet theory. The purpose of all these designs is to combine RSGW with convolutional neural networks to form a new deep CNN driven by RSGW.

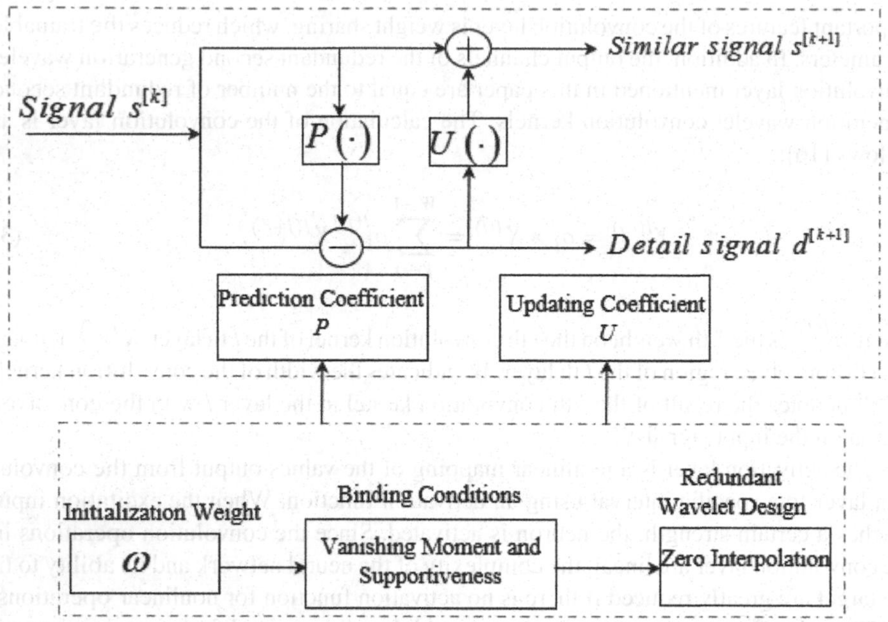

Fig. 2. Design of RSGW convolution layer

3.1 Design of RSGW Convolution Kernel

The Construction of the Initial Predictor *P* and Updater *U*

Suppose that the coefficient of the predictor P with length N is $P = [p_1, p_2, \cdots, p_{N/2}, p_{N/2+1}, \cdots, p_N]$, and the coefficient of the updater U with length $\tilde{N}$ is $U = [u_1, u_2, \cdots, u_{\tilde{N}}]$. Claypoole uses the equivalent filter method to obtain the prediction operator P and updater U, that is, the specific coefficients of P and U can be obtained by solving the linear equations [18]. If the order of the prediction polynomial of constraint P is M. For a linear predictor P with a length of N, only the polynomial $M < N$ order of predictor P is required to be constrained, and the remaining $N - M$ degrees of freedom will be updated by the back-propagation algorithm to adapt to the signal characteristics. The relationship between RSGW equivalent high-pass filter $\tilde{H}$ and prediction operator P is

$$\tilde{H} = \left[-p_1 \; 0 \; -p_2 \; 0 \cdots -p_{N/2} \; 1 \; -p_{N/2+1} \; 0 \cdots 0 \; -p_N \right]. \tag{4}$$

The specific expression of the constraint predictor coefficient polynomial of order M is

$$\sum_{k=-N+1}^{N-1} k^q \tilde{H}_k = 0, 0 \leqslant q < M. \tag{5}$$

The relationship between RSGW reconstruction equivalent high-pass filter H and predictor P and updater U is

$$H_{2l-1} = \begin{cases} 1 - \sum_{m=1}^{N} Ps_m Us_{l-m+1} & l = (N+\tilde{N})/2 \\ \sum_{m=1}^{N} Ps_m Us_{l-m+1} & l \neq (N+\tilde{N})/2 \end{cases} . \tag{6}$$

$$H_{2l+N-2} = Us_l \qquad l = 1, 2, \ldots, \tilde{N}$$

Equation (6) can be simplified as follows:

$$\tilde{V}H = 0. \tag{7}$$

where $\tilde{V}$ is a matrix of size $\tilde{N} \times \left[2 \times \left(N + \tilde{N}\right) - 1\right]$, whose elements are represented as follows:

$$[\tilde{V}]_{m,n} = n^m. \tag{8}$$

where $n = -N-\tilde{N}+2, -N-\tilde{N}+3, \cdots, N+\tilde{N}-3, N+\tilde{N}-2, m = 0, 1, \cdots, \tilde{N}-1$. Since the coefficient P is determined by the order of the constraint coefficient polynomial and the adaptive signal, Eq. (7) is a linear system of equations containing only the coefficients of U, and then U can be determined by the least square method. Duan proved through experiments that when $N - M \leq 2$ can make the predictor adapts to the signal best [12]. Therefore, the order of the predictor coefficient polynomial of the RSGW selected in this paper is $N - 2$, and the remaining degrees of freedom are determined by the neural network by fitting the input signal.

The Construction of Redundant Predictor $P^{[k]}$ and Updater $U^{[k]}$

Based on the initial predictor, the coefficient $p_r^{[k]}$ of the k th RSGW decomposition predictor is calculated as follows 15:

When $r - 1$ can be divisible by 2^k,

$$p_r^{[k]} = p_{(r-1)/2^k}. \tag{9}$$

When $r - 1$ can't be divisible by 2^k,

$$p_r^{[k]} = 0. \tag{10}$$

Then we can get the redundant predictor $P^{[k]} = \left\{p_r^k, r = 1, 2, \cdots, 2^r N\right\}$ in the decomposition of the layer k.

Based on the initial updater U, the coefficient $u_l^{[k]}$ of the redundant updater decomposed by layer k is designed as follows:

When $l - 1$ can be divisible by 2^k,

$$u_l^{[k]} = u_{(l-1)/2^k}. \tag{11}$$

When $l-1$ can't be divisible by 2^k,

$$u_l^{[k]} = 0. \tag{12}$$

Then the redundant updater $U^{[k]} = \left\{u_l^k, l = 1, 2, \cdots, 2^l\tilde{N}\right\}$ of the k th layer decomposition can be obtained.

3.2 Deep CNN Driven by Multiscale RSGW Kernels

The main structure of RW-Net includes an RSGW convolution layer (Conv1), 1D convolution layers, adaptive maximum pooling layers, and fully connected layers, as shown in Fig. 3. Two RSGW decompositions are performed in the Conv1, which can better extract the input signal features without losing the useful information in the signal. The RSGW convolution kernel is a, and then the initial prediction operator b and the updating operator c are obtained according to the wavelet vanishing moment and the equivalent filter method, and their lengths are 10. The longer the length of P and U, the more the waveform changes of the RSGW, and the stronger the ability of the RSGW to adapt to the signal. However, the longer the length P and U will increase the training time of the network, which is considered as 10. When performing RSGW transform, the initial P and U will be interpolated according to the number of scale transformations to increase the length. The length relationship between the redundant prediction operator P and the updating operator U of the k th scale transformation is shown in Eq. (13). Therefore, the length of the redundant prediction operator $P^{[1]}$ and updating operator $U^{[1]}$ of the first RSGW transform is 20, and the second is 40. Subsequently, too many RSGW convolution kernels will increase the training time of the network, and too few will affect

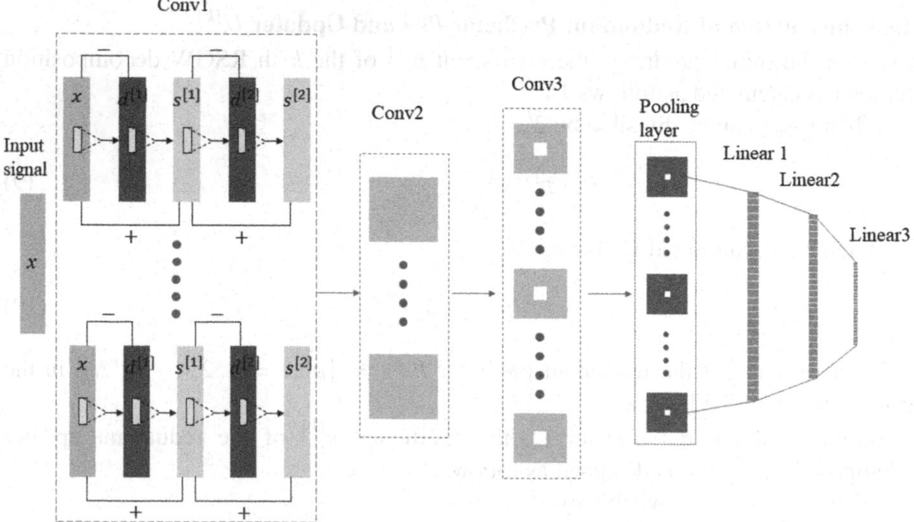

Fig. 3. RW-Net network structure diagram

the network performance. The number of kernels in Conv1 is set to 6.

$$\begin{aligned} P^{[k]} &= 2^k P \\ U^{[k]} &= 2^k U \end{aligned} \cdot \tag{13}$$

The specific parameters of RW-Net are shown in Table 1.

Table 1. Specific parameters of RW-Net

Network layers	Kernel size	Channels	Output	Padding	Activation function
Conv1	2	6	1*6*1024	Yes	Tanh
Conv2	5	16	1*16*1020	No	Tanh
Conv3	25	32	1*32*996	No	Tanh
Pooling	–	32	1*32*32	–	–
Linear1	–	–	1*1*120	–	Tanh
Linear2	–	–	1*1*72	–	Tanh
Linear3	–	–	m	–	Softmax

4 Experimental Verification

4.1 Case1: CWRU

4.1.1 Dataset Description

In this paper, the bearing fault dataset of CWRU is selected as the experimental object [19]. The motor speed is 1730 RPM and the sampling frequency is 12000 Hz. The bearing label categories to be identified are shown in Table 2. There are 10 kinds of labels, including 0.007, 0.014, and 0.021 inch fault diameter bearing with inner ring, roller, and outer ring fault and healthy bearing. The total number of samples in the bearing signal dataset is 2000, and the training and test samples are divided in a ratio of 3:1. That is, the training data sample is 1500, and the test data sample is 500. In order to reflect the advantages of this method for short sample length (fewer sample points), the input signal length is 512.

4.1.2 Selection of Activation Function

The activation function is very important for neural network nonlinearity and diagnostic ability. The ReLU activation function allows eigenvalues greater than zero in the signal to pass through, and values less than zero are treated as zero, while the Tanh activation function can map eigenvalues to the interval $[-1, 1]$. In order to explore the influence of ReLU and Tanh activation functions on the diagnostic performance of RW-Net, RW-Net with different activation functions is utilized to do comparative experiments on CWRU

Table 2. Label information

	0 Inch	0.007 Inch	0.014 Inch	0.021 Inch	Motor speed
Normal	Label1	–	–	–	1797 RPM
Inner ring	–	Label2	Label3	Label4	
Roller	–	Label5	Label6	Label7	
Outer ring	–	Label8	Label9	Label	

bearing fault dataset. In order to guarantee the objectivity of the results, each network is trained 5 times. Figure 4 and Table 3 show the influence of ReLU and Tanh activation functions on the accuracy of the network. It can be seen from Fig. 4 that the Tanh activation function can make the RW-Net diagnostic performance better than the ReLU activation function, and its maximum, minimum, and average accuracy are the highest. Table 3 records the specific values of the correct rate 99.6%, 99.2% and 99.4%. In this paper, Tanh is selected as the activation function of RW-Net.

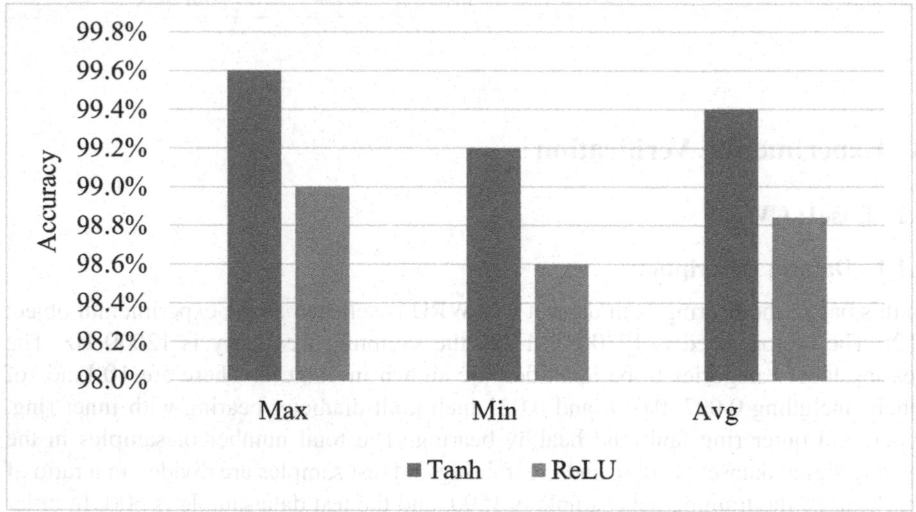

Fig. 4. The effect of ReLU and Tanh activation functions on the accuracy of RW-Net

Table 3. The statistical results of the test under different activation functions

Activation function	Maximum accuracy	Minimum accuracy	Average accuracy
Tanh	**99.6%**	**99.2%**	**99.40%**
ReLU	99.0%	98.6%	98.84%

4.1.3 Experimental Contrast Analyses

In order to explore the influence of the number of RSGW transform in the Conv1 on the overall performance of RW-Net, this paper will perform an RSGW transform in the convolution layer, called RW-Net1. Figure 5 and Fig. 6 show the change in loss rate and accuracy rate of RW-Net and RW-Net1 during training on the CWRU dataset. It can be seen that the loss rate of RW-Net in training decreases faster than that of RW-Net1, and the loss has been close to 0 as the number of iterations increases. In terms of accuracy, RW-Net can also reach 100% quickly in training and remain stable. All these indicate that RW-Net is better than RW-Net1 in instability and convergence in training.

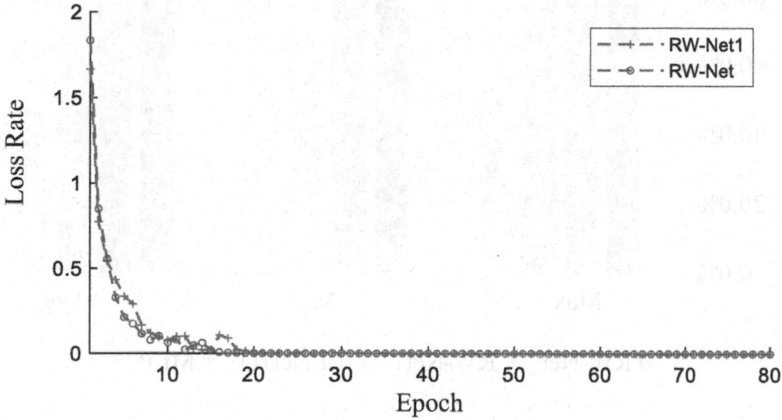

Fig. 5. The loss rate of RW-Net and RW-Net1 during training

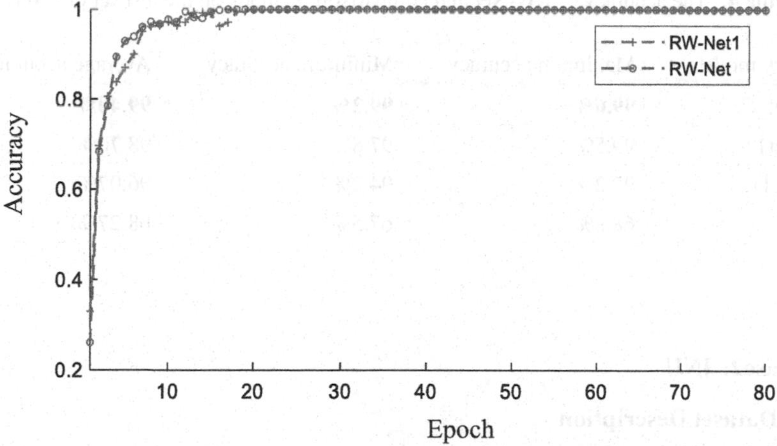

Fig. 6. The accuracy of RW-Net and RW-Net1 during training

Compared with the current mainstream intelligent diagnosis models LeNet1D and MLP. Figure 7 and Table 4 show the test results of RW-Net and comparison methods on the CWRU bearing dataset, respectively. It can be seen that RW-Net has achieved the best results. Although RW-Net1 also has good diagnostic ability on this dataset, the diagnostic accuracy is always less than RW-Net. By analyzing the experimental results, it can be concluded that: 1) RW-Net and RW-Net1 have better recognition ability for CWRU bearing dataset than comparison methods; 2) Two RSGW transforms in the Conv1 can improve the network diagnostic ability more than only once.

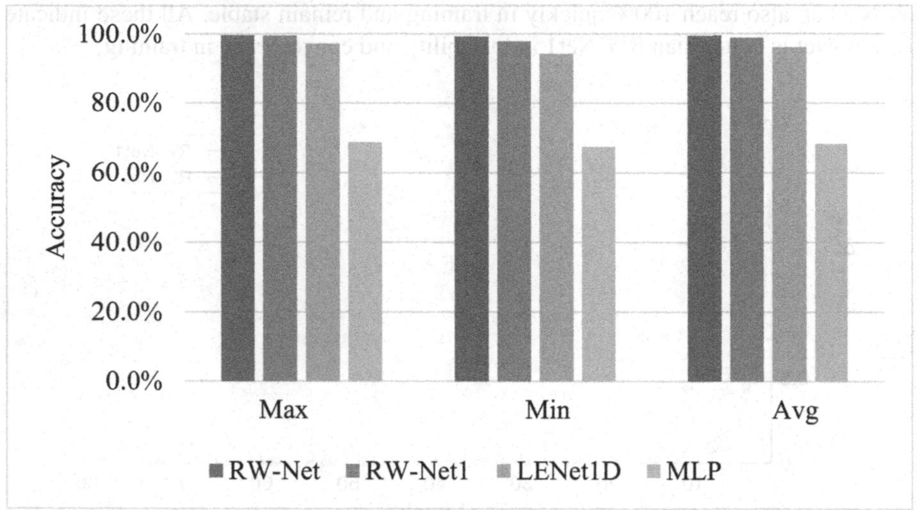

Fig. 7. The accuracy of RW-Net and comparison methods on the test set of CWRU

Table 4. The accuracy of RW-Net and comparison methods on the test set of CWRU

Network models	Maximum accuracy	Minimum accuracy	Average accuracy
RW-Net	**99.6%**	**99.2%**	**99.40%**
RW-Net1	99.5%	97.8%	98.70%
LENet1D	97.2%	94.2%	96.07%
MLP	68.8%	67.5%	68.27%

4.2 Case2: JNU

4.2.1 Dataset Description

Jiangnan University (JNU) bearing datasets were provided by Jiangnan University [20, 21] The JNU datasets are composed of three bearing vibration datasets with different rotating speeds, and the data acquisition frequency is 50 kHz. In this experiment, the 1000 RPM bearing fault dataset is adopted. Its vibration signals are shown in Fig. 8. This dataset contains one health state and three fault patterns, including inner ring fault, outer ring fault and rolling element fault. Each category of health condition takes 400 samples, and each sample contains 512 data points. The dataset contains a total of 1200 samples. The samples corresponding to each label were assigned to the training and test sets in a ratio of 3:1, respectively.

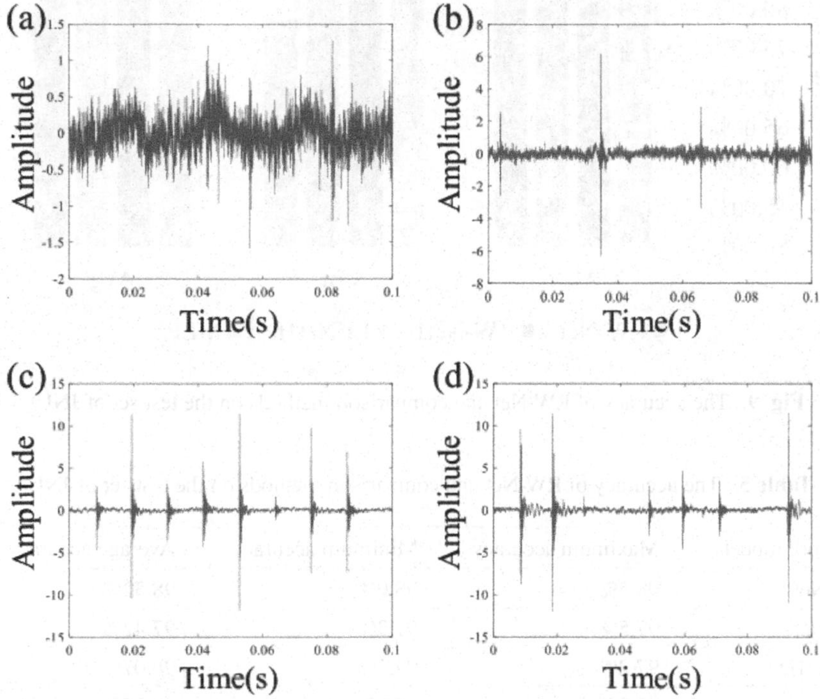

Fig. 8. (a) Health state (b) Inner ring (c) Outer ring (d) Rolling element

4.2.2 Experimental Contrast Analyses

Likewise, the experiment is used as a comparison with the current mainstream intelligent diagnostic models LeNet1D and MLP. Figure 9 and Table 5 show the test results of RW-Net and comparison methods on the CWRU bearing dataset, respectively. It can be seen that RW-Net has achieved the best results. The experiments demonstrate not only the superior fault identification capability of RW-Net, but also its robustness and generalization capability.

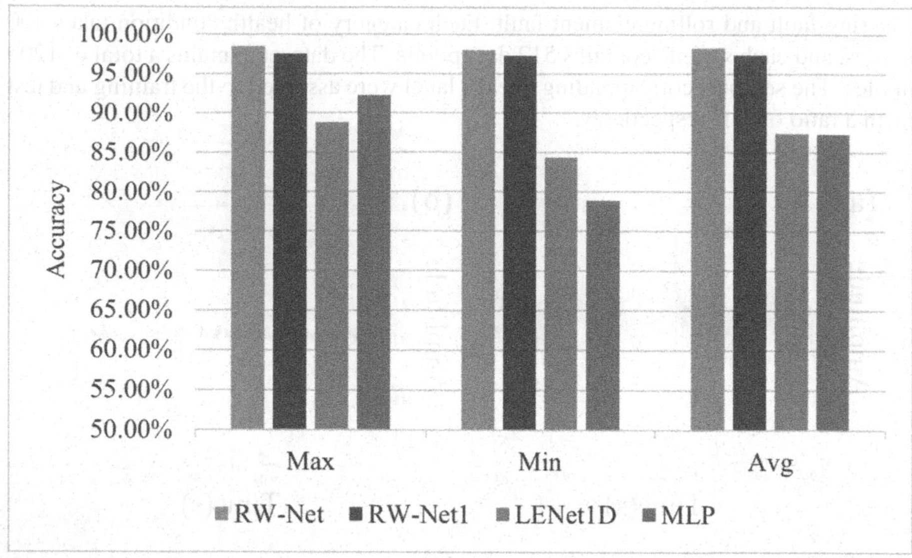

Fig. 9. The accuracy of RW-Net and comparison methods on the test set of JNU

Table 5. The accuracy of RW-Net and comparison methods on the test set of JNU

Network models	Maximum accuracy	Minimum accuracy	Average accuracy
RW-Net	98.5%	98.0%	98.32%
RW-Net1	97.5%	97.3%	97.42%
LENet1D	97.2%	94.2%	96.07%
MLP	68.8%	67.5%	68.27%

5 Conclusion

This paper proposes an improved CNN based on RSGW theories, called RW-Net. The shallow layer of RW-Net performs RSGW transform on time-domain signals. This layer inherits the advantages of RSGW in signal processing to extract signal features. RW-Net takes time domain signal as input. The RSGW layers are used as a multi-channel filter to simultaneously extract multiple fault features, and then the features extracted by the RSGW layers are fused as the input of the pooling layer. By enhancing the feature extraction ability of the shallow layer of the proposed method, the fault features can be accurately extracted by using small sample datasets, and the network training parameters are reduced. In this paper, the feasibility of RW-Net is verified by the CWRU and JNU bearing fault datasets. From the analysis of the experimental results, it can be seen that the average accuracy of RW-Net reaches 99.4% and 98.32%, which is better than other comparison methods. The feasibility and effectiveness of the proposed method are fully illustrated.

In this paper, we use the time-domain signal as the input signal of the network and each labeled data segment is independent of each other when segmenting the data set, i.e., the correlation between data segments is not considered. We think the correlation between data segments can be the next research direction. The graph convolutional neural network may be a good method.

References

1. Huang, Z., Lei, Z., Huang, X., et al.: A multisource dense adaptation adversarial network for fault diagnosis of machinery. IEEE Trans. Ind. Electron. **69**(6), 6298–6307 (2021)
2. Jiang, Y., Yin, S.: Recursive total principle component regression based fault detection and its application to vehicular cyber-physical systems. IEEE Trans. Ind. Inf. **14**(4), 1415–1423 (2017)
3. Yuan, J., Cao, S., Ren, G., et al.: LW-Net: an interpretable network with smart lifting wavelet kernel for mechanical feature extraction and fault diagnosis. Neural Comput. Appl. **34**(18), 15661–15672 (2022)
4. Yuan, J., He, Z., Zi, Y., et al.: Adaptive multiwavelets via two-scale similarity transforms for rotating machinery fault diagnosis. Mech. Syst. Sig. Process. **23**(5), 1490–1508 (2009)
5. Shao, H., Jiang, H., Zhang, H., et al.: Rolling bearing fault feature learning using improved convolutional deep belief network with compressed sensing. Mech. Syst. Sig. Process. **100**, 743–765 (2018)
6. Li, Z., Chen, J., Pan, J.: Independence-oriented VMD to identify fault feature for wheelset bearing fault diagnosis of high-speed locomotive. Mech. Syst. Sig. Process. **85**, 512–529 (2016)
7. Chen, J., Li, Z., Chen, G., et al.: Wavelet transform based on inner product in fault diagnosis of rotating machinery: a review. Mech. Syst. Sig. Process. **70**, 1–35 (2016)
8. Ming, A., Qin, Z., Zhang, W., et al.: Spectrum auto-correlation analysis and its application to fault diagnosis of rolling element bearings. Mech. Syst. Sig. Process. **41**(1–2), 141–154 (2013)
9. Wang, P., Song, L., Guo, X., et al.: A high-stability diagnosis model based on a multiscale feature fusion convolutional neural network. IEEE Trans. Instrum. Meas. **70**, 1–9 (2021)
10. Lund, D., MacGillivray, C., Turner, V., et al.: Worldwide and regional internet of things (IoT) 2014–2020 forecast: a virtuous circle of proven value and demand. International Data Corporation (IDC), Technical Report, vol. 1, no. 1, p. 9 (2014)
11. Chen, Z., Li, W.: Multisensor feature fusion for bearing fault diagnosis using sparse autoencoder and deep belief network. IEEE Trans. Instrum. Meas. **66**(7), 1693–1702 (2017)
12. Duan, C., Li, L., He, Z.: Application of second generation wavelet transform to fault diagnosis of rotating machinery. Mech. Sci. Technol. **23**(2), 224–226 (2004)
13. Duan, C., He, Z.: Second generation wavelet denoising and its application in machinery monitoring and diagnosis. Mini-Micro Syst. **25**(7), 1341–1343 (2004)
14. Gao, L., Tang, W., et al.: Noise reduction technology based on redundant second generation wavelet. J. Beijing Univ. Technol. **34**(12), 1233–1237 (2008)
15. Jiang, H., He, Z., Duan, C.: Construction of redundant second generation wavelet and mechanical signal feature extraction. J. Xi'an Jiaotong Univ. **38**(11), 1140–1142 (2004)
16. Zhang, W.: Study on Bearing Fault Diagnosis Algorithm Based on Convolutional Neural Network. Harbin University of Science and Technology (2017)
17. Chen, X., Xiang, S., Liu, C., et al.: Vehicle detection in satellite images by hybrid deep convolutional neural networks. IEEE Geosci. Remote Sens. Lett. **11**(10), 1797–1801 (2014)

18. Claypoole, R.L., Baraniuk, R.G., et al.: Adaptive wavelet transforms via lifting. In: Proceedings of the 1998 IEEE International Conference on Acoustics, Speech and Signal Processing (ICASSP), vol. 3, pp. 1513–1516 (1998)
19. Bearing Data Center, Case Western Reserve University, Cleve land, OH, USA (2004). http://csegroups.case.edu/bearing datacenter/home
20. Li, K.: School of Mechanical Engineering, Jiangnan University (2019)
21. Li, K., Ping, X., Wang, H., Chen, P., Cao, Y.: Sequential fuzzy diagnosis method for motor roller bearing in variable operating conditions based on vibration analysis. Sensors **13**(6), 8013–8041 (2013)

Unsupervised Deep Transfer Learning Model for Tool Wear States Recognition

Qixin Lan[1], Binqiang Chen[1(✉)], Bin Yao[1], and Wangpeng He[2]

[1] School of Aerospace Engineering, Xiamen University, Xiamen, China
{cbq,yaobin}@xmu.edu.cn

[2] School of Aerospace Science and Technology, Xidian University, Xian, China
hewp@xidian.edu.cn

Abstract. Heavy worn tools can cause severe cutting vibrations, leading to a decrease in the surface quality of the workpiece. It is important to monitor tool states and replace the worn tool in time. The traditional tool wear states monitoring methods are mainly based on machine learning and features engineering for the specific cutting condition. In this paper, a novel tool wear states monitoring method is proposed for the multi working conditions monitoring task. The similarity of tools wear process is used to realize the transformation of the priori knowledge from the labeled source domain to the unlabeled target domain. An unsupervised deep transfer learning model is built for the tools wear states recognition, based on neural networks. The network is composed of one-dimensional (1D) convolutional neural network (CNN) and multi-layer perceptron (MLP). There is a domain adaptation unit in the penultimate layer to achieve the deep features alignment. Our experiments demonstrate that the proposed model can achieve a classification accuracy of higher than 80% in the target domain.

Keywords: Tool wear states recognition · Transfer learning · 1D convolutional Neural network · Domain adaptation

1 Introduction

The tool wear states monitoring system embedded in the production line can sense the tool wear states in real time, which is of great significant to ensure production safety and cutting quality. Traditional machine learning methods are mainly based on feature engineering [1–3]. These methods extract the features of monitoring signals based on the complex signal analysis. The wear states of new tool can be predicted based on the feature laws. With the development of machine learning technology, tool wear monitoring methods based on machine learning model have achieved better performance than the traditional methods. Li et al. proposed a data driven monitoring approach based on radar map feature fusion for tool wear recognition [4]. Bazi et al. built a tool wear prediction model, using 1D convolutional neural network and bidirectional long short-term memory network [5]. The models are always employed in the specific cutting condition. However, in the production environment, the cutting conditions changes frequently,

H. Zhang et al. (Eds.): NCAA 2023, CCIS 1870, pp. 279–290, 2023.
https://doi.org/10.1007/978-981-99-5847-4_20

which lead to the failure of the recognition models. This is mainly because the signal features extracted by these methods vary with the cutting conditions. The features from the new cutting conditions don't obey the proposed laws. Tool wear monitoring methods under multiple working conditions has better value.

Researchers have proposed various methods to solve the problem of tool wear monitoring under multiple working conditions. Feature engineering methods can extract signal features across working conditions. Li et al. proposed a time-frequency intrinsic feature-extraction strategy of acoustic emission signal, which is cutting condition independent [6]. Pan et al. used instantaneous cutting force model to extract the milling force coefficients which are independent of milling parameters [7]. Obviously, these methods need complex feature extraction work and enough prior knowledge.

Transfer learning is another monitoring method to solve the problem of states recognition under multiple working conditions. Feuz et al. transferred knowledge between domains with different feature spaces, using a new transfer learning technique, called Feature-Space Remapping [8]. Jiang et al. proposed a multi-label metric transfer learning algorithms jointly considering instance space and label space distribution divergence [9]. Liao et al. proposed a dynamic distribution adaptation algorithm, which can estimate the influences of marginal and conditional distribution at the same time [10].

Transfer learning can transfer existing knowledge from one domain to similar domains [11]. For tool wear states monitoring, different working conditions can be regarded as different domains. The signal characteristics related to cutting parameters vary under different cutting conditions. But for the same tool, under the same wear state, the contact state between the tool and the workpiece does not change with the working condition. The same contact state results in the same spectral characteristics of the monitoring signal. For source domain with labeled samples, machine learning based tool wear states recognition models can achieve good performance. Therefore, transfer learning can transfer the label knowledge of the source cutting conditions to the target domain by using the invariable characteristics of the spectrum characteristics, so as to achieve tool wear condition monitoring under multiple working conditions. At present, some researchers have applied transfer learning technology to health monitoring under multiple working conditions. Lee et al. used a newly devised multi-objective instance weight to decrease domain discrepancy [12]. Li et al. used maximum mean square discrepancy method to evaluate the similarity of the historical tool and new tool features [13].

For supervised transfer learning, the target domain data should have a small number of labels, which can be used to fine tune the source domain model. For tool wear monitoring, the observation of cutting tool wear states needs to be involved in the production process, and the measurement of tool wear is very difficult. Therefore, the monitoring data of tool wear under multiple working conditions usually have no labels. Unsupervised transfer learning method is used to solve the situation where the target domain lacks labels. It is more suitable for tool wear monitoring under multiple working conditions. Zhu et al. proposed a unsupervised Dual-Regression Domain Adversarial Adaptation network for tool wear prediction in multi-working conditions [14]. Lu et al. used transfer component analysis algorithm to minimize the distance between the marginal distributions of the source and target domains in an unsupervised intelligent fault diagnosis system [15]. In the cutting process of small batch and multi variety products, the cutting conditions change frequently, and most of the monitoring data are lack of labels. At present, there are few researches on tool wear condition monitoring for frequent changes of working conditions.

Aiming at the problem of tool wear states monitoring under multiple working conditions, this study proposes a tool wear state monitoring method based on unsupervised deep transfer learning method. The main contributions of this paper are: (a) a tool wear states monitoring method based on unsupervised transfer learning model is proposed; (b) a deep convolutional neural network is built for tool wear states classification with a domain adaptation unit embedded in the perceptron network; (c) based on the classification network, we propose an unsupervised deep transfer learning model, which can realize the tool wear states prediction of target domain samples.

2 Background and Preliminaries

2.1 Data Distribution Adaptation of Deep Transfer Learning

Deep transfer learning models can directly extract features from sample data, which meets the end-to-end requirements in practical applications. In particular, it uses the adaptive layer embedded in the multi-layer neural network to complete the deep features adaptation of the source and target domain data. The historical working condition data can be regarded as the source domain $\mathcal{D}_{\mathrm{s}} = \{(\boldsymbol{x}_i, y_i)\}_{i=1}^{N_{\mathrm{s}}}$, and the new working condition data can be regarded as the target domain $\mathcal{D}_{\mathrm{t}} = \{\boldsymbol{x}_j\}_{j=1}^{N_{\mathrm{t}}}$. The two domains have the same feature space and label space, namely $\mathcal{X}_{\mathrm{s}} = \mathcal{X}_{\mathrm{t}}$, $\mathcal{Y}_{\mathrm{s}} = \mathcal{Y}_{\mathrm{t}}$. The joint probability distribution of the two fields is different, namely $P_{\mathrm{s}}(\boldsymbol{x}, y) \neq P_{\mathrm{t}}(\boldsymbol{x}, y)$. The unified learning objective of the deep transfer learning method can be expressed as Eq. (1), where $\boldsymbol{v} \in \mathbb{R}^{N_{\mathrm{s}}}$ is the source domain sample weight, $v_i \in [0, 1]$; B is the number of samples of a batch of the training dataset; $R(\cdot, \cdot)$ is the transfer regularization term, which is used to represent the data distribution discrepancy between the source domain and the target domain; λ is the weight parameter of the two sub-goals; $f(\cdot)$ is the mapping function from sample $\boldsymbol{x}_i$ to label y_i; $\ell(\cdot, \cdot)$ is the loss function; $\mathcal{B}$ represents samples of a batch.

$$f^* = \arg\min_{f \in \mathcal{I}} \frac{1}{B} \sum_{i=1}^{B} \ell(f(v_i \boldsymbol{x}_i), y_i) + \lambda R(\mathcal{B}_{\mathrm{s}}, \mathcal{B}_{\mathrm{t}}) \tag{1}$$

2.2 Data Distribution Alignment

The maximum mean discrepancy is a nonparametric measure, which is used to measure the distance of two distributions based on kernel embedding in the reproducing kernel Hilbert space (RHKS) [16]. $\boldsymbol{X}_{\mathrm{s}} = \{\boldsymbol{x}_i\}_{i=1}^{N_{\mathrm{s}}} \subset \mathbb{R}^d$ represents the feature vector set sampling from the marginal distribution P_{s} of source domain; $\mathbf{X}_{\mathrm{t}} = \{\boldsymbol{x}_i\}_{i=1}^{N_{\mathrm{t}}} \subset \mathbb{R}^d$ represents the feature vector set sampling from the marginal distribution P_{t} of target domain. $\mathbb{E}_{x\ P_{\mathrm{s}}}[g(\boldsymbol{x})]$ and $\mathbb{E}_{x\ P_{\mathrm{t}}}[g(\boldsymbol{x})]$ represent the mathematical expectation of $g(\boldsymbol{x})$.

The maximum mean discrepancy between P_s and P_t is defined in Eq. (2), where H is the unit sphere H in the RHKS, $\phi(x)$ is a feature mapping from sample space to RKHS space, which makes $\mathbb{E}_{x} P[g(x)] = \langle \mathbb{E}_{x} P\phi(x), g \rangle_{\mathcal{H}}$, $\langle \cdot, \cdot \rangle$ represents the inner product operation.

$$D_{\mathcal{H}}(P_s, P_t) \triangleq \left\| \mathbb{E}_{\mathbf{x} \sim P_s}[\phi(\mathbf{x})] - \mathbb{E}_{\mathbf{x} \sim P_t}[\phi(\mathbf{x})] \right\|_{\mathcal{H}}^2 \tag{2}$$

3 Proposed Methodology

Aiming at the problem of tool wear states monitoring under multiple working conditions, this paper proposes a tool wear states recognition method based on unsupervised deep transfer learning. The tool states classification neural network is composed of 1D convolutional neural network and multi-layer perceptron. The frequency spectrum of machine center spindle vibration signal is used to train the network.

3.1 Monitoring Data Processing Method

In actual cutting environment, the monitoring system should intervene in the processing as little as possible. Compared with cutting force signal, acoustic emission signal, the vibration signal is easier to obtain, and the measuring equipment is cheaper. The frequency spectrum of vibration signal can describe the contact states between tool and workpiece. Therefore, the vibration signal is selected as the signal for end mill wear monitoring in this study. In order to measure the tool ware, the images of tool flank are obtained through a camera. The vibration signal and images of tool processing method in this paper is shown in Fig. 1.

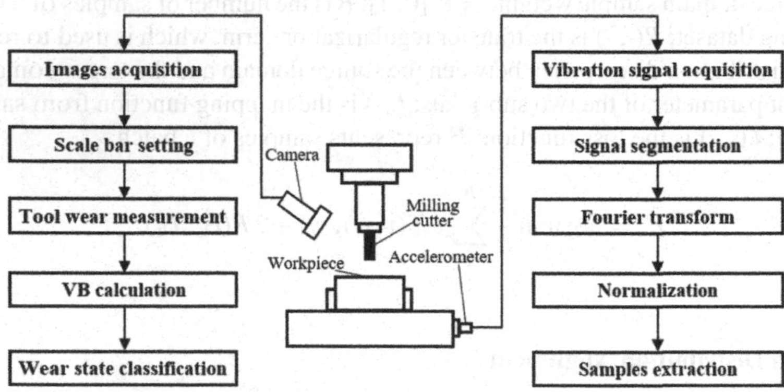

Fig. 1. Monitoring data preprocessing method

3.2 Tool Wear States Recognition Neural Network

Convolutional networks are widely used for feature extraction of multidimensional samples. Compared with the perceptron network, the convolution network has fewer parameters and the parameter training is simpler. The sample data in this paper is one-dimensional spectrum data. Therefore, this paper uses one-dimensional convolution to construct the first half of the tool wear states recognition network for feature extraction. After convolution network, the perceptron network is used to further extract the deep features of vibration spectrum data. The classification of tool wear states is realized in the last layer network. The tool wear states recognition neural network in this paper is shown in Fig. 2. The network is composed of 1D convolution layers (Conv layers) and full connection layers (Dense layers). There are three convolution layers and three dense layers. The convolutional network and the perceptron network are connected using a flatten layer. The last layer of deep feature is domain adaptive feature. The layer parameters and activation function of each layer are shown in Table 1.

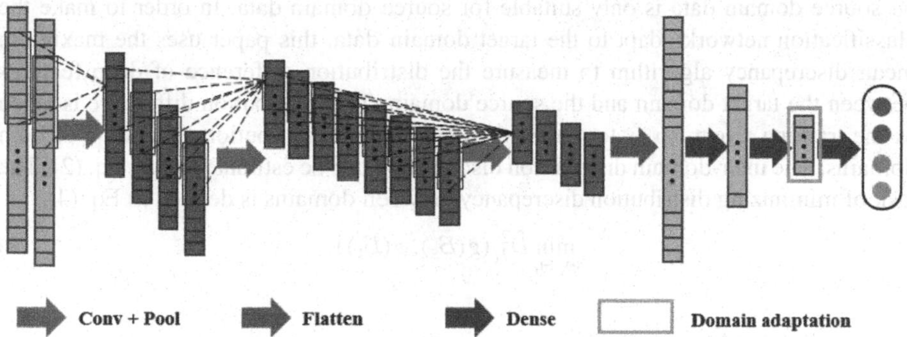

Fig. 2. Tool wear states recognition neural network

Table 1. The parameters of the proposed model

Layers	Parameters			
	Filters or Units	Kernel Size	Stride	Activation
Conv1	4	30	5	Leaky ReLU
Conv2	8	10	3	Leaky ReLU
Conv3	4	3	1	Leaky ReLU
Dense1	32	/	/	Leaky ReLU
Dense2	16			Leaky ReLU
Dense3	8			Softmax

3.3 Model Learning Task

Source Domain Classification Task: From the perspective of sample data mapping, tool wear states recognition network is divided into feature extractor g and classifier f. $g : \mathcal{X} \rightarrow \mathcal{Z}$ represents the mapping from sample data to deep feature vector. $f : \mathcal{Z} \rightarrow \mathcal{Y}$ represents the mapping from deep feature vector to prediction label. Θ_g and Θ_f is used to represent the parameter set of feature extractor and classifier respectively. The optimization objective of the source domain classification task can be expressed as Eq. (3), where $\boldsymbol{x}_s$ is the data of source domain, y_s is the one-hot code of the label of source domain, $\mathcal{L}_{en}(\cdot, \cdot)$ represents the categorical cross entropy loss function.

$$\min_{\Theta_g \Theta_f} \frac{1}{N_b} \sum_{i=1}^{N_b} \mathcal{L}_{en}(f(g(\boldsymbol{x}_s^i)), y_s^i) \tag{3}$$

Inter-Domain Distribution Discrepancy Minimization Task: Due to the influence of cutting parameters, the distribution of deep features in different domain is different under the same tool wear state. The wear states classification network trained based on source domain data is only suitable for source domain data. In order to make the classification network adapt to the target domain data, this paper uses the maximum mean discrepancy algorithm to measure the distribution difference of deep features between the target domain and the source domain. The distribution difference is taken as the training target, so as to realize the deep feature distribution adaptation of both domains. The inter-domain distribution discrepancy can be estimated using Eq. (2). The task of minimizing distribution discrepancy between domains is defined as Eq. (4).

$$\min_{\Theta_g \Theta_f} \hat{D}_{\mathcal{H}}^2(g(\mathcal{B}_s), g(\mathcal{B}_t)) \tag{4}$$

3.4 Model Training Method

The main goal of the model training in this paper is to make the tool wear states recognition model reduce the difference of deep features distribution between source domain and target domain, so as to realize the transfer of source domain label knowledge. The trained model can realize the tool wear states classification of target domain data. The learning strategy of the tool wear states recognition model is:

Model pre-training: the tool wear states recognition model is pre-trained by the labeled source domain samples, so that the model has the basic source domain classification ability. The optimization objective is shown in Eq. (3).

Transfer learning training: the goal of transfer learning is to make the tool wear states recognition model adapt to the data of source domain and target domain at the same time. Transfer learning training can improve the classification accuracy of the model in the target domain while maintaining the performance in the source domain. The optimization target is set as Eq. (5), where α is a weight parameter used to control the learning process.

$$\min_{\Theta_g \Theta_f} \frac{1}{N_b} \sum_{i=1}^{N_b} \mathcal{L}_{en}(f(g(\boldsymbol{x}_s^i)), y_s^i) + \alpha \hat{D}_{\mathcal{H}}^2(g(\mathcal{B}_s), g(\mathcal{B}_t)) \tag{5}$$

4 Experiment and Discussion

4.1 Milling Experiment Setup

A cutting experimental platform is built on a CNC machining center. The platform is composed of CNC machining center, cutting vibration data acquisition system and tool image acquisition system. This platform can be used to carry out cutting experiments in the whole life cycle of the tool, and realize the acquisition of cutting vibration data and tool images. In order to verify the proposed method, we designed a full-life cutting experiment for a quick-feed milling cutter with four inserts. The workpiece material is annealed Cr12MoV steel. Three vibration sensors are installed in three mutually perpendicular directions of the machining center spindle. The sampling frequency is set as 12.8 kHz. The experimental platform is shown in Fig. 3.

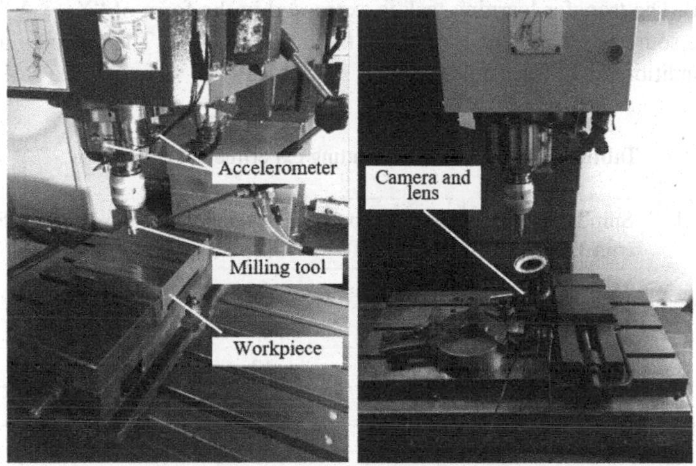

Fig. 3. Cutting experimental platform

The full file cutting experiment takes plane milling as processing method, which is in the form of alternating up and down milling. The size of the milling plane is 200 mm * 200 mm. There are 54 cutting conditions in the experiment, where the spindle speed, feed per tooth, milling direction and radial cut-ting depth is changeable. Various processing parameters are shown in Table 2. The milling area is 10000 mm^2 for every kind of cutting condition. The processing under one cutting condition is regarded as one step. A cutting cycle has 54 steps. The full life cutting has 65 cutting cycles. During the experiment, the vibration signal is collected uninterruptedly, and the image of flank is collected once per cycle.

Table 2. Machining parameters of milling experiment

Spindle speed (rpm)	Feed per tooth (mm)	Radial cutting depth (mm)	Milling direction
1500, 2000, 2500	0.1, 0.2, 0.3	6, 12, 18	Up, Down

4.2 Data Preprocessing

In this study, the sample data of 6 cutting conditions are selected to train and test the proposed tool wear states recognition model. The parameters of 6 working conditions are shown in Table 3. The vibration signal and tool images are processed according to the method described in Sect. 3.1 to obtain the dataset for the model training. The sample data is the first 4000 numbers of the signal spectrum. It is folded into the format of 2×2000. The transfer learning task is expressed in the form of "X → Y", where X refers to the source domain and Y refers to the target domain. The sample data of each working condition constitutes a domain.

Table 3. Parameters of six cutting experimental conditions

Domain label	Spindle speed (rpm)	Feed per tooth (mm)	Radial cutting depth (mm)	Milling direction
A	1500	0.1	6	Up
B	1500	0.2	12	Up
C	2000	0.2	12	Up
D	2000	0.3	18	Down
E	2500	0.1	6	Down
F	2500	0.3	18	Down

In this study, according to the tool image processing method in Sect. 3.1, the average VB value of four inserts in 65 cutting processes was obtained. This study divides the wear states of tool into four states: early wear, stable wear, early severe wear, and severe wear. According to the trend of the VB value curve, the wear state of each cutting cycle is determined. The cutting cycle number corresponding to the four wear states is: 1–10, 11–44, 45–54, 55–65.

4.3 Model Performance Analysis

Task Weight Parameters Optimization. Referring to the normal training methods of neural networks in the literature, the pre-trained tool wear states classification network easily got a classification accuracy of 95% on the test samples of source domain. But in transfer learning training, the model needs to complete the classification task and the domain adaptation task at the same time. The two tasks have different effects on the performance of the model. Hyperparameters α are used to control the optimization direction of the model. In this study, three transfer learning tasks (A $\rightarrow$ C, B $\rightarrow$ E, D $\rightarrow$ F) are selected to study the optimization of α.

When $\alpha = 0$, the model is trained only under the source domain classification task. The source domain classification cross entropy (SDCE) and the maximum mean discrepancy (MMD) is showed in Fig. 4. As the number of training rounds increases, the source domain classification cross entropy decreases, which shows that the classification ability of the model for source domain data is getting better and better. The maximum mean discrepancy is very small at the beginning of training. However, as the adaptability of the model to the source domain data increases, the maximum mean difference increases rapidly. This shows that the distribution difference between the deep features of source domain and target domain becomes larger. The deep features distribution of target domain and source domain is very different, which leads to the low classification accuracy of the trained model for the target domain samples. This proves that it is very necessary to carry out transfer learning on the model to improve the adaptability of the model to the target domain data.

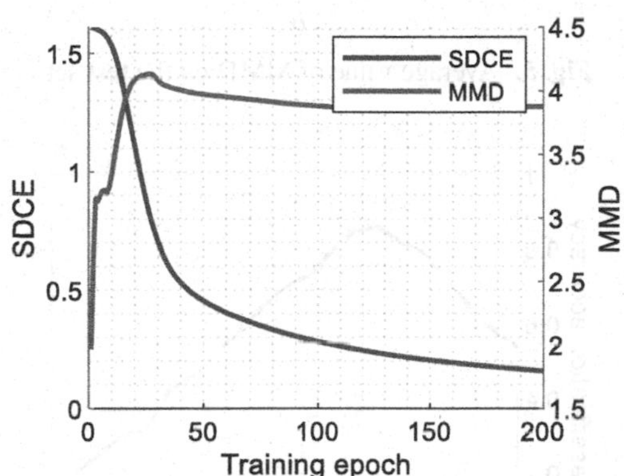

Fig. 4. The model training loss change process when $\alpha = 0$

α will affect the optimization direction of model parameters, and ultimately affect the classification accuracy of the model in the source domain and target domain. Our training goal is to make the model have better performance in both source domain and target domain. If weight parameter α is too small, the model can't align the distribution of deep features; If the weight parameter is too large, the deep features of each category of data may be mixed with each other, resulting in a decline in classification accuracy. Therefore, this study conducts a grid search on α to find the optimal value. Figure 5 shows the relationship between the MMD mean value and α. Compared with Fig. 4, it can be seen that MMD plays a very important role in model optimization. The MMD value is 1.4 when $\alpha = 0.3$, which is nearly 3 times less than that when $\alpha = 0$. With the increase of α, the alignment effect of the model on deep features becomes stronger and stronger, and the MMD value also gradually decreases. Figure 6 shows the value of the acc of the model on the target domain test set as α changes. It can be seen from the figure that as the weight parameter increases, the classification accuracy of the model in the target domain first increases and then decreases, with a maximum value in the middle. When the weight parameter is greater than 0.6, the classification accuracy of the model is less than that when the weight parameter is equal to zero. This proves that too large value of α will cause the features of each category to gather. At about $\alpha = 0.3$, the model achieves the highest accuracy of 86%.

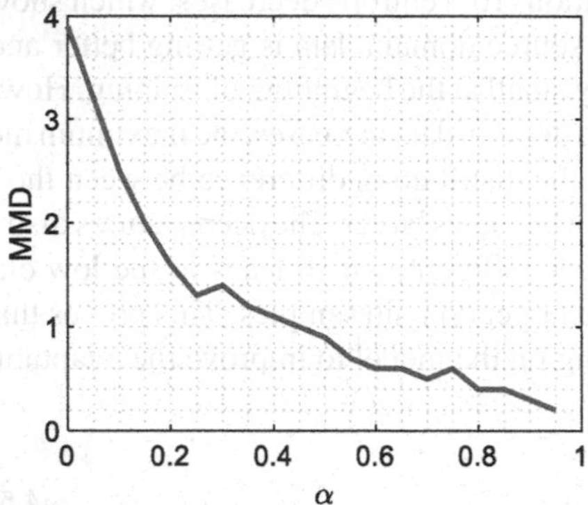

Fig. 5. Average value of MMD on the test set

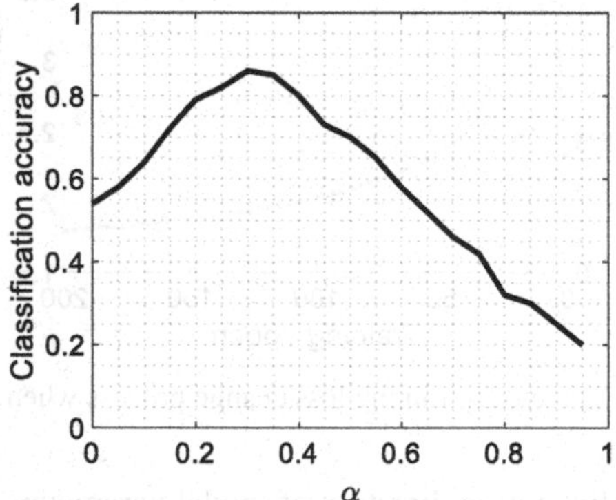

Fig. 6. The acc on the target domain test set

Model Performance Comparison. In order to further verify the performance of the proposed unsupervised transfer learning model for tool wear states recognition, five unsupervised transfer learning tasks are built, using the domain listed in Table 3. The weight parameter α is set as 0.3. 1D convolutional neural network (1DCNN) without transfer learning and two transfer learning methods disclosed in documents are used to compare with our proposed model. The two transfer learning methods are: Dual-path dynamic adversarial domain adaptation (DP-DADA) [17], Pretrained one-dimensional convolutional neural network based transfer learning model (PCNNTLM) [18].

Table 4. Classification accuracy on different transfer learning tasks

Task	Model			
	1DCNN	DP-DADA	PCNNTLM	Proposed model
A → E	0.72	0.77	0.86	0.84
A → C	0.46	0.70	0.82	0.83
E → F	0.85	0.88	0.88	0.92
B → C	0.44	0.69	0.80	0.83
B → F	0.45	0.77	0.76	0.81
C → E	0.40	0.74	0.75	0.79

Table 4 shows the average classification accuracy of four models in the target domain test set. The benchmark model 1DCNN is only trained on the source domain, which gets the lowest classification accuracy on the target domain. It can't be used for tool wear states recognition. Compared with the baseline, the classification accuracy of the three transfer learning models in the target domain has been significantly improved. Our proposed model achieves the best test accuracy on three transfer learning tasks. The accuracy on the other two transfer learning tasks is more than 80%. This proves that the model in this paper has better comprehensive performance than the other two models.

5 Conclusion

The severely worn tool will affect the cutting stability and the surface quality of the workpiece. It is necessary to carry out active tool wear states monitoring to ensure the quality of cutting. In view of the cutting scenes in which the cutting conditions change frequently, this paper proposes a tool wear monitoring method based on unsupervised transfer learning. This method constructs a tool wear states classification deep learning network based on 1D convolutional neural network and multi-layer perceptron network. The tool wear state recognition model based on the proposed network realizes the distribution adaptation of deep features of two conditions monitoring data through transfer learning. The experimental results in this paper show that the proposed model achieves better performance than the methods in the open literature.

References

1. Stavropoulos, P., Papacharalampopoulos, A., Vasiliadis, E., Chryssolouris, G.: Tool wear predictability estimation in milling based on multi-sensorial data. Int. J. Adv. Manuf. Technol. **82**(1–4), 509–521 (2016)
2. Liao, X.P., Zhou, G., Zhang, Z.K., Lu, J., Ma, J.Y.: Tool wear state recognition based on GWO-SVM with feature selection of genetic algorithm. Int. J. Adv. Manuf. Technol. **104**(1–4), 1051–1063 (2019)

3. Chen, N., Hao, B.J., Guo, Y.L., Li, L., Khan, M.A., He, N.: Research on tool wear monitoring in drilling process based on APSO-LS-SVM approach. Int. J. Adv. Manuf. Technol. **108**(7–8), 2091–2101 (2020)
4. Li, X.B., et al.: A data-driven approach for tool wear recognition and quantitative prediction based on radar map feature fusion. Measurement **185** (2021)
5. Bazi, R., Benkedjouh, T., Habbouche, H., Rechak, S., Zerhouni, N.: A hybrid CNN-BiLSTM approach-based variational mode decomposition for tool wear monitoring. Int. J. Adv. Manuf. Technol. **119**(5–6), 3803–3817 (2022)
6. Li, Z.M., Zhong, W., Shi, Y.G., Yu, M., Zhao, J., Wang, G.F.: Unsupervised tool wear monitoring in the corner milling of a titanium alloy based on a cutting condition-independent method. Machines **10**(8) (2022)
7. Pan, T.H., Zhang, J., Zhang, X., Zhao, W.H., Zhang, H.J., Lu, B.H.: Milling force coefficients-based tool wear monitoring for variable parameter milling. Int. J. Adv. Manuf. Technol. **120**(7–8), 4565–4580 (2022)
8. Feuz, K.D., Cook, D.J.: Transfer learning across feature-rich heterogeneous feature spaces via feature-space remapping (FSR). ACM Trans. Intell. Syst. Technol. **6**(1) (2015)
9. Jiang, S.Y., et al.: Multi-label metric transfer learning jointly considering instance space and label space distribution divergence. IEEE Access **7**, 10362–10373 (2019)
10. Liao, Y.X., Huang, R.Y., Li, J.P., Chen, Z.Y., Li, W.H.: Dynamic distribution adaptation based transfer network for cross domain bearing fault diagnosis. Chin. J. Mech. Eng. **34**(1), 52 (2021)
11. Li, C., Zhang, S.H., Qin, Y., Estupinan, E.: A systematic review of deep transfer learning for machinery fault diagnosis. Neurocomputing **407**, 121–135 (2020)
12. Lee, K., et al.: Multi-objective instance weighting-based deep transfer learning network for intelligent fault diagnosis. Appl. Sci.-Basel **11**(5) (2021)
13. Li, J.B., Lu, J., Chen, C.Y., Ma, J.Y., Liao, X.P.: Tool wear state prediction based on feature-based transfer learning. Int. J. Adv. Manuf. Technol. **113**(11–12), 3283–3301 (2021)
14. Zhu, Y.M., Zi, Y.Y., Xu, J., Li, J.: An unsupervised dual-regression domain adversarial adaption network for tool wear prediction in multi-working conditions. Measurement **200** (2022)
15. Lu, N.N., Wang, S.C., Xiao, H.H.: An unsupervised intelligent fault diagnosis system based on feature transfer. Math. Probl. Eng. **2021** (2021)
16. Borgwardt, K.M., Gretton, A., Rasch, M.J., Kriegel, H.P., Scholkopf, B., Smola, A.J.: Integrating structured biological data by kernel maximum mean discrepancy. Bioinformatics **22**(14), E49–E57 (2006)
17. Li, K., Chen, M.S., Lin, Y.C., Li, Z., Jia, X.S., Li, B.: A novel adversarial domain adaptation transfer learning method for tool wear state prediction. Knowl.-Based Syst. **254** (2022)
18. Bahador, A., Du, C.L., Ng, H.P., Dzulqarnain, N.A., Ho, C.L.: Cost-effective classification of tool wear with transfer learning based on tool vibration for hard turning processes. Measurement **201** (2022)

Fault Diagnosis of High-Voltage Circuit Breakers via Hybrid Classifier by DS Evidence Fusion Algorithm

Xiaofeng Li, Liangwu Yu, Hantao Chen, Yue Zhang, and Tao Zhang(✉)

College of Power Engineering, Naval University of Engineering, Wuhan 430072, China
xiaofengli@whu.edu.cn, 927495831@qq.com

Abstract. Accurate and timely fault diagnosis is of significance for the stability of high-voltage circuit breaker (HVCB), which plays an important role in ensuring the safety of the power system. Current fault diagnosis techniques generally depend on an individual classifier. In this study, combining support vector machine (SVM) with extreme learning machine (ELM) by Dempster-Shafer (DS) evidence fusion algorithm, we propose DS_SE, a hybrid classifier for clearance joint fault diagnosis of HVCB. At first, through variational mode decomposition (VMD), the energy distribution is extracted as the feature value of vibration signals. Then DS evidence fusion algorithm is proposed for fusing analysis of evidence from different sub-classifiers and sensors. Extensive evaluation of DS_SE based on a real Zn12 HVCB indicates that it can achieve better performance by fusing conflicting evidence.

Keywords: High-voltage circuit breakers · Mechanical Fault diagnosis · Hybrid classifier · Evidence fusion

1 Introduction

Affected by mechanical, electrical, and chemical stresses, the mechanical properties of HVCB are prone to failure [1]. According to international survey, almost 50% of major failures are of mechanical origin, like joint clearance fault led to by corrosion and wear effect [2–4]. However, conventional scheduled maintenance of HVCB is usually highly dependent on prior knowledge about diagnostic expertise, which is inefficient and difficult to ensure reliability. Besides, since a certain degree of disassembly is necessary for that process, secondary damage might also be caused to HVCB [5, 6].

In the past decades, intelligent fault diagnosis and online state-monitoring of HVCB have attracted wide attention [7, 8]. For example, a hybrid classifier is proposed by Huang et al. [9], in which support vector data description (SVDD) and fuzzy c-means (FCM) clustering method is applied for outlier detection and specific fault diagnosis, respectively. Thus both known and unknown mechanical faults can be detected. Ma et al. [10, 11] built the diagnosis model by random forest (RF) and proposed a strategy to reduce the non-essential feature component. As HVCB is always being in a long-term standstill, it's unrealistic to obtain sufficient samples for diagnosis model training and

H. Zhang et al. (Eds.): NCAA 2023, CCIS 1870, pp. 291–302, 2023.
https://doi.org/10.1007/978-981-99-5847-4_21

testing. In that case, the fault diagnosis model built by SVM is reported to have little dependence on the amount of data [12]. Besides, after the kernel and penalty parameters optimization by annealing, genetic, as well as particle swarm algorithm, SVM can further improve its fault diagnosis performance than others [13]. Therefore, one class [14, 15], and multi-class [16] type of SVM are investigated for mechanical fault diagnosis of HVCB, by which the looseness of base screws, electromagnet stuck and overtravel are diagnosed successfully.

Compared with other feedforward neural networks, ELM randomly predetermines the weights and offsets between the input and hidden layer and calculates the network parameter between the hidden and output layer by Moore-Penrose generalized inverse [17], which makes it more efficient and has therefore been widely applied in the field of fault diagnosis [18, 19]. For instance, integrating multiple independently trained one-class ELM, a multi-class classifier for spring stress, and the installation error of the drive shaft of HVCB was developed by Chen et al. [20]. Gao et al. [21] composed multiple binary ELM, then assigned the multi-class diagnosis task to different level's classifiers. Similar to the multi-layer filter, the input samples corresponding to various types of faults were filtered out one by one by different binary classifiers. Beyond that, with the advantage of storing complex mapping relationships, the deep neural network is also increasingly popular in the fault diagnosis field [22]. Classifiers of convolutional neural networks (CNN) [23, 24] were reported recently, which might imply an important trend for the development of mechanical fault diagnosis of HVCB. The abovementioned mechanical fault diagnosis technologies contribute to improving the service reliability of HVCB to a certain extent.

Overall, feature extraction and fault diagnosis are mainly involved in current fault diagnosis investigations. For the feature extraction section, sound [23], contact travel curve [25, 26], electromagnet coil current [27, 28], and vibration [18–20] are typical signals for fault diagnosis of HVCB. In the past decades, a larger percentage of fault diagnosis models of HVCB are building by vibration signal process, as rich state-related information is contained in vibration signal [29]. Thanks to the previous valuable works, empirical mode decomposition (EMD) [14, 18], local mean decomposition (LMD) [9], empirical wavelet packet decomposition (WPD) [10], empirical wavelet transform (EWT) [30] and variational mode decomposition (VMD) [15, 20]. Afterward, different types of amplitude- and frequency-based features of vibration, such as time-frequency entropy [10], permutation entropy [20], singular entropy [18], and energy entropy [9] have been extracted for fault diagnosis. Among these signal-processing methods, VMD has good noise robustness, and can effectively alleviate the end effect and modal aliasing problems, which makes it perform better at processing non-stationary and non-linear vibration signals. Thence, vibration is considered as the input signal and VMD is utilized to decompose the signal in this study.

In the study, we report a novel synthesis model named DS_SE that combines different individual classifiers via the DS evidence fusion algorithm for mechanical fault diagnosis of HVCB. In detail, the energy distributions of vibration in IMFs by VMD are first calculated as the feature vector for diagnostic model building. Besides, multiple SVM and ELM classifiers are trained by features of different sensors and each one is employed as the sub-classifier. In the end, the DS evidence fusion algorithm is adapted for joint clearance fault diagnosis of HVCB.

The rest of our paper is structured as follows. Section 2 outlines the technical framework. Section 3 describes the theoretical background of the paper. Section 4 introduces the experiment setup and analyzes the feature extraction of vibration by VMD method. Section 5 presents the effectiveness of the DS evidence fusion algorithm by comparing the fault diagnosis accuracy of our DS_SE with those of single classifiers. Conclusions are given in Sect. 6.

2 Technical Framework

In the paper, the energy distribution in different IMFs of VMD is adopted as a vibration signal feature to build a hybrid classifier for fault diagnosis of HVCB, in which SVM and ELM are employed as sub-classifier and fused through the DS evidence algorithm. For demonstration, the main steps of fault diagnosis in the paper is given as Fig. 1

At first, switching experiments of a real ZN12 HVCB in various mechanical states are conducted. In addition to normal conditions, clearance joints fault at three locations are constructed, each type of mechanical state is carried out by 100 times, and a total of 400 groups of vibration signals of two measuring points under four mechanical states are collected.

Next, VMD is used to decompose the obtained vibration signal, and the energy distribution in different IMF is calculated as the mechanical state-related feature of HVCB. Then, as shown in Table 1, the obtained feature is divided into two datasets for

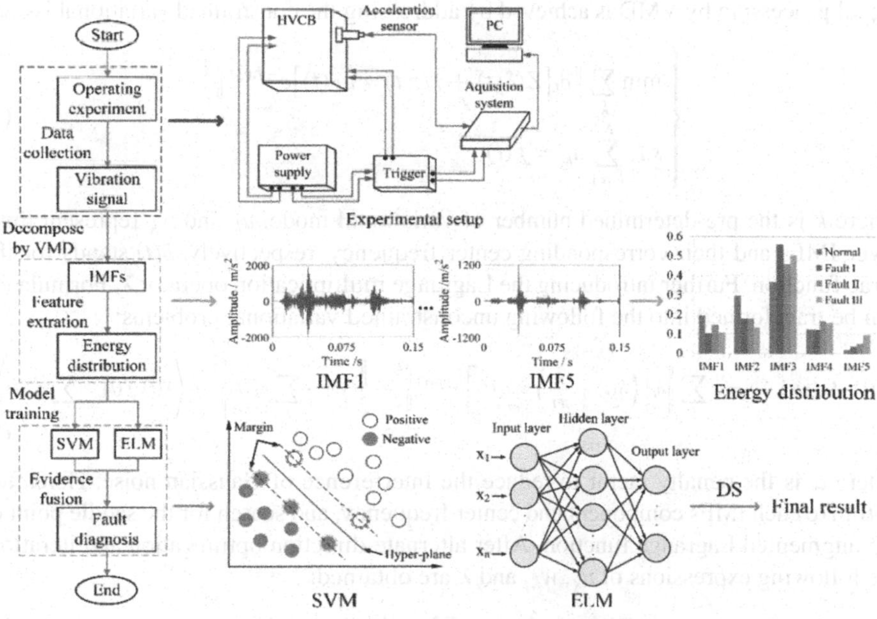

Fig. 1. Overall flow chart of fault diagnosis of HVCB

model building. It should be noted that the testing data should not be included in the training dataset; otherwise, over-optimistic problems could emerge in the testing result.

Table 1. Dataset for model building

Fault type		Normal		Fault I		Fault II		Fault III	
Clearance size (mm)		< 0.04	< 0.04	0.25	0.75	0.25	0.75	0.25	0.75
Dataset A	training	40	40	40	40	40	40	40	40
	testing	10	10	10	10	10	10	10	10
Dataset B	training	50		50		50		50	
	testing		50		50		50		50

Finally, SVM and ELM are considered as the sub-classifier, and their modeling training and testing are implemented by the above dataset. In the fault diagnosis process of the hybrid model, the output of SVM and ELM are considered as original evidence and perform fusion analysis by the DS evidence algorithm.

3 Theoretical Background

3.1 Variational Mode Decomposition

In the theory of VMD, the raw signal *f(t)* can be divided into a series of IMFs with different center frequencies and limited bandwidth. The basic implementation of the signal processing by VMD is achieved by addressing the constrained variational issues:

$$\begin{cases} \min \sum_{k} \left\| \partial_t \left[Z(\delta(t) + j/\pi t) * u_k(t) \right] e^{-jw_k t} \right\|_2^2 \\ s.t. \sum_{k=1}^{K} u_k = f(t) \end{cases} \tag{1}$$

where k is the pre-determined number of variational mode, u_k and w_k represent the k layer IMFs and their corresponding center frequency, respectively. $\delta(t)$ stands for the dirac function. Further introducing the Lagrange multiplication operator λ, Formula (1) can be transformed into the following unconstrained variational problems:

$$L(\{u_k\}, \{w_k\}, \lambda) = \alpha \sum_k \left\| \partial_t \left[\left(\delta(t) + \frac{j}{\pi t} \right) * u_k(t) \right] e^{-jw_k t} \right\|_2^2 + \left\| f(t) - \sum_k u_k(t) \right\|_2^2 + \left\langle \lambda(t), f(t) - \sum_k u_k(t) \right\rangle \tag{2}$$

where α is the penalty factor to reduce the interference of Gaussian noise. Hereafter, optimize each IMFs component and center frequency, and search for the saddle point of the augmented Lagrange function. After alternate direction optimization and iteration, the following expressions of u_k, w_k, and λ are obtained:

$$\hat{u}_k^{n+1}(w) = \frac{\hat{f}(w) - \sum_{i \neq k} \hat{u}_i(w) + \hat{\lambda}(w)/2}{1 + 2\alpha (w - w_k)^2} \tag{3}$$

$$w_k^{n+1} = \frac{\int_0^\infty w\left|\hat{u}_k^{n+1}(w)\right|^2 dw}{\int_0^\infty \left|\hat{u}_k^{n+1}(w)\right|^2 dw} \tag{4}$$

$$\hat{\lambda}^{n+1}(w) = \hat{\lambda}^n(w) + \gamma(\hat{f}(w) - \sum_k \hat{u}_k^{n+1}(w)) \tag{5}$$

where Υ is the noise tolerance, $\hat{u}_k^{n+1}(w)$, $\hat{u}_i(w)$, $\hat{f}(w)$ and $\hat{\lambda}(w)$ correspond to the Fourier transform results of $u_k^{n+1}(t)$, $u_i(t)$, $f(t)$ and $\lambda(t)$, respectively. For signal processing by VMD method, the main iterative step can be summarized as follows:

Step 1: Initialize $\hat{u}_k^1$, $\hat{w}_k^1$, λ^1 and the maximum number of iterations number N.

Step 2: Calculate each IMFs and center frequency $\hat{u}_k$, w_k by Formula (3) and (4).

Step 3: Update Lagrange multiplication operator $\hat{\lambda}$ by Formula (5).

Step 4: Judge whether $\sum_k \left\|\hat{u}_k^{n+1} - \hat{u}_k^n\right\|_2^2 / \left\|\hat{u}_k^n\right\|_2^2 < \varepsilon$ (ε is the accuracy convergence criterion). or the maximum number of iterations is reached (n < N). If it is satisfied, the iteration is completed, then obtain $\hat{u}_k$ and w_k, otherwise return to step 2.

3.2 DS Evidence Fusion Algorithm

Information from multiple aspects can be comprehensively analyzed by the DS evidence fusion algorithm to reach a synthesis decision. Taking a set comprised of N elements as an example, it can be named the discernment frame.

$$\Theta = \{A_1, A_2, ...A_i, ..., A_N\} \tag{6}$$

In real fault diagnosis scenarios, element A can represent specific working states. Define $m(A)$ as the basic probability assignment (BPA), which is subject to the following limitations:

$$\begin{cases} m(\phi) = 0 \\ 0 \leqslant m(A) \leqslant 1, \ \forall A \subset \Theta \\ \sum_{A\subset\Theta} m(A) = 1 \end{cases} \tag{7}$$

For fault diagnosis issue, as the possible working state is limited, there is finite BFA function $m_1, m_2, ..., m_n$, and the evidence fusion law is defined as below:

$$(m_1 \oplus m_2 ... \oplus m_n)(A) = \frac{1}{1-k} \sum_{A_1\cap A_2 ... \cap A_n = A} m_1(A_1) * m_2(A_2) ... * m_n(A_n) \tag{8}$$

where k represents the conflict coefficient that is applied to describe the degree of evidence conflict and can further be expressed as follows:

$$k = \sum_{A_1\cap A_2 ... \cap A_n = \emptyset} m_1(A_1) * ... m_n(A_n) = 1 - \sum_{A_1\cap A_2 ... \cap A_n \neq \emptyset} m_1(A_1) * ... m_n(A_n) \tag{9}$$

DS evidence fusion algorithm has a strong capability to synthesize information from different sources. For fault diagnosis of HVCB, the raw diagnostic evidence usually come from different sensors and classifiers. DS evidence fusion provides a comprehensive analysis approach.

4 Experimental Application

4.1 Experiment Setup

Limited by the insufficiency of fault data, the mechanical fault type that has been diagnosed is relatively simple, most of which are obvious faults such as the looseness of base screws, electromagnet stuck, overtravel, and spring fatigue [10–16], rarely involve the clearance joints fault of the mechanical system. Joint clearance is inevitable in articulated mechanical systems such as HVCB. Meanwhile, affected by a manufacturing error, wear, and corrosion, clearance joint fault is a common cause of mechanical failure of HVCB (Fig. 2).

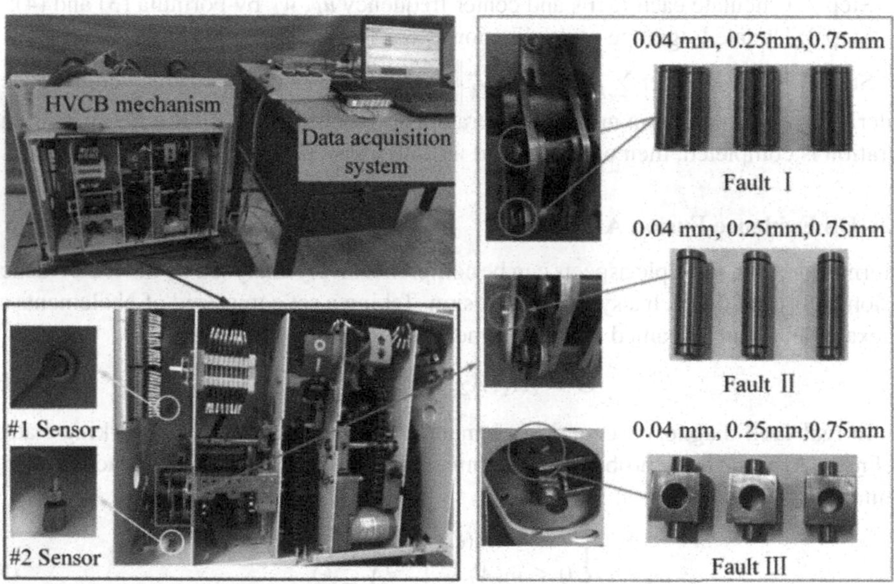

Fig. 2. Experimental setup

Compared with spring fatigue and overtravel, the state-related feature of clearance joint-caused fault is weaker, which makes it more difficult to diagnose. To verify the effectiveness of the proposed fault diagnosis model, an experimental setup based on a real ZN12 HVCB is built. Except for the normal working state, three joint fault working states are constructed by adjusting joint clearance size (fault size of 0.25 mm, 0.75 mm). For measurement Two CCLD/IEPE acceleration sensors (coming from Brüel & Kjær, type 8339) are installed in the HVCB for the vibration recording of the closing operation. For our accelerometer, its measurement range, sensitivity, and upper cut-off frequency are ± 10000 m/s^2, 0.25 mv/g, and 20 kHz, respectively. Besides, a signal collection card (coming from Brüel & Kjær, type 3053-B-120) is applied for vibration acquisition. It should be noted that, in order to avoid interference between two adjacent switching operation tests of HVCB, A 3-min interval for each switching operation test is necessary. Meanwhile, the experimental trials prove the measuring point of vibration is crucial for related fault diagnosis analysis. For ensuring that the extracted vibration signal contains abundant state-related information, the two accelerometers should be installed near the joint-caused fault. As the axis line of the two accelerometers, it is better to be perpendicular to each other.

In the specific experimental process, set the signal acquisition frequency of the signal collection card to 65536 Hz with a sampling time of 1 s. For the normal and three types of joint clearance faults working states, there are 400 groups of vibration signals from two acceleration sensors.

4.2 Feature Extraction

The vibration energy distribution in the different frequency bands of an articulated mechanism would change along with its structural health status [14]. Therefore, the signal energy of multiple IMFs by VMD is calculated as the feature. In general, the center frequencies interval of adjacent IMFs would gradually shrink with the increase of IMFs number, and further lead to the over-decomposition issue. Meanwhile, too small predetermined IMFs number might lead to the lack of center frequency of IMFs, which lower the degree of discrimination for different working condition. In the study, multiple signal process circles by VMD method are tried for the analysis of optimal IMFs number, the final optimal IMFs number is set to 5. Subsequently, employed as the state-related feature of the HVCB. Calculate the energy distribution of the vibration signal as follows:

$$P_i = \frac{E_i}{E} \tag{10}$$

$$\begin{cases} E_i = \int_{t_0}^{t_i} |A(t)|^2 dt \\ E = \sum_{1}^{5} E_i \end{cases} \tag{11}$$

where i represents the serial number of IMFs. t_i and t_0 stand the end and start time of the analyzed vibration signal. E_i and $A(t)$ denote the energy of each IMF and amplitude at different time points.

5 Result Analysis

In this section, we compared the performance of the DS_SE with its sub-classifiers. Dataset A in Table 1 is first considered, wherein 40 groups of the signal feature of each joint clearance size are randomly selected as training samples. Thus for each type of condition, the training set contained 80 samples and the remaining 20 samples are grouped into the testing set. On this basis, the average result of 10 fault diagnosis tries is shown in Table 2.

Table 2. Fault diagnosis comparison of dataset A

Classifier	Normal	Fault I	Fault II	Fault III	Ave
SVM_P1	85.0%	85.0%	80.0%	85.0%	83.8%
SVM_P2	85.0%	55.0%	100.0%	70.0%	77.5%
ELM_P1	90.0%	85.0%	80.0%	85.0%	85.0%
ELM_P2	85.0%	65.0%	100.0%	80.0%	82.5%
DS_SE	85.0%	85.0%	85.0%	90.0%	86.3%

In Table 2, SVM_P1 and SVM_P2 represent the fault diagnosis of SVM by the #1 accelerometer and the #2 accelerometer, respectively. Through comparison, we can find some interesting observations. First, the average accuracy of the traditional single classifier under a single sensor is within 85%, lower than previous research. This can be attributed to the fact that compared with other mechanical faults like the looseness of base screws, the representative features characterizing the clearance joint fault in measured signals are weaker. Second, both SVM and ELM perform better under the #1 accelerometer than that under the #2 accelerometer, which indicates that the vibration signal of the #1 accelerometer contains more state-related information. As the two sensors in this paper are installed in different positions with mutually perpendicular axes, different information is carried out, and improved fault diagnosis performance can be obtained by combining different sensor information. After evidence fusion analysis, the average diagnostic accuracy of DS_SE can reach 86.3%.

To further demonstrate how our DS_SE improves the performance of fault diagnosis, the confusion matrices of the sub-classifiers and hybrid classifiers are given in Fig. 3, Fig. 4, and Fig. 5. The lateral axis and vertical axis represent the predicted labels and true labels, respectively. The number of predicted samples for each fault kind is listed in the matrices. For instance, 20 groups of normal samples in Fig. 3(a) are predicted as 17 normal, 2 fault II and 1 fault III. For the #2 accelerometer, the prediction accuracy of fault I is worse, and the error diagnosis between normal and fault I is higher than that of the #1 accelerometer. Specifically, both SVM_P2 and ELM_P2 misdiagnose 3 normal samples as fault I. On the other hand, for the #1 accelerometer, the fault diagnosis performance between normal and fault I is obviously better, but more misdiagnosis cases emerge between fault II and fault III. It indicates that different information in different sensors features a certain degree of complementarity. Additionally, SVM_P1 and ELM_P1 achieve the same average prediction accuracy of 85% for normal and fault II but with different misdiagnosis distributions. Taking fault III as an instance, 3 samples is diagnosed as 1 normal and 2 fault II by mistake, but the result of ELM_P1 is just the opposite. It shows that there are some contradictions in the diagnosis results of different sub-classifier. Therefore, fusing the results of different sub-classifier with different sensor information can improve the prediction accuracy. This can be explained that the DS_SE model having a stronger ability to address paradoxical evidence.

To test the robustness and adaptability of our model, we trained each fault diagnosis model by dataset B in Table 1, and related testing results are given in Table 3 and

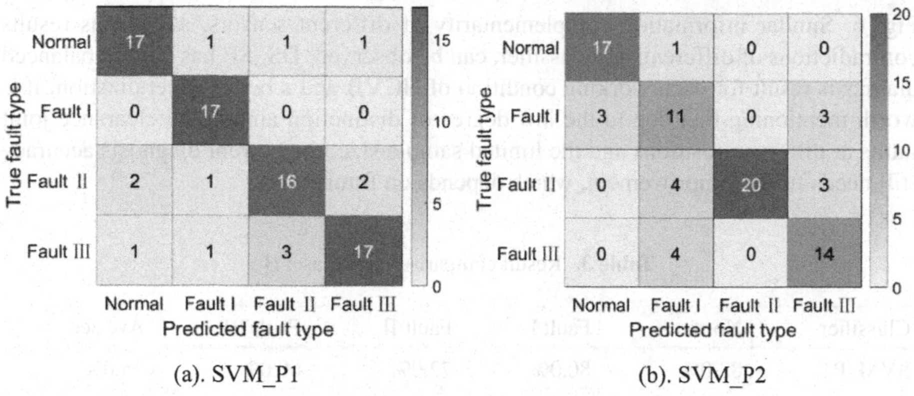

(a). SVM_P1 (b). SVM_P2

Fig. 3. Result comparison of ELM classifier

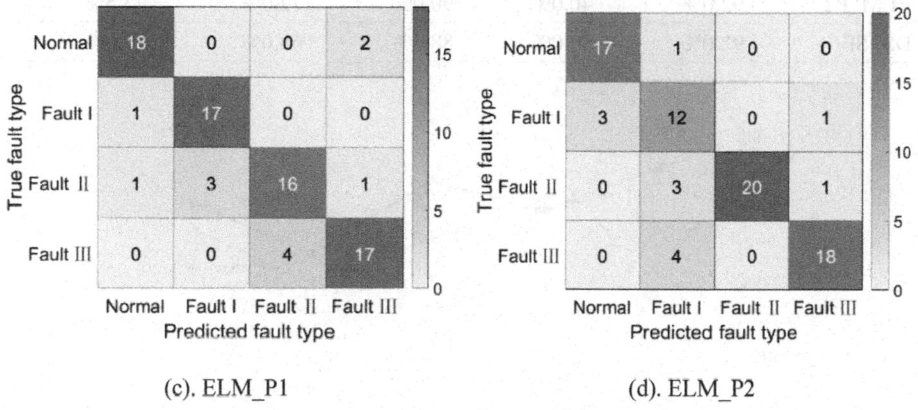

(c). ELM_P1 (d). ELM_P2

Fig. 4. Result comparison of ELM classifier

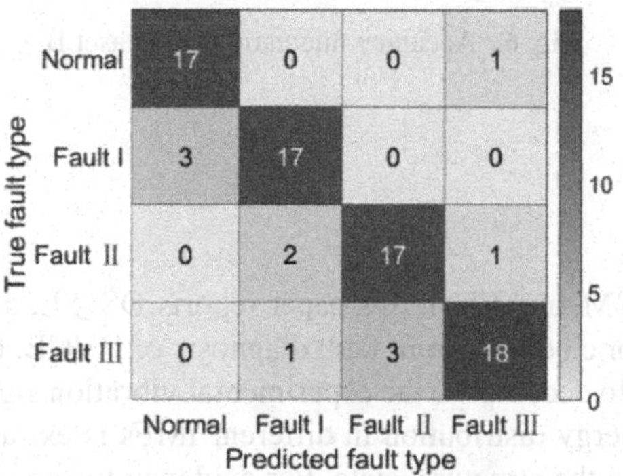

Fig. 5. Result comparison of hybrid classifier

Fig. 6. Similar information complementarity of different sensors, as well as results contradictions of different sub-classifier, can be observed. DS_SE has a more balanced diagnosis result for each working condition of HCVB and a better generalization. It is worth mentioning that due to the low degree of distinction among the clearance joint faults at different positions and the limited sample size, the current diagnosis accuracy still needs further improvement, which depends on future work.

Table 3. Result comparison of dataset B

Classifier	Normal	Fault I	Fault II	Fault III	Ave acc
SVM_P1	88.0%	80.0%	72.0%	60.0%	75.0%
SVM_P2	78.0%	24.0%	86.0%	66.0%	63.5%
ELM_P1	86.0%	84.0%	62.0%	84.0%	79.0%
ELM_P2	92.0%	40.0%	90.0%	72.0%	73.5%
DS_SE	92.0%	80.0%	88.0%	68.0%	82.0%

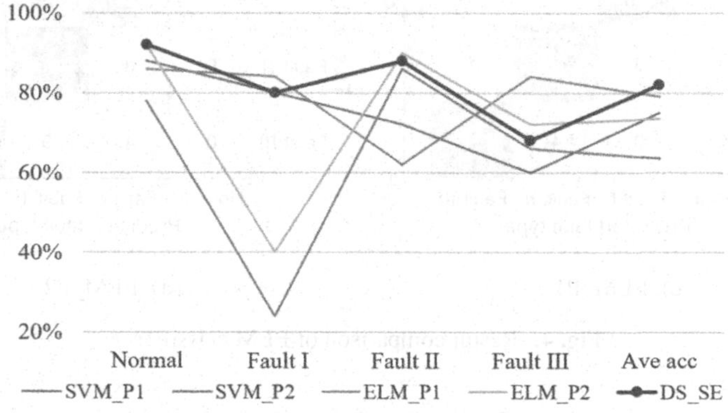

Fig. 6. Accuracy fluctuation of dataset B

6 Conclusion

Combining VMD, SVM, and ELM, this paper reports DS_SE, a hybrid classifier with DS evidence fusion for clearance joint fault diagnosis of HVCB. For feature extraction, VMD is first adopted to decompose the experimental vibration signal of two accelerometers, and then the energy distribution in different IMFs is extracted as representative features characterizing the structural state. For evidence fusion of SVM and ELM, we propose a fusion method based on the DS evidence fusion algorithm, which helps to deal with its paradoxical evidence fusion problems. Comparative results show that the proposed hybrid classifier can achieve better performance than traditional single classifiers due to the consideration of information complementarity of different sensors. It has a stronger ability than the traditional method of a single classifier. In further work, hybrid state-related features and model performance in case of unbalanced input would be evaluated by the reported methodology. Some advanced optimization algorithms will also be adopted to enhance the model performance and promote the application in other emerging applications.

References

1. Liu, Y., Zhang, G., Zhao, C., Qin, H., Yang, J.: Influence of mechanical faults on electrical resistance in high voltage circuit breaker. Int. J. Electr. Power **129** (2021)
2. Li, X., Zhang, T., Guo, W., Wang, S.: Multi-layer integrated extreme learning machine for mechanical fault diagnosis of high-voltage circuit breaker. In: Zhang, H., et al. (eds.) NCAA 2022. CCIS, vol. 1638, pp. 287–301. Springer, Singapore (2022). https://doi.org/10.1007/978-981-19-6135-9_22
3. Zhang, X., Gockenbach, E., Liu, Z., Chen, H., Yang, L.: Reliability estimation of high voltage SF6 circuit breakers by statistical analysis on the basis of the field data. Electr. Power Syst. Res. **103**, 105–1013 (2013)
4. Razi-Kazemi, A.A., Niayesh, K.: Condition monitoring of high voltage circuit breakers: past to future. IEEE Trans. Power Deliv. **36**, 740–750 (2021)
5. Ramentol, E., et al.: Fuzzy-rough imbalanced learning for the diagnosis of High Voltage Circuit Breaker maintenance: the SMOTE-FRST-2T algorithm. Eng. Appl. Artif. Intell. **48**, 134–139 (2016)
6. Vianna, E.A.L., Abaide, A.R., Canha, L.N., Miranda, V.: Substations SF6 circuit breakers: Reliability evaluation based on equipment condition. Electr. Power Syst. Res. **142**, 36–46 (2017)
7. Rudsari, F.N., Kazemi, A., Shoorehdeli, M.A.: Fault analysis of high voltage circuit breakers based on coil current and contact travel waveforms through modified SVM classifier. IEEE Trans. Power Deliv. **34**, 1608–1618 (2019)
8. Geng, S., Wang, X.: Research on data-driven method for circuit breaker condition assessment based on back propagation neural network. Comput. Electr. Eng. **86**, 106732 (2020)
9. Huang, N., Fang, L., Cai, G., Xu, D., Chen, H., Nie, Y.: Mechanical fault diagnosis of high voltage circuit breakers with unknown fault type using hybrid classifier based on LMD and time segmentation energy entropy. Entropy **18** (2016)
10. Ma, S., Chen, M., Wu, J., Wang, Y., Jia, B., Yuan, J.: Intelligent fault diagnosis of HVCB with feature space optimization-based random forest. Sensors **18**, 1221 (2018)
11. Ma, S., Chen, M., Wu, J., Wang, Y., Jia, B., Jiang, Y.: High-voltage circuit breaker fault diagnosis using a hybrid feature transformation approach based on random forest and stacked auto-encoder. IEEE Trans. Ind. Electron. **66**, 9777–9788 (2018)
12. Lin, L., Wang, B., Qi, J., Chen, L., Huang, N.: A novel mechanical fault feature selection and diagnosis approach for high-voltage circuit breakers using features extracted without signal processing. Sensors **19** (2019)
13. Yin, Z., Hou, J.: Recent advances on SVM based fault diagnosis and process monitoring in complicated industrial processes. Neurocomputing **174**, 643–650 (2016)
14. Jian, H., Hu, X., Fan, Y.: Support vector machine with genetic algorithm for machinery fault diagnosis of high voltage circuit breaker. Measurement **44**, 1018–1027 (2011)
15. Li, X., Wu, S., Li, X., Yuan, H., Zhao, D.: Particle swarm optimization-support vector machine model for machinery fault diagnoses in high-voltage circuit breakers. Chin. J. Mech. Eng. **33**, 6 (2020)

16. Huang, N., Chen, H., Cai, G., Fang, L., Wang, Y.: Mechanical fault diagnosis of high voltage circuit breakers based on variational mode decomposition and multi-layer classifier. Sensors (Basel, Switzerland) **16** (2014)
17. Huang, G.B., Zhu, Q.Y., Siew, C.K.: Extreme learning machine: theory and applications. Neurocomputing **70**, 489–501 (2006)
18. Gao, W., Wai, R.J., Qiao, S.P., Guo, M.F.: Mechanical faults diagnosis of high-voltage circuit breaker via hybrid features and integrated extreme learning machine. IEEE Access **7**, 60091–60103 (2019)
19. Wan, S., Chen, L.: Fault diagnosis of high-voltage circuit breakers using mechanism action time and hybrid classifier. IEEE Access **7**, 85146–85157 (2019)
20. Chen, L.: Mechanical fault diagnosis of high-voltage circuit breakers using multi-segment permutation entropy and a density-weighted one-class extreme learning machine. Meas. Sci. Technol. **31**, 85107–85118 (2020)
21. Gao, W., Qiao, S.P., Wai, R.J., Guo, M.F.: A newly designed diagnostic method for mechanical faults of high-voltage circuit breakers via SSAE and IELM. IEEE Trans. Instrum. Meas. **1** (2020)
22. Feng, J., Lei, Y., Jing, L., Xin, Z., Na, L.: Deep neural networks: a promising tool for fault characteristic mining and intelligent diagnosis of rotating machinery with massive data. Mech. Syst. Sig. Process. **72–73**, 303–315 (2016)
23. Zhao, S., Wang, E., Hao, J.: Fault diagnosis method for energy storage mechanism of high voltage circuit breaker based on CNN characteristic matrix constructed by sound-vibration signal. J. Vibroeng.**21** (2019)
24. Yang, Q., Ruan, J., Zhuang, Z., Huang, D.: Condition evaluation for opening damper of spring operated high-voltage circuit breaker using vibration time-frequency image. IEEE Sens. J. **19**, 8116–8126 (2019)
25. Niu, W., Liang, G., Yuan, H., Li, B.: A fault diagnosis method of high voltage circuit breaker based on moving contact motion trajectory and ELM. Math. Probl. Eng. 1–10 (2016)
26. Ali, F., Akbar, A.A., Ali, N.G.: Model-based fault analysis of a high-voltage circuit breaker operating mechanism. Turk. J. Electr. Eng. Comput. Sci. **25**, 2349–2362 (2017)
27. Mei, F., Pan, Y., Zhu, K., Zheng, J.: On-line hybrid fault diagnosis method for high voltage circuit breaker. J. Intell. Fuzzy Syst. **33**, 2763–2774 (2017)
28. Mei, F., Mei, J., Zheng, J., Wang, Y.: Development and application of distributed multilayer on-line monitoring system for high voltage vacuum circuit breaker. J. Electr. Eng. Technol. **8**, 813–823 (2013)
29. Li, X., Zheng, X., Zhang, T., Guo, W., Wu, Z.: Robust fault diagnosis of a high-voltage circuit breaker via an ensemble echo state network with evidence fusion. Complex Intell. Syst. (2023)
30. Li, B., Liu, M., Guo, Z., Ji, Y.: Mechanical fault diagnosis of high voltage circuit breakers utilizing EWT-improved time frequency entropy and optimal GRNN classifier. Entropy **20**, 448 (2018)

Multi-Feature Fusion and Reinforcement Model for High-Speed Train Axle Box Bearing Fault Diagnosis Under Variable Speed Domain

Yuyan Li[1], Jingsong Xie[1(✉)], Tiantian Wang[2], Jinsong Yang[1], and Buyao Yang[2]

[1] Central South University, Changsha 410000, China
{liyuyan,yangjs}@csu.ecu.cn, jingsongxie@foxmail.com
[2] Hunan University, Changsha 410000, China
{wangtt,yangbuyaoyby}@hnu.edu.cn

Abstract. The bogie of high-speed trains is the key component of the high-speed train. The fault diagnosis of bogie axle box bearing can ensure the safe operation of the whole vehicle. The signal obtained from the axle box bearing has the characteristics of variable speed and multiple distributions. To solve the above problems, we designed a new multi-scale feature fusion network, which is combined with a feature reinforcement mechanism. The proposed SE-MSFNet creatively combines signal processing methods with neural networks to construct different multiscale branches. Cascade convolutions with odd-even mixed dilated rates are proposed to expand the receptive field and reduce grid effects. A weight unit combined with the channel focus mechanism is designed to increase the channel feature weight conducive to classification and improve its sensitivity to fault features. The proposed method improves the feature extraction ability and robustness of the model on multiple distributed data through multi-scale feature fusion and enhancement. Compared with other advanced feature fusion methods, this method achieves better classification performance in both the full speed domain and the variable speed domain of the bogie.

Keywords: Fault diagnosis · High-speed train · Axle box bearing · Variable speed domain

1 Introduction

With the rapid development of high-speed trains (HST), higher requirements are put forward for comfort, safety, and reliability [1, 2]. The bogie is a key part to ensure the riding quality and operation security of a train. Its main function is to support the train and transmit the load between the train body and the wheelset. As a vulnerable part of the bogie, the health of the axle box bearing is of vital significance for the security of the bogie and even the whole vehicle [3]. To monitor train operation in real-time and avoid sudden faults, it is an effective fault diagnosis method to install sensors on the axle box and analyze them based on vibration signals.

H. Zhang et al. (Eds.): NCAA 2023, CCIS 1870, pp. 303–317, 2023.
https://doi.org/10.1007/978-981-99-5847-4_22

Existing fault detection and diagnosis methods of HST axle box bearing usually focus on signal-based methods that analyze sensor signals by extracting fault-related characteristics. As the monitoring equipment is featured with a big scale, high sampling frequency, and long-term collection, the generated large-volume data have brought new challenges to fault diagnosis. Machine learning is a great way to deal with the above problems. Traditional machine learning based on artificial feature extraction, such as SVM [4], extreme learning machine [5], sparse representation [6], and fuzzy inference [7] has been widely used in fault diagnosis of mechanical equipment. However, feature extraction requires extensive domain expertise and prior experience, and the quality of extracted features will directly affect the fault diagnosis performance. Especially for fault diagnosis of the HST axle box bearing whose internal structure, transmission path, and dynamic characteristics are complex, the collected vibration signals are non-stationary and contain a lot of noise, making it more dependent on the experience of researchers.

Compared with shallow neural networks, intelligent diagnosis methods based on deep neural networks do not need to design handcrafted features, which can directly form end-to-end frameworks. The end-to-end network is fed with raw data and directly outputs classification results without intermediate processes. Common deep-learning methods include DAE [8], DBN [9], CAE [10], and CNN [11]. Among these methods, CNN is highly valued by researchers because of weight sharing, good robustness, and strong supervised learning ability, which has been widely used in various fields such as image recognition and fault diagnosis. For end-to-end bearing fault diagnosis model. Zhang et al. [12] proposed a novel method named WDCNN, which uses wide kernels in the first convolutional layer for extracting features and suppressing high-frequency noise. Li et al. [13] combined Resnet and the squeeze operation to get the optimal combination of channels. Ye et al. [14] proposed a data-driven method, which combines multiscale permutation entropy and linear local tangent space alignment to diagnose the faults of vehicle suspension systems. Man et al. [15] use a graph convolutional network and attention mechanism to classify train bogie faults. It provides a new idea for bearing fault diagnosis based on graph theory. Kou et al. [16] fused multi-sensor data to achieve end-to-end detection of bearing faults in high-speed trains. Sun et al. [17] combine SSVM, multi-domain features, and feature selection to propose a noise-robust fault diagnosis method, which is successfully applied to HST bearing vibration data set. Jia et al. [18] proposed a model integrating DTW and depth separable convolution, which can classify axle box bearing faults of high-speed trains at different speeds, but has not been verified with real vehicle data.

For the diagnosis of bearing faults at stable speeds, the above-mentioned neural networks can effectively solve the problem. However, under the actual operating conditions of high-speed trains, the operating speed often varies within a certain range with the vehicle types and lines. Therefore, it is very important to improve the accuracy of axle box bearing fault identification in a variable speed environment. This issue includes two aspects: the first is that the model has good robustness for full-speed domain fault diagnosis for each speed interval from low to high speed. Second, the model has good classification performance for variable speed domains containing a small range of speed changes. However, speed variations will cause changes in the response of fault features on both time and spatial scales, meaning that the same fault is characterized differently at different speeds. At the same time, the mixed data set formed by vibration data of different speeds has uncertainty in its sample distribution, which makes the classification of neural networks difficult. To solve the above problems, a multi-scale feature fusion network (SE-MFSNet) is proposed for axle box bearing fault diagnosis in the full-speed domain. The main contributions are as follows:

(a) A multi-branch network combined with signal processing methods is proposed. Each branch automatically learns the features of different speed scales from low speed to high speed in parallel, which improves the feature extraction ability of the model for signals in variable speed domain. Generally, the network consists of three consecutive stages: multi-scale feature extraction stage, feature enhancement stage (attention mechanism) and feature classification stage.
(b) A new parity mixed dilated convolution combination is proposed to expand the receptive field and weaken the "grid effect" [19] to obtain more fault information. Parity mixed convolution combined with residual structure can fuse shallow and deep features, suppress the impact of noise and prevent over-fitting.
(c) In the feature reinforcement stage, a channel focus mechanism applicable to the 1-D fused feature is designed to filter and reinforce the fused features. Specifically, the weighting unit combined with the channel focus mechanism is designed to increase the weight of channel features conducive to classification, and improve their sensitivity to fault features.

The model was validated on the bogie simulation test bed dataset and the real train dataset. The results show that the proposed method obtains high prediction accuracy in the full-speed domain and speed fluctuation conditions, and has significant advantages over other advanced models.

The rest of this paper is organized as follows. Section 2 presents the proposed method in this paper. Section 3 validates the effectiveness of the proposed method through experiments on both public and real vehicle datasets. Finally, Sect. 4 concludes this article.

2 Method

2.1 Multi-Scale Feature Extraction Network

The network structure proposed in this paper is shown in Fig. 1, which is mainly composed of two parts, the multi-scale feature fusion module, and the channel attention mechanism module. In this model, we first design a large-scale convolution kernel with the size of 64×1, which is similar to the short-time Fourier transform and can extract the short-time characteristics of the signal. The difference is that the window function of the short-time Fourier transform is a sine function, and the weight of the large convolution kernel is obtained through the optimization algorithm and backpropagation, so the weight oriented to fault classification can be automatically learned. After the large-scale convolution layer, there is a pooling layer with a convolution stride of 2. The purpose of this layer is to reduce the size of the feature to half of the original size, which is conducive to reducing the computation of the subsequent feature fusion network and improving the training speed. Next is the feature fusion network, which consists of three branches.

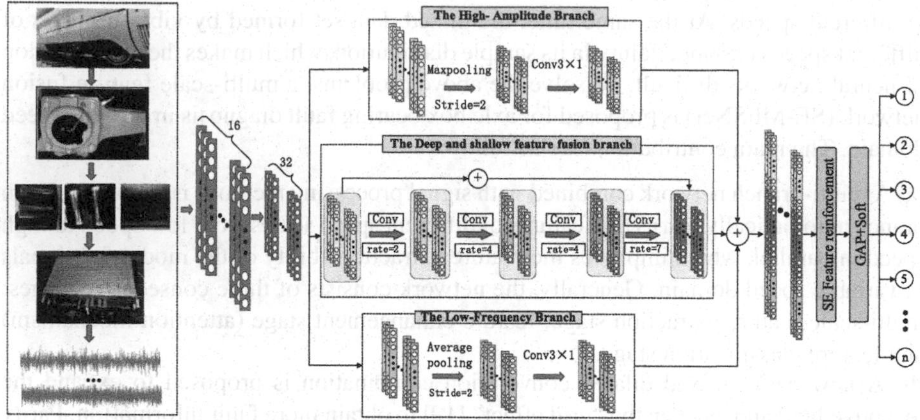

Fig. 1. The structure of SE-MSFNet

The High-Amplitude Branch. The fault point on the bearing usually brings sudden changes in the signal amplitude. The high-amplitude branch is a max pooling layer with a stride of 2. The maximum pooling layer is operated to obtain the maximum value point in the kernel acceptance domain. For one-dimensional vibration signals, it can reduce information redundancy and noise effects, and retain the main feature information. Then, after the max pooling layer is a 1 × 1 convolution, used for dimension promotion.

Suppose that the input signal of the branch is $x = \{x_1, x_2 ... x_N\}$. The output sequence after the maximum pooling operation is $x^H = \{x_1^h, x_2^h, ... x_N^h\}$, where x_i^H is the value at the i-node of x^H and represented as

$$x_i^h = Max(\sum_{i}^{i+m-1} x_i), 1 \leq i + m \leq N \tag{1}$$

where m is the kernel size. The final output of this branch is:

$$y^H = Conv(x^H) \tag{2}$$

The Low-Frequency Branch. The purpose of this branch is to extract more fault features from the low-frequency component of the original signal. For bearing vibration signals, the fault signal is usually characterized by the low-frequency component, so the low-frequency band tends to contain more discriminative fault features. The low-frequency information of the vibration signal is obtained by the average pooling layer and then using a convolution layer to learn the fault features of the low-frequency components. This operation is similar to a moving average filter in signal processing. The signal input to this branch is $x = \{x_1, x_2 \ldots x_N\}$. And the new time series obtained by averaging the pooling layer is $x^L = \{x_1^l, x_2^l, ... x_N^l\}$, where x_i^L is the value at the i-node of and represented as

$$x_i^l = \frac{1}{m}(\sum_{i}^{i+m-1} x_i), 1 \leq i + m \leq N \tag{3}$$

where m is the kernel size. The final output of this branch is:

$$y^L = Conv(x^L) \tag{4}$$

The Deep and Shallow Feature Fusion Branch. The deep and shallow feature fusion branch is a series dilated convolution module. Four dilated convolution kernels are used to obtain an expanded receptive field while acquiring low and high-dimensional information about the data. In a neural network, the larger the receptive field of a feature, the more it responds to the characteristic information of the original signal or the more fault information it contains. To increase the receptive field of a feature, there are usually two approaches. One is to make the convolution step larger than 1, but this will reduce the size of features, resulting in a lot of information loss. The other is to use a larger convolution kernel, but this will increase the number of parameters and computation of the model. Therefore, dilated convolution is proposed to solve the above problems. Like the second method, it uses larger convolutional kernels to obtain a larger receptive field, but the difference is that it does not add additional computation.

Dilated convolution can create sparse similar filters by inserting different numbers of zeros into the convolution kernel. For 1-dimensional features, n is the size of ordinary convolution kernel; r is the dilated rate of convolution kernel; W is the feature vector input size; S is the convolution step size; P is the number of zero filling layers. And the calculation formulas for the dilated convolution kernel N and the eigenvector size M are as follows:

$$N = [n + (n-1) \times (r-1)] \tag{5}$$

$$M = \frac{W + 2P - n}{S} + 1 \tag{6}$$

The receptive field is the size of the area mapped on the original signal by one sample point on each layer. The formation of receptive field size is equivalent to the process of inverting the input features from the output features. Assuming that zero filling operation is not considered, the number of convolution layers is k, the receptive field of this layer is l_k, the receptive field of the upper layer is l_{k-1}, and the size of the convolution kernel of the upper layer is n_{k-1}. The formula for calculating the receptive field size in this layer is:

$$l_k = l_{k-1} + n_{k-1} \prod_{i=1}^{k-1} S \tag{7}$$

Since convolution is a continuous process with each layer of convolution superimposed, the receptive field will increase step by step. Combined with (5) and (6), the formula for calculating dilated convolutional receptive field is derived as follows:

$$l_k = l_{k-1} + [rate(n-1) - 2] \times \prod_{i=1}^{k-1} S \tag{8}$$

To obtain a larger receptive field, it is necessary to use multiple convolutions. The maximum perceptual field with minimum information loss is guaranteed when the setting of the null convolution rate matches. Continuous dilated convolution will produce a "grid effect": dilated convolution can only cover data information in the form of a grid. The gradual increase of the high-level dilated rate will cause the input sampled data to become more and more sparse, resulting in local information loss and information irrelevance. To suppress the "grid effect" caused by continuous dilated convolution, this paper proposes parity mixed dilated convolutions. Figure 2 shows the changes in the feature receptive field under a three-layer 3 × 1 normal convolution. Figure 3 shows the changes in the receptive field of the features in the three convolutional layers when the dilated rate is (2, 4, 8). The receptive fields from convolution layer 3 to convolution layer 1 are 17 × 17, 25 × 25, and 29 × 29. At the same time, it can be seen that if the dilation rate is in the same proportion or the dilation rate has the same common divisor, the dilated convolution will produce the "grid effect". The strategy proposed in this paper is to expand the convolution kernel by using the parity mixed dilated rate shown in Fig. 4. The dilated convolution of (2, 4, 4, 7) can form the receptive field characteristics from dense center to gradually sparse edge. The size of the receptive field from convolution layer 4 to convolution layer 1 is 15 × 15, 23 × 23, 31 × 31, and 35 × 35. Our dilated rate method can reduce the "grid effect" while expanding the receptive field.

At the same time, the batch normalization algorithm [20] and Relu [21] activation function is added to each convolution. The output data of each layer is normalized (normalized to mean value 0 and standard deviation 1) and nonlinear activated to stabilize the data distribution of the next layer and prevent overfitting, to accelerate network convergence and improve training speed.

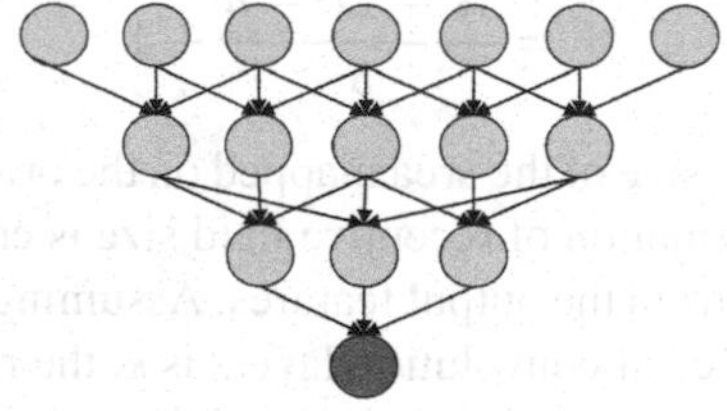

Fig. 2. Change of receptive field with normal convolution

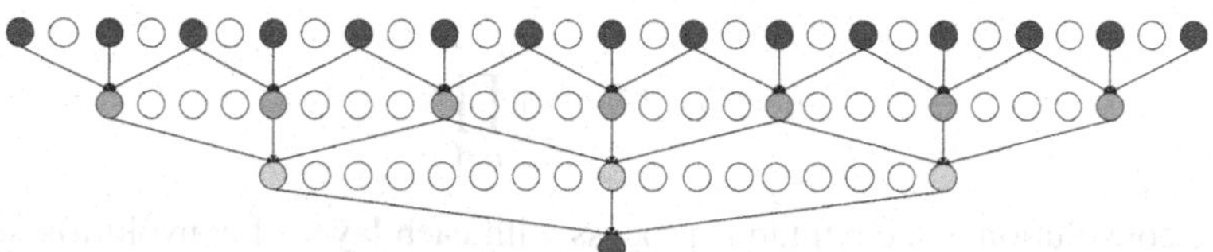

Fig. 3. Change of receptive field with (2,4,8) dilated convolution

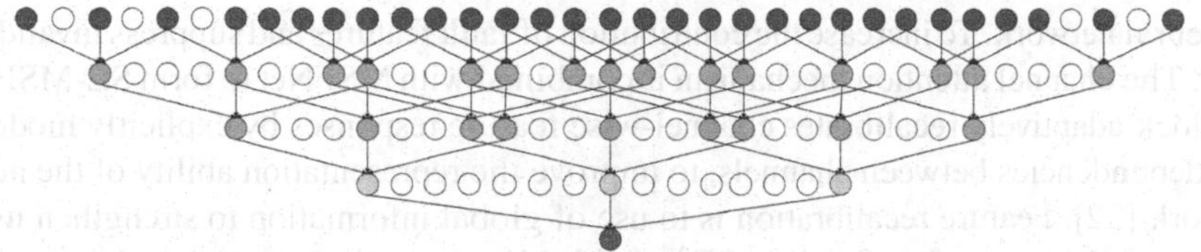

Fig. 4. Change of receptive field with (2,4,4,7) dilated convolution

2.2 Multi-Scale Feature Fusion Mechanism

As shown in Fig. 2, feature fusion includes two steps. The first step is to fuse the input features and the high-dimensional features obtained by a series of dilated convolutions. This fusion does not change the dimensionality and size of the features, but simply adds up the deep and shallow features at the same position in the same channel. Low-dimensional features mainly contain local information, while with the increase of the receptive field, high-dimensional features can contain more global information. The advantage of this method is that it can ensure that the output features of this branch have a good extraction function for fault features at different speeds. Assume that the signal input to this branch is, and the output of the first layer is. Suppose that the signal input to this branch is $x = \{x_1, x_2...x_N\}$, the output of the first layer $y_1 = Conv_{rate}(x)$. The output of each dilated convolute layer can be deduced as:

$$y_n = Conv_{rate}(y_{n-1}) \tag{9}$$

The final output of this branch is:

$$y^F = x + y_n \tag{10}$$

The second step is the feature fusion of the three branch networks. This fusion does not change the number of channels, but increases the length of the fused features, which we call "splicing". This has the advantage of making full use of the features obtained from each branch while minimizing the number of model operations. This provides a good basis for subsequent feature selection and enhancement. In this way, the multi-scale features obtained from the three branches are fused here. The output high amplitude features, low-pass filtering features, and deep-shallow fusion feature respectively. According to Formula (2), (4), and (10), the output characteristics after fusion are:

$$y^M = \{y^H, y^F, y^L\} \tag{11}$$

2.3 Feature Reinforcement Mechanism

Although the multi-scale feature fusion network can obtain signal features of different dimensions and scales, these features not only include fault features but also include non-related features. These non-related features will affect the classification accuracy of the neural network. To increase the contribution of fault features and suppress invalid features, The channel attention mechanism is combined with MSFNet to form SE-MSFNet. SE block adaptively recalibrates channel-wise feature responses by explicitly modeling interdependencies between channels, to improve the representation ability of the neural network [22]. Feature recalibration is to use of global information to strengthen useful features and dilute useless features. SE is divided into two steps: squeeze and then excitation. "Squeeze" compresses the features of each channel as the weight description of the channel by using global average pooling. In fact, "Squeeze" turns each feature channel into a real number, which has a global receptive field to a certain extent. For a 1-D feature with $L \times C$, its feature can be expressed as $F_{tr} : \mathrm{X} \rightarrow \mathrm{U},\ \mathrm{X} \in \mathbb{R}^{L' \times C'},\ \mathrm{U} \in \mathbb{R}^{L \times C}$. The convolution operation can be expressed as:

$$\mathrm{u}_c = \mathrm{v}_c * \mathrm{X} = \sum_{s=1}^{C'} \mathrm{v}_c^s * \mathrm{x}^s \tag{12}$$

Here v_c refers to the parameters of the c filter, $*$ denotes convolution, and u_c is the input of F_{tr}. "Squeeze" can be expressed as:

$$Zc = \mathrm{F}_{sq}(\mathrm{u}_c) = \frac{1}{L}\sum_{i=1}^{L} u_c(i) \tag{13}$$

The "exception" operation is similar to the mechanism of the gate in the recurrent neural network. The weight of each channel is learned through two fully connection layers, and the activation function is selected as Relu and Sigmoid [23] in turn. The first layer $\mathrm{W}_1 \in \mathbb{R}^{\frac{C}{r} \times C}$ has the function of decreasing dimension, compressing the characteristics of C dimension, and fully capturing the relationship between channels. The second layer $\mathrm{W}_2 \in \mathbb{R}^{C \times \frac{C}{r}}$ can restore dimensions. So "Excitation" can be expressed as:

$$s = \mathrm{F}_{ex}(\mathrm{z}, \mathrm{W}) = \sigma(g(\mathrm{z}, \mathrm{W})) = \sigma(\mathrm{W}_2\delta(\mathrm{W}_1\mathrm{z})) \tag{14}$$

where z is the output of "squeeze"; δ refers to the ReLu function; σ refers to the Sigmoid function.

Finally, there is a reweight operation. SE regards the weight of the output of the "exception" as the importance of each feature channel after feature selection and then weights the previous features channel by channel through multiplication to complete the recalibration of the original features in the channel dimension. The output of SE can be expressed as:

$$\tilde{\mathrm{x}}_c = \mathrm{F}_{scale}(\mathrm{u}_c, s_c) = s_c \cdot \mathrm{u}_c \tag{15}$$

The implementation of this paper is shown in Fig. 5. First, the features are converted from 32 to 1 channel by using global average pooling. This 1-D feature contains 32 neurons, corresponding to the feature of 32 channels. And then, it was reduced to 8 neurons and then increased to 32 neurons. Through the above operations, the weights with channel correlation are obtained. The weights are multiplied with the original 32-channel fusion features in a one-to-one correspondence. And then the multiplied features are converted to 1D features by global average pooling. Through the above operations, the channel weight including fault features will gradually increase, which improves the sensitivity of the network to fault features. Meanwhile, it can suppress the impact of noise and accelerate network convergence.

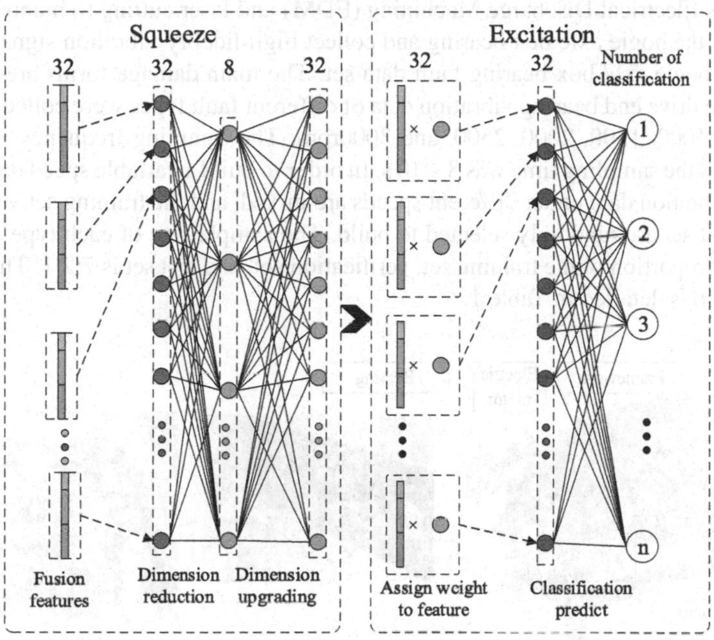

Fig. 5. SE feature reinforcement block

3 Experiment Verification

SE-MSFNet is implemented in the Keras library under Tensorflow2.6. The network was trained and tested on a workstation with a Windows 10 operating system and a GTX 3090Ti GPU. During training, we used a cross-entropy loss function and the Adam optimization algorithm with a learning rate of 0.0001 and a batch size of 128. The number of iterations is 200.

To verify the performance of SE-MSFNet, three classic end-to-end models: Inception [24], Resnet18 [25], WDCNN [12], and three state-of-the-art feature fusion networks: MSCNN [26], MBSCNN [27], MK-ResCNN [28] are used as comparison networks.

3.1 Bogie Axlebox Bearing Fault Simulation Test Bench

In order to study the multi fault diagnosis of bogie axlebox bearings in the variable speed range. The bogie axlebox bearing failure simulation test bench was designed to obtain vibration data of more types of faulty bearings in variable speed domain. The main structure of the test bench is shown in Fig. 6, which consists of the gearbox, motor, base, structural frame, sliding bearing, spindle, swing arm, and other structures, which is the same as the suspension and transmission structure of the real bogie and uses sliding bearing to simulate the wheels, which can better simulate the dynamic response of the real bogie. The ratio of the bogie test bench to the real bogie is 1:2.

To simulate the real fault sample characteristics and sample distribution, this experiment uses Electrical Discharge Machining (EDM) and laser cutting to manually inject faults into the bogie axle box bearing and collect high-fidelity vibration signals to construct the bogie axle box bearing fault data set. The main damage forms are shown in Fig. 7. The drive end bearing vibration data of different fault types were collected at five speeds of 1000, 1500, 2000, 2500, and 3000rpm. The sampling frequency was set to 12 kHz and the sampling time was 8 ~ 10 s. In order to build a variable speed data set, the bearing vibration data under different speeds are mixed, and the training set verification set and test set are randomly selected to build. The sample size of each type of fault is 500. The proportion of the training set, verification set, and test set is 7:2:1. The specific information is detailed in Table 1.

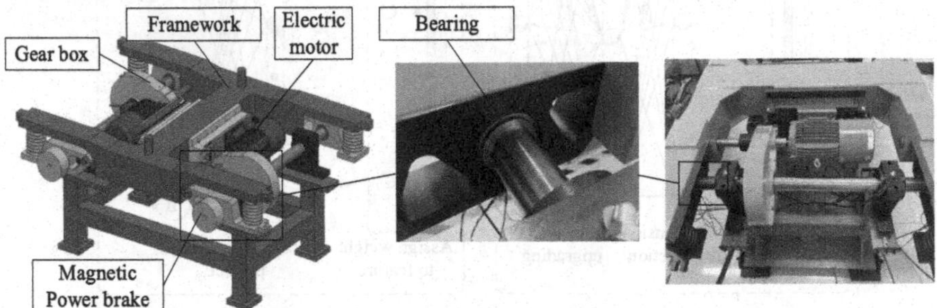

Fig. 6. Bogie fault simulation experiment system

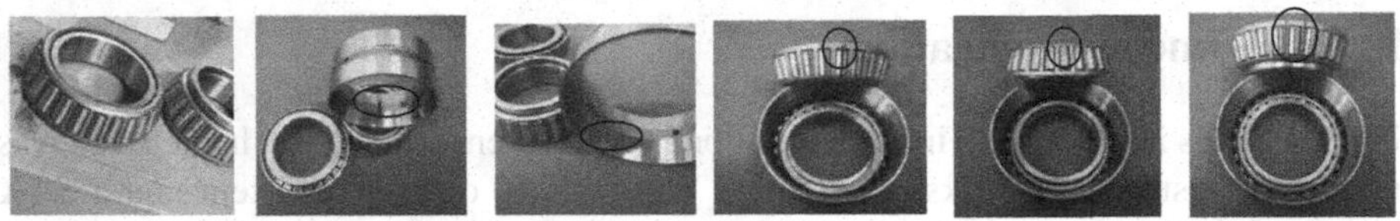

(a)Normal (b) Outer crack (c)Outer pitting (d)Outer crack (d)Roller pitting (e) Cage crack

Fig. 7. Fault bearings

3.2 Classification Comparison and Analysis

As shown in Table 2, the ten models were trained on steady speed datasets of 1000rpm, 1500rpm, 2000rpm, 2500rpm, and 3000rpm and variable speed domain of 1000–1500rpm, 1500–2000rpm, 2000–2500rpm, and 2500–3000. The speeds we set have

Table 1. Bogie axle box bearing datasets

Class label	Fault	Speed condition/rpm
B0	Normal	1000/1500/2000/2500/3000
B1	Cage crack	
B2	Outer crack (0.5mm)	
B3	Outer crack (1.0mm)	
B4	Outer crack (2.0mm)	
B5	Outer pitting (light)	
B6	Outer pitting (moderate)	
B7	Outer pitting (severe)	
B8	Roller crack (0.4mm)	
B9	Roller crack (0.8mm)	
B10	Roller crack (1.2mm)	
B11	Roller pitting (light)	
B12	Roller pitting (moderate)	
B13	Roller pitting (severe)	

practical engineering applications. 1000–1500rpm corresponds to 160–240 km/h and 1500–2000rpm corresponds to 240–320 km/h. Considering economic and route factors, these two intervals are the actual operating speeds of most commercial high-speed trains in China. 2500rpm corresponds to 400 km/h, which is the highest design speed for commercial high-speed railways in China and the next generation of high-speed railways. The 400 km/h is also the operating speed of the next generation high-speed railroad. 3000rpm corresponds to 480 km/h, which is the maximum design speed required to reach 400km/h. It can be seen that our method can achieve 96.42% accuracy in the range of 320–400 km/h and 92.58% accuracy in the speed range of 400–480 km/h, which is at least 7% better than other methods. Therefore, the method in this paper is not only applicable to the fault diagnosis of current commercial high-speed trains but also has good adaptability to the future next generation high-speed trains bogie axlebox bearing fault diagnosis.

The experimental results show that the classification accuracy of most models in the steady speed domain is higher than that in the variable speed domain. For example, the precision of MBSCNN is 97.5% on the 1000rpm dataset, 96.86% on the 1500rpm dataset, and only 91.4% on the 1000-1500rpm dataset. This is because the sample distribution of the same fault containing two speeds is not unique, which makes the classification difficult. The proposed model is less disturbed by it, and the classification accuracy varies only about 2% on the two types of datasets. Therefore, the proposed model is more robust to the variable speed domain.

It can be seen from the confusion matrix in Fig. 8 that in the speed range of 320–400 km/h, MSCNN has classification errors in 14 categories, MK- ResCNN has classification errors in 10 categories, MBSCNN and SE-MSFNet have classification errors in 9 categories. However, the number of MBSCNN errors is 94 and the number of SE MSFNet errors is 54. Therefore, the proposed model is more robust to multi-classification tasks.

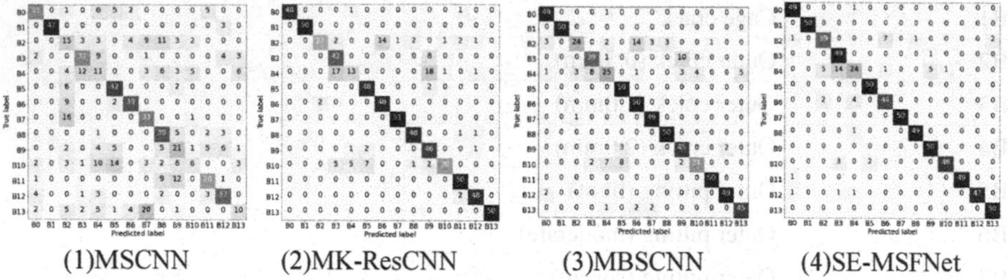

(1)MSCNN (2)MK-ResCNN (3)MBSCNN (4)SE-MSFNet

Fig. 8. Classification confusion matrix of four feature fusion methods in 320–400 km/h

Table 2. The test results on bogie axlebox bearing fault datasets

Method	160 km/h	160-240 km/h	240 km/h	240-320 km/h	320 km/h	320-400 km/h	400 km/h	400-480 km/h	480 km/h
WDCNN	97.50%	86.07%	96.00%	83.92%	89.58%	83.21%	88.87%	78.60%	77.46%
Resnet18	91.07%	84.28%	94.15%	66.78%	85.30%	66.07%	77.46%	49.64%	59.20%
SE-Resnet18	90.71%	82.85%	94.29%	76.78%	77.31%	58.21%	75.32%	56.77%	64.76%
Inception	87.14%	87.50%	94.57%	84.64%	88.15%	70.35%	82.88%	72.61%	79.45%
SE-Inception	91.07%	84.28%	96.86%	88.21%	91.44%	77.14%	84.16%	75.46%	81.88%
MSCNN	55.00%	55.00%	91.86%	54.64%	71.75%	49.10%	70.47%	53.63%	66.33%
MK-ResCNN	98.21%	96.42%	98.57%	90.97%	94.86%	85.57%	93.86%	85.16%	90.58%
MBSCNN	97.50%	91.40%	96.86%	87.64%	96.71%	87.14%	92.01%	84.45%	87.01%
MSFNet	**99.64%**	96.07%	99.57%	94.06%	**99.14%**	92.14%	**98.28%**	92.15%	93.43%
SE-MSFNet	**99.28%**	**99.28%**	**99.71%**	**96.91%**	**99.00%**	**96.42%**	97.00%	**92.58%**	**95.29%**

The performance of each model on the dataset is analyzed to illustrate the advantages of the proposed model in functional design. WDCNN and SE-MSFNet both have the first layer of large scale convolution kernel, and the classification accuracy of WDCNN is not as good as that of SE-MSFNet. This shows that although the first layer of large-scale convolution is conducive to fault classification, the multi-scale feature fusion and attention mechanism proposed in this paper are the key to further improving the classification accuracy; The number of network layers of Resnet18 is far greater than that of SE-MSFNet, but its accuracy is not as good as that of SE-MSFNet, which indicates that deepening the model depth cannot effectively improve the classification effect under variable speed conditions. SE-MSFNet only uses 4 layers of convolution in the deep and shallow feature fusion branch, and achieves good classification results under the condition of greatly optimizing model parameters and training speed; MSCNN does not use convolution kernels of different sizes in each branch and achieves the worst classification effect among all methods. It shows that convolution kernels of different sizes have a favorable effect on multi fault classification under variable speed conditions; Although MBSCNN uses convolution of different sizes to obtain multi-scale features, it does not integrate the concept of traditional filtering. In addition to using convolution kernels of different sizes, SE-Inception only superimposes each channel on feature fusion (the number of channels after fusion is the sum of the number of branches). MBSCNN and SE-Inception are not as accurate as SE-MSFNet in classification accuracy, which shows the effectiveness of SE-MSFNet's combination of traditional filtering concepts, such as high amplitude branch, low-pass filter branch, and splicing strategy (the number of channels remains unchanged). MK ResCNN uses different convolution cores and residual networks to fuse high-dimensional information, but the classification accuracy is not as good as SE-MSFNet. In addition to proving the effectiveness of high-dimensional branches and low-pass filtering branches, it further proves the effectiveness of parity mixed convolution to obtain high-dimensional features. Finally, MSFNet is not as accurate as SE-MSFNet, which proves the effectiveness of the SE feature reinforcement block in improving the classification effect.

4 Conclusion

Focusing on the multi fault diagnosis problem of bogie axle box bearings in the full-speed domain and variable-speed domain of high-speed trains, this paper proposes a fault diagnosis method based on the SE-MSFNet network. The network uses multi branch structure combined with signal processing methods and parity mixed dilated convolution to obtain multi-scale features. Different scale features of the original signal are spliced together, making thc single channel contain more abundant features. The feature reinforcement mechanism can filter and strengthen the features after fusion, which is conducive to improving the efficiency and utilization of feature fusion. The bearing fault data in variable speed domains are obtained on the bogie simulation test bench. It is demonstrated that SE-MSFNet achieves better classification results than other state-of-the-art methods on different speed domains. The proposed strategy of multi-scale feature fusion and feature reinforcement mechanism is proved to have a positive effect on bearing fault diagnosis in the full speed domain and variable speed domain. More importantly, the proposed method is applicable to the speed range of the next generation of high-speed trains, which has important reference significance for fault diagnosis of the next generation of high-speed trains.

Future work includes further revealing the fault mechanism of bogie axle box bearings in the variable speed range, aiming to obtain more accurate and effective fault characteristics. Meanwhile, it is difficult to obtain the fault bearing of real vehicles, and the next step is to improve the generalization of the neural network model by using simulation data.

References

1. Jiang, Y., et al.: Safety-assured model-driven design of the multifunction vehicle bus controller. IEEE Trans. Intell. Transp. Syst. **19**(10), 3320–3333 (2018)

2. Jo, O., Kim, Y.K., Kim, J.: Internet of things for smart railway: feasibility and applications. IEEE Internet Things J. **5**(2), 482–490 (2017)
3. Chen, H., Jiang, B.: A review of fault detection and diagnosis for the traction system in high-speed trains. IEEE Trans. Intell. Transp. Syst. **21**(2), 450–465 (2019)
4. Ding, S., Hao, M., Cui, Z., Wang, Y., Hang, J., Li, X.: Application of multi-SVM classifier and hybrid GSAPSO algorithm for fault diagnosis of electrical machine drive system. ISA Trans. **133**, 529–538 (2023)
5. Chen, Z., Gryllias, K., Li, W.: Mechanical fault diagnosis using convolutional neural networks and extreme learning machine. Mech. Syst. Signal Process. **133**, 106272 (2019)
6. Wang, C., Gan, M., Zhu, C.A.: A supervised sparsity-based wavelet feature for bearing fault diagnosis. J. Intell. Manuf. **30**, 229–239 (2019)
7. Li, C., De Oliveira, J.V., Cerrada, M., Cabrera, D., Sánchez, R.V., Zurita, G.: A systematic review of fuzzy formalisms for bearing fault diagnosis. IEEE Trans. Fuzzy Syst. **27**(7), 1362–1382 (2018)
8. Mao, W., Feng, W., Liu, Y., Zhang, D., Liang, X.: A new deep auto-encoder method with fusing discriminant information for bearing fault diagnosis. Mech. Syst. Signal Process. **150**, 107233 (2021)
9. Jin, Z., He, D., Wei, Z.: Intelligent fault diagnosis of train axle box bearing based on parameter optimization VMD and improved DBN. Eng. Appl. Artif. Intell. **110**, 104713 (2022)
10. Ma, M., Sun, C., Chen, X.: Deep coupling autoencoder for fault diagnosis with multimodal sensory data. IEEE Trans. Industr. Inf. **14**(3), 1137–1145 (2018)
11. Zhang, W., Li, C., Peng, G., Chen, Y., Zhang, Z.: A deep convolutional neural network with new training methods for bearing fault diagnosis under noisy environment and different working load. Mech. Syst. Signal Process. **100**, 439–453 (2018)
12. Zhang, W., Peng, G., Li, C., Chen, Y., Zhang, Z.: A new deep learning model for fault diagnosis with good anti-noise and domain adaptation ability on raw vibration signals. Sensors **17**(2), 425 (2017)
13. Su, L., Ma, L., Qin, N., Huang, D., Kemp, A.H.: Fault diagnosis of high-speed train bogie by residual-squeeze net. IEEE Trans. Industr. Inf. **15**(7), 3856–3863 (2019)
14. Ye, Y., Zhang, Y., Wang, Q., Wang, Z., Teng, Z., Zhang, H.: Fault diagnosis of high-speed train suspension systems using multiscale permutation entropy and linear local tangent space alignment. Mech. Syst. Signal Process. **138**, 106565 (2020)
15. Man, J., Dong, H., Jia, L., Qin, Y.: AttGGCN model: a novel multi-sensor fault diagnosis method for high-speed train bogie. IEEE Trans. Intell. Transp. Syst. **23**(10), 19511–19522 (2022)
16. Kou, L., Qin, Y., Zhao, X.: A multi-dimension end-to-end CNN model for rotating devices fault diagnosis on high-speed train bogie. IEEE Trans. Veh. Technol. **69**(3), 2513–2524 (2019)
17. Sun, B., Liu, X.F.: Significance support vector machine for high-speed train bearing fault diagnosis. IEEE Sens. J. **23**(5), 4638–4646 (2023)
18. Jia, X., Qin, N., Huang, D., Zhang, Y., Du, J.: A clustered blueprint separable convolutional neural network with high precision for high-speed train bogie fault diagnosis. Neurocomputing **500**, 422–433 (2022)
19. Wang, P., Chen, P., Yuan, Y., Liu, D., Huang, Z., Hou, X., Cottrell, G.: Understanding convolution for semantic segmentation. In: 2018 IEEE Winter Conference on Applications of Computer Vision (WACV), pp. 1451–1460. IEEE (2018)
20. Ioffe, S., Szegedy, C.: Batch normalization: accelerating deep network training by reducing internal covariate shift. In: International Conference on Machine Learning, pp. 448–456. PMLR (2015)
21. Wang, J., Zhuang, J., Duan, L., Cheng, W.: A multi-scale convolution neural network for featureless fault diagnosis. In: 2016 International Symposium on Flexible Automation (ISFA), pp. 65–70. IEEE (2016)

22. Hu, J., Shen, L., Sun, G.: Squeeze-and-excitation networks. In: Proceedings of the IEEE Conference on Computer Vision and Pattern Recognition, pp. 7132–7141. IEEE (2018)
23. Zheng, J., Jiang, Z., Pan, H.: Sigmoid-based refined composite multiscale fuzzy entropy and t-SNE based fault diagnosis approach for rolling bearing. Measurement **129**, 332–342 (2018)
24. Szegedy, C., Liu, W., Jia, Y., Sermanet, P., Reed, S., Anguelov, D., Rabinovich, A.: Going deeper with convolutions. In: Proceedings of the IEEE Conference on Computer Vision and Pattern Recognition, pp. 1–9. IEEE (2015)
25. He, K., Zhang, X., Ren, S., Sun, J.: Deep residual learning for image recognition. In Proceedings of the IEEE Conference on Computer Vision and Pattern Recognition, pp. 770–778. IEEE (2016)
26. Jiang, G., He, H., Yan, J., Xie, P.: Multiscale convolutional neural networks for fault diagnosis of wind turbine gearbox. IEEE Trans. Industr. Electron. **66**(4), 3196–3207 (2018)
27. Peng, D., Wang, H., Liu, Z., Zhang, W., Zuo, M.J., Chen, J.: Multibranch and multiscale CNN for fault diagnosis of wheelset bearings under strong noise and variable load condition. IEEE Trans. Industr. Inf. **16**(7), 4949–4960 (2020)
28. Liu, R., Wang, F., Yang, B., Qin, S.J.: Multiscale kernel based residual convolutional neural network for motor fault diagnosis under nonstationary conditions. IEEE Trans. Industr. Inf. **16**(6), 3797–3806 (2019)

Degradation Modelling and Remaining Useful Life Prediction Methods Based on Time Series Generative Prediction Networks

Xusheng Chen, Wanjun Hou, and Yizhen Peng(✉)

College of Mechanical and Vehicle Engineering, Chongqing University, Chongqing 400044, China
{202207131317,202107131346}@stu.cqu.edu.cn, pengyz@cqu.edu.cn

Abstract. Currently, in the field of prediction and health management, there is a proliferation of deep learning approaches to degradation modeling and prediction. However, deep learning networks not only have a large number of parameters, but also their construction often relies on a large substantial quantity of historical data for each state. In practical work, it is extremely difficult to collect full lifecycle data of critical components, so lifecycle data for degradation modeling and prediction is scarce. Therefore, this paper proposes a time-series generation and prediction network based on GRU-GAN to solve the problem of sample enhancement prediction under minor degradation conditions. This method was applied to degradation modeling and lifespan prediction of various critical components. The results indicate that the method proposed in this paper enhances the accuracy of predicting the lifespan of bearings and lithium batteries. Therefore, it can be concluded that the approach is a reliable and effective predictive method.

Keywords: Degradation · Remaining useful life · Deep learning · GAN · GRU

1 Introduction

Key components are crucial to the normal operation of machines, and whether a machine can operate properly always depends on their health status. During operation, the performance of key components such as rolling bearings and lithium batteries tends to deteriorate due to the influence of external environment and internal factors. In such cases, if the key components are not repaired or replaced in time, their deterioration will eventually make them unable to perform their duties as expected, leading to significant economic losses [1]. Therefore, accurate prediction of their remaining useful life (RUL) is essential to elevate the reliability and safety of mechanical systems.

Based on data-rich fault diagnosis, nearly perfect predictive performance is demonstrated in most public datasets. However, in many cases, fully labeled data or an adequate quantity of training data are rare, which limits the systems that may be studied in the context of deep learning [2]. In addition, the required amount of statistics for training deep neural networks is not only huge but also costly [3]. During the operation of the machine, most components are in normal working condition, and the collected

H. Zhang et al. (Eds.): NCAA 2023, CCIS 1870, pp. 318–329, 2023.
https://doi.org/10.1007/978-981-99-5847-4_23

data mostly reflect the working status of key components under healthy conditions. For safety considerations, machines should not be allowed to operate for too long when they break down. So as a result, the amount of fault data that can be collected is very limited. Therefore, the life cycle data available for component life prediction is extremely small. The problem of monitoring big data and small sample failures or degradation greatly limits the progress and application of key equipment component life prediction approaches. At present, most data used for training and testing models is very limited and extremely imbalanced, resulting in certain biases in the model's decision boundaries [4]. Consequently, there is a great need for some data generation methods in order to increase scarce data and improve the accuracy of predictions. Generative adversarial networks (GAN) as an emerging data generation way is a pioneering deep learning tool first proposed by Goodfellow et al. [5]. It has been used by many researchers to generate abundant types of normal or faulty signals for addressing imbalanced datasets in fault diagnosis and improving the accuracy of other prediction means [6]. The use of GAN-based techniques has gained significant popularity in various domains, including speech recognition [7], image generation [8], sentiment analysis [9], and so on. This is because GAN has the ability to learn the underlying distribution of the training data and generate new data that are different but similar to the original distribution. Gao et al. [10] generated fault samples using the GAN method and used the data for finite element simulation of fault diagnosis, improving the predictive performance. Li et al. [11] presented a cross-domain fault diagnosis approach that based on GAN and provided effective diagnostic results. Gao et al. [12] suggested a technique of data augmentation that utilizes Wasserstein generative adversarial networks and redesigned the loss function of WGAN. This approach can significantly enhance the stability of the model and enhance the quality of sample generation. Zhou et al. [13] improved the GAN method, redesigned a new generator and discriminator, and optimized the approach for discriminating fault samples. Meanwhile, more novel methods based on GAN have been proposed, making GAN flourish in the field of fault diagnosis. However, some of these studies are limited to structural innovations of GAN and only predict experimental data without applying the GAN method to the actual fault diagnosis of key components.

Various data augmentation methods based on GAN have become increasingly popular in the fields of prediction and health management. However, most of these methods only expand the data of failed stages, without enhancing the degraded time series data. Meanwhile, in degradation modeling and life prediction, the problem of a lack of small sample strongly stands out due to the dual constraints of time cost and economic cost, making the collection of full-lifespan degradation data extremely difficult. Due to the swift progress of data-driven methods, using deep learning methods for fault diagnosis of key components is currently a hot topic in research. In this paper, a GAN-based data generation approach is put forward to generate time-series degraded data via GAN and predict lifespans through GRU. It effectively solves the problem of insufficient data for fault diagnosis of key components and improves prediction accuracy through training models on multiple types of data. Firstly, the temporal features of key component lifetimes are obtained as the training data for the GRU-GAN network, and approximately real data is generated based on the trained model to construct a key component lifespan prediction model. Secondly, the GRU neural network is used to construct a multi-data prediction model built upon real, generated, and mixed data. Finally, a fusion prediction is conducted with multiple datasets to obtain the predicted results for key components. The proposed method is utilized in the degradation modeling and lifespan prediction of key components, achieving good prediction results for the remaining useful life of bearings and high-accuracy predictions for lithium-ion battery remaining life, demonstrating the effectiveness of the proposed approach.

2 The Proposed Degradation Trend Prediction Model

2.1 GAN Method

GAN is a deep learning model which comprises a generator network G and a discriminator network D. The objective of G is to generate samples that are similar to the real distribution, while D aims to differentiate between whether these samples are real or fake. To achieve this, GAN uses random input vectors z from a pre-defined distribution to generate synthetic data while optimizing the generator to make it challenging for D to distinguish between raw and generated data. During training, the discriminator and generator are both alternately trained to attain the final Nash equilibrium state, resulting in the synthetic data being comparable to the samples in the training set [14]. The following figure shows the training process of GAN (Fig. 1).

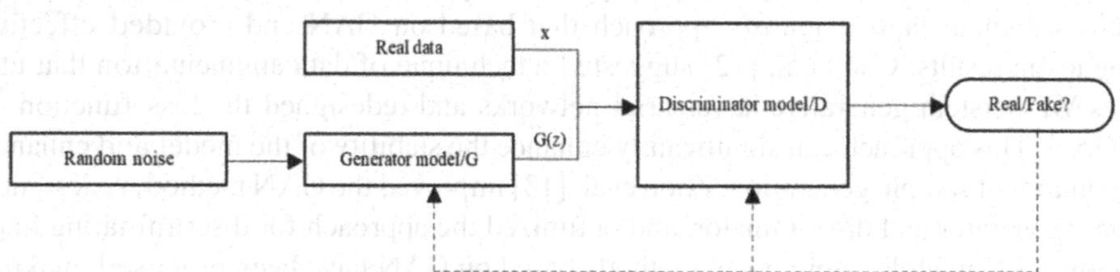

Fig. 1. The training process of GAN

The objective function of GAN during training is shown below:

$$L_{GAN}(G, D) = \mathbb{E}_{x\sim p_{data}(x)}[\log D(x)] + \mathbb{E}_{x\sim p_z(z)}[\log(1 - D(G(z))] \quad (1)$$

The GAN equation involves variables such as x, z, $G(Z)$, $D(Z)$, $p_{data}(x)$, and $P_z(z)$, Specifically, x represents a real sample from the true data distribution, while z denotes the input noise in the network sampled, $G(Z)$ represents the synthesized sample generated by the generator network G, and $D(Z)$ denotes the probability that the discriminator network D correctly determines whether $G(Z)$ is real or fake data, $p_{data}(x)$ represents the real sample data distribution, and $p_z(z)$ represents the input noise distribution.

We can think of this equation as two functions: First, we consider any given generator G and an optimal discriminator D. Given a random vector z, and then the optimal generator G is expressed as:

$$\max_D V(G, D) = \mathbb{E}_{x\sim p_{data}(x)}[\log D(x)] + \mathbb{E}_{x\sim p_z(z)}[\log(1 - D(G(z)))] \quad (2)$$

$$\min_G V(G, D) = \mathbb{E}_{x\sim p_z(z)}[\log(1 - D(G(z)))] \quad (3)$$

2.2 GRU Network

Recurrent Neural Network (RNN) is often used to process continuous data, especially for time series data with excellent predictive performance [15]. Therefore, many scholars have applied RNNs to the fields of speech recognition, machine translation, video analysis, image description generation, and achieved good results [16, 17]. However, RNNs also have their limitations. Due to problems such as vanishing or exploding gradients and short-term memory, RNNs cannot handle longer time series datasets.

In recent years, some experts and scholars have addressed the problem of long-term dependency in neural networks and LSTM neural network, which allows for the memorization of long-term information. In addition, researchers have applied LSTM to plenty of fields [18], such as human speech recognition, machine translation [19], and road traffic-related prediction [20]. In recent years, many variants of LSTM have evolved according to different needs. To achieve better prediction performance and faster computation, researchers have improved LSTM and proposed a new gated recurrent unit (GRU) neural network, which balances input and forgotten information by introducing gate mechanisms [21]. GRU has fewer parameters than LSTM, with better predictive performance.

Because GRU was developed by researchers based on LSTM, the internal structure of GRU is similar to that of LSTM. However, the improvement of GRU lies in combining two gates of LSTM into one gate, that is, replacing the input gate and forget gate with an update gate. In this case, GRU has fewer parameters, reducing computation time and improving predictive performance compared to LSTM. The mathematical expression of GRU is as follows:

$$r_t = \sigma W_r \cdot h_{t-1}, x_t \tag{4}$$

$$z_t = \sigma W_z \cdot h_{t-1}, x_t \tag{5}$$

$$\tilde{h}_t = \tanh W_{\tilde{h}} r_t * h_{t-1}, x_t \tag{6}$$

$$h_t = 1 - z_t * h_{t-1} + z_t * \tilde{h}_t \tag{7}$$

This equation includes the current state h_t, candidate state $\tilde{h}_t$, and update gate. The role of the update gate z_t is to control how much information the current state h_t should retain from the previous state and how much new information it should accept from the candidate state $\tilde{h}_t$. The reset gate r_t determines how much past information needs to be forgotten. Figure 2 illustrates the composition of the GRU unit.

2.3 The GAN-GRU Network

The process of the time series generation and prediction network based on GRU-GAN is as follows: First, we extract feature data of relevant components and train GAN generator and discriminator to obtain time series data that closely resemble real feature data. Then, the training data consists of a combination of both the original dataset and the generated dataset. Finally, the remaining life prediction and degradation of the components are predicted based on the GRU neural network. Our proposed method's process is displayed in Fig. 3.

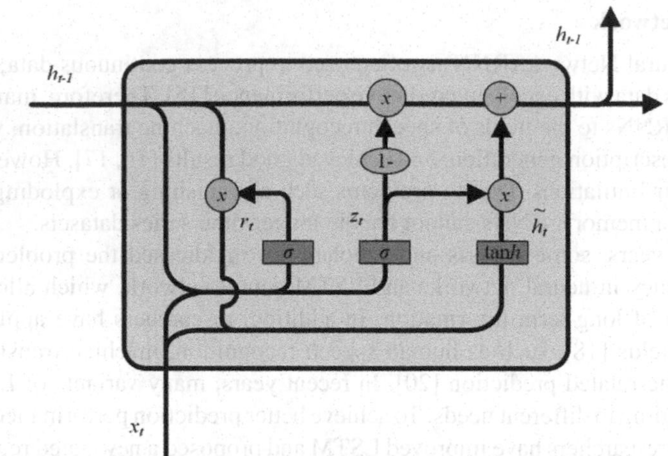

Fig. 2. The Structure of GRU

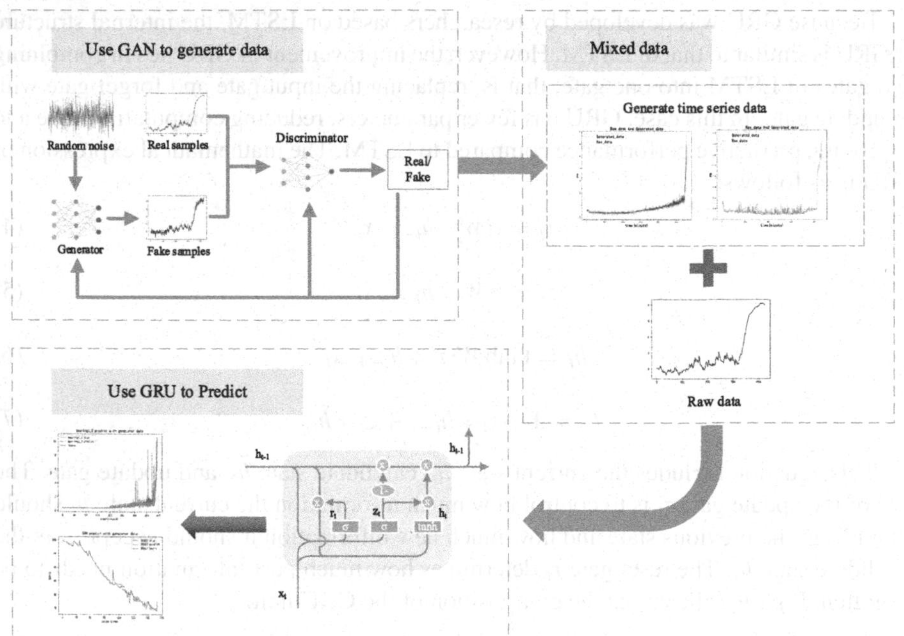

Fig. 3. Flowchart of key components RUL prediction

3 Experimental Verification and Analysis

In order to verify the effectiveness of the proposed GAN-GRU model in this article, we conducted experiments on two datasets: the bearing dataset from the 2012 IEEE PHM Conference Predictive Maintenance Challenge held in Denver, USA, and the lithium-ion battery dataset from NASA.

3.1 Using GAN to Generate Data

In this section, we generate simulation data based on the raw data. In obtaining; After generating the required data, we compare it with actual data to confirm data feasibility (Fig. 4).

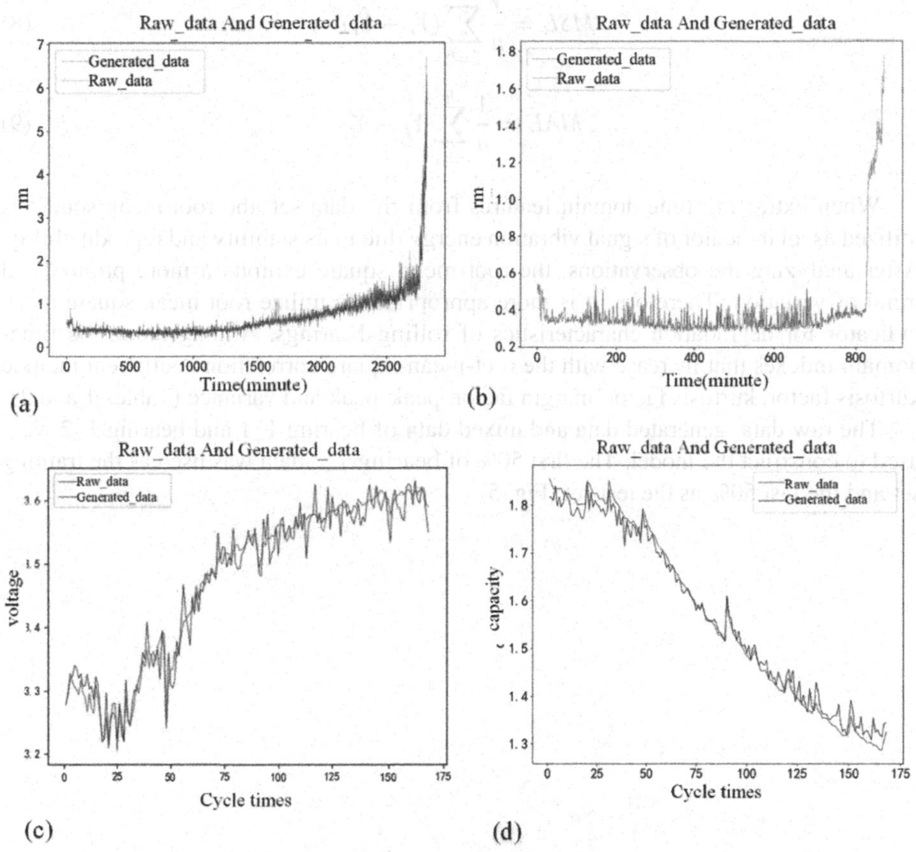

Fig. 4. Raw data and generated data

In the above figures, we can see the results of generating bearing and battery data, where (a) and (b) show the generated results of bearing data, while (c) and (d) show the generated results of battery data. The red line represents the original data, while the blue line represents the data generated using GAN.

It can be observed from the figure above that the generated data bears a strong resemblance to the original data, but not the same data and there are still some differences in some places. However, the deviation is very small, so that the data generated can be used as data for future research.

3.2 Prediction Results of Bearing Data Set PHM-2012

Here, MSE and MAE are used as error calculation indexes.

$$MSE = \frac{1}{n}\sum_{i=1}^{n}(Y_i - \hat{Y}_i)^2 \tag{8}$$

$$MAE = \frac{1}{n}\sum_{i=1}^{n} Y_i - \overline{Y}_i \tag{9}$$

When extracting time domain features from the data set, the root mean square is utilized as an indicator of signal vibration energy due to its stability and reproducibility. After analyzing the observations, the root mean square exhibits a more pronounced trend of variation. Therefore, It is more appropriate to utilize root mean square as an indicator for degradation characteristics of rolling bearings. Among them, the time-domain indexes that increase with the root-mean-square correlation coefficient include kurtosis factor, kurtosis factor, margin factor, peak-peak and variance (Tables 1 and 2).

The raw data, generated data and mixed data of bearing 1_1 and bearing 1_2 were used to construct the model. The first 50% of bearing 1_3 data was used as the training set and the last 50% as the test set (Fig. 5).

The predicted results are shown in the figures:

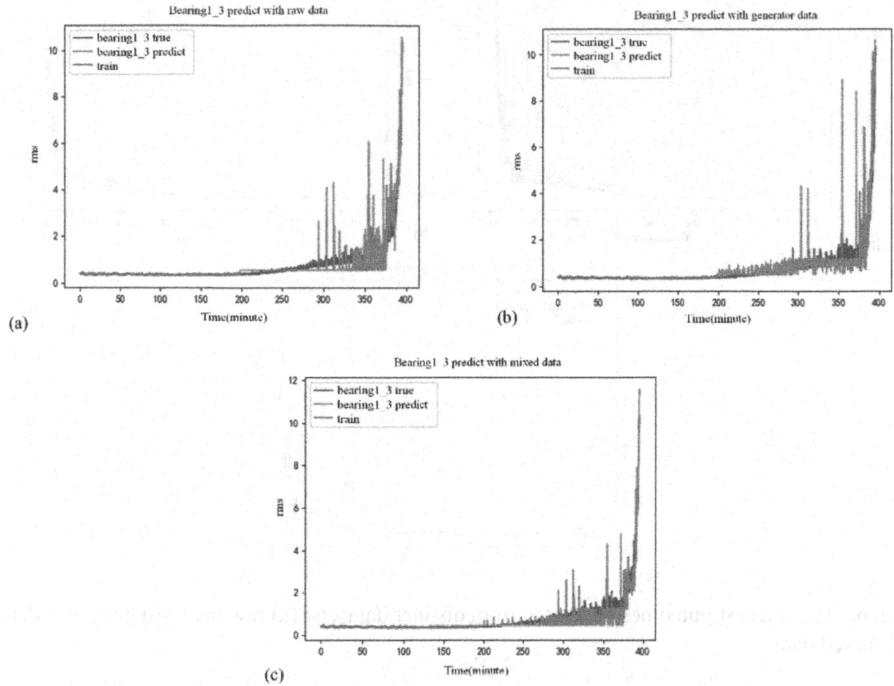

(a) (b) (c)

Fig. 5. The forecast outcomes generated from distinct datasets: (a) raw data; (b) generator data; (c) mixed data

Table 1. MAE, MSE of bearing1_3(50%)

Bearing1_3(50%)	MAE	MSE
Raw data	0.3362	0.3689
Generator data	0.2566	0.316
Mixed data	0.2248	0.1073

The experimental results in the figure and table above show that the prediction error of the model constructed with mixed data is much smaller than that of the single model constructed with raw data or generated data (Fig. 6).

Then, 60% of the data of bearing 1_3 is used to predict, and the results are shown in the following figures:

The experimental results show that GAN-GRU model has stronger predictive ability than GRU when less historical data is used for training. However, when more historical data are used for prediction, the prediction results of the model have little difference. Therefore, when bearing data is scarce and the collected historical data is small, the GAN-GRU method can be used for prediction with high accuracy (Fig. 7).

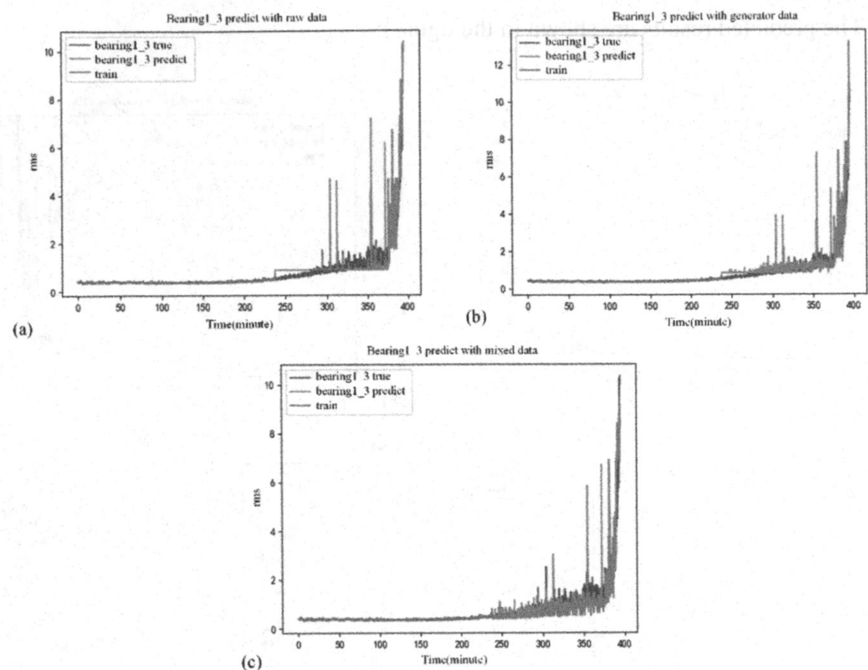

Fig. 6. The forecast outcomes generated from distinct datasets: (a) raw data; (b) generator data; (c) mixed data

Table 2. MAE, MSE of bearing1_3(60%)

Bearing1_3(50%)	MAE	MSE
Raw data	0.3082	0.2434
Generator data	0.2929	0.1984
Mixed data	0.2324	0.1443

Next, it was tested on NASA lithium batteries. Data of B0005 and B0006 are used to predict B0007. The prediction results are shown in the following figures:

According to the experimental results, we can find that the hybrid model exhibits significantly lower prediction errors in comparison to the single model. In addition, when the data is scarce and the historical data collected is small, the GAN-GRU method yields better prediction results than others (Table 3).

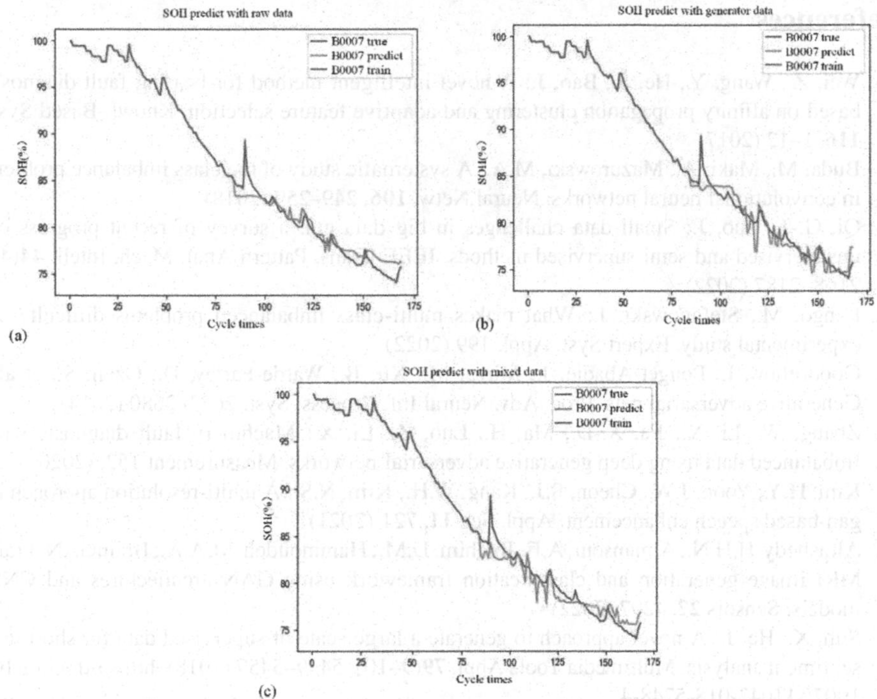

Fig. 7. The forecast outcomes generated from distinct datasets: (a) raw data; (b) generator data; (c) mixed data

Table 3. MAE, MSE of NASA battery

Bearing1_3(50%)	MAE	MSE
Raw data	0.8476	1.3614
Generator data	0.9446	1.794
Mixed data	0.7748	1.104

4 Conclusion

In order to solve the problem that neural network models have difficulty in prediction and poor performance under conditions of sparse data, this paper introduces a technique for generating data that utilizes GRU-GAN and also demonstrates how GRUs can be employed for lifespan forecasting. This method can build a prediction model based on the degradation process of rolling bearings. When bearing data is scarce and historical data collected is small, the proposed method can be used to predict with high accuracy, and the uncertainty of the model is taken into account. The proposed method is also effective in lithium battery prediction.

References

1. Wei, Z., Wang, Y., He, S., Bao, J.: A novel intelligent method for bearing fault diagnosis based on affinity propagation clustering and adaptive feature selection. Knowl.-Based Syst. **116**, 1–12 (2017)
2. Buda, M., Maki, A., Mazurowski, M.A.: A systematic study of the class imbalance problem in convolutional neural networks. Neural Netw. **106**, 249–259 (2018)
3. Qi, G.-J., Luo, J.: Small data challenges in big data era: a survey of recent progress on unsupervised and semi-supervised methods. IEEE Trans. Pattern Anal. Mach. Intell. **44**(4), 2168–2187 (2022)
4. Lango, M., Stefanowski, J.: What makes multi-class imbalanced problems difficult? an experimental study. Expert Syst. Appl. 199 (2022)
5. Goodfellow, I., Pouget-Abadie, J., Mirza, M., Xu, B., Warde-Farley, D., Ozair, S., et al.: Generative adversarial nets. Proc. Adv. Neural Inf. Process. Syst. 2672–2680 (2014)
6. Zhang, W., Li, X., Jia, X.-D., Ma, H., Luo, Z., Li, X.: Machinery fault diagnosis with Imbalanced data using deep generative adversarial networks. Measurement 152, (2020)
7. Kim, H.Y., Yoon, J.W., Cheon, S.J., Kang, W.H., Kim, N.S.: A multi-resolution approach to gan-based speech enhancement. Appl. Sci. **11**, 721 (2021)
8. Alrashedy, H.H.N., Almansour, A.F., Ibrahim, D.M., Hammoudeh, M.A.A.: BrainGAN: brain MRI image generation and classification framework using GAN architectures and CNN models. Sensors **22**, 4297 (2022)
9. Sun, X., He, J.: A novel approach to generate a large scale of supervised data for short text sentiment analysis. Multimedia Tools Appl. **79**(9–10), 5439–5459 (2018). https://doi.org/10.1007/s11042-018-5748-4
10. Gao, Y., Liu, X., Xiang, J.: FEM simulation-based generative adversarial networks to detect bearing faults. IEEE Trans. Ind. Inform. **16**(7), 4961–4971 (2020)
11. Li, X., Zhang, W., Ding, Q.: Cross-domain fault diagnosis of rolling element bearings using deep generative neural networks. IEEE Trans. Ind. Electron. **66**(7), 5525–5534 (2019)
12. Gao, X., Deng, F., Yue, X.: Data augmentation in fault diagnosis based on the Wasserstein generative adversarial network with gradient penalty. Neurocomputing **396**, 487–494 (2020)
13. Zhou, F., Yang, S., Fujita, H., Chen, D., Wen, C.: Deep learning fault diagnosis method based on global optimization GAN for unbalanced data. Knowl.-Based Syst. 187 (2020)
14. Ratliff, L.J., Burden, S.A., Sastry, S.S.: Characterization and computation of local Nash equilibria in continuous games. In: 51st Annual Allerton Conference on Communication, Control, and Computing (Allerton), pp. 917–924 (2013)
15. Zaremba, W., Sutskever, I., Vinyals, O.: Recurrent neural network regularization. arXiv preprint arXiv:1409.2329 (2014)
16. Palangi, H., Deng, L., Shen, Y., et al.: Deep sentence embedding using long short-term memory networks: analysis and application to information retrieval. IEEE/ACM Trans. Audio, Speech Lang. Process. **24**(4), 694–707 (2016)
17. Mikolov, T., Sutskever, I., Chen, K., et al.: Distributed representations of words and phrases and their compositionality. In: Proceedings of the 26th International Conference on Neural Information Processing Systems, vol. 2, pp. 3111–3119 (2013)
18. Du, X.B., et al.: An efficient LSTM network for emotion recognition from multichannel EEG signals. IEEE Trans. Affect. Comput. **13**, 1528–1540 (2020)
19. Srivastava, N., Mansimov, E., Salakhutdinov, R.: Unsupervised learning of video representations using LSTMs. In: Proceedings of the 32nd International Conference on Machine Learning, pp. 843–852 (2015)

20. Donahue, J., Hendricks, L.A., Rohrbach, M., et al.: Longterm recurrent convolutional networks for visual recognition and description. IEEE Trans. Pattern Anal. Mach. Intell. **39**(4), 677–691 (2015)
21. Jiao, R., Huang, X., Ma, X., et al.: A model combining stacked auto encoder and back propagation algorithm for short-term wind power forecasting. IEEE Access **6**, 17851–17858 (2018)

Sequence Learning for Spreading Dynamics, Forecasting, and Intelligent Techniques Against Epidemic Spreading (2)

Data-Model Intergrowth Makes Better Time Series Prediction

Lixian Chen(✉), Hongda Liu, Chongqi Sun, Yi Wang, and Yongheng Hu

College of Intelligent Systems Science and Engineering, Harbin Engineering University, Harbin 150001, China
chenlixian@hrbeu.edu.cn

Abstract. In common prediction problems, the open-loop prediction framework is popular because of its fast prediction process and easy implementation. However, time series prediction is a special prediction problem because of the temporal nature of the time series, in which missing values are not only the target of the time series prediction, but also the flexible decision variables in the training sets of the prediction model, which indirectly control the predicted values, allowing the prediction process to be considered as a closed loop. This paper proves that the closed-loop prediction framework performs better on prediction accuracy than the open-loop framework. In this paper, the missing value prediction of the time series is modeled as a special optimization problem, in which the missing values and prediction models are decision variables constraining each other. The prediction models keep being trained on the training set containing the predicted values through iteration. The objective function is the virtual accuracy of the target dimension prediction model. When the objective function reaches its minimum, the best missing values and prediction models are given, which can solve any time series prediction problem. Therefore, the method is a general closed-loop prediction framework for time series prediction. In the experiment, 5 typical machine learning models in the closed-loop framework perform better than ones in the open-loop framework on 4 datasets with different frequency characteristics.

Keywords: Time-series prediction · Missing value padding · Machine learning

1 Introduction

Data is the basis of various advanced technologies in the era of big data. Time series data is a common data type. The temporal nature of the time series data provides the possibility to predict the futural states of the real dynamic systems. The time series prediction has been applied to make decisions in lots of engineering fields, such as sensor network monitoring [1], energy management for smart grid [2], economics and finance [3] and disease transmission analysis

H. Zhang et al. (Eds.): NCAA 2023, CCIS 1870, pp. 333–344, 2023.
https://doi.org/10.1007/978-981-99-5847-4_24

[4]. Unfortunately, the datasets in the real world always miss values, caused by sensor faults, communication faults or privacy problems, directly affecting the accuracy of downstream tasks, such as classification and regression [5]. Therefore, missing value processing plays an important role in the whole process.

The missing value processing methodology for time series consists of direct deletion, statistical interpolation, machine learning and generative model. The direct deletion method deletes all samples containing missing values, which not only ignores their potentially important information [6], but also leads to a smaller sample set and fallacies in the conclusion [7]. Statistical interpolation methods, including mean interpolation, regression interpolation, Lagrange interpolation and Chebyshev interpolation, has limited ability to fit high dimensional nonlinear time series. Machine learning and generative models are modern time series prediction methods, using prediction models to fit the relationship among sequential values in time series. The structure design of prediction models has attracted great attention. Its mathematical essence is to design functions with learnable parameters with stronger fitting ability for specific data, such as regression models, neural networks, ensemble decision trees, attention models. For example, Bania *et al.* [8] used the K-nearest neighbor (KNN) interpolation method to pad the missing values, which improved the accuracy of medical data classification. Zhuang *et al.* [9] proposed a traffic data interpolation method, which converts the original data into spatiotemporal images and develops a context encoder based on convolutional neural network (CNN) to estimate the missing data. Besides, the generative models have been improved in learning high-dimensional and complex data distribution on images and the similarities between the missing value padding and the data generation allow generative methods to be applied in missing value padding. Qu *et al.* [10] proposed a multi-optimization data interpolation method based on generating adversarial networks to pad the missing values in wind turbine data. Gondara *et al.* [11] proposed a multi-interpolation model based on over-complete deep denoising autoencoder, which can handle different data types, missing modes, missing proportions and missing distributions.

Missing value padding is not only a typical task in time series prediction but also a part of the data preprocessing for model training. In the time series prediction, each value in the dataset always appears in multiple samples for training, leading to strong dependence of prediction performance on the missing value processing. The ideal missing value processing is to pad the missing values with the actual values, which are unfortunately hard to know. The difficulty of the missing value padding is to design a robust assessment criterion without knowing actual values. The existing researches propose methods and test the error between the predicted values and the actual values, providing more and better choices to pad the missing values. The related works seem to be blind attempts without essential explanations about why their methods improve the prediction performance. In this paper, the accessible training error is proved to orient the missing value optimization as the similar direction as the inaccessible testing error.

The greater complexity and more mature learning mechanism of machine learning models lead to their better fitting ability, which indeed provides more choices for the prediction problem. From another perspective, this paper focuses on improving the prediction framework to improve the prediction accuracy without building any new models. This paper proposes a closed-loop prediction framework of the time series missing value padding. In this paper, the time series missing value padding problem is modeled as a special optimization problem. Its decision variables are missing values and the prediction models, which are constrained by each other. The missing values are predicted by the models while the models are trained on the training set with the predicted missing values. The objective function is only related to the current padded dataset and the target prediction model. The proposed framework can be equipped with any prediction model, of which the adaptive ability can be fully utilized by iterating the missing values until the best fitting between the model and the dataset. The proposed framework is tested on 4 time series with different frequency characteristics equipped with 5 classical prediction models. In most cases, the closed-loop framework makes better prediction than the open-loop framework.

2 Methodology

In this paper, the missing value padding problem of a time series dataset is modeled as an optimization problem. The missing values and the prediction models are taken as decision variables constrained by each other. To pick out the best iteration in the optimization process, an objective function only related to the current dataset and the current target prediction model is designed. The prediction procedure is shown in Algorithm 1.

2.1 Problem Modeling

The methodology of time series prediction is to use machine learning models to fit the relationship among the sequential terms in the time series. Mathematically, a multivariate time series is regarded as a multidimensional sequence $\{b_t = (a_t^0, a_t^1, ..., a_t^n)\}$, in which each dimension $\{a_t^i\}$ has the property of recursive sequence because of the temporal nature of the time series.

$$a_t^i = f_1\left(a_{t-1}^i, a_{t-2}^i, ..., a_{t-m}^i\right) = f_1\left(\left\{a_{t-\tau}^i \middle| 1 \leq \tau \leq m, \tau \in N\right\}\right) \tag{1}$$

in which $m+1$ is the window size of the time series.

Besides, the dimensions influence each other.

$$a_t^i = f_2\left(\left\{a_t^j \middle| j \neq i\right\}\right) \tag{2}$$

Algorithm 1: Data-Model Intergrowth

Input: dataset $\{b_t = (a_t^0, a_t^1, ..., a_t^n)\}$ with missing values, target dimension $*$
Output: $\{b_t\}$

begin
  Initialization: $\{f_\theta^i | i \leq n, i \in N\}$, $\{b_t\}$ with $a_t^i \leftarrow a_{t-1}^i$ if a_t^i is missing
  repeat
    for i **do**
      $S^i = \{S_t^i = (i_t^i, o_t^i) | t \in N\}$
      in which $\begin{cases} i_t^i = \{a_{t-\tau}^i | \tau \leq m, \tau \in N, i \leq n, i \in N\} - \{a_t^i\} \\ o_t^i = a_t^i \end{cases}$
      $T^i = \{s_t^i = (i_t^i, o_t^i) | \text{if } o_t^i \text{ is actual}\} \subset S^i$
      Update $\{b_t\}$ with $o_t^i = f_{\theta(T^i)}^i (i_t^i)$ if o_t^i is missing
    end
  until $min\, RMSE\left(\left\{o_t^* \middle| (i_t^*, o_t^*) \in T_{0.3}^*\right\}, \left\{f_{\theta(T_{0.7}^*)}^*(i_t^*) \middle| (i_t^*, o_t^*) \in T_{0.3}^*\right\}\right)$
end
return $\{b_t\}$,$\{f_\theta^i\}$

Combining (1) and (2), the influence function of each dimension is given.

$$\begin{aligned} a_t^i &= f\left(\{a_{t-\tau}^i | 1 \leq \tau \leq m, \tau \in N\} \cup \left\{a_t^j \middle| j \neq i\right\}\right) \\ &= f\left(\left\{a_{t-\tau}^j \middle| 1 \leq \tau \leq m, \tau \in N, j \leq n, j \in N\right\} \cup \left\{a_t^j \middle| j \neq i\right\}\right) \\ &= f\left(\left\{a_{t-\tau}^j \middle| \tau \leq m, \tau \in N, j \leq n, j \in N\right\} - \{a_t^i\}\right) \end{aligned} \quad (3)$$

The most important dimension in a time series is called the target dimension $\{a_t^*\}$, while others are feature dimensions.

Time series prediction with machine learning fits the influence function of the target dimension by training its prediction model on its samples. A prediction model, such as the regression model, neural network, ensemble decision tree and attention model, is a special function $f_\theta^i(i_t^i) = o_t^i$ with learnable parameters θ trained on its training set $S^i = \{s_t^i = (i_t^i, o_t^i) | t \in N\}$, in which the sample s_t^i consists of its input i_t^i and its output o_t^i for the tth moment.

In practice, the time series prediction is a typical optimization problem, aiming to pad the missing values with some predicted values closest to the unknown actual values. However, its objective function contains the unknown actual values, leading to failure of using optimization methods to solve prediction problems.

This paper models the time series prediction as a special optimization problem. The decision variables consist of the missing values and prediction models for each dimension. The main constraint is that the missing values must be predicted by corresponding prediction models. The objective function is the statistical error of prediction, such as mean absolute error (MAE), mean relative error (MRE) and root mean square error (RMSE), on the part of training set instead of the testing set.

$$MAE = \frac{1}{N}\sum_{i=1}^{N} |\hat{y}_i - y_i| \tag{4}$$

$$MRE = \frac{1}{N}\sum_{i=1}^{N} \left|\frac{\hat{y}_i - y_i}{y_i}\right| \tag{5}$$

$$RMSE = \sqrt{\frac{1}{N}\sum_{i=1}^{N} (\hat{y}_i - y_i)^2} \tag{6}$$

Different from the popular open-loop perdition framework, in which the model predicts the missing values once after training, this paper proposes a closed-loop prediction framework, in which the models keep predicting and training alternatively during iterations until a balance between the missing values and the models. The proposed framework is called data-model intergrowth (DMI) because the models and the dataset with missing values are improved synchronically.

2.2 Initialization

Mark Missing Values. The missing values are decision variables, being updated through iterations. Missing values appear in the input or output of the sample, resulting in different usages of the sample in the DMI framework. Only samples with actual outputs are worth learning and testing.

Initialize Missing Values. As an optimization framework, DMI requires initial missing values for the first iteration. A conservative padding method, such as the forward padding or the linear padding, is likely to improve the convergence of the DMI optimization.

Initialize Prediction Models. DMI uses prediction models to fit the influence functions for each dimension in the time series. Although the existing researches have provided various models capable to fit complex functions, it is difficult to choose a best model from them for a specific dataset. To solve a real problem, the engineers always suffer from comparing different model structures and optimizing the hyperparameters, which greatly influence the prediction performance. As an optimization framework for the dataset, DMI has the potential to narrow the performance difference among different models, because different models preform more similarly on the purer and cleaner dataset. A simpler model is likely to perform better in the DMI framework because of its smaller chance to overfit.

2.3 Iteration Process

Construct Sample Sets. Each prediction model requires a different sample set S^i, as mentioned in 2.1. The sample sets are restructured from the dataset with missing values. Therefore, the missing values appear in the inputs and outputs of multiple samples. A sample with its output being actual is worth learning and testing.

Extract Training Sets. The training set for each model T^i consists of the samples with actual outputs, because the model is supposed to be less sensitive to the partly mistakes in high-dimensional input and more confident in its output.

$$T^i = \left\{ s_t^i = \left(i_t^i, o_t^i \right) \middle| \text{if } o_t^i \text{ is actual} \right\} \subset S^i \tag{7}$$

Train Prediction Model. Each prediction model f_θ^i learns from its training set T^i to adjust its learnable parameters $\theta\left(T^i\right)$. The training of models is always the most time-consuming part in the prediction process. In the open-loop framework, the prediction model is trained only once. On the contrary, in the DMI framework, models are trained on the updated training set in each iteration, leading to more time cost than in the open-loop framework. In order to reduce the time cost of DMI iterations, it is suggested to use the models with fewer parameters and faster training mechanism.

Update Missing Values. In each dimension, missing values $\{o_t^i\}$ are predicted by its corresponding model f_θ^i. The inputs of the models are containing padded missing values in the last iteration, which are believed to get closer to the actual values through iterations. In a word, DMI improves the prediction performance of the target model by updating missing values in its training set and its inputs in the real prediction problem.

$$o_t^i = f_{\theta(T^i)}^i \left(i_t^i \right) \text{ if } o_t^i \text{ is missing} \tag{8}$$

Calculate the Objective Function. The goal of the real prediction problem is to minimize the statistical error between the predicted values and the actual values. However, in the actual prediction problem, the prediction should be completed before the actual values come, resulting in a lack of direction for the optimization methods. In the DMI framework, an objective function without actual missing values is calculated to assess the virtual accuracy of the missing values in each iteration.

The training set of the target dimension T^* is randomly divided by 7 to 3 into a sub training set $T_{0.7}^*$ and a sub test set $T_{0.3}^*$. An initialized target model is trained on the sub training set and tested on the sub test set to calculate the RMSE. The objective function is the mean value of the test RMSEs in multiple times of random division and test.

The proposed objective function, regarded as the virtual accuracy, is believed to be close to the actual accuracy, which keeps orienting the optimization of the missing values and the target prediction model to a better place.

$$i_t^* = \{a_{t-\tau}^i | \tau \leq m, \tau \in N, i \leq n, i \in N\} - \{a_t^*\} \tag{9}$$

$$o_t^* = a_t^* \tag{10}$$

$$T^* = \{(i_t^*, o_t^*) | \text{if } o_t^* \text{ is actual}\} \tag{11}$$

$$T_{0.7}^* + T_{0.3}^* = T^* \tag{12}$$

$$\min RMSE\left(\left\{o_t^* \middle| (i_t^*, o_t^*) \in T_{0.3}^*\right\}, \left\{f_{\theta(T_{0.7}^*)}^*(i_t^*) \middle| (i_t^*, o_t^*) \in T_{0.3}^*\right\}\right) \tag{13}$$

End with Convergence. The goal of the DMI optimization is to find the best predicted missing values and the best prediction models, which minimize the objective function. The objective function through iterations is supposed to decline to a minimum, when a balance is achieved between the models and the missing values. The optimization ends when the objective function no longer has new minimum value in the next few iterations. The best missing values and prediction models are given from the iteration with the minimum objective function. Besides, the objective function has a certain randomness because of the random division of the test set during each calculation. The modest randomness brings more iterations and makes the optimization more robust.

3 Experiment

3.1 Datasets

The 4 datasets in the experiment were selected from the NCAA2022 time series prediction competition datasets, which are generated from coronavirus (COVID-19) pandemic data via a prediction problem generation process composed of the finite impulse response filter-based approach and problem setting module [12]. The 4 generated time series datasets have different frequency characteristics, including low-pass, high-pass, band-pass and band stop. Each dataset has 4 dimensions, in which the target dimension has 3 feature dimensions and each dimension has missing values. Before the experiment, each dataset is preprocessed by min-max normalization.

3.2 Results

The 5 typical machine learning models, including BP [13], CNN [14], LSTM [15], XGBoost [16] and LightGBM [17], are tested in both the open-loop and DMI prediction framework to predict the missing values in the 4 datasets. The predicted and actual values in the time series are compared in Fig. 1. MAE, MRE

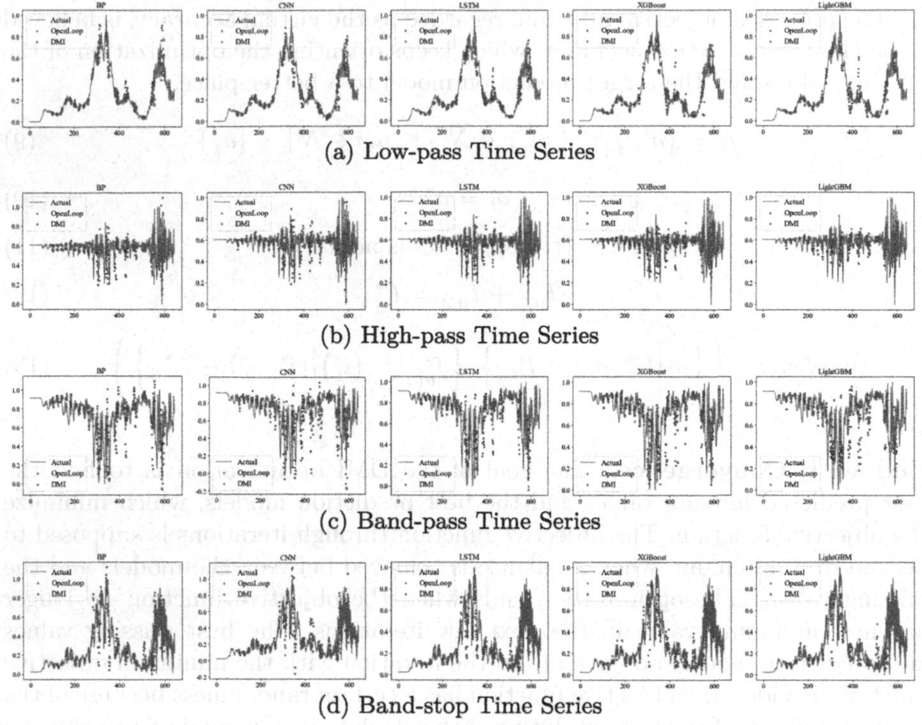

(a) Low-pass Time Series

(b) High-pass Time Series

(c) Band-pass Time Series

(d) Band-stop Time Series

Fig. 1. Prediction Performance

and RMSE of the target dimension are calculated as the evaluation metrics, shown in Table 1.

In the open-loop prediction framework, each dataset has a different best model, BP for low-pass, LSTM for high-pass and band-pass, and LightGBM for band-stop. However, in the DMI prediction framework, BP performs best for all the 4 datasets. In most cases, a model performs better in the DMI framework than in the open-loop framework and takes more time. Especially on the low-pass dataset, the RMSE of XGBoost in the DMI framework reaches half of that in the open-loop framework, and the time cost is 200 times of that in open-loop framework. On the high-pass dataset, the DMI framework narrows the performance gap between different models.

3.3 Discussion

The most time-consuming part in the time series prediction is the training of the prediction model. The open-loop prediction framework uses the trained model to predict the missing values, in which the model is trained only once, while the prediction models in the DMI framework need to be trained several times, resulting in more time costs.

Table 1. Model Performance

Dataset	Model	Frame	Time/s	MAE	MRE	RMSE
Low-pass	BP	OpenLoop	**14.2**	0.0526	0.1712	0.0919
		DMI	6387.43	**0.0395**	**0.133**	**0.0615**
	CNN	OpenLoop	**12.51**	0.0666	0.2078	0.1097
		DMI	9401.52	**0.0541**	**0.1801**	**0.0952**
	LSTM	OpenLoop	**27.5**	0.0596	0.2048	0.0932
		DMI	3319.25	**0.0493**	**0.1425**	**0.0814**
	XGBoost	OpenLoop	**0.12**	0.0749	0.2931	0.12
		DMI	24.07	**0.0422**	**0.1774**	**0.0653**
	LightGBM	OpenLoop	**0.16**	0.0658	0.2549	0.1064
		DMI	34.94	**0.0443**	**0.1868**	**0.0694**
High-pass	BP	OpenLoop	**12.82**	0.0949	0.1772	0.1403
		DMI	7642.15	**0.0541**	**0.111**	**0.0911**
	CNN	OpenLoop	**12.55**	0.085	0.165	0.1355
		DMI	11187.5	**0.0601**	**0.1244**	**0.0954**
	LSTM	OpenLoop	**28.28**	0.0723	0.1364	0.1155
		DMI	2919.1	**0.0606**	**0.1247**	**0.0956**
	XGBoost	OpenLoop	**0.13**	0.0828	0.1498	0.1309
		DMI	14.07	**0.0594**	**0.1233**	**0.1025**
	LightGBM	OpenLoop	**0.2**	0.0739	0.1444	0.1126
		DMI	15.68	**0.0596**	**0.1249**	**0.0995**
Band-pass	BP	OpenLoop	**8.56**	0.1306	0.55	0.1869
		DMI	6940.82	**0.0737**	**0.4636**	**0.1009**
	CNN	OpenLoop	**11.87**	0.1496	**0.5149**	0.2263
		DMI	7561.03	**0.1117**	0.6613	**0.1602**
	LSTM	OpenLoop	**27.26**	0.1029	0.496	0.1564
		DMI	3315.28	**0.0758**	**0.4318**	**0.1041**
	XGBoost	OpenLoop	**0.27**	0.1392	**0.5274**	0.2083
		DMI	28.98	**0.101**	0.8456	**0.1646**
	LightGBM	OpenLoop	**0.36**	0.1154	**0.5533**	0.1706
		DMI	52.21	**0.1011**	0.5948	**0.151**
Band-stop	BP	OpenLoop	**11.22**	0.114	0.5109	0.1806
		DMI	7896.97	**0.0749**	**0.267**	**0.1122**
	CNN	OpenLoop	**12.05**	0.1165	0.4965	0.1824
		DMI	7129.95	**0.1023**	**0.4377**	**0.1564**
	LSTM	OpenLoop	**27.67**	0.1123	0.5186	0.1707
		DMI	3551.28	**0.0759**	**0.2517**	**0.1295**
	XGBoost	OpenLoop	**0.17**	0.113	0.5626	0.1624
		DMI	15.89	**0.1056**	**0.4721**	**0.1554**
	LightGBM	OpenLoop	**0.18**	0.1055	0.5298	0.1565
		DMI	18	**0.0895**	**0.3258**	**0.1425**

Each value in the time series dataset appears in multiple samples for the prediction model, leading to a strong dependance of the prediction performance on the missing value processing of the training set. The open-loop framework regards the training set as the one-time material of the prediction model with only once processing, while the DMI framework optimizes the training set through iterations, contributing to the higher accuracy.

The 5 models tested in this paper have totally different structures. Although they are all capable to fit the complex influence function for each dimension in the time series, they differ in their sensitivity to the wrong samples, resulting in different prediction performances. The open-loop framework trains the 5 models on the same training set, while the DMI framework allows models to rewrite their own training sets to correct the wrong samples, leading to the smaller performance gap among them.

BP is the least complex model among the tested models but performs best in the DMI framework. In the DMI framework, less complex models are likely to perform better because the DMI framework focuses on correcting the training set, leading to simpler influence functions to fit.

4 Conclusion

As one of the common time series prediction tasks, predicting missing values in the time series is the main problem in this paper. In traditional machine learning methods, the solution is to optimize a prediction model to fit the influence function of the target dimension, including data cleaning of the training set, model structure design and hyperparameter optimization. However, the temporal nature of the time series leads to multiple appearances of a same value in different samples, especially in their inputs. The samples containing missing values are too many to ignore. Therefore, padding the missing values is a necessary part of the data cleaning of the training set. Meanwhile, the time series prediction methods can predict not only the unknown futural values, but also the missing values in the historical dataset. In this paper, the missing value prediction is regarded as both a problem and its solution, which is the basis of the proposed methodology.

This paper proposes an optimization modeling method to predict missing values in the time series. In the optimization problem, the decision variables are missing values and the prediction models, which are constrained by each other. The missing values are predicted by the models while the models are trained on the training set with the predicted missing values. The objective function is only related to the current padded dataset and the target prediction model. The method is called Data-Model Intergrowth because the dataset and the models are keeping updated by each other during the iterative process.

DMI is far beyond a missing value padding method for time series datasets. The DMI framework optimizes both the missing values and their prediction models, with which any time series prediction problem can be solved. Therefore, DMI can be regarded as a general closed-loop prediction framework for time series prediction problems as well.

The popular open-loop prediction process is short because the fast prediction and the slow training of the model are separated as independent parts and the training has been finished in advance. However, DMI repeats the training and prediction of the model in the prediction process, without a pretraining process, sacrificing more time for better accuracy. Rather than improving the structure and hyperparameters of the prediction model, DMI focuses on cleaning the training set to improve the learnable parameters of the model.

The popular framework uses the prediction model to fit the map from the input to the output in each old sample, then uses the model to predict the output from the new input. Just like taking a closed-book exam, students firstly learn an inference model to deduce the answer from the question in old exams, then answer new questions with the model without looking into the old exams again. However, the time series prediction is more like an open-book exam, always allowing the model to look into the dataset. Compared with the data-to-model training plus the model-to-result prediction, DMI combines the dataset and the prediction model as an evolutionary unity, which is a data-to-result prediction method, making the most of the dataset.

After comparing performances of five different typical prediction models on four datasets with different frequency characteristics in open-loop and DMI prediction framework, it is turned out that DMI helps the improvement of prediction accuracy on missing values in time series.

References

1. Bochenek, B., Ustrnul, Z.: Machine learning in weather prediction and climate analyses-applications and perspectives. Atmosphere **13** (2022)
2. Kathirgamanathan, A., Rosa, M.D., Mangina, E., Finn, D.P.: Data-driven predictive control for unlocking building energy flexibility: a review. Renew. Sustain. Energy Rev. **135**, 110120 (2021)
3. Cao, J., Li, Z., Li, J.: Financial time series forecasting model based on CEEMDAN and LSTM. Physica A **519**, 127–139 (2019)
4. Nadim, S.S., Ghosh, I., Chattopadhyay, J.: Short-term predictions and prevention strategies for COVID-19: a model-based study. Appl. Math. Comput. **404**(2), 126251 (2021)
5. Lin, Y., Wang, Y.: Multivariate time series imputation with bidirectional temporal attention-based convolutional network. In: Zhang, H., et al. (eds.) NCAA 2022, Part II. CCIS, vol. 1638, pp. 494–508. Springer, Singapore (2022). https://doi.org/10.1007/978-981-19-6135-9_37
6. Little, R.J.A., Rubin, D.B.: Statistical Analysis with Missing Data, vol. 793. Wiley, Hoboken (2019)
7. Mohamed, A.K., Nelwamondo, F.V., Marwala, T.: Estimating missing data using neural network techniques, principal component analysis and genetic algorithms (2007)
8. Rubul Kumar Bania and Anindya Halder: R-ensembler: a greedy rough set based ensemble attribute selection algorithm with KNN imputation for classification of medical data. Comput. Methods Programs Biomed. **184**, 105122 (2020)

9. Zhuang, Y., Ke, R., Wang, Y.: R-ensembler: a greedy rough set based ensemble attribute selection algorithm with KNN imputation for classification of medical data. Intell. Transp. Syst. IET **13**(4), 605–613 (2019)
10. Fq, B., Jla, B., Ym, C., Dong, Z.B., Mf, B.: A novel wind turbine data imputation method with multiple optimizations based on GANs. Mech. Syst. Sig. Process. **139** (2020)
11. Gondara, L., Wang, K.: Mida: multiple imputation using denoising autoencoders. arXiv e-prints (2017)
12. Wu, Z., Jiang, R.: Time-series benchmarks based on frequency features for fair comparative evaluation. In: Neural Computing and Applications (2023)
13. Buscema, M.: Back propagation neural networks. Substance Use Misuse **33**(2), 233–270 (1998)
14. Hoseinzade, E., Haratizadeh, S.: CNNpred: CNN-based stock market prediction using a diverse set of variables. Expert Syst. Appl. **129**, 273–285 (2019)
15. Hochreiter, S., Schmidhuber, J.: Long short-term memory. Neural Comput. **9**(8), 1735–1780 (1997)
16. Chen, T., Guestrin, C.: XGBoost: a scalable tree boosting system. In: Proceedings of the 22nd ACM SIGKDD International Conference on Knowledge Discovery and Data Mining, pp. 785–794 (2016)
17. Ke, G., et al.: LightGBM: a highly efficient gradient boosting decision tree. In: Advances in Neural Information Processing Systems, vol. 30 (2017)

Spatial-Temporal Electric Vehicle Charging Demand Forecasting: A GTrans Approach

Liangliang Zhu[1,3], Yangyang Ge[2,4], Kui Wang[1,3], Yongzheng Fan[5], Xintong Ma[2,4], and Lijun Zhang[5](✉)

[1] State Grid Electric Power Research Institute Co., Ltd., Nanjing 210000, China
[2] State Grid Liaoning Electric Power Co., Ltd., Liaoning 110004, China
[3] State Grid Electric Power Research Institute Wuhan Energy Efficiency Evaluation Co., Ltd., Wuhan 430074, China
[4] State Grid Liaoning Electric Power Co., Ltd. Electric Power Research Institute, Shenyang 110000, China
[5] School of Artificial Intelligence and Automation, Huazhong University of Science and Technology, Wuhan 430074, China
{yongzheng_fan,lijunzhang}@hust.edu.cn

Abstract. Accurate forecasting of electric vehicle charging demand is vital for city planning and power system scheduling. Existing studies on this topic fall short in terms of exploiting the spatial locations of the charging stations and utilizing long-term temporal dependencies. In this study, we combine a graph convolutional network (GCN) and a Transformer network to address the aforementioned challenges. The resulting forecasting network termed "GTrans" extracts the spatial couplings among charging stations using the GCN and achieves charging demand forecast over time dimension by learning the long-term correlations encoded in the historical data. A real-world dataset from the city of Palo Alto was used for the model training and performance validation. Comparison results with recent works show that GTrans has better accuracy on both short and long-term forecasting tasks.

Keywords: Charging demand forecasting · Electric vehicle · Graph convolutional network · Transformer

1 Introduction

Electric vehicles (EVs) produce less carbon dioxide to the environment in comparison to gasoline-powered vehicles [14]. However, a larger number of charging stations are required for the proper operation of EVs. For the planning [3] and operational scheduling of these charging stations, it is essential to forecast the distribution of EV charging demand over both time and space dimensions.

There are two research directions on the EV charging station (EVCS) demand forecasting problem. The first direction treats the forecasting task of EVCS as a pure time-series problem. The second direction considers the same task as

H. Zhang et al. (Eds.): NCAA 2023, CCIS 1870, pp. 345–358, 2023.
https://doi.org/10.1007/978-981-99-5847-4_25

a spatio-temporal distribution problem. Application of methods ranging from queuing theory to statistical machine learning [9] and deep learning techniques [7] have been reported on this task.

In particular, the main challenge in the first direction is to improve the forecasting accuracy of the multi-series model. The majority of studies in this line are based on statistic machine learning techniques like auto-regressive integrated moving average (ARIMA) and its variants [1,10]. For instance, [2] presented a chance-constrained day-ahead scheduling problem for power systems utilizing an ARIMA model for EVCS demand forecasting.

Taking a different approach, the core question of the second line of studies is: "how to deal with complex spatio-temporal relationships of the charging demand?". As the underlying principles are still unclear, many studies resorted to deep learning models to solve the spatio-temporal problem [8,21]. Many use the recurrent neural network (RNN) [7] and its variant long short-term memory network (LSTM) [7] for temporal feature modelling. LSTM is particularly popular because of its simplicity and fast training speed [16]. Borrowing ideas from the local feature extraction ability of convolution neural networks (CNN) in imaging processing [12], some researchers attempted to model the spatial couplings amongst EVCSs by dividing an area containing multiple EVCSs into a raster map [20]. CNN can then be applied to the resulting map to extract spatial features of the spatio-temporal problem [17]. Consequently, CNN and LSTM are combined for the spatial-temporal demand forecasting of the EVCS demand [8,15]. There are also studies that reported the use of graph convolutional network (GCN) [11,20] instead of CNN combined with LSTM for the same purpose [8].

It is intuitively desirable to use a graph topology for the spatial feature extraction and a model that can capture the long-term dependence for temporal forecasting. As such, a combination of GCN and Transformer, which was originally reported in [18] and shown to have superior performance on forecasting tasks [13,19], provides a new pathway to tackle the spatial-temporal EVCS demand forecasting problem. The design and application of such a GCN-Transformer combination on the said problem have not yet been investigated in the literature.

Therefore, this paper proposes to combine the GCN and Transformer(GTrans) to deal with the complex topological structure of the EVCS demand forecasting task. The GTrans model utilizes GCN and Transformer to capture the spatial and temporal dynamics of the charging demand, respectively. We focus on GTrans approaches to deal the spatial temporal EVCS forecasting where the data is provided by a publically available dataset. We hope to showcase the advantages of the proposed model in comparison to existing methods. As a result, both the geographical correlation among the stations and the long-term temporal dependence of the charging demand patterns can be exploited to form accurate forecasts.

In this paper, We define the EVCS forecasting problem (Sect. 2.1) and provide the model structure with the spatial modeling (Sect. 2.2) and temporal modeling

(Sect. 2.2). We introduce the public dataset (Sect. 3.1) used in the model and the evaluation methods (Sect. 3.2). After that, We describe the hyperparameters of our model (Sect. 3.3) and compare and analysis the experiment results (Sect. 3.4) to verify the effectiveness of our proposed model.

2 Methodology

The problem statement and its corresponding mathematical modelling using the proposed GTrans are provided in this section.

2.1 Problem Statement

The problem focuses on forecasting the electrical charging power demand at multiple EVCS located near each other making use of past charging demand and geographical correlations.

This paper defines a temporal signal $X_t = [x_t^1, x_t^2, ..., x_t^s]$ which contains the historical daily energy demand for each node s at time t on the topology G, which represents the spatial distribution of the EVCS. The spatial-temporal charging demand forecasting of the EVCS can be defined as acquiring a function f defined as follows:

$$[X_{t+1}, X_{t+2}, ...X_{t+T}] = f(G; (X_{t-n}, X_{t-n+1}, ..., X_t)), \tag{1}$$

where n is the length (number of days) of the historical data and T is the forecast horizon.

2.2 Modeling

The charging demand forecasting problem considered intrinsically has spatial and temporal components. To this end, we propose to model the dynamics of these two dimensions separately and then form a complete forecast through combination. First, a GCN is employed to extract the spatial features representing coupling among the EVCS. The output of the GCN model is then cascaded to a Transformer, which is chosen to encode the long-term temporal correlations in the charging demand. This GCN-Transformer (GTrans) architecture is shown in Fig. 1 and detailed in the following.

Spatial Modelling. Existing studies on spatial demand forecasting made use of raster maps to represent the geographical couplings, which however has difficulties in determining the appropriate grid size and capturing the exact topological connections among the EVCS [17]. To overcome the aforementioned issues, this study takes inspiration from [8], which captures the spatial correlations of the EVCS by an undirected weighted graph $G = (V, E)$, where $V = \{v_1, v_2, ..., v_N\}$ is the set of charging stations in the entire area, N is the total number of charging stations, and the E represent edges. To facilitate spatial feature extraction, this

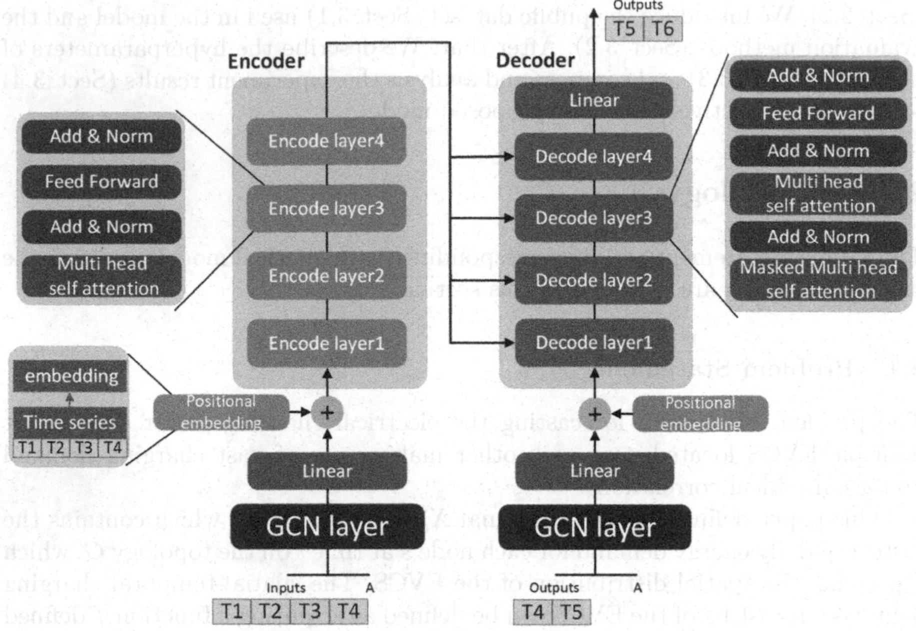

Fig. 1. Architecture of the GTrans for EVCS demand forecasting

weighted graph is further denoted by its adjacency matrix $A = \{a_{ij}; i, j \in V\}$ are defined by:

$$a_{ij} = \begin{cases} exp(-h(x_i, x_j)) & \text{if } h(x_i, x_j) < 2.5, \\ 0 & \text{otherwise,} \end{cases} \tag{2}$$

where x_i and x_j are the locations of the i-th and j-th charging stations and $h(x_i, x_j)$ is the Haversine distance of the corresponding stations measured in km. We think that the two stations are independent of each other when their distance is more than 2.5 km. This explains the zero value of $h(x_i, x_j)$ in (2).

Apart from its intuitiveness and versatility for representing geographical relationships, a graph in Fig. 2.b allows the adoption of GCN to extract the spatial characteristics of the demand forecasting problem with graph convolutions, which constructs a filter in the Fourier domain. In this study, we use the 2-layered GCN model described by:

$$f(A, X) = relu(\hat{A} relu(\hat{A} X W_0) W_1), \tag{3}$$

where X is the feature matrix, A is the adjacency matrix of the graph G, with $\hat{A} = \tilde{D}^{-\frac{1}{2}} \tilde{A} \tilde{D}^{-\frac{1}{2}}$ representing its symmetric normalization. $\tilde{A} = A + I$ means all nodes in the graph are neighbours of themselves. $\tilde{D} = \sum_j \tilde{A}_{ij}$ is the degree matrix of the $\tilde{A}$. W_0 and W_1 are the weight matrices of the first and second layers of the GCN, respectively. The *relu* is the chosen activate function to facilitate model training.

The aggregated features from the output of the GCN can then be readily made available for a subsequent temporal model to form a combined spatial-temporal forecasting scheme.

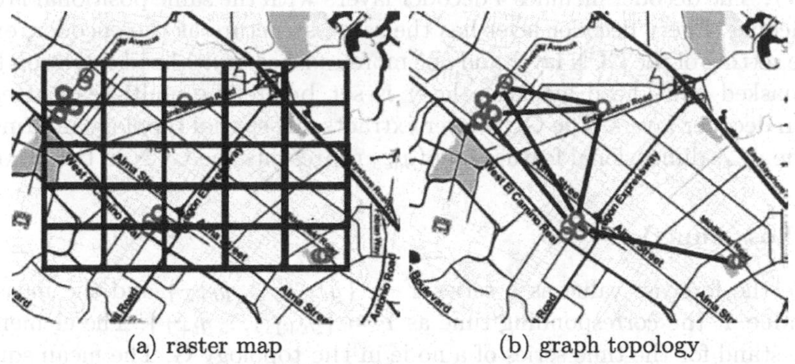

(a) raster map (b) graph topology

Fig. 2. Two topologies of the same area.

Temporal Modelling. The Transformer network originated from [18] is employed for the temporal forecasting task. The attention mechanism of the Transformer allows it to model long-distance correlations in its input sequences. In particular, the attention mechanism utilizes the following equation to calculate its output from three linear transforms of its input, namely, the query Q, the key K, and the value V:

$$Z = softmax(\frac{QK^T}{d_{model}})V, \tag{4}$$

where Q, K, V are 3-dimensional matrices, with indices {Batch, Sequence, d_{model}}, the d_{model} is the number of the feature dimension for the forecasting assignment. Q is the query vector used to represent the information that needs to be focused on. K represents the key vector used to represent the importance or relevance of each input signal. V represents the value vector, which represents the input signal itself. To obtain the attention weights, the softmax function normalizes the results of QK^T. The resulting attention output is denoted by Z.

Encoder. The encoder includes four encoder layers, a positional encoding layer, with a GCN layer. Every encoder layer is composed of a multi-head attention layer and a feed forward layer, each sub-layer of the encoder layer is followed by a normalization layer with a residual connection. For the positional encoding layer, the embedding function is used to get the sequential information of the continuous numbers into the d_{model} features. The GCN layer extracts the spatial topology information into a g_{enci}-dimensional feature map(g_{enc} represents the

GCN in the encoder, the subscript i represents the GCN layer number), then the last feature map is converted to a d_{model}-dimensional feature map by the fully connected layer.

Decoder. The decoder includes 4 decoder layers with the same positional layer as the encoder. Every decoder layer has the same structure as the encoder, except for the setting of the GCN layer and one more masked multi-head attention layer. The masked multi-head attention layer is set before the multi-head attention layer in decoder layers. The GCN layer extracts the spatial topology information into the g_{deci}-dimensional feature map(g_{dec} represents the GCN in the decoder).

2.3 Loss Function

Denote the forecast value as a series $\hat{Y} = \{\hat{y}_{t+1}, \ldots, \hat{y}_{t+T}\}$ and the measured true value at the corresponding time as $Y = \{y_{t+1}, \ldots, y_{t+T}\}$. The elements y_t and $\hat{y}_t$ stand for the time series of a node in the topology G. The mean squared error with ℓ_2 regularization is used as the loss function in model training. As a result, the following loss function is defined:

$$L = \frac{1}{n}\sum_{i=1}^{n}(Y - \hat{Y})^2 + \frac{\lambda}{2n}\sum_{j=1}^{n} w_j^2, \tag{5}$$

where λ is a hyperparameter which can be used to tune the strength of regularization, n is the total number of samples, and w_j represents the parameters of the jth model.

3 Experiments and Data

The proposed model is implemented in Python utilizing the PyTorch stellar graph libraries [5] with a NVIDIA GeForce RTX 3060 GPU device. To validate its performance over multi-period forecasts, the proposed GTrans model is tested with 7-day and 30-day forecast horizons. Further, the results are compared with four existing approaches reported in the literature (listed in Table 1). The data we use comes from publicly available online datasets [4] for better comparison.

3.1 Data

In order to ensure the comparability of the experimental results, this paper uses a public dataset to test the performance of the proposed model. The dataset used includes the charging demand information of charging stations in Palo Alto, USA [4] from 2011 to 2020, as depicted in Fig. 2.b. The dataset contains information such as the longitude and latitude of the station, and charging time information (e.g., charging time, total energy consumption). This dataset provides sufficient data with different temporal and spatial resolutions to facilitate model training and results validation.

Table 1. Models used for comparison analysis

	Spatial model	Temporal model
CNN-LSTM	CNN with raster map	LSTM
T-GCN	GCN with graph	LSTM
CNN-Tran	CNN with raster map	Transformer
Transformer	–	Transformer
GTrans	GCN with graph	Transformer

3.2 Evaluation Function

In order to verify the fitting accuracy of the model. This paper uses three different evaluation metrics in terms of the absolute and relative errors, namely, R_p, root-mean-squared error (RMSE), and mean absolute percentage error. These metrics are defined by:

$$R_p = \frac{\sum_{i=1}^{n} |y_i - \hat{y}_i|}{\sum_{i=1}^{n} |y_i|}, \tag{6}$$

$$RMSE = \sqrt{\frac{1}{n}\sum_{i=1}^{n}(\hat{y}_i - y_i)^2}, \tag{7}$$

$$MAPE = \frac{1}{n}\sum_{i=1}^{n}|\frac{\hat{y}_i - y_i}{y_i}|, \tag{8}$$

where the meaning of y and $\hat{y}$ is set to the one-day charging demand of the forecast days in this paper. The closer the evaluation metrics are to 0, the better the model performs on the test dataset.

3.3 Model Settings

The hyperparameters of the GTrans model mainly include learning rate lr, training epochs, ℓ_2 regulation parameter λ, rate of dropout, the number of encoder layers and decoder layers, the feature dimensions of GCN layers g_{enci} and g_{deci}, the dimension of positional embedding, and dimensions of the encoder layer and decoder layer input features d_{model}. To compare the effectiveness of CNN and GCN on spatial feature extraction, we constructed a 5×5 raster map in the geographical area because CNN cannot handle the graph structure. Figure 2.a shows this topology.

In the 7-day forecast, 30-day historical time-series data is used as the input to the model. In the GTrans model, the hyperparameters g_{enc1} and g_{enc2} of the encoder GCN are set to 64 and 30, respectively. For the decoder GCN, $g_{dec1} = 32$, $g_{dec2} = 7$. The GTrans model uses the four encoder layers, four decoder layers, eight heads, 100-dimension positional embedding, and $d_{model} = 128$. Other hyperparameters of the models used for comparison are set as follows.

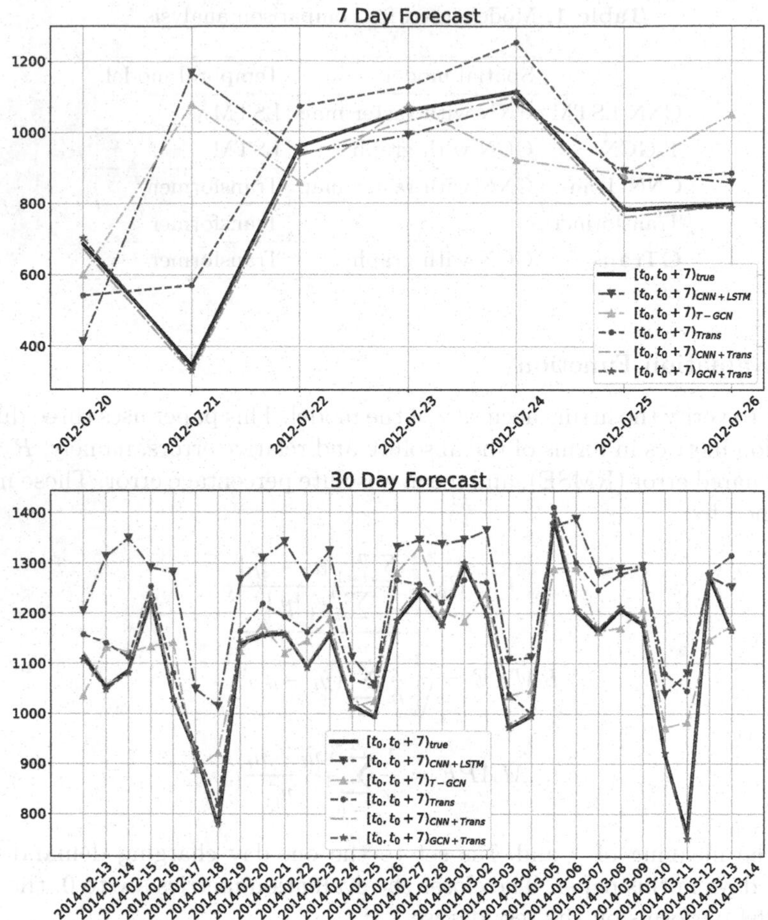

Fig. 3. Forecast results for both 7-day and 30-day horizons

CNN uses 16 filters with 3×3 kernels, LSTM uses 100 hidden units and ten layers, and the positional embedding of Trans in the Experiment 3 is set to 256.

In the 30-day forecast, 120-day historical time-series data is used as the input to the model. In the GTrans model, the hyperparameters are the same as those used for the 7-day forecast except the dimension of positional embedding is changed to 200. The hyperparamerters of GCN are set to: $g_{enc1} = 64, g_{enc2} = 120, g_{dec1} = 32, g_{dec2} = 7$. The CNN, LSTM and Trans models used for comparison are kept the same as in the 7-day forecast case.

The learning rate and other parameters are shown in the Table 2. They remain constant across different forecasting missions. While training, the data is divided into training dataset (80% of the overall data) and test dataset (20% of the overall data) randomly. The batch size is set to 64. All models are fitted using the loss function in (5) with the ADAM optimizer [6] and tested using the evaluation metrics given in Sect. 3.2.

Table 2. Learning rate lr, training epochs, λ and dropout in different models

model	lr	training epochs	λ	dropout
CNN-LSTM	$1e^{-5}$	10000	$1e^{-8}$	0.1
T-GCN	$1e^{-4}$	10000	$1e^{-8}$	0.1
Transformer	$1e^{-5}$	10000	$1e^{-8}$	0.1
CNN-Trans	$1e^{-5}$	10000	$1e^{-9}$	0.1
GTrans	$1e^{-4}$	10000	$1e^{-9}$	0.1

3.4 Results and Analysis

Three comparison experiments are conducted among the models shown in Table 1. We evaluate the models with the R_p, RMSE and MAPE on the different forecast horizons. The experimental results are shown in Tables 3 and 4, and Fig. 3 and Fig. 4.

Experiment 1. The capabilities of LSTM and Transformer on the temporal modelling are analysed.

This is achieved by comparing models with the same spatial component but different temporal networks. As Table 3 and Table 4 show, CNN-Trans and GTrans have superior results than CNN-LSTM and T-GCN. Their training metric MSE and evaluation metrics (Rp, RMSE, MAPE) have obvious advantages. What's more, GTrans and CNN-Trans have lower volatility in terms of training metrics and evaluation metrics as the history sequence length increases, which indicates that Transformer has better stability and accuracy than LSTM on time modelling.

As shown in Fig. 3, the CNN-Trans and GTrans are all better than the CNN-LSTM and T-GCN. GTrans exhibits the best forecasting accuracy for both the 7-day and 30-day forecasts. In Fig. 4, Transformer combined with the spatial feature extraction model demonstrates better predictive performance than LSTM at each node (grid or station). This indicates that the Transformer performs better than the LSTM in multi-dimensional time-series forecasting.

Experiment 2. The capabilities of CNN and GCN on spatial modelling are compared. As can be seen from Tables 3 and 4, During training, T-GCN has a higher training loss than the CNN-LSTM, and GTrans has a higher training loss than CNN-Trans for both 7-Day and 30-Day forecast. This indicates that CNN is easier to train than GCN. In terms of the spatial feature extraction, although the models demonstrate comparable performance in the 7-Day forecast, GTrans and T-GCN show better evaluation metrics than CNN-Trans and CNN-LSTM in the 30-Day forecast. This suggests as the time dimension increases, GCN can better extract the spatial feature than CNN.

As shown in Fig. 3, in the 7-day forecast, the prediction results of the models compared are similar. In the 30-Day forecast, however, T-GCN has significantly

Table 3. Model comparison over 7-day forecasting

Model 7-Day	MSE	R_p	RMSE	MAPE
CNN-LSTM	$5.67 * 10^{-4}$	0.132	133.805	0.246
T-GCN	$3.03 * 10{-}3$	0.140	142.720	0.218
Transformer	$9 * 10^{-5}$	0.136	147.556	0.226
CNN-Trans	$\mathbf{1.7 * 10^{-5}}$	0.0245	29.245	0.055
GTrans	$5 * 10^{-5}$	**0.0059**	**7.373**	**0.012**

better prediction results than CNN-LSTM. This indicates that GCN can obtain a more stable feature than CNN as the historical time dimension increases.

Experiment 3. The necessity of a spatial feature extraction is demonstrated. All spatial models combined with Transformer have lower training metrics and evaluation metrics than the Transformer alone (see Tables 3 and 4), indicating that the inclusion of spatial features is meaningful for the EVCS forecasting problem. As shown in Fig. 3, the CNN-Trans and GTrans both show better performance than the Transformer. This indicates that the results from the Fig. 3 are consistent with those in Table 3, 4.

Table 4. Model comparison over 30-day forecasting

Model 30 Days	MSE	R_p	RMSE	MAPE
CNN-LSTM	$1.29 * 10^{-3}$	0.226	180.472	0.558
T-GCN	$4.4 * 10^{-3}$	0.109	117.457	0.167
Transformer	$8.4 * 10^{-5}$	0.089	115.256	0.111
CNN-Trans	$\mathbf{1.2 * 10^{-5}}$	0.0201	30.844	0.0667
GTrans	$4.7 * 10^{-5}$	**0.0051**	**5.212**	**0.0102**

In this paper, we use the Transformer for the temporal problem, which uses the attention sores on the time series. The GTrans attention score is shown in Fig. 5 and Fig. 6. Figure 5 displays the connection of the 7-day decoder and the encoder input features, which indicates that GTrans finds some small correlation between the forecast data and the historical data. As the time series becomes longer, this characteristic becomes more easily identifiable. Figure 6 shows a better trend of this, suggesting that the Transformer can handle long time-series data and has stronger resistance to forecasting than the LSTM as the sequence length increases.

Nevertheless, the proposed model also has some limitations. First, although a graph instead of a raster map is used to represent the geographical coupling, it is still defined subjectively according to the relative distances among the stations.

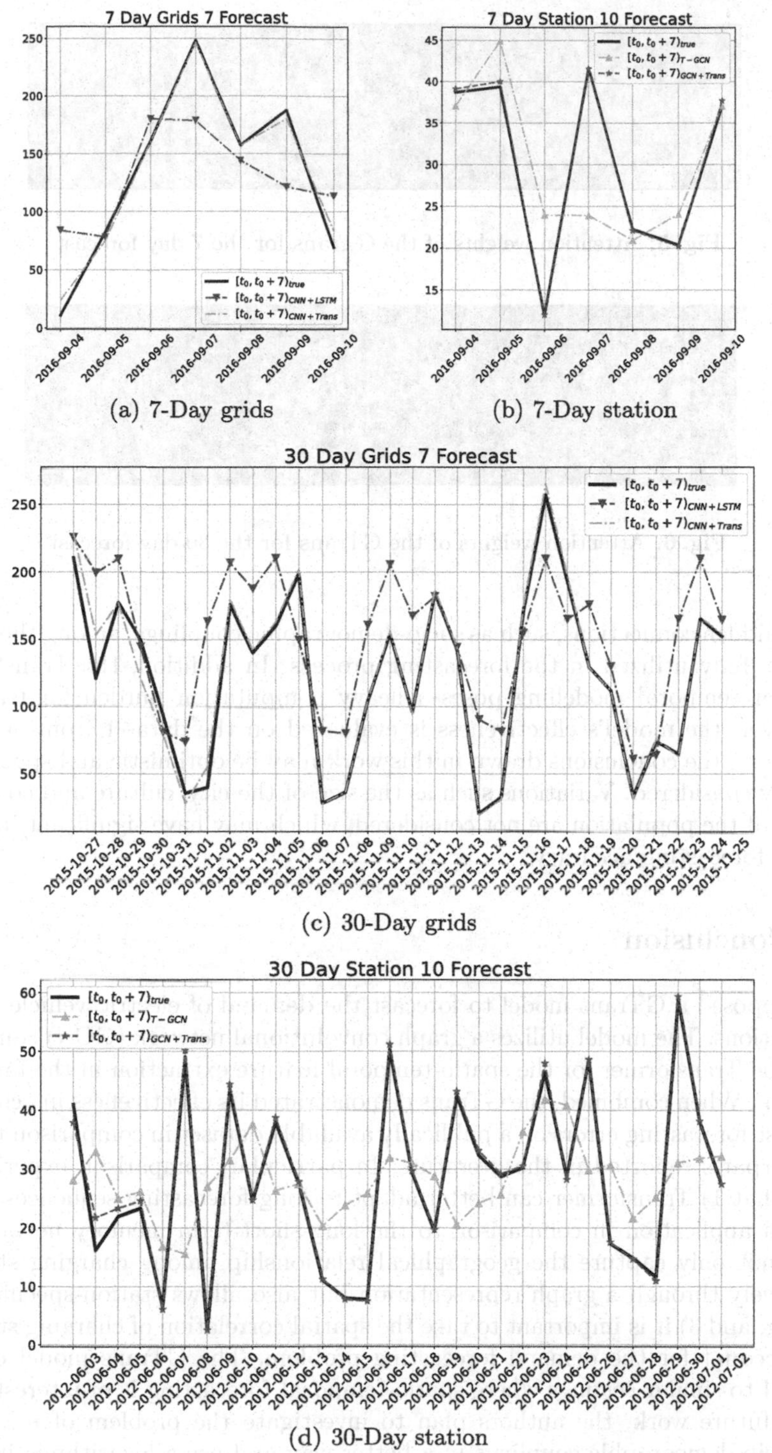

(a) 7-Day grids (b) 7-Day station

(c) 30-Day grids

(d) 30-Day station

Fig. 4. 7-Day and 30-Day forecast at the grid and station

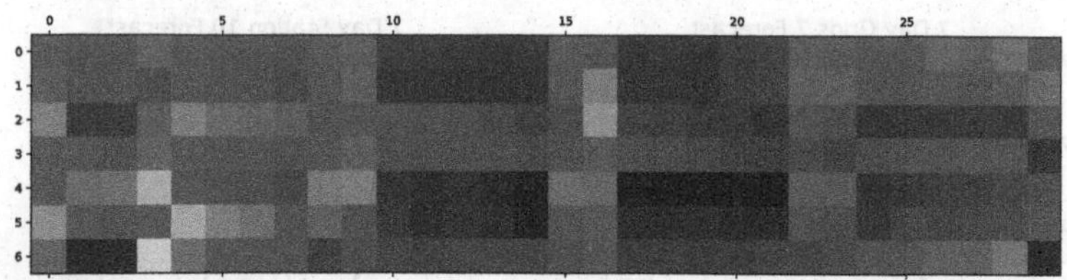

Fig. 5. Attention weights of the GTrans for the 7-day forecast

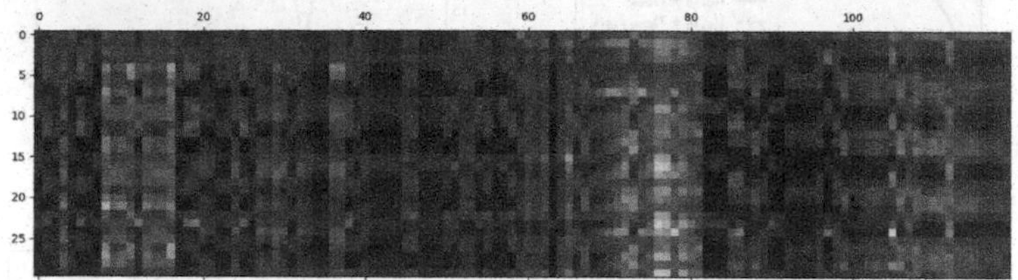

Fig. 6. Attention weights of the GTrans for the 30-day forecast

Some hidden connections, such as socio-demographic couplings, among the nodes are not fully utilized in the forecasting process. In addition, the Transformer used for temporal modelling poses a heavy computation burden for training. Moreover, the model's effectiveness is evaluated on the dataset from one city. Therefore, the conclusions drawn in this work may be optimistic and specific for the city considered. Variations such as the size of the city, culture and commute habits of the population are not considered, which may have significant impacts on the forecasting accuracy.

4 Conclusion

We proposed a GTrans model to forecast the demand of electric vehicle charging stations. The model utilizes a graph convolutional network (GCN) combined with the Transformer for the spatio-temporal feature extraction in the forecasting task. When combined, the GTrans demonstrated its effectiveness in achieving the best forecasting errors on a publically available dataset in comparison to four counterparts reported in the literature. In particular, comparison experiments show that 1) Transformer can better adapt to long forecasting sequences in the studied application in comparison to the long-short-term memory network; 2) GCN not only capture the geographical relationship among charging stations intuitively through a graph representation but also allows station-specific forecasting; and 3) it is important to take the spatial correlation of charging stations into account for the demand forecasting problem. The GTrans model can be applied to similar settings where spatio-temporal forecasting is of interest.

In future work, the authors plan to investigate the problem of modelling the socio-demographic couplings in a better way and use a logarithmic interval attention mechanism to reduce the model's computational cost. We will consider more features of the urban environment to make the model more explanatory, and select some simplified Transformer models to improve our model's computational speed. We can also add more city datasets to make the model training more robust.

Acknowledgements. This work was supported by Science and Technology Project of SGCC (No.5400-202228170A-1-1-ZN).

References

1. Amara-Ouali, Y., Goude, Y., Massart, P., Poggi, J.M., Yan, H.: A review of electric vehicle load open data and models. Energies **14**(8) (2021)
2. Amini, M.H., Kargarian, A., Karabasoglu, O.: ARIMA-based decoupled time series forecasting of electric vehicle charging demand for stochastic power system operation. Electric Power Syst. Res. **140**, 378–390 (2016)
3. Cheon, S., Kang, S.J.: An electric power consumption analysis system for the installation of electric vehicle charging stations. Energies **10**(10) (2017)
4. City of Palo Alto: Electric vehicle charging station usage in Palo Alto, CA (2020). https://data.cityofpaloalto.org/dataviews/257812/electric-vehicle-charging-station-usage-july-2011-dec-2020/
5. Data61, C.: Stellargraph machine learning library. https://github.com/stellargraph/stellargraph
6. Diederik, P., Kingma, J.B.: Adam: a method for stochastic optimization. CoRR **9**(4) (2014)
7. Hochreiter, S., Schmidhuber, J.: Long short-term memory. Neural Comput. **9**(8), 1735–1780 (1997)
8. Hüttel, F.B., Peled, I., Rodrigues, F., Pereira, F.C.: Deep spatio-temporal forecasting of electrical vehicle charging demand. CoRR abs/2106.10940 (2021)
9. Khoo, Y.B., Wang, C.H., Paevere, P., Higgins, A.: Statistical modeling of Electric Vehicle electricity consumption in the Victorian EV Trial, Australia. Transp. Res. Part D: Transp. Environ. **32**, 263–277 (2014)
10. Kim, Y., Kim, S.: Forecasting charging demand of electric vehicles using time-series models. Energies **14**(5) (2021)
11. Kipf, T.N., Welling, M.: Semi-supervised classification with graph convolutional networks. CoRR abs/1609.02907 (2016)
12. LeCun, Y., et al.: Backpropagation applied to handwritten zip code recognition. Neural Comput. **1**(4), 541–551 (1989)
13. Li, S., et al.: Enhancing the locality and breaking the memory bottleneck of transformer on time series forecasting. In: Wallach, H., Larochelle, H., Beygelzimer, A., d'Alché-Buc, F., Fox, E., Garnett, R. (eds.) Advances in Neural Information Processing Systems, vol. 32. Curran Associates, Inc. (2019)
14. Miotti, M., Supran, G.J., Kim, E.J., Trancik, J.E.: Personal vehicles evaluated against climate change mitigation targets. Environ. Sci. Technol. **50**(20), 10795–10804 (2016)
15. Rodrigues, F., Pereira, F.C.: Beyond expectation: deep joint mean and quantile regression for spatiotemporal problems. IEEE Trans. Neural Netw. Learn. Syst. **31**(12), 5377–5389 (2020)

16. Sherstinsky, A.: Fundamentals of recurrent neural network (RNN) and long short-term memory (LSTM) network. Physica D **404**, 132306 (2020)
17. SHI, X., Chen, Z., Wang, H., Yeung, D.Y., Wong, W.K., WOO, W.C.: Convolutional LSTM network: a machine learning approach for precipitation nowcasting. In: Cortes, C., Lawrence, N., Lee, D., Sugiyama, M., Garnett, R. (eds.) Advances in Neural Information Processing Systems. Curran Associates Inc (2015)
18. Vaswani, A., et al.: Attention is all you need. In: Guyon, I., et al. (eds.) Advances in Neural Information Processing Systems, vol. 30. Curran Associates, Inc. (2017)
19. Wu, N., Green, B., Ben, X., O'Banion, S.: Deep transformer models for time series forecasting: the influenza prevalence case. CoRR abs/2001.08317 (2020)
20. Zhao, L., et al.: T-GCN: a temporal graph convolutional network for traffic prediction. IEEE Trans. Intell. Transp. Syst. **21**(9), 3848–3858 (2020)
21. Ziat, A., Delasalles, E., Denoyer, L., Gallinari, P.: Spatio-temporal neural networks for space-time series forecasting and relations discovery. In: 2017 IEEE International Conference on Data Mining (ICDM), pp. 705–714 (2017)

A Long Short-term Memory Model for COVID-19 Forecasting Using High-efficiency Feature Representation

Zhengyang Hou[1], Jingeng Fang[1], Yao Huang[1], Lulu Sun[1(✉)], Choujun Zhan[2], and Kim-Fung Tsang[1]

[1] School of Electrical and Computer Engineering, Nanfang College Guangzhou, Guangzhou 510970, China
jenqyanghou@gmail.com, {fangjg200082,huangy200065,sunll}@nfu.edu.cn, ee330015@cityu.edu.hk

[2] School of Computer Science, South China Normal University, Guangzhou 510641, China

Abstract. COVID-19 is a large family of viruses that are spreading around the world at an unprecedented rate. Due to the large number of infected people, medical institutions around the world are overwhelmed. Therefore, an accurate forecast of the number of cases and deaths can help governments and other organizations formulate response strategies in advance and ease the pressure on the public health system. Deep learning offers a new perspective to combat novel coronavirus outbreaks. In this research, we propose a unique hybrid framework that integrates empirical wavelet transform (EWT), variational mode decomposition (VMD), and long short-term memory (LSTM) network for COVID-19 prediction. More specifically, to fully exploit COVID-19 series information, we developed an efficient feature representation method by integrating EWT and VMD. By this method, the original time series, which exhibits intricate variations, can be decomposed into subsequences with more discernible change characteristics.. LSTM-based models leverage memory and gating mechanisms to capture both short-term and long-term dependencies within time series data. To demonstrate the predictive performance of the proposed model, we perform short, medium, and long-term forecasts of weekly new confirmed cases and deaths, respectively. Experimental results show that our model outperforms the other seven baseline models and the hybrid artificial neural network (ANN) models used for comparison in terms of root mean squared logarithmic error (RMSLE), mean absolute percentage error (MAPE), coefficient of determination (R^2) and pearson correlation coefficient (PCC).

Keywords: COVID-19 · Hybrid deep learning · Empirical wavelet transform · Variational mode decomposition · Long short-term memory

H. Zhang et al. (Eds.): NCAA 2023, CCIS 1870, pp. 359–374, 2023.
https://doi.org/10.1007/978-981-99-5847-4_26

1 Introduction

In December 2019, an outbreak of acute atypical respiratory illnesses emerged in Wuhan, China, and rapidly disseminated globally. The illness is attributed to a novel coronavirus known as severe acute respiratory syndrome coronavirus-2 (SARS-CoV-2). [22]. The WHO has named the pandemic coronavirus disease 2019 (COVID-19). As of February 8, 2023, there have been a total of 755,041,562 cases of infection and 6,830,867 deaths worldwide[1]. The huge number of infected people has created an unprecedented shock and challenge to the healthcare system and government decision-making. It is very important to use epidemic prediction to help the government formulate long-term strategies and allocate resources reasonably to relieve the pressure on the medical system. Through the research on COVID-19, many methods and models have been developed to make predictions, which can be roughly divided into the following three categories:

The epidemiological dynamics model is a fundamental tool for studying the dynamic behavior of the disease and estimating the trend of disease transmission. Yang et al. [21] integrated the COVID-19 data into the susceptible-exposed-infectious-removed (SEIR) model to obtain a dynamic model to predict the peak of COVID-19. Chen et al. [6] proposed the susceptible-exposed-infected-unreported-removed (SEIUR) model and the segmental SEIUR model, which the segmented SEIUR model can update the β and γ values at the beginning of each cycle to make the model more perfect. Although these models provide effective explanations for the outbreak and growth of COVID-19, they can only incorporate a limited number of variables and parameters, making it difficult to guarantee the accuracy of predictions.

Statistical models for COVID-19 prediction mainly are mainly autoregressive models. Monllor et al. [14] used the ARMA [3] model to predict the transmission speed and the number of confirmed cases of COVID-19 in China, Italy, and Spain using existing data. Yang et al. [20] constructed the ARIMA [4] model by combining the daily confirmed cases and deaths during the period from outbreak to zero in Wuhan. The model was then used to predict the epidemic situation in Italy. Nevertheless, relying solely on fixed mathematical formulas, most of these models are inadequate in capturing the complete essence of time series characteristics. Furthermore, the real-world time series data exhibit intricate and fluctuating patterns, posing challenges for these models to achieve accurate prediction outcomes.

As an important part of the AI field, traditional machine learning has also been used to predict the spread of the epidemic. Such as support vector regression (SVR) Parbat et al. [15], random forest (RF) Zhan et al. [23], XGBoost Mehta et al. [13]. Machine learning methods with nonlinear mapping capabilities can achieve better results than linear statistical models. But the nonlinear mapping ability of traditional machine learning methods is limited. It also relies on feature engineering and parameter tuning, resulting in poor performance. Deep learning is a more intelligent method than traditional machine learning because of its

[1] https://covid19.who.int/.

powerful nonlinear feature extraction ability and generalization performance. Phba et al. [17] proposed a multiple layer perceptron (MLP)-based model that can effectively describe and predict the 6-day number of COVID-19 infections and deaths. Rauf et al. [18] used long short-term memory (LSTM), recurrent neural network (RNN), and gated recurrent unit (GRU) to predict the number of infections in Pakistan, Afghanistan, India, and Bangladesh for the next 10 days. Predictive models based on deep learning can effectively help develop coping strategies for COVID-19 [12] [16] [2] and achieve better results than traditional machine learning. However, the deep learning method has high requirements for the data set. COVID-19 data are nonlinear and non-stationary time series data, which are influenced by various factors, and the data set that can be used for deep learning training is small. Therefore, it is still difficult to fully extract sequence features by using only deep learning methods, thus unable to achieve the ideal effect.

Data augmentation is an excellent and efficient feature representation method that can capture hidden features of time series. As a method of data augmentation, signal decomposition has been widely used in nonlinear and non-stationary time series analysis. Empirical wavelet transform (EWT) and variational mode decomposition (VMD) are two of them. EWT uses data-driven frequency band division rules to extract amplitude modulated and frequency modulated (AM-FM) from the original sequence. VMD divides the original sequence into a set of intrinsic mode functions (IMFs) components. These components possess simpler frequency components and stronger correlations, rendering them more amenable to accurate predictions. As a variant of RNN, LSTM controls information transmission through a gating mechanism, which can effectively solve the problem of gradient disappearance and explosion during long sequence training, and is more suitable for long sequence prediction. Therefore, this study proposes an EWT-VMD-LSTM hybrid model to predict the weekly new confirmed cases and deaths of COVID-19. Our primary contributions can be summarized as follows:

(1) We validate the ability of the proposed hybrid model to predict the COVID-19 time series on a small dataset. It is demonstrated that data augmentation can achieve more efficient feature representation.
(2) Through EWT and VMD, the original time series, which exhibits intricate variations, can be decomposed into subsequences with more discernible change characteristics. LSTM networks are able to fully learn these features. The obtained results demonstrate that our model outperforms both the baseline and comparison models, thereby showcasing its superior performance.
(3) The proposed EWT-VMD-LSTM hybrid model can predict the weekly new confirmed cases and deaths of COVID-19 accurately in the short, medium, and long-term, respectively. The experimental results can provide early warning of the epidemic and help governments and other regulatory agencies allocate resources in a timely and reasonable manner.

2 Related Works

2.1 Empirical Wavelet Transform

The EWT, introduced by Gilles et al. [10] in 2013, is an innovative adaptive signal processing technique. This method combines the adaptability of empirical mode decomposition (EMD) and the theoretical framework of the wavelet transform. In this method, the original signal is regarded as the sum of a group of AM-FM components, and empirical wavelets can be adaptively constructed according to different signal contents. The basic steps of EWT are as following:

Step 1: The Fourier spectrum of the original signals is partitioned into N consecutive segments. Determine the spectrum division boundary ω_n according to the frequencies of two adjacent main maxima in the spectrum.

Step 2: Construct empirical scaling function $\hat{\phi}(\omega)$ and empirical wavelet function $\hat{\psi}(\omega)$:

$$\hat{\phi}_n(\omega) = \begin{cases} 1, |\omega| \le (1-\gamma)\omega_n \\ \cos\left[\frac{\pi}{2}\beta\left(\frac{1}{2\gamma\omega_n}(|\omega| - (1-\gamma)\omega_n)\right)\right], (1-\gamma)\omega_n \le |\omega| \le (1+\gamma)\omega_n \\ 0, otherwise \end{cases} \tag{1}$$

$$\hat{\psi}_n(\omega) = \begin{cases} 1, (1+\gamma)\omega_{n+1} \le |\omega| \le (1-\gamma)\omega_n \\ \cos\left[\frac{\pi}{2}\beta\left(\frac{1}{2\gamma\omega_{n+1}}(|\omega| - (1-\gamma)\omega_{n+1})\right)\right], \\ (1-\gamma)\omega_{n-1} \le |\omega| \le (1+\gamma)\omega_{n+1} \\ \sin\left[\frac{\pi}{2}\beta\left(\frac{1}{2\gamma\omega_n}(|\omega| - (1-\gamma)\omega_n)\right)\right], (1-\gamma)\omega_n \le |\omega| \le (1+\gamma)\omega_n \\ 0, otherwise \end{cases} \tag{2}$$

where:

$$\gamma < \min_n\left(\frac{\omega_{n+1} - \omega_n}{\omega_{n+1} + \omega_n}\right) \tag{3}$$

$$\beta(x) = \begin{cases} x^4\left(35 - 84x + 70x^2 - 20x^3\right), 0 < x < 1 \\ 0, x \le 0 \\ 1, x \ge 1 \end{cases} \tag{4}$$

Step 3: Finally, the empirical wavelet transform is implemented on $f(t)$. The operation process can be expressed as:

$$W_f^{\varepsilon}(0,t) = \langle f, \phi_1 \rangle = \int f(\tau)\bar{\phi}_1(\tau - t)\mathrm{d}\tau = F^{-1}\left(\hat{f}(\omega)\hat{\phi}_1(\omega)\right) \tag{5}$$

$$W_f^{\varepsilon}(n,t) = \langle f, \psi_n \rangle = \int f(\tau)\bar{\psi}_n(\tau - t)\mathrm{d}\tau = F^{-1}\left(\hat{f}(\omega)\hat{\psi}_n(\omega)\right) \tag{6}$$

where $\bar{\phi}_1(t)$ and $\bar{\psi}_n(t)$ are the complex conjugates of $\phi_1(t)$ and $\psi_n(t)$, respectively. $\hat{\phi}_1(\omega)$ and $\hat{\psi}_n(\omega)$ are the Fourier transforms of $\phi_1(t)$ and $\psi_n(t)$, respectively. $F^{-1}[\cdot]$ is defined as the inverse Fourier transform. At this point, the original signal can be reconstructed according to the following formula:

$$f(t) = W_f^{\varepsilon}(0,t) * \phi_1(t) + \sum_{n=1}^{N} W_f^{\varepsilon}(n,t) * \psi_n(t) =$$
$$F^{-1}\left[\left(\hat{W}_f^{\varepsilon}(0,\omega) * \hat{\phi}_1(\omega) + \sum_{n=1}^{N} \hat{W}_f^{\varepsilon}(n,\omega) * \hat{\psi}_n(\omega)\right)\right] \tag{7}$$

where * represents convolution operation. The empirical mode function $c_k(t)$ obtained by decomposing the signal $f(t)$ is defined as:

$$c_0(t) = W_f^{\varepsilon}(0,t)^* \phi_1(t) \tag{8}$$
$$c_k(t) = W_f^{\varepsilon}(k,t)^* \psi_k(t) \tag{9}$$

2.2 Variational Mode Decomposition

VMD is a non-recursive signal decomposition method that builds upon the variational model initially proposed by Dragomiretskiy et al. [8], which is an extension of the EMD technique. The method decomposes the original sequence into IMFs with specific sparse properties. By employing the classical Wiener filter, we employ a fully intrinsic and adaptive variational approach to determine the center frequency and bandwidth limits. These parameters are obtained by solving a constrained variational problem. Subsequently, we identify the corresponding effective components in the frequency domain that facilitate the separation of modes. The basic steps of VMD are as following:

Step 1: Dragomiretskiy et al. made a more stringent definition of the IMF based on the modulation criterion:

$$u_k(t) = A_k(t)\cos(\phi_k(t)) \tag{10}$$

where $A_k(t)$ is the envelope amplitude of $u_k(t)$, $\phi_k(t)$ is the instantaneous phase, instantaneous frequency $\omega_k(t) := \phi_k^{'}(t)$.

Step 2: Firstly, the analytical signal of $u_k(t)$ is obtained by the Hilbert transform, and its unilateral spectrum is calculated. By multiplying with the operator $e^{-j^{\omega}kt}$, the central bandwidth of $u_k(t)$ is modulated to the corresponding baseband. Secondly, calculate the square norm L^2 of the demodulation gradient to estimate the bandwidth of each module component, and obtain the constrained variational problem as follows:

$$\min_{\{u_k\},\{\omega_k\}} \left\{ \sum_k \left\| \partial_t \left[\left(\delta(t) + \frac{j}{\pi t} \right) * u_k(t) \right] e^{-j\omega_k t} \right\|_2^2 \right\}$$
$$s.t. \quad \sum_k u_k = f \tag{11}$$

where k is the number of modes, $\{u_k\} := \{u_1, \ldots, u_K\}$ is a set of modal functions, $\{\omega_k\} := \{\omega_1, \ldots, \omega_K\}$ is the center frequency set.

Step 3: Introducing the quadratic penalty term α and the Lagrangian multiplier λ turns the constrained variational problem into an unconstrained variational problem, and obtains the augmented Lagrangian $\mathcal{L}$:

$$\mathcal{L}(\{u_k\}, \{\omega_k\}, \lambda) := \alpha \sum_k \left\| \partial_t \left[\left(\delta(t) + \frac{j}{\pi t} \right) * u_k(t) \right] e^{-j\omega_k t} \right\|_2^2 + \left\| f(t) - \sum_k u_k(t) \right\|_2^2 + \left\langle \lambda(t), f(t) - \sum_k u_k(t) \right\rangle \tag{12}$$

Step 4: The alternate direction multiplier method (ADMM) is used to iteratively find the saddle point of $\mathcal{L}$, which is the optimal solution of the variational constrained problem. The iteration formula of all components is:

$$\hat{u}_k^{n+1}(\omega) = \frac{\hat{f}(\omega) - \sum_{i<k} \hat{u}_i^{n+1}(\omega) - \sum_{i>k} \hat{u}_i^n(\omega) + \frac{\hat{\lambda}^n(\omega)}{2}}{1 + 2\alpha \left(\omega - \omega_k^n\right)^2} \tag{13}$$

$$\omega_k^{n+1} = \frac{\int_0^\infty \omega \left| \hat{u}_k^{n+1}(\omega) \right|^2 d\omega}{\int_0^\infty \left| \hat{u}_k^{n+1}(\omega) \right|^2 d\omega} \tag{14}$$

$$\hat{\lambda}^{n+1}(\omega) = \hat{\lambda}^n(\omega) + \tau \left(\hat{f}(\omega) - \sum_k \hat{u}_k^{n+1}(\omega) \right) \tag{15}$$

where $f(\omega)$, $\lambda(\omega)$, $u_i(\omega)$, $u_k^{n+1}(\omega)$ denote Fourier transform of $f(t)$, $\lambda(t)$, $u_i(t)$, $u_k^{n+1}(t)$ respectively. Continue to iterate until $\sum_k \left\| \hat{u}_k^{n+1} - \hat{u}_k^n \right\|_2^2 / \left\| \hat{u}_k^n \right\|_2^2 < \epsilon$ is satisfied, where n represent the number of iterations, τ denotes the update coefficient.

2.3 Long Short-Term Memory

Accurate time series forecasting necessitates the consideration of not only the most recent data but also the historical data. The recurrent neural network (RNN) exhibits advantages in addressing long-term forecasting challenges due to its self-feedback mechanism within the hidden layer. LSTM, an enhanced variant of RNN, was introduced by Hochreiter et al. [11] in 1997, further enhancing the model's performance in handling long-term dependencies. The model is jointly controlled by the forget gate, input gate, internal memory unit, and output gate. The equations governing the operation of an LSTM cell are presented below:

$$f_t = \sigma\left(W_f^T \times h_{t-1} + U_f^T \times x_t + b_f\right) \tag{16}$$

$$i_t = \sigma\left(W_i^T \times h_{t-1} + U_i^T \times x_t + b_i\right) \tag{17}$$

$$\widetilde{c}_t = \tanh\left(W_c^T \times h_{t-1} + U_c^T \times x_t + b_c\right) \tag{18}$$

$$c_t = f_t \times c_{t-1} + i_t \times \widetilde{c}_t \tag{19}$$

$$h_t = o_t \times \tanh(c_t) \tag{20}$$

$$o_t = \sigma\left(W_o^T \times h_{t-1} + U_o^T \times x_t + b_o\right) \tag{21}$$

where f_t is the output of the forget gate for time t, h_{t-1} represents the output of the module at time $t-1$, i_t represents the input at time t, o_t represents the input at time t, c_t represents the output from the cell unit. W and U are weight matrices. b is the bias matrix. σ and $tanh$ are activation functions. Figure 1 illustrates the comprehensive structure of the LSTM cell, incorporating all the gates and variables.

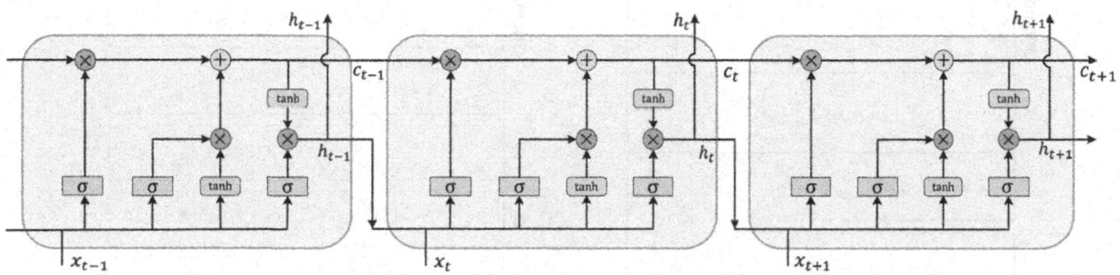

Fig. 1. Architecture of the LSTM cell.

3 Methodologies

3.1 Problem Formulation

Forecasting COVID-19 data is actually a time series forecasting problem. A time series refers to a collection of random variables that are ordered chronologically, typically obtained by observing a specific underlying process at regular intervals with a fixed sampling rate. COVID-19 data can be represented as:

$$\mathbf{x} = \left(x^1, x^2, \cdots, x^N\right) \in \mathbb{R} \tag{22}$$

where N represents the length of the time series. Time series prediction is to calculate the predicted value $y_{T\hat{+}H}$ at the time $T+H$ according to the observed data at the previous time T, where H refers to the horizon of the prediction. The input sequence and target sequence of the model can be expressed as:

$$\mathbf{X} = \left(\mathbf{x}_1, \mathbf{x}_2, \cdots, \mathbf{x}_{N-(T+H-1)}\right) \in \mathbb{R}^{(N-T-H+1)\times T} \tag{23}$$

$$\mathbf{y} = \left(y_1, y_2, \cdots, y_{N-(T+H-1)}\right)^\top \in \mathbb{R}^{N-(T+H-1)} \tag{24}$$

Our goal is to establish a model F that can learn an accurate mapping from past values to future values:

$$y_{T\hat{+}H} = F\left(y_1, \cdots, y_{N-(T+H-1)}, \mathbf{x}_1, \mathbf{x}_2, \cdots, \mathbf{x}_{N-(T+H-1)}\right) \tag{25}$$

3.2 Data Augmentation

EWT adaptively divides the signal spectrum based on the frequency-domain extreme point distribution and constructs a suitable wavelet filter bank according to the frequency-domain segmentation results. The EWT is applied in the divided frequency domain to separate the AM-FM component with a tight support spectrum from the original signal. Fig 2 shows the decomposition results of EWT. The number of decompositions is obtained through repeated experiments. The decomposed components exhibit a progressive arrangement of characteristics, ranging from low frequency to high frequency.

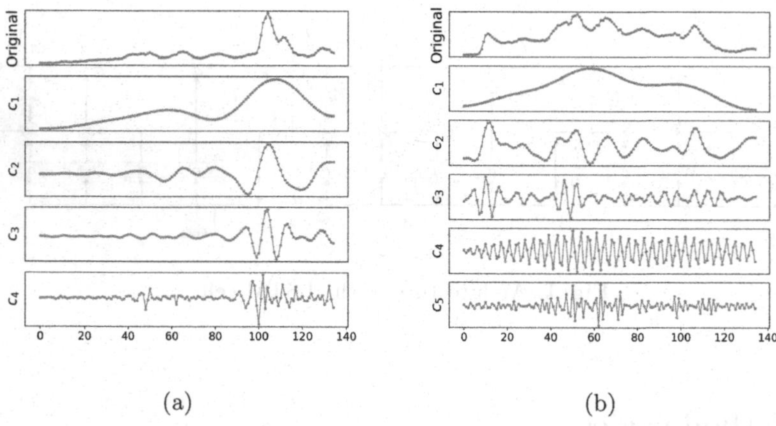

Fig. 2. (a) EWT for the weekly new confirmed cases. (b) EWT for the weekly new deaths.

VMD converts data from the time domain to the frequency domain for decomposition, which can not only capture the nonlinear characteristics of time series data well, but also avoid variable information overlap, and its decomposition process has strong robustness. Fig 3 shows the decomposition results of VMD. The number of decompositions is obtained through repeated experiments. The weekly confirmed cases sequence is decomposed into 4 IMFs and 1 residual, and the weekly deaths sequence is decomposed into 3 IMFs and 1 residual.

Compared to the original data, the decomposed subsequences possess a simpler structure, increased stability, and enhanced regularity. These characteristics contribute to improved accuracy in terms of fitting and forecasting.

3.3 Proposed EWT-VMD-LSTM Model

The proposed hybrid model initially decomposes the time series into several components, which are then trained in parallel to provide reliable forecasts. Through data augmentation, the model is able to learn an accurate mapping between historical data and future data. Figure 4 showcases the architectural

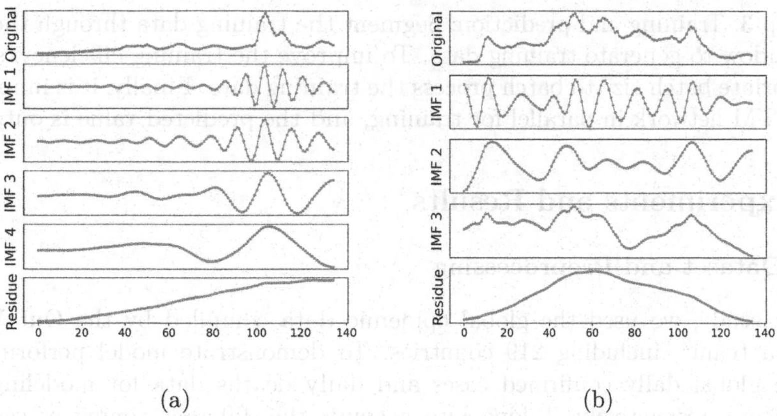

Fig. 3. (a) VMD for the weekly new confirmed cases. (b) VMD for the weekly new deaths.

design of the proposed hybrid framework. The process of forecasting using a hybrid model is as follows:

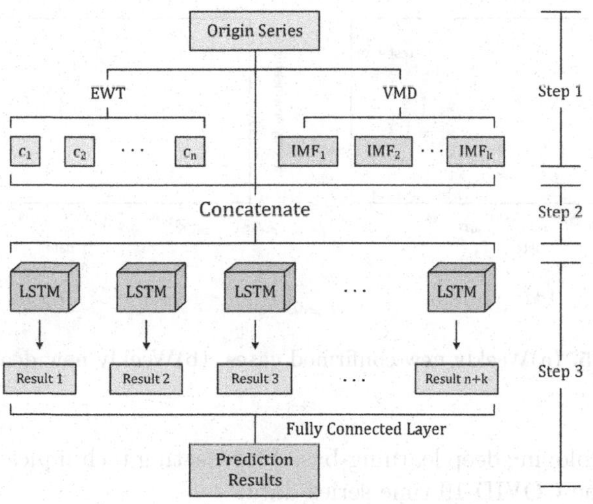

Fig. 4. Framework of EWT-VMD-LSTM model.

Step 1: Data augmentation: Use the decomposition algorithm to decompose the normalized input sequence, and decompose the original sequence into 10 subsequences.

Step 2: Mode vector merge: Merge the subsequences together with the original sequence so that it can be fed into the LSTM network in parallel.

Step 3: Training and prediction: Segment the training data through the sliding window to generate training data. To improve the training efficiency, set an appropriate batch size to batch process the training data. Finally, it is input into the LSTM network in parallel for training, and the predicted value is output.

4 Experiments and Results

4.1 Dataset and Preprocessing

In this study, we used the global epidemic data compiled by the Our World in Data team[2], including 219 countries. To demonstrate model performance, we use global daily confirmed cases and daily deaths data for modeling and forecasting, respectively. Taking into account the different timing of cases in different countries, we use data from January 22, 2020, to August 24, 2022. Through further sorting and processing according to the number of new people added every week, a total of 135 sample points can be obtained, as shown in Fig 5. The first 95 sample points are used as the training data, and the remaining 40 sample points are used as the test data. The ratio of the training set to the test set is 7:3.

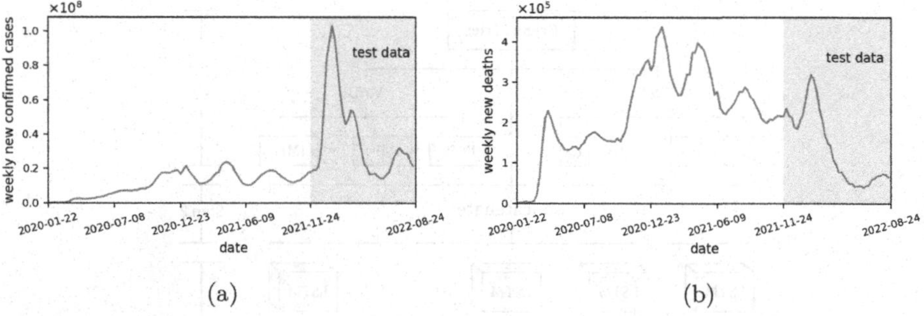

Fig. 5. (a)Weekly new confirmed cases. (b)Weekly new deaths.

Prior to employing deep learning-based forecasting techniques, it is necessary to normalize the COVID-19 time series data:

$$\widetilde{x} = \frac{(x - x_{\min})}{(x_{\max} - x_{\min})} \tag{26}$$

where $x_{\max}$ and $x_{\min}$ represent the maximum and minimum values of the training data, respectively. $x_{\max}$ and $x_{\min}$ will be used to normalize training and test data.

[2] https://github.com/owid/covid-19-data.

4.2 Experiment Settings and Evaluation Metrics

We set the sliding window size $T = 21$ for the proposed EWT-VMD-LSTM model, and the forecasting horizon H is set to 1, 2, and 3 weeks (representing short-term, medium-term, and long-term forecasts, respectively). Our LSTM network contains one hidden layer. The hidden size is 128 for short-term and medium-term forecasting. The hidden size is 256 for long-term prediction. At the same time, we set batch size as 16, epoch number as 300, learning rate as 0.3 and Adam optimizer as optimizer.

In this work, four performance factors were analyzed for a fair comparison. Root mean squared logarithmic error (RMSLE), mean absolute percentage error (MAPE), coefficient of determination (R^2), and pearson correlation coefficient (PCC) can be calculated by using Eq 27-30, respectively.

$$RMSLE = \sqrt{\frac{1}{n}\sum_{i=1}^{n}(log(y_i + 1) - log(\hat{y}_i + 1))^2} \tag{27}$$

$$MAPE = \frac{100}{n}\sum_{i=1}^{n}\left|\frac{y_i - \hat{y}_i}{y_i}\right| \tag{28}$$

$$R^2 = 1 - \frac{\sum_{i=1}^{n}(y_i - \hat{y}_i)^2}{\sum_{i=1}^{n}(y_i - y)^2} \tag{29}$$

$$PCC = \frac{\sum_{i=1}^{n}(\bar{y}_i - \bar{\hat{y}})(y_i - \bar{y})}{\sqrt{\sum_{i=1}^{n}(\bar{y}_i - \bar{\hat{y}})^2}\sqrt{\sum_{i=1}^{n}(y_i - \bar{y})^2}} \tag{30}$$

4.3 Forecasting Performance

After the data preprocessing, the reconstructed data series was processed into the input of the EWT-VMD-LSTM. Table 1-4 show the prediction performance of the proposed hybrid model in terms of the number of confirmed cases and deaths. Especially in terms of long-term forecasting, our model shows excellent performance.

When predicting the number of confirmed cases, low RMSLE and MAPE values of 0.0828 and 6.8412, high R^2, and PCC values of 0.8997 and 0.9626 were obtained by the proposed constructor for long-term prediction. When predicting the number of deaths, low RMSLE and MAPE values of 0.0664 and 5.5018, high R^2, and PCC values of 0.9027 and 0.9710 were obtained by the proposed constructor for long-term prediction. Through data augmentation for more efficient feature representation, our model shows clear advantages in predicting both confirmed and death cases. It is also robust and can predict future data stably and accurately. The fitting effects can be shown in Fig 6.

Table 1. Comparisons with single models for weekly new confirmed cases.

H	Metrics	ARIMA	XGBoost	MLP	RNN	LSTM	GRU	TCN	Ours
1	RMSLE	0.1962	0.1461	0.0914	0.0983	0.0863	0.0969	0.0908	**0.0534**
	MAPE	15.8391	12.0921	7.2224	7.2374	7.3049	8.1452	7.0170	**4.5043**
	R^2	0.5568	0.5532	0.8933	0.8925	0.9057	0.8970	0.8973	**0.9655**
	PCC	0.8685	0.8240	0.9483	0.9478	0.9523	0.9473	0.9492	**0.9838**
2	RMSLE	0.2929	0.2201	0.1453	0.1374	0.1391	0.1457	0.1575	**0.0747**
	MAPE	21.5573	18.9186	12.4120	11.1518	11.6664	11.4657	12.3477	**6.3555**
	R^2	0.2249	0.1762	0.7160	0.7803	0.7702	0.7034	0.7173	**0.9399**
	PCC	0.8005	0.6595	0.8505	0.8906	0.8857	0.8795	0.8768	**0.9715**
3	RMSLE	0.4278	0.2974	0.2122	0.1585	0.2121	0.2010	0.2045	**0.0828**
	MAPE	29.3659	24.2864	18.5412	11.7947	18.5044	16.3553	15.2078	**6.8412**
	R^2	-0.5219	-0.3176	0.3784	0.7341	0.5409	0.5431	0.5669	**0.8997**
	PCC	0.6322	0.4371	0.6202	0.8788	0.7694	0.7595	0.8116	**0.9626**

Table 2. Comparisons with single models for weekly new deaths.

H	Metrics	ARIMA	XGBoost	MLP	RNN	LSTM	GRU	TCN	Ours
1	RMSLE	0.1190	0.3047	0.1705	0.1048	0.1049	0.0889	0.0945	**0.0420**
	MAPE	10.1782	27.4840	14.7653	8.3438	9.4350	7.4481	7.1057	**3.0300**
	R^2	0.7931	-1.4424	0.6033	0.8734	0.8488	0.8768	0.8818	**0.9761**
	PCC	0.9040	0.4992	0.8552	0.9447	0.9253	0.9505	0.9419	**0.9894**
2	RMSLE	0.1767	0.3560	0.1707	0.1509	0.1720	0.1718	0.1432	**0.0644**
	MAPE	14.2010	34.6548	14.1792	12.4823	13.6737	13.8485	12.2998	**5.0375**
	R^2	0.4774	-2.4931	0.4983	0.6002	0.5424	0.3374	0.6041	**0.9147**
	PCC	0.8155	0.2863	0.7954	0.8480	0.8305	0.8707	0.8098	**0.9772**
3	RMSLE	0.2383	0.4111	0.2122	0.1793	0.1575	0.2177	0.1830	**0.0664**
	MAPE	18.9110	42.2521	18.5412	13.4934	12.2927	16.1537	14.5805	**5.5018**
	R^2	-0.0377	-4.2968	0.3784	0.3319	0.5430	0.0751	0.3216	**0.9027**
	PCC	0.6672	-0.0322	0.6202	0.7311	0.8255	0.8147	0.6353	**0.9710**

4.4 Comparisons and Ablation Study

We will compare with seven baseline models, including ARIMA [4], XGBoost [5], MLP [19], RNN [9], LSTM [11], GRU [7] and temporal convolutional network (TCN) [1]. Baseline models use only original data as input and also make predictions on all three horizons. Tables 1 and 2 display the evaluation metrics for each model, enabling a comprehensive assessment of their overall performance. The forecasting results strongly indicate that the hybrid models outperform the baseline models that were considered in this study. When predicting the number of confirmed cases, the RMSLE of the suggested hybrid model is 72.78% lower than ARIMA, 63.45% lower than XGBoost, 41.58% lower than MLP, 45.68% lower than RNN, 38.12% lower than LSTM, 44.89% lower than GRU, 41.19% lower than TCN of short-term prediction. MAPE, R^2, and PCC are also significantly better than baseline models. When predicting the number of deaths, the RMSLE of the suggested hybrid model is 64.71% lower than ARIMA, 86.22% lower than XGBoost, 75.37% lower than MLP, 59.92% lower than RNN, 59.96% lower than LSTM, 52.76% lower than GRU, 55.56% lower than TCN of short-term prediction. MAPE, R^2, and PCC are also significantly better than baseline models. Upon applying the VMD and EWT decomposition techniques, the original time series, characterized by intricate and complex variations, is effectively decomposed into subsequences that exhibit more discernible and simplified change characteristics. The LSTM network possesses robust prediction capabilities when it comes to time series forecasting tasks. Therefore, the model proposed in this paper has a more accurate prediction effect than the other seven baseline models.

Table 3. Ablation experiments for weekly new confirmed cases.

model	Forecasting horizon(H)											
	1				2				3			
	RMSLE	MAPE	R^2	PCC	RMSLE	MAPE	R^2	PCC	RMSLE	MAPE	R^2	PCC
EWT-RNN	0.0556	3.7469	0.9633	0.9824	0.0820	6.8034	0.9054	0.9586	0.1044	8.3365	0.8693	0.9417
EWT-LSTM	0.0531	3.7848	0.9660	0.9867	0.0735	6.5811	0.9139	0.9698	0.0931	7.7744	0.8902	0.9582
EWT-GRU	0.0506	4.1825	0.9653	0.9848	0.1146	8.6230	0.8707	0.9344	0.1155	8.6739	0.7859	0.9306
EWT-TCN	0.0640	4.5104	0.9571	0.9785	0.0845	7.6362	0.9040	0.9555	0.1127	8.5631	0.8488	0.9327
VMD-RNN	0.0588	4.6242	0.9517	0.9786	0.0992	8.1575	0.9021	0.9636	0.1183	9.6449	0.8223	0.9381
VMD-LSTM	0.0629	4.8918	0.9480	0.9752	0.0743	5.9942	0.9155	0.9574	0.1227	9.6776	0.8249	0.9513
VMD-GRU	0.0602	4.1919	0.9608	0.9829	0.0987	7.5676	0.9053	0.9740	0.1332	10.4525	0.8166	0.9434
VMD-TCN	0.0603	4.5080	0.9595	0.9813	0.0617	4.7617	0.9368	0.9813	0.1137	9.4773	0.8399	0.9277
EWT-VMD-RNN	0.0540	4.1721	0.9673	0.9837	0.0853	7.0559	0.9131	0.9700	0.1367	8.5798	0.8705	0.9689
EWT-VMD-GRU	0.0704	5.6451	0.9476	0.9753	0.1029	7.2386	0.9084	0.9676	0.0977	7.6493	0.8952	0.9628
EWT-VMD-TCN	0.0595	4.2776	0.9642	0.9848	0.0807	6.3171	0.9270	0.9666	0.1049	8.1095	0.8981	0.9502
Ours	**0.0534**	**4.5043**	**0.9655**	**0.9838**	**0.0747**	**6.3555**	**0.9399**	**0.9715**	**0.0828**	**6.8412**	**0.8997**	**0.9626**

Table 4. Ablation experiments for weekly new deaths.

model	Forecasting horizon(H)											
	1				2				3			
	RMSLE	MAPE	R^2	PCC	RMSLE	MAPE	R^2	PCC	RMSLE	MAPE	R^2	PCC
EWT-RNN	0.0369	2.8074	0.9843	0.9942	0.1181	9.4921	0.7677	0.8793	0.1201	9.6938	0.7461	0.8961
EWT-LSTM	0.0318	2.3077	0.9834	0.9937	0.8520	7.2050	0.8600	0.9290	0.1687	16.6175	0.2438	0.8116
EWT-GRU	0.0972	8.9931	0.8041	0.9505	0.0024	6.9241	0.8556	0.9304	0.2060	19.8121	0.1300	0.7680
EWT-TCN	0.0405	3.0167	0.9827	0.9927	0.1030	8.3561	0.8195	0.9210	0.1956	18.1353	0.2555	0.7780
VMD-RNN	0.0551	4.6046	0.9550	0.9803	0.0937	7.4652	0.8587	0.9636	0.1472	11.8541	0.6401	0.9307
VMD-LSTM	0.0704	5.1035	0.9216	0.9678	0.0974	7.9621	0.8416	0.9404	0.1445	12.4962	0.5924	0.8578
VMD-GRU	0.0603	4.8635	0.9440	0.9718	0.1007	8.9429	0.7936	0.9401	0.1428	12.0895	0.6357	0.9263
VMD-TCN	0.0600	5.8687	0.9202	0.9680	0.0957	7.8054	0.8252	0.9099	0.1096	8.5550	0.7425	0.8849
EWT-VMD-RNN	0.0429	3.5390	0.9740	0.9919	0.0903	6.4201	0.8940	0.9734	0.1030	8.8634	0.7317	0.9426
EWT-VMD-GRU	0.0411	3.2377	0.9732	0.9917	0.0670	5.1007	0.9150	0.9694	0.1244	10.5703	0.7381	0.8700
EWT-VMD-TCN	0.0421	3.3048	0.9779	0.9917	0.0812	6.7211	0.8686	0.9384	0.1541	11.3813	0.5427	0.8768
Ours	**0.0420**	**3.0300**	**0.9761**	**0.9894**	**0.0644**	**5.0375**	**0.9147**	**0.9772**	**0.0664**	**5.5018**	**0.9027**	**0.9710**

At the same time, ablation experiments are performed to compare with EWT-ANN (RNN, LSTM, GRU, and TCN, respectively), VMD-ANN, and EWT-VMD-ANN. It can be seen from Table 3 and Table 4 that our proposed model also exhibits superior performance than other models, especially in long-term forecasting. When predicting the number of confirmed cases, the RMSLE of the proposed hybrid model is 11.06% lower than EWT+LSTM, and 32.52% lower than VMD+LSTM of long-term prediction. MAPE, R^2, and PCC are also significantly better than other comparable models. When predicting the number of deaths, the RMSLE of the suggested hybrid model is 60.64% lower than EWT+LSTM, and 54.05% lower than VMD+LSTM of long-term prediction. MAPE, R^2, and PCC are also significantly better than other comparable models. EWT constructs wavelet filters based on component boundary frequencies, while VMD constructs Wiener filters based on component center frequencies. By integrating both VMD and EWT decomposition with LSTM-based models, deep learning models are capable of capturing time-variant properties and significant patterns within historical data. LSTM-based models leverage memory and gating mechanisms to effectively capture both short-term and long-term dependencies within time series data. Additionally, LSTM takes into account past data to generate their output. This structure makes it more suitable for modeling temporal dependencies than traditional feed-forward and temporal convolutional architecture. Furthermore, experiments show that our model can provide satisfactory prediction results even when the amount of data is relatively small.

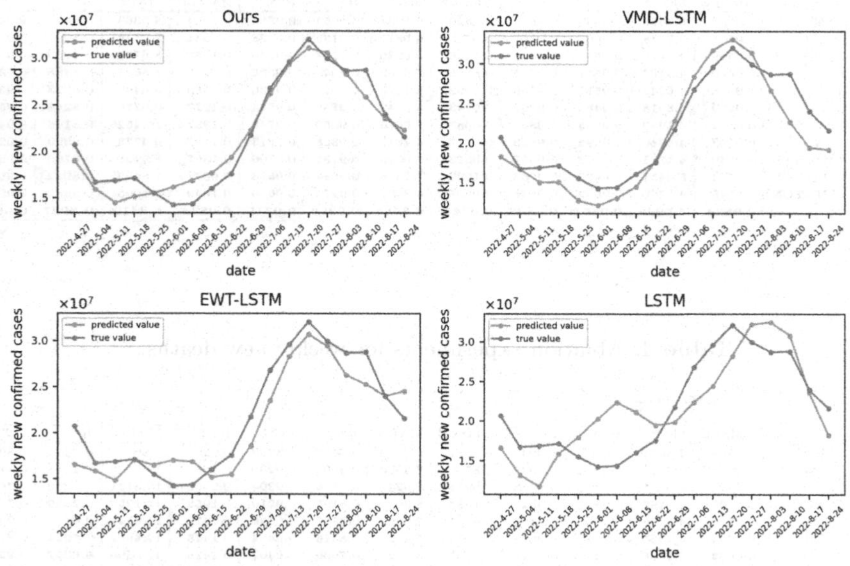

Fig. 6. Prediction results for weekly new confirmed cases in long-term forecasting.

5 Conclusion

In this study, we propose an EWT-VMD-LSTM hybrid model for COVID-19 prediction. By applying EWT and VMD techniques, the complex change characteristics of the original time series can be decomposed into subsequences exhibiting more apparent changes. Additionally, LSTM-based models utilize memory and gating mechanisms to effectively capture dependencies, both short-term and long-term, within time series data. Combining the signal decomposition algorithms and the deep learning model to build a hybrid model can get more accurate predictions. Experiments show that our model using high-efficiency feature representation outperforms other baseline models and ANN models used for comparison in predicting the number of weekly new confirmed cases and deaths.

At present, our method can only be used to predict a single variable, that is, to predict a single time series. However, since multiple variables of the time series are correlated, forecasting multiple target series at the same time is also beneficial to improve forecasting accuracy. In addition, since the decomposition and forecast stages are separate, parameter selection depends on manual selection. In future work, we will further investigate methods for multivariate forecasting and find a better way to implement hybrid models in deep learning frameworks.

Acknowledgement. This work was supported by, the Natural Science Foundation of Guangdong Province, China (2023A1515011618), Key Scientific Research Platform and Project of Guangdong Provincial Education Department (2022ZDZX1040) and Educational Science Planning Topic of Guangdong Province (2022GXJK382), Higher Education Teaching Reform Project of Nanfang College Guangzhou (XJJG2220).

References

1. Bai, S., Kolter, J.Z., Koltun, V.: An empirical evaluation of generic convolutional and recurrent networks for sequence modeling. arXiv preprint arXiv:1803.01271 (2018)
2. Boccaletti, S., Ditto, W., Mindlin, G., Atangana, A.: Modeling and forecasting of epidemic spreading: the case of COVID-19 and beyond. Chaos, Solitons Fractals **135**, 109794 (2020)
3. Box, G.E.: GM Jenkins Time Series Analysis: Forecasting and Control. Holdan-Day, San Francisco (1970)
4. Box, G.E., Jenkins, G.M., Reinsel, G.C., Ljung, G.M.: Time Series Analysis: Forecasting and Control. John Wiley & Sons, New York (2015)
5. Chen, T., Guestrin, C.: XGBoost: a scalable tree boosting system. In: Proceedings of the 22nd ACM SIGKDD International Conference on Knowledge Discovery and Data Mining, pp. 785–794 (2016)
6. Chen, Z., Feng, L., Lay, H.A., Jr., Furati, K., Khaliq, A.: SEIR model with unreported infected population and dynamic parameters for the spread of COVID-19. Math. Comput. Simul. **198**, 31–46 (2022)
7. Cho, K., Van Merriënboer, B., Bahdanau, D., Bengio, Y.: On the properties of neural machine translation: Encoder-decoder approaches. arXiv preprint arXiv:1409.1259 (2014)
8. Dragomiretskiy, K., Zosso, D.: Variational mode decomposition. IEEE Trans. Signal Process. **62**(3), 531–544 (2013)
9. Elman, J.L.: Finding structure in time. Cogn. Sci. **14**(2), 179–211 (1990)
10. Gilles, J.: Empirical wavelet transform. IEEE Trans. Signal Process. **61**(16), 3999–4010 (2013)
11. Hochreiter, S., Schmidhuber, J.: Long short-term memory. Neural Comput. **9**(8), 1735–1780 (1997)
12. Hussain, A.A., Bouachir, O., Al-Turjman, F., Aloqaily, M.: Notice of retraction: AI techniques for COVID-19. IEEE Access **8**, 128776–128795 (2020)
13. Mehta, M., Julaiti, J., Griffin, P., Kumara, S., et al.: Early stage machine learning-based prediction of us county vulnerability to the COVID-19 pandemic: machine learning approach. JMIR Public Health Surveill. **6**(3), e19446 (2020)

14. Monllor, P., Su, Z., Gabrielli, L., Taltavull de La Paz, P.: COVID-19 infection process in Italy and Spain: are data talking? Evidence from arma and vector autoregression models. Front. Public Health. **8**, 550602 (2020)
15. Parbat, D., Chakraborty, M.: A python based support vector regression model for prediction of COVID19 cases in India. Chaos, Solitons, Fractals **138**, 109942 (2020)
16. Pham, Q.V., Nguyen, D.C., Huynh-The, T., Hwang, W.J., Pathirana, P.N.: Artificial intelligence (AI) and big data for coronavirus (COVID-19) pandemic: a survey on the state-of-the-arts. IEEE Access **8**, 130820–130839 (2020)
17. Phba, B., Oz, A., Jpt, A.: A COVID-19 time series forecasting model based on MLP ANN. Proc. Comput. Sci. **181**, 940–947 (2021)
18. Rauf, H.T., et al.: Time series forecasting of COVID-19 transmission in Asia pacific countries using deep neural networks. Pers. Ubiquit. Comput. 1–18 (2021)
19. Rosenblatt, F.: The perceptron, a perceiving and recognizing automaton Project Para. Cornell Aeronautical Laboratory (1957)
20. Yang, Q., Wang, J., Ma, H., Wang, X.: Research on COVID-19 based on Arima modelδ-taking Hubei, China as an example to see the epidemic in Italy. J. Infect. Public Health **13**(10), 1415–1418 (2020)
21. Yang, Z., et al.: Modified SEIR and AI prediction of the epidemics trend of COVID-19 in china under public health interventions. J. Thorac. Dis. **12**(3), 165 (2020)
22. Yuki, K., Fujiogi, M., Koutsogiannaki, S.: COVID-19 pathophysiology: a review. Clin. Immunol. **215**, 108427 (2020)
23. Zhan, C., Zheng, Y., Zhang, H., Wen, Q.: Random-forest-bagging broad learning system with applications for COVID-19 pandemic. IEEE Internet Things J. **8**(21), 15906–15918 (2021)

Stepwise Fusion Transformer for Affective Video Content Analysis

Zeyu Chen(✉), Xiaohong Xiang, Xin Deng, and Qi Wang

Chongqing University of Posts and Telecommunications, Nan'an District, Chongqing 400065, China

{s210231023,S200201065}@stu.cqupt.edu.cn, {xiangxh,dengxin}@cqupt.edu.cn

Abstract. In the field of video content analysis, affective video content analysis is an important part. This paper presents an efficient multimodal multilevel Transformer derivation model based on standard self-attention mechanism and cross-attention mechanism by multilevel stepwise fusion of features. We also used the loss function for the first time to constrain the learning of tokens in the transformer, and achieved good results. The model begins by combining the global and local features of each modality. The model then uses the cross-attention module to combine data from the three modalities and then uses the self-attention module to integrate the data from each modality. In classification and regression experiments, we achieved better results in previous papers. Compared to the state-of-the-art results in recent years [19], we have improved the Valence and Arousal correct rates in the classification dataset by 4.267% and 0.924%, respectively. On the regression dataset, Valence results improved by 0.007 and 0.08 on the MSE and PCC metrics, respectively; Arousal correspondingly improved by 0.117 and 0.057.

Keywords: Affective Video Content Analysis · LIRIS-ACCEDE · Transformer · Multi-level structure

1 Introduction

In recent years, Affective Video Content Analysis has received a lot of attention for its application in various fields. For example, the identification of bloody and violent videos [18], human-computer interaction [24], emotion-based tailored content recommendation [4], personalized video recommendation [25], and so on. Although today's sentiment video content analysis has improved compared to past practice, it is still a big challenge to analyze the exact sentiment of videos by program algorithms.

Affective Video Content Analysis is aimed at analyzing the emotion of a video. The emotional content of a video is defined as the intensity and type of emotion expected to arise when people watch the video. In general, video is divided into audio content and video content. The visual and audio representations are first extracted by features, and then the multimodal information is

H. Zhang et al. (Eds.): NCAA 2023, CCIS 1870, pp. 375–386, 2023.
https://doi.org/10.1007/978-981-99-5847-4_27

merged by a series of alignment and fusion for classification or regression, which is the mainstream approach for sentiment video content analysis nowadays. The biggest problems faced using these fusion techniques are the differences between different modal features and the problems in modal alignment. These mainly include intra-modal fusion, inter-modal fusion and the integration of temporal information. The intra-modal fusion takes into account the relationship between the whole and the local, and the inter-modal fusion mainly involves the alignment of information from different modalities at different moments. To solve this problem, we propose a multi-layer multimodal Transformer-based fusion model. By multi-level fusion and mutual constraints between modalities to improve the efficiency of model alignment of features rather than a single same attention to synthesize the information of all modalities. Compared with the standard Transformer approach of [16], which is based on linear mapping of features and then using a self-attention mechanism, we add global and temporal information and constraint loss between different modalities to the features before doing the attention so that the model learns multimodal information more efficiently, which is more suitable for the characteristics of video tasks. The idea of graded fusion is added to make the fusion of features more delicate.

Transformer has been widely used in various fields, and there are experiments to prove that today's transformer structure surpasses the traditional convolutional network to some extent, but there are only a few papers that use transformer structure in affective video content analysis. Most of the models used for sentiment content analysis based on arousal and valence metrics, such as the paper [16,23], are very traditional and out of touch with the rapid development of transformer today. We present this paper inspired by the multi-loss functions and FPN [11] networks for the image classification task [2] for reasons such as the above. The idea of using multilevel progressive fusion modalities to improve the efficiency of arousal and valence metrics tasks.

The contributions of this paper mainly lie in the following points: 1. An efficient new structure for multimodal multilevel fusion based on a pure transformer architecture is proposed 2. A global local structure is used in the transformer and a constraint method on the transformer structure is proposed to improve the correctness of the model 3. This paper combines the transformer method to a new level for arousal-valence based tasks and achieves promising results in, LIRIS-ACCEDE [1], MediaEval 2015 [15] and MediaEval 2016 [7] datasets.

2 Related Work

2.1 Transformer

Since the Transformer architecture has made a splash in NLP, it has gradually extended to other domains. Due to its competitive modeling capabilities, Transformer has achieved impressive performance improvements over convolutional neural networks (CNNs) in multiple benchmark tests on a variety of tasks.

Transformer [17] is an encoder-decoder architecture in which both the encoder and decoder are superimposed with multi-headed self-attention modules. It is the decoder architecture that has been used most in all fields of applications. The Transformer architecture used in this paper refers to the decoder in Transformer in the article. The encoder consists of multiple identical self-attention modules stacked on top of each other, each with two sub-layers, the first sub-layer is a multi-headed self-attention layer, and the second sub-layer is with a feed-forward neural network and each sub-layer uses residual linking. The encoder structure first represents the input sequence with word embedding and adds location information. After that, the input sequence of the current encoder layer is put into a multi-headed self-attention layer to generate new vectors. Specifically, in computing the encoder's self-attention, the queries, keys and values are derived from the output of the previous encoder layer. Then the output of the multi-headed self-attention layer is connected to the input of the current encoder layer as a residual and the result is layer normalized. Finally, the result after layer normalization is put into a fully connected layer (feed forward) layer. The role of this layer is to transform all the location representations output from the self-attention layer, so it is called a location-based feed forward neural network. It is worth mentioning that the above steps are executed N times in a loop.

2.2 Affective Video Content Analysis

Affective content analysis aims to recognize automatically the emotions triggered by a video. Psychologists use two primary methods to measure affect: the discrete method and the dimensional method [20]. The discrete approach classifies emotions into different categories. According to Ekman [9], six different categories, namely happiness, sadness, surprise, disgust, anger, and fear, were used to classify emotions. The dimensional approach was first proposed by Russell [14] of affect]. It divides emotions into a 2D continuous space: arousal and potency. Some works discretize dimensional descriptions into different categories, such as positive and negative valence and high or low arousal.

In recent years, many researchers have worked on developing efficient and concise multimodal structures to recognize video emotions. For example, Yi et al. [21] used a dual-stream network (TSN) to extract temporal features and physical signs of objects, manually extracted MKT features to depict motion features, and used selected EmoBase10 features to depict audio features. The strategy of early fusion used, by SVM and SVR to derive the results. Chen et al. [6] explored the dependence between visual elements and emotion using picture brightness, picture color, and clip tempo, and used low-level elements to improve the predicted phase rate of emotional videos.

3 Proposed Method

The overview of our proposed method is shown in Figure 1. First we extract the feature vectors of three modalities from the given video, one audio modality and

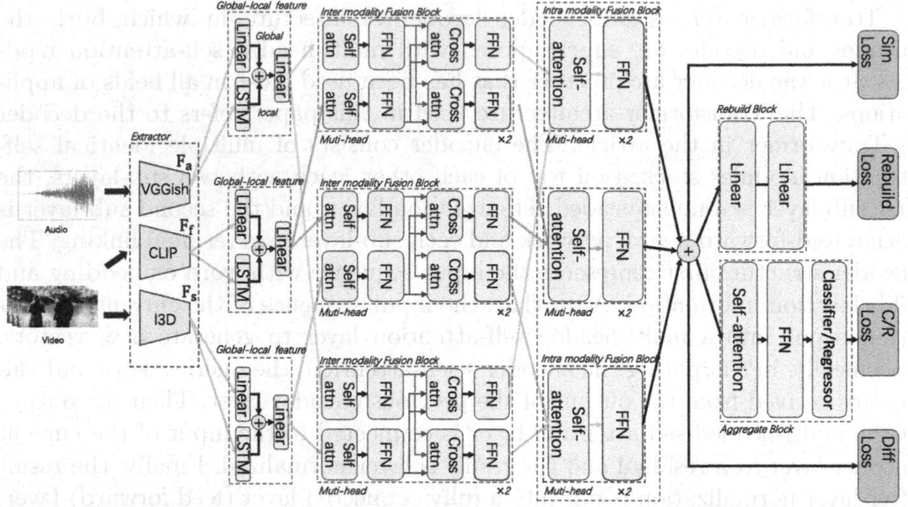

Fig. 1. Overview of the model. The activation function is omitted in the figure.

two visual modalities, corresponding to the extraction module in Fig. 1. We then separate the corresponding global and local features from the obtained feature vectors of each modality separately and stitch them together, which corresponds to the global-local feature module in Figure 1. Then we send them two-by-two to the Inter modality Fusion Block for cross-modality information interaction. Finally, we pass the obtained features through the branches with different loss functions to obtain the final experimental results.

3.1 Features

Generally speaking, the two most used modalities in video sentiment analysis are audio and visual. We further decompose the visual modality into frame modality and segment modality. The frame modality captures information about the objects present in a single frame of the video, and the segment modality captures information about the motion of the objects over a period of time. The audio modality is to capture the emotional ups and downs in the audio. Based on our results, we found that the fusion of the three modalities based on ViT's [8] Clip model [13], I3D [3] model and VGGish [10] model achieved the best results for the experiment. The specific extraction process will be described in Chap. 4, Experimental Configuration.

3.2 Model

We use the corresponding feature extractor to do the processing of the three modalities and get three feature vectors with different shapes. We define the extracted modal feature set as F_i and the shape of the extracted features as

$(n \times c)$, $i \in (a, f, s)$,$F_i \in \mathbb{R}^{n \times c}$. Where n denotes the number of a certain feature, c denotes the size of the corresponding modal extraction, and (a, f, s) denotes three feature frames, segments (multiple frames) and audio, respectively. Next, we will present each of these modules in the figure. For the process of specific attention, we follow the process in the paper [17].

Then we further integrate the modal information in the respective modalities in the Intra modality Fusion Block module. We also use Sim, Diff and rebuild loss as constraints to improve the efficiency of the model.

Global-Local Feature Block is used to fuse global-local features. After extracting the video or audio features, we can get the feature vector of shape $(b \times n \times d)$, where b is the batch size, n is the number of video or audio clips, and d is the dimension of each video or audio clip after extraction. The dimensionality of d varies for different modalities. For local features, we send each segment vector accordingly to the LSTM and map it to a dimensional size of D. The final layer of the result is taken as a local feature of shape $(b \times 1 \times D)$, and the role of the local vector is to describe how the feature changes throughout the video. For global features, we map d dimensions to D dimensions and average over n dimensions to obtain global features with the same shape as the local features, describing the feature tone of the whole video. The final global local vectors all have the shape $(b \times 1 \times D)$. Where H_i^G is the global feature of mode i and H_i^L represents the local feature of mode i. Finally we splice the two features together to get E, and the formula is expressed as follows.

$$E_i(F_i) = Linear([H_i^G; H_i^L]) \tag{1}$$

Inter Modality Fusion Block is used to exchange information between modalities. As shown in the Fig. 1, we feed the obtained combination of global local vectors of each modality into the inter-modality fusion block two by two to transfer the inter-modality information two by two. The inter-modal fusion block uses the standard self-attention structure for initial integration of global and local features, and the cross-attention structure for cross-modal information transfer.

Intra Modality Fusion Block is used to integrate the sub-modalities of the results obtained from the previous processing. Since the feature vectors of one modality must interact with the feature vectors of the other two modalities, the number of features will be doubled after the Inter modality Fusion Block. In this module, we simply add the feature vectors of the corresponding modality to fuse the information of different Tokens in the same modality to reduce the computational effort, and finally integrate them in their own branch of the self-attention mechanism. We have combined the formulas for the Inter Modality Fusion Block and the Intra Modality Fusion Block and listed them together below.

$$m_{i,(i,j)}, m_{j,(i,j)} = CA_{i,j}(SA_i(E_i(X_i)), SA_j(E_j(X_j))),\\ i,j \in (a,f,s),\\ i \neq j \tag{2}$$

where $CA_{i,j}$ and SA_i are Cross-Attention and Self-Attention operations. i, j represent the incoming vectors from different branches. It is worth mentioning that we feed the same vector into different branches and finally add up the corresponding vectors from these two different branches to get the final module output. The expression is described as:

$$M_i = MSA(\sum_i m_{i,(*,*)}) \tag{3}$$

where MSA stands for Intra modality Fusion Block module, which uses a self-attention structure.

Aggregate Block is used to extract the final multimodal features, which are composed of self-attention. After the previous operations, we extract the final fused features by stitching all the obtained vectors together as the input of this module. This completes the hierarchical feature fusion.

Rebuild Block serves to reconstruct the loss. The feature vector of each modality is mapped back to the shape of the original modality in order to preserve all the feature details of the original feature and prevent the model from learning irrelevant trivial knowledge. The mathematical operations are as follows.

$$f_i = Linear(act(Linear(M_i^G + M_i^L))) \tag{4}$$

where Linear() is the linear layer and act() is the activation function.

Classifier/Regressor consists of linear layers that are used to obtain the final classification or regression prediction results. The expressions are as follows.

$$R = Head(SA(M_a, M_f, M_s)) \tag{5}$$

where $Head()$ is a linear layer of two dimensions and R denotes the final result.

3.3 Loss Functions

For the loss functions we use Difference Loss, Similarity Loss loss and Rebuild Loss.

Similarity Loss is a constraint on the result of the Intra Modality Fusion Block. Since the Transformer does not change the shape of the input feature vector, F_i and M_i have the same shape and correspond to each other. We compute the two-by-two similarity of all global features in M_i in order to get more similar global features to learn the common points in different modalities and learn the modality invariant representation. The specific formula is shown as follows.

$$\mathcal{L}_{sim} = \frac{s^2(M_i^G - M_j^G)}{n^2} \tag{6}$$

where n represents the total number of elements and $s()$ represents the summation function of all elements. M_i^G represents the global eigenvector corresponding to mode i.

Difference Loss. If we just use the above loss function to capture the common features of the modalities, we will lose the specific features within each modality. We use Difference Loss to avoid this situation by ensuring that the local feature vectors within individual branches learn different information from the global feature vectors. The specific formula is shown as follows.

$$\mathcal{L}_{diff} = \sum_{i \in \{a,f,s\}} ||(M_i^G)^{\mathrm{T}} M_i^L||_F^2 \tag{7}$$

where $|| * ||_F^2$ is the squared Frobenius norm.

Rebuild Loss is used to retain the original details in the extracted features of the extracted model, so as to avoid learning useless features after applying Similarity and Difference Loss, instead of just following the unrepresentative vectors of loss learning. The formula is as follows.

$$\mathcal{L}_{re} = (\frac{\sum^{(a,f,s)_i} ||F_i - f_i||^2}{3}) \tag{8}$$

Task Loss is the loss function of the task itself. For the classification task it is the standard cross-entropy loss function and for the regression task it is the squared loss function. The specific formula is as follows

$$\begin{aligned} \mathcal{L}_{classification} &= -\frac{1}{N}\sum_{i=0}^{N} y_i \cdot log\hat{y}_i \\ \mathcal{L}_{regression} &= \frac{1}{N}\sum_{i=0}^{N} ||y_i - \hat{y}_i||_2^2 \end{aligned} \tag{9}$$

where y_i is the label of sample i and $\hat{y}_i$ is the prediction of sample i.

Finally the total loss function of the model in total is:

$$\mathscr{L}_{Loss} = a\mathscr{L}_{sim} + b\mathscr{L}_{diff} + c\mathscr{L}_{re} + d\mathscr{L}_{task} \tag{10}$$

a,b,c,d are scaling factors. And $a + b + c + d = 1$.

4 Experiments

4.1 Datasets

As our main dataset, we make use of LIRIS-ACCEDE. 9,800 clips from 160 feature-length and short films make up LIRIS-ACCEDE. It is the largest video dataset on the market, annotated with induced emotion tags by a wide and representative community. The 9800 clips from the LIRIS-ACCEDE dataset and an additional 1100 clips make up the categorization dataset known as mediaEval 2015. A regression dataset called mediaEval 2016 is made up of 9800 clips from the LIRIS-ACCEDE dataset and an additional 1200 clips that forecast awakening and value.

4.2 Implementation Details

All experiments were performed on an NVIDIA RTX A5000 GPU using the pytorch framework. The learning rate of the model is set to 0.000045. Batchsize is set to 256. Task loss, reconstruction loss, similarity loss, different loss scaling factor is 0.7 0.1 0.1 0.1. inter modality Fusion Block iterates once, Intra modality Fusion Block iterates twice, Aggregate Block iterates twice. The number of Transformer heads is set to 64 and the length of the hidden dimension is set to 1024. Relu is used as the loss function in the Rebuild Block layer and silu is used as the loss function in all the remaining linear layers in the FFN module.

For frame-level features, we use a CLIP pre-training model based on ViT training to extract them. An average of 18 frames are extracted from the video and fed to the extractor to obtain a sequence feature with a hidden dimension of 512 and a number of 18. For segment features we use the I3D model with the original frame rate of the video and a 2048-dimensional feature vector for extraction. For audio features, we use the VGG model trained on audioset. Each segmented one-dimensional audio signal is converted into a two-dimensional spectrum, from which 128-dimensional embedded feature vectors are extracted (Table 1).

4.3 Experimental Results

As we can see from the results in Table I, our method is ahead of recent methods for sentiment video content analysis in all metrics. Our top results show that the proposed method is able to facilitate the understanding of the data distribution. Collectively, our method achieves better results in both classification and regression tasks. We have improved the Valence and Arousal correct rates in the classification dataset by 3.55% and 0.713%, respectively. On the regression dataset, Valence results improved by 0.015 and 0.082 on the MSE and PCC metrics, respectively; Arousal correspondingly improved by 0.015 and 0.047.

Table 1. Comparison of different methods.

	AIMT15		EIMT16			
	Valence(ACC %)	Arousal(ACC %)	Valence		arousal	
			MSE	PCC	MSE	PCC
ours	**52.817**	**59.294**	**0.167**	**0.590**	**0.405**	**0.581**
RUC [5]	-	-	0.201	0.419	1.479	0.467
Yi et al. [21]	46.22	57.40	0.198	0.399	1.173	0.446
Ou et al. [12]	46.57	57.51	0.194	0.445	1.077	0.491
Yi et al. [22]	48.61	58.22	0.193	0.468	0.542	0.522
Thao et al. [16]	-	-	0.185	0.467	0.742	0.503
Yi et al. [23]	-	-	0.176	-	0.441	-
Wang et al. [19]	48.55	58.37	0.174	0.510	0.522	0.524

4.4 Ablation Study

Table 2. The ablation study of Local/Global features.

	AIMT15		EIMT16			
	Valence(ACC %)	Arousal(ACC %)	Valence		arousal	
			MSE	PCC	MSE	PCC
baseline	52.817	59.294	0.167	0.590	0.405	0.581
(-)local feature	50.694	59.062	0.166	0.593	0.468	0.561
(-)global feature	50.147	59.104	0.164	0.577	0.542	0.511

Table 3. The ablation study of Modality.

	AIMT15		EIMT16			
	Valence(ACC %)	Arousal(ACC %)	Valence		arousal	
			MSE	PCC	MSE	PCC
baseline	52.817	59.294	0.167	0.590	0.405	0.581
(-)Vggilsh	50.568	58.368	0.167	0.558	0.448	0.511
(-)I3D	51.808	59.209	0.161	0.587	0.507	0.580
(-)Clip	47.246	57.675	0.187	0.456	0.621	0.485

Table 4. The ablation study of Blocks.

	AIMT15		EIMT16			
	Valence(ACC %)	Arousal(ACC %)	Valence		arousal	
			MSE	PCC	MSE	PCC
baseline	52.817	59.294	0.167	0.590	0.405	0.581
(-)Inter Modality Block	52.082	59.041	0.164	0.596	0.465	0.582
(-)Intra Modality Block	51.745	59.272	0.166	0.586	0.556	0.563

Local and Global Ablation Study. The results of the experiment are shown in Table 2. In this experiment we did ablation experiments on the global local variables in the model. In order to keep the overall structure of the model unchanged, we replace the global or local feature vectors with all-0 vectors to overwrite the information in the original feature vectors. In the table, (-) means to remove the information of certain feature vectors, for example, "(-) Local" means to cover all local vectors. Comparing the experimental results in Table 2, we find that the simultaneous use of the two modalities has an enhanced effect on the performance of the model. Removing either feature causes a decrease in the metrics. The two have a complementary effect on each other.

Modality Ablation Study. The experimental results as shown in Table 3. The leftmost column lists the modalities used in the model. Specifically, the main logic of the model with the bimodal feature is the same as that of the model with the trimodal feature, with the difference being the removal of the branched pair of removed modes.

We find that the best results in the EIMT16 dataset are achieved only when all three modalities are used together, and the results are balanced in the AIMT15 dataset and higher than in recent papers. The modal effects on the model are, in descending order, *clip>vgg>i3d*. The results show that although the i3d modality has a very slight negative effect on the EIMT15 dataset, it is in a completely acceptable range compared to the model improvement effect of this modality. All three modalities have a boosting effect on the sentiment video analysis task.

Blocks Ablation Study. The results of the experiment are shown in Figure 4. In this experiment we did ablation experiments for each module in the model. The ablation experiments were performed by directly removing the modules we wanted to perform the ablation experiments on. We found that the absence of either module has a negative impact on the model results.

Loss Ablation Study. The last one is the loss function ablation experiment, and the experimental results are shown in Fig. 5. We simply remove all additional loss functions in the experiment to make a comparison. In the experiments, we found that the loss function has different effects on two metrics of the AIMT15 dataset. There is a slight effect on the Arousal accuracy but a significant increase on Valenc accuracy. For the EIMT16 dataset, removing the loss function has a large impact on the larger loss function, and the MSE results under this dataset have a significant decrease.

Table 5. Loss ablation study.

	AIMT15		EIMT16			
	Valence(ACC %)	Arousal(ACC %)	Valence		arousal	
			MSE	PCC	MSE	PCC
baseline	52.817	59.294	0.167	0.590	0.405	0.581
(-)Exrta loss function	51.493	59.651	0.167	0.568	0.505	0.573

5 Conclusion

In this paper we propose a Transformer structure with multi-level fusion of features and compatible with the popular multi-loss constraint approach. The model progressively fuses features from small to large scales in different intervals, which improves the fusion efficiency and increases the robustness of the model. Finally the mode-invariant properties and mode-specific peculiarities in the four loss-function constrained model are used to achieve better prediction results at once. In conclusion, we emphasize the importance of asymptotic fusion and multimodal constraints, and demonstrate the effectiveness of the model experimentally. In the future we will try to further improve it by combining more fusion schemes.

Acknowledgement. This work was supported in part by the Natural Science Foundation of Chongqing under Grant cstc2020jcyj-msxmX0284; in part by the Scientific and Technological Research Program of Chongqing Municipal Education Commission under Grant KJQN202000625.

References

1. Baveye, Y., Dellandrea, E., Chamaret, C., Chen, L.: LIRIS-accede: a video database for affective content analysis. IEEE Trans. Affect. Comput. **6**(1), 43–55 (2015)
2. Bousmalis, K., Trigeorgis, G., Silberman, N., Krishnan, D., Erhan, D.: Domain separation networks. In: Advances in Neural Information Processing Systems, vol. 29 (2016)
3. Carreira, J., Zisserman, A.: Quo Vadis, action recognition? a new model and the kinetics dataset. In: proceedings of the IEEE Conference on Computer Vision and Pattern Recognition, pp. 6299–6308 (2017)
4. Chan, C.H., Jones, G.J.: Affect-based indexing and retrieval of films. In: Proceedings of the 13th Annual ACM International Conference on Multimedia, pp. 427–430 (2005)

5. Chen, S., Jin, Q.: RUC at mediaeval 2016 emotional impact of movies task: fusion of multimodal features. In: MediaEval, vol. 1739 (2016)
6. Chen, T., Wang, Y., Wang, S., Chen, S.: Exploring domain knowledge for affective video content analyses. In: Proceedings of the 25th ACM International Conference on Multimedia, pp. 769–776 (2017)
7. Dellandréa, E., Chen, L., Baveye, Y., Sjöberg, M.V., Chamaret, C.: The mediaeval 2016 emotional impact of movies task. In: CEUR Workshop Proceedings (2016)
8. Dosovitskiy, A., et al.: An image is worth 16×16 words: transformers for image recognition at scale. arXiv preprint arXiv:2010.11929 (2020)
9. Ekman, P.: Basic emotions. Handbook of Cognition And Emotion **98**(45–60), 16 (1999)
10. Hershey, S., et al.: CNN architectures for large-scale audio classification. In: 2017 IEEE International Conference on Acoustics, Speech and Signal Processing (ICASSP), pp. 131–135. IEEE (2017)
11. Lin, T.Y., Dollár, P., Girshick, R., He, K., Hariharan, B., Belongie, S.: Feature pyramid networks for object detection. In: Proceedings of the IEEE Conference on Computer Vision and Pattern Recognition, pp. 2117–2125 (2017)
12. Ou, Y., Chen, Z., Wu, F.: Multimodal local-global attention network for affective video content analysis. IEEE Trans. Circuits Syst. Video Technol. **31**(5), 1901–1914 (2020)
13. Radford, A., et al.: Learning transferable visual models from natural language supervision. In: International Conference on Machine Learning, pp. 8748–8763. PMLR (2021)
14. Russell, J.A.: A circumplex model of affect. J. Pers. Soc. Psychol. **39**(6), 1161 (1980)
15. Sjöberg, M., et al.: The mediaeval 2015 affective impact of movies task. In: MediaEval, vol. 1436 (2015)
16. Thao, H.T.P., Balamurali, B., Roig, G., Herremans, D.: Attendaffectnet-emotion prediction of movie viewers using multimodal fusion with self-attention. Sensors **21**(24), 8356 (2021)
17. Vaswani, A., et al.: Attention is all you need. In: Advances in Neural Information Processing Systems, vol. 30 (2017)
18. Wang, J., Li, B., Hu, W., Wu, O.: Horror video scene recognition via multiple-instance learning. In: 2011 IEEE International Conference on Acoustics, Speech and Signal Processing (ICASSP), pp. 1325–1328. IEEE (2011)
19. Wang, Q., Xiang, X., Zhao, J., Deng, X.: P2SL: private-shared subspaces learning for affective video content analysis. In: 2022 IEEE International Conference on Multimedia and Expo (ICME), pp. 1–6. IEEE (2022)
20. Wang, S., Ji, Q.: Video affective content analysis: a survey of state-of-the-art methods. IEEE Trans. Affect. Comput. **6**(4), 410–430 (2015)
21. Yi, Y., Wang, H.: Multi-modal learning for affective content analysis in movies. Multimed. Tools App. **78**(10), 13331–13350 (2019)
22. Yi, Y., Wang, H., Li, Q.: Affective video content analysis with adaptive fusion recurrent network. IEEE Trans. Multimed. **22**(9), 2454–2466 (2019)
23. Yi, Y., Wang, H., Tang, P.: Unified multi-stage fusion network for affective video content analysis. SSRN 4080629
24. Zeng, Z., Tu, J., Liu, M., Huang, T.S., Pianfetti, B., Roth, D., Levinson, S.: Audio-visual affect recognition. IEEE Trans. Multimedia **9**(2), 424–428 (2007)
25. Zhao, S., Yao, H., Sun, X., Xu, P., Liu, X., Ji, R.: Video indexing and recommendation based on affective analysis of viewers. In: Proceedings of the 19th ACM International Conference on Multimedia, pp. 1473–1476 (2011)

An Adaptive Clustering Approach for Efficient Data Dissemination in IoV

Weiyang Chen[1,2], Yuhao Liu[1], Yang Lu[1], Weizhen Han[1(✉)], and Bingyi Liu[1,2]

[1] School of Computer Science and Artificial Intelligence, Wuhan University of Technology, Wuhan 430070, China
{weiyangchen,yhliu,ly_work,hanweizhen,byliu}@whut.edu.cn

[2] Chongqing Research Institute, Wuhan University of Technology, Chongqing 401135, China

Abstract. Due to the inherent characteristics of Vehicular Ad-hoc Networks (VANETs), such as uneven distribution and high mobility, establishing and maintaining efficient routes for data dissemination is a significant and challenging issue. To enhance communication efficiency, many cluster-based protocols have been designed to reduce data redundancy and the number of control messages by integrating vehicles into manageable groups headed by a superior vehicle, known as the cluster head (CH). Nevertheless, most existing protocols are unable to adaptively adjust the cluster, resulting in a significant network burden and message transmission delay. To address this issue, we propose a cluster-based routing method empowered by Vehicle Fog Computing(VFC), which takes advantage of the clustering architecture to reduce the overhead of routing discovery and maintenance. Specifically, based on data transmission requirements and vehicle environment status, the proposed method can adaptively adjust the cluster structure and the number of CHs to reduce data redundancy and transmission load in concurrent scenarios of massive data transmission. Moreover, cooperating with the adaptive clustering method, a routing method is proposed to improve the efficiency of data transmission. Lastly, we conducted extensive experiments to verify the cluster-based routing scheme based on VFC. Our experimental results demonstrate that the proposed routing protocol is feasible and performs well compared to existing methods.

Keywords: Routing protocol · Vehicular Ad-hoc Networks · Reinforcement learning · Clustering

1 Introduction

Vehicular Ad-hoc Network (VANET) is a crucial part of Intelligent Transport Systems (ITS), which provide flexible and fast data transmission for potential risk warnings and safe vehicle movement. However, the high mobility of vehicles and the randomness of driving trajectories make the topology of VANETs

H. Zhang et al. (Eds.): NCAA 2023, CCIS 1870, pp. 387–402, 2023.
https://doi.org/10.1007/978-981-99-5847-4_28

unstable and volatile, resulting in broadcast storms, especially when vehicle density is high [1,2]. Moreover, the variability of road driving conditions brings about an uneven distribution of network nodes, which poses a severe challenge for establishing reliable routing in VANETs [3].

Edge computing, which extends computing capabilities to the network's edge, has emerged as a solution to address the increasing need for processing and storing data closer to the source in the Internet of Things (IoT) context [4–6]. For example, [7] proposed an optimization method for heterogeneous task replication in Mobile Edge Computing (MEC) networks, and an MEC-based architecture for adaptive bitrate(ABR)-based multimedia streaming in VANET was proposed by Penglin Dai [8]. MEC leverages the proximity and mobility of mobile devices for efficient data processing and storage and has the potential to support applications such as real-time traffic management and efficient message delivery in vehicular networks [9,10].

The routing algorithm aims to minimize algorithm overhead, increase data packet delivery rate, and reduce packet transmission delay while satisfying all requirements. Cluster-based routing protocols are an effective approach to address the challenges of the dynamic nature of VANETs, which divide the network into manageable subparts by grouping vehicles into clusters to reduce communication overhead [11]. CHs are management nodes that facilitate network topology management and improve scalability, simplifying the handling of topology changes caused by vehicle nodes' addition and relocation [12]. The clustering structure can dynamically adapt to meet specific requirements such as communication load balance and service quality, improving routing efficiency, and enhancing data transmission performance.

Most existing cluster-based routing methods lack the ability to adjust the cluster adaptively, resulting in a significant network burden and delay of messages. Research on VANET network routing algorithms has traditionally focused on common vehicles. However, their unpredictable trajectories and variable speeds in urban environments cause their network structure to change frequently, leading to a significant increase in VANET routing overhead. On the other hand, buses follow fixed routes and maintain a relatively constant speed, making them a more stable component of VANETs. In this context, the bus can act as the cluster coordinator in the proposed cluster-based routing method, enabling efficient management of cluster members and simplifying network topology control.

The main contributions of this paper can be summarized as follows:

1. We design an edge-intelligent-enabled routing protocol based on a three-layer architecture, which includes an end-user layer, an edge layer, and a cloud computing layer.
2. We propose an adaptive clustering algorithm where buses act as CHs to adjust the cluster size in real time based on the connectivity and stability of the vehicles.

3. We propose a cluster-based routing protocol named Bp-ADOV, which utilizes reinforcement learning (RL) and considers buses as fog nodes. Bp-ADOV constructs routing paths by connecting multi-hop links between adjacent CHs.
4. The experiment shows that the proposed approach can significantly enhance packet delivery rates while reducing transmission load and delay. It highlights the potential of edge intelligence in improving communication performance in urban traffic scenarios.

2 Related Work

2.1 Routing Algorithms in VANET

Various routing algorithms have been proposed for VANETs, including geographic, opportunistic, topology-based, RL-based, and VFC-based routing protocols. The Ad hoc On-Demand Distance Vector (AODV) protocol, a reactive topology-based routing protocol, was first proposed in 1999 and is widely used in the Internet of Vehicles [13]. To improve AODV's performance in VANETs, UrbanAODV (U-AODV) [14] was introduced, which includes a new route discovery and selection phase. For networks with up to 60 nodes, U-AODV has demonstrated lower overhead than AODV, although analysis for networks with more nodes is not included. AODV-MEC, proposed in [15], is an MEC-based routing protocol that utilizes a new clustering approach combined with Q-learning to select better intermediate nodes. The experimental results indicate that AODV-MEC is capable of reducing the average End-to-End (E2E) delay and topological network overhead in highway VANETs with up to 220 nodes.

The classical Greedy Perimeter Stateless Routing (GPSR) is a geolocation-based routing protocol that uses the location information of neighboring nodes to select the next-hop relay node. However, GPSR has a locally optimal solution problem and is not reconfigurable [16]. The authors of [17] proposed an innovative clustering control scheme based on fuzzy logic to achieve scalability, improve network topology stability, incentivize spectrum sharing by owners, and enable efficient and cost-effective use of spectrum.

2.2 Cluster Algorithms in VANET

Due to vehicular networks' high mobility and dynamic topology, most algorithms suffer from low packet delivery ratios and high E2E delays. To address these challenges, a clustering architecture has been designed to improve data dissemination efficiency and reduce the overhead of route discovery and maintenance. Kayis et al. [18] proposed one of the earliest vehicle clustering algorithms, the passive clustering algorithm, where clustering is only based on the predefined speed interval. However, more than velocity intervals are required to capture similarities in mobility. In [19], researchers proposed a multipath routing protocol that utilizes the probabilities of the street and path consistency.

Ant colony optimization (ACO)-based clustering introduces a new relay-bus selection mechanism to enhance packet forwarding. The relay bus is a critical component of the proposed protocol, as it is responsible for efficiently transmitting packets to the next forwarding relay. In [20], Tseng et al. provided the stable clustering algorithm CATRB. It applies bus traffic rules and considers how mobile vehicles are in terms of their position, speed, and direction. An equilateral triangle's centroid, combustion center, and circumference center are simultaneously located. CATRB uses this concept to choose the best CH for VANET. He et al. [21] proposed a two-level routing protocol that leverages vehicle attribute information. To maintain the stability of the cluster structure, the protocol selects CHs based on vehicle type, total distance, vehicle speed change, and number of neighbors in the first stage.

Cluster-based routing protocols are distributed algorithms where each node only requires information about its cluster rather than the entire network topology. This approach reduces the burden on the network's central node, enhancing scalability and reliability. However, there are still several issues with the existing protocols:

- Most current research focuses on establishing stable and efficient communication paths for the given VANET clustering structure, focusing less on the specific clustering and cluster maintenance processes.
- Existing clustering management methods typically assume a fixed cluster size and lack adaptive transmission range adjustment based on data transmission requirements.
- Traditional cluster-based routing protocols dynamically adjust routing paths based on vehicle mobility and routing quality metrics without considering the stability of the network structure.
- While cluster-based routing protocols increasingly incorporate edge intelligence, most still rely on expensive infrastructure such as base stations and roadside units (RSUs).

3 Clustering Management Scheme

In this section, we first propose the system model. Next, we describe a clustering construction scheme based on data transmission requirements and vehicle environment, including clustering construction, aggregation, and decomposition.

3.1 System Model

We present a three-tier cloud-edge-end architecture based on edge intelligence to model packet transmission, as depicted in Fig. 1. The layers of the system model are as follows:

- End-User Layer: In this layer, vehicles on the road periodically report their surrounding conditions to the bus, which serves two purposes. Firstly, these parameters are used in the training process of the RL framework. Secondly,

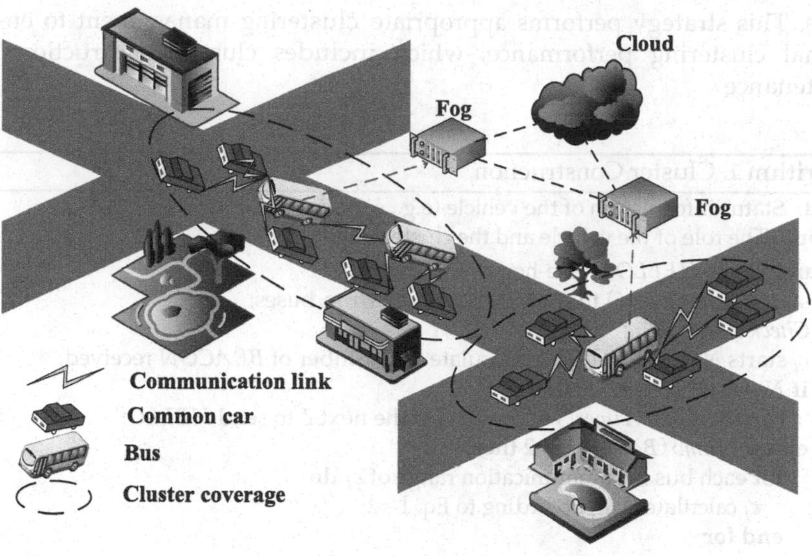

Fig. 1. System model based on *VFC*.

vehicles exchange packets for cluster construction, merging, and decomposition, which helps establish stable routing transmissions in a variable traffic environment.

- Edge Layer: The edge layer deploys edge servers with computing resources at the network's edge to meet specific task requirements, reduce the computing pressure on cloud servers, and minimize backhaul consumption time and communication latency. Buses can act as virtual fog nodes without requiring additional deployment since they better sense their surroundings. They can gather data about the perceived condition of the road and nearby vehicles and send it to the cloud layer via the uplink.
- Cloud Layer: The top layer in the VFC model has powerful computing capacity and abundant data resources. It can gather data from the fog nodes and conduct data analysis. In the packet routing scenario, all the local traffic information uploaded by the bus is aggregated into a global traffic environment in the cloud for RL training.

To model message dissemination, we use a simplified representation of the traffic scenario with a set of road segments, $L = \{l_1, l_2, \cdots, l_n\}$, and a set of vehicles $C = \{c_1, c_2, \cdots, c_m\}$. The communication links between the vehicles are modeled by an undirected graph $G = (C, E)$, where E is the set of edges representing to single-hop links.

3.2 Clustering Management Strategy

In this section, we propose a comprehensive cluster-based collaborative operations strategy that monitors metrics such as network connectivity and vehicle status. This strategy performs appropriate clustering management to ensure optimal clustering performance, which includes cluster construction and maintenance.

Algorithm 1. Cluster Construction

Input: Status information of the vehicle (e.g., speed and direction).
Output: The role of the vehicle and the cluster it belongs to.

Car c_i sends *HELLO* to one-hope vehicle;
c_i starts *CheckTimer*() to wait for *BEACON* from buses;
if *CheckTimer*() $\leq \xi$ **then**
  c_i starts *NumofBeacon*() to calculate the number of *BEACON* received;
  if *NumofBeacon*() = 0 **then**
    c_i resets *CheckTimer*() and waits for the next ξ to send *HELLO*;
  else if *NumofBeacon*() $\leq$ 2 **then**
    for each bus in communication range of c_i **do**
      c_i calculates CR_i according to Eq. 1 - 2;
    end for
    c_i sends *JOIN* to the bus with the highest CR_i;
  end if
  else
    c_i sends *JOIN* to the bus;
end if

Cluster Construction. Cluster construction is initiated when a vehicle joins the VANETs based on mobility and stability metrics similarity, as shown in Algorithm 1. Communication reliability (CR) is defined as the probability of a direct communication link between vehicles c_i and b_i being available for a given time t, refer to [22]. Communication reliability assumes that vehicle speeds follow the normal distribution $N(\mu, \sigma^2)$ and predicts the duration of link availability by considering the vehicle's transmission range and geographical location. Equation 1 is the probability density function of the link duration T between c_i and b_i :

$$f(CR) = \frac{R_{bc}}{\sigma_{\Delta v_{bc}} \cdot \sqrt{2\pi T^2}} \exp\left[-\frac{\left(\frac{R_{bc}}{T} - X_{\Delta v_{bc}}\right)^2}{\left(2\sigma_{\Delta v_{bc}}\right)^2}\right] \tag{1}$$

where R_{bc} represents the relative transmission range, which varies based on the direction and speed of the vehicle. The expected mean of velocity v_{bc} is denoted by $X_{\Delta v_{bc}}$, $\sigma_{\Delta v_{bc}}$ is the deviation in vehicle b and c velocities.

The probability that communication will be available for $\triangle T$ duration is computed as follows, which means the reliability of communication between b and c at time t.

$$CR_i = \int_{t+\Delta T}^{t} f(CR)dt \tag{2}$$

Algorithm 2. Cluster Maintenance

Input: Status information and role of the vehicle.
Output: The role of the vehicle and the cluster it belongs to.

While Bus b_i receives packet on two consecutive ξ periods **do**
  b_i determines the type of packet and the status of the sender;
  if the type is *BEACON* **and** the distance between buses$(b_i,b_j) \leq 150m$ **then**
    if the cluster sizes of $b_i \leq$ the cluster sizes of b_j **then**
      CMs of b_i become CMs of b_j;
      b_i becomes the CM of b_j;
    end if
  end if
end while
While Car c_i approach an intersection **do**
  if the direction of c_i is different from CH **then**
    c_i leaves the cluster;
  else
    c_i calculates *DD* according to Eq.3;
    if $DD > 0.9$ **then**
      c_i leaves the cluster;
    end if
  end if
end while

where ΔT refers to the estimated availability time, which be calculated by $\Delta T = \frac{R_{bc}}{\Delta v_{bc}}$.

Cluster Maintenance. Traffic dynamics may cause an increase in cluster overlap, leading to decreased managerial effectiveness and increased transmission overhead in VANET. When the distance between two buses is under fifty percent of the communication distance, cluster aggregation is performed. The bus with fewer members relinquishes its cluster-head status, reducing unnecessary communication overhead due to overlap.

The cluster maintenance algorithm is described in Algorithm 2. To mitigate the impact of unstable connections caused by vehicles moving out of communication range within the same cluster, we propose a decomposition degree (*DD*) metric in Eq. 3:

$$DD = \frac{1 - e^{-|\Delta d_{ij}|}}{1 + e^{-|\Delta v_{ij}|}} \tag{3}$$

where Δd_{ij} and Δv_{ij} represent the relative distance and speed between the cluster member (CM) and the CH, respectively. The *DD* metric determines when a CM should leave its current cluster and join another one to improve network connectivity.

4 Bp-ADOV Routing Algorithm

In existing Q-learning-based algorithms, the selection of the optimal next-hop node is typically based solely on routing performance metrics, without considering critical factors such as network structure stability and connectivity. In this section, we propose an adaptive clustering routing algorithm called Bp-AODV to addresses this problem.

4.1 Reference Factors for Routing Algorithm

In VANET routing algorithms like AODV [13] and IoDSCF [23], packet forwarding between vehicle nodes follows a "store-carry-forward" approach. However, due to frequent link breaks in real-world scenarios, updating the routing table constantly becomes one of the major challenges faced by VANET routing algorithms. To tackle this challenge, we define the packet transmission process as a Markov Decision Process (MDP), represented by a tuple (N, S, A, R, γ).

Definition 1: An MDP is comprised of the following components:

- **Agent**: The VANET is the environment. We consider the packet as agents and the set of agents denoted as $I = \{1, ..., N\}$.
- **State**: The state s represents the cluster structure where the packet is located. For any packet, its state space S includes all clusters in the road. Under each state s, the agent needs to define the relationship between the possible next state s_{next} and the corresponding action, denoted as $< s|s_{\text{next}}, a >$.
- **Action**: The action set A is the collection of neighboring vehicles within one hop of the node. State transition involves forwarding packets from the current node to the neighboring cluster structure. Once the forwarding is completed, instant rewards R are obtained. We denote the action space of the agent as A, where $a_i \in A$.
- **Policy**: The state transition function $P(s|s_{\text{next}}, a)$ represents the probability of transitioning to state s_{next} after taking action a in state s. The agent utilizes this function to decide the next action to be taken.
- **Reward**: After executing each action, the agent receives a reward or punishment based on the defined reward function. We define the reward function as $reward(s, a)$, representing the reward for acting a in state s.

To ensure uniformity among neighboring nodes and optimize routing efficiency, we propose the use of two key metrics: the Difference Ratio of Nodal Degrees ($DRNR$) and the Link Effective Time Ratio (LTR). The $DRNR$ measure is designed to address the negative impacts of having excessive or insufficient neighboring nodes, which can lead to issues such as broadcast storms or inefficient use of bandwidth resources. It calculates the difference ratio of nodal degrees for a given node x.

$$DRNR(x) = e^{-|n_x - \bar{n}|} \tag{4}$$

where n_x is the node degree of x, $\bar{n}$ is the average node degree, i.e. the ratio of the sum of the node degrees to the total number of nodes in the network.

The Link Effective Time Ratio (LTR) is a metric that quantifies the duration of data communication between two nodes in terms of the time taken for a common vehicle to travel from one cluster to the boundary of another cluster. It is computed based on the geographic coordinate information, velocities, and directions of the CH node a $(x_a, y_a, v_a, \theta_a)$ and the CM node b $(x_b, y_b, v_b, \theta_b)$. The formula for calculating the LTR is given as Eq. 5:

$$LTR(a,b) = \frac{-(eg+fh) + \sqrt{(v_a - v_b)^2 r^2 - (eh - fg)^2}}{2(v_a - v_b)^2 r} \tag{5}$$

where e and f represent the relative differences in velocity between the nodes in the horizontal and vertical axes, respectively. g and h represent the relative distances between the nodes in the horizontal and vertical directions, in that order, and r represents the transmission radius of the CH.

4.2 Adaptive Clustering Routing Algorithm Combined with Reinforcement Learning

The traditional Q-learning algorithm used in VANET typically employs fixed learning and discount factors, which may limit its adaptability to different VANET scenarios. In order to enhance adaptiveness, we propose incorporating the bandwidth factor as the learning rate. Initially, when the Q-table is empty, the learning rate is set to 1, allowing the algorithm to fully adopt new knowledge. As the number of successfully forwarded packets increases, more records are added to the Q-table. The algorithm dynamically adjusts the convergence rate based on real-time bandwidth conditions to prevent network congestion.The improved Q-learning formula is shown is Eq. 6:

$$Q_p(q,m) = (1 - BWF)Q_p(q,m) + BWF\left[R(p,m) + \gamma \max_{n \in S(m)} Q_m(q,n)\right] \tag{6}$$

where $Q_p(q,m)$ represents the Q-value of the current node p, which sends the packet to the next node m and then forwards it to the destination node q one after another. The bandwidth factor BWF is taken as the Q-learning algorithm's learning rate, and the discount rate γ is set to 0.8. Node m is a neighbor of current node p, and $S(m)$ represents the set of neighboring nodes of node m.

$R(p,m)$ represents the instant reward that node p obtains for forwarding the packet to node m. The calculation formula for $R(p,m)$ is shown in Eq. 7:

$$R(p,m) = \begin{cases} 1 & d = m \\ \omega_1 \times LTR(p,m) + \omega_2 \times \mathrm{DRNR}(m) & \text{otherwise} \end{cases} \tag{7}$$

where $LTR(p,m)$ is the link effective time ratio of nodes p and m, $DRNR(m)$ is the difference ratio of nodal degrees of node m. ω_1 and ω_2 are constants, and ω_1+ω_2=1.

During packet transmission, intermediate vehicle nodes apply Eq. 7 to determine the reward function R of different next-hop nodes based on node connectivity and vehicle stability. They then update the Q-value using Eq. 6 until the best node has been selected, optimizing the selection of intermediate nodes. The routing link is created when the ultimate destination vehicle node receives RREQ and responds to the source vehicle node following the established reverse route.

4.3 Message Propagation Scheme

During the route discovery phase of the AODV routing protocol, when a source node has to forward a packet to the destination node, it queries the routing table. Suppose that the corresponding path is not in the routing table. In that case, the source node broadcasts an RREQ packet, and intermediate nodes that receive it establish a reverse route until the destination node receives the RREQ. The destination node replies to the source node with an RREP through the previously established reverse route. However, in VANET environments where vehicle nodes move quickly, the network topology changes frequently, and route discovery is often triggered. This increases the amount of RREQs in the network, causing congestion and excessive overhead [24].

To address the above issue, Bp-AODV routing protocol utilizes the density of neighbor nodes to dynamically adjust the route discovery probability. VANET is comprised of high-density and low-density areas. On the one hand, in high-density networks, there are many routes with similar routing indicators, so choosing the best route is unnecessary. On the other hand, dense networks consist of many nodes, which can quickly generate numerous unnecessary $RREQs$ in the local neighborhood [14]. Therefore, reducing the number of $RREQs$ during the route discovery process in dense networks can help not only maintain the selection of high-quality routes but also decrease the routing overhead.

Establish Multi-hop Links Between Adjacent CHs. Using the local vehicle information uploaded by the fog nodes, the Bp-AODV protocol routing trains a RL model to obtain the optimal path between neighboring CHs. Each CH maintains a Q-table that stores the optimal path to the adjacent CH. If a link connection breaks, CH initiates route discovery based on the Q-table to find the following best path. This approach reduces the number of unnecessary $RREQs$ and selects high-quality routes, thereby improving the overall performance of the VANET.

During route discovery between adjacent CHs, the CH will send out $RREQs$ to all reachable nodes, causing a high load on the network. We propose a probabilistic forwarding mechanism based on neighbor density to address this issue. When CM receives the $RREQ$ from CH, it computes the probability p according to Eq. 8, where Nb_i is the number of neighbor nodes of CH and k is a constant. According to the simulation results, we chose the best value of $k = 6$.

$$p_i = \begin{cases} 1 & k \geq Nb_i \\ \frac{k}{Nb_i} & \text{otherwise} \end{cases} \tag{8}$$

To determine whether to forward the *RREQ*, the node generates a uniformly distributed random value p_r in the range of (0,1). If p_r exceeds the probability p_i, the *RREQ* will not be forwarded again. The above steps are repeated until the *RREQ* reaches the destination.

Establish Multi-hop Links Between Nodes. When the source node transmits a packet, it first checks the Q-table to see whether the optimal path to the destination node is documented.

If the next-hop node is not found, the source vehicle starts route discovery by sending a *RREQ* to the CH. The CH checks whether the destination vehicle is inside the cluster based on the membership table. The CH starts the route discovery to the destination node and updates the Q-table if the destination vehicle is within the cluster.

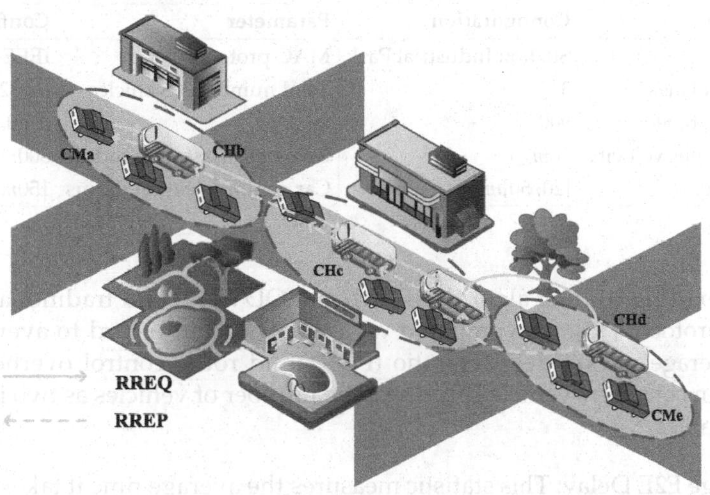

Fig. 2. Establishing multi-hop links across clusters.

As shown in Fig. 2, if the destination node is outside the cluster, the transmission path from CM_a to CM_e can be viewed as several multi-hop links stitched together, including from CM_a to CH_b, from CH_b to CH_c, from CH_c to CH_d, and from CH_d to CM_e. The optimal transmission path between two CHs is established by the optimal transmission paths between neighboring CHs, which have been trained in the cloud. According to the pre-trained optimal path between adjacent CHs, the source CH forwards RREQ to the neighboring CH. If the destination node is not within the cluster, RREQ is forwarded to the next neighboring CH until it reaches the destination CH. Then the destination CH initiates the route discovery to the destination node and updates the Q-table. During packets' transmission, Bp-AODV requires fewer transmission beacons to establish multi-hop links and select routes, thus reducing the chance of channel conflicts.

5 Simulations

5.1 Bp-AODV Simulation Settings

The performance evaluation of the proposed Bp-AODV protocol is conducted in a real-world urban scenario in Suzhou Industrial Park. The real-world road network is obtained from OpenStreetMap and is used to establish a traffic simulation environment. The simulation parameters for the communication and traffic model are summarized in Table 1.

Table 1. Simulation parameters.

Parameter	Configuration	Parameter	Configuration
Scenario	Suzhou Industrial Park	MAC protocol	IEEE 802.11p
Number of lanes	3	Total number of vehicles	[60, 200]
Simulation time	900s	ω_1 ω_2	0.5 0.5
Maximum bus velocity	30m/s	Bus communication radius	300m
Car velocity	[20, 50]m/s	Car communication radius	150m

Two existing routing algorithms, Urban-AODV [14] and traditional AODV routing protocol [13], are compared with Bp-AODV in regard to average E2E delay, average packet delivery ratio (PDR), and route control overhead. The simulation considers vehicle speed and the number of vehicles as two independent variables.

- Average E2E Delay: This statistic measures the average time it takes for each data packet to be delivered from the source to the destination.
- Average Packet Delivery Ratio (PDR): This metric represents the percentage of packets transmitted without errors and successfully received at their destination.
- Route Control Overhead: This metric represents the ratio of total routing control messages transmitted in the network to the total number of data packets delivered successfully.

5.2 Analysis of Experimental Results

Average E2E Delay. Figures 3(a) and 3(b) depict the average E2E delay variation for varying numbers and velocities of vehicles. As the number of vehicle nodes increases, the average E2E delay for all three protocols decreases. This is because more nodes are involved in route forwarding, resulting in reduced E2E delay. Bp-AODV further reduces the delay by pre-training optimal transmission paths between adjacent CHs.

Average Packet Delivery Ratio. Figures 3(c) and 3(d) depicts the PDR variation under different numbers and velocities of vehicles, respectively. The stable cluster structure constructed by Bp-AODV reduces the probability of link disconnection due to the fast-changing topology. High-speed node movements at high vehicle node speeds can cause frequent topology changes, which increase the probability of link disconnection, and thus harm communication quality.

Routing Control Overhead. Figure 4 illustrates the average ratio of routing control overhead across various numbers and velocities of vehicles. The three

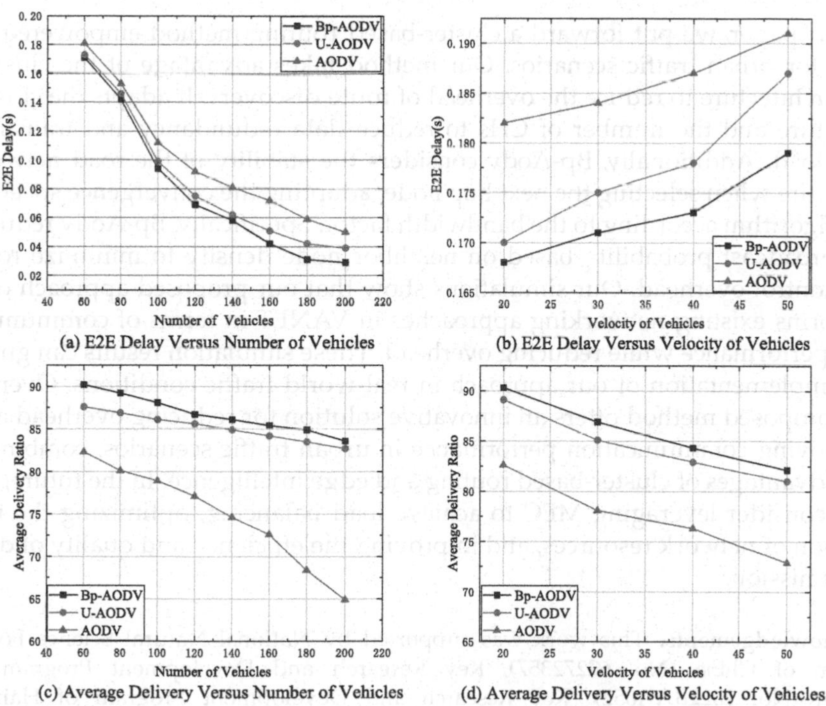

Fig. 3. Average E2E delay and average delivery rate under different numbers and velocity of vehicles

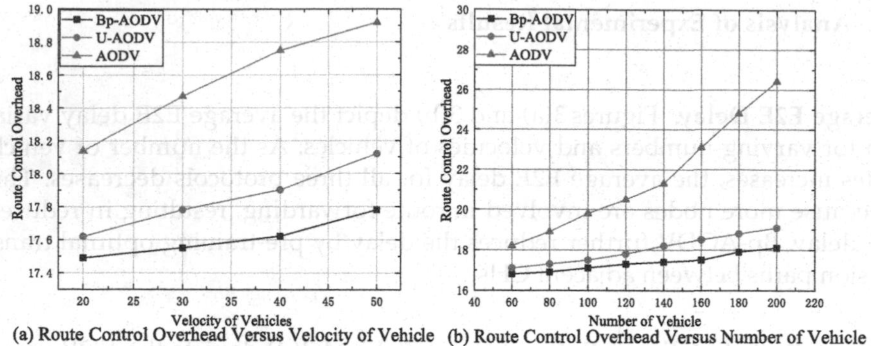

(a) Route Control Overhead Versus Velocity of Vehicle (b) Route Control Overhead Versus Number of Vehicle

Fig. 4. Route control overhead under different numbers and velocity of vehicle

routing protocols are impacted by alterations in the network topology. However, Bp-AODV incurs lower routing control overhead than U-AODV and AODV by using fewer routing control messages to establish a more dependable route.

6 Conclusion

In this paper, we put forward a cluster-based routing method empowered by VFC for urban traffic scenarios. Our method takes advantage of the clustering architecture to reduce the overhead of route discovery. It adapts the cluster structure and the number of CHs to reduce data redundancy and transmission load. Additionally, Bp-Aodv considers the stability of the road network structure when selecting the next hop node, adapting the convergence speed of the algorithm according to the bandwidth factor. Specifically, Bp-Aodv reduces the broadcast probability based on neighbor node density to minimize routing control overhead. Our simulations show that our proposed approach outperforms existing networking approaches in VANET in terms of communication performance while reducing overhead. These simulation results can guide the implementation of our approach in real-world traffic conditions. Overall, our proposed method offers an innovative solution for reducing overhead and improving communication performance in urban traffic scenarios, combining the advantages of cluster-based routing and edge intelligence. In the future, we will consider leveraging MEC to achieve load balancing, optimizing the utilization of network resources, and improving the efficiency and quality of data transmission.

Acknowledgements. This work was supported by National Natural Science Foundation of China (No. 62272357), Key Research and Development Program of Hubei (No. 2022BAA052), Key Research and Development Program of Hainan (No. ZDYF2021GXJS014), Science Foundation of Chongqing of China (cstc2021jcyj-msxm4262), and Research Project of Chongqing Research Institute of Wuhan University of Technology (ZD2021-04, ZL2021-05).

References

1. Arif, M., Wang, G., Bhuiyan, M.Z.A., Wang, T., Chen, J.: A survey on security attacks in VANETs: communication, applications and challenges. Veh. Commun. **19**, 100179 (2019)
2. Liu, B., et al.: Collaborative intelligence enabled routing in green IOV: a grid and vehicle density prediction-based protocol. IEEE Trans. Green Commun. Netw. **7**(2), 1012–1022 (2023)
3. Hussein, N.H., Yaw, C.T., Koh, S.P., Tiong, S.K., Chong, K.H.: A comprehensive survey on vehicular networking: communications, applications, challenges, and upcoming research directions. IEEE Access **10**, 86127–86180 (2022)
4. Dai, P., Hu, K., Wu, X., Xing, H., Teng, F., Yu, Z.: A probabilistic approach for cooperative computation offloading in MEC-assisted vehicular networks. IEEE Trans. Intell. Transp. Syst. **23**(2), 899–911 (2020)
5. Shao, X., Hasegawa, G., Dong, M., Liu, Z., Masui, H., Ji, Y.: An online orchestration mechanism for general-purpose edge computing. IEEE Trans. Serv. Comput. **16**(2), 927–940 (2023)
6. Liu, B., et al.: Multi-agent attention double actor-critic framework for intelligent traffic light control in urban scenarios with hybrid traffic. IEEE Trans. Mobile Comput. 1–13 (2023)
7. Dai, P., Han, B., Wu, X., Xing, H., Liu, B., Liu, K.: Distributed convex relaxation for heterogeneous task replication in mobile edge computing. IEEE Trans. Mobile Comput. 1–16 (2022)
8. Dai, P., Song, F., Liu, K., Dai, Y., Zhou, P., Guo, S.: Edge intelligence for adaptive multimedia streaming in heterogeneous internet of vehicles. IEEE Trans. Mobile Comput. (2021)
9. Liu, K., Xiao, K., Dai, P., Lee, V.C., Guo, S., Cao, J.: Fog computing empowered data dissemination in software defined heterogeneous VANETs. IEEE Trans. Mob. Comput. **20**(11), 3181–3193 (2021)
10. Liu, B., et al.: A novel V2V-based temporary warning network for safety message dissemination in urban environments. IEEE Internet Things J. **9**(24), 25136–25149 (2022)
11. Liu, B., et al.: A region-based collaborative management scheme for dynamic clustering in green VANET. IEEE Trans. Green Commun. Netw. **6**(3), 1276–1287 (2022)
12. Ayyub, M., Oracevic, A., Hussain, R., Khan, A.A., Zhang, Z.: A comprehensive survey on clustering in vehicular networks: current solutions and future challenges. Ad Hoc Networks (2021)
13. Perkins, C., Belding-Royer, E., Das, S.: Ad hoc on-demand distance vector (AODV) routing. Technical report (2003)
14. Mubarek, F.S., Aliesawi, S.A., Alheeti, K.M.A., Alfahad, N.M.: Urban-AODV: an improved AODV protocol for vehicular ad-hoc networks in urban environment (2018)
15. Zhang, D., Gong, C., Zhang, T., Zhang, J., Piao, M.: A new algorithm of clustering AODV based on edge computing strategy in IOV. Wirel. Netw. **27**(4), 2891–2908 (2021). https://doi.org/10.1007/s11276-021-02624-z
16. Karp, B.: GPSR: greedy perimeter stateless routing for wireless networks. In: ACM Mobicom (2000)
17. Alsarhan, A., Kilani, Y., Al-Dubai, A., Zomaya, A.Y., Hussain, A.: Novel fuzzy and game theory based clustering and decision making for VANETs. IEEE Trans. Veh. Technol. **69**(2), 1568–1581 (2020)

18. Kayis, O., Acarman, T.: Clustering formation for inter-vehicle communication. In: Intelligent Transportation Systems Conference (2007)
19. Khan, Z., Fang, S., Koubaa, A., Fan, P., Farman, H.: Street-centric routing scheme using ant colony optimization-based clustering for bus-based vehicular ad-hoc network. Comput. Elect. Eng. **86**(1), 106736 (2020)
20. Tseng, H.W., Wu, R.Y., Lo, C.W.: A stable clustering algorithm using the traffic regularity of buses in urban VANET scenarios. Wirel. Netw. **26**(1), 2665–2679 (2020)
21. He, C., Qu, G., Ye, L., Wei, S.: A two-level communication routing algorithm based on vehicle attribute information for vehicular ad hoc network. Wirel. Commun. Mob. Comput. **2021**, 6692741:1–6692741:14 (2021)
22. Khan, Z., Fan, P., Fang, S., Abbas, F.: An unsupervised cluster-based VANET-oriented evolving graph (CVoEG) model and associated reliable routing scheme. IEEE Trans. Intell. Transp. Syst. **20**(10), 3844–3859 (2019)
23. Bine, L.M.S., Boukerche, A., Ruiz, L.B., Loureiro, A.A.F.: IoDSCF: a store-carry-forward routing protocol for joint bus networks and internet of drones. In: 2022 IEEE 42nd International Conference on Distributed Computing Systems (ICDCS), pp. 950–960 (2022)
24. Malnar, M., Jevtic, N.: An improvement of AODV protocol for the overhead reduction in scalable dynamic wireless ad hoc networks. Wirel. Netw. **28**(3), 1039–1051 (2022)

Strategy Determination for Multiple USVs: A Min-max Q-learning Approach

Le Hong and Weicheng Cui(✉)

Key Laboratory of Coastal Environment and Resources of Zhejiang Province (KLaCER), School of Engineering, Westlake University, Hangzhou 310024, Zhejiang Province, China
{hongle,cuiweicheng}@westlake.edu.cn

Abstract. The application of Unmanned Surface Vehicles (USVs) has gained significant momentum in various domains, including surveillance operations and security enforcement. Efficient coordination among multiple USVs is crucial for successful pursuit and defense in these applications. This paper employs a specific min-max Q-learning method to handle multiple pursuit-USVs and their interactions with a moving evasion-USV. Firstly, to reflect the cooperation and competence mechanism among the USVs, the proposed multi-USV pursuit-evasion problem is formulated by a sequence of Zero-Sum Pursuit-Evasion Games (ZSPEGs). During the games, pursuit-USVs aim to intercept the evasion-USV and prevent it from reaching its intended location. To solve the proposed model efficiently, a specific min-max Q-learning method is well designed to improve the model's robustness in a complex environment. The simulation results indicate that the designed min-max Q-learning approach provides a reliable defending strategy for the pursuit-USVs. It is more effective than the conventional solution for the matrix games, with a higher success rate of capture and faster computation time.

Keywords: Cooperative strategy determination · Surface Pursuit-Evasion · Zero-Sum Games · Min-max Q-learning

1 Introduction

Unmanned Surface Vehicles (USVs), equipped with advanced sensing and navigation capabilities, offer unique advantages in the surface Pursuit-Evasion (PE) scenarios [1,2]. The surface PE scenario involves multiple USVs, where some USVs act as pursuers and others as evaders, each aiming to achieve their own objectives in a complex environment [3]. Coordinating the actions of multiple USVs engaged in PE poses significant challenges. The pursuers should strategize to capture the evaders efficiently, while the evaders seek to elude capture by employing evasion tactics [4]. The optimal strategies for each pursuit-USV depend on a multitude of factors, such as their individual capabilities, the evaders' behavior, and the dynamics of the game. Traditional approaches based on heuristics or rule-based methods often struggle to handle the complexity and uncertainty inherent in such scenarios [5–7].

H. Zhang et al. (Eds.): NCAA 2023, CCIS 1870, pp. 403–417, 2023.
https://doi.org/10.1007/978-981-99-5847-4_29

To effectively address the complexities and uncertainties inherent in such scenarios, the integration of game theory has emerged as a valuable approach [8]. In particular, Zero-Sum Game (ZSG) theory provides a powerful framework to model and analyze competitive scenarios where the objectives of the pursuers and evaders are directly opposed. The formed Zero-Sum Pursuit-Evasion Games (ZSPEG) has been extensively studied [9–11]. By adopting the ZSG theory, the PE problem can be recast as a competitive game, where the interests of the pursuers and evaders are diametrically opposed. This framework allows for a deeper analysis of the strategic interactions and enables the development of optimal strategies for both sides [12,13]. In the early stage, researchers focus on how to build the accurate objective function in the multi-agent PE model. A divide-and-conquer approach is used for a multi-player PE problem where the pursuers have a twofold goal combining caption and an allocated task [14]. A multi-model adaptive estimator has been implemented to identify unknown sets of incoming attackers in the offline designed policy sets [15]. However, these ZSPEG-based models are mainly established from the known input-output relationship, which makes the model depend on the chosen parameters, the constructed goal functions, and the known environment.

Nowadays, reinforcement learning has gained popularity as effective tools for selecting optimal strategies in decision-making problems, particularly in uncertain environments involving either single or multiple participants [16,17]. In the context of the two-player ZSG, Q-learning methods have been successfully applied to solve simple problems with discrete states and action spaces [16]. Besides, an integral Q-learning algorithm incorporating continuous differential countermeasures is designed to address the allocation problem within ZSGs [17]. Furthermore, a min-max Q-learning algorithm has been employed to refine the performance of the evader's strategy throughout the normal PE scenario [18]. The challenges and key issues of reinforcement learning for multi-agent problems have been well-discussed, shedding light on the complexity of such scenarios [19]. However, despite the advancements in reinforcement learning methods, their application in specific PE scenarios, such as assisting USV systems in capturing invading vehicles, still lacks exploration. Recently, a novel approach has been proposed to leverage multi-agent deep reinforcement learning for offering multiple pursuit-USVs an efficient pursuit planning [20]. However, under the surface communication conditions where computational resources may be limited, deep reinforcement learning may not be a suitable choice for the current in-service USVs. In contrast with the deep reinforcement learning methods, a lightweight multi-agent reinforcement learning method, named min-max Q-learning, may offer a more practical and feasible solution. Min-max Q-learning allows for faster training and decision-making processes [21]. Besides, it does not require explicit knowledge of the underlying dynamics or accurate models of the environment. These characteristics makes min-max Q-learning applicable in real-world USV PE scenario, in which obtaining accurate models is challenging.

Inspired by the above discussions, this paper proposes a novel strategy determination approach that applies ZSG theory and min-max Q-learning for multiple USVs engaged in the PE scenario. The main contributions of this paper lie in the following three aspects.

1) Enhanced understanding of multi-USV PE scenarios: By formulating the multi-USV PE problem into a sequence of ZSPEGs, this paper offers valuable insights into the dynamics of cooperative defense strategies among the multiple pursuit-USVs.
2) A specific min-max Q-learning method for the proposed PE model: The designed min-max Q-learning method closely mirrors real-world USV PE scenarios. It introduces a tailored learning-state space and reward function, which take into account crucial factors such as the time-to-capture and the invading capability of the evasion-USV.
3) Empirical evaluation on the proposed model and approach: The simulations utilize parameters derived from an actual USV. By comparing the performance of our proposed approach with the classic solution for the matrix game, concrete evidences of the effectiveness and superiority of our method are provided.

The rest of this paper is organized as follows. Section 2 gives the mathematical description of the problem to be explored. In Sect. 3, based on the multi-USV ZSPEG model, a specific min-max Q-learning is designed. Two simulation cases are conducted in Sect. 4. Finally, conclusions are drawn in Sect. 5.

2 Preliminaries and Problem Statement

A schematic diagram of applying multiple pursuit-USVs to capture the invading USV is shown in Fig. 1. At the time t_1, the evasion-USV invades the protected marine field. Then, when the time is t_1, the marine-monitoring equipment would discover the evasion-USV and transfer its position to the signal receptor for the pursuit team. Once the pursuit team is informed that the evasion-USV has invaded the protected field, at t_3, the pursuit team starts to capture the evasion-USV. At the same time, the evasion-USV expects to reach its target position to finish its task. The 2D-planar map about how the participant-USV moves at t_3 are presented at the bottom right of Fig. 1.

Within the framework of multiple pursuit-USVs coordinating to intercept the evasion-USV, the success of the pursuit team depends on how the pursuit system determines the pursuit strategies for each pursuit-USV from the time $t3$. Considering that the evasion-USV can escape consciously, this paper treats the pursuit process as multiple pursuit-USVs pursuit-evasion games. Then, we apply the relevant approaches in the pursuit-evasion game to provide the optimal defending and attacking strategy for the pursuit system. The motion models and the problem formulation for the pursuit-USV system are introduced as follows.

2.1 Motions for the USV

First of all, several notations are given. The set of real numbers, n-dimensional real vectors, and $m \times n$ dimensional matrices are denoted by $\mathbb{R}$, $\mathbb{R}^n$, and $\mathbb{R}^{m\times n}$, respectively. During the pursuit-evasion process, the motions of the studied participant-USV are described discretely.

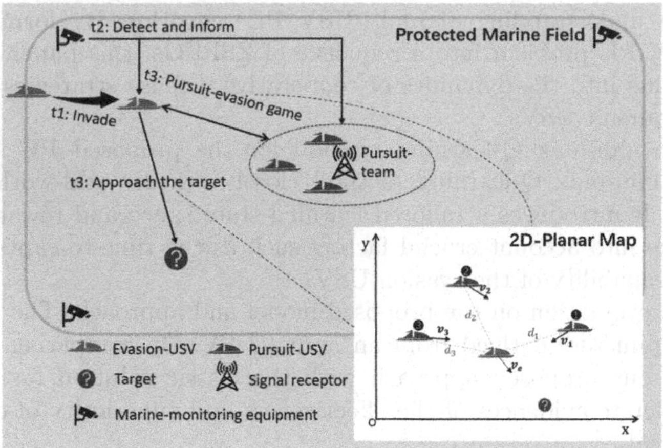

Fig. 1. The schematic of the pursuit-evasion scenario in a pursuit system.

From the 2D-Planar map shown in the Fig. 1, we assume that one participant USV has the initial position of $\mathbf{x(t)} := (x(t), y(t))$. The corresponding velocity is v, and the acceleration is $\mathbf{a(t)}$. Here, $\mathbf{a(t)}$ could be equivalent to the control inputs for the discrete-time motion model. The value of $\mathbf{a(t)}$ is assumed in the set $\mathcal{U} := \{\|\mathbf{a}\| = 1 \text{ or } \|\mathbf{a}\| = 0\}$. Then, the dynamic equation for the participant-USV is shown in Eq.(1).

$$\dot{\mathbf{x}}(\mathbf{t}) = v \times t \times \mathbf{a(t)} \tag{1}$$

where the time $t \in [0, T_f]$, T_f is the maximum continuous moving time of the participant-USV, and v is the speed of this USV. After one discrete time period T, the position of the USV is changed to $\mathbf{x(t+T)} := (x(t+T), y(t+T))$. To fit the application of the game theory, the USV's motion is expected to in its discrete state space. Therefore, it is assumed that the game consists of K stages at most. $\triangle t$ is chosen for $T_f := \triangle t \times K$. The following discrete-time state-space model for the participant-USV at stage $k \in \{1, 2, \ldots, K-1\}$ could be obtained.

$$\mathbf{x_d(k+1)} = \mathbf{x_d(k)} + v \times \triangle t \times \mathbf{a_d(k)} \tag{2}$$

where $\mathbf{x_d(k)}$ and $\mathbf{a_d(k)}$ denote the position and the acceleration for the USV at stage $k \in \{1, 2, \ldots, K-1\}$. Under the assumption that $\mathbf{a}(\tau)$ is equivalent to $\mathbf{a_d(k)}$ for all $\tau \in [k \triangle t, (k+1) \triangle t)]$, and for all $k \in \{1, 2, \ldots, K-1\}$, the value of $\mathbf{x_d(k)}$ is subsequently equal to $\mathbf{x}(\tau)$.

2.2 Problem Formulation

The 2D plannar map in Fig. 1 shows how the multiple USVs move in the pursuit-evasion scenario. Considering that N pursuit-USVs take the responsibility for the caption, when $i \in \{1, 2, \ldots, N\}$, P_i stands the ith pursuit-USV. v_i denotes the velocity of P_i. d_i represents the distance from the P_i to the evasion-USV

E. As for the evasion-USV E, d_i denotes the distance from the E to the target position. Besides, $\mathbf{x_i} \in \mathbb{R}^2$ and $\mathbf{x_e} \in \mathbb{R}^2$ represent the current positions of the ith pursuit-USV and the evasion-USV, respectively. Based on the above illustration, the multiple pursuit-USV pursuit-evasion problem is formulated in Problem 1.

Problem 1. The goal for the pursuit-USV team is to capture the evasion-USV and prevent it from reaching the target $\mathbf{x_T} \in \mathbb{R}^2$ as soon as possible. At the stage $K_{cap} \in \{1, 2, \ldots, K-1\}$, if there exists the ith pursuit-USV, its distance to the evasion-USV is less than the capture distance d_{cap}. At the same time, the distance from the evasion-USV to the target is larger than the destroyed distance d_{de}. Then, the pursuit system is judged to have successfully captured the evasion-USV. Therefore, the final successful captured situation at the stage of $K_{cap} \in \{1, 2, \ldots, K-1\}$ can be expressed in Eq. (3).

$$\exists\ i \in \{1, 2, \ldots, N\} : d_i(K_{cap}) \leq d_{cap},\ \ d_e(K_{cap}) > d_{reach} \tag{3}$$

3 The Specific Min-max Q-learning

In this section, the discrete-time game model with $N+1$ USVs is reformulated as a multiple pursuit-USVs ZSPEG with two players. The steps are as follows. The pursuit-USV team is treated as a single team $P := \{P_1, P_2, \ldots, P_N\}$. Another player is the evasion-USV E. To avoid the emergence of a dimensional explosion, we enable the pursuit team to accurately deploy one of the pursuers at a given stage. This chosen strategic approach is named relay-pursuit strategy, which is explained in Definition 1. As a result, the dimensions of the strategy set among the pursuer group are greatly reduced at a given stage.

Definition 1. Relay-pursuit strategy
In the relay pursuit, only one pursuit-USV is active, while the others are stationary. The distribution of the active pursuit-USVs changes over time, which depends on the outcome of each game. At a certain stage, if the active pursuit-USV has been determined and designated by the index i^*, the feedback strategy of the ith pursuer could be obtained in Eq. (4).

$$\mathbf{a}_i := \begin{cases} \frac{\mathbf{r}_i}{\|\mathbf{r}_i\|}, ifi = i^* \\ 0, others \end{cases} \tag{4}$$

where $\mathbf{r}_i := x_e - x_i$ is the relative position vector of the evasion-USV from the ith pursuit-USV. Since every pursuit-USV wants to remain spatially localized and conserve the relevant resources, adopting the relay pursuit is a suitable strategy choice for a group of pursuit-USVs.

At every stage, the pursuit team has total N strategies. When $i \in \{1, 2, \ldots, N\}$, the ith strategy for the pursuit-team is about the situation that the ith pursuit-USV actively pursues the evasion-USV. Similarly, E's restricted decision-making space consists of $N+1$ actions. The first N actions represent evading the corresponding pursuers. To be specific, when $j \in \{1, 2, \ldots, N\}$, the jth action is about evading the jth pursuit-USV. The $N+1$ action is the goal-seeking behavior, which means that the evasion-USV towards the target. Therefore, at the kth stage, $M_k \in \mathbb{R}^{N\times(N+1)}$ denotes the payoff matrix for the pursuit-team P. Considering that $N = 2$, the formulation, and entries of the payoff matrix M_k are presented in Table 1. In this table, the two rows represent the choice of the active pursuit-USV. The first two columns represent evasion-USV's strategy to evade one pursuit-USV. The last column indicates evasion-USV moves to the target position directly. When $i \in \{1, 2\}$ and $j \in \{1, 2, 3\}$, $M_{(i,j)}$ represents the payoff given to the pursuit-USV team.

Table 1. Payoff matrix for the proposed USV system at the kth stage.

P	Payoff function		
	E		
	Evading P_1	Evading P_2	Towards the target
P_1 in pursuit	$M_{k(1,1)}$	$M_{k(1,2)}$	$M_{k(1,3)}$
P_2 in pursuit	$M_{k(2,1)}$	$M_{k(2,2)}$	$M_{k(2,3)}$

To relax the model's dependency on the parameters of the USVs and enhance the model's transferability, a specific min-max Q-learning structure is utilized to help generating the payoff matrix, which can is demonstrated in Fig. 2.

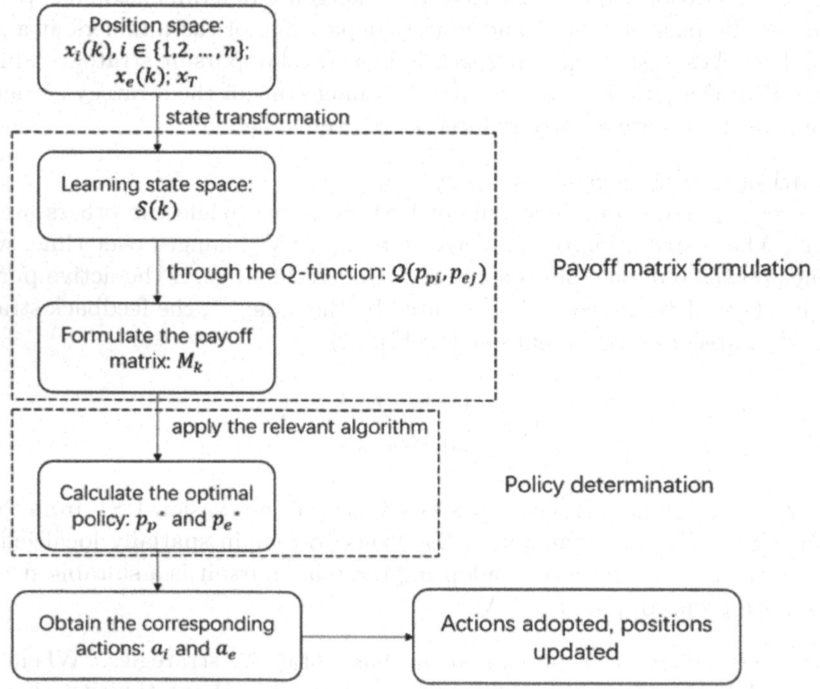

Fig. 2. Min-max Q-learning structure for the proposed model.

3.1 Construction of the Learning State Space

Calculation for the Q-function could be divided into two parts. The first is the construction of the learning state space. The other one is the approximation of the Q-function based on the learning state space. To reduce the computational expense, the learning state space S is assumed to consist of four variables at all stages, $S := [S_1, S_2, S_3, S_4]$. The four variables are all dependent on the velocity and the positions of every USVs at a given game stage. First of all, the index about the minimum time-to-intercept $T_I(p_{p_i}, p_{e_j}) := T_I(\mathbf{a_{p_i}}, \mathbf{a_{e_j}})$ is introduced and solved by Eq. (5).

$$(v_e^2 - v_i^2)T^2 + 2(v_e \mathbf{a_{e_j}}^\top \mathbf{r_i} - d_{cap} v_i)T + \|r_i\|^2 - \|d_{cap}\|^2 = 0 \tag{5}$$

where the minimum positive solution $T_0(\mathbf{x_e}, \mathbf{x_i})$ corresponds to the minimum time-to-intercept $T_I(\mathbf{a_{p_i}}, \mathbf{a_{e_j}})$, which is the ith defending USV should cost to capture the invading USV [22]. To solve the $T_0(\mathbf{x_e}, \mathbf{x_i})$, $\mathbf{r_i} := \mathbf{x_e} - \mathbf{x_i}$ and $\mathbf{a_{e_j}} := (\mathbf{x_e} - \mathbf{x_i}) / \|\mathbf{x_e} - \mathbf{x_i}\|$. v_i and v_e denote the velocity of the ith defending USV and the invading USV, respectively.

To calculate the time-to-intercept for the defending USV team in two different situations, Eq. (6) and Eq. (7) are rewritten from Eq. (5) to formulate the subsequent four variables in the S. When the invading USV is moving in a randomly chosen direction, the minimum positive solution of the Eq.(6) is the minimum time-to-intercept $T_{0r}(\mathbf{x_e}, \mathbf{x_i}, \mathbf{a_{er}})$.

$$(v_e^2 - v_i^2)T^2 + 2(v_e \mathbf{a}_{\mathbf{e_r}}^\top \mathbf{r_i} - d_{cap} v_i)T + \|r_i\|^2 - \|d_{cap}\|^2 = 0 \tag{6}$$

where $\mathbf{a_{e_r}}$ is randomly determined. The chosen of the other parameters and the variables are unchanged. The second situation is the invading USV towards the target point. The corresponding time-to-intercept $T_{0d}(\mathbf{x_e}, \mathbf{x_i}, \mathbf{x_T})$ is solved by Eq. (7).

$$(v_e^2 - v_i^2)T^2 + 2(v_e \mathbf{a}_{\mathbf{e_T}}^\top \mathbf{r_i} - d_{cap} v_i)T + \|r_i\|^2 - \|d_{cap}\|^2 = 0 \tag{7}$$

where $\mathbf{a_{e_T}} := (\mathbf{x_T} - \mathbf{x_e}) / \|\mathbf{x_T} - \mathbf{x_e}\|$ and the minimum positive solution is $T_{0d}(\mathbf{x_e}, \mathbf{x_i}, \mathbf{x_T})$. Besides, the minimum time $T_{0T}(\mathbf{x_e}, \mathbf{x_T})$ for the invading USV to reach the target is also introduced. It is calculated by: $T_{0T}(\mathbf{x_e}, \mathbf{x_T}) := \frac{\|\mathbf{x_T} - \mathbf{x_e}\|}{v_e}$.

After the introduction of the above three time-metrics related to the participant-USVs, the corresponding four overall metrics would be obtained from a comprehensive view, which are (i) the least minimum time-to-intercept T_{Il} when the invading USV is in the evasion mode, (ii) the least minimum time-to-intercept T_{rl} when the invading USV is in the random mode, (iii) the least minimum time-to-intercept T_{dl} when the invading USV is in the target-oriented mode, and (iv) the minimum time to reach the target T_T when the invading USV is in the target-oriented mode. These four metrics are presented in Eq. (8) - Eq.(11), respectively.

$$T_{Il} = \min_i T_0(\mathbf{x_e}, \mathbf{x_i}) \tag{8}$$

$$T_{rl} = \min_i T_{0r}(\mathbf{x_e}, \mathbf{x_i}, \mathbf{a_e}) \tag{9}$$

$$T_{dl} = \min_i T_{0d}(\mathbf{x_e}, \mathbf{x_i}, \mathbf{x_T}) \tag{10}$$

$$T_T = \min T_{0T}(\mathbf{x_e}, \mathbf{x_T}) \tag{11}$$

The proposed ZSPEG model wishes to maximize the time for the invading USV to reach the target and minimize the time for the pursuit-USV team to intercept the invading USV. Therefore, based on the above metrics, at the kth stage, the learning state space is constructed as follows. $\mathcal{S}$ denotes the sigmoid function.

$$S(k) = [\mathcal{S}(\frac{T_T}{T_{Il}}), \mathcal{S}(\frac{T_T}{T_{rl}}), \mathcal{S}(\frac{T_T}{T_{dl}}), \frac{1}{T_T}] \tag{12}$$

3.2 Calculation of the Q-function

The relationship between the Q-function and the payoff function is shown in Eq. (13).

$$M_k(p_{p_i}, p_{e_j}) := Q(S(k), p_{p_i}, p_{e_j}) \tag{13}$$

where the $M_k(p_{p_i}, p_{e_j})$ denotes the payoff for the defending team when the system adopt the policy set (p_{p_i}, p_{e_j}) . It stands the game situation that P chooses the ith strategy and E selects its jth strategy. As for the construction of the Q-function $Q(S(k), p_{p_i}, p_{e_j})$, it is related to the learning state at this stage, the policies that two types of USVs take, and the dynamics of the chosen key factors. Thus, at the kth stage, the Q-function for the policy set (p_{p_i}, p_{e_j}) is initially constructed in Eq.(14).

$$Q(S(k), p_{p_i}, p_{e_j}) = q_{(4)}^{\top} \times \zeta_{(4)}(S(k), p_{p_i}, p_{e_j}) \tag{14}$$

where the $\zeta_{(4)}(S(k), p_{p_i}, p_{e_j})$ is assumed to including four variables to satisfy the requirement for the Q-function. $q_{(4)}$ is the measurement of these four variables in the Q-function, which is to be determined in the training process. The chosen of the four variables would be illustrated as follows. For the current state $S(k)$, if the system adopts the policy set (p_{p_i}, p_{e_j}) . The positions of USVs would be transferred to a new position state. Then, the learning state space would be subsequently changed. The new learning state space for the system is shown in Eq. (15).

$$S(k+1; p_{p_i}, p_{e_j}) = S(k \mid p_{p_i}, p_{e_j}) \tag{15}$$

As for the dynamics, D_{T_T} , the changes of the time to reach the target for the invading USV is chosen to form the $\zeta_{(4)}(S(k), p_{p_i}, p_{e_j})$. This is because that to protect the target point is the most important thing that the defending USV team must do. Based on the Eq. (11) for the T_T, D_{T_T} is to be calculated by the Eq. (16).

$$D_{T_T}(S(k), p_{e_j}, \mathbf{x_e}; \mathbf{x_T}) = \frac{\mathrm{d}T_T(x_e(t), x_T)}{\mathrm{d}t \mid (t = k\triangle)} \tag{16}$$

Based on the above introduction and construction of the $S(k+1; p_{p_i}, p_{e_j})$ and D_{T_T}. $\zeta_{(4)}(S(k), p_{p_i}, p_{e_j})$ could be formulated as follows.

$$\zeta_{(4)}(S(k), p_{p_i}, p_{e_j}) := \begin{cases} D_{T_T}(S(k), p_{e_j}, \mathbf{x_e}; \mathbf{x_T}) \\ S(k+1; p_{p_i}, p_{e_j})_1 \\ S(k+1; p_{p_i}, p_{e_j})_2 \\ S(k+1; p_{p_i}, p_{e_j})_3 \end{cases} \tag{17}$$

3.3 Reward Function

To estimate the Q-function needs to train the model to learn the best measurement $q_{(4)}$. The best $q_{(4)}$ relies on the Q-function update according to the designed reward function and the update principles of the min-max Q-learning approach. Here, the reward function $R(S(k), p_{p_i}, p_{e_j})$ would be designed based on the target-protected reward $R_{T_T}(S(k), p_{p_i}, p_{e_j})$, state-transition reward $R_S(S(k), p_{p_i}, p_{e_j})$, and the terminal reward $R_F(S(k))$. The target-protected reward R_{T_T} is designed to be the changes of the time to reach the target D_{T_T} for the invading USV, which is shown in Eq. (18).

$$R_{T_T}(S(k), p_{p_i}, p_{e_j}) := D_{T_T}(S(k), p_{e_j}, \mathbf{x_e}; \mathbf{x_T}) \tag{18}$$

To positively reward the decrease in the time-to-intercept for the defending USV team and the increase in the time to reach the target for the invading USV, the state-transition reward is calculated by Eq. (19).

$$R_S(S(k), p_{p_i}, p_{e_j}) := S(k+1; p_{p_i}, p_{e_j}) - S(k)_1 \tag{19}$$

The terminal reward aims to positively reward the defending USV team capturing the invading USV and negatively reward the invading USV's reaching the target. Thus, the terminal reward is constructed in Eq. (20). Under different situations, its values are also different.

$$R_F(S(k) = \begin{cases} +1 & , \frac{1}{S(k)_1} = \frac{1}{S(k)_2} = \frac{1}{S(k)_3} = 0 \\ -1 & , \frac{1}{S(k)_3} = 0 \\ 0 & , \text{otherwise} \end{cases} \tag{20}$$

Based on the introduction of the above three rewards, they associate with each other to form the final reward for the defending USV team. Finally, the reward function is given by Eq. (21).

$$R = R_{T_T} + R_S + R_F \tag{21}$$

So far, the preliminary requirement for the subsequent step for the implementation of the min-max Q-learning method have been accomplished. The detailed process to calculate the Q-function in the proposed model is presented in Algorithm 1.

Algorithm 1 Min-max Q-function

procedure MINMAXQ($S(k)$, p_{p_i}, p_{e_j})
  while $n \leq N_{tr}$ or $||\delta_{\omega_Q}|| > \delta_{tol}$ **do**
    randomly initialize $\mathbf{x_i}^0$, $\mathbf{x_e}^0$
    while $k \leq K_{max}$ **do**
      calculate $S(k)$
      for $i \leftarrow 1, N_P$ **do**
        for $j \leftarrow 1, N_E$ **do**
          $M_{i,j} = Q(S(k), p_i, e_j)$
        end for
      end for
      Obtain the mixed strategy-sets $\mathbf{p_P}$ and $\mathbf{q_E}$
      $V(S(k)) = min_i(Q \cdot \mathbf{p_P})$
      based on $\mathbf{p_P}$ and $\mathbf{p_E}$, choose the actions p^* and e^* for **P** and **E**
      $\triangle = \mathcal{R}_Q(S(k), p^*, e^*) + \gamma V(S(k+1)) - Q(S(k), p^*, e^*)$
      $\delta_{\omega_Q} = \partial \cdot \triangle \cdot \zeta(S(k), p^*, e^*)$
      $\omega_Q := \omega_Q + \delta_{\omega_Q}$
      update the position of every USV
      $k := k + 1$
    end while
    $\partial := \partial \cdot \delta_\partial$
    $\beta := \beta - \delta_\beta$
    $n := n + 1$
  end while
end procedure

4 Simulations

USV systems with defending and attacking ability have been attached great importance all over the world. According to the open data of the latest USV, MANTAST-12, developed by the US military, the parameters of the participant-USVs in this paper are assumed as follows. The velocity of the pursuit-USV is assumed to be 2.0 m/s. All USV's lengths are set to be 4.3 m. Then, the assumed velocity for the evasion-USV is in the range of [1.6 m/s-1.8 m/s]. Therefore, the safe distance between the two pursuit- USVs is set to be 10.32 m by the equation $d_{\text{safe}} = 2.4 \times \ell$. Besides, the successful captured distance for the pursuit-USV is set to be d_{cap} =12 m. The destroyed distances for the evasion-USV to attack the target point is assumed as d_{de} =6 m.

4.1 Case 1

The codes are written by Python. Classic solution refers to the linear programming for the matrix game problem. In the classic solution, to solve the linear programming problem, Pulp package is used to solved the matrix game. Parameters for the proposed min-max Q-learning approach are presented in Table 2.

In Case 1, a benchmark simulation environment is constructed within 40 m×40 m square-size protected area. The relevant parameters setting is given

Table 2. Parameters for the min-max Q-learning.

Learning rate	Discount factor	Training episode
0.01	0.9	30000

in Table 3. The effective time $K\Delta t$ for the two defending USVs to intercept the invading USV is set to be 30 s, where the time-period Δt between two discrete stages is 1s. $\mathbf{x_e}^0$ and $\mathbf{x_T}$ is randomly chosen within the protected square. With different $\mathbf{x_e}^0$ and $\mathbf{x_T}$, the simulation results are different. To let the pursuit-evasion process easy to be understood, a simulation group with $\mathbf{x_e}^0 := (5, 40)$ and $\mathbf{x_T} := (35, 35)$ is chosen to be simulated under the two algorithms. Simulation results are shown in Fig. 3.

Table 3. Parameters in case 1.

$\mathbf{x_1}^0$	$\mathbf{x_2}^0$	Protected area	Effective time ($K \times \Delta t$)
$(10, 10)$	$(30, 30)$	40 m×40 m	30 s

From the illustration tables in Fig. 3, we can clearly know how the participant-USVs choose strategy at different game stages. Solved by the min-max Q-learning, the pursuit-USV in green captures the invading USV at t =5 s. Under the classic solution, the pursuit-USV in purple captures the evasion-USV at t =12 s. Within this simulation group, it is evident that the min-max Q-learning algorithm performs better in terms of time-to-intercept compared to the classic solution.

To make a comprehensive comparison, a benchmark group consisting of 1000 simulations is conducted for each algorithm. In these simulations, $\mathbf{x_e}^0$ and $\mathbf{x_T}$ is randomly chosen within the protected square. The success rate of capture and the calculation time are obtained and presented in Table 4. The data clearly demonstrates that the min-max Q-learning algorithm exhibits faster calculation speed. In terms of success rate of capture, the classic solution significantly lags behind the min-max Q-learning method. Based on the aforementioned analysis, it can be concluded that the min-max Q-learning approach outperforms the classic solution in addressing the proposed model.

Table 4. Comparison among the two solutions in the benchmark group.

Solution	Total calculation time (s)	Successful captured rate (%)
Min-max Q	**156.58**	**93.0**
Classic solution	675.25	76.2

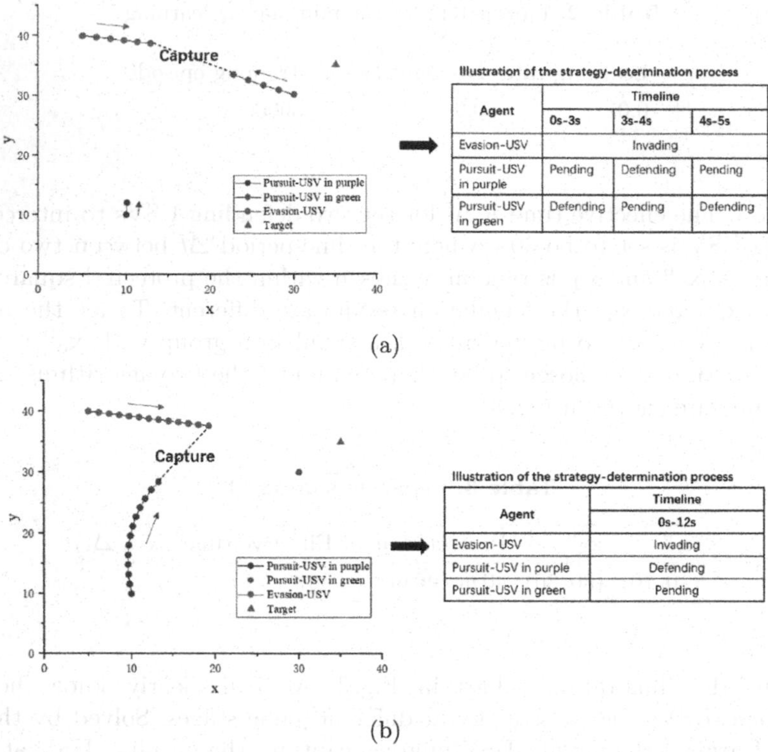

Fig. 3. The waypoints and the strategy determination illustrations of the participant-USVs under: (a) the min-max Q-learning method, and (b) the classic solution.

4.2 Case 2

In this case, by conducting simulations with three pursuit-USVs, we aim to explore the scalability of the proposed model. In comparison to the benchmark group in Case 1, the only change made is in the initial positions of the pursuit-USVs. In Case 2, the initial positions for the pursuit-USVs are set as follows: $\mathbf{x_1}^0 = (0, 0)$, $\mathbf{x_2}^0 = (20, 20)$, and $\mathbf{x_3}^0 = (40, 40)$.

Under the modified initial positions, another set of 1000 simulations is conducted. Various types of data are collected during these simulations, including the success rate of capturing the evasion-USV and the failure rate, which are presented in Fig. 4. The failure rate consists of two components: the rate of exceeding the expected capture time and the rate of the target being destroyed by the evasion-USV.

The simulations yield data for both the min-max Q-learning algorithm and the classic solution, considering two scenarios involving two pursuit-USVs and three pursuit-USVs. These results are presented in an informative sunburst chart in Fig. 4. By examining the chart horizontally, we observe that the system with three pursuit-USVs achieves a higher success rate of capture. Besides, the vertical comparison within the three pursuit-USVs highlights that the min-max Q-learning algorithm still outperforms the classic solution in the proposed model.

Overall, the findings suggest that increasing the number of pursuit-USVs in the team can enhance the success rate of capture, and the min-max Q-learning algorithm remains superior to the classic solution in the given model.

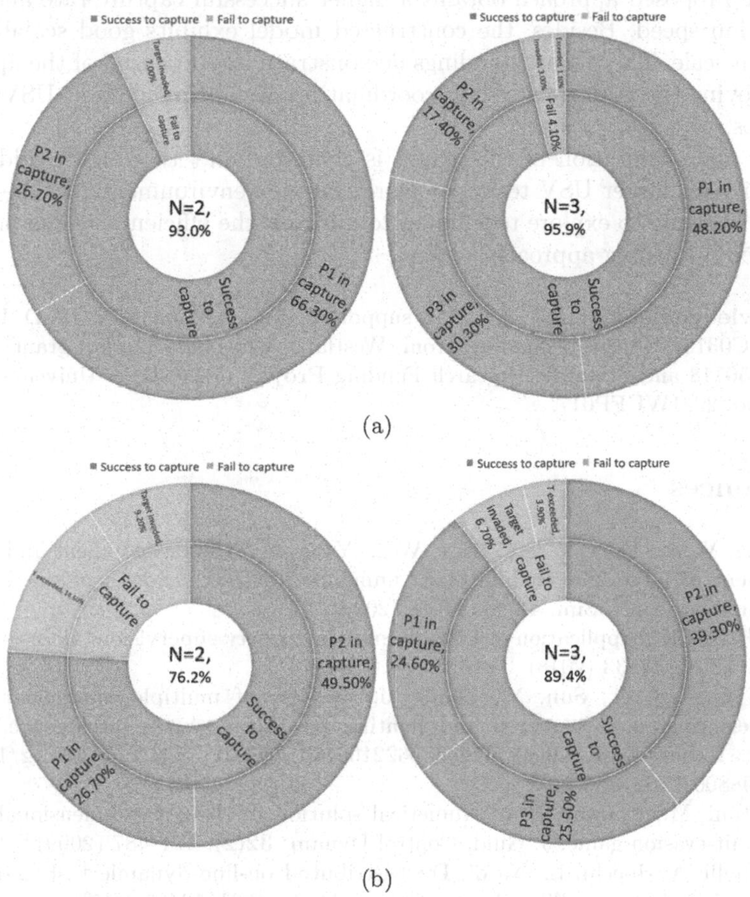

(a)

(b)

Fig. 4. The obtained data solved by: (a) the min-max Q-learning and (b) the classic solution.

5 Conclusions

This paper has presented a novel strategy determination approach using a min-max Q-learning framework for multiple USVs in the ZSPEG. The objective is to enhance the coordination and decision-making capabilities of the USVs in dynamic and adversarial environments. Based on the constructed multi-USV ZSPEG-based model, a specific min-max Q-learning approach is designed and utilized to solve the proposed model. The adoption of min-max Q-learning allows for the approximation of the payoff function during the game process.

After conducting two simulation cases, the effectiveness of the min-max Q-learning approach is verified. Compared to the classic solution, it is observed that the proposed approach obtains a higher successful capture rate and faster calculation speed. Besides, the constructed model exhibits good scalability at the team scale. The research findings demonstrate the potential of the approach in improving the performance and coordination of multiple pursuit-USVs in PE scenarios.

The main limitation of this paper is that it didn't thoroughly address its scalability for larger USV teams or more complex environments. In the future, we will continue to explore techniques to improve the efficiency of the proposed min-max Q-learning approach.

Acknowledgement. This work was supported by Zhejiang Key R&D Program No.2021C03157, start-up funding from Westlake University under grant number 041030150118 and Scientific Research Funding Project of Westlake University under Grant No. 2021WUFP017.

References

1. Kong, W.W., Feng, W.Q., Zhuge, W.Z., Yang, X.: The development and enlightenment of large and medium-sized unmanned surface vehicles of the US navy. Command Contr. Sim. **44**(5), 14–18 (2022)
2. Li, F., Yi, H.: Application of USV to maritime safety supervision. Chinese J. Ship Res. **13**(6), 27–33 (2018)
3. Wu, G., Xu, T., Sun, Y., Zhang, J.: Review of multiple unmanned surface vessels collaborative search and hunting based on swarm intelligence. Int. J. Adv. Robot. Syst. **19**(2), 17298806221091885 (2022). https://doi.org/10.1177/17298806221091885
4. Pontani, M., Conway, B.A.: Numerical solution of the three-dimensional orbital pursuit-evasion game. J. Guid. Control Dynam. **32**(2), 474–487 (2009)
5. Farinelli, A., Iocchi, L., Nardi, D.: Distributed on-line dynamic task assignment for multi-robot patrolling. Auton. Robot. **41**(6), 1321–1345 (2017)
6. Ge, J.H., Tang, L., Reimann, J., Vachtsevanos, G.: Suboptimal approaches to multiplayer pursuit-evasion differential games. In: AIAA Guidance, Navigation, and Control Conference, pp. 5272–5278 (2006). https://doi.org/10.2514/6.2006-6786
7. Pehlivanoglu, Y.V.: A new vibrational genetic algorithm enhanced with a Voronoi diagram for path planning of autonomous UAV. Aerosp. Sci. Technol. **16**(1), 47–55 (2012)
8. Wishart, D.: Differential games: a mathematical theory with applications to warfare and pursuit, control and optimization. Phys. Bull. **17**(2), 60 (1966). https://doi.org/10.1088/0031-9112/17/2/009
9. Kothari, M., Manathara, J.G., Postlethwaite, I.: Cooperative multiple pursuers against a single evader. J. Intell. Robot. Syst. **86**(3), 551–567 (2017)

10. Shaferman, V., Shima, T.: A Cooperative differential game for imposing a relative intercept angle. In: AIAA Guidance, Navigation, and Control Conference, pp. 1015–1039 (2017)
11. Yan, R., Shi, Z., Zhong, Y.: Reach-avoid games with two defenders and one attacker: an analytical approach. IEEE Trans. Cybern. **49**(3), 1035–1046 (2019)
12. Garcia, E., Fuchs, Z.E., Milutinovic, D., Casbeer, D.W., Pachter, M.: A geometric approach for the cooperative two-pursuer one-evader differential game. In: the 20th World Congress of the International-Federation-of-Automatic-Control (IFAC), Toulouse (2017). https://doi.org/10.1016/j.ifacol.2017.08.2366
13. Huang, H.M., Zhang, W., Ding, J., Stipanovic, D.M., Tomlin, C.J.: Guaranteed Decentralized Pursuit-Evasion in the Plane with Multiple Pursuers. In: the 50th IEEE Conference of Decision and Control (CDC), Orlando (2011). https://doi.org/10.1109/CDC.2011.6161237
14. Makkapati, V.R., Tsiotras, P.: Optimal evading strategies and task allocation in multi-player pursuit-evasion problems. Dyn. Games Appl. **9**(4), 1168–1187 (2019)
15. Shaferman, V., Shima, T.: Cooperative multiple-model adaptive guidance for an aircraft defending missile. J. Guid. Control Dynam. **33**(6), 1801–1813 (2010)
16. Frenay, B., Saerens, M.: QL2, a simple reinforcement learning' scheme for two-player zero-sum Markov games. Neurocomputing **72**(7), 1494–1507 (2009)
17. Hu, Y., Gao, Y., An, B.: Multiagent reinforcement learning with unshared value functions. IEEE Trans. Cybern. **45**(4), 647–662 (2015)
18. Selvakumar, J., Bakolas, E.: Min-Max Q-learning for multi-player pursuit-evasion games. Neurocomputing **475**, 1–14 (2022)
19. Wang, Y., He, H., Sun, C.: Learning to navigate through complex dynamic environment with modular deep reinforcement learning. IEEE Trans. Games **10**(4), 400–412 (2018)
20. Qu, X., Gan, W., Song, D., Zhou, L.: Pursuit-evasion game strategy of USV based on deep reinforcement learning in complex multi-obstacle environment. Ocean Eng. **273**, 114016 (2023)
21. Wu, G., Tan, G., Deng, J., Jiang, D.: Distributed reinforcement learning algorithm of operator service slice competition prediction based on zero-sum markov game. Neurocomputing **439**, 212–222 (2021). https://doi.org/10.1016/j.neucom.2021.01.061
22. Selvakumar, J., Bakolas, E.: Feedback strategies for a reach-avoid game with a single evader and multiple pursuers. IEEE Trans. Cybern. **51**(2), 696–707 (2021)

COVID-19 Urban Emergency Logistics Planning with Multi-objective Optimization Model

Baiming Zeng, Yanfen Mao, Dongyang Li(✉), and Weian Guo(✉)

Tongji University, 4800 Cao'an Road, Shanghai, China
{1952019,maoyanfen,1710333,guoweian}@tongji.edu.cn

Abstract. In response to the negative effects of epidemics on society, it is crucial to examine urban emergency supplies, particularly during the COVID-19 pandemic. To meet the material needs of urban residents and reduce the risk of infection, it is necessary to devise a vehicle scheduling system that maximizes transportation efficiency and minimizes direct human contact. This study proposes an all-encompassing strategy that takes into account transportation time, distance, and economic benefits. The analytic hierarchy process determines the input's total weight, whereas the efficiency coefficient method calculates the total efficiency coefficient. The proposed method determines the optimal distribution center for the thirteen sub-districts within Chengdu's second ring road. The vehicle routing problem is solved by an enhanced genetic algorithm with time as the primary constraint. The results of the calculations indicate that the developed algorithm effectively reduces distribution costs and improves the complete load rate of vehicles. This method takes into account the urgency of the need for emergency supplies and provides an optimal solution for the distribution of vehicle routes during urban epidemic outbreaks. These findings contribute considerably to the optimization of emergency supply distribution during crises.

Keywords: emergency supplies analytic hierarchy process (AHP) · Pre-location · genetic algorithm · vehicle routing problem

1 Literature Review

1.1 Traditional Emergency Logistics

Emergency logistics is a specialized logistics activity that seeks to provide emergency relief materials as quickly and efficiently as possible in response to unexpected events, while minimizing logistics costs. Unlike general logistics, which prioritizes both efficiency and benefits, emergency logistics prioritizes efficiency to minimize event-related losses [1]. In addition to the six basic elements of general logistics systems, namely flow, carrier, direction, volume, process, and velocity, emergency logistics also incorporates a unique time element due to the typically urgent time constraints present in emergency logistics processes. In contrast to general logistics, which emphasizes both logistics efficiency and logistics benefits, emergency logistics frequently accomplishes its logistics benefits by way of logistics efficiency [2]. As a general rule, logistics emphasizes both

H. Zhang et al. (Eds.): NCAA 2023, CCIS 1870, pp. 418–433, 2023.
https://doi.org/10.1007/978-981-99-5847-4_30

efficiency and benefits, with logistics efficiency being achieved through the optimization of logistics processes and logistics benefits being realized through the effective management of logistics costs. As such, emergency logistics often places a greater emphasis on logistics efficiency than on logistics benefits. To guarantee necessary emergency supplies to ensure the handling of emergencies, scientists have designed many solutions and methods with optimize algorithms, in order to achieve following goals:

1. Quick response and minimalize delay. The rapid response measures the supply speed of emergency logistics, and it is related to the ability of the emergency logistics system to meet the needs of the demand side in a timely manner. Barbarosoglu [3] investigates the transportation plan of helicopters, to get the minimize transport time in a model with multiple limits. Fiedrich [4] has developed an optimal resource allocation decision support system suitable for search and rescue activities after earthquakes. Knott's [5] linear programming model with the objective of minimizing costs for a single distribution center cannot adequately meet the requirement for a multi-origin distribution approach necessary to satisfy the enormous material demands of modern cities. Lucas [6] used heuristic algorithms and Ghannadpor [7] used differential evolution algorithm to solve vehicle routing optimization.
2. Reduction of logistics cost. Like general commercial logistics, emergency logistics also aims to reduce logistics costs. This is because the government spends a large amount of national financial resources in organizing emergency relief operations. To quickly and efficiently respond to the complex and urgent supply of relief materials in natural disasters, it is necessary to address the related logistical issues. Nickel et al. [8]. Considered the total cost of installation and maintenance facilities and discussed the location and quantity of ambulances and their bases in a certain area. Sudtachat et al. [9] proposed a relocation strategy to improve the performance of the Emergency Medical Service (EMS) system and demonstrated its effectiveness using an example.
3. Meeting the requirements. Emergency logistics supply of goods must meet the needs of emergency response in terms of variety and quantity. Caunhye [10] studied single emergency supplies scheduling problems. Zhang [11] conducted a study on the distribution of emergency supplies when the required quantities were not met at demand points. Heung-Suk Hwang [12, 13] used sector clustering analysis to maximize meeting the needs of disaster victims under the premise of insufficient total resources. Zhu Jianming [14] proposed a distribution method with the aim of fairness using ant colony algorithm under the theory of unfairness aversion, and verified the model through examples.
4. Minimum variance. The concept refers to how a system can deviate from its expected behavior in logistics. When natural disasters or crises happen, the rescue supply chain has to work in very chaotic conditions, and infrastructure like roads and airports may be damaged, limiting transportation capacity. Additionally, the quantity of supplies required in a short period of time can be enormous [15]. Zheng and Ling [16] studied the combinatorial optimization problem of vehicle dispatching for multiple rescue points in the event of a large-scale disaster. Kolobov [17] discussed the shortest path search problem in a directed, acyclic network with independently distributed arc lengths.

1.2 Recent Studies Focusing on Emergency Logistics Under Pandemic

In the past, the research focus in the field of emergency logistics transportation has been mainly on the distribution of supplies in humanitarian aid and the search for reliable paths in natural disaster situations [1]. The COVID-19 pandemic has highlighted the importance of efficient distribution of emergency supplies in urban areas. In response, there has been increased research on urban transportation during public health emergencies. Distributing emergency supplies during lockdowns can be complex due to population differences and urgency. Additionally, social media makes it difficult to distinguish between true and false information, which can lead to panic [18]. In addition, urban residents under epidemic lockdown are different from disaster victims, with higher psychological expectations and material needs. Therefore, providing a variety of abundant residential supplies is of great and positive significance for reducing epidemic panic, increasing public cooperation, and improving lockdown efforts to realize the fastest control of epidemic.

Yang et al. [19, 20] analyzed the impact of distance, risk area distribution, and capacity on emergency logistics in Wuhan City during the COVID-19 outbreak, used a multi-dimensional robust optimization (MRO) model and a non-dominated sorting genetic algorithm. Zhou et al. [21] conducted a study on the unconventional epidemic prevention strategy for urban public transportation during the COVID-19 outbreak. The study conducted by Liu et al. [22] aimed to address the negative impacts of epidemic outbreaks on human society by investigating the vehicle routing problem (VRP) of urban emergency supplies in response to such outbreaks. However, some limitations may exist, such as insufficient corresponding indicators within the indicator layer, which could affect the accuracy and effectiveness of the proposed solutions.

1.3 Multi-Objective Optimization Problem

Multi-objective optimization refers to the process of optimizing multiple objectives simultaneously. This task can be challenging when the objectives have conflicting characteristics, meaning that the optimal solution for one objective function may differ from that of another. When solving such problems, with or without constraints, a set of trade-off optimal solutions is generated, which are commonly referred to as Pareto-optimal solutions [23]. Mirjalili et al. [24] proposed a Multi-objective ant lion algorithm to solve engineering problems and proved its prove the effectiveness. Li Xue et al. [25] employed the fuzzy comprehensive evaluation method in their study on urban transportation efficiency. Deng et al. [26] use developed Data Envelopment Analysis to quantify the comprehensive efficiency, technical efficiency, scale efficiency, input redundancy and output insufficient of urban road network resources, so as to verify the validity and rationality of regional road network structure.

Genetic algorithms (GA) have gained popularity in solving multi-objective optimization problems due to their simplicity, ability to obtain multiple solutions, implicit parallel search in various regions of the search space, and the absence of assumptions about the problem being optimized [27]. Konak et al. [28] studied the general solution to use genetic algorithms for problems with multiple objectives. Asadi et al. [29] proposed a multi-objective optimization model using genetic algorithm and artificial neural network to quantitatively assess technology choices in a building retrofit project. Ko et al. [30] enhanced the genetic algorithm and applied it to the scheduling of emergency supplies. Dai Min et al. [31] applied enhanced genetic algorithm is developed to solve the nonlinear programming problem in resource flexibility and complex constraints in flexible manufacturing system. Thus, genetic algorithm is chosen later to develop the optimize model, more detail will be introduced in following chapter.

2 Model Establishment

2.1 Weight Calculation

In an epidemic outbreak, the distribution of emergency supplies to residents is constrained by the limited availability of supplies, transportation capacity, and resources of distribution companies. Therefore, it is crucial to consider the urgency of demand for different emergency supplies when distributing supplies. In order to improve the corresponding indicators in the demand urgency evaluation index system, this paper builds upon the "wounded urgency evaluation index system" established by Cheraghi et al. [32]. by incorporating Time of delivery, Capacity of the distribution center, and cost of delivery into the evaluation system. The resulting evaluation index system is shown in Table 1, and it represents a more scientifically and reasonably approach for assessing the demand urgency of emergency supplies demand points.

Analytic Hierarchy Process (AHP) is a decision-making method that decomposes a problem into hierarchies, such as goals, criteria, and alternatives, and combines qualitative and quantitative analysis. Bencure et al. [33] used AHP to build an evaluation model for Land value and reached good result. The purpose module has two main features. The first is that the goal of the plan is strong and clear, which is to ensure timely and adequate distribution of supplies to residents under lockdown. This goal is given top priority, and other factors such as economic considerations are secondary. The second feature is that the plan relies less on feedback from results, which allows for quick development of optimized plans in emergency situations and makes it highly reproducible for other scenarios.

Table 1. Efficiency evaluation index system of distribution center

Target layer A	Criterion layer B	Index layer C
Efficiency of distribution centers relative to each delivery point	Time of delivery B_1	Expected delivery time C_{11}
	Capacity of the distribution center B_2	Warehouse capacity C_{21}
		Fully loaded volume C_{22}
	Cost of delivery B_3	Delivery distance from center to point C_{31}
		Transportation type cost C_{32}

Table 1 shows the evaluation index system for the urgency of emergency supplies demand points. It has three levels, with five different indices: expected delivery time, warehouse capacity, fully loaded volume, delivery distance from center to demand point, and transportation type cost. Table 2 lists all the symbols used in the model.

Table 2. Symbol description of the evaluation module

Symbol	Definition	Symbol	Definition
M_{nn}	Comparison and Judgement Matrix with *n* columns and *n* rows and n columns	CI	Consistency index refers to the rough level of consistency
m_{xy}	The importance grade of index *x* compared to index *y*	CR	Consistency ratio refer to the precise level of consistency
V_n	Coefficient vector refers to the weight value of every index	RI	Random consistency index
$norm_{xy}$	Normalization result of the index *y*	x_i	The original data value of index *i*
w_y	Weight value of index *y*	x_i^h	The maximum original data value of index *i*
B_x	The *x*. layer of Criterion layers	x_i^s	The minimum original data value of index *i*
C_{xy}	The *y*. index of the *x*. criterion layer	f_i	Standardized efficacy coefficient of index *i*
λ	Eigenvalue of V_n	F_i	Comprehensive efficacy coefficient of index *i*
$\lambda_{\max}$	Maximum eigenvalue among V_n	G	Overall efficacy coefficient score

The evaluation model for emergency supply demand points has been improved to make it more scientific and reasonable. However, some assumptions are necessary to ensure the model's feasibility and reliability. These assumptions include: (1) it is assumed that each distribution center can provide all types of resources needed. (2) The model assumes a specific emergency scenario. (3) The distribution center's capacity can meet the total nominal demands of demand points. (4) Each demand point must be covered

Table 3. Construction standard of Comparison and Judgement Matrix

Grade	Meaning of grades indicating difference between indices
1	Represents that two factors have equal importance
3	Represents that one factor is slightly more important than another factor
5	Represents that one factor is clearly more important than another factor
7	Represents that one factor is strongly more important than another factor
9	Represents that one factor is extremely more important than another factor
2, 4, 6, 8	Represents the median value between two adjacent judgments

by at least one distribution center. (5) The emergency supply demand points are fixed and do not change during the planning horizon.

Based on the 1–9 scale shown in Table 3, the importance of each element in the evaluation system is judged, and then the indicators are quantified. After scoring comparison judgment matrix $\mathrm{M} = (m_{XY})_{5\times5}$ is constructed as shown in Eq. (1). In Matrix M_{nn}, each value m_{xy} is the relative weight, which reflects the relationship of mutual importance and influence between the elements No. x and No. y within the same level.

$$\mathrm{M}_{5\times5} = \left[m_{xy}\right]_{5\times5} = \begin{bmatrix} 1 & 5 & 4 & 3 & 5 \\ 0.2 & 1 & 0.8 & 0.6 & 1 \\ 0.25 & 1.25 & 1 & 0.75 & 1.25 \\ 0.33 & 1.67 & 1.33 & 1 & 1.67 \\ 0.2 & 1 & 0.8 & 0.6 & 1 \end{bmatrix} \tag{1}$$

In order to transform the comparison judgment matrix into the coefficient vector V_n, relative weight should be normalized with each column and dividing according to $norm_{xy} = \frac{m_{xy}}{\sum_{x=1}^{n} m_{xy}}$ and $w_y = \frac{\sum_{y=1}^{n} norm_{xy}}{n}$. The results are presented in Table 4. Thus, the value of coefficient vector V_n is $[0.504, 0.101, 0.126, 0.168, 0.101]^{\mathrm{T}}$.

Table 4. Analytic hierarchy process results

Index	Weight value
C_{11}	0.504
C_{21}	0.101
C_{22}	0.126
C_{31}	0.168
C_{32}	0.101

Expert scoring is subjective and can be biased by the personal opinions of experts. Therefore, it is important to test the consistency of the judgment matrix to minimize this influence and ensure that the urgency of demand is consistent with the actual situation of epidemic relief. λ is adopted to represent the eigenvalue of V_n, with the maximum eigenvalue being denoted by $\lambda_{\max}$. n is the order of the judgment matrix. The calculation of the consistency index CI is shown in Eq. (2). After introducing the random consistency index RI, the method for computing CR is illustrated in Eq. (3). It is worth noting that a smaller value of CI indicates greater consistency of the judgment matrix M. Specifically, when $CR < 0.1$, it suggests that matrix M meets the consistency test and its level of inconsistency falls within an acceptable range. It should be emphasized that the use of CR is crucial in minimizing the impact of expert subjectivity on the evaluation of the urgency of demand. As a result of Table 4, the consistency ratio CR value is $0.0016 < 0.1$, indicating that the judgment matrix satisfies the consistency test, and the calculated weights are consistent.

$$CI = \frac{\lambda_{\max} - n}{n - 1} \tag{2}$$

$$CR = \frac{CI}{RI} \tag{3}$$

2.2 Application of Efficacy Coefficient Method

The Data Envelopment Analysis (DEA) is a method used to evaluate decision-making units by considering multiple indicators. DEA quantifies the degree to which a scheme deviates from the production frontier principle. This method helps to identify the most effective decision-making units, which are those with the highest efficiency coefficient. To use DEA, input and output indicators must be identified, and the input and output values for each decision-making unit are entered into a linear programming model. The efficiency coefficient for each decision-making unit is then calculated and ranked. The normalization or standardization of the indices is necessary to ensure that each index has equal weightage in the evaluation process. The Min-Max method is used in this study to normalize the data into a range of 0 to 1. The normalized data is then used as inputs for subsequent analysis, ensuring that each index is treated equally and with-out bias. Equation (4) is used in the normalization process.

$$f_i = \frac{x_i - x_i^s}{x_i^h - x_i^s} \tag{4}$$

where f_i indicates the standardized efficacy coefficient of index i. x_i^h and x_i^s are the maximum and minimum values of the original data among this index layer. The lowest value $f_{i\min}$ takes the worst evaluated index value as benchmark, and equal to 0. The highest value $f_{i\max}$ is 1, which refers to the best evaluated index among all the layer.

In this model economic costs through different transportation method are 0–1 integer variables. Considering that air resources should be allocated preferentially to emergency supplies such as medical supplies and the price of freight, the indicator for the airport as a distribution center is taken as 0, while the train station is taken as 1.

The fitness level of the distribution center to the demand point considering all indices is then obtained by Eq. (5). f_{ji} is the standardized efficacy coefficient of demand point j's evaluation index i. One thing should be notice, that all the value f_{ji} conforms to the limitation Eq. (6) Evaluation result F presents the comprehensive fitness degree of considering all the indices, the higher the value is, the better the distribution center fit the demand point.

$$F = \sum_{i=1}^{n} f_{ji} \times u_i \tag{5}$$

$$st. \qquad f_i \leq 1 \tag{6}$$

In order to give a clearer and better reflection of the evaluation result, a scoring method is introduced to present the fitness level of the distribution center to the demand point with overall efficacy coefficient score G, which comes from Eq. (7).

$$G = c + F \times d \tag{7}$$

where c and d are both constant for deformation of index F. Constant d is set as 40, which plays the role of zooming in or zooming out the calculated value. Constant c is set as 60, which shift the value changed by constant c. So, the overall efficacy coefficient score G construct to the final result of evaluation of all index, which is limited with $G \in [60, 100]$. The whole process of the model evaluation is presented as flow chart in Fig. 1.

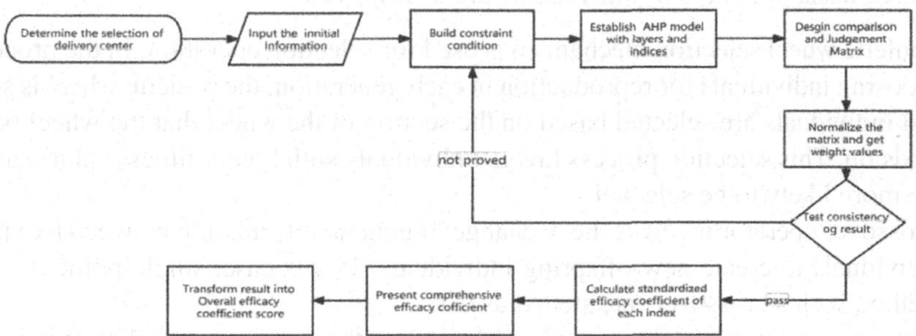

Fig. 1. The flow chart of evaluation model

3 Optimal Vehicle Routing Planning

3.1 Problem Statement

The Vehicle Routing Problem (VRP) is a well-known optimization problem that is classified as NP-hard. It was first introduced by Dantzig and Ramser [34]. Who applied it to solve petrol delivery problems. The VRP is a combinatorial optimization problem that involves finding the optimal set of routes for a fleet of vehicles to serve a set of customers, while minimizing the total cost or distance traveled. Due to its computational complexity, the VRP has been a popular research topic in the fields of operations research and computer science for several decades [35].

The vehicle routing model is subject to several assumptions, which include the following: (1) Demand points are concentrated at one point for simplicity. (2) The model is designed to address a specific emergency scenario, as many emergency situations are too complex to generalize. (3) The distribution center's capacity is assumed to be sufficient to meet the total nominal demands of the demand points. (4) Total delivery cost only includes fixed expenses of open facilities, transportation costs, penalty costs for violating time constraints, and psychological trauma to the public caused by the incident. (5) Euclidean distance is used as the distance metric in this model. (6) Every demand point must be completely covered. (7) The position coordinates of candidate points are known.

3.2 Vehicle Routing Planning Based on Non-Dominated Sorting Genetic Algorithm

The genetic algorithm (GA) belongs to a class of evolutionary algorithms that employs the fundamental principles of biological evolution, such as natural selection, crossover, and mutation, to search for and optimize solutions to problems through selection and variation. This algorithm possesses strong features of adaptability, self-organization, and intelligence, among others, which have attracted significant attention from the scientific community and found broad applications in various fields [36]. By continuously applying selection, crossover, and mutation to the population, the genetic algorithm can quickly identify the optimal or suboptimal solutions to problems, and it excels in handling complex problems with high dimensions, nonlinearity, multimodality, and so on.

Three basic operators of this method are as follows:

1. Roulette wheel selection mechanism is used for selection operation. In the process of choosing individuals for reproduction in each generation, the roulette wheel is spined and individuals are selected based on the section of the wheel that the wheel pointer lands on. This selection process favors individuals with higher fitness values, as they are more likely to be selected.
2. Crossover operator involves the exchange of genetic information between two parent individuals to create new offspring individuals. In this case, single-point crossover method with a crossover rate stetted as 0.7.
3. Mutation operator introduces random changes in the genetic material of an individual to produce a new candidate solution. The mutation operator is important for maintaining diversity in the population and preventing the GA from getting stuck in local optima, but a too high mutation probability may reduce the quality of the solutions by introducing too much randomness. In this paper, the mutation rate is set as 0.3 after multiple experiment.
4. Stop criterion the condition that is used to determine when a search or GA should stop iterating and return the best solution found so far. After experiment, the result is certain after 200 generation.

The fitness function is shown in Eq. (8), which considers the expected delivering time t_n and delivering distance d_n along the routing s_n, which can be later accurately predicted in chapter 4.1. The population of each generation contains four members. These two parameters are added up after normalizing to reflect the multi-objective programming. The flow of genetic algorithm is shown in Fig. 2

$$\text{cost}_n = \frac{\exp(-t_n)}{\sum_n^4 \exp(-t_n)} + \frac{\exp(-d_n)}{\sum_n^4 \exp(-d_n)} \tag{8}$$

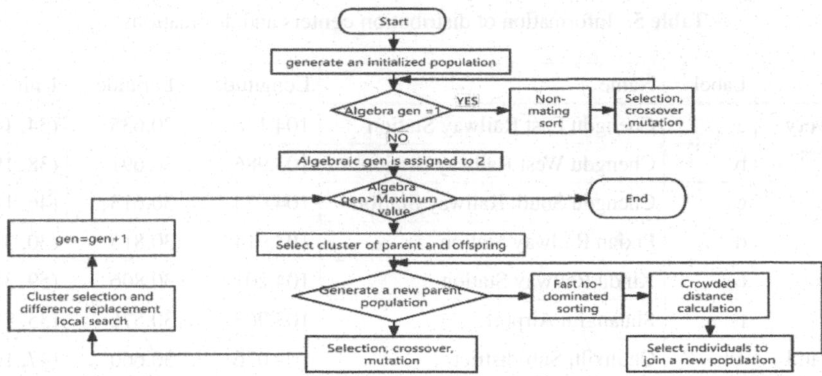

Fig. 2. Process of genetic algorithms searching vehicle routing plan

4 Simulation and Scheme of Route Plan

4.1 Data Analyses

GPS data from taxis has allowed for extensive research on predicting travel times for taxis. This research has important implications for transportation planning and management, given the significance of taxis as a mode of transportation. The data is currently used to study traffic behavior, conditions, and travel predictions, and has strong value as a reference for transportation research [37].

This article used data from ride-hailing orders in April 2016 to predict travel time in a rectangular area of 5222 km^2 between two locations in Chengdu. The area was divided into unit cells of equal size in $1km^2$, , and the predicted travel time was calculated as the average of all orders between two cells within the last thirty days. The study focused on 1,197,786 residents within the Second Ring Road of Chengdu, divided into 13 sub-districts. The distribution center was selected to be the main railway and aviation hub in Chengdu. Other data was obtained from the official website of Chengdu Municipal Government(http://www.chengdu.gov.cn/).

According to the multi-criteria evaluation model established in Sect. 2.1, the geographic and temporal information processed was input into Equations to calculate the total efficiency coefficient score for each distribution center and destination, as shown in Table 5. (Chengdu East Railway Station is under construction until 2026, so is out of consideration) The total efficiency coefficient scores are presented in Table 6. Based on the total efficiency scores, the best distribution center for each destination can be determined, as shown in Table 7. D, E, and F have no corresponding optimal distribution center, so only A, B, and C are considered as distribution centers in the subsequent route planning.

In a transportation scenario with high demand for a single type of goods, a point-to-point transportation model can be used to move goods between the distribution center and the community, based on information in Table 7. However, when a vehicle can fulfill multiple points starting from the distribution center, the vehicle routing problem is converted to the Traveling Salesman Problem and can be solved using the genetic algorithm model developed in Chapter 3.2.

Table 5. Information of distribution centers and destinations

	Label	Name	Longitude	Latitude	Unit
Delivery centers	a	Chengdu East Railway Station	104.148	30.635	(54, 14)
	b	Chengdu West Railway Station	103.986	30.691	(38, 19)
	c	Chengdu South Railway Station	104.074	30.613	(46, 11)
	d	Pixian Railway Station	103.914	30.815	(30, 31)
	e	Xindu Railway Station	104.201	30.806	(59, 31)
	f	Shuangliu Airport	103.963	30.576	(35, 8)
Demand points	1	Chunxilu Sub-district	104.076	30.660	(47, 16)
	2	Shuyuanlu Sub-district	104.094	30.664	(48, 16)
	3	Hejiangting Sub-district	104.086	30.652	(48, 15)
	4	Niushikou Sub-district	104.108	30.649	(50, 15)
	5	Taishenlu Sub-district	104.085	30.672	(47, 17)
	6	Caotang Sub-district	104.048	30.664	(43, 16)
	7	Funan Sub-district	104.033	30.681	(42, 18)
	8	Caoshi Sub-district	104.082	30.678	(47, 17)
	9	Simaqiao Sub-district	104.094	30.689	(48, 18)
	10	Fuqin Sub-district	104.051	30.692	(44, 19)
	11	Hehuachi Sub-district	104.083	30.695	(47, 19)
	12	Shuangqiaozi Sub-district	104.113	30.654	(51, 16)
	13	Mengzhuiwan Sub-district	104.103	30.668	(49, 16)

Finally, the prioritized sequence of temporary distribution centers for emergency supplies is determined by the stable solution after multiple experiments, and is obtained as follows and in Fig. 3 on map: b → 7 → 6 → 8 → 11 → 10 → b; a → 12 → 3 → 2 → 13 → 5 → 9 → a; c → 1 → 4 → c.

Table 6. Scoring table of total efficacy coefficient

	1	2	3	4	5	6	7
a	98.04	94.22	98.59	64.12	71.88	82.42	98.04
b	99.08	96.47	95.17	64.12	82.61	84.71	99.08
c	99.28	94.56	98.51	64.12	72.02	76.46	99.28

(continued)

Table 6. (*continued*)

	1	2	3	4	5	6	7
d	86.36	99.01	93.51	68.80	79.51	80.00	86.36
e	97.53	97.42	97.17	64.12	68.57	82.38	97.53
f	92.69	100.00	96.97	79.72	64.12	84.54	92.69
	8	9	10	11	12	13	
a	94.22	98.59	64.12	71.88	82.42	98.04	
b	96.47	95.17	64.12	82.61	84.71	99.08	
c	94.56	98.51	64.12	72.02	76.46	99.28	
d	99.01	93.51	68.80	79.51	80.00	86.36	
e	97.42	97.17	64.12	68.57	82.38	97.53	
f	100.00	96.97	79.72	64.12	84.54	92.69	

Table 7. Best distribution centers

Demand points	1	2	3	4	5	6	7	8	9	10	11	12	13
Optimal Delivery center	c	a	a	c	a	b	b	b	a	b	b	a	a

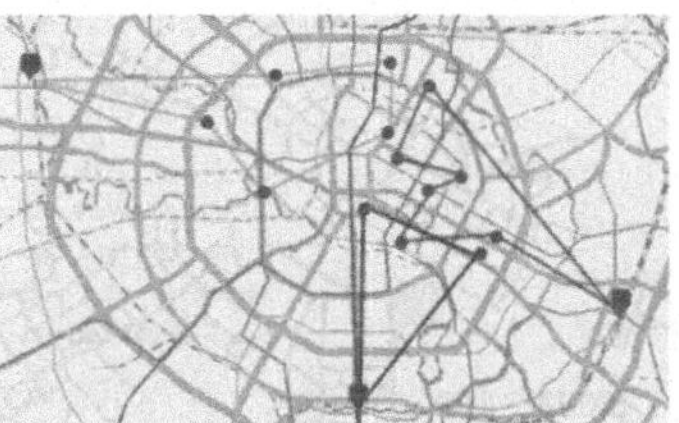

Fig. 3. Best route of distribution

4.2 Results Analysis and Comparison

Based on the algorithm described above, the routes starting from only points a, b, and c with a single distribution center model were obtained as shown in Fig. 4, and the time, distance, and number of people served for each route were calculated, as shown in Table 8. According to the ratio of served residents, the mean predicted values of time and distance for the optimization method were 6607.273 s and 20.119 km. Comparing the parallel optimization method of delivering from each optimal delivery endpoint to the respective delivery center and the single distribution center model from points a, b, and c, it can be seen from Figs. 5 and 6 that the optimization method has advantages in both time and distance, achieving the goal of minimizing input and maximizing vehicle utilization.

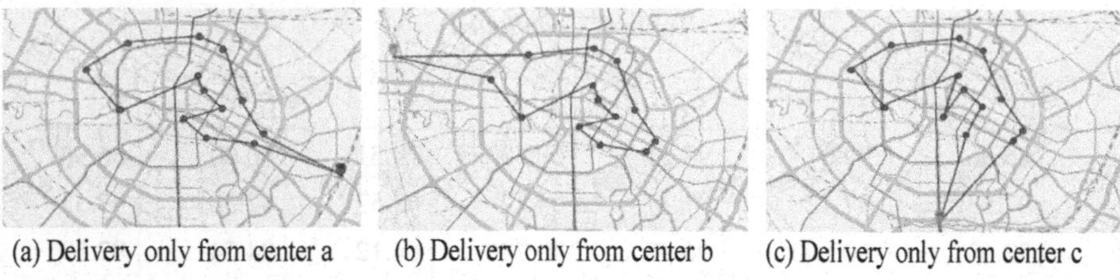

(a) Delivery only from center a (b) Delivery only from center b (c) Delivery only from center c

Fig. 4. Best routes in single distribution center method

Table 8. Information of delivery routes

Route scheme	Predict delivery time(s)	Routing distance(km)	Served residents
[a, 4, 3, 2, 5, 8, 1, 6, 7, 10, 11, 9, 13, 12]	10420.7	33.60	1197786
[b, 10, 11, 9, 13, 12, 4, 3, 2, 5, 8, 1, 6, 7]	11498.1	32.21	1197786
[c, 3, 2, 5, 8, 1, 6, 7, 10, 11, 9, 13, 12, 4]	10278.3	31.80	1197786
[b,7, 6, 8, 11, 10]	5964.4	21.71	557465
[a, 12, 3, 2, 13, 5, 9]	8471.6	20.31	474244
[c,1,4]	3355.9	16.30	166077

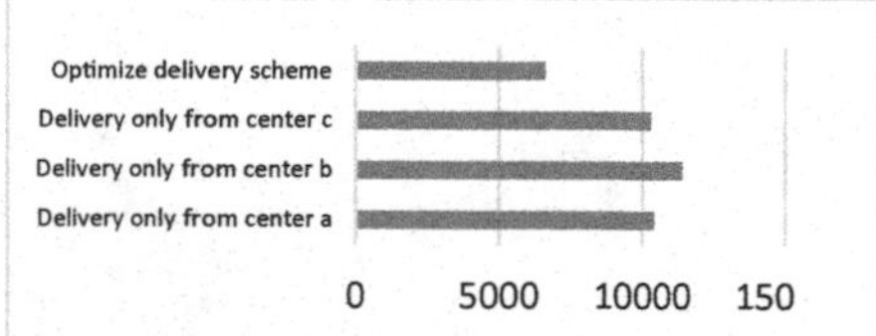

Fig. 5. Predicted delivery time(s)

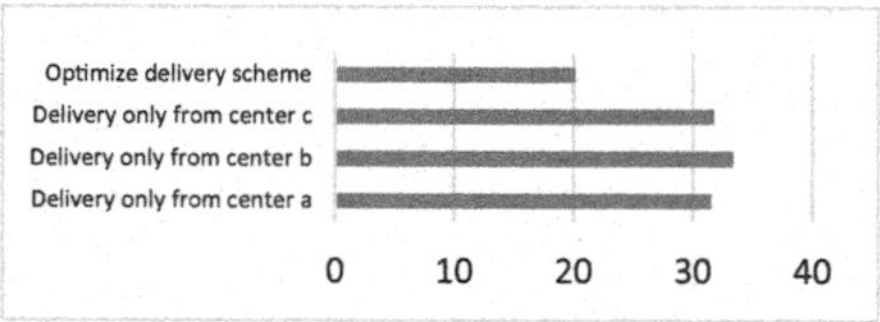

Fig. 6. Predicted delivery distance(km)

5 Conclusions and Future Work

This study developed a model for distributing emergency supplies in urban areas during an epidemic outbreak, which makes several contributions to urban humanitarian logistics in response to an epidemic outbreak: (1) the mathematical model considers the demands of residents and emergency supplies carriers during an outbreak, providing a scientific and flexible evaluation of the distribution center selection, which can be reliably tested. (2) the non-dominated sorting genetic algorithm effectively solves the vehicle routing problem with consistent results across multiple experiments. (3) the research demonstrates that the developed model can be effectively applied to the distribution of urban emergency supplies in response to epidemic outbreaks.

In the future work, the researchers suggest using fuzzy AHP for a more objective result and including more secondary indicators to refine the hierarchy structure of the analytic hierarchy process.

Acknowledgements. This work is supported by National Key R&D Program of China under Grant Number 2022YFB2602200; the National Natural Science Foundation of China under Grant Number 62273263, 72171172 and 71771176; Shanghai Municipal Science and Technology Major Project (2022-5-YB-09); Natural Science Foundation of Shanghai under Grant Number 23ZR1465400; Natural Science Foundation of Shanghai, China under Grant Number 19ZR1479000.

References

1. Cheng, M.: The Research of Transportation Optimization Problem of Emergency Logistic under Public Emergency, Tongji University (2007)
2. Sheu, J.-B.: An emergency logistics distribution approach for quick response to urgent relief demand in disasters. Transp. Res. Part E: Logistics Transp. Rev. **43**(6), 687–709 (2007)
3. Barbarosoğlu, G., Özdamar, L., Çevik, A.: An interactive approach for hierarchical analysis of helicopter logistics in disaster relief operations. Eur. J. Oper. Res. **140**(1), 118–133 (2002)
4. Fiedrich, F., Gehbauer, F., Rickers, U.: Optimized resource allocation for emergency response after earthquake disasters. Saf. Sci. **35**(1), 41–57 (2000)
5. Knott, R.: The logistics of bulk relief supplies. Disasters **11**(2), 113–115 (1987)
6. Arnold, F., Sörensen, K.: What makes a VRP solution good? The generation of problem-specific knowledge for heuristics, Computers and Operations Research **106**, 280–288 (2019)
7. Ghannadpour, S.F., Zandiyeh, F.: An adapted multi-objective genetic algorithm for solving the cash in transit vehicle routing problem with vulnerability estimation for risk quantification. Eng. Appl. Artif. Intell. **96**, 103964 (2020)
8. Nickel, S., Reuter-Oppermann, M., Saldanha-da-Gama, F.: Ambulance location under stochastic demand: a sampling approach. Oper. Res. Health Care **8**, 24–32 (2016)
9. Sudtachat, K., Mayorga, M.E., McLay, L.A.: A nested-compliance table policy for emergency medical service systems under relocation. Omega **58**, 154–168 (2016)
10. Caunhye, A.M., Zhang, Y., Li, M., Nie, X.: A location-routing model for prepositioning and distributing emergency supplies. Transp. Res. Part E: Logistics Transp. Rev. **90**, 161–176 (2016)
11. Zhang, Z., Qin, H., Li, Y.: Multi-objective optimization for the vehicle routing problem with outsourcing and profit balancing. IEEE Trans. Intell. Transp. Syst. **21**(5), 1987–2001 (2020)
12. Hwang, H.S.: A food distribution model for famine relief. Comput. Ind. Eng. **37**(1–2), 335–338 (1999)
13. Hwang, H.S.: An improved model for vehicle routing problem with time constraint based on genetic algorithm. Comput. Ind. Eng. **42**(2–4), 361–369 (2002)
14. Zhu, J., Wang, R.: Study on multi-stage distribution of emergency materials in disaster rescue based on people's psychological perception. J. Saf. Sci. Technol. **16**(2), 5–10 (2020)
15. Thomas, A.: Humanitarian Logistics, Enabling Disaster Response, The Fritz Institute, San Francisco (2003)
16. Zheng, Y.-J., Ling, H.-F.: Emergency transportation planning in disaster relief supply chain management: a cooperative fuzzy optimization approach. Soft. Comput. **17**(7), 1301–1314 (2013)

17. Kolobov, A., Mausam, M., Weld, D., Geffner, H.: Heuristic search for generalized stochastic shortest path MDPs. Proceedings of the International Conference on Automated Planning and Scheduling **21**(1), 130–137 (2011)
18. Xu, Z., Liu, Y., Zhang, H., Luo, X., Mei, L., Hu, C.: Building the multi-modal storytelling of urban emergency events based on crowdsensing of social media analytics. Mob. Netw. Appl. **22**(2), 218–227 (2016). https://doi.org/10.1007/s11036-016-0789-2
19. Yang, Y., Ma, C., Ling, G.: Pre-location for temporary distribution station of urban emergency materials considering priority under COVID-19: A case study of Wuhan City. China, Physica A: Stat. Mech. Appl. **597**, 127291 (2022)
20. Yang, Y., Ma, C., Zhou, J., Dong, S., Ling, G., Li, J.: A multi-dimensional robust optimization approach for cold-chain emergency medical materials dispatch under COVID-19: a case study of Hubei Province. J. Traffic Transp. Eng. (English Ed.) **9**(1), 1–20 (2022)
21. Zhou, J., Ma, C., Dong, S., Zhang, M.-J.: Unconventional prevention strategies for urban public transport in the COVID-19 epidemic: taking Ningbo City as a case study, China J. Highway Transport, **33**(12), 1–10 (2020)
22. Liu, H., Sun, Y., Pan, N., Li, Y., An, Y., Pan, D.: Study on the optimization of urban emergency supplies distribution paths for epidemic outbreaks. Comput. Oper. Res. **146**, 105912 (2022)
23. Deb, K.: Multi-objective optimisation using evolutionary algorithms: an introduction, Springer (2011)
24. Mirjalili, S., Jangir, P., Saremi, S.: Multi-objective ant lion optimizer: a multi-objective optimization algorithm for solving engineering problems. Appl. Intell. **46**(1), 79–95 (2016). https://doi.org/10.1007/s10489-016-0825-8
25. Li, X., Wu, F., Zuo, J.: The evaluation and analysis of Lanzhou traffic transportation efficiency. Technol. Econ. Areas Commun. **3**, 92–94 (2007)
26. Qichun, D., Meng, L., Liang, Z., Helai, H.: Evaluating transport efficiency of road network in Changsha-Zhuzhou-Xiangtan urban agglomeration. J. Railway Sci. Eng. **13**(2), 388–393 (2016)
27. Jara, E.C.: Multi-objective optimization by using evolutionary algorithms: the p-optimality criteria. IEEE Trans. Evol. Comput. **18**(2), 167–179 (2014)
28. Konak, A., Coit, D.W., Smith, A.E.: Multi-objective optimization using genetic algorithms: a tutorial. Reliab. Eng. Syst. Saf. **91**(9), 992–1007 (2006)
29. Asadi, E., Silva, M.G.D., Antunes, C.H., Dias, L., Glicksman, L.: Multi-objective optimization for building retrofit: a model using genetic algorithm and artificial neural network and an application. Energy Build. **81**, 444-456 (2014)
30. Ko, Y.D., Song, B.D., Hwang, H.: Location, capacity and capability design of emergency medical centers with multiple emergency diseases. Comput. Ind. Eng. **101**, 10–20 (2016)
31. Dai, M., Tang, D., Giret, A., Salido, M.A.: Multi-objective optimization for energy-efficient flexible job shop scheduling problem with transportation constraints. Robot. Comput.-Integrat. Manuf. **59**, 143–157 (2019)
32. Cheraghi, S., Hosseini-Motlagh, S.-M.: Responsive and reliable injured-oriented blood supply chain for disaster relief: a real case study. Ann. Oper. Res. **291**(1–2), 129–167 (2018). https://doi.org/10.1007/s10479-018-3050-5
33. Bencure, J.C., Tripathi, N.K., Miyazaki, H., Ninsawat, S., Kim, S.M.: Development of an Innovative Land Valuation Model (iLVM) for Mass Appraisal Application in Sub-Urban Areas Using AHP: An Integration of Theoretical and Practical Approaches, Sustainability (2019)
34. Dantzig, G.B., Ramser, J.H.: The truck dispatching problem. Manage. Sci. **6**(1), 80–91 (1959)
35. Tiwari, K.V., Sharma, S.K.: An optimization model for vehicle routing problem in last-mile delivery. Expert Syst. Appl. **222**, 119789 (2023)

36. Livak, K.J., Schmittgen, T.D.: Analysis of relative gene expression data using real-time quantitative PCR and the 2(T)(-Delta Delta C) method. Methods **25**(4), 402–408 (2001)
37. Sheng, Z., et al.: Taxi travel time prediction based on fusion of traffic condition features. Comput. Electr. Eng. **105**, 108530 (2023)

Joint Data Routing and Service Migration via Evolutionary Multitasking Optimization in Vehicular Networks

Yangkai Zhou[1], Hualing Ren[1], Ke Xiao[2(✉)], and Kai Liu[1]

[1] College of Computer Science, Chongqing University, Chongqing, China
{angytree,renharlin,liukai0807}@cqu.edu.cn

[2] College of Computer and Information Science, Chongqing Normal University, Chongqing, China
kexiao0625@gmail.com

Abstract. The growing Internet of Vehicles and intelligent transportation systems pose challenges in meeting real-time application demands due to increased computation costs and problem complexity. This work embarks on the first study exploring the potential relationship between data routing and service migration in vehicular networks and aims to realize efficient joint optimization of multiple tasks. We consider a scenario where vehicles request data routing tasks and service migration tasks, which would be served via V2V/V2I communications. We propose an edge-based model that formulates the joint optimization problem for data routing and service migration. This model considers the heterogeneous transmission and computation resources of edge nodes and vehicles, the mobility of vehicles, aiming at maximizing both the completion rate of data routing tasks and service migration tasks. Furthermore, we propose a novel Location Mapping based Evolutionary Multitasking (LM-EMT) algorithm. This algorithm uses different integer-based coding schemes for each of the two problems, and utilizes an explicit knowledge transfer strategy to exploit problem dependence for accelerated solving. We design a two-stage transferring strategy to mitigate the negative effects between solutions. Finally, we build a simulation model and conduct a comprehensive performance evaluation to verify the superiority of the proposed algorithm.

Keywords: data routing · service migration · evolutionary multitasking optimization · vehicular network

1 Introduction

Recent advances in wireless communication have paved the way for the development of the Internet of Vehicles (IoV). Many intelligent transportation applications, such as autonomous driving, intersection control, and vehicular argument reality, have been developed and attracted great attention in both the industry and the academia [1–3]. These applications typically require a substantial

H. Zhang et al. (Eds.): NCAA 2023, CCIS 1870, pp. 434–449, 2023.
https://doi.org/10.1007/978-981-99-5847-4_31

amount of input data and computing resources. In order to enable such applications, edge computing has emerged as a state-of-the-art paradigm, offering promising support for low-latency and high-reliability applications through task offloading [4,5]. However, the high mobility of vehicles and limited communication ranges pose significant challenges to effective service migration [6].

Currently, a significant amount of research has been focused on data routing in vehicular networks [7]. Taek et al. [8] addressed the routing problem in "unstructured" ad-hoc vehicular networks, proposing a flooding algorithm to fulfill on-demand routing in dense environments. Murugeswari et al. [9] designed a set of heuristic algorithms such as evolutionary algorithms to solve the multicast routing problem, aimed at meeting the low latency requirements of data transmission. Sedigheh et al. [10] established a communication substructure among cluster heads using hierarchical routing and data aggregation, which eliminates the need for the route discovery procedure. On the other hand, extensive studies have been conducted on service migration in vehicular networks. Ning et al. [11] constructed an energy-efficient resource scheduling framework based on mobile edge computing in IoV, and a cross-node task scheduling algorithm was designed to minimize the energy consumption of edge nodes. Liang et al. [12] proposed a relaxation-and-rounding-based solution to migrate services to nearby idle servers, thus reducing the burden on overloaded nodes. [13] formulated the service migration problem as an integer programming problem and proposed the Seq-Greedy heuristic and the particle swarm optimization algorithm to minimize migration costs. However, none of these studies attempt to leverage the valuable information between the routing problem and the migration problem for collaborative optimization, resulting in an implicit waste of computational resources.

Evolutionary multitasking (EMT) is a novel paradigm that aims to explore the potential correlations among problems in order to optimize multiple tasks simultaneously [14]. In comparison to traditional genetic algorithms [15,16], EMT continuously transfers valuable knowledge between distinct but potentially related optimization problems until optimal solutions for all problems are discovered. This unique characteristic of EMT can expedite the evolutionary search process and reduce the overall computational cost [17,18].

To address the aforementioned issue, this study introduces a novel task optimization model for vehicular networks based on EMT. The model aims to jointly optimize the data routing and service migration problems by incorporating an edge-based scheduling architecture, formulating the two problems, and designing the EMT algorithm. To the best of our knowledge, this is the first attempt to tackle the service migration and data routing problems in vehicular networks simultaneously using EMT. The main contributions of this paper are outlined below:

- We establish an edge-based model for joint data routing and service migration in vehicular networks, which enables edge-based scheduling with a global view of vehicles within its coverage. Vehicles periodically transmit their status, including location, data size, bandwidth, and task requests, to the edge nodes. Based on this collected information, the edge node is responsible for simultaneously determining the migration nodes for the service migration problem and the relay nodes for the data routing problem.

- We formulate two problems in the heterogeneous vehicular network: the data routing (DR) problem and the service migration (SM) problem. The DR problem focuses on optimizing data routing paths to maximize the data routing ratio. It considers the characteristics of the data and the heterogeneous transmission resources to model the routing delay. The SM problem aims to maximize the service migration ratio by migrating services to different edge nodes based on vehicle mobility. It considers both the features of the services and the resources to model the service delay. These two problems are fundamental yet crucial in vehicular networks, and solving them simultaneously can significantly enhance the communication and service efficiency of vehicular networks.
- To solve the DR and SM problems simultaneously, we propose a Location Mapping based Evolutionary Multitasking (LM-EMT) algorithm. Firstly, we introduce two integer-based representations to encode the solutions for the DR and SM problems. Next, we design specific crossover and mutation operators to search for optimal solutions independently for each problem. Notably, considering the significant impact of physical location distribution, we design a mapping strategy for solution transfer based on the location information associated with the tasks.

The remaining sections of this paper is organized as follows. Section 2 presents the system architecture. Section 3 formulates the joint data routing and service migration optimization problem. Section 4 proposes the EMT algorithm. Section 5 builds the simulation model and evaluates the performance. Finally, Sect. 6 concludes the paper.

2 System Architecture

In this section, we propose a system architecture for joint service migration and data routing optimization in vehicular edge computing (VEC). The architecture is illustrated in Fig. 1, which showcases multiple edge nodes (e.g., e_1, e_2) deployed along the roadside, such as Roadside Units (RSUs) and Micro base stations. Each edge node offers robust communication and computation services to support vehicles with limited resources in scheduling and executing their applications. These edge nodes exhibit heterogeneity in terms of bandwidth, communication coverage, and computation capabilities. To accommodate the heterogeneous network environment, we assume that all vehicles are equipped with both cellular and Dedicated Short-Range Communication(DSRC) interfaces. When vehicles are within the coverage area of an edge node, they can communicate with it through vehicle-to-infrastructure (V2I) communication. Additionally, vehicles can communicate with each other through vehicle-to-vehicle (V2V) communication. Specifically, the edge nodes have access to global information, including submitted requirements (i.e., data routing, service migration), as well as vehicle status (i.e., locations, driving directions, available resources) within its coverage. This information enables the edge node to make informed scheduling decisions, such as determining the routing paths for data transmission and identifying suitable migration nodes for service migration.

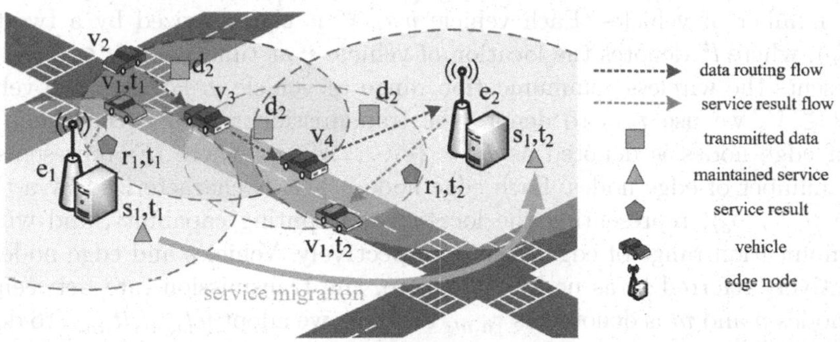

Fig. 1. System Architecture

The system architecture primarily considers two types of tasks: 1) Data Routing (DR). The DR task consists of three main components: the requested vehicle (i.e., v_2), the objective node (i.e., e_2), and the relay nodes (i.e., v_3, v_4). When vehicle requests a data to be delivered, it sends the request to the dispatching center for routing scheduling. The vehicle would forward the data packet to relay nodes via V2V communications following the routing decisions. The goal of the DR Task is to find a reasonable communication link that allows data to be transferred from the requested vehicle to the objective node within a certain time constraint. 2) Service Migration (SM). Due to the high mobility of vehicles and the limited coverage of edge nodes, vehicles often need to hand over their services from one edge node to another (i.e., v_1 moves from e_1 to e_2). To ensure the completion of services, the system needs to migrate services based on the mobility of vehicles. The goal of the SM task is to identify the appropriate migration edge node and the transmission link for the service, enabling the vehicle to obtain the service result within a given time constraint.

Although the DR and SM seem totally different, we observe that certain implicit knowledge can be transferred through EMT. Firstly, both DR and SM shares common geographical distribution and mobility feature of vehicles. And the edge node is responsible to schedule both the service demands of passing vehicles. This naturally constrains the solution of two problems in similar spaces. Secondly, the SM task needs to download data from edge nodes, leading to the dependency between SM and DR. When the vehicle moves out of coverage, the result would return to the requested vehicle via data routing. These features motivate us to explore the EMT algorithm to optimize SM and DR simultaneously.

3 Problem Formulation

3.1 Preliminary

In the proposed system, we define the set of time slots as $T = \{t_1, \ldots, t_{|T|}\}$. The set of vehicles is denoted as $V = \{v_1, \ldots, v_{|V|}\}$, where $|V|$ represents the total number of vehicles. Each vehicle $v \in V$ is characterized by a two-tuple $\langle l_v^t, \delta_v \rangle$, where l_v^t denotes the location of vehicle v at time slot t ($t \in T$), and δ_v represents the wireless communication range of vehicle v. For any two vehicles v, $v\prime \in V$, we use $\tau_{v,v\prime}$ to denote the transmission rate between them. The set of edge nodes is denoted as $E = \{e_1, \ldots, e_{|E|}\}$, where $|E|$ represents the total number of edge nodes. Each edge node $e \in E$ is characterized by a three-tuple $\langle l_e, c_e, \delta_e \rangle$, representing the location, computing capability, and wireless communication range of edge node e, respectively. Vehicles and edge nodes are collectively referred to as nodes. Therefore, the transmission rate between any two nodes n and $n\prime$ is denoted as $\tau_{n,n\prime}$. Further, we adopt $[et_{n,n\prime}, lt_{n,n\prime}]$ to denote the dwelling interval of node n in the coverage of $n\prime, whereet_{n,n\prime}$ and $lt_{n,n\prime}$ are the enter and left time, respectively.

We define the set of data routing tasks as $D = \{d_1, \ldots, d_{|D|}\}$. Each routing task $d_i \in D$ is represented by a four-tuple $\langle \sigma_{d_i}, ddl_{d_i}, v_{d_i}, ob_{d_i} \rangle$, where σ_{d_i} denotes the data size, ddl_{d_i} denotes the routing deadline, v_{d_i} denotes the index of task requested vehicle and ob_{d_i} denotes the objective node of routing task d_i. Moreover, the set of services to be migrated is denoted as $S = \{s_1, \ldots, s_{|R|}\}$. Each service $s_j \in S$ is associated with a six-tuple $\left\langle \alpha_{s_j}, \upsilon_{s_j}, \sigma_{s_j}, ddl_{s_j}, v_{s_j}, or_{s_j} \right\rangle$, where α_{s_j} denotes the computing requirement, υ_{s_j} denotes the size of service, σ_{s_j} denotes the size of download data, ddl_{s_j} denotes the service deadline, v_{s_j} denotes the index of service requested vehicle and or_{s_j} denotes the original node that maintains service s_j.

Table 1 summarizes the primary notations.

3.2 Data Routing Problem

For routing task d_i, we denote the routing path between the requested vehicle v_{d_i} and the objective node ob_{d_i} as $U_{d_i} = \left\langle v_{i,1}, ..., v_{i,|Udi|} \right\rangle$, where $|U_{d_i}|$ is the number of relay nodes on the routing path.

To calculate the routing time of d_i, we consider two factors: transmitting time and waiting time. (1) Transmitting time: For arbitrary two relay nodes $v_{i,j}$ and $v_{i,j+1}$ on the routing path, we calculate the time required for data transmitting. Given the transmission rate $\tau_{v_{i,j},v_{i,j+1}}$ and data size σ_{d_i}, the transmitting time between $v_{i,j}$ and $v_{i,j+1}$ is denoted as

$$t_{v_{i,j},v_{i,j+1}}^{trans} = \frac{\sigma_{d_i}}{\tau_{v_{i,j},v_{i,j+1}} / \theta_{v_{i,j},v_{i,j+1}}} \tag{1}$$

Where $\theta_{v_j,v_{j+1}}$ represents the number of tasks transmitted between $v_{i,j}$ and $v_{i,j+1}$.

Table 1. Summary of primary notations

Notations	Descriptions
V	the vehicles set
l_v^t	the coordinate information of v at time slot t
δ_v	the communication range of v
$r_{n,n\prime}$	the transmission rate between n and $n\prime$
l_e	the coordinate information of e
c_e	the computing capacity of e
δ_e	the communication range of e
$N_t^{n_i}$	neighboring nodes of n_i at time slot t
D	the data routing task set
σ_{d_i}	transmitted packet size of d_i
ddl_{d_i}	the maximum task tolerance time of d_i
v_{d_i}	the index of task requested task d_i's vehicle
ob_{d_i}	the objective node of routing task d_i
S	the service set
α_{s_j}	the computation required by s_j
v_{s_j}	the size of service s_j
σ_{s_j}	the size of the packet that need to transmit by s_j
ddl_{s_j}	the maximum task tolerance time of s_j
v_{s_j}	the index of service s_j's requested vehicle
or_{s_j}	the original edge node that maintains the service s_j

(2) Waiting time: The data transmitted by v_{d_i} are queued based on the ascending order of deadlines. The data with the earliest deadline is transmitted first. We denote the pending queue of v_{d_i} as $D_{v_{d_i}}$. For each data $d_{v_{d_i},k} \in D_{v_{d_i}}$, we denote its data size as $\sigma_{v_{d_i},k}$. The waiting time is calculated as

$$t_{v_{i,j},v_{i,j+1}}^{wait} = \sum_{k=1}^{j-1} t_{v_{k,j},v_{k,j+1}}^{trans} \tag{2}$$

The routing time of d_i can then be calculated as follows:

$$t_{d_i} = \sum_{j=1}^{|U_{d_i}|-1} t_{v_{i,j},v_{i,j+1}}^{trans} + t_{v_{i,j},v_{i,j+1}}^{wait} \tag{3}$$

The data routing task can be successfully completed only if it satisfies the following constraints: (1) Objective node e_{d_i} receives the data before deadline ddl_{d_i}, .i.e., $t_{d_i} \leq ddl_{d_i}$. (2) The transmission interval between two adjacent nodes $v_{i,j}$ and $v_{i,j+1}$ should within its communicable interval. i.e.,

$\left[st_{i,j},\ st_{i,j}+t^{trans}_{v_{i,j},v_{i,j+1}}\right] \subseteq [et_{v_j,v_{j+1}},\ lt_{v_j,v_{j+1}}]$, where $st_{i,j}$ is the start transmission time of $v_{i,j}$.

Finally, the objective of DR is to find set of routes of data routing tasks to maximize the completion ratio. We denote W_D as the number of satisfied tasks, data routing problem can be expressed as:

$$
\begin{aligned}
&P_1 : \max_{U_{d_i}} f_1 = \frac{W_D}{|D|} \\
&s.t.\quad t_{d_i} \leq ddl_{d_i}, \forall d_i \epsilon D \\
&\qquad \left[st_{i,j},\ st_{i,j}+t^{trans}_{v_{i,j},v_{i,j+1}}\right] \subseteq \left[et_{v_j,v_{j+1}},\ lt_{v_j,v_{j+1}}\right], \forall d_i \epsilon D,\ \forall d_j \epsilon V \cup E
\end{aligned}
$$

3.3 Service Migration Problem

For migration task s_j, we introduce a binary variable $x_{j,m,m'}$ to indicate whether s_j is migrated from edge node e_m to $e_{m'}$. If $x_{j,m,m'} = 1$, it means that s_j is migrated from e_m to $e_{m'}$; otherwise, s_j remains at e_m. Given the computation requirement α_{s_j}, migration size v_{s_j}, and download data size σ_{s_j} of s_j, the total service time can be divided into four parts: migrating time, waiting time, computing time, and downloading time.

(1) Migrating time: This is the time taken to transmit service s_j from e_m to $e_{m'}$. It is calculated as

$$t^{mig}_{s_j} = \frac{v_{s_j}}{r_{e_m,e_{m'}}} \tag{4}$$

(2) Waiting time: Since the service computing follows the Shortest-Deadline-First-Served policy at the edge node, the waiting time is defined as the duration of services pending at the edge node until they are being computed. We denote the pending queue of $e_{m'}$ as $S_{e_{m'}} = s_{e_{m'}}.k|s_{e_{m'}}.k \in S_{e_{m'}}$. The waiting time can be calculated as

$$t^{wait}_{s_j} = \sum_{k=1}^{j-1}(t^{mig}_{e_{m'},k} + t^{cal}_{s_{e_{m'}}.k}) \tag{5}$$

(3) Computing time: This is the duration for computing service s_j at edge node $e_{m'}$ and is calculated as

$$t^{cal}_{s_j} = \frac{\alpha_{s_j}}{c_{e_{m'}}} \tag{6}$$

(4) Downloading time: This is the duration for $e_{m'}$ to return the computing result to the requested vehicle v_{s_j} and is calculated as

$$t^{trans}_{s_j} = \frac{\sigma_{s_j}}{r_{e_{m'},v_{s_j}}} \tag{7}$$

When v_{s_j} moves out of the communication coverage of $e_{m'}$, the downloading time will be equal to the routing time defined in Sect. 3.2.

The service time of migration task s_j can be calculated as follows:

$$t_{s_j} = t_{s_j}^{mig} + t_{s_j}^{wait} + t_{s_j}^{cal} + t_{s_j}^{trans} \tag{8}$$

We denote W_S as the number of satisfied services, the objective of SM is to maximize the service ratio by migrating service to appropriate edge nodes, which is formulated as:

$$\begin{aligned} P_2: \ & \max_{e_{m'}} f_2 = \frac{W_S}{|S|} \\ s.t. \ & x_{j,m,m'} \in \{0,\ 1\}, \forall s_j \in S,\ e_m \in E \\ & \textstyle\sum_{m'=1}^{|E|} x_{j,m,m'} = 1, \forall s_j \in S,\ e_m \in E \\ & t_{s_j} \leq ddl_{s_j},\ \forall s_j \in S \end{aligned}$$

Above constraints guarantee that each service can be migrated to at most one node in E, and the migration procedure has to be finished before its deadline.

4 Proposed Solution

In this section, the details of the proposed Location Mapping based Evolutionary Multitasking (LM-EMT) are presented. Specifically, as shown in Fig. 2, given two task groups DR and SM, the first step is to establish mappings across task groups. Based on the physical location associated with any specific task, the corresponding task in another task group is found, and the mapping relationship is established between all tasks in the two task groups (i.e., $M_{sm \to dr}$ from SM to DR and $M_{dr \to sm}$ from DR to SM). This process ensures that the task can transfer the solution to the mapping task close to the physical location. Next, two independent evolutionary solvers are used to optimize the DR Problem and SM problem, respectively, to obtain the max completion rate of DR tasks and SM tasks. We assume that knowledge transfer occurs at fixed intervals and propose a two-stage knowledge transfer method to mitigate the negative transfer effect.

To facilitate knowledge transfer across task groups and enable the sharing of positive knowledge related to communication links and node load distribution, we focus on establishing the correct mappings between tasks. Since the number of tasks differs between the DR Task and the SM task, creating one-to-one mappings between tasks is impractical. Instead, we aim to map similar tasks from different task groups. Comparing tasks directly is challenging due to the distinct feature spaces of the task groups. To address this, we leverage the significant impact of physical location on communication and load. We artificially extract low-dimensional projections of the task groups by using location information to map tasks. This approach allows us to capture the spatial relationship between tasks and find corresponding tasks with similar physical locations.

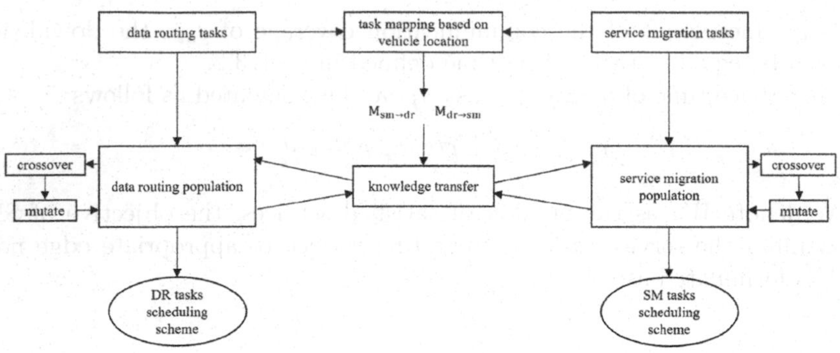

Fig. 2. Illustration of solving DR and SM via EMT algorithm

Using the example of mapping from SM tasks to DR tasks, the following steps can be followed:

First, we denote the set of request vehicles of DR tasks as $V^{dr} \subseteq V$, and denote the set of request vehicles of SM tasks as $V^{sm} \subseteq V$. The distance between v_i and v_j is represented as $dist\left(v_i, v_j\right)$. For v_i^{dr}, we select vehicle v_j^{sm} from V^{sm} as its mapping vehicle according to the distance relation. v_j^{sm} is the vehicle closest to v_i^{dr} in the V^{sm}. For v_i^{sm}, we select vehicle v_j^{dr} from V^{dr} as its mapping vehicle according to the distance relation. v_j^{dr} is the vehicle closest to v_i^{sm} in the V^{dr}. Thus, we determine the mapping vehicle for each vehicle.

Then, establish the mapping relationship between the tasks of the vehicle and the tasks of the mapping vehicle. We denote the task set of vehicle v_i^{dr} as $D_i^{dr} = \{d_{i,1}^{dr}, \ldots, d_{i,|D_i^{dr}|}^{dr}\}$, and the task set of mapping vehicle v_j^{sm} as $S_j^{sm} = \{s_{j,1}^{sm}, \ldots, s_{j,|S_j^{sm}|}^{sm}\}$. Our goal is to transform S_j^{sm} into D_i^{dr}. When $\left|D_i^{dr}\right| \leq \left|S_j^{sm}\right|$, $d_{i,k}^{dr} = s_{j,k}^{sm}$. Otherwise, $[d_{i,k}^{dr}] = \begin{cases} s_{j,k}^{sm}, & k \leq \left|S_j^{sm}\right|, \\ s_{j,rand(1,|S_j^{sm}|)}^{sm}, & \left|S_j^{sm}\right| < k \leq \left|D_i^{dr}\right| \end{cases}$, where $rand(1, |S_j^{sm}|)$ represents taking random integers in $[1, |S_j^{sm}|]$. Thus, we identified the SM task as the source of transformation for each DR Task and established the mapping relationship between the SM problem and the DR problem. The method of mapping the DR problem to the SM problem is the same as above.

4.1 Evolutionary Solver Setting

Encoding Strategy: In the proposed approach, a real number coding strategy is adopted to represent the solutions of the DR and SM tasks. N is the maximum routing hop constraint. Each element in a chromosome is represented by an integer, which implies the index of routing nodes in the condition that all nodes are ordered according to their indexes. The element will be set to 0 when the index of the objective node has been obtained in its previous genes. Specifically, for DR, the chromosome is encoded as a $|D| \times N$ two-dimensional array, where $|D|$ is the number of DR tasks. For SM, we design the chromosome to represent the migration node as well as routing decisions for vehicle moves out of communication coverage. Specifically, the chromosome is encoded as a two-dimensional array of $|S| \times N$, where $|S|$ is the number of SM tasks.

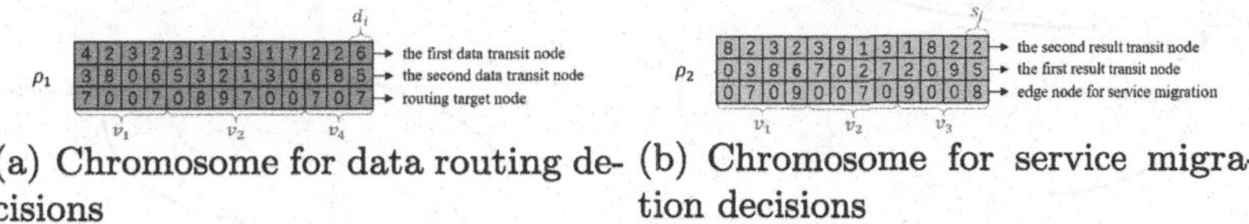

(a) Chromosome for data routing decisions (b) Chromosome for service migration decisions

Fig. 3. Example of solution presentation

Crossover: Since each chromosome corresponds to the solution of tasks requested by multiple vehicles, the crossover operation is conducted based on the unit of genes belonging to the same vehicle. Let $V\prime$ be the set of requested vehicles. For $\forall v_i \in V\prime$, denote sp_i and ep_i as the start position and end position of genes corresponding to the solution of v_i, respectively, we can get the crossover range $[sp_i, ep_i]$. On this basis, the Two-points crossover operation is conducted based on crossover probability p_c.

Mutation: We traverse the whole chromosome and randomly decide whether conduct a mutation operation based on mutation probability p_m. Considering the mutation operators may generate infeasible solutions, we find the mutation solution in the range of the neighboring nodes of its corresponding vehicle. Let n be the index of the element to be mutated, the new value of n is randomly generated from the neighboring nodes of the previous hop node.

4.2 Knowledge Transfer Strategy

To achieve knowledge transfer, we periodically select top-Q excellent individuals in the population according to the transfer frequency and add them to another population through mapping. But too many same migrants can decrease the diversity of the population, making it difficult to converge. To solve the such challenge, we design a novel knowledge transfer strategy.

As shown in Fig. 4, we propose a two-stage strategy to explicitly transfer knowledge between two populations by selecting and exchanging individuals between population Pop_1 and population Pop_2, where Pop_1 is the set of feasible solutions of DR, Pop_2 is the set of solutions of SM. Take the exchange from Pop_1 to Pop_2 as an example. 1) In the first stage, the top-Q best individuals are selected from Pop_1 and transformed into SM individuals. Those transformed individuals are stored at the temporary population Pop_t. In the second stage, $\lambda \times Q$ individuals are randomly selected from Pop_2 to crossover with the individuals in Pop_t. Then mutation operation is performed based on the remaining $(1 - \lambda) \times Q$ individuals. The newly generated individuals in Pop_t would be transferred into Pop_2. By employing this two-stage strategy, the individuals

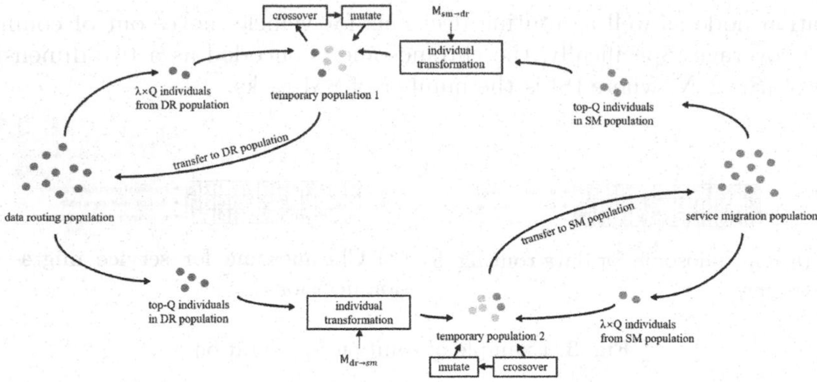

Fig. 4. Illustration of the knowledge transfer section

transferred into Pop_2 are similar to the selected individuals but not identical. This approach mitigates the negative effects of over-transfer since consistent individuals no longer flood the population, preserving population diversity.

5 Experiment

5.1 Set up

We build the simulation model based on the architecture proposed in Sect. 3. A 1.5 km × 1.5 km area is extracted from Chengdu downtown, China, and 180 vehicles are simulated. The algorithm is implemented by MATLAB.

By default, there are 10 edge nodes simulated within the concerned area. Each edge node's computing capacity is randomly generated within the range of 50 to 60 cycles/s. The transmission rate between any two edge nodes is randomly generated from 10 to 15 units/s, and the transmission rate between an edge node and a vehicle is generated within the range of 5 to 10 units/s. Regarding vehicles, the transmission rate between any two vehicles is set between 2 and 5 units/s. The number of DR tasks requested by a vehicle is randomly generated within the range of 1 to 10, while the number of SM tasks requested by a vehicle is randomly generated within the range of 1 to 8. Specifically, for each DR task, the data size and deadline are generated from intervals of 1 to 6 units and 160 to 280 s, respectively. For each SM task, the computing requirement, size, and deadline are generated from intervals of 2 to 12 cycles, 2 to 8 units, and 200 to 360 s, respectively. In the EMT algorithm, the maximum number of iterations is set to 1000, the population size is set to 100, and the maximum hop constraint is set to 2. The crossover probability is set to 0.95, and the mutation probability is set to 0.01. Additionally, explicit transferring of individuals occurs at a frequency of 40 individuals per generation.

For performance evaluation, three competitive algorithms were implemented in addition to the proposed LM-EMT algorithm. Here are brief descriptions of these algorithms:

- MFEA (Multifactorial Evolutionary Algorithm) [19]: It is a classical implicit multitasking algorithm that encodes the possible solutions of all tasks into the unified chromosome to crossover and mutation. Then the solutions to these tasks can be obtained via decoding optimal chromosomes.
- EEMTA (Explicit Evolutionary Multitasking Algorithm) [20]: It is an EMT algorithm that transfers knowledge explicitly among all tasks, where all tasks have their chromosome for evolving. The chromosomes belonging to different tasks are established mappings based on a manifold alignment strategy.
- SOEA (Single Objective Evolutionary Algorithm): It is a traditional evolutionary algorithm for single-task optimization, which solves two tasks independently.

Besides the DRR (Data Routing Ratio) and SMR (Service Migration Ratio), which is the primary objective of DR and SM, respectively, we define another metric for further discussion, which is the Number of Positive Transferred Individuals (NPTI). The NPTI is defined as the number of transferred individuals between populations which survive into the next generation, the higher value of NPTI indicates more effective transferring strategy, that is, the knowledge learned from a population can play a good role in another population.

5.2 Simulation Results

Convergence Tends: Figure 5(a) and Fig. 5(b) show the convergence trend of the DR problem with 900 routing tasks and the SM problem with 700 migration tasks, respectively. It is clear that our proposed LM-EMT coverages fast than other algorithms in both DR and SM problems. And our algorithm always achieves the highest service rate of DR and SM. The comparison results between LM-EMT and SOO indicates that our algorithm can transfer valid knowledge between different chromosomes adaptively. By comparing the performance between LM-EMT and MFEA, we note that our explicit knowledge transfer can significantly improve the scheduling performance. Moreover, the result of LM-EMT and EEMTA indicates the efficiency of our proposed mapping strategy.

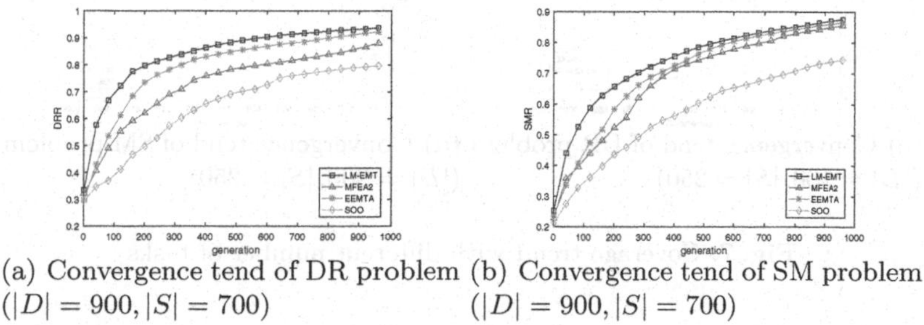

(a) Convergence tend of DR problem ($|D| = 900, |S| = 700$)

(b) Convergence tend of SM problem ($|D| = 900, |S| = 700$)

Fig. 5. Convergence trend of DR and SM problem

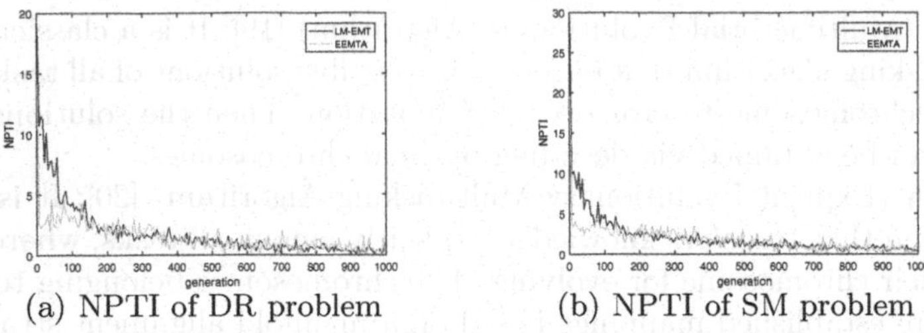

(a) NPTI of DR problem (b) NPTI of SM problem

Fig. 6. NPTI of DR and SM problem

Survival of Transferred Individuals: By analyzing Fig. 6 (a) and Fig. 6 (b), it can be seen that at the early stage of iteration, the number of positive transferred individuals is large, which indicates the knowledge learned between DR and SM can effectively improve the service ratio. With an increase of generations, the NPTI becomes lower, but it doesn't go to zero. This is because the two problems have already found their own good solutions, and the benefits of knowledge transfer for similar problems become limited. This tendency indicates that our proposed transfer strategy can provide significant effective knowledge at an early stage, and will not converge to the suboptimal solution in all ranges.

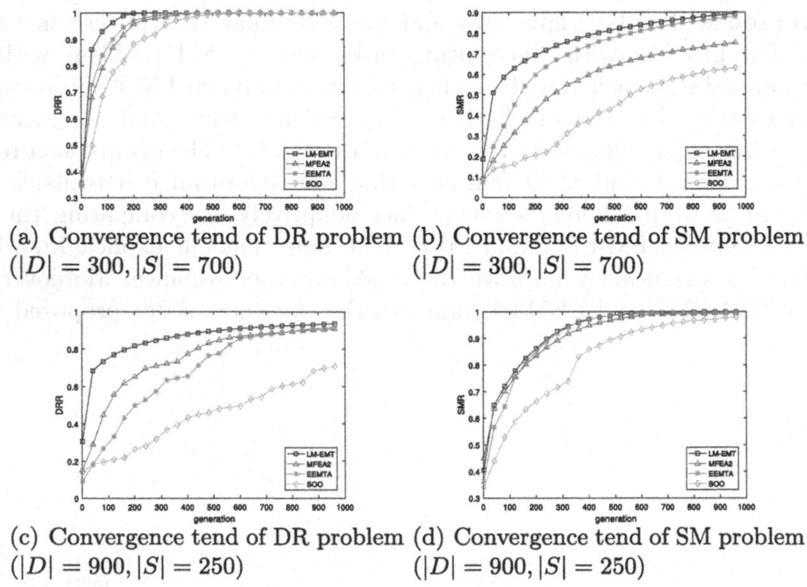

(a) Convergence tend of DR problem ($|D| = 300, |S| = 700$) (b) Convergence tend of SM problem ($|D| = 300, |S| = 700$)

(c) Convergence tend of DR problem ($|D| = 900, |S| = 250$) (d) Convergence tend of SM problem ($|D| = 900, |S| = 250$)

Fig. 7. Coverage trend with different number of tasks

Affection of the Number of Tasks: Figure 7 reflects the impact of various algorithms when the number of tasks is not balanced. Figure 7(a) and Fig. 7(d) show that all algorithms converge in a short time when the number of tasks is reduced, which is due to the smaller burden of fewer tasks on nodes. Figure 7(b) and Fig. 7(c) show that even if the number of source tasks for knowledge transfer is significantly smaller, knowledge transfer can still play a very helpful role in solving. Our algorithm shows good robustness in all cases, especially its ability to mine useful knowledge in fewer tasks is superior to other algorithms. This indicates our algorithm can be suitable for different scheduling scenarios.

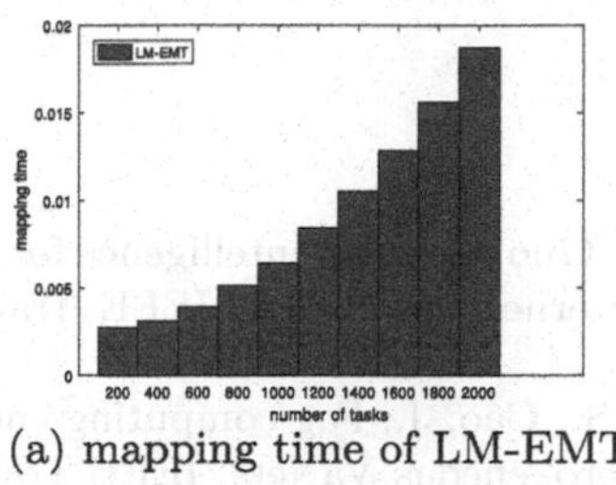

(a) mapping time of LM-EMT

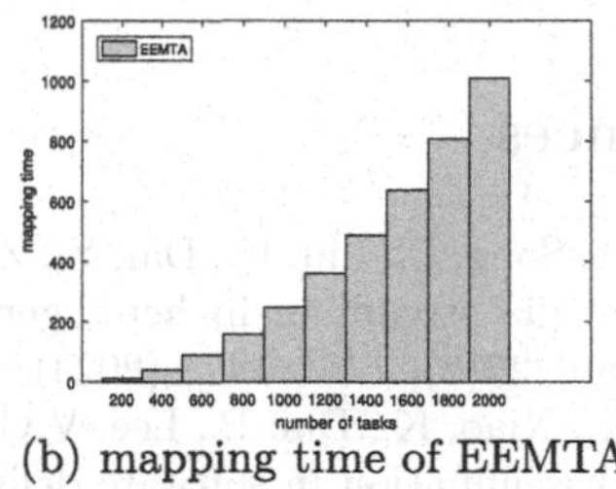

(b) mapping time of EEMTA

Fig. 8. NPTI of DR and SM problem

Elapsed Time to Establish a Mapping: Figure 8 (a) and (b) show the alignment time consumed by LM-EMT and EEMTA with the different number of tasks, respectively. The number of DR is equal to SM, and then gradually increases the total number of tasks from 200 to 2000. With the increase in the number of tasks, the growth trend of the time-consuming of the two methods increases, because the search space of finding the mapping task for a single task becomes larger. However, our method is significantly better than EEMTA in terms of mapping time, because LM-EMT adopts the method of direct dimensionality reduction to the low dimensional space, without the need to do a complex linear transformation of the manifold to project to the common subspace.

6 Conclusion

In this paper, we presented a joint data routing and service migration operation architecture based on edge computing in vehicular networks, where edge nodes and vehicles have heterogeneous transmission and computation resources. In particular, the edges node can collect information on all vehicles within its coverage to make routing and migration decisions simultaneously. Targeting to maximize the service ratio for successful transmission, we proposed an explicit EMT algorithm with the following critical components. We designed to integer encoding strategy to represent the solution to DR and SM problems. On this basis, we designed a knowledge transfer strategy to exchange solutions between two populations, where a mapping strategy was designed to unify the form of different solutions. Finally, we built the simulation model and gave a comprehensive simulation. The results demonstrated the effectiveness of our algorithm in jointly optimizing DR and SM problems. In time-sensitive vehicular network scenarios, the algorithm exhibits significant advantages, enabling efficient decision-making for data routing and service migration.

Acknowledgements. This work was supported in part by the National Natural Science Foundation of China under Grant No. 62172064, the Science and Technology Research Program of Chongqing Municipal Education Commission (Grant No. KJQN202200503), and the Chongqing Young-Talent Program (Project No. cstc2022ycjh-bgzxm0039).

References

1. Dai, P., Song, F., Liu, K., Dai, Y., Zhou, P., Guo, S.: Edge intelligence for adaptive multimedia streaming in heterogeneous internet of vehicles. IEEE Trans. Mob. Comput. **22**(3), 1464–1478 (2023)
2. Liu, K., Xiao, K., Dai, P., Lee, V.C., Guo, S., Cao, J.: Fog computing empowered data dissemination in software defined heterogeneous VaNets. IEEE Trans. Mob. Comput. **20**(11), 3181–3193 (2021)
3. Liu, K., Xu, X., Chen, M., Liu, B., Wu, L., Lee, V.C.S.: A hierarchical architecture for the future internet of vehicles. IEEE Commun. Mag. **57**(7), 41–47 (2019)
4. Ren, H., Liu, K., Yan, G., Li, Y., Zhan, C., Guo, S.: A memetic algorithm for cooperative complex task offloading in heterogeneous vehicular networks. IEEE Trans. Network Sci. Eng. **10**(1), 189–204 (2023)
5. Liu, C., Liu, K., Ren, H., Xu, X., Xie, R., Cao, J.: RTDs: real-time distributed strategy for multi-period task offloading in vehicular edge computing environment. Neural Comput. Appl. **35**, 12373–12387 (2021)
6. Wang, S., Xu, J., Zhang, N., Liu, Y.: A survey on service migration in mobile edge computing. IEEE Access **6**, 23511–23528 (2018)
7. Togou, M.A., Hafid, A., Khoukhi, L.: SCRP: stable CDS-based routing protocol for urban vehicular ad hoc networks. IEEE Trans. Intell. Transp. Syst. **17**(5), 1298–1307 (2016)
8. Kwon, T.J., Gerla, M., Varma, V., Barton, M., Hsing, T.: Efficient flooding with passive clustering-an overhead-free selective forward mechanism for ad hoc/sensor networks. Proc. IEEE **91**(8), 1210–1220 (2003)
9. Murugeswari, R., Kumar, K.A., Alagarsamy, S.: An improved hybrid discrete PSO with GA for efficient QoS multicast routing. In: 2021 5th International Conference on Electronics, Communication and Aerospace Technology (ICECA), pp. 609–614 (2021)
10. Sharifi, S.S., Barati, H.: A method for routing and data aggregating in cluster-based wireless sensor networks. Int. J. Commun Syst **34**, e4754 (2021)
11. Ning, Z., Huang, J., Wang, X., Rodrigues, J.J.P.C., Guo, L.: Mobile edge computing-enabled internet of vehicles: toward energy-efficient scheduling. IEEE Network **33**(5), 198–205 (2019)
12. Liang, Z., Liu, Y., Lok, T.M., Huang, K.: Multi-cell mobile edge computing: joint service migration and resource allocation. IEEE Trans. Wireless Commun. **20**(9), 5898–5912 (2021)
13. Kim, T., et al.: Modems: optimizing edge computing migrations for user mobility. In: IEEE INFOCOM 2022 - IEEE Conference on Computer Communications, pp. 1159–1168 (2022)

14. Feng, L., et al.: Evolutionary multitasking via explicit autoencoding. IEEE Trans. Cybern. **49**(9), 3457–3470 (2019)
15. Deb, K., Jain, H.: An evolutionary many-objective optimization algorithm using reference-point-based nondominated sorting approach, part I: Solving problems with box constraints. IEEE Trans. Evol. Comput. **18**(4), 577–601 (2014)
16. Blank, J., Deb, K., Roy, P.C.: Investigating the normalization procedure of NSGA-III. In: International Conference on Evolutionary Multi-Criterion Optimization (2019)
17. Martinez, A.D., Del Ser, J., Osaba, E., Herrera, F.: Adaptive multifactorial evolutionary optimization for multitask reinforcement learning. IEEE Trans. Evol. Comput. **26**(2), 233–247 (2022)
18. Wang, D., Liu, K., Feng, L., Dai, P., Wu, W., Guo, S.: Evolutionary multitasking for cross-domain task optimization via vehicular edge computing. In: 2021 IEEE Global Communications Conference (GLOBECOM), pp. 1–6 (2021)
19. Gupta, A., Ong, Y.S., Feng, L.: Multifactorial evolution: toward evolutionary multitasking. IEEE Trans. Evol. Comput. **20**(3), 343–357 (2016)
20. Feng, L., et al.: Explicit evolutionary multitasking for combinatorial optimization: a case study on capacitated vehicle routing problem. IEEE Trans. Cybern. **51**(6), 3143–3156 (2021)

Applications of Data Mining, Machine Learning and Neural Computing in Language Studies

Teaching Pre-editing for Chinese-to-English MT: An Experiment with Controlled Chinese Rules

Ying Zheng[1,2](✉)

[1] School of Foreign Languages, Chaohu University, Hefei 238024, China
783893009@qq.com

[2] Centre for Digital Humanities and Intercultural Communication, Chaohu University, Hefei 238024, China

Abstract. This paper aims to explore whether the training in pre-editing based on controlled Chinese rules can help students better cooperate with machine translation (MT) to deliver more accurate and effective English MT output. Three main controlled Chinese rules are put forward for web-based product description text. One rule, which requires removing redundant expressions, deals with extra-linguistic aspect. The other two rules, which require adding missing subjects and making sentences short, deal with linguistic aspect. Fifty-one students were trained in the controlled Chinese rules and then participated in a test to evaluate their grasp of relevant pre-editing methods. The results show that 39% of students can significantly improve Chinese-to-English MT output by pre-editing and over 90% of them believe pre-editing based on controlled Chinese rules can help them improve MT quality and translation efficiency. However, students are still deeply influenced by the Chinese thinking pattern and writing customs. They find it hard to determine which parts are redundant; what kind of subject to add and where to add such subjects; and how to divide long Chinese sentences. Students need more time and more practice to grasp the differences in sentence structure between Chinese and English, to master the differences in writing style between Chinese and English business texts, so as to better leverage controlled Chinese rules in pre-editing. This teaching experiment on pre-editing based on controlled Chinese rules for C-E translation is a new attempt and can provide insights for future C-E translation pre-editing teaching. With the machine translation becoming increasingly intelligent and powerful, our translation training maybe should pay more attention to the aspect of writing style, so that we can better cooperate with MT to produce effective output.

Keywords: Pre-editing for Chinese-to-English Translation · Controlled Language · Controlled Chinese Rules · Pre-editing Teaching

1 Introduction

Neural machine translation (NMT) has improved significantly in the aspects of translation accuracy and fluency. But for Chinese-to-English (C-E) translation, due to the great differences in their language structure, text style, and social culture, MT (specifically

H. Zhang et al. (Eds.): NCAA 2023, CCIS 1870, pp. 453–464, 2023.
https://doi.org/10.1007/978-981-99-5847-4_32

referring to NMT in this paper) cannot yet deliver high-quality output for operative texts, for instance, web-based business text. Students often use MT in Translation/English classes for the simple fact that MT could provide some clues or even deliver adequate translation versions. Instead of prohibiting students from using free online MT in translation-learning classes, the author proposes the pre-editing module to help students better cooperate with MT to produce more effective output.

The author currently teaches students majoring in Business English the course of business translation between Chinese and English in a local university in China. The students would need to translate Chinese business texts into English in their future work, and they would probably use free online MT for help. So, learning how to use pre-editing to improve C-E MT might be useful for their future translation practice. Based on controlled Chinese rules deduced from comparative texts in Chinese and English, students were trained in controlled Chinese and pre-editing for three weeks and then took a test to find out if the pre-editing training could help them improve MT output. The contribution of the paper includes:

1) The test results and analysis can reveal the difficulties students (or most Chinese) may encounter in their interaction with MT and provide information for future translation and pre-editing training.
2) The teaching module can help students better cooperate with MT to produce relatively effective output, which can reduce the workload of post-translation editing and improve translation efficiency.
3) The module will enable students to have a better comparative knowledge of the language structures and styles of Chinese and English, and thus improving their grasp of both languages.

2 Pre-editing and Controlled Chinese Rules

2.1 Literature Review on Pre-editing and Controlled Language

Pre-editing is "the process of preparing a source language (SL) for translation by a machine translation system" [1]. Carbonell & Tomita pointed out the aim of pre-editing is to "eliminate complex grammatical structures, ambiguous words and problematic nuances" [2]. Pre-editing is often based on a controlled language (CL). Controlled language refers to "subset of natural languages whose grammars and dictionaries have been restricted in order to reduce or eliminate both ambiguity and complexity" [3]. Since 1970s, controlled language has been applied in aviation and other technology industries to improve the readability and translatability of technical texts. Studies on controlled translation (controlled language combined with machine translation) have found that controlled language can help improve the overall quality of the output from earlier MT approaches [4, 5]. But for NMT, Shaimaa Marzouk found that "neural MT delivered distinctly better results than earlier MT approaches, … which indicates that neural MT offers a promising solution that no longer requires CL rules for improving the MT output" [6].

Marzouk's study focused on German-to-English translation of technical texts. German and English belong to the same Indo-European language family, and they have closely-related linguistic forms, thinking patterns, and cultural habits, which might make it easier for MT to deliver accurate output. But for Chinese-to-English translation of business texts, the MT output can often be inappropriate due to different thinking patterns behind the two languages (linguistic factors) and different writing styles of business texts (extra-linguistic factors). Thus, pre-editing based on controlled Chinese is quite necessary to help neural MT deliver effective as well as accurate output. Studies on pre-editing have mainly focused on the aspect of accuracy; a recent study in 2021 by Chung-ling Shih talked about the communicative aspect from the perspective of Grice's Cooperative Maxims [7]. In this paper, both aspects have been taken into account in the pre-editing training. Based on controlled Chinese rules, students were trained in how to improve MT output in both accuracy and communicative effect with pre-editing. Before the training module, it was necessary to compare Chinese and English business texts of the same genre and establish controlled Chinese rules for pre-editing.

2.2 Controlled Chinese Rules for Web-based Business Text

In the author's previous study [8], controlled Chinese rules have been designed for product description text, a typical web-based business text. The controlled Chinese rules are based on the analysis of comparable bilingual texts, the principles of modern business writing, and the general differences between Chinese and English languages. The comparable texts were: the English text from Apple's official website; the Chinese text from Xiaomi's official website (a successful Chinese electronic brand); and the English translation text, which is the English version of the Chinese text on Xiaomi's UK official website. Here we have chosen three main controlled Chinese rules for the teaching experiment, which can be classified into the linguistic aspect and extra-linguistic aspect.

For the linguistic aspect, one rule is that every sentence should have an explicit subject (**Rule 1**), and the corresponding pre-editing method is adding a subject. This rule is derived from the differences in sentence structure between Chinese and English. The Chinese grammar permits a null subject when the subject can be inferred from the context, so sentences with no subject are very common in Chinese language. But with the English language—a subject-prominent language, the subject is the start of a sentence and is rarely omitted in writing. When MT deals with Chinese sentences with no explicit subjects, if the subject is obvious from the context, the MT may add the right subject; but if the subject is not so obvious, the machine might translate the sentence into the passive voice, or add the impersonal pronoun "it" as the subject. The passive voice does not conform to the business writing principle of adopting the active voice, and "it" as the subject can sometimes be confusing in meaning. Therefore, adding a subject in the pre-editing of null-subject sentences can help MT recognize the correct subject and adopt the active voice, which is preferred in business writing [9].

Another rule for the linguistic aspect is that sentences should be short (**Rule 2**), and the relevant pre-editing method is making sentences short. A Chinese sentence can be long and consist of several comma-connected short clauses, which is not permitted in English grammar. The short clauses connected by commas, and the hidden logic, are easy for Chinese readers to comprehend. However, if such long sentences are machine-translated into English, the MT output might be long-winded sentences, joined by "and"

and other coordinating conjunctions, if the machine cannot recognize the logic to automatically cut the sentences. The long sentences can slow down the reading speed and lose readers. The main factors affecting the reading difficulty of Flesch Reading Ease are word length and sentence length, which indicates that to attract more readers, we should choose short words and cut sentences to reduce the reading difficulty when translating such texts into English. So, making sentences short in the pre-editing step can help MT deliver short sentences, which are considered to be "the best" for business writing [9].

For the extra-linguistic aspect, the rule is that there should be no repetitive complimentary expressions (**Rule 3**), and the pre-editing method is removing redundant expressions. The English text rarely has complimentary expressions with similar meanings, while the Chinese text adopts a variety of complimentary expressions with similar meanings to enhance the advertising effect and highlight the advantages of the product. Redundant expressions in English translation text will not in fact better present the product advantage and do not include more information. According to Grice's cooperative maxim of quantity, a wordy text is often boring to the reader, and many feel it is a waste of time to read a long text for information scanning or skimming [10]. In Shih's research, the modified Grice's maxim of quality is that "For web audiences, instead of genuine, reliable information, the important and interesting information is what they are more concerned about." [7] So, to make a text less wordy and keep only the important and interesting information, redundant complimentary expressions should be removed in pre-editing. This rule also concurs with the business writing principle of briefness [9].

3 The Teaching Experiment

3.1 Participants

The participants were 51 juniors majoring in Business English from a local university in China. Before the teaching experiment, they had already taken courses on basic English skills and business translation. They have varying language proficiency levels and learning motivations. 71% of them (36) passed TEM-4 (Test for English Majors Band 4, a national English proficiency test for English majors in China), and all of them passed CET-6 (College English Test Band 6, a national English test for all college students). Therefore, they have basic translation knowledge and skills.

3.2 Procedure

Students learned controlled Chinese rules and pre-editing methods in class as part of a three-week module. The teacher explained the controlled Chinese rules and pre-editing methods, and then gave students exercise texts for pre-editing and MT. After students finished the exercises, the teacher guided students to compare the MT output of pre-edited text with MT output without pre-editing, so that they could see the effect of pre-editing at first hand. Then the teacher commented on students' performance and gave suggestions for improving their pre-editing skills. At the end of the module, they participated in a pre-editing test to see if they can improve C-E translation by pre-editing. After the test, a survey was conducted to find out students' opinions on the effectiveness of pre-editing based on controlled Chinese rules. The translation engine selected is Sogou Translate based in China. Sogou Translate is one of the mainstream generic C-E translation engines and has received good comments from users. According to Cai Xinjie and Wen Bing's case study on MT of publicity texts, the comprehensive performance of Sogou Translate in C-E translation is better than that of Youdao Translate, Baidu Translate and Google Translate [11]. The detailed procedure is shown in Fig. 1.

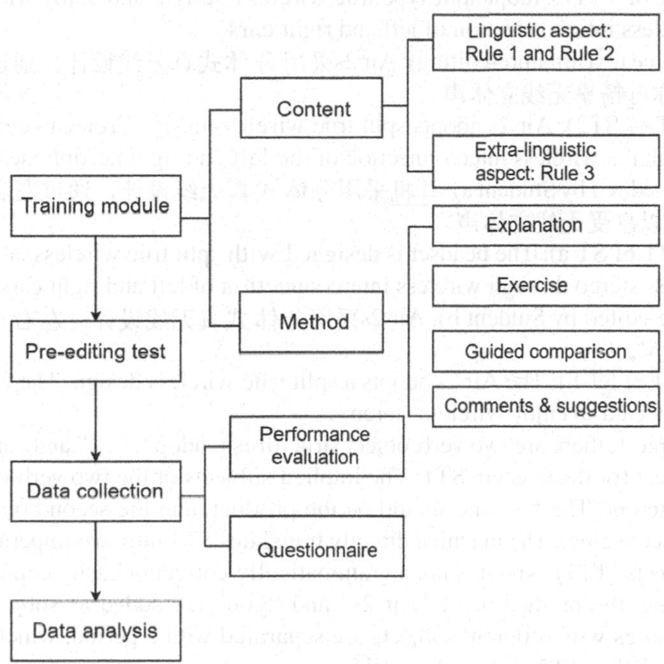

Fig. 1. Research procedure

4 Data Analysis and Discussion

20 students (39%) have significantly improved C-E performance by pre-editing; 19 students (37%) did show that they can use controlled Chinese rules on certain scenarios, but they did not improve overall MT performance significantly; 12 students (24%) didn't improve MT performance or even reduced MT performance by inappropriately revising the source text. Since students have only learned controlled Chinese rules for three weeks, this experiment can reveal the difficult points for students, and which areas we should pay attention to in future pre-editing teaching. The application of each controlled Chinese rule by students in the pre-editing test will be analyzed in detail with examples. The reasons as well as countermeasures will be talked about.

4.1 Analysis on Students' Application of Rule 1—Every Sentence Should Have an Explicit Subject

Example 1

ST1 (source text without pre-editing): 采用分体式真无线设计, 通过左右耳无线互联, 畅享无线立体声。

TT1 (MT of TT1): Adopt split-type true wireless design, and enjoy wireless stereo through wireless interconnection of left and right ears.

ST2 (source text after pre-editing): Air 2s采用分体式真无线设计。通过左右耳机无线互联, 你可畅享无线立体声。

TT2 (MT of ST2): Air 2s adopts split true wireless design. You can enjoy wireless stereo through the wireless interconnection of the left and right headphones.

ST a (pre-edited by Student a): 耳机采用分体式真无线设计，通过左右耳无线互联，用户可以享受无线立体声。

TT a ((MT of ST a): The headset is designed with split true wireless, and users can enjoy wireless stereo through wireless interconnection of left and right ears.

ST b (pre-edited by Student b): Air 2s采用分体式真无线设计。左右耳互联，畅享无线立体声。

TT b (MT of ST b): The Air 2s adopts a split true wireless design. The left and right ears are connected to enjoy wireless stereo.

In Example 1, there are two verb-object structures ("adopt ……" and "enjoy ……") with no subject (or the doer) in ST1; The implied subjects of the two verb-object structures are different. The first one should be the product, and the second one should be "you" or the consumer. The machine directly translates ST1 into two imperative clauses with no subjects (TT1), which is not a grammatically correct or easily comprehensible. By pre-editing, the product name "Air 2s" and "you" are added as subjects, and the two short clauses with different subjects are separated with a period, which makes the machine translation TT2 clear and unambiguous.

Most students can add a subject in the test, but there are two problems in practice. The first problem is that only 17 students (33%) could add the subjects *你/您* (you) and Air 2s, and the majority still prefer *用户* (users) to *你* (you), ***耳机*** (headphones) to the specific product name "Air 2s" (see ST a). *用户* (users) is from third person point of view and ***耳机*** (headphones) is a generic term. When talking about product features on product description page, the Chinese language tends to adopt generic terms and the perspective of third person, as opposed to specific terms and the first person and second person point of view preferred by English business writing style. Although students have been taught to adopt a conversational tone and use specific expressions in pre-editing, their pre-editing performance shows it is still hard for them to adapt their Chinese thinking pattern. So, in future pre-editing teaching, we should pay attention to the fact that students are deeply influenced by Chinese language custom, and should be advised to use first person (we)/second person (you) and specific terms for product description translation. The second problem also stems from influence by Chinese language custom. The Chinese language often omits the subject if the subject can be inferred from the context, as shown by the source text. If the subject at the beginning of a sentence is omitted, most students can detect the problem and add the omitted subject. But if a subject is omitted in the middle of a sentence, 61% of students could not see the lack of a subject (see ST b). In Example 1, 86% students can add a subject at the sentence beginning, but only 20 students (39%) added a subject before *畅享*(enjoy), which as a verb clearly needs a subject when translated into English. We can see from TT b that without the subject, the second clause is in passive voice and confusing in meaning. In future pre-editing training, we should suggest students pay attention to subject omission not only at the beginning of Chinese sentence, but also in the middle of a sentence, so that they can add subjects properly to make the semantic meaning of source text complete and clear for MT.

4.2 Analysis on Students' Application of Rule 2—Sentences Should be Short

Example 2

ST1: 小米真无线蓝牙耳机 Air 2s升级双核芯片,采用全新双耳同步传输技术,将声音信号同时传递到左右耳机，有效降低声音延迟，同时具有强抗干扰性，左右耳机连接更稳定。

TT1: Xiaomi's true wireless Bluetooth headset Air 2s upgrades dual-core chip, adopts new binaural synchronous transmission technology, and transmits sound signals to the left and right headphones at the same time, effectively reducing sound delay, <u>and at the same time,</u> it has strong anti-interference performance, and the connection between the left and right headphones is more stable.

ST2: <u>我们</u>升级 Air2s双核芯片，采用全新双耳同步传输技术，将声音信号同时传递到左右耳机，有效降低声音延迟<u>。</u>同时<u>Air 2s</u>具有强抗干扰性，左右耳机连接更稳定。

TT2: We upgraded Air 2s dual-core chip, and adopted a new binaural synchronous transmission technology to transmit the sound signal to the left and right headphones at the same time, effectively reducing the sound delay. At the same time, Air 2s has strong anti-interference performance, and the connection between left and right headphones is more stable.

ST c (Student c): Air 2s升级双核芯片，采用全新双耳同步传输技术，将声音同时传递到左右耳机，有效降低声音延迟，同时具有强抗干扰性，左右耳机连接更稳定。

TT c: Air 2s upgrades dual-core chip and adopts new binaural synchronous transmission technology to transmit sound to the left and right headphones at the same time, which effectively reduces the sound delay, and has strong anti-interference and more stable connection between the left and right headphones.

ST1 in Example 2 is a long Chinese sentence composed of several comma-connected short clauses. The sentence describes two advantages of Air 2s—reducing sound delay and strong anti-interference. However, there is no period or semicolon between these two advantages, and a comma is used instead. Although the Chinese writing habit of connecting several semantically related short clauses with comma is quite common, this loose structure cannot present the sentence logic clearly. Thus, the machine translates the whole sentence into several short clauses connected by commas or "and". The translated sentence is long-winded and the logical relationship is not clear enough, which will slow down the reading speed of customers and lose their interest. If the long sentence is reasonably cut short with a period and subjects are added at proper positions (see ST2), the machine-translated sentence can be relatively shorter. In addition, with shortened sentences, the machine can recognize and deliver the sentence logic more clearly and accurately (see TT2).

Students' pre-editing performance shows that 29 students (57%) were able to cut sentences short reasonably. However, 22 students (43%) failed to break the sentence effectively, but rewrote the sentence by removing the expressions that they thought were redundant (such as ST c). Although the removal makes TT c more concise to some extent, the logic of TT c is still unclear. Such pre-editing is not only time-consuming, but also ineffective. The reason why some students can't break sentences effectively is that their Chinese reading and writing habits are engrained: they pay little attention to punctuation marks when reading and tend to understand sentences based on context; their writings often adopt a lot of commas, and the habit of using punctuation marks carefully has not been formed. This means that in future pre-editing and translation teaching, students should be guided to pay attention to the usage of punctuation marks and the logical relationship between sentences. They should grasp the differences between Chinese and English punctuation rules and strengthen their editing practice in Chinese long comma-connected sentences.

4.3 Analysis on Students' Application of Rule 3—There Should be no Repetitive Complimentary Expressions

Example 3

ST1: 大尺寸动圈单元 听感浑厚自然

采用 14.2mm 大尺寸的复合振膜动圈单元，声音更加丰满浑厚，听感自然，还原更多真实质感和细节。同时支持 AAC 音频编解码技术，带来听得到的音质提升与体验，尽享动听。

TT1: Large-scale moving coil unit has a rich and natural listening sense.

Using 14.2 mm large-size compound diaphragm moving coil unit, the sound is fuller and more natural, and more real texture and details are restored. At the same time, it supports AAC audio coding and decoding technology, which brings audible sound quality improvement and experience and enjoys the sound.

ST2: 高品质声音

14.2 mm 大尺寸的复合振膜动圈单元让声音更加丰满。同时 Air 2支持 AAC 音频 编解码技术，带来更好音质。

TT2: High quality sound.

14.2 mm large composite diaphragm moving coil unit makes the sound fuller. At the same time, Air 2 supports AAC audio coding and decoding technology, which brings better sound quality.

ST d (Student d): 大尺寸动圈单元让声音更加浑厚自然

耳机采用 14.2mm 大尺寸的复合振膜动圈单元，让声音更加丰满浑厚，听感自然，还原更多真实质感和细节。同时支持 AAC 音频编解码技术，带来听得到的音质提升，尽享动听音乐。

TT d: Large-size moving coil unit makes the sound more rich and natural.

Headphones use 14.2 mm large-size composite diaphragm moving coil unit, which makes the sound fuller and more natural, and restores more real texture and details. At the same time, it supports AAC audio coding and decoding technology, which improves the audible sound quality and enjoys beautiful music.

ST1 of Example 3 contains a subheading and a detailed explanation. As for the subheading, its first phrase is the advanced product technology, and its second phrase is the effect brought by that technology. According to the comparable corpus (the product description pages of mainstream wireless headphones on Amazon website), English subheadings are generally 3–4 words, rarely exceeding 5 words. There are obviously too many words in the Chinese subheading, so ST2 rewrites the subheading and shortens it into one phrase. In the explanation part, *声音更加丰满浑厚，听感自然，还原更多真实质感和细节* and *带来听得到的音质提升与体验，尽享动听* are similar complimentary expressions indicating one meaning—high-quality sound. According to Shih's research, online texts should be as short as possible to speed up reading and reduce cognitive load, and only key and "intriguing" information should be kept to attract readers' attention [7]. Therefore, ST2 removes repetitive content and only keeps key information. TT2 is correspondingly more concise.

Most students show their effort to eliminate redundant expressions, but they have different opinions as for which expressions are redundant. For the subheading, only 20 students removed the second part of the subheading; other students either explained the subheading into a longer sentence, which does not comply with the shortness requirement for subheadings and may cost too much time, or they changed several words to make the subheading sound more reasonable in Chinese, but didn't significantly improve MT performance (see ST d and TT d); and a few students didn't change any part of the subheading. For the repetitive expressions, 20 students removed the redundant expressions; 18 students deleted the expressions they considered as "talkative" and revised ST1 based on instincts, but the revised version couldn't help improve MT performance (see ST d and TT d). We can conclude that students may find it hard to determine which parts should be removed, and which parts shouldn't, and they may be doing pre-editing based on instinct, not rules. In future pre-editing training, students should be guided to make a comparative analysis of comparable texts, so that they can conclude by themselves what expressions are considered redundant in English (including repetitive expressions, unnecessary compliments, or exaggerations), although such expressions may be perceived as necessary in Chinese writing. By comparative learning, they can also have a better understanding of the differences in writing styles between Chinese and English. The removal of certain expressions might result in translation unfaithfulness at the superficial level, so students are advised to explain the omissions to their patrons and decide whether the eliminations are appropriate and acceptable to their patrons.

In addition to the application of controlled Chinese rules, 13 students (25%) paraphrased the whole passage, but didn't significantly improve overall MT performance. Paraphrasing the whole text may cost a lot of time and cannot significantly help improve MT quality. Therefore, it is suggested that students only pre-edit the parts that need revision. As the standard document ISO18587:2017 suggested, when translators post-edit, they should revise the inappropriate expressions and try to use the original output as much as possible, to increase efficiency and also to keep accuracy delivered by MT

[12]. Students should also abide by this rule in pre-editing and try to avoid unnecessary revisions.

4.4 Students' Feedback

In the survey questionnaire, students were asked to evaluate pre-editing and specific controlled Chinese rules on a five-point scale. Table 1 describes the mean values of students' ratings for each question. We can see that the ratings are quite positive, and the scores of "I believe pre-editing can help me improve MT quality"(4.529) and "Evaluate the rule: Every sentence should have an explicit subject"(4.510) are higher than the scores of other questions. 97% students (giving scores of 4 or 5 points) believe pre-editing could help them improve MT output quality, and 90% think pre-editing could help them improve translation efficiency—the lower proportion indicating students need more practice to grasp pre-editing and use this tool efficiently. As for specific rules, 97% consider the pre-editing method of adding a subject as effective, 93% consider the method of removing redundant expressions as effective, and only 87% of students believe the method of making sentences short is effective, which means a small percent of students don't recognize the importance of cutting short long-winded Chinese sentences.

Table 1. Descriptive statistics of students' evaluation of pre-editing and controlled Chinese rules (n = 51).

Question item	Statistical counting					Mean	Std. Deviation
	5	4	3	2	1		
1. I believe pre-editing can help me improve MT quality	29	20	2	0	0	4.529	0.578
2. I believe pre-editing can help me improve my translation efficiency	24	22	5	0	0	4.373	0.662
3. In my future translation work, I will use pre-editing to help improve MT quality	26	19	6	0	0	4.392	0.695
4. Evaluate the rule: Every sentence should have an explicit subject	29	20	1	1	0	4.510	0.644
5. Evaluate the rule: Sentences should be short	26	19	6	0	0	4.392	0.695
6. There should be no repetitive complimentary expressions	27	21	1	2	0	4.431	0.728

The last question in the questionnaire was an open question: "Do you have any suggestions for pre-editing based on controlled Chinese rules?" Thirty-one students answered the question. Ten students commented that pre-editing should eliminate ambiguities, pay attention to logic as well as details. Three students said we should pay attention to Chinese words with multiple meanings or implied meanings at the pre-editing step. Two students emphasized cultural differences and pointed out adaptations should be made in certain contexts. Two students said that the pre-editing should not change the original meaning of the source text. One student pointed out we should diversify the expressions in different contexts, which is an interesting and valuable idea. Three students wrote they need more practice to carry out effective pre-editing, and three wrote that they need to improve their Chinese language skills. Five students repeated that they think the pre-editing approach is very "useful". Students' comments are very enlightening and can provide reference for future research work.

5 Conclusion

Although the training module, due to time restriction, didn't last long, it is a new attempt at teaching MT pre-editing based on controlled Chinese rules, and can offer advice for future C-E translation training. From the analysis of students' application of controlled Chinese rules, we can see that students are deeply influenced by Chinese thinking patterns and Chinese writing customs. For the pre-editing method of adding a subject, students might neglect the need for a subject in the middle of a Chinese sentence, and they might forget to adopt the conversational tone; For the pre-editing method of making sentences short, most students showed the tendency to cut Chinese sentences short, but some students, influenced by their long-existing reading and writing habits, couldn't break Chinese long sentences effectively. For the pre-editing method of removing redundant expressions, they may find it hard to determine which parts are redundant. Although students generally respond positively to the effectiveness of controlled Chinese rules, they need more time and more practice to effectively apply controlled Chinese rules to interaction with MT. In future research, we could establish more controlled Chinese rules and expand students' training in interaction with MT. Efforts could also be made to explore the possibility of embedding controlled Chinese rules into neural machine translation system. With machine translation, especially ChatGPT, becoming increasingly intelligent and powerful, our translation training maybe should pay more attention to the aspect of writing style and communicative effect, so that students can better cooperate with MT to produce effective output.

Acknowledgements. The work was substantially supported by The Anhui Education Bureau Social Science Fund for Higher Education, Project No. 2022AH051699: Research on the Application of Controlled Chinese Rules in the Pre-editing of Business Texts.

References

1. Shuttleworth, M., Cowie, M.: Dictionary of Translation Studies. Routledge, London (1997)
2. Carbonell, J.G., Tomita, M.: Knowledge-based machine translation, the CMU approaches. In: Nirenburg, S. (ed.) Machine translation: theoretical and methodological issues, pp. 68–89. Cambridge University Press (1987)
3. Yuan, Y.: Controlled language and machine translation. Shanghai J. Translators Sci. Technol. **3**, 77–80 (2003)
4. Bernth, A., Gdaniec, C.: MTranslatability. Mach. Transl. **16**(3), 175–218 (2001)

5. Reuther, U.: Two in one – can it work? readability and translatability by means of controlled language. In: EAMT Workshop: Improving MT through other language technology tools: resources and tools for building MT (2003)
6. Marzouk, S.: An in-depth analysis of the individual impact of controlled language rules on machine translation output: a mixed-methods approach. Mach. Transl. **35**(2), 167–203 (2021). https://doi.org/10.1007/s10590-021-09266-0
7. Shih, C.: How to empower machine-translation-to-web pre-editing from the perspective of Grice's Cooperative Maxims. Theor. Practice Lang. Stud. **11**(12), 1554–1561 (2021)
8. Zheng, Y.: Designing controlled Chinese rules for MT pre-editing of product description text. Int. J. Translation, Interpretation, Appl. Linguistics **4**(2), 1–13 (2022)
9. Taylor, S.: Model business letters, e-mails & other business documents. 7th edn. (Y. Lu & R. Bai, Trans.) Foreign Language Teaching and Research Press, Beijing(2014). (Original work published 2012)
10. Grice, H.P.: Studies in the Way of Words. Harvard University Press, Cambridge (1989)
11. Cai, X., Wen, B.: Statistical analysis on types of C-E machine translation errors: a case study on C-E translation of publicity text. J. Zhejiang Sci-Tech Univ. **46**(2), 162–169 (2021)
12. International Organization for Standardization. (2017). Translation services—post-editing of machine translation output—requirements (ISO Standard No. 18587:2017). https://www.iso.org/standard/62970.html

Research on the Application of Computer Aided Corrective Feedback in Foreign Language Grammar Teaching

Jie Xu[1,2](✉)

[1] School of Foreign Languages, Chaohu University, Hefei 238024, China
maggiexu0915@sina.com

[2] Centre for Digital Humanities and Intercultural Communication, Chaohu University, Hefei 238024, China

Abstract. This paper aims to explore the effective mode of computer-assisted corrective feedback applied to foreign language teaching. Through experimental design, it focuses on the corrective effects of different feedback methods, including immediate effects and delayed effects. The results of statistical analysis show that written corrective feedback is generally effective, indicating that corrective feedback as a teaching method can promote foreign language learning; and that the effect of metalinguistic corrective feedback is the best, indicating that the effect of corrective feedback is basically positively correlated the degree of dominance of the feedback method; and that the effect of test time on the effectiveness of corrective feedback is negatively correlated with the degree of dominance of the feedback method. The methods and findings of this study can provide a reference for further research on second language grammar acquisition in the computer communication environment.

Keywords: written corrective feedback · foreign language teaching · computer-aided · Grammar acquisition · non-predicate verbs

1 Introduction

Under the impact of the COVID-19 epidemic, the traditional teaching mode has been greatly affected, while it has also promoted the rapid development of online teaching. Benefiting from the diversification of Internet resources, online education has unique advantages and has become the trend of future development. Meanwhile, how to make good use of this new model and how to optimize and integrate it organically with traditional teaching is a problem that needs to be explored first.

In recent years, with the continuous deepening of the cognition of foreign language teaching by domestic foreign language experts, scholars and front-line teachers, the foreign language classroom is also undergoing changes. The traditional teacher-delivered classroom is being transformed into a communicatively competent, interactive, experiential learning environment. In class, taking the language ability of students as the goal,

H. Zhang et al. (Eds.): NCAA 2023, CCIS 1870, pp. 465–477, 2023.
https://doi.org/10.1007/978-981-99-5847-4_33

teachers tend to introduce more flexible and diverse teaching modes such as heuristic teaching and interactive teaching to give full play to the initiative and creativity of students. Meanwhile, teachers also pay more attention to the promotion of critical thinking and communicative skills.

According to Long's interactive hypothesis [1], interactive teaching is favorable to the second language acquisition. Corrective feedback, as an important form of classroom interaction between teachers and students, is undoubtedly conducive to improving the results of foreign language classroom teaching. This article analyzed the importance and key issues of corrective feedback from two different perspectives of teaching method and second language acquisition, and explored the use strategies of corrective feedback in foreign language classroom teaching to improve the teaching effect.

2 Literature Review

Corrective feedback refers to the feedback that language learners receive from others when they use the second language incorrectly for oral or written expression, including oral corrective feedback and written corrective feedback, which means that the teacher or classmates give corrective feedback when students make mistakes in oral or written expressions in foreign language teaching. The current research on corrective feedback can be summarized in three areas:

First, the effectiveness of corrective feedback research. Throughout the history of second language acquisition research, corrective feedback has been considered to be a positive promotion for students' language learning. Some studies have proved that written corrective feedback can improve learners' control over the target structure [2, 3]. Written corrective feedback provides learners with the opportunity to compare manuscripts and revise manuscripts, and promotes internalizing grammar knowledge [4, 5]. Written feedback is a kind of positive feedback in second language acquisition teaching, which can effectively promote the development of second language ability [6]. However, Truscott believes that written feedback is not only ineffective, but also harmful, hence error correction feedback should be stopped [7]. Due to the limited scope of current research, it is difficult to determine the extent of corrective feedback effect. Moreover, because only a few language forms (for example, articles, general past tense, etc.) have been studied, so that it is difficult to be sure that corrective feedback is effective for all language forms. Therefore, more complex language forms and structures need to be further studied. For example, Yu et al. [8] suggests a need for longitudinal naturalistic studies adopting mixed methods and some theoretical frameworks to better explain learners' dynamic engagement in response to teacher written feedback.

Second, what type of corrective feedback is more effective. A large number of studies have examined different types of corrective feedback methods, and the findings vary. First, most studies compare the effects of direct versus indirect corrective feedback. Robb et al. [9] concluded that there was no significant difference in the effects of the two approaches. Kang et al. [10] demonstrated that direct corrective feedback is a more effective form of corrective feedback and better facilitates learners' grammar acquisition. Indirect corrective feedback is more effective for advanced language learners. Direct corrective feedback is more helpful for less proficient learners. Zhang et al. [11] revealed a tendency for learners to prefer more explicit types of written corrective feedback (i.e., metalinguistic explanation and overt correction) for most error types. Guo et al. [12] proved that the form-focused instruction prompt group outperformed the form-focused instruction with recast group and the control group on the immediate post-test; the form-focused instruction prompt group also achieved significantly higher scores than the other groups on the delayed post-test in the written test. Secondly, the contrast between the effect of focus and non-focus corrective feedback. Ellis et al. [13] proves that focused corrective feedback is more effective for low-level learners because such feedback requires less language processing ability of learners. Yang et al. [14] also found that focused corrective feedback is more effective than non-focused corrective feedback. However, due to the limitation of the experimental design, the research results need to be further verified.

Third, the research on the influencing factors of corrective feedback effect. Due to the inconsistent results of many empirical studies on the effect of corrective feedback, the influencing factors of corrective feedback effect have attracted more and more attention. Sheen [2] uses individual differences (learners' learning background) as a regulating factor to explore the impact of corrective feedback on learners' article acquisition. Zhang et al. [15] found that learners' corrective feedback beliefs can modulate their language accuracy. Tsao et al. [16] examine how individual differences in terms of factors such as writing anxiety and motivation predict learners' self-evaluative judgments of both teacher-corrected and peer-corrected feedback. Cheng et al. [17] examine how teachers' language and sociocultural backgrounds mediate the performance of written feedback from three aspects: feedback scope, feedback focus, and feedback strategy. In addition, cognitive and affective factors such as working memory [18], verbal analysis ability [19], and cognitive engagement [20] have been included as moderating variables or influencing factors in the framework of written corrective feedback studies. However, there are many factors that affect the effect of corrective feedback (for example, individual learners, target language factors, etc.), and such studies need to be further improved.

In short, the existing research on corrective feedback is not comprehensive and in-depth in grammar scope, feedback methods and influencing factors, which leads to disagreement in the role and effect of corrective feedback. This study will design experiments to examine the effects of written corrective feedback on English grammar acquisition from these three aspects.

3 Research Design

3.1 Research Topics

This paper mainly discusses the influence of different written corrective feedback methods on the acquisition of English target structure----non-predicate verbs, under the computer-aided communication environment. Non-predicate verb is a complex grammatical phenomenon. There is no corresponding language form in Chinese. It is a verb form that acts as various sentence components except predicates in a sentence. It has many characteristics of verbs, and it is a corresponding component to predicate verbs in the sentence, where they are consistent with tenses. It is one of the most difficult grammar to be mastered. This study carried out the test at three time levels: pre-test, instant post-test and delayed post-test, in order to investigate whether there are significant differences in grammar acquisition among college students in different corrective feedback and tests at different times.

3.2 Research Participants

The subjects were from the 2021 undergraduate students of a university in Hefei, a total of 42 people, aged between 19 and 22 years old. The subjects were randomly divided into three groups: the first was control group, consisting of 14 people, who received the feedback being merely correct or not, without further explanation; the second group was the direct feedback group, consisting of 14 people, who were given feedback on whether it was correct or not, and also the correct form for the error; the third was metalinguistic feedback group, consisting of 14 people, who were given feedback on whether it was correct or not, the correct form for the error, and also further explanation of the cause of the error.

3.3 Test Materials

The non-predicate verb structures examined in this study include: (1) doing, (2) done, (3) to do, (4) to be done, (5) having done, (6) having been done, (7) being done.

This experiment was carried out during the epidemic period and the test and corrective feedback were carried out with the help of online teaching methods. The tests included one pre-test and two post-tests. The inspection point was only the usage of non-predicate verbs. Three test papers with equivalent difficulty were issued at the beginning of the research, for pre-test, real-time and delayed post-test respectively. Each paper contains 21 Chinese sentences, every 3 of which refer to each type of target structure. The participants were requested to quickly translate them into English within 40 min. The tests and feedback were both conducted online with computer-aided methods.

3.4 Research Methods

Type of Corrective Feedback

According to the explicit degree of feedback, written corrective feedback can be divided into indirect feedback, direct feedback and metalinguistic feedback. Indirect feedback refers to pointing out errors to students without correcting them; direct feedback refers to providing the correct form for errors without explanation; metalinguistic feedback refers to further clarifying the nature of the error or the rules involved on the basis of pointing out or correcting the error [21].

Tencent Document (https://docs.qq.com) supports multiple editors, so all parties can communicate in real time. People can use the discussion forum to communicate instantly about changes to the text, and the changes to the text can be saved so that the process of making changes can be understood at a glance. Through the Tencent document acting as an instant communication platform, students could send their test paper to the teacher, and discuss and feedback with the teacher using the computer.

Test Time

In order to check the promotion effect of corrective feedback on language acquisition (including immediate effect and long-term effect), the present study also examined the time factor as a variable. The subjects were tested at 3 time points to investigate changes in the mastery of the target structure before feedback (pre-test), just after receiving feedback (instant post-test) and approximately 4 weeks after receiving feedback (delayed post-test).

Experimental Design

The experiment lasted for 6 weeks, and the pre- and post-test was performed 3 times. Three sets of test papers were designed in advance to investigate the usage of non-predicate verbs. A pre-test was administered in the first week, and the results of the pre-test revealed no significant differences between the three groups. And then teachers intervened in the next week. The teacher sent back the graded pre-test papers to the students. According to different groups, the corresponding corrective feedback to the errors involving the target structure was made along with the paper respectively. After complete comprehension of corresponding corrective feedback, instant post-test could be conducted for the participants with another set of papers in this week. And the delayed post-test was conducted in the 6th week with the 3rd set of paper.

Each type of non-predicative verb structure was designed to be translated from Chinese to English in 3 questions, a total of 21 questions. The test questions were only for the target grammatical structure, and the rest were as simple as possible to ensure that in formal experiments, other factors such as vocabulary and other grammatical structures will not become translation obstacles. Only the target structure was scored, with 1 point for correctness and 0 points for wrongness. The full score is 21 points each time.

3.5 Data Analysis

After scoring, the score of each type of target grammatical structure is counted. Scores reflect the effectiveness of corrective feedback. The R statistical software was used for statistical analysis. And the data of each group were compared to determine the significant difference in the effect value of each group.

In statistics, students' test scores are approximately obedience to normal distribution. However, when the number of samples is small, it is difficult to reliably determine the normality of the data, so it cannot be tested using analysis of variance. Fundamentally, non-parametric tests do not require normality of the data. Therefore, in the case of a small sample size, the more robust Kruskal-Wallis test should be used for statistical inference.

4 Results

The purpose of this study was to explore the effects of two factors, feedback mode and test time, on the acquisition of English grammar. Table 1 shows the overall profile of the three tests. The mean values of the scores of the target structure for the three groups of subjects in each experiment over the three tests were reported. Figure 1 shows the data dispersion in a box plot, reflecting the characteristics of the original data distribution, and also allows for comparison of the distribution characteristics of multiple data sets. The vertical coordinate is the test score, the horizontal coordinate is the test time, and the scores of the 3 test groups are obtained for each test time point. Box plot are used to reflect the central location and spread of groups of continuous quantitative data distributions. In a box plot, the line in the middle of the box represents the median of the data; the top and bottom of the box are the upper (Q3) and lower quartiles (Q1) of the data, respectively; thus the box contains 50% of the data; the upper and lower edges represent the maximum and minimum values of the set of data. Sometimes there are some points outside the box, which mean outliers in the data. Therefore, the height of the box reflects the fluctuation degree of the data to a certain extent, and the flatter the box, the more concentrated the data. It can be seen that in the pre-test, the scores of the three groups of subjects are relatively close; in the immediate post-test, the scores of the direct feedback group and the metalinguistic feedback group have improved, and the metalinguistic feedback group has the highest score; in the delayed post-test, the scores of the direct feedback group and the metalinguistic feedback group decreased compared with the immediate post-test, and the direct feedback group decreased more significantly, but still scored higher than the control group.

Table 1. Mean value of the effect of different written corrective feedback at different test times

Groups	Pre-test		Instant post-test		Delayed post-test	
	Average score	Standard deviation	Average score	Standard deviation	Average score	Standard deviation
Control group	8.36	0.99	8.14	0.64	8.30	0.85
Direct feedback group	8.79	1.00	14.76	2.02	12.74	1.44
Metalinguistic feedback group	8.21	0.95	19.68	1.08	18.08	2.24

Table 2 shows the results of the Kruskal-Wallis H-test using the R language, to evaluate whether different feedback methods have different effects on students' acquisition of English grammar. The Kruskal-Wallis H test is a rank-based nonparametric test method that can be used to test whether there are differences in continuous or ordered categorical variables between multiple groups (or two groups). Kruskal-Wallis test does not require samples' normality and homogeneity of variance. Its null hypothesis is that each sample follows a probability distribution with the same median. A rejected null hypothesis means that the median of the probability distribution of at least one sample is different from the other samples. If the differences are large, the groups are not from the same population at a particular significance level (e.g. $p < 0.05$). From the results, basing on the level of $\alpha = 0.05$, in the pre-test, $p = 0.3455$, it can be considered that there is no statistical difference in the scores of the three groups; as to the instant post-test and the

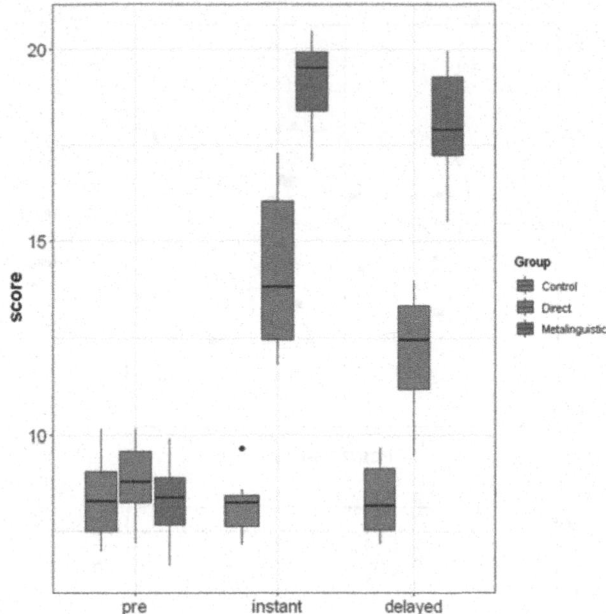

Fig. 1. Distribution of scores of different written corrective feedbacks at different test times

Table 2. Results of Kruskal-Wallis analysis of the error correction effect of different feedback methods

Parameters	Pre-test	Instant post-test	Delayed post-test
chi-squared	2.1253	36.28	35.037
Df	2	2	2
p-value	0.3455	1.324e–08	2.464e–08

delayed post-test, the p-value is much smaller than the α-value, and it can be considered that the difference in the scores of the three groups is statistically significant.

The statistical results show that the control group exhibits no obvious real-time and delay effects; corrective feedback is generally effective and the trends of changes tend to be consistent, but the delayed effect is worse than the immediate effect; the immediate effect of direct feedback is more obvious; the effect of metalinguistic feedback is the most significant. The result suggests that the effect of corrective feedback is proportional to the degree of dominance of the feedback method; The negative impact of test time on the effectiveness of corrective feedback is more pronounced for direct feedback; or in other words, the long-term effect of metalinguistic feedback on errors is better than direct feedback, indicating that it is more helpful for students to master grammar rules.

PCA analysis was applied to the scores of each part of participants using R 4.2.1 to extract data features. The result is shown in Fig. 2. PCA analysis (Principal Components

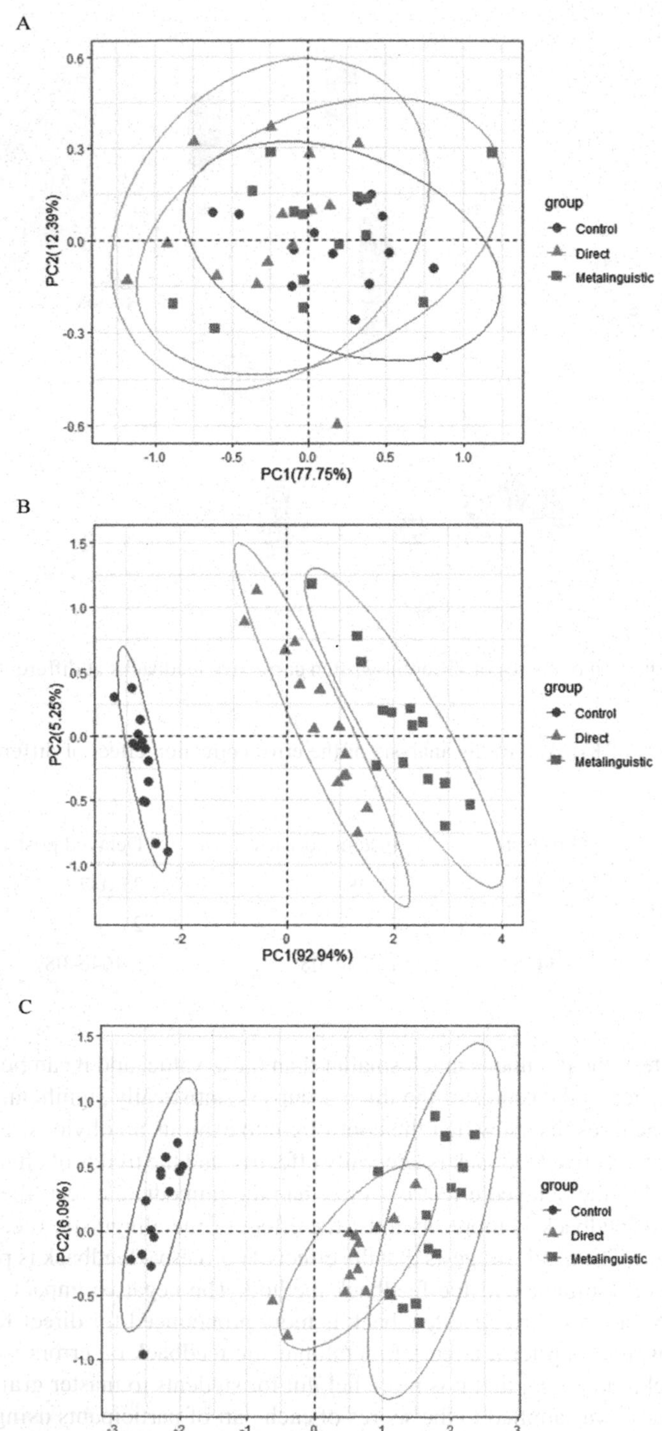

Fig. 2. PCA analysis of different written corrective feedback at different test times (A) Pre-test (B) Instant post-test (C) Delayed post-test.

Analysis) is a common method for data dimensionality reduction, which is often used to extract the main feature components of high-dimensional data. PCA is an unsupervised algorithm that reduces the dimensionality of data without the need for labels. This method uses principal components to retain the main and identifiable features in the original data to the greatest extent, reduces data dimensions, and facilitates better identification and calculation.

PCA can initially reflect the distribution of different groupings of samples that may exhibit dispersion and aggregation, so that it can determine whether the composition of samples with the same conditions is similar. The horizontal coordinates indicate the first principal component, and the percentages in parentheses indicate the contribution of the first principal component to the sample variance; The vertical coordinate indicates the second principal component, and the percentage in parentheses indicates the contribution of the second principal component to the sample variance. Different colors represent samples belonging to different groupings, and the farther the distance between two points, the greater the difference between the scores of the two samples. As can be seen from Fig. 2, the distribution of the scores of the three groups of participants in the pre-test cannot be distinguished from each other, and the three groups in the instant post-test and delayed post-test can be separated more clearly, indicating that there are more significant differences.

5 Discussion

Basis of Experimental Design

The input hypothesis holds that in order to improve the overall learning effect of second language learners, it should be improved from the two levels of input content and quantity [22]. On the one hand, the input should be consistent with the overall level of the student. If the content is relatively difficult, students' attention and interest will be reduced to some extent, which will discourage students and affect the effectiveness of their learning. If the input content is too popular and simple, it is not conducive to the improvement of students' motivation and attention, and the effect of learning is greatly reduced. On the other hand, it is important to ensure that the amount of input is sufficient to meet the requirements. If the amount of input content is not sufficient to meet the specific requirements, it may affect the overall results achieved by the students. If the amount of input content is too large, it is easy to cause students' learning pressure, which is not conducive to improving the overall learning effect of students. Practice has proved that only by ensuring sufficient input can students' comprehension ability and the generation of language meaning be effectively enhanced. The output hypothesis holds that the grammar learning of second language acquisition should also have sufficient output content. The acquired grammatical knowledge is further processed to form an output based on understanding, which helps learners to notice the distance between their own learning ability and the target learning in the process of learning the language, thus increasing the final effect of their learning. The scheme of this experiment was designed based on the above considerations.

Analysis of Results

The direct feedback group and the metalinguistic feedback group maintained a significant advantage over the control group in both immediate and delayed post-tests, indicating that error correction in a computer environment can effectively improve the accurate use of target structures. According to the noticing hypothesis [23], corrective feedback can highlight the grammatical features to be learned to learners, help learners focus first on form, and then solidify explicit structure through the combination of form and meaning. Moreover, the corrective feedback in the computer online communication mode strengthens the prominence of the feedback, enhances the learner's attention, and achieves the necessary condition for the acquisition -- noticing [24], as a result, is conducive to the learning effect. In addition, conditional hypothetical sentences are complex structures [25]. Corrective feedback on complex grammatical structures has a strong degree of salience and may promote learning effectiveness.

The results of the direct feedback group and the meta-linguistic feedback group have essentially the same trend. From the pre-test to the instant post-test and then to the delayed post-test, the scores go up and then down. For most foreign language learners, the basic knowledge of foreign language skills can only come mainly from explicit knowledge. Explicit knowledge is non-procedural knowledge, which has the characteristics of insufficient stability and easy fluctuation. It can be said that the success of foreign language learning depends to a large extent on whether the right explicit knowledge can be provided / obtained in an appropriate way [26]. The decrease of the scores in post-test may suggest, to some extent, that the improvement in grammatical accuracy may be influenced by direct or explicit feedback, and that explicit feedback has a positive effect on the development of learners' explicit grammatical knowledge. Polio [27] has also stated that direct written corrective feedback can increase learners' explicit knowledge.

Although, the effects of direct feedback and meta-linguistic feedback in the computerized communication environment were all significantly superior to the control group and both decreased in the delayed post-test, they showed significant differences between them in the post-test scores. The effect of corrective feedback is directly proportional to the degree of dominance of the feedback method. The reason is there are also two types of explicit knowledge, one for memorization and the other for comprehension. To ensure good learning outcomes from errors-correcting, learners need to not only "pay attention" to the corrective information provided by the teacher, but also "understand" the intent of the correction [28]. In this study, metalinguistic feedback exhibits the best effect probably because it provides not only memory knowledge, but also comprehension knowledge, so that the subjects know what it is and why it is, hence is more conducive to memory or acquisition. Whereas direct feedback provides only isolated memory knowledge, it may work well for grasping errors that have been learned but are caused by forgetfulness or negligence. As for the errors caused by the lack of grammatical knowledge or the lack of understanding of grammatical rules in this experiment, it is obviously not enough for memory or full acquisition just knowing what it is and not knowing why it is. In the control group, the subjects were only informed whether their answer is correct or not, and often still do not know why the answer is correct or wrong, so that have no idea about how to correct the errors. Zhang et al. [29] found that explicit feedback can help learners understand the nature of their errors and eliminate confusion, thus improving the effectiveness of corrective feedback. Diab [30] proved that the more direct and adequate information provided by the feedback, the better the results tend to be.

Computer-assisted corrective feedback has a significant impact on the development of explicit knowledge [4]. Written corrective feedback provides an interactive platform for both parties to understand language and an important method for accurately understanding language. Its purpose is to better promote language understanding and acceptance. In the process of correction, teacher's corrective feedback left on the interactive computer interface is key information. This kind of feedback not only has the characteristics of real-time (for example, having some of the characteristics of face-to-face oral communication), but also has a certain communication delay. On the one hand, such feedback could provide negotiation on meaning and form, and create a situation similar to face-to-face oral communication; and on the other hand, it could reduce the sense of time constraint and anxiety of face-to-face communication [31], prolong the processing time for learners to understand and correct formal errors, and extend working memory. In the process of combining these two advantages, more opportunities for negotiation of form and meaning can be produced, which is conducive to the development of acquisition.

6 Conclusion

To sum up, the corrective feedback of college students in the computer communication environment has a significant positive effect on grammar acquisition. Further research also showed that target grammar scores were positively correlated with the degree of dominance of the feedback method; the degree of decrease of the score in the delayed post-test was negatively correlated with the dominant degree of feedback, but the performance was still better than in the pre-test. These results throw light on the effects of written corrective feedback on students' acquisition of grammar of non-predicate verbs. The research methods used in this study can be applied to similar studies. However, the research method of this study is relatively simple, relying only on quantitative analysis, and qualitative research is relatively scarce.

With the development of internet and computer technology, its combination with foreign language teaching will have more space for language learning. However, modern technological means are only tools that serve education. The way and process of using technology, and how technology makes changes in teaching and learning are the key points that need to be paid attention to in the research of computer-aided teaching. How to bring more opportunities for learners to negotiate in meaning and form, how to highlight the content of acquisition and thus facilitate acquisition development from a technological perspective, and how to choose the types of tasks to exploit the features of new technologies to facilitate grammar acquisition, still need further exploration.

Acknowledgments. The work was substantially supported by The Anhui Education Bureau 2023 Social Science Fund for Higher Education, Lingjiatan Culture Translation and Communication Strategy Research from the Perspective of Convergence Media and by The Anhui Education

Bureau 2021 Education and Teaching Research Project, A Study on the Development of Intercultural Foreign Language Teaching Competence of Applied Undergraduate Teachers Based on Community of Practice.

References

1. Long, M.H.: Native speaker/non-native speaker conversation and the negotiation of comprehensible input. Appl. Linguis. **4**(2), 126–141 (1983)
2. Heen, Y.: The effect of focused written corrective feedback and language aptitude on ESL learners' acquisition of articles. TESOL Q. **41**, 255–283 (2007)
3. Bitchener, J., Knoch, U.: The contribution of written corrective feedback to language development: a ten-month investigation. Appl. Linguis. **31**, 193–214 (2009)
4. Shintani, N., Aubrey, S.: The effectiveness of synchronous and asynchronous written corrective feedback on grammatical accuracy in a computer-mediated environment. Mod. Lang. J. **100**(1), 296–319 (2016)
5. Shintani, N., Ellis, R., Suzuki, W.: Effects of written feedback and revision on learners' accuracy in using two English grammatical structures. Lang. Learn. **64**(1), 103–131 (2014)
6. Stefanou, C., Revesz, A.: Direct written corrective feedback, learner differences, and the acquisition of second language article use for generic and specific plural reference. Mod. Lang. J. **2**, 263–282 (2015)
7. Truscott, J., Hsu, A.Y.: Error correction, revision, and learning. J. Second. Lang. Writ. **17**(4), 292–305 (2008)
8. Yu, R., Yang, L.: ESL/EFL learners' responses to teacher written feedback: reviewing a recent decade of empirical studies. Front Psychol. **12**, 735101 (2021)
9. Robb, T.L., Ross, S.M., Shortreed, I.: Salience of feedback on error and its effect on EFL writing quality. TESOL Q. **20**(1), 83–96 (2012)
10. Kang, E.Y., Han, Z.: The efficacy of written corrective feedback in improving L2 written accuracy: a meta-analysis. Mod. Lang. J. **99**(1), 1–18 (2015)
11. Zhang, T., Chen, X., Hu, J., Ketwan, P.: EFL students' preferences for written corrective feedback: do error types, language proficiency, and foreign language enjoyment matter? Front Psychol. **12**, 660564 (2021)
12. Guo, X., Yang, Y.: Effects of corrective feedback on EFL learners' acquisition of third-person singular form and the mediating role of cognitive style. J. Psycholinguist Res. **47**(4), 841–858 (2018)
13. Ellis, N.C.: At the interface: how explicit knowledge affects implicit language learning. Stud. Second. Lang. Acquis. **27**, 305–352 (2005)
14. Yang, Y., Lyster, R.: Effects of form-focused practice and feedback on Chinese EFL learners' acquisition of regular and irregular past tense forms. Stud. Second. Lang. Acquis. **32**(2), 235–263 (2010)
15. Zhang, J., Cao, X., Zheng, N.: How learners' corrective feedback beliefs modulate their oral accuracy: a comparative study on high - and low-accuracy learners of Chinese as a second language. Front Psychol. **13**, 869468 (2022)
16. Tsao, J.J., Tseng, W.T., Wang, C.: The effects of writing anxiety and motivation on EFL college students' self-evaluative judgments of corrective feedback. Psychol Rep. **120**(2), 219–241 (2017)
17. Cheng, X., Zhang, L.J.: Teacher written feedback on English as a foreign language learners' writing: examining native and nonnative English-Speaking teachers' practices in feedback provision. Front Psychol. **12**, 629921 (2021)

18. Li, S., Roshan, S.: The associations between working memory and the effects of four different types of written corrective feedback. J. Second. Lang. Writ. **45**, 1–15 (2019)
19. Benson, S., DeKeyser, R.: Effects of written corrective feedback and language aptitude on verb tense accuracy. Lang. Teach. Res. **23**(6), 702–726 (2019)
20. Zheng, Y., Yu, S.: Student engagement with teacher written corrective feedback in EFL writing: a case study of Chinese lower-proficiency students. Assess. Writ. **37**, 13–24 (2018)
21. Ellis, R.: A typology of written corrective feedback types. ELT J. **63**(2), 97–107 (2008)
22. Latif, A., Muhammad, M.: Toward a new process-based indicator for measuring writing fluency: evidence from L2 writers' think-aloud protocols. Can. Mod. Lang. Rev. / La revue ca. **4**, 99–105 (2009)
23. Schmidt, R.: The role of consciousness in second language learning. Appl. Linguis. **11**, 129–158 (1990)
24. Schmidt, R.: Awareness and second language acquisition. Annu. Rev. Appl. Linguist. **13**, 206–226 (1993)
25. Izumi, S., Bigelow, M., Fujiwara, M., Fearnow, S.: Testing the output hypothesis: effects of output on noticing and second language acquisition. Stud. Second. Lang. Acquis. **21**(3), 421–452 (1999)
26. Robert, M.: DeKeyser: the robustness of critical period effects in second language acquisition. Stud. Second. Lang. Acquis. **22**(4), 499–533 (2000)
27. Polio, C.: The relevance of second language acquisition theory to the written error correction debate. J. Second. Lang. Writ. **21**(4), 375–389 (2012)
28. Schmidt, R.: Attention and Awareness in Foreign Language Learning. University of Hawai'i Press, Honolulu (1995)
29. Zhang, K., Li, B., Chen, K.Q.: The effectiveness of teachers' written corrective feedback for L2 development: an explanatory sequential mixed-method study. Mod. Foreign Lang. **42**(3), 363–373 (2019)
30. Diab, N.M.: Effectiveness of written corrective feedback: does type of error and type of correction matter? Assess. Writ. **24**, 16–34 (2015)
31. Lv, M.L., Guan, Y.P., Tang, J.M.: An empirical study on foreign language learners' anxiety under synchronous computer-mediated communication mode. J. PLA Univ. Foreign Lang. **39**(3), 28–36 (2016)

Student-centered Education in Metaverse: Transforming the Language Listening Curriculum

Xiaoming Lu[1(✉)], Guangrong Dai[1], Lu Tian[1], and Wenjian Liu[2]

[1] School of Interpreting and Translation Studies, Guangdong University of Foreign Studies, Guangzhou, China
andylau@cityu.mo

[2] Faculty of Data Science, City University of Macau, Macao, China

Abstract. The paper explores transforming the English Listening curriculum for university freshmen with the aid of technology. The extension of Computer-assisted Language Learning (CALL) courses goes beyond the aspect of soft-ware and hardware, to the concept of the metaverse. Surveys have been conducted to collect students' requirements and responses to probe the extent of their technology acceptance as well as expectations for the course. The surveyed 150 university freshmen are involved as research objects. The result based on the SPSS software analysis reveals that students' language proficiency is not necessarily related to the fulfillment of instructors' requirements. Despite the integration of multimedia technology for language learning in the class, university students usually have their own learning techniques. Teachers need to respect their methods and provide guidance based on their nature. Data retrieved from frequently used online learning platforms is analyzed to find out that freshmen students are highly dependent on offline interaction with the instructor. The paper concludes that teachers' involvement in terms of on-site content delivery is a fundamental source of students' motivation. Authors are also seeking a way of combing the advantages of both traditional classrooms and modern technology. Mataverse may be an appropriate setting to accommodate students' proactive participation and language acquisition.

Keywords: Computer-assisted Language Learning (CALL) · Metaverse · Teacher's involvement · Motivation

1 Introduction

Today, listening, rather than being considered a passive skill, is usually associated with speaking, reading, and writing. It requires L2 learners' interaction with other native English-speaking objects to fully acquire a non-mother tongue. This enables the integration of technology into traditional language teaching classes. Instructors need to analyze listening skills and engage students in the communication of English context (Greenleaf 2011).

Technology-aided language teaching and learning create an effective and enjoyable process for learners. It drives the changes that occur in the linguistic field with innovation.

H. Zhang et al. (Eds.): NCAA 2023, CCIS 1870, pp. 478–491, 2023.
https://doi.org/10.1007/978-981-99-5847-4_34

It helps create realistic English contexts for students who get involved with interests and learn with a clear goal after class. Therefore, language teachers should create a favorable environment based on the availability of multi-modal resources. Provided with multimedia technology, students would be an incentive to socialize in a second language. The whole process is in this sense more student-centered and effective through multimedia technology (Warschauer 2000).

The year 2021 could be considered the turning point where virtual reality (VR) became recognised to access the metaverse during the COVID-19 period. Since the world's social media giant Facebook was renamed to Meta, such transformation indicates a reform of redefining learning in the future. VR is likely to take control of interactive learning environments, offering fresh and incredibly versatile prospects for immersive experiences and educational settings (Rospigliosi 2022). Educators need to notice how students learn the visual presentation with consciousness and intention to help them immerse in and explore the metaverse (Han 2020).

Previously research touches on the analysis of students' cognition, education methodology and technology, and metaverse as a newly emerged concept. However, scholars have rarely explored the futher application in education driven by research which links to metaverse in the future. The study focuses on how technology-driven approaches can be applied to the classroom from the perspective of both teachers and students. It investigates two questions:

1. What technical aspects should the instructor consider while designing the CALL course?
2. What factors may affect freshmen students' satisfaction and performance during their learning?

2 Theoretical Literature Review

The era witnessed a more compact integration of technology and education. However, lacking effective interaction between teachers and students causes the failure of tool implementation. The curriculum design guided by the educational theory does not always appear well-organized and explicit. Educational theories can function as guidelines informing how the technology would be used and when it would be used in its interest. Particularly, certain uses of technology may be more useful than inefficient integration. In this way, the use of different technological approaches can be evaluated scientifically (Greenleaf 2011; Sandars et al. 2015).

In addition to this, before using any education approach, instructors need to learn about the learner's level and prerequisite knowledge (Vandergrift 2003). The pace of developing cognitive processes recognizes a positive correlation with the extent of the experience or knowledge we gain (Moskovsky et al. 2014). Then we become well-trained listeners with much quicker responses to the second language. It requires teachers to step in to help students connect with what lies underneath the surface of the language presentation. Besides, in the metaverse, technology is the medium of transmitting culture (Smith-Shank 2007). Our experience shapes knowledge, therefore, during the cognitive process, it is the foremost concern (Efland 2002). Our cultural or social backgrounds affect the experience (Brown 2007).

The bottom layer is schema knowledge. On this level, students can achieve in a self-contained way (Ahlin 2020), which means that the help from the instructor and the interactions with peers are excluded. At this time, the intervention of traditional technology or VR steps in. There are three goals within this layer: (1) extract critical information and develop inferences, trends, and generalizations; (2) analyze the contexts or connections between ideas; (3) remember and understand facts and concepts related to the topics of listening tasks. The second layer is grammar knowledge, which can be acquired by traditional technology or VR as well. There are two goals within this layer: (1) complete online tests, quizzes or exams where students may encounter unfamiliar vocabularies and collocations, short-answer or multiple-choice questions, in a way either online or offline; (2) create and share their presentations in the class. The third layer is logical thinking, which should be trained under the guidance of teachers. There are also three goals in this process: (1) use technology to complete specific tasks (i.e., compare machine translation results generated by different MT engines); (2) provide comments or feedback for students' presentations, particularly focusing on students' speech structure; (3) teachers can introduce the materials following an organized structure, provide a clear outline, select and filter data, then encourage individual presentation and group discussion.

3 Data Analysis

There are around 150 freshmen students (No. of female > No. of male) involved in the data collection process. They have been enrolled in the course where they learn both the English listening and propel tools that facilitate their studies). These students have excellent academic performances in the university entrance exam in China. Data was gathered from the following platforms and was analyzed qualitatively (Tables 1 and 2).

Table 1. CALL Online APPs

Platform	Function(s)	Purpose
DingDing (钉钉打卡)	Automatic clock-in	Students transcribed a piece of English recording after class
Duifenyi (对分易)	Class management, homework submission, curriculum resources upload, etc	Teachers can check students' class attendance, assign, and score homework and share teaching resources
Wenjuanxing (问卷星)	Survey and vote	Teachers can view students' anonymous feedback and students feel involved
Yikao (易考)	Online Tests	Students are required to complete the online tests sourced from IELTS, TOEFL and textbooks

Table 2. Comparison of different teaching approaches

Teaching approach	Teacher-centered	Student-centered	Test-driven
Online Instruction	√		
Online Test			√
Student Presentation		√	
On-site Lecture	√		

What students aspire most from the course include applying (81.82%), remembering, understanding (66.67%), and creating (51.52%). They put language mastery ahead of technology proficiency. Regarding the most frequently downloaded item on the class management platform Duifenyi, the data shows that students are much more interested in downloading/previewing the outline or the requirements of the course, as well as the materials they fail to access online, rather than instructional slides provided each class by the teacher.

4 Instructional Methods

4.1 Teaching Resources

It is common to use multimedia resources as instructional materials to restore the authentic language environment. In the university where authors work, there have been certain laboratories provided with professional auditory equipment. However, at its first launch, though PowerPoint slides, Microsoft Word documents with scattered knowledge points and somehow disorganized video/audio materials are provided, students may find it difficult to assess whether they have mastery of course content with a coursebook that helps them understand and review this course. Among students who received its first launch, one of the survey questions is "Would you suggest that a digital/paper coursebook facilitates your learning experience for this course?". Approximately 70% of students agreed with this proposal. Even without textbooks for the first time, enough online language-learning resources are available to utilize. 85% considered that they have received sufficient course materials before/after class.

In the course's second launch, a textbook named LISTENING FOR SUCCESS: MINI-LECTURES has been used as the coursebook for instruction. Nearly 85% of surveyed students considered the book as proper tutorial material and nearly 94% of them thought they have received sufficient course materials from the class. Particularly, compared to the first group of students, the second group has more learning of technical skills such as online search skills and grammar correction platforms. 97% of students in the second group recognize the helpfulness of the teaching of CALL tools (i.e., Wordcloud, use of Microsoft PowerPoint Master slide, Paratrans/DeepL, text-to-speech platform).

4.2 Teacher Education and Course Management

Multimedia technology brings up the combination of audio and visual effects to display textual materials vividly. This process enables students to comprehend the newly learnt knowledge. But their critical thinking may be activated and stimulated merely by using PowerPoint slides to display teaching content. Students should be encouraged to think independently and explicitly express themselves. Teachers need to prepare various forms of interactive activities and thought-provoking content to guide the student-centred teaching process. The instructor's teaching leads to how students structure their thinking and are motivated. The technological maneuver changes the instructor's role in the classroom, less of monotony but more of enjoyment and stimulation. For instance, teachers require students to do presentations by using PowerPoint slides, which motivates students to speak, think, comprehend, and communicate in English. Language learning is transformed into capacity cultivation and the construction of knowledge systems (Warschauer 2000; Rönnberg 2019).

Both teachers and students need to know how to use, adapt, and integrate technologies in the teaching process to support their classes, and consider each student's learning style (visual, auditory, or kinesthetic/physical), easing the teacher-student interaction. Instructors should notice new advancements and updates in technology, as well as its applications in language education. Such multimodal teaching is a development of teaching patterns driven/aided by technology, expectedly minimizing the gap between field-dependent (FD) and field-independent (FI) students. Thus, it will be a gradual transformation from multimedia technology developing into a visual reality world (metaverse).

Successful implementation of technology in the classroom depends upon the use of pedagogical activities rather than other learning settings (Salaberry 2001). It is imperative to instruct teachers on how to optically use technologically advanced materials in class activities. Proper training for teachers allows the effective input offered to L2 learners.

Thus, the use of technology is not sufficient for the completion of teaching tasks (Hughes, Thomas and Scharber 2006). The combination of technology and teaching pedagogy could remain a debated issue. To record the combination, a teaching plan for every class may be of great help. Multimedia technology as an addition of stimulation and diversity must not replace the instructor's role and restrain the advantages of traditional classrooms. When using multimedia technology, teachers should take the lead, rather than leave the task to students themselves without strict guidance and supervision. More importantly, the teacher's role should not be replaced by technological devices.

Teachers can provide keywords and sentences during a class to remind students of the class foci which are elements students need to articulate. Research shows that asking students to recall and reorganize the class content they received during the last few minutes exerts a positive effect on the mastery of knowledge (Lyle and Crawford 2011). Besides, through displaying students the instruction, they may be exposed to both answering questions and understanding basic concepts and key facts.

It is not the whole picture if only the linguistics part of English studies were delivered, including lexicon, grammar, and syntax. Computer-assisted helps to satisfy organizational requirements of FD students and provides the autonomy learning mode to FI ones. Both parties, in this way, enhance the learning outcome with the aid of technology display. This project-based instruction is the consideration of replacing traditional teaching with the implementation of tailoring different instructional strategies. Hypermedia instruction engages students in a combination of autonomy, interaction, instant stimulus, and feedback. It would be critical for teachers to link cognition with thought patterns in multimedia environments and consider what aspects should be included in the design of instructional strategies (Stash and De Bra 2004; Chen and Liu 2008; Tinajero et al. 2011) (Table 3).

Table 3. Student Learning Styles

Visual	53.7%
Auditory	27.78%
Kinesthetic/Tactile	12.96%
Mixed/Unsure	5.56%

As we learn from the chart, students are sensitive to visual representations, which affects the project-driven PowerPoint presentation tasks. It is interesting to learn that two-thirds of students preferred such a task as well as considering it an effective way of peer learning. Over 98% of students value the feedback and assessment from the instructor after their presentations have been given. In their opinion, teachers are more professional and educated, whose comments and advice would become an indispensable part of language teaching and learning. 73% expect to receive both peer evaluation and the teacher's evaluation.

Teachers should also focus on the handling of students' feedback towards teaching and their learning motivations. Students' motivational incentives usually reveal the lack of traditional classroom settings. Certain needs have not been covered for students who have various cognitive styles. For instance, FD students are more likely to thrive in a competitive environment where they may have a better performance. Students of such a cognitive type may be more sensitive to the giving and receiving of feedback. They see feedback and comment as guidance as well as a source of motivation. They usually have selective attention to social information. From the authors' observation, the presentation tasks assigned to students remind them of language practice as well as give them enough freedom and autonomy to create at their will.

After each individual/group presentation, the student audiences are required to fill in the assessment form to evaluate their classmates' presentation performance based on the Oral Presentation Evaluation Rubric. Moreover, the classroom setup should also be considered. In this course, a specific multimedia classroom accommodating 60 people was used, when it is offline. The issue relevant to class size online herein will not be discussed. It is worth mentioning that the form online or offline does not matter much for students if the teacher's involvement and interaction with them are presented during class hours (Tables 4 and 5).

Speaking of the multimedia classroom, the equipment includes an instructor's desktop computer with dual monitors, an audio amplifier, and students' screens that the teacher can broadcast. Students themselves do not have keyboards and hardware in front of them. The projector is also unavailable in this circumstance. Even so, students can still view the simulated visual content from their display screen, in a clearer manner than that from the projector and big screen. Multimodal technology makes the class lively

Table 4. Teacher's involvement

The teacher's involvement is part of the motivation for studying this course	
Yes	76.79%
No	19.64%
Never mind	3.57%

Table 5. Interaction between the teacher and students

It is necessary to increase the interaction between the teacher and the individual during this course	
Yes	54.55%
No	27.27%
Never mind	18.18%

and improves students' willingness to more active participation. Students in this case acquire enough language information. Besides, students can resort to teachers online after class.

In the test-driven class, the long-term memory of learnt knowledge is better enhanced and retained. Types of tests, be it short answers or multiple choices would not affect the effectiveness of retrieving information from the memory. One example is the formative assessment, which enables students to progress through repeated training. During the recorded teaching activities, an online test given every week revolves around a specific topic (i.e., tourism, political meeting, electronic product, marketing, aerospace, etc.), sourcing from IELTS, TOFEL, or chosen coursebooks. However, teachers should pay attention to the negative effect of information cramming and the difficulty level setting of tests. Once students are overloaded with the content of both language and technology, according to the feedback from surveyed students, they find it difficult to assimilate, time-consuming and less useful.

To present test-driven teaching in class, the advantages in this regard should be considered. First, it is useful to include different types of quick quizzes (i.e., short-answer or fill-in-the-blank) from time to time during the teaching. The form could be either online or face-to-face. The subsequent feedback is also practical. During the testing phase, students are at the remembering stage of the pyramid model reviewing unfamiliar vocabularies and their collocations (Huang and Tsao 2021).

Students should be informed beforehand about the testing effect. Instructors can provide the positive outcome to improve students' metacognition. By informing students that practicing quizzes regularly helps and that there are effective types of quizzes available, they may add such techniques to their learning repertoire (Stanger-Hall et al. 2011).

Incorporating pre-testing can emphasize students' self-domination over their learning, for instance, by utilizing exams as a means. The effect may not be satisfying if they are asked to simply take a daily check of their knowledge or undertake regular dictation training without intervention from the instructor (Brame and Biel 2015). For instance, the result is not as useful and effective as expected of clocking in with records of students' after-class listening exercises automatically updated on the Ding Ding platform (钉钉) (Table 7).

Table 6. After-class self-training and exam score

Variable	Sample no	Median	Mean	Standard deviation	Skewness	Kurtosis	Shapiro-Wilk	Kolmogorov-Smirnova
Clock-in Rate	59	0.21	0.348	0.36	0.585	−1.169	0.841(0.000***)	0.195(0.019**)
Final Exam Score	59	88	87.271	4.405	−1.429	2.608	0.886(0.000***)	0.204(0.012**)

Note: ***, **, and * respectively represent significance levels of 1%, 5%, and 10%. As we can learn from the results within (Table 6), the data does not respond to strict normal distribution. However, considering that the absolute value of skewness is smaller than 3 and kurtosis smaller than 10, combined with its P-P plot and Q-Q plot, it fits with the normal distribution

Table 7. Correlation between attendance and exam score

	Clock-in rate	Final exam score
Clock-in rate	1.000(0.000***)	−0.001(0.992)
Final exam score	−0.001(0.992)	1.000(0.000***)

As $p = 0.992 > 0.05$, there is no correlation between the clock-in rate and final exam score. Students stated the main reasons for not sticking to the habit of clocking in: ① Practicing English listening every day is time-consuming and of less fun; ② They are unwilling to impress other classmates by being diligent.

5 Student Perceptions

5.1 Student Literacy

Teachers support students' learning experiences through the discrete selection of topics, texts, activities, and resources. Nevertheless, the importance of students' self-learning activities after class is often neglected and underestimated. Students practice English beyond classrooms. Their language literacy is built beyond texts, PPT slides and teachers' instruction. They also interact with peers who are also language enthusiasts. They learn by being exposed to the L2 environment in relevant websites, social networks, and online gaming. The purpose of socialization motivates students' literacy (Paleeri 2021), which is usually related to the completion of tasks in some time and certain places. The extent of helpfulness and effectiveness of such self-learning methods leaves much room for researchers and teachers to explore. Therefore, it can be inferred that it obtains positive correlations between students' existing literacy in English with their language acquisition. Being Generation Z, students receive literacy training based on the Internet mainly without intervention from teachers.

5.2 Student Motivation

It is a well-known principle from teaching experience that instructional methods and textbook materials are less important than students' exploration and contributions. Students should be inspired and guided to discover and memorize things by their curiosity aroused, interest stimulated, and inquiry dealt with. While learning should come from inside, motivational strategies can bring to students' performance in the classroom. The way teachers educate usually leads to students' motivation in the process of language learning. Students' oral production is something that should be studied from the motivational field through focus groups, questionnaires, classroom observations, and interviews. Students are usually more enthusiastic towards the subject that they are well versed in.

A response from each participant is considered important during the in-class language conversation (Brown 2007). Clear instructions for cooperative learning are also key strategies (Peterson 2001). Students may not be aware of the benefits of group discussion and peer interaction. In this case, they should be informed of the communication strategies delivered in L2 and the benefits of doing so. Without interaction activities, students are less spontaneous to practice on their own. The data of students' Ding Ding Clock-in performance which records their daily completion of transcribing an English news text reveals that the standard deviation of completion percentage reaches as high as 36%. This means an extreme distribution that some students are so diligent that they stick to the task every day while some never set off.

The conventional classroom teaching techniques become inferior to the use of multimedia technology that combines elements such as audio, visual, and animation, by which students can acquire English proficiency effectively. Through the presentation, students learn by doing. Technology provides the visual effect that affects our judgement of education. In the second language teaching, the instructor should emphasize both the development of students' language skills and their interaction capability across the culture. The consideration of different cultures during communication is weighted in computer-aided language learning. The extra-linguistic factors include worldviews, rituals, customs, gestures, taboos, images, and symbols. Features such as different ways of addressing, expressing as well as idioms are applied in communication to distinguish the uniqueness of culture (Makhmudov 2020).

5.3 Student Technology Acceptance

New technologies become a norm in our daily life, particularly NLP techniques with the advent of ChatGPT by OpenAI. According to students' feedback, they expect that part of the after-class assignment is about examining their use of the technology (Fig. 1).

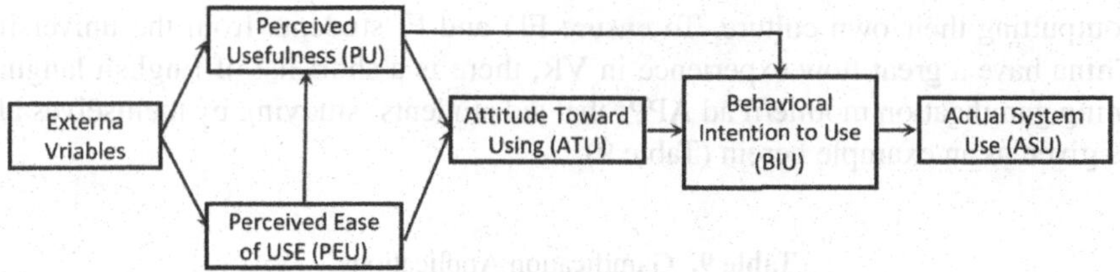

Fig. 1. TAM model

The Technology Acceptance Model (TAM) (Davis et al. 1989) model indicated that (1) PU and PEU, to the greatest extent, affect users' technological receptions; (2) the model exemplifies users' e-learning acceptance; and (3) the reliability and validity of TAM depend on individuals as well as the online medium (Murillo et al. 2021). It is essential to teach students how to use technology to the maximum benefit of improving their English mastery in the first year of university. The questionnaire result shows that 97% of surveyed students attach significance to tools that facilitate their English learning (eg. Word Cloud Creator, use of Microsoft PowerPoint master slide, Paratrans/DeepL, Text to Speech Converter, Grammarly.com, ReWriter (改写匠), Wordnet, Related Words generator, etc.). Among these tools, what first-year freshmen considered most useful is (Table 8):

Table 8. English learning tools

Use of Microsoft PowerPoint master slide	6.06%
Machine Translation platforms (Paratrans/DeepL)	21.21%
Text to Speech Converter	3.03%
Grammarly.com/ReWriter (改写匠)	69.7%
Wordnet	0
Word Cloud creator	0
Others	0

The prerequisite that students are prepared for and embrace the integration of technological updates in studies would help teachers plan how to support them better. Moreover, beyond the traditional technology scope and furthering educational uses, research has been done to explore and accentuate how the use of AR and VR intrigues better performances and enthusiasm from students (Jang et al. 2021).

6 Schema Knowledge via Metaverse

In an immersive metaverse, various types of cultural memory are captured, where cultures are narrated and reproduced in pixels and virtual texts. Users coming from different countries and ethnic backgrounds gather and meet in such a virtual environment. Education could be a competitive sector in the metaverse that aggregates potential talents for outputting their own culture. To ensure FD and FI students from the universities in China have a great flow experience in VR, there is a short list of English language learning gamification mobile/iPad APPs that aid students' studying by themselves after class given as an example herein (Table 9).

Table 9. Gamification Applications

English Mobile Name	Game Type	Price
Life is Strange	Choice-based narrative game	Free
Scribblenauts	Adventure and puzzle-solving game	Free
PRG Ruinverse	First-person role-play game	USD 7.99
Trails of Heroes	Role-play game	Free
InMind VR (Cardboard)	Adventure action/arcade VR game	Free
Anne Frank House	Museum Visitor Guide	Free (unlock all for USD 2.99)
Mount Everest Story	Mountain climbing simulation game	USD 3.99
Rec Room	Multiplayer adventure video game	Free

Metaverse provides the function of simulating an authentic teaching environment, where students imagine them being instructed in the classroom and feel the atmosphere of listening and interacting. Traditional classrooms fail to transcend limitations (e.g., classrooms and textbooks) of content delivery. Meanwhile, Massive Open Online Courses (MOOCs) achieves unrestrained online sharing of educational resources. However, it may lack the creation of an engaging environment where a crowd of students on site are acquiring academic update led by the instructor's presence. Such revolutionized educational attempt still confronts many unsatisfying obstacles, such as on-site tutoring and supervision, the lack of knowledge concretization, and the unavailability of space and time (Lin et al. 2022). Metaverse may be a combination of traditional scenarios and newly emerged technology explorations.

7 Conclusions

FI learners could develop their internal referents and restructure their knowledge while FD learners relied on external structure and guidance (Yang and Chen 2023). The present objective is to find out the connection between instructional dimensions and cognitive style, as well as employ useful and appropriate pedagogical strategies. The cognitive style in education covers two facets of consideration: instructional methods and student modality. To date, computer-assisted teaching with effective instruction for individuals and groups should be attended with more focus.

Regarding instructional methods, how to organize or structure a class with proper tools is a teaching process to be considered at the time of course design. FD students often struggle with logical structures. In this case, the teaching should be presented to them in an organized way to guide them to notice key information. Students would be motivated or stimulated to extract key information, then summarize and conclude the newly learnt knowledge into a knowledge system. Teachers need to give them guidelines and training in this regard (Sternberg, Grigorenko and Zahn 2008). Both motivations and feedback should be given the same amount of attention by teachers. FD students are becoming inferior without supportive resources. The highlight of teaching should be targeted at the consideration of students' cognitive styles. FD students may be aware of their self-motivated learning strategies (Trawick and Corno 1995 and Wolters 2003) while FI students should focus on their communication skills.

Beyond assessment methods, teachers and researchers also need to contemplate the fact that certain exams are disadvantageous for students with a particular cognitive style. It may work better if teachers abide by formative evaluation to avoid the interference of different cognitive styles and help students find out their weaknesses in the evaluation process and receive corresponding training (Vanderheiden and Donavan 2007). These methods may be categorized to be part of adaptive teaching (Corno 2008). Instructors need to respond to FD students and give out support when necessary. As the paper has learnt from the analysis of questionnaire results, students' course satisfaction is highly dependent on the instructor's on-site elaborated guidance. Multimedia technology motivates students with a portfolio of audio, visual and textual effects. However, interactions among different parties may decrease. Students may have less time for thinking, speaking and communication. They are considered only as viewers passively receiving what is presented to them rather than actively participating (Warschauer 2000).

Even with the aid of technology, the quality teaching that brings students a sense of achievement is not necessarily proven (Healey et al. 2011). It is a mistake that teachers fully rely on available technologies, without revealing and displaying the essence of teaching, which is effective knowledge transfer. Multimedia technology may attract students' attention temporarily, they may miss the chance of cultivating other language skills by just looking at the screen and being inactive. This is exactly what traditional tutoring and lecturing can achieve. Practically, it is a test of teachers' knowledge foundation and teaching competence in the classrooms, which could benefit students. Multimedia technology should be an addition to the teaching process without replacing the teacher's dominant role. The technology-based teaching environment should combine teachers' presentations with their academic capacity and teaching experience. Teachers should keep in mind students' process of acquiring knowledge and shift dedication to individuals' academic growth (Warschauer 2000). Metaverse could be a worth-trying step in the right direction.

Acknowledgement. The work reported in this paper was supported by a 2022 Teaching Quality and Reform Project for Undergraduate Education in Guangdong Province, "Learning Journal Based Reform and Practice of Critical Thinking and Value Education in College English Teaching".

References

Ahlin, E.M.: A mixed-methods evaluation of a hybrid course modality to increase student engagement and mastery of course content in undergraduate research methods classes. J. Crim. Justice Educ. **32**(1), 22–41 (2020)

Brame, C.J., Biel, R.: Test-enhanced learning: the potential for testing to promote greater learning in undergraduate science courses. CBE—Life Sci. Educ. **14**(2), es4 (2015)

Brown, M.R.: Educating all students: creating culturally responsive teachers, classrooms, and schools. Interv. Sch. Clin. **43**(1), 57–62 (2007)

Chen, S.Y., Liu, X.: An integrated approach for modeling learning patterns of students in web-based instruction: a cognitive style perspective. ACM Trans. Comput.-Hum. Interact. (TOCHI) **15**(1), 1–28 (2008)

Corno, L.Y.N.: On teaching adaptively. Educational psychologist **43**(3), 161–173 (2008)

Davis, F.D., Bagozzi, R.P., Warshaw, P.R.: User acceptance of computer technology: a comparison of two theoretical models. Manage. Sci. **35**(8), 982–1003 (1989)

Efland, A.: Art and Cognition: Integrating the Visual Arts in the Curriculum. Teachers College Press (2002)

Greenleaf, J.G.: Implementing Computer-Assisted Language Learning in the Teaching of Second Language Listening Skills. Minnesota State University, Mankato (2011)

Han, H.C.S.: From visual culture in the immersive metaverse to visual cognition in education. In: Cognitive and Affective Perspectives on Immersive Technology in Education, pp. 67–84. IGI Global (2020)

Healey, D., Smith, E.H., Hubbard, P., Ioannou-Georgiou, S., Kessler, G., Ware, P.: TESOL technology standards. TESOL, Alexandria Virginia [in English] (2011)

Huang, P.Y., Tsao, N.L.: Using collocation clusters to detect and correct English L2 learners' collocation errors. Comput. Assist. Lang. Learn. **34**(3), 270–296 (2021)

Hughes, J., Thomas, R., Scharber, C.: Assessing technology integration: the RAT–replacement, amplification, and transformation-framework. In: Society for Information Technology and Teacher Education International Conference, pp. 1616–1620. Association for the Advancement of Computing in Education (AACE), March 2006

Jang, J., Ko, Y., Shin, W.S., Han, I.: Augmented reality and virtual reality for learning: an examination using an extended technology acceptance model. IEEE Access **9**, 6798–6809 (2021)

Lin, H., Wan, S., Gan, W., Chen, J., Chao, H.C.: Metaverse in education: vision, opportunities, and challenges. arXiv preprint arXiv:2211.14951 (2022)

Lyle, K.B., Crawford, N.A.: Retrieving essential material at the end of lectures improves performance on statistics exams. Teach. Psychol. **38**(2), 94–97 (2011)

Makhmudov, K.: Ways of forming intercultural communication in foreign language teaching. Sci. Educ. **1**(4), 84–89 (2020)

Moskovsky, C., Jiang, G., Libert, A., Fagan, S.: Bottom-up or top-down: English as a foreign language vocabulary instruction for Chinese university students. TESOL Q. **49**(2), 256–277 (2014)

Murillo, G.G., Novoa-Hernández, P., Rodriguez, R.S.: Technology acceptance model and Moodle: a systematic mapping study. Inf. Dev. **37**(4), 617–632 (2021)

Paleeri, S. (n.d.): Contributions of Noam Chomsky for Educational Psychology. www.academia.edu. https://www.academia.edu/12688648/Contributions_of_Noam_Chomsky_for_Educational_Psychology. Accessed 30 May 2021

Peterson, R.A.: On the use of college students in social science research: insights from a second-order meta-analysis. J. Cons. Res. **28**(3), 450–461 (2001)

Rönnberg, J., Holmer, E., Rudner, M.: Cognitive hearing science and ease of language understanding. Int. J. Audiol. **58**(5), 247–261 (2019)

Rospigliosi, P.A.: Metaverse or Simulacra? Roblox, Minecraft, Meta and the turn to virtual reality for education, socialisation and work. Interact. Learn. Environ. **30**(1), 1–3 (2022)

Salaberry, M.R.: The use of technology for second language learning and teaching: a retrospective. Mod. Lang. J. **85**(1), 39–56 (2001)

Sandars, J., Patel, R.S., Goh, P.S., Kokatailo, P.K., Lafferty, N.: The importance of educational theories for facilitating learning when using technology in medical education. Med. Teach. **37**(11), 1039–1042 (2015)

Smith-Shank, D.L.: Reflections on semiotics, visual culture, and pedagogy (2007)

Stanger-Hall, K.F., Shockley, F.W., Wilson, R.E.: Teaching students how to study: a workshop on information processing and self-testing helps students learn. CBE—Life Sci. Educ. **10**(2), 187–198 (2011)

Stash, N., De Bra, P.: Incorporating cognitive styles in AHA! (The adaptive hypermedia architecture). In: Proceedings of the IASTED International Conference Web-Based Education, pp. 378–383, February 2004

Sternberg, R.J., Grigorenko, Zahn, L.F.: Styles of learning and thinking matter in instruction and assessment. Perspect. Psychol. Sci. **3**, 486–506 (2008)

Trawick, L., Corno, L.: Expanding the volitional resources of urban community college students. New Dir. Teach. Learn. **1995**(63), 57–70 (1995)

Tinajero, C., Castelo, A., Guisande, A., Páramo, F.: Adaptive teaching and field dependence-independence: Instructional implications. Revista Latinoamericana de Psicología **43**(3), 497–510 (2011)

Vandergrift, L.: Orchestrating strategy use: toward a model of the skilled second language listener. Lang. Learn. **53**(3), 463–496 (2003)

Vanderheiden, S., Donavan, J.: Connecting student outcomes to exam preparation strategies: promoting self-reflective learning. Paper Presented at Annual Meeting of the Midwest Political Science Association, Chicago, Illinois (2007)

Warschauer, M., Meskill, C.: Technology and second language teaching. In: Handbook of Undergraduate Second Language Education, vol. 303, no.18 (2000)

Wolters, C.A.: Regulation of motivation: evaluating an underemphasized aspect of self-regulated learning. Educ. Psychol. **38**(4), 189–205 (2003)

Yang, T.C., Chen, S.Y.: Investigating students' online learning behavior with a learning analytic approach: field dependence/independence vs. holism/serialism. Interact. Learn. Environ. **31**(2), 1041–1059 (2023)

A Positive-Negative Dual-View Model for Knowledge Tracing

Qiqi Xu[1], Guanhua Chen[1], Lan Shuai[2], Shi Pu[2], Hai Liu[1](✉), and Tianyong Hao[1]

[1] School of Computer Science, South China Normal University, Guangzhou, China
{xuqiqi,2021023267,haoty}@m.scnu.edu.cn, namelh@gmail.com

[2] Educational Testing Service, Princeton, USA
spu@ets.org

Abstract. Knowledge Tracing (KT) aims to accurately trace the states of evolving knowledge of students and reliably predict students' performances on future exercises. This task has been widely studied, leading to fast promotion on the development of online education. However, KT still faces two problems. First, most of previous work directly assigned an embedding for each question, which ignores semantic information contained in the questions. Secondly, students may learn differently from correct and incorrect answers to a question. Therefore, the embedding of a question should change based on the correctness of a student's answer. In this paper, we propose a positive-negative dual-view model named PDNV for knowledge tracing. Firstly, we leverage two Graph Convolutional Networks to learn question embeddings from both positive and negative perspectives. Secondly, an information filtering module is designed based on students' answers to selectively enhance positive or negative information in question embeddings. Experiment results based on three widely-used datasets demonstrate that our model outperforms state-of-the-art baseline models.

Keywords: Knowledge Tracing · Dual-view Model · Graph Convolutional Network

1 Introduction

Knowledge Tracing (KT) is an essential task in online education, which aims to trace the evolving knowledge states of students [1]. Specifically, KT can dynamically simulate a student's mastery level of knowledge concepts related to a certain question according to his/her learning interaction history and predict the probability that student will be able to answer correctly on the next question [1]. KT devotes to helping online education systems to develop personalized learning paths for students and recommending suitable learning resources to students, which can effectively improve their motivations and greatly enhance their learning outcomes [2]. Since the booming development of online education has brought massive educational data, which provides a solid research basis for KT. Recent years, KT has gained considerable attentions, and numerous researchers have devoted themselves to the field of KT [1, 3–5].

H. Zhang et al. (Eds.): NCAA 2023, CCIS 1870, pp. 492–504, 2023.
https://doi.org/10.1007/978-981-99-5847-4_35

Existing methods [6–8] generally build predictive models by distinguishing between the questions themselves. However, these methods face the problems of data sparsity because of the great difference between the number of questions and the number of questions that students actually attempt on. To solve this problem, Yang et al. [5] leveraged Graph Convolutional Network (GCN) to extract high-order relations between questions and skills. All of them directly assign an embedding for each question. However, students might learn differently from correct and incorrect answers to a question. Therefore, the embedding of a question should change based on the correctness of a student's answer. For example, Grimaldi et al. [9] found that a failed attempt might enhance a student's subsequent encoding and learning. Kang et al. [10] proposed an error correction theory, which explained the principle that failed attempts could enhance the knowledge encoding of subsequent learning. These learning theories also confirm that it is important to focus on both correct and incorrect answers, and learn question embeddings from both. For the rest of the paper, we refer to a correct answer to a question as a positive response, and an incorrect answer to a question as a negative response.

In this paper, we propose a dual-view model to fully exploit positive and negative responses to questions. Specifically, we leverage two GCNs to generate positive and negative question embeddings. Then, we use an information filtering module to selectively activate the positive or negative question embedding based on the correctness of the student's answer. This model architecture has several advantages. Firstly, a GCN that aggregates skill embeddings and question embeddings enables it to extract high-order skill-question relations [5]. Secondly, learning the question embeddings from different dimensions can mine more profound feature information of questions and then obtain a more fine-grained representation. At last, introducing student's responses as supervised signals can selectively enhance positive or negative information in question embeddings.

To sum up, this paper proposes a Positive-Negative Dual-View (PNDV) model for knowledge tracing. Our contributions are three-fold as follows:

1) Introducing a new dual-view model PNDV to represent questions from positive and negative perspective by capturing positive or negative information.
2) Utilizing an information filtering module to selectively enhance positive or negative information in question embeddings.
3) Conducting extensive experiments based on three real-world datasets to demonstrate the effectiveness of PNDV.

2 Related Work

KT models students' knowledge states based on their previous learning histories to accurately predict performance in future exercises. On top of this, online learning platforms can provide students with timely guidance and appropriate learning resources [2]. According to the modeling methods, existing KT methods can be divided into two groups: traditional methods and deep learning based methods.

Traditional methods consist of two representative methods, Bayesian Knowledge Tracing (BKT) and Additive Factor Model (AFM). BKT [1] regards students' knowledge states as a tuple of binary values and updates the values by a hidden Markov model. Many variations were developed to extend the applications of BKT, for example, extending students' knowledge states [11], integrating forget and slip probabilities [12] and adding item difficulty into model [13]. Based on the Item Response theory (IRT), the AFM model [14] introduced learning rate and practiced opportunities for modeling the dynamic learning process. Further extensions of the AFM model include, introducing constraints on parameters values [15], dividing learning opportunities into positive and negative [16], etc.

With the development of deep learning techniques, researchers have leveraged deep learning methods to automatically capture relationships between skills. DKT [3] first introduced deep neural networks into KT, which exploit LSTM to model a student's learning process. Zhang et al. [17] enriched input features at the problem-level by observing more combinations of input features to extend DKT. In order to model a student's forgetting behavior during learning process, Nagatani et al. [18] incorporated more information related to forgetting behavior into DKT. Pandey et al. [19] introduced a transformer to calculate the attention weights for previously answered questions, enhancing relevant information and weakening irrelevant ones. Pu et al. [20] modified the Transformer structure to leverage between-question relations and simulated forgetting behavior in the learning process.

For tracing a student's mastery level of each skill, Dynamic Key-Value Memory Networks (DKVMN) [4] exploited a static matrix to store skills with a dynamic matrix storing and updating mastery levels of corresponding skills. Dynamic student classification on memory networks (DSCMN) [21] assumed that learning ability changes over time and captured it at each time interval for dynamic student classification.

Considering that skills are interrelated, Graph-based Knowledge Tracing (GKT) [22] leveraged graph neural network for KT based on a graph-based structure of skills. Inspired by GCN which updated graph representations by aggregating information from neighbors, GIKT [5] exploited GCN to aggregate skill embeddings and question embeddings for extracting high-order skill-question relations. In order to leverage spatial associations and complex structures of nodes in "exercise-to-exercise" (E2E) relational subgraph, Bi-Graph Contrastive Learning based Knowledge Tracing (Bi-CLKT) [23] leveraged contrastive learning to learn representation of exercise at global and local levels.

Inspired by Grimaldi et al. [9], we believe that it is important to focus on both positive and negative responses. In our paper, we propose a dual-view model to learn question embeddings from positive and negative responses by leveraging two GCNs. In addition, we introduce an information filtering module to selectively activate corresponding question embeddings.

3 Preliminary

3.1 Problem Definition

For each student, the learning interactions are denoted as M-length sequence $X = \{X_1, X_2, \ldots, X_t, \ldots, X_M\}$, where each X_t consists of (q_t, s_t, r_t).$q_t \in Q$ represents the question answered by the student at time step t, and Q represents the set of questions. $s_t \in S$ denotes the skill examined by the question q_t and S represents the set of skills. $r_t \in \{0, 1\}$ represents the student's response at time step t, in which 1 represents a positive response and 0 represents a negative response. The KT task aims to predict a student's performance in questions $P = \{p_1, p_2, \ldots, p_t, \ldots, p_M\}$.

3.2 Heterogeneous Information Network

A heterogeneous information network $H = (V, E)$ is exploited to model the relation of skills and questions, where V and E represent the set of nodes and edges in the network, respectively. There are two types of nodes in the network: skill nodes and question nodes. The relationship between skills and questions are represented by edges between nodes.

A question usually involves multiple skills and one skill may relate to multiple questions. Taking Fig. 1 as an example, there are three edges between *question* 1 and {*skill* 1, *skill* 2, *skill* 7}, representing that *question* 1 examines three skills. Similarly, there are two edges between *skill* 3 and {*question* 2, *question* 3}, representing that *skill* 3 is related to two questions.

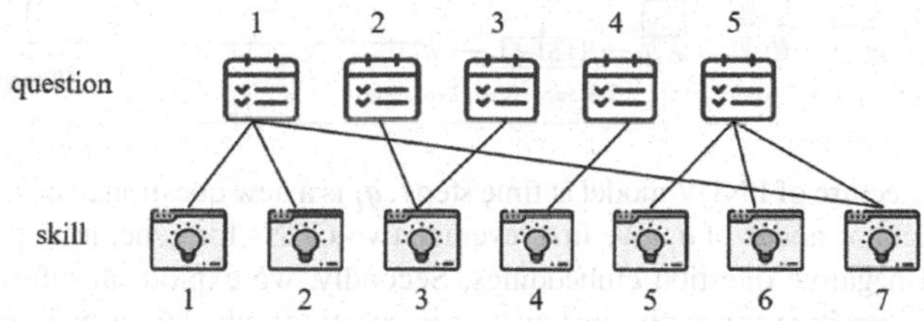

Fig. 1. An example of a skill-question heterogeneous information network

4 The Approach

This section introduces the Positive-Negative Dual-View (PNDV) model in detail. The overall model architecture is presented in Fig. 2. Specifically, the PNDV consists of four parts: two GCNs for learning question embeddings and skill embeddings in skill-question network, an information filtering module for selectively enhancing positive or negative information in question embeddings, a LSTM for modeling the evolving knowledge states of student and an interaction module for prediction by capturing long term dependency and extracting history information. Below, we denote vectors with bold italic small letters.

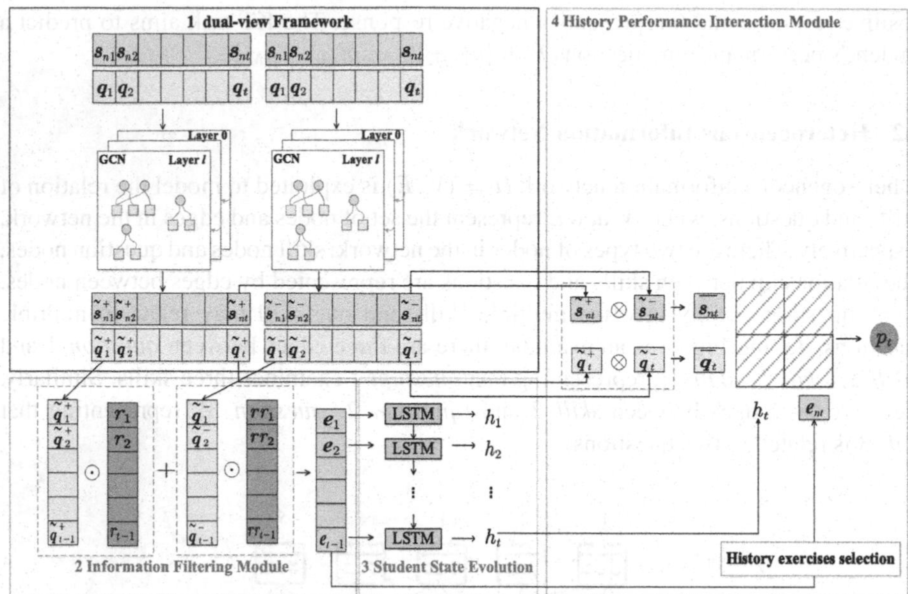

Fig. 2. The architecture of PNDV model at time step t, $\boldsymbol{q}_t$ is a new question to be predicted and $\boldsymbol{s}_{nt}$ are the set of neighbor nodes of $\boldsymbol{q}_t$. We first leverage two GCNs for generating positive question embeddings and negative question embeddings. Secondly, we exploit an information filtering module to enhance positive or negative information in question embeddings by introducing student responses as supervised signal. Thirdly, we use LSTM to model the evolving knowledge states of students. Finally, we use the history performance interaction module to make prediction p_t.

4.1 Input Representation

We randomly initialize $\boldsymbol{q}_t \in R^d$ as the embedding of question $\boldsymbol{q}_t$. Thus the set of questions can be represented as an embedding matrix $\boldsymbol{Q} \in R^{|Q| \times d}$, where $|Q|$ is the number of questions, d is the number of dimensions. Similarly, we denote $\boldsymbol{s}_t \in R^d$ as the embedding of skill $\boldsymbol{s}_t$, and the set of skills can be represented as an embedding matrix $\boldsymbol{S} \in R^{|S| \times d}$. Since responses are binary values, the two rows in embedding matrix $\boldsymbol{R} \in R^{2 \times d}$ represent the positive and negative responses, respectively.

4.2 Dual-View Model

We propose a dual-view model to express questions from a positive perspective and a negative perspective, separately. With different parameters, the GCNs are exploited to generate positive question embeddings and negative question embeddings, respectively. The skill-question network convolution performs information transfer and node feature aggregation through the relations of skills and questions.

In the convolution process, the 1st hop neighbor nodes are the skill nodes corresponding to the question nodes and the 2nd hop neighbor nodes are the question nodes related to the skill nodes. We denote the representation of node v in the network as $\boldsymbol{v}^{(0)}$, which expresses skill embedding s or question embedding q. The propagation process of the embedding can be expressed as follows:

$$v^{(l+1)} = \sigma\left(\left(\frac{1}{|N_{(v)}|}\sum_{u\in N_{(v)}} u^l \oplus v^l\right)W^{l+} + b^{l+}\right), v^{(0)} \in R^d, \tag{1}$$

$$v^{(l+1)} = \sigma\left(\left(\frac{1}{|N_{(v)}|}\sum_{u\in N_{(v)}} u^l \oplus v^l\right)W^{l-} + b^{l-}\right), v^{(0)} \in R^d, \tag{2}$$

where $N_{(v)}$ is the set of the neighboring nodes of node v, $\oplus$ is the operation of adding two vectors, σ is ReLU function, W^{l+} and W^{l-} are the weight matrix in the l-th layer in GCNs respectively. b^{l+} and b^{l-} are the biases in the l-th layer in GCNs, respectively.

After the propagation process, we obtain the aggregate embedding of questions and skills from the positive and negative perspectives and use $\tilde{q}^+$, $\tilde{q}^-$, $\tilde{s}^+$, $\tilde{s}^-$ to express them. In addition, we fix the number of neighbors in each batch for better parallelization.

4.3 Information Filtering Module

For each question, we introduce an information filtering module based on student responses for selectively enhancing positive and negative information in question embeddings. If the student response is positive, the information filtering nodule will activate positive question embedding, otherwise, it activates negative question embedding. Firstly, we use the reverse operation in responses and denotes them as rr. Then, for each time step t, we leverage the response as a supervised signal to activate positive or negative question embedding. This process can be expressed as,

$$rr_t = \sim r_t, \tag{3}$$

$$e_t = \tilde{q}_t^+ \times r_t + \tilde{q}_t^- \times rr_t, \quad 1 \le t \le M-1. \tag{4}$$

After that, the two question embeddings tend to be associated with corresponding responses and continue to enhance themselves during the training process.

4.4 Student State Evolution

Whether the next question can be answered correctly depends not only on the current knowledge state, but also on the student's historical performance. Recurrent neural networks are proposed to solve this problem. Inspired by [12], the phenomenon of forgetting exists in the student's learning process, therefore, we use the forgetting gate in LSTM [24] to model the forgetting behavior of students. This process can be expressed as,

$$f_t = \sigma([h_{t-1}, e_t]W_f + b_f), \tag{5}$$

$$i_t = \sigma([h_{t-1}, e_t]W_i + b_i), \tag{6}$$

$$\tilde{C}_t = tanh([h_{t-1}, e_t]W_C + b_C), \tag{7}$$

$$C_t = f_t \times C_{t-1} + i_t \times \tilde{C}_t, \tag{8}$$

$$o_t = \sigma([h_{t-1}, e_t]W_o + b_o), \tag{9}$$

$$h_t = o_t \times tanh(C_t), \tag{10}$$

where f is the decay coefficient of the memory, i_t is the decay coefficient of the currently learned memory, $\tilde{C}_t$ is the current state learned memory, C_t is the current memory state at time step t, o_t is the coefficient of the output gate, and h_t is the output result. h_t is a summary of the student's knowledge mastery from time step *1* to time step *t*, representing the student's knowledge state at time step *t*.

4.5 History Performance Interaction Module

Students are always susceptible to the influence of similar questions in history when completing the current question [19]. We introduce an interaction module for leveraging useful information in history states in the following two aspects: (i) In order to model a student's mastery degree at question and skill level respectively, we use the student's knowledge state h for interaction. (ii) Students tend to give similar responses when answering questions on the same topic, thus we select the relevant history exercises based on the methods of sharing the same skills with the new question. In addition, we generate the positive and negative embeddings of question and skill in the previous module, and use fusion coefficients to fuse the two embeddings respectively.

$$e_{nt} = \{e_i | N_{q_i} = N_{q_t}, 1 \le i < t\}, \tag{11}$$

$$\overline{q}_t = \lambda \times \tilde{q}_t^+ + (1 - \lambda) \times \tilde{q}_t^-, \tag{12}$$

$$\overline{s}_{nt} = \lambda \times \widetilde{s_{nt}}^+ + (1 - \lambda) \times \tilde{s}_{nt}^-, \tag{13}$$

$$I_e = [\overline{q}_t, \overline{s}_{nt}], \tag{14}$$

$$I_h = [e_{nt}, h_t], \tag{15}$$

$$I = I_e \times I_h, \tag{16}$$

where e_{nt} represents the relevant history exercises of q_t, λ is an adjustable parameter. We denote the neighbors of q_t as s_{nt}. [,] represents the vector concatenation. After the operation of multiplication, we use I to represent the result after a two-by-two interaction.

In order to calculate the correlation between the interactions, we introduce an attention network to calculate the bi-attention weights between interactions and compute the weighted sum as the final prediction score.

$$\alpha = Softmax(\sigma[\boldsymbol{I}_e, \boldsymbol{I}_h]), \tag{17}$$

$$p_t = \alpha \times \boldsymbol{I}. \tag{18}$$

4.6 Optimization

To update the trainable parameters in our model and optimize the model predicted results, we choose the cross entropy loss between true score r_t and predicted score p_t as the objective function.

$$L = -\sum_t (r_t \log p_t + (1 - r_t)\log(1 - p_t)). \tag{19}$$

5 Experiments and Results

5.1 Datasets

We evaluate our model on three real-world educational datasets, the statistics of which are shown in Table 1.

ASSIST09 dataset [25] was collected from the ASSISTments online platform from 2009 to 2010. It contains 123 skills, 17,737 questions, and involves 282,619 learning interactions from 3,852 students.

ASSIST12 dataset [5] also came from the ASSISTments online platform, covering the period of 2012. It includes 265 skills, 53,065 questions, and involves 2,709,436 learning interactions from 27,485 students.

EdNet dataset [26] was collected by Santa. Due to the overwhelming amount of data, we randomly selected 676,974 learning interactions generated by 5,000 students, which contained 189 skills and 12,161 questions.

Table 1. The statistics of datasets

Statistics	ASSIST09	ASSIST12	EdNet
#skills	123	265	189
#questions	17,737	53,065	12,161
#students	3,852	27,485	5,000
#interactions	282,619	2,709,436	676,974

In the implementation, we split all the datasets into training and testing sets according to the ratio of 80% and 20%. Following previous work, we use the Area Under the Curve (AUC) to evaluate the performance of the proposed model.

5.2 Baselines

We compare our proposed method PNDV with the following state-of-the-art KT methods.

BKT [1] uses a hidden Markov model to update the student's knowledge state, considered to be a tuple of binary values.

DKT [3] first introduces deep learning methods into the KT field and makes good progress. It exploits LSTM to model the changing of a student's knowledge state.

DKVMN [4] draws on the idea of Memory-Augmented Neural Network and creatively uses two matrices to accomplish static storage and dynamic update operations.

GIKT [5] exploits the GCN to aggregate information from neighbors based on the relations of skills and questions.

5.3 Implementation Details

The embedding matrices of skills, questions and responses are randomly initialized and updated during the training process, and the dimension of embedding matrices is set as 100. Following the setting of GIKT [5], the number of propagation layers and sampled neighbors are respectively set as 3 and 4 in the implementation of GCN. Similarly, the number of related exercises of the new question is set as 3. We use a two-layer stacked LSTM, where the sizes of the memory cells are 200 and 100 respectively. The model is optimized by Adam with a learning rate of 0.001 and a batch size of 32. The maximum length of sequence is 200 and the minimum length is 3. The results of model performance are averaged over ten runs.

5.4 Results

We report experiment results comparing PNDV with the other baseline models in Table 2, where the best results are highlighted in bold. Our model performs better than all the baseline models on the three datasets. Specifically, compared with the best baseline GIKT model, PNDV improves AUC on ASSIST09 by 0.20%, on ASSIST12 by 0.17%, on EdNet by 0.19%, which confirms the effectiveness of our model.

PNDV performances the best on ASSIST09 dataset than on the other two datasets. The reason may be that negative responses account for the largest proportion of total responses in the ASSIST09 dataset among the three datasets, thus the proposed positive-negative dual-view model facilitates the model to make full use of negative information.

Table 2. The AUC results of comparison methods

Dataset	Results (AUC)				
	BKT	DKT	DKVMN	GIKT	PNDV
ASSIST09	0.6571	0.7561	0.7550	0.7883	**0.7903(+0.0020)**
ASSIST12	0.6204	0.7286	0.7283	0.7732	**0.7749(+0.0017)**
EdNet	0.6027	0.6822	0.6967	0.7411	**0.7430(+0.0019)**

The amount of positive information expressed in the question embeddings is controlled by the hypo parameter λ (as shown in Eq. (11)). Figure 3 shows the AUC results of different values of λ setting on the three datasets. As the λ value increases, the proportion of positive information in the question embeddings increases. When $\lambda = 0.7,\ 0.3,\ 0.5$, our model achieves the best performance on the three datasets, respectively, which shows that the inclusion of negative information can help to obtain more fine-grained questions embeddings and improve model performance.

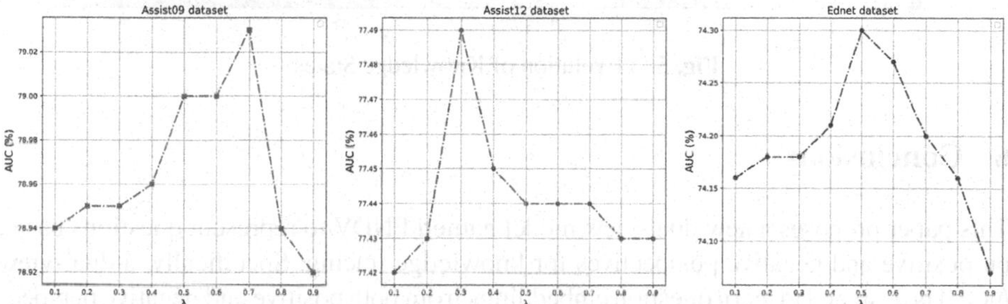

Fig. 3. The AUC results with different configuration of λ on three datasets

Figure 4 and Fig. 5 compare the evolution of knowledge states of six students traced by GIKT and PNDV. From these cases, we can observe several important findings. Firstly, when predicting performance on a coming question, GIKT refers to performance on other questions that share the same skills as that question in the student's history exercises. However, this may provide a false reference for the prediction. For example, when predicting student performance on *question* 7939, GIKT incorrectly refers to the performance on *question* 7923, resulting in a wrong prediction. Our proposed PNDV method can reduce the interference of incorrect reference and improve the accuracy of prediction (as shown in Fig. 4). Secondly, in Fig. 5, we note that GIKT argues the knowledge state will decline if a student gives an incorrect response to the question, which is clearly not in line with the educational cognitive theory. Students can learn from both positive and negative experiences when facing the same question. Our model corrects for this faulty predicted behavior and models the student's learning process very well. Moreover, in Fig. 5, we can observe that the improvement of a student's knowledge states after he/she answers a question incorrectly is less than that after he/she answers the question correctly.

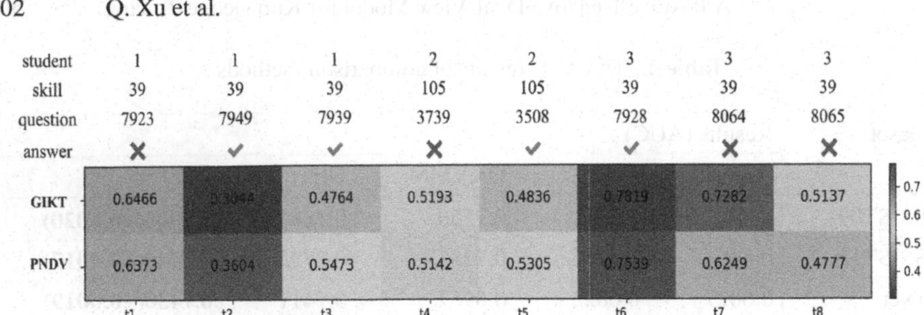

Fig. 4. . Evolution of Knowledge States

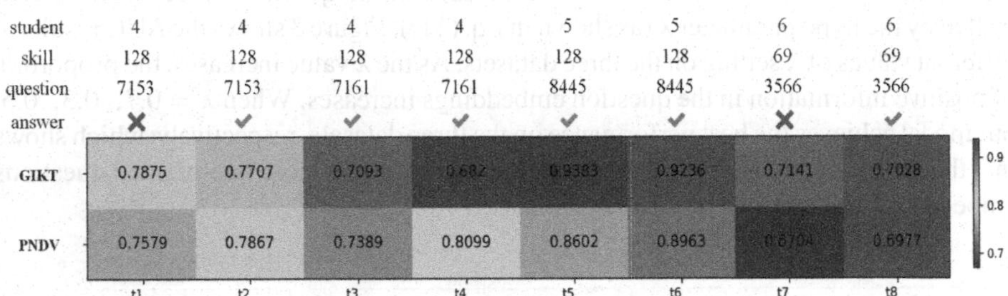

Fig. 5. . Evolution of Knowledge States

6 Conclusions

This paper proposes a new dual-view model named PNDV to represent questions based on positive and negative perspectives for knowledge tracing. Specifically, a dual-view model is designed to learn question embeddings from both positive and negative perspectives to fully utilize semantic information contained in questions. Then an information filtering module is leveraged to selectively activate positive or negative question embeddings. Experiment results based on three educational datasets prove that PNDV can accurately model students' learning process and achieve better performance than state-of-the-art baselines. In future work, we plan to optimize model architecture and improve interpretability of model. For example, introducing attention mechanism to identify the relationship between questions and skills, and exploring latent knowledge structure information. Besides, we can further optimize the structure of the History Performance Interaction Module to better exploit historical information from learning interactions, thereby improving the predictive performance of model.

References

1. Corbett, A.T., Anderson, J.R.: Knowledge tracing: modeling the acquisition of procedural knowledge. User Model. User-Adap. Inter. **4**, 253–278 (1994)
2. Wu, Z., Tang, Y., Liu, H.: Survey of personalized learning recommendation. J. Front. Comput. Sci. Technol. **16**, 21 (2022)
3. Piech, C., et al.: Deep knowledge tracing. Adv. Neural. Inf. Process. Syst. **28**, 505–513 (2015)

4. Zhang, J., Shi, X., King, I., Yeung, D.Y.: Dynamic key-value memory networks for knowledge tracing. In: Proceedings of the 26th international conference on World Wide Web, pp. 765–774. The International World Wide Web Conferences Steering Committee, Republic and Canton of Geneva, CHE (2017)
5. Yang, Y., et al.: Gikt: a graph-based interaction model for knowledge tracing. In: Hutter, F., Kersting, K., Lijffijt, J., Valera, I. (eds.) ECML PKDD 2020. LNCS (LNAI), vol. 12457, pp. 299–315. Springer, Cham (2021). https://doi.org/10.1007/978-3-030-67658-2_18
6. Choi, Y., et al.: Towards an appropriate query, key, and value computation for knowledge tracing. In: Proceedings of the Seventh ACM Conference on Learning@ Scale, pp. 341–344. Association for Computing Machinery, New York (2020)
7. Shin, D., Shim, Y., Yu, H., Lee, S., Kim, B., Choi, Y.: Saint+: integrating temporal features for ednet correctness prediction. In: LAK21: 11th International Learning Analytics and Knowledge Conference, pp. 490–496. Association for Computing Machinery, New York (2021)
8. Shen, S., et al.: Learning process-consistent knowledge tracing. In: Proceedings of the 27th ACM SIGKDD Conference on Knowledge Discovery & Data Mining, pp. 1452–1460. Association for Computing Machinery, New York (2021)
9. Grimaldi, P.J., Karpicke, J.D.: When and why do retrieval attempts enhance subsequent encoding? Mem. Cogn. **40**, 505–513 (2012)
10. Kang, S.H., Pashler, H., Cepeda, N.J., Rohrer, D., Carpenter, S.K., Mozer, M.C.: Does incorrect guessing impair fact learning? J. Educ. Psychol. **103**, 48 (2011)
11. Zhang, K., Yao, Y.: A three learning states bayesian knowledge tracing model. Knowl.-Based Syst. **148**, 189–201 (2018)
12. Qiu, Y., Qi, Y., Lu, H., Pardos, Z.A., Heffernan, N.T.: Does time matter? modeling the effect of time with bayesian knowledge tracing. In: EDM 2011 - Proceedings of the 4th International Conference on Educational Data Mining, pp. 139–148 (2011)
13. Pardos, Z.A., Heffernan, N.T.: Kt-idem: Introducing item difficulty to the knowledge tracing model. In: Konstan, J.A., Conejo, R., Marzo, J.L., Oliver, N. (eds.) UMAP 2011. LNCS, vol. 6787, pp. 243–254. Springer, Heidelberg (2011). https://doi.org/10.1007/978-3-642-22362-4_21
14. Cen, H.: Generalized learning factors analysis: improving cognitive models with machine learning. Carnegie Mellon University (2009)
15. Cen, H., Koedinger, K., Junker, B.: Comparing two irt models for conjunctive skills. In: Woolf, B.P., Aïmeur, E., Nkambou, R., Lajoie, S. (eds.) ITS 2008. LNCS, vol. 5091, pp. 796–798. Springer, Heidelberg (2008). https://doi.org/10.1007/978-3-540-69132-7_111
16. Pavlik Jr, P.I., Cen, H., Koedinger, K.R.: Performance factors analysis–a new alternative to knowledge tracing. In: 14th International Conference on Artificial Intelligence in Education (AIED 2009), Frontiers in Artificial Intelligence and Applications, p. 531+. IOS press, Amsterdam (2009)
17. Zhang, L., Xiong, X., Zhao, S., Botelho, A., Heffernan, N.T.: Incorporating rich features into deep knowledge tracing. In: Proceedings of the Fourth (2017) ACM Conference on Learning@ Scale, pp. 169–172. Association Computing Machinery, New York (2017)
18. Nagatani, K., Zhang, Q., Sato, M., Chen, Y.Y., Chen, F., Ohkuma, T.: Augmenting knowledge tracing by considering forgetting behavior. In: The World Wide Web Conference, pp. 3101–3107. Association for Computing Machinery, New York (2019)
19. Pandey, S., Karypis, G.: A self-attentive model for knowledge tracing. In: EDM 2019 - Proceedings of the 12th International Conference on Educational Data Mining, pp. 384–389. International Educational Data Mining Society (2019)
20. Shi, P., Michael Yudelson, L.O., Huang, Y.: Deep knowledge tracing with transformers. In: Bittencourt, I., Cukurova, M., Muldner, K., Luckin, R., Millán, E. (eds.) AIED 2020. LNCS

(LNAI), vol. 12164, pp. 252–256. Springer, Cham (2020). https://doi.org/10.1007/978-3-030-52240-7_46
21. Minn, S., Desmarais, M.C., Zhu, F., Xiao, J., Wang, J.: Dynamic student classiffication on memory networks for knowledge tracing. In: Yang, Q., Zhou, Z.-H., Gong, Z., Zhang, M.-L., Huang, S.-J. (eds.) PAKDD 2019. LNCS (LNAI), vol. 11440, pp. 163–174. Springer, Cham (2019). https://doi.org/10.1007/978-3-030-16145-3_13
22. Nakagawa, H., Iwasawa, Y., Matsuo, Y.: Graph-based knowledge tracing: modeling student proficiency using graph neural network. In: IEEE/WIC/ACM International Conference on Web Intelligence, pp. 156–163. Association for Computing Machinery, New York (2019)
23. Song, X., Li, J., Lei, Q., Zhao, W., Chen, Y., Mian, A.: Bi-clkt: bi-graph contrastive learning based knowledge tracing. Knowl.-Based Syst. **241**, 108274 (2022)
24. Hochreiter, S., Schmidhuber, J.: Long short-term memory. Neural Comput. **9**, 1735–1780 (1997)
25. Feng, M., Heffernan, N., Koedinger, K.: Addressing the assessment challenge with an online system that tutors as it assesses. User Model. User-Adap. Inter. **19**, 243–266 (2009)
26. Choi, Y., et al.: Ednet: a large-scale hierarchical dataset in education. In: Bittencourt, I., Cukurova, M., Muldner, K., Luckin, R., Millán, E. (eds.) AIED 2020. LNCS (LNAI), vol. 12164, pp. 69–73. Springer, Cham (2020). https://doi.org/10.1007/978-3-030-52240-7_13

A Study of Chinese-English Translation Teaching Based on Data Mining

Leiming Kang[1,2], Wenhui Zhang[1,2](✉), and Mengfan Guo[1]

[1] School of Foreign Studies, Hefei University of Technology, Hefei, China
1592634228@qq.com
[2] Center for Literature and Translation Studies, Hefei University of Technology, Hefei, China

Abstract. The paper presents specific applications of data mining methods in the analysis of the data obtained from the translation process. The authors take knowledge nodes involved in Textwells platform as the analysis object, importing the data into Gephi, one of the data mining tools, to generate and visualize the relevance graph of all related knowledge nodes and analyze the hidden characteristics. This effort concludes that there are top four communities in the knowledge-based network, namely, semantic matching, syntactic matching, word-class alteration, rank alteration. In these four communities, the co-occurrence of the knowledge nodes is more prominent, which definitely will help teachers with their translation teaching. In addition, the authors review the existing Chinese-English translation courses on the Chinese University MOOCs platform, study the teaching syllabuses carefully and suggest redesigning and modifying the current translation courses based on the findings above so as to improve the teaching effectiveness and efficiency.

Keywords: Translation Teaching · Data Mining · Knowledge Nodes

1 Introduction

Due to globalization, translation teaching has become increasingly important. This may bring us to think about the keys to improve translation ability effectively and efficiently. The ultimate goal of translation teaching, whether online or offline, college or social institution training, is to improve learners' translation competence and translation quality. Translation competence consists of a series of related components such as language competence, text competence, theme competence, culture competence and transfer competence summarized by Albrecht Neubert [1]. According to Miao, translation competence plays a vital role, which includes bilingual competence, transfer competence, and world/subject knowledge. Language competence is the foundation of translation competence [2].

Over the years the traditional translation teaching is closely related to language teaching. Nevertheless, the training of translation methods and skills is often based on impressionistic experience teaching while a systematic and theoretically supported text analysis is absent. What's more, little attention has been paid to the importance

H. Zhang et al. (Eds.): NCAA 2023, CCIS 1870, pp. 505–515, 2023.
https://doi.org/10.1007/978-981-99-5847-4_36

and relevance of knowledge nodes in the process of translation, including language phenomena, bilingual competence and transfer competence [3].

Therefore, translation teaching is supposed to involve the training of skills or methods and pay more attention to the construction of a knowledge network combining multiple knowledge modules as well [3]. In this way we can be clear about the obvious characteristics and the hidden relationship or close relevance between different knowledge nodes in the process of translation.

The development of science and technology in the 21st century has made the impact of data mining more and more indispensable. Data analysis and data mining technology have attracted the attention and interest from analysts and scholars of different fields like finance, healthcare, manufacturing, engineering design, media, education, etc. Because of its powerful data organization and visualization ability, it is applied more and more frequently in the field of education.

The paper presents specific applications of data mining methods in the analysis/investigation of the data obtained from the translation process [4]. The authors take knowledge nodes involved in Textwells as the analysis object, importing the data into Gephi, one of the data mining tools, to generate and visualize the relevance graph of all related knowledge nodes and analyze the hidden characteristics. This effort would help teachers to redesign and modify the current translation courses so as to improve the teaching effectiveness and efficiency.

2 Literature Review

The term "data mining" was not coined until the 1990s. As an interdisciplinary subfield of computer science and statistics, data mining can be treated as a process of discovering patterns in large data sets [5]. It is used to help uncover hidden patterns, bursts of activity, correlations and laws of data because of its powerful data organization and visualization ability [6].

Early applications of data mining were mostly employed in financial field [7]. The application of data mining in educational field is a relatively new discipline, which is called Educational Data Mining, combining the techniques of data mining with educational data in order to provide students, instructors, and researchers with knowledge that can benefit academic processes [8].

According to the review of the existing research efforts published concerning the current practices in Educational Data Mining applications performed with the database of Web of Science and CNKI (China National Knowledge Infrastructure), it's found that the current research of data mining in foreign language learning is mainly to predict the performance of students, to check learners' motivation, and to provide feedback for instructors [8]. Wang and Liao (2011) constructed an Adaptive Learning in Teaching English as a Second Language (TESL) for e-learning (AL-TESL-e-learning system) and explored the learning performance of various students using a data mining technique when three different levels of content (grammar, vocabulary, and reading) were set in the system [9]. Alina A. von Davier 2012 made an overview of the use of data mining for monitoring scaled scores [10].

In the past few years, data mining research in fields like manufacturing has increased rapidly. In terms of translation studies, various approaches have been addressed [4].

Taking Zhiwang or CNKI as literature sources, we search for CSSCI core journals with the key concept "translation", and the retrieving results suggest that the number of the papers published is 191,400. When "translation teaching" or "translation studies" is used as the key concept, researches are mostly focused on areas like translation theory, interdisciplinary research, machine translation, corpus translation studies, translation of various text types, translation standards, translation norms, translation techniques, etc. However, there are still few researches on data mining-based translation studies, and even fewer data-based and in-depth studies in this field. Overall, researches on data mining application to translation teaching are still at an initial stage and call for further development.

This paper attempts to analyze the use of data mining in the field of education, the teaching of translation in particular, targeting the advantages of data mining in teaching practices, optimization ideas, and so on, in order to achieve the positive significance of enriching and improving the quality of translation courses in universities and colleges.

3 Textwells Platform

Data used for the empirical case study of the paper is from Textwells Platform, a knowledge-network-based system for online translation teaching/learning, which is an interdisciplinary outcome covering computer science, translation studies, language acquisition, functional linguistics, corpus linguistics, stylistics and discourse studies. It features a computable network with "tag-words" as the knowledge nodes to form a roadmap of navigation to introduce theory-informed annotations [11, 12].

On the Platform, there are 236 bilingual texts collected including the subject domains like cultural texts, financial texts, legal documents, entertainment texts, literary texts, public affairs documents, scientific texts, news reports, etc. The data is annotated for the identification and explanation of language phenomena or translation techniques by using specific knowledge nodes which are classified into six fields, word level, sentence level, information, text, rhetoric, and intertextual associations.

Although the size of its database is not amazing, it is the first attempt of this kind in the field. What is recognized in its field is the knowledge-based, theoretically-informed delicacy and relevance of annotations and its teacher/learner-friendly data management [12]. There are some papers published introducing and demonstrating the feasibility and validity of this platform for teaching translation online. This data-based and computer-aided translation training paradigm with text annotation and knowledge nodes can help present the common features and the relevance of translation knowledge in the process of translating real texts, optimize learning resources and alleviate the pressure of staff shortage/labor-intensive teaching [11].

4 The Relevance Visualization of Translation Knowledge Nodes

4.1 Gephi-a Network Visualization Software

Gephi is a general-purpose, open-source social network software for data application and analysis. It is jointly developed by engineers and scientists from various countries. It was put into use in France in 2008. Gephi Alliance, a non-profit organization, was established to support, protect and promote the Gephi project [13].

It is the leading visualization and exploration software for all kinds of graphs and networks, a paradigm appeared in the Visual Analytics field of research [14], which is widely used in various disciplines like social network analysis, biology, genomics, etc.

Some prominent and powerful features lie in its layout algorithms and efficient real-time visualization. It helps the analysis of data and exploration and understanding of graphs by providing various layout algorithms to present network and transform it into a map. The goal is to help data analysts make hypothesis, intuitively discover patterns, and isolate structure singularities or faults during data sourcing [14] so that the hidden patterns can be revealed.

4.2 Translation Knowledge Nodes Relevance Network Visualization

We are authorized to mine the data in Textwells. The relevance between the knowledge nodes in the knowledge network is visually presented in the form of graphs. Gephi tool in this paper is used for link analysis of the knowledge nodes deriving from various bilingual texts, and it is used to help find the hidden relevance between them by applying layout algorithms of the graphs.

Translation knowledge nodes are collected from Textwells with Web crawler-like tools. And we have done the necessary data preprocessing operations such as data preparation, data cleaning, data selection, data integration, data reduction, data filtering, data transformation, incorporation of appropriate prior knowledge. With the help of Gephi text analysis software and data mining association rule technology, the data of text annotations of Textwells is mined and processed, and subsequently the text network is displayed through Gephi's visualization function. The text network transforms the abstract data into a graph that is easier for researchers to understand and observe, vividly describing the internal relationship of the data and revealing the important relevance hidden in the data, so as to display the results of data analysis more clearly and effectively. It is shown in Fig. 1.

It can be clearly seen from Fig. 1 that there are many knowledge nodes involved in the translation process, and the internal correlation between knowledge points is active. But the data nodes are too complex, which is not conducive to observing the distribution characteristics and internal relationship. In order to facilitate the visual analysis of the co-occurrence characteristics of translation knowledge nodes and enhance the indicative performance and visual effect of the graph, we have reset the data extraction rules. And the data extraction scope is limited to the texts with 20 or more knowledge nodes annotated in the text, a total of 465 data. As a result, Fig. 2 is obtained. It is helpful to facilitate the observation and discovery of the relationship between translation nodes, and facilitate the better realization of the purpose of translation research.

Relevance analysis is carried out by using Gephi statistical method, which is to visualize the relevance between translation knowledge nodes, including graph density, modularity, average degree, average clustering coefficient and average path length, etc., as is shown in Table 1.

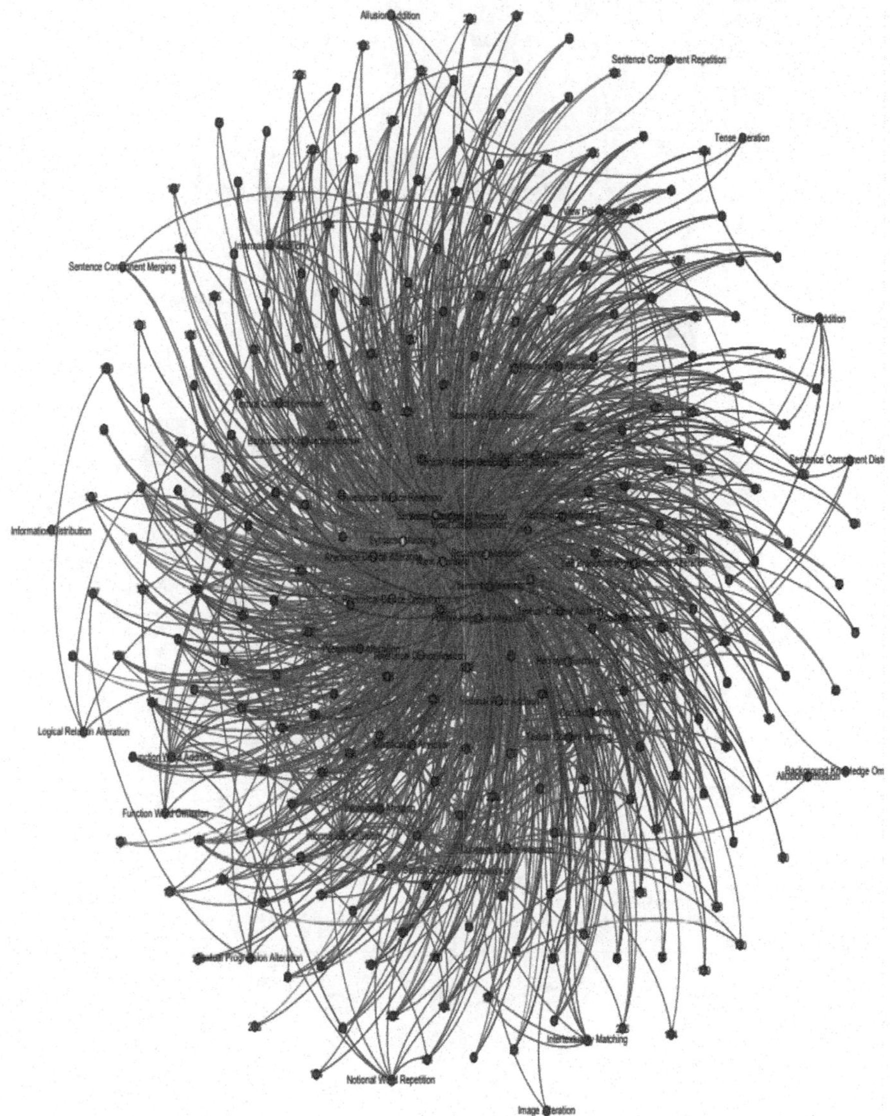

Fig. 1. Translation knowledge nodes co-occurrence relevance network

4.3 Translation Knowledge Nodes Relevance Analysis

Data mining tool like Gephi can extract and visualize relevant information that is commonly hidden or difficult to observe for us. The organic combination of data mining and visualization technology can make up for the defects of traditional data mining process and strengthen the processing ability of data mining. The visualization method makes the application of data mining technology more vivid and intuitive [15].

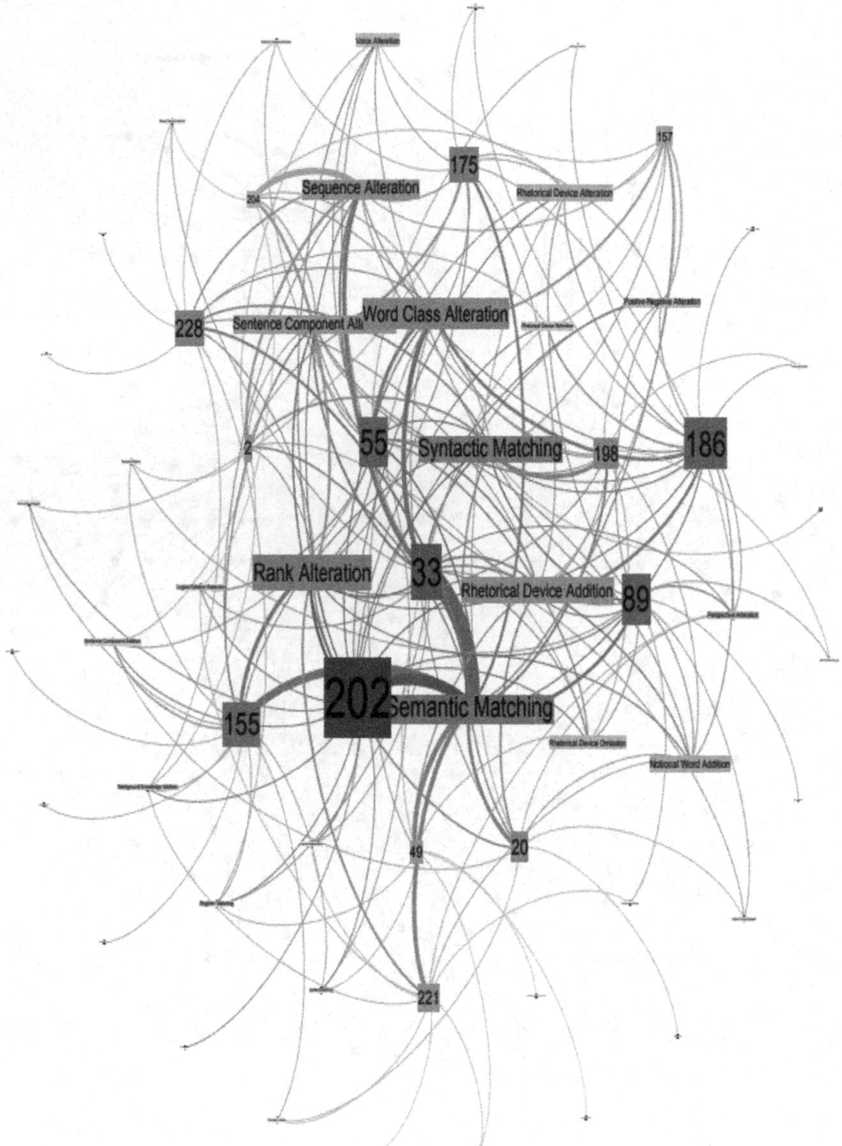

Fig. 2. Extracted translation knowledge nodes co-occurrence relevance network

The line between the two nodes in Figs. 1 and 2 is used to present the relationship between the two nodes. It is a visual representation of all the related data. The line here represents the co-occurrence relationship between the keywords. The thicker the line between the two node marks, the higher the co-occurrence frequency of the two keywords [13].

Table 1. Translation knowledge nodes co-occurrence relevance characteristics value

Characteristics	Graph density	Modularity	Average clustering coefficient	Average path length	Average degree
Value	0.057	0.235	0	1	3.350

The characteristics value of the co-occurrence relationship of translation knowledge nodes in Table 1 shows that the graph density is 0.057, the modularity is 0.235, the average clustering coefficient is 0, the average path length is 1, and the average degree is 3.350. Graph density is used to describe the intimacy of all translation knowledge nodes, that is, the degree in which each knowledge node is closely connected. The higher the density, the closer the nodes in the graph are connected. The graph density presented in this study is 0.057, which shows that not all nodes are directly related while the key nodes are related. Based on the combination and correlation between the knowledge nodes, an interconnected knowledge-network-based system is constructed with the knowledge nodes extracted from the text, cultural phenomena and translation methods itself [3]. As is shown in Figs. 1 and 2, the connection of each knowledge node is relatively close as a whole, but there are still some relatively evacuated connections.

The size distribution of the community is shown in Fig. 3, in which one community is represented by one red dot. The smaller the modularity is, the more important communities there are. The modularity presented in this study is 0.235, and it shows that there are not many related communities in the knowledge network. This is mainly because the number of knowledge nodes such as translation phenomena and translation skills involved in the translation process is relatively limited, which is different from other disciplines like engineering science and technology, statistics, etc.

According to modularity optimization algorithm of Gephi, it is found that there are 4 communities in the network, namely, semantic matching, syntactic matching, word-class alteration, rank alteration. In these four communities, the co-occurrence of the knowledge nodes is more prominent. Because of this co-occurrence relationship, the relevance between knowledge nodes is more prominent. This kind of research belongs to the innovation of knowledge mining. Only by observing the co-occurrence of knowledge nodes in the bilingual texts can knowledge nodes be presented more objectively and clearly. Around these four co-occurrence relations, there are some other knowledge nodes that can be presented, reflecting comprehensively and objectively the knowledge nodes involved in the translation process or the accumulation of knowledge nodes.

5 Reflections of the Teaching Design of the Current Translation Courses

Good command of language related and the translating techniques have been regarded as highly valued features, because the competences like those may greatly help the outcome of translation with high quality. In translation practice, many factors like translator identity, ideology, cultural awareness, contexts, the use of translation techniques, etc.,

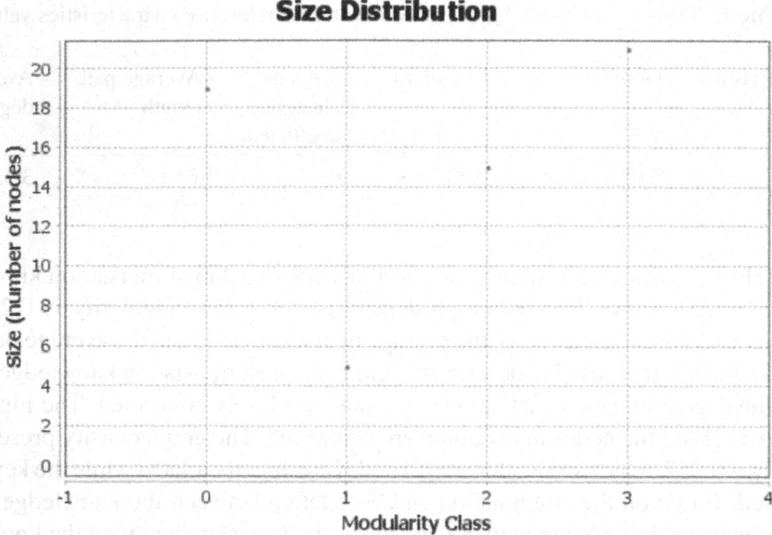

Fig. 3. Communities of translation knowledge nodes

may affect the translation process and quality. Undoubtedly, the basic and most important factors lie in the translation techniques like semantic matching, syntactic matching, word-class alteration, rank alteration, etc.

Chinese University MOOCs, co-produced by iCourse and Netease Cloud Classroom, is a platform featuring courses offered by renowned higher education institutions, where you can access informative and interesting content, improve your ability and skills, and even receive an e-certificate signed by the teacher via application.

The authors have conducted an extensive survey with Chinese University MOOCs as sources by searching for courses with the key concept "Chinese-English translation", and this process has yielded about 25 courses to study. The retrieving results suggest that these courses can be classified into three types: Chinese- English translation basics; translation studies and practices in specific fields such as research papers, international business, medicine, science and technology, economy and trade, etc. translation exercises in post- graduate entrance examination. The survey reflects that quite a number of courses put emphasis on translation theories and translation practices of different text types while neglecting how the translation knowledge-nodes and their relationship affect the translation process. A typical translation course offered by Shanghai International Studies University is chosen here to illustrate.

According to the syllabus shown in Fig. 4, nine chapters are designed to help students understand the basic issues, theory, history, skills of translation and translation practices of different text types as well. The course seems rich in content and logical in design. However, as an elementary course of translation, students are more eager to learn how to achieve faithfulness or equivalence by putting Chinese into English in word level, syntactic level, rhetoric level, text level, etc. Pitifully, there is only one chapter (chapter three) dealing with translation skills.

01 General issues of translation (翻译概论)
课时
1.1 What is translation?
1.2 What are the types of translation?
1.3 Why does translation matter?
1.4 How can one become a qualified translator?
02 The history of translation (翻译史)
课时
2.1 Chinese translation history (I)
2.2 Chinese translation history (II)
2.3 Western translation history
03 Translation skills (翻译的技巧)
课时
3.1 Shift of perspective
3.2 Sentence division/combination
3.3 Transposition
04 How to avoid Chinglish (避免中式英语)
课时
4.1 Unnecessary words
4.2 Padding
4.3 Repetition
4.4 Some spoken English issues
05 Literary translation (文学翻译)
课时
5.1 Literature and non-literature
5.2 Translatable figures of speech
5.3 Untranslatable figures of speech
5.4 Conversation and character
06 Translation of Chinese classics (典籍翻译)
课时
6.1 Translation of The Analects
6.2 Translation of classical poems
6.3 Translation of the Four Great Classical Novels
6.4 Biography translation
07 Translating for cultural promotion (文化宣传翻译)
课时
7.1 Cultural promotional translation
7.2 Relationship with the source text
7.3 Re-creating the source culture
7.4 Exploring translation acceptability
08 Translating academic texts (学术文献翻译)
课时
8.1 Purposes
8.2 Basic rules
8.3 Useful tools
8.4 Examples
09 Translating song lyrics (歌词翻译)
课时
9.1 Song lyrics and poetry
9.2 Rules for translating singable songs
9.3 Untranslatability of song lyrics
9.4 Translating Wo he wode zuguo (《我和我的祖国》)

Fig. 4. Syllabus of the Chinese-English translation course

According to our Gephi-based data mining and data visualization analysis, knowledge nodes do not exist in isolation during the translation process, and the translation output does not only point to the role of a knowledge node. A certain number of knowledge nodes can always be seen in the translation, which can be clearly displayed on the Textwells platform. Gephi visualization software enables us to present the relationship between translation knowledge nodes objectively and scientifically. The translated work is the result of the close connection and integration of multiple translation knowledge nodes under the guidance of translation theory. From the perspective of qualitative analysis, these common knowledge nodes presented by data mining and analysis are what we think should be emphasized in translation teaching. This is a conclusion based on data rather than the traditional judgment based on teachers' personal experience. We believe that the four knowledge nodes are more important: semantic matching, syntactic matching, word-class alteration and rank alteration. The co-occurrence in the text is more prominent.

No doubt, the findings of the Gephi-based data mining and data visualization analysis point out clearly the direction of translation teaching. First of all, translation teaching is supposed to vary with the level of students. As an introductory course of translation, focus should be taken on the knowledge nodes and translation skills. What teachers attempt to do is help students master some basic translating skills so as to achieve semantic matching, syntactic matching, etc. When it comes to advanced course of translation, different text types' translation teaching is a preferred choice according to students' major and interests. Also, in translation teaching design, how to arrange the knowledge nodes and their proper translation skills is another key problem. According to Figs. 1 and 2, the top four communities are semantic matching, syntactic matching, word-class alteration and rank alteration, which can be arranged in translation teaching design in descending order of importance. How to achieve the translation knowledge nodes through various translation skills or approaches in each chapter should be presented, explained and practiced in detail. Moreover, in these four communities, the co-occurrence of the knowledge nodes is more prominent, which calls for various translation skills or combination of them.

6 Conclusion

Teachers are constantly exploring effective approaches to improve translation teaching in their own educational practice. The results of the data analysis in educational field can help us have a basic knowledge about what to teach and how to teach.

Compared to traditional methods, data mining techniques perform fairly better in extracting hidden data and transforming reams of data into useful information without explicit assumptions. Gephi can directly display them in forms of graphs, thus making up for the defects of traditional methods. It is a complementary tool to traditional statistics, as visual thinking with interactive interfaces is now recognized to facilitate reasoning [14].

From the perspective of core methodology, knowledge nodes are used for mining. It is theoretically feasible to put knowledge nodes in Gephi software and let it generate such mining results. The results of mining are not subjective assumptions based on personal experiences, but data-based and computer-assisted data mining applications. It shows that data mining can be applied in teaching and is also valuable for translation teaching. The endeavor will help to search for the hidden relationships and relevance of the knowledge nodes in the process of translation. What's more, the network connection between knowledge nodes is also open. With the continuous expansion of corpus examples and the continuous deepening of text analysis, the knowledge network will become richer and more new connections will emerge, forming a dynamic connection system [3]. This paper seeks to gain insight as to how data mining is being used to benefit the teaching of translation, and more in-depth and data-based research in this field is urgently needed.

Acknowledgements. The work was supported by Teaching and Research Project of Anhui Province (2020jyxm1519), Postgraduate Teaching and Research Project of Anhui Province (2022jyjxggyj041), and Postgraduate Teaching and Research Project of Hefei University of Technology (2021YJG006).

References

1. Neubert, A.: Competence in language, in languages and in translation. In: Schäffner, C., Adab, B. (eds.) Developing Translation Competence (06), pp. 3–18. John Benjamins, Amsterdam (2000)

2. Miao, J.: Research on translation competence – the basis of constructing translation teaching model. Foreign Lang. Teach. **04**, 47–50 (2007)
3. Zhu, C., Mu, Y.: Towards a textual accountability-driven mode of teaching and (Self-) learning for translation and bilingual writing: with special reference to a city U On-line teaching platform. Chin. Trans. J. (02), 56–62+127 (2013)
4. Nowakowska, M., Beben, K., Pajecki, M.: Use of data mining in a two-step process of profiling student preferences in relation to the enhancement of English as a foreign language teaching. Stat Anal Data Min: ASA Data SCI J. 1–17 (2020)
5. Rupashi, K.: Overview of data mining. Int. J. Trend Sci. Res. Dev. (IJTSRD)(4), 1333–1336 (2020)
6. Yang, J., Cheng, C., Shen, S., Yang, S.: Comparison of complex network analysis software: Citespace, SCI2 and Gephi. In: 2017 IEEE 2nd International Conference on Big Data Analysis, pp.169–172. IEEE (2017)
7. Phyu, A.P., Wai, K.K.: To development manufacturing and education using data mining: a review. Int. J. Trend Sci. Res. Dev. (IJTSRD) **3**, 2168–2173 (2019)
8. Bravo-Agapito, J., Bonilla, C.F., Seoane, I.: Data mining in foreign language learning. WIREs Data Min. Knowl. Discov. e1287 (2018)
9. Wang, Y.H., Liao, H.C.: Data mining for adaptive learning in a TESL-based e-learning system. Expert Syst. Appl. **38**, 6480–6485 (2011)
10. von Davier, A.A.: The use of quality control and data mining techniques for monitoring scaled scores: an overview. Research Report ETS RR-12-20. ETS, Princeton, New Jersey (2012)
11. Mu, Y., Tian, L., Yang, W.: Towards a knowledge management model for online translation learning. In: Hao, T., Chen, W., Xie, H., Nadee, W., Lau, R. (eds.) Emerging Technologies for Education. SETE 2018. Lecture Notes in Computer Science(), vol. 11284. Springer, Cham (2018). https://doi.org/10.1007/978-3-030-03580-8_21
12. Mu, Y., Yang, W.: A teaching experiment on a knowledge-network-based online translation learning platform. In: Popescu, E., Hao, T., Hsu, TC., Xie, H., Temperini, M., Chen, W. (eds.) Emerging Technologies for Education. SETE 2019. Lecture Notes in Computer Science, vol. 11984. Springer, Cham (2020). https://doi.org/10.1007/978-3-030-38778-5_35
13. Deng, J., Ma, X., Bi, Q.: A comparative study of the social network analysis tools—Ucinet and Gephi. Inf. Theory Appl. (ITA) **08**, 133–138 (2014)
14. Gephi Homepage. https://gephi.org. Accessed 16 Feb 2023
15. Sun, Q., Rao, Y.: Survey of network data visualization technology based on association analysis. Comput. Sci. **6A**, 484–488 (2015)

Computational Intelligent Fault Diagnosis and Fault-Tolerant Control, and Their Engineering Applications

Novel TD-based Adaptive Control for Nonlinearly Parameterized Stochastic Systems

Yanli Liu(✉), Yihua Sun, and Li-Ying Hao

School of Marine Electrical Engineering, Dalian Maritime University, Dalian 116026, Liaoning, China
liuyanlizl@163.com

Abstract. In this article, the adaptive tracking control of nonlinearly parameterized stochastic systems is developed via the nonlinear tracking differentiator technique. First, the Barrier Lyapunov function (BLF) is brought to handle the constraint. Secondly, for the "explosion of complexity" issue, the tracking differentiator (TD) technique is brought in to make up the above issue and improve the tracking performance by integrating the error compensation signal technology into the control design process. The whole signals are bounded under the designed control tactic, the convergence of tracking error is ensured within the constrained set. And the availability of the above tactic is illustrated through the examples.

Keywords: Stochastic nonlinear systems · Nonlinear parameterization · Tracking differentiator · Barrier Lyapunov function

1 Introduction

As the widespread application of control in real industry, the requirements for control quality is becoming increasingly high. Due to the fact that most systems in the real world are affected by random factors or environmental noise, thus the research on stochastic systems have received widespread attention and many results have been published in [1,2]. Owing to the fact that "randomness" is one of the focuses in control theory research, for the research of existing stochastic systems, different Lyapunov functions has been studied to design the control schemes, which can be divided into two categories: quadratic and quartic Lyapunov functions. Furthermore, by combining the quartic Lyapunov function and the neural networks technique, an output feedback tracking control issue was studied for stochastic switched nonlinear systems in the strict-feedback form with input saturation in [3].

In practical engineering control problems, for a practical system, it is impossible to establish a precise mathematical model since that the controlled system inevitably involves the uncertainty. When this uncertainty is the parameter uncertainty, the system with parameter uncertainty in the equation is called a parameterized system. Over the past period of time, scholars have conducted research on

H. Zhang et al. (Eds.): NCAA 2023, CCIS 1870, pp. 519–531, 2023.
https://doi.org/10.1007/978-981-99-5847-4_37

the nonlinear parameterized systems and achieved many beneficial results in [4–8]. For a type of nonlinear parameterized systems, by applying the integral Lyapunov function to adaptive repetitive learning control, the singularity problem of the controller in [4] has been discussed. In [5], for control issue of nonlinear parameterized uncertain system, an adaptive learning control law was developed.

As everyone knows, constraints are common in practical engineering applications. The system states are generally required to satisfy a certain set to realize safety and performance owing to the physical factors and safety issues. To address the constraints of Brunovsky standard form systems, Ngo et al. constructed the prototype of arctangent and logarithmic barrier functions, and used constraint intervals as the domain of definition in [9]. Subsequently, Tee et al. carried on more research and formatted the Lyapunov function with obstacle characteristics as the BLF in [10]. Subsequently, many achievements were made by using BLFs of various forms to discuss various constraints in [11,12].

A plentiful of practical systems in the real world are nonlinear, and no system is even absolutely linear, so the research of nonlinear systems has drawn growing concern [13–17]. The backstepping technique has solved an awful lot of adaptive control design issues over the past decades for nonlinear systems with uncertainties in [18–20]. Due to the fact that virtual controllers contain many derivatives about variables–"explosion of complexity" (EOC) phenomenon, which inevitably appears when the backstepping technique is used. To handle the EOC problem, many methods have been studied, such as dynamic surface control technology, command filter and tracking differentiator, which was led in the design process to make the control design better [21].

Based on the aforementioned discussions, in this paper, the nonlinear TD and the BLF are utilized to control the nonlinearly parameterized stochastic systems. Different from other methods of dealing with "explosion of complexity" problem, the nonlinear TD has a good tracking ability for input signals, which can filter out noise and extract differential signals. The features are outlined as follows:

i) The nonlinear TD technique is blended into adaptive backstepping control to clear off the EOC issue for nonlinearly parameterized stochastic systems.
ii) The error compensation signals are constructed to improve the tracking performance and get a higher quality tracking effect, thus, the developed control tactic is more easily to apply in engineering.

2 Mathematical Preliminaries

2.1 System Description

Consider stochastic systems

$$\begin{aligned} dx_i =&(x_{i+1} + f_i(\bar{x}_i, \theta))dt + h_i^T(\bar{x}_i, \theta)d\omega \\ dx_n =&(u + f_n(x, \theta))dt + h_n^T(x, \theta)d\omega \\ y =&x_1, \quad i = 1, 2 \cdots, n \end{aligned} \tag{1}$$

in which, $\bar{x}_i = [x_1, x_2, \cdots, x_i]^T, i = 1, 2, \cdots, n, u$ and y denote the system states, input and output. $\theta \in R^p$ means the unknown parameter. Additionally, $f_i(\bar{x}_i, \theta)$ and $h_i(\bar{x}_i, \theta)$ denote the unknown nonlinear functions. ω denotes a standard Brownian motion satisfying $E\{d\omega(t)\} = 0$.

Control Purpose: Designing an adaptive control tactic in light of the nonlinear tracking differentiator technique, such as the tracking error can converge to the given constrained set and all signals are bounded in probability, simultaneously.

2.2 Preliminaries

Assumption 1. [20] *The smooth functions $f_i(\bar{x}_i, \theta)$ and $h_i(\bar{x}_i, \theta)$ are unknown but bounded, and satisfy $|f_i(\bar{x}_i, \theta)| \le a_i \bar{f}_i$, $|h_i(\bar{x}_i, \theta)| \le b_i \bar{h}_i$ with constant $a_i > 0, b_i > 0$.*

Assumption 2. [21] *Tracking signal y_d and $\dot{y}_d$ are bounded and smooth.*

Lemma 1. [12] *Given $\rho > 0$, open sets $\Omega_* : |\chi_1| < \rho$ and $\mathcal{N} := R^l \times \Omega_* \subset R^{l+1}$. For system $d\eta = f(t, \eta)dt + h(t, \eta)d\omega$, in which, $\eta := [\iota, \chi_1]^T \in \mathcal{N}$ being the state, $f(\cdot), h(\cdot) := R_+ \times \mathcal{N} \to R^{l+1}$ are partly Lipschitz in χ_1, uniformly in t and piecewise continuous in t, on $R_+ \times \mathcal{N}$. For positive-definite and continuously differentiable functions $U : R^l \to R_+$ and $V_1 : \Omega_* \to R_+$*

$$V_1(\chi_1) \to \infty \quad as \quad \chi_1 \to \pm\rho \tag{2}$$

$$\underline{\gamma}(\|\iota\|) \le U(\iota) \le \bar{\gamma}(\|\iota\|) \tag{3}$$

hold, in which, $\underline{\gamma}, \bar{\gamma}$ are class $\mathcal{K}_\infty$ functions. Define $V(\eta) = V_1(\chi_1) + U(\iota)$, constants $a_0, \quad b_0 > 0$, for $\chi_1(0) \in \Omega_*$, if

$$\mathcal{L}V = \frac{\partial V}{\partial \eta} f + \frac{1}{2} Tr\{h^T \frac{\partial^2 V}{\partial \eta^2} h\} \le -a_0 V + b_0, \quad \eta \in \mathcal{N} \tag{4}$$

then ι is bounded, and $\chi_1(t)$ stays in the set $\Omega_*, \forall t \ge 0$.

Lemma 2. [12] *For any an integer $\kappa > 0$, a constant $\rho > 0$, any $\chi_1 \in R$ satisfies $|\chi_1| < \rho$, the inequality*

$$\log \frac{\rho^{2\kappa}}{\rho^{2\kappa} - z^{2\kappa}} \le \frac{z^{2\kappa}}{\rho^{2\kappa} - z^{2\kappa}} \tag{5}$$

holds.

2.3 Main Controller Design Process

Devise the controller in view of the coordinate changes

$$z_1 = y - y_d \tag{6}$$

$$z_i = x_i - \alpha_{id} \tag{7}$$

where y_d is the smooth and bounded tracking signal.

To remedy the EOC problem, the following tracking differentiator is introduced

$$\dot{\alpha}_{id} = \mu_i$$
$$\mu_i = -\Upsilon_i^2(a_{i1}\sinh(l_{i1}(\alpha_{id} - \alpha_{i-1})) + a_{i2}\sinh(\frac{l_{i2}x_i}{\Upsilon_i})) \tag{8}$$

in which $\Upsilon_i, a_{i1}, a_{i2}, l_{i1}, l_{i2} (i = 2, 3, \cdots, n)$ present positive parameters; α_{id} means the estimate of α_{i-1}, then $s_{id} = \alpha_{id} - \alpha_{i-1}$ means the estimation errors. It is easily can be obtained that there exist scalars $s_{iM} > 0$ such that $|s_{id}| \leq s_{iM} (i = 2, 3, \ldots, n)$ when the suitable parameters are selected.

Remark 1. Different from the common dynamic surface control, the TD technique with the error compensation signals is introduced to handle the EOC issue, in which the filter errors are considered and discussed in light of the error compensation signals. It can be seen that the developed control method can remedy the EOC issue well and improve the control effect.

To improve the tracking performance, the novel tracking error signals yield

$$\chi_i = z_i - \pi_i, \quad 1 \leq i \leq n \tag{9}$$

where π_i is the designed error compensation signals later.

Step 1. Consider

$$V_1 = \frac{1}{4}\log\frac{\rho^4}{\rho^4 - \chi_1^4} + \frac{1}{2r_1}\tilde{\vartheta}_1^2 \tag{10}$$

where constant $r_1 > 0$, $\vartheta_1 = \tilde{\vartheta}_1 + \hat{\vartheta}_1$, $\hat{\vartheta}_1$ denotes the estimation value of unknown parameter ϑ_1, $\tilde{\vartheta}_1$ means the estimation error.

Furthermore, according to the tracking constrained condition $|\chi_1| < \rho$, V_1 means an efficient Lyapunov function within the constraints.

In the light of (9), it gets

$$d\chi_1 = (\chi_2 + \pi_2 + \alpha_1 + s_{2d} + f_1 - \dot{y}_d - \dot{\pi}_1)dt + h_1^T dw \tag{11}$$

then

$$\begin{aligned}\mathcal{L}V_1 =& \frac{1}{\rho^4 - \chi_1^4}\chi_1^3(\chi_2 + \pi_2 + \alpha_1 + s_{2d} + f_1 - \dot{y}_d - \dot{\pi}_1) \\ &+ \frac{1}{2(\rho^4 - \chi_1^4)^2}\chi_1^2(3\rho^4 + \chi_1^4)h_1^T h_1 - \frac{1}{r_1}\tilde{\vartheta}_1\dot{\hat{\vartheta}}_1\end{aligned} \tag{12}$$

By Young's inequality and Assumption 1, we get

$$\begin{aligned}\frac{1}{\rho^4 - \chi_1^4}\chi_1^3\chi_2 \leq& \frac{3}{4}\frac{\rho^4}{(\rho^4 - \chi_1^4)^{\frac{4}{3}}}\chi_1^4 + \frac{1}{4}\chi_2^4 \\ \frac{1}{\rho^4 - \chi_1^4}\chi_1^3 f_1 \leq& \frac{3}{4}\frac{1}{(\rho^4 - \chi_1^4)^{\frac{4}{3}}}a_1^{\frac{4}{3}}\bar{f}_1^{\frac{4}{3}}\chi_1^4 + \frac{1}{4} \\ \frac{1}{2(\rho^4 - \chi_1^4)^2}\chi_1^2(3\rho^4 + \chi_1^4)h_1^T h_1 \leq& \frac{1}{4}\frac{1}{(\rho^4 - \chi_1^4)^4}\chi_1^4(3\rho^4 + \chi_1^4)^2 b_1^4\|\bar{h}_1\|^4 + \frac{1}{4}\end{aligned} \tag{13}$$

Substituting (13) into (12) leads to

$$\mathcal{L}V_1 \le \frac{1}{\rho^4 - \chi_1^4}\chi_1^3(\pi_2 + \alpha_1 + s_{2d} - y_d - \dot{\pi}_1 + \frac{3}{4}\frac{1}{(\rho^4 - \chi_1^4)^{\frac{1}{3}}}\chi_1(1 + a_1^{\frac{4}{3}}\bar{f}_1^{\frac{4}{3}})$$
$$+ \frac{1}{4}\frac{1}{(\rho^4 - \chi_1^4)^3}\chi_1(3\rho^4 + \chi_1^4)^2 b_1^4\|\bar{h}_1\|^4) + \frac{1}{4}\chi_2^4 + \frac{2}{4} - \frac{1}{r_1}\tilde{\vartheta}_1\dot{\hat{\vartheta}}_1 \tag{14}$$

Choose $\vartheta_1 = \max\{1, a_1^{\frac{4}{3}}, b_1^4\}$ as a virtual parameter, $\phi_1 = \frac{3}{4}\frac{1}{(\rho^4-\chi_1^4)^{\frac{1}{3}}}\chi_1(1+\bar{f}_1^{\frac{4}{3}})+\frac{1}{4}\frac{1}{(\rho^4-\chi_1^4)^3}\chi_1(3\rho^4 + \chi_1^4)^2\|\bar{h}_1\|^4)$ as a computable scalar function, and substitute them to

$$\mathcal{L}V_1 \le \frac{1}{\rho^4 - \chi_1^4}\chi_1^3(\pi_2 + \alpha_1 + s_{2d} - y_d - \dot{\pi}_1 + \vartheta_1\phi_1\chi_1)$$
$$+ \frac{1}{4}\chi_2^4 + \frac{2}{4} - \frac{1}{r_1}\tilde{\vartheta}_1\dot{\hat{\vartheta}}_1 \tag{15}$$

Choose $\dot{\pi}_1$ and α_1 as

$$\dot{\pi}_1 = -c_1\pi_1 + \pi_2 + s_{2d} \tag{16}$$
$$\alpha_1 = -c_1 z_1 - \hat{\vartheta}_1\phi_1\chi_1 + y_d \tag{17}$$

where $c_1 > 0$ is the design constant, then

$$\mathcal{L}V_1 \le \frac{1}{\rho^4 - \chi_1^4}\chi_1^3(-c_1 z_1 - \hat{\vartheta}_1\phi_1\chi_1 + c_1\pi_1 + \vartheta_1\phi_1\chi_1)$$
$$+ \frac{1}{4}\chi_2^4 + \frac{2}{4} - \frac{1}{r_1}\tilde{\vartheta}_1\dot{\hat{\vartheta}}_1$$
$$\le -\frac{1}{\rho^4 - \chi_1^4}c_1\chi_1^4 + \frac{2}{4} - \frac{1}{r_1}\tilde{\vartheta}_1\dot{\hat{\vartheta}}_1 - \frac{r_1}{\rho^4 - \chi_1^4}\chi_1^4\phi_1) \tag{18}$$

Choose $\dot{\hat{\vartheta}}_1$

$$\dot{\hat{\vartheta}}_1 = \frac{r_1}{\rho^4 - \chi_1^4}\chi_1^4\phi_1 - \sigma_1\hat{\vartheta}_1 \tag{19}$$

where $\sigma_1 > 0$ is the design constant and we consider Lemma 2, thus

$$\mathcal{L}V_1 \le -c_1 \log\frac{\rho^4}{\rho^4 - \chi_1^4} - \frac{\sigma_1}{2r_1}\tilde{\vartheta}_1^2 + d_1 \tag{20}$$

where $d_1 = \frac{1}{2} + \frac{\sigma_1}{2r_1}\vartheta_1^2$.

Step i. $(2 \le i \le n-1)$ V_2 is designed as

$$V_i = \frac{1}{4}\chi_i^4 + \frac{1}{2r_i}\tilde{\vartheta}_i^2 + V_{i-1} \tag{21}$$

with

$$d\chi_i = (\chi_{i+1} + \pi_{i+1} + \alpha_i + s_{(i+1)d} + f_i - \dot{\alpha}_{id} - \dot{\pi}_i)dt + h_i^T dw \tag{22}$$

then

$$\begin{aligned}\mathcal{L}V_i = &\mathcal{L}V_{i-1} + \chi_i^3(\chi_{i+1} + \pi_{i+1} + \alpha_i + s_{(i+1)d} + f_i - \dot{\alpha}_{id} - \dot{\pi}_i) \\ &+ \frac{3}{2}\chi_i^2 h_i^T h_i - \frac{1}{r_i}\tilde{\vartheta}_i\dot{\hat{\vartheta}}_i\end{aligned} \tag{23}$$

Similarly, we get

$$\begin{aligned}\chi_i^3\chi_{i+1} \le &\frac{3}{4}\chi_i^4 + \frac{1}{4}\chi_{i+1}^4 \\ \chi_i^3 f_i \le &\frac{3}{4}a_i^{\frac{4}{3}}\bar{f}_i^{\frac{4}{3}}\chi_i^4 + \frac{1}{4} \\ \frac{3}{2}\chi_i^2 h_i^T h_i \le &\frac{3}{4}\chi_i^4 b_i^4\|\bar{h}_i\|^4 + \frac{3}{4}\end{aligned} \tag{24}$$

Substituting (24) into (23) yields

$$\begin{aligned}\mathcal{L}V_i \le &- c_1 \log\frac{\rho^4}{\rho^4 - \chi_1^4} - \sum_{j=1}^{i-1} c_j\chi_j^4 - \sum_{j=2}^{i-1}\frac{\sigma_j}{2r_j}\tilde{\vartheta}_j^2 + d_{i-1} + \chi_i^3(\pi_{i+1} + \alpha_i \\ &+ s_{(i+1)d} + \frac{1}{4}\chi_i - \dot{\pi}_i - \dot{\alpha}_{id} + \frac{3}{4}\chi_i(1 + a_i^{\frac{4}{3}}\bar{f}_i^{\frac{4}{3}}) + \frac{3}{4}\chi_i b_i^4\|\bar{h}_i\|^4) \\ &+ \frac{1}{4}\chi_{i+1}^4 + 1 - \frac{1}{r_i}\tilde{\vartheta}_i\dot{\hat{\vartheta}}_i\end{aligned} \tag{25}$$

Choose $\vartheta_i = \max\{1, a_i^{\frac{4}{3}}, b_i^4\}$ as a virtual parameter, $\phi_i = \frac{3}{4}(1 + \bar{f}_i^{\frac{4}{3}}) + \frac{3}{4}\|\bar{h}_i\|^4$ as a computable scalar function, then

$$\begin{aligned}\mathcal{L}V_i \le &- c_1 \log\frac{\rho^4}{\rho^4 - \chi_1^4} - \sum_{j=2}^{i-1} c_j\chi_j^4 - \sum_{j=1}^{i-1}\frac{\sigma_j}{2r_j}\tilde{\vartheta}_j^2 + d_{i-1} + \chi_1^3(\pi_{i+1} + \alpha_i \\ &+ s_{(i+1)d} + \frac{1}{4}\chi_i - \dot{\pi}_i - \dot{\alpha}_{id} + \vartheta_i\phi_1\chi_i) + \frac{1}{4}\chi_{i+1}^4 + 1 - \frac{1}{r_i}\tilde{\vartheta}_i\dot{\hat{\vartheta}}_i\end{aligned} \tag{26}$$

Choose $\dot{\pi}_i$ and α_i as

$$\dot{\pi}_i = - c_i\pi_i + \pi_{i+1} + s_{(i+1)d} \tag{27}$$

$$\alpha_i = - c_i z_i - \hat{\vartheta}_i\phi_i\chi_i - \frac{1}{4}\chi_i + \dot{\alpha}_{id} \tag{28}$$

where $c_i > 0$ denotes the design constant, we get

$$\begin{aligned}\mathcal{L}V_i \le &- c_1 \log\frac{\rho^4}{\rho^4 - \chi_1^4} - \sum_{j=2}^{i-1} c_j\chi_j^4 - \sum_{j=1}^{i-1}\frac{\sigma_j}{2r_j}\tilde{\vartheta}_j^2 + d_{i-1} \\ &- c_i\chi_i^4 + \frac{1}{4}\chi_{i+1}^4 + 1 - \frac{1}{r_i}\tilde{\vartheta}_i(\dot{\hat{\vartheta}}_i - r_i\chi_i^4\phi_i)\end{aligned} \tag{29}$$

Choose the adaptation law as

$$\dot{\hat{\vartheta}}_i = r_i \chi_i^4 \phi_i - \sigma_i \hat{\vartheta}_i \tag{30}$$

where $\sigma_i > 0$ is the design constant, thus

$$\mathcal{L}V_i \leq -c_1 \log \frac{\rho^4}{\rho^4 - \chi_1^4} - \sum_{j=2}^{i} c_j \chi_j^4 - \sum_{j=2}^{i} \frac{\sigma_j}{2r_j} \tilde{\vartheta}_j^2 + d_i + \frac{1}{4}\chi_{i+1}^4 \tag{31}$$

where $d_i = d_{i-1} + 1 + \frac{\sigma_i}{2r_i}\vartheta_i^2$.

Step n. Consider the Lyapunov function

$$V_n = \frac{1}{4}\chi_n^4 + \frac{1}{2r_n}\tilde{\vartheta}_n^2 + V_{n-1} \tag{32}$$

with

$$d\chi_n = (u + f_n - \dot{\alpha}_{nd} - \dot{\pi}_n)dt + h_n^T dw \tag{33}$$

then

$$\begin{aligned}\mathcal{L}V_n =& \mathcal{L}V_{n-1} + \chi_n^3(u + f_n - \dot{\alpha}_{nd} - \dot{\pi}_n) \\ &+ \frac{3}{2}\chi_n^2 h_n^T h_n - \frac{1}{r_n}\tilde{\vartheta}_n \dot{\hat{\vartheta}}_n\end{aligned} \tag{34}$$

In like manner, we get

$$\begin{aligned}\mathcal{L}V_n \leq& -c_1 \log \frac{\rho^4}{\rho^4 - \chi_1^4} - \sum_{j=2}^{n-1} c_j \chi_j^4 - \sum_{j=1}^{n-1} \frac{\sigma_j}{2r_j}\tilde{\vartheta}_j^2 + d_{n-1} + \chi_n^4(u + \frac{1}{4}\chi_n \\ &- \dot{\pi}_n - \dot{\alpha}_{nd} + \frac{3}{4}\chi_n a_n^{\frac{4}{3}} \bar{f}_n^{\frac{4}{3}} + \frac{3}{4}\chi_n b_n^4 \|\bar{h}_n\|^4) + 1 - \frac{1}{r_n}\tilde{\vartheta}_n \dot{\hat{\vartheta}}_n\end{aligned} \tag{35}$$

Choose $\vartheta_n = \max\{1, a_n^{\frac{4}{3}}, b_n^4\}$ as a virtual parameter, $\phi_n = \frac{3}{4}(1 + \bar{f}_n^{\frac{4}{3}}) + \frac{3}{4}\|\bar{h}_n\|^4$ as a computable scalar function, and substitute them to

$$\begin{aligned}\mathcal{L}V_n \leq& -c_1 \log \frac{\rho^4}{\rho^4 - \chi_1^4} - \sum_{j=2}^{n-1} c_j \chi_j^4 - \sum_{j=1}^{n-1} \frac{\sigma_j}{2r_j}\tilde{\vartheta}_j^2 + d_{n-1} + \chi_n^4(u + \frac{1}{4}\chi_n \\ &- \dot{\pi}_n - \dot{\alpha}_{nd} + \vartheta_n \phi_1 \chi_n) + 1 - \frac{1}{r_n}\tilde{\vartheta}_n \dot{\hat{\vartheta}}_n\end{aligned} \tag{36}$$

Constructing u, $\dot{\pi}_n$ and $\dot{\hat{\vartheta}}_n$ as

$$u = -c_n z_n - \hat{\vartheta}_n \phi_1 \chi_n - \frac{1}{4}\chi_n + \dot{\alpha}_{nd} \tag{37}$$

$$\dot{\pi}_n = -c_n \pi_n \tag{38}$$

$$\dot{\hat{\vartheta}}_n = r_n \chi_n^4 \phi_n - \sigma_n \hat{\vartheta}_n \tag{39}$$

where $c_n, \sigma_n > 0$ are design constants. Furthermore

$$\mathcal{L}V_n \leq -c_1 \log \frac{\rho^4}{\rho^4 - \chi_1^4} - \sum_{j=2}^{n} c_j \chi_j^4 - \sum_{j=2}^{n} \frac{\sigma_j}{2r_j} \tilde{\vartheta}_j^2 + d_n \tag{40}$$

where $d_n = d_{n-1} + 1 + \frac{\sigma_n}{2r_n} \vartheta_n^2$.

Theorem 1. *For the stochastic nonlinearly parameterized system (1). Through the tracking differentiator method to construct control scheme (19), (31),(41) and (43), which make all signals keep bounded, the tracking error converges to nearby zero within the constrained set.*

Prior to proving Theorem 1, the following Lemma 3 of the constructed error compensation signals is established obviously.

Lemma 3. [21] *For the error compensation signals described in (16), (27) and (42), we can get* $\lim_{t\to\infty} \|\vartheta(t)\| \leq \frac{s_M}{2c_0}$, *where* $c_0 = \frac{1}{2}\min_i(c_i - 1)$ *with* $s_M = \max_i(s_{iM})$.

Proof. As $a_0 = \min\{4c_j, \sigma_j | j = 1, 2, \cdots, n\}, b_0 = d_n$, we have

$$\mathcal{L}V \leq -a_0 + b_0 \tag{41}$$

Furthermore, we get

$$0 \leq EV(t) \leq V(0) + \frac{b_0}{a_0} \tag{42}$$

then, we can get

$$|\chi_1| \leq \rho^4 \sqrt{(1 - e^{-4\frac{b_0}{a_0}})}, \quad |\chi_i| \leq \sqrt[4]{4\frac{b_0}{a_0}}, i = 2, 3, \cdots, n \tag{43}$$

which ends the proof.

3 Simulation Example

We will prove the validity of the developed tactic in this section. Consider the stochastic nonlinear system

$$\begin{aligned} dx_1 &= (x_2 + f_1(x_1, \theta))dt + h_1^T(x_1, \theta)d\omega \\ dx_2 &= (u + f_2(x_1, x_2, \theta))dt + h_2^T(x_1, x_2, \theta)d\omega \\ y &= x_1 \end{aligned} \tag{44}$$

where $f_1 = 0.5x_1^2, f_2 = -x_1 - 2x_2, h_1^T = 2x_1^2, h_2^T = 2x_1^2 + 0.5x_2^2$. The desired tracking signal is defined as $y_d = 0.2\sin(0.5t)$. The tracking error χ_1 is expected to stay in $|\chi_1| \leq \rho = 0.02$.

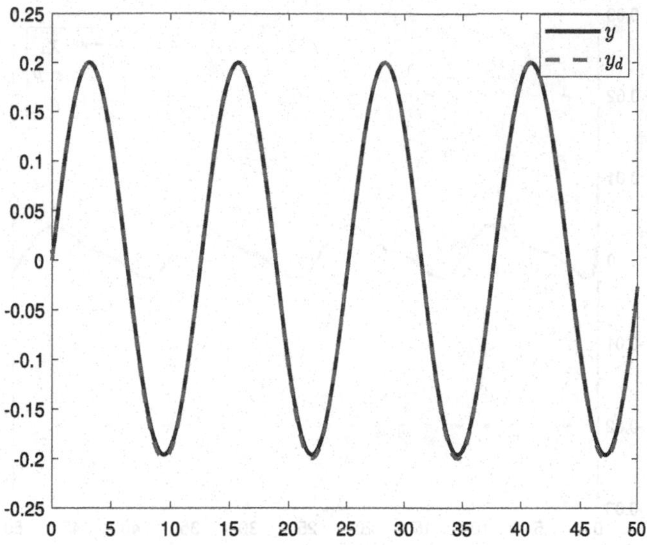

Fig. 1. Tracking performance.

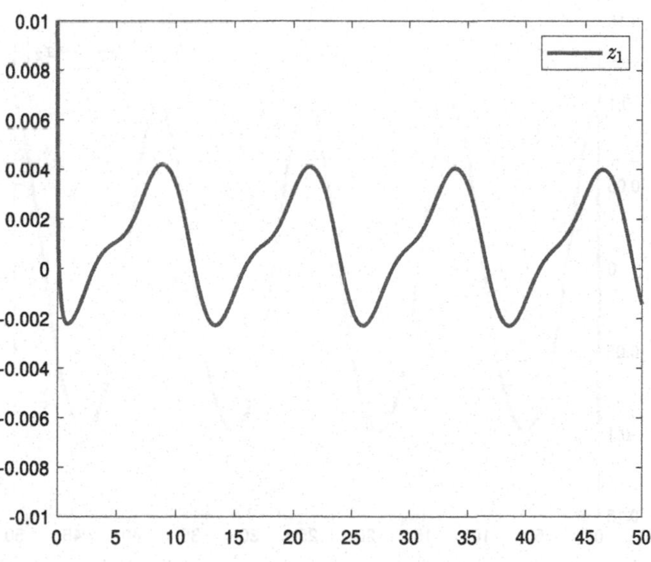

Fig. 2. Response of z_1.

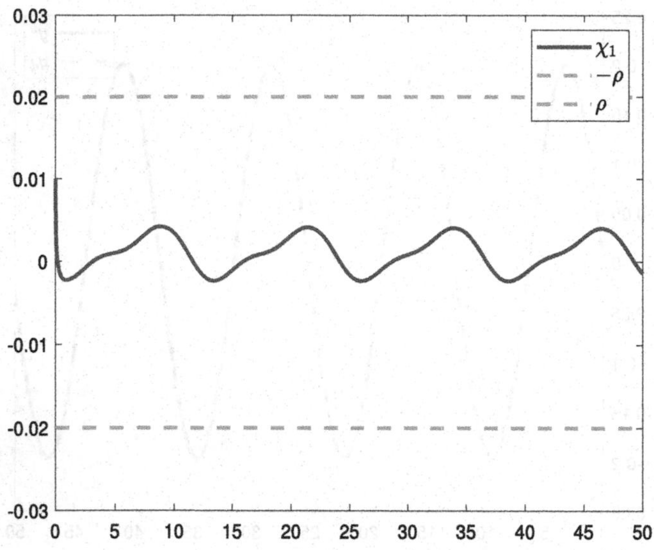

Fig. 3. Response of χ_1.

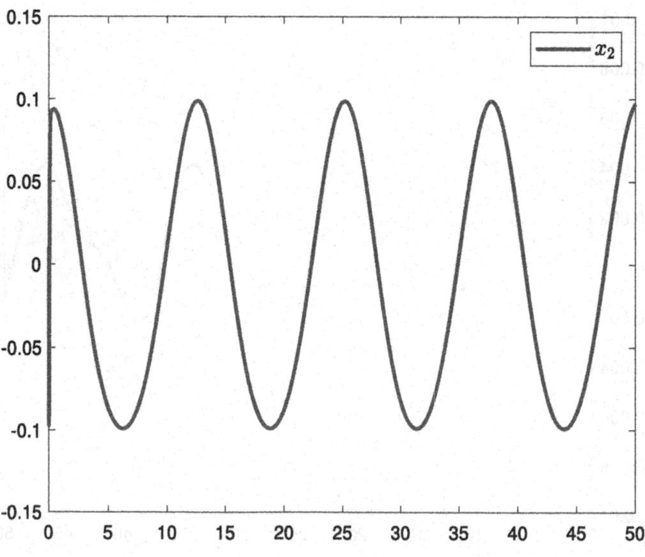

Fig. 4. Response of x_2.

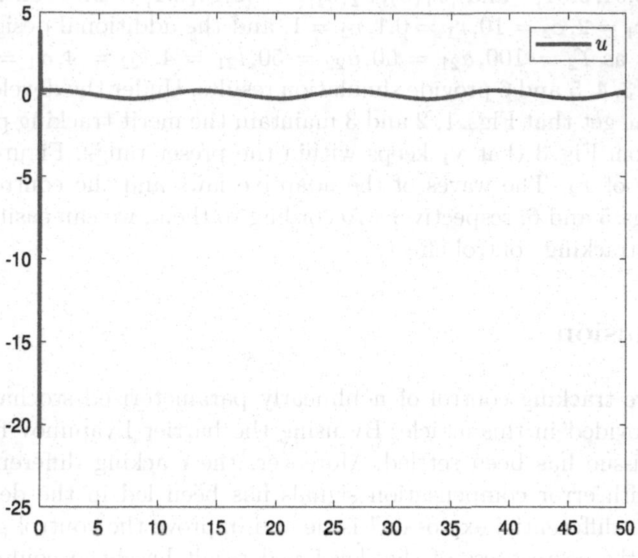

Fig. 5. Response of u.

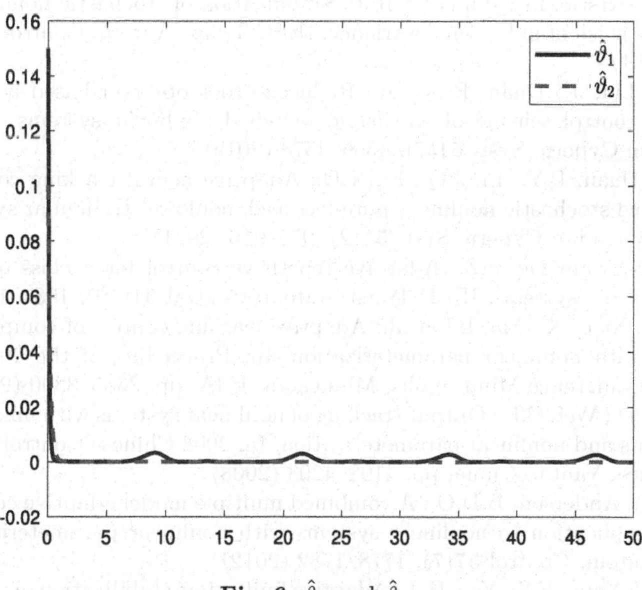

Fig. 6. $\hat{\vartheta}_1$ and $\hat{\vartheta}_2$.

The simulation experiment is ran under the initial conditions $[x_1(0), x_2(0)]^T = [0.01, 0.01]^T$ and $[\hat{\vartheta}_1(0), \hat{\vartheta}_2(0)]^T = [0.15, 0.1]^T$. The control gains are selected as $c_1 = 2, c_2 = 10, r_1 = 0.1, r_2 = 1$, and the additional designed parameters are set as $\Upsilon_2 = 100, a_{21} = 50, a_{22} = 50, l_{21} = 4, l_{22} = 4, \sigma_1 = 10, \sigma_2 = 5$. Figures 1, 2, 3, 4, 5 and 6 provide simulation results. Under the developed control tactic, we can get that Figs. 1, 2 and 3 maintain the merit tracking performance. It follows from Fig. 3 that χ_1 keeps within the preset range. Figure 4 gives the performance of x_2. The waves of the adaptive laws and the control signal are shown in Figs. 5 and 6, respectively. According to them, we can easily get a good constrained tracking control effect.

4 Conclusion

The adaptive tracking control of nonlinearly parameterized stochastic systems has been provided in this article. By using the barrier Lyapunov function, the constraints issue has been settled. Moreover, the tracking differentiator (TD) technique with error compensation signals has been led in the design process to deal with "differential explosion" issue and improve the control performance. In the end, the correctness of the developed result has been confirmed on the simulations.

References

1. Deng, H., Krstić, M., Williams, R.J.: Stabilization of stochastic nonlinear systems driven by noise of unknown covariance. IEEE Trans. Autom. Control **46**(8), 1237–1253 (2001)
2. Niu, B., Liu, J., Duan, P., et al.: Reduced-order observer-based adaptive fuzzy tracking control scheme of stochastic switched nonlinear systems. IEEE Trans. Syst. Man Cybern. Syst. **51**(7), 4566–4578 (2019)
3. Niu, B., Duan, P.Y., Li, J.Q., Li, X.D.: Adaptive neural tracking control scheme of switched stochastic nonlinear pure-feedback nonlower triangular systems. IEEE Trans. Syst. Man Cybern. Syst. **51**(2), 975–986 (2021)
4. Sun, M.X., Sam Ge, S.Z.: Adaptive repetitive control for a class of nonlinearly parameterized systems. IEEE Trans. Autom. Control **51**(10), 1684–1688 (2006)
5. Fang, Y., Xiao, X., Ma, B., et al.: Adaptive learning control of complex uncertain systems with nonlinear parameterization. In: Proceeding of the 2006 American Control Conference Minneapolis, Minneaota, USA, pp. 3385–3390 (2006)
6. Wang, Q.D., Wei, C.L.: Output tracking of nonlinear systems with unknown control coefficients and nonlinear parameterization. In: 2008 Chinese Control and Decision Conference, Yantai, China, pp. 4198–4203 (2008)
7. Chen, W., Anderson, B.D.O.: A combined multiple model adaptive control scheme and its application to nonlinear systems with nonlinear parameterization. IEEE Trans. Autom. Control **57**(7), 1778–1782 (2012)
8. Gao, F.Z., Yuan, F.S., Yao, H.J.: Adaptive finite-time stabilization of nonholonomic systems with nonlinear parameterization. In: 2010 Chinese Control and Decision Conference, Xuzhou, China, pp. 2132–2137 (2010)

9. Ngo, K.B., Mahony, R., Jiang, Z.P.: Integrator backstepping using barrier functions for systems with multiple state constraints. In: Proceedings of the 44th IEEE Conference on Decision and Control, Seville, Spain, pp. 8306–8312. IEEE (2005)
10. Tee, K.P., Ge, S.S.: Control of nonlinear systems with partial state constraints using a barrier Lyapunov function. Int. J. Control **84**(12), 2008–2023 (2011)
11. Kim, B.S., Yoo, S.J.: Approximation-based adaptive tracking control of nonlinear pure-feedback systems with time-varying output constraints. Int. J. Control Autom. Syst. **13**(2), 257–265 (2015). https://doi.org/10.1007/s12555-014-0084-6
12. Liu, Y.L., Ma, H.J.: Adaptive tracking control of stochastic switched nonlinear systems with unknown dead-zone output. Int. J. Robust Nonlinear Control **31**(10), 4511–4530 (2021)
13. Zhu, J., Zhang, P., Hou, Y.: Optimal control of nonlinear time-delay systems with input constraints using reinforcement learning. Neural Comput. Adv. Appl. **1265**, 332–344 (2020)
14. Jia, Q., Wang, J.Y.: Aperiodic sampling based event-triggering for synchronization of nonlinear systems. Neural Comput. Adv. Appl. **163**, 234–246 (2022)
15. Zhang, Y., Tao, G., Chen, M., et al.: A matrix decomposition based adaptive control scheme for a class of MIMO non-canonical approximation systems. Automatica **103**, 490–502 (2019)
16. Liu, L., Liu, Y.J., Chen, A., et al.: Integral barrier Lyapunov function-based adaptive control for switched nonlinear systems. Sci. China Inf. Sci. **63**(3), 1–14 (2020)
17. Zhao, L., Yu, J.P., Lin, C.: Command filter based adaptive fuzzy bipartite output consensus tracking of nonlinear coopetition multi-agent systems with input saturation. ISA Trans. **80**, 187–194 (2018)
18. Song, Y., Huang, X., Wen, C.: Tracking control for a class of unknown nonsquare MIMO nonaffine systems: a deep-rooted information based robust adaptive approach. IEEE Trans. Autom. Control **61**(10), 3227–3233 (2015)
19. Qian, C., Lin, W.: A continuous feedback approach to global strong stabilization of nonlinear systems. IEEE Trans. Autom. Control **46**(7), 1061–1079 (2001)
20. Zhou, S., Song, Y., Luo, X.: Fault-tolerant tracking control with guaranteed performance for nonlinearly parameterized systems under uncertain initial conditions. J. Franklin Inst. **357**, 6805–6823 (2020)
21. Liu, Y.L., Wang, R.Z., Hao, L.Y.: Adaptive TD control of full-state-constrained nonlinear stochastic switched systems. Appl. Math. Comput. **427**, 127165 (2022)

A Data-driven Intermediate Estimator-based Approach for Collaborative Fault-tolerant Tracking Control of Multi-agent Systems

Lianghuan Ying, Junwei Zhu(✉), and Yasi Pan

Institute of Cyberspace Security, Zhejiang University of Technology, Hangzhou 310023, China
junweizhu1001@zjut.edu.cn

Abstract. This paper presents a data-driven distributed intermediate estimator-based cooperative tracking control method for multi-agent systems with actuator faults and unknown model parameters. A residual generator is constructed using the process input and output data of the system with unknown model parameters, and a new error state variable is constructed. Moreover, a distributed intermediate estimator is designed to estimate the combined unknown input signal of the followers. Based on this estimation signal, the control law is designed to compensate for the unknown inputs to the followers. Finally, the experimental results verify the effectiveness of the proposed method.

Keywords: Data-driven distributed intermediate estimator · Cooperative tracking control · Residual generator · Multi-agent systems

1 Introduction

In recent years, multi-agent systems have been widely used in many fields, such as power grids, unmanned aerial vehicles, satellite formations, and mobile robots. Multi-agent systems can exchange information with other agents through communication topologies to accomplish some complex tasks, such as cooperative tracking control and coherent control, etc. However, as the dimensions of the system increases, the probability of fault increases significantly. Faults [1] are defined as an unacceptable deviation of not less than one characteristic property or parameter of a system from an acceptable standard condition. In general, faults can be classified as actuator faults [2], sensor faults [3], and process faults [4]. Due to the presence of these types of faults, the control actions of the controller on the device may be interrupted and a large number of measurement errors may be generated, which results in a drastic reduction of system performance and leads to the fault and collapse of the system. As a result, distributed control of multi-agent systems and the issues of safety and reliability have become research topics, and numerous research results have been produced.

H. Zhang et al. (Eds.): NCAA 2023, CCIS 1870, pp. 532–544, 2023.
https://doi.org/10.1007/978-981-99-5847-4_38

Among the existing model-based fault diagnosis schemes, the so-called observer-based [5] techniques have received much attention since the 1990s. For example, Xiong et al. [6] investigated the problem of signal quantization fault detection for uncertain linear systems by designing a fault detection filter. Li et al. [7] proposed a real-time weighted diagnostic observer-based fault detection method for T-S fuzzy systems, which improved fault detectability. Wang et al. [8] proposed a fuzzy state observer based on time and space sampling. Li et al. [9] proposed a new control method based on dilation observation for non-integral chain systems with mismatch uncertainty by reasonably selecting the disturbance compensation gain. In addition, Zhu et al. [10] designed an intermediate estimator that is capable of estimating system states and faults simultaneously. Most fault-tolerant control (FTC) methods of multi-agent systems are observer-based methods. The FTC methods [11] are investigated by scholars to improve the resistance of control systems to faults, which can be divided into two categories, one is model-based. For example, Zhou and Ren [12] proposed a new feedback controller architecture, which designed the controller separately to ensure both system performance and robustness. Another category of fault-tolerance approach is based on data-driven. Currently, there are some data-driven FTC methods, which are only for systems where process faults have occurred. For example, Hua et al. [13] designed an optimal fault-tolerant compensation controllers based on the value iteration-based reinforcement learning algorithm to reduce the impact of system parameter changes on system performance.

So far, various types of multi-agent fault observers have been proposed by numerous researchers. For example, a distributed extended state observer with $H\infty$ performance metrics based on adjustable parameters was designed in the [14]. A set of sliding mode observers based on global relative output information was designed in the [15] to handle the fault estimation problem for multi-agent systems with undirected graphs. Adaptive observers [16] were also extended to estimate node faults in multi-agent systems. Besides, intermediate estimators have also evolved into distributed structures to solve tracking control problems with disturbances of multi-agent systems [17]. In addition, the [18] used distributed intermediate estimators to study the cooperative fault-tolerant tracking control problem for linear multi-agent systems with faults and mismatched disturbances. It is important to note that all the above-mentioned FTC methods are model-based and require accurate system model parameters while designing the observer. In this paper, a comprehensive data-driven estimation algorithm that is capable of detecting and estimating the actuator fault is proposed. Moreover, the proposed method has good performances in transient time, steady-state time and computation complexity. Therefore, for some multi-agent systems with unknown model parameters, the data-driven approach can be more effective in solving the cooperative fault-tolerant tracking control problem.

2 Preliminaries and Problem Statement

2.1 Graph Theory

Assuming that there is N agent in the multi-agent systems, the directed graph between the individual nodes can be represented as $G = (v, \varepsilon)$, where $v = \{1, 2, \ldots, N\}$ stands for a node collection in the graph, ε represents an edge collection in the graph, and the edge $\varepsilon_{ij} = (i, j) \in \varepsilon$ directed path setting out from node i and reaches node j, $\mathcal{A} = [a_{ij}] \in \Re^{N \times N}$ is the adjacent matrix of G, where $a_{ii} = 0$. If the information of the agent i can be obtained by the agent j, $a_{ij} = 1$, else $a_{ij} = 0$. If a leader exists, the involved pinning matrix is indicated as $G = diag\{g_i\} \in \Re^{N \times N}$, where $g_i = 1$ if node i observes the leader, $g_i = 0$ otherwise. The Laplacian matrix in the directed graph can be represented as:

$$\mathcal{L}_{ij} = \begin{cases} \sum_{j=1}^{N} \mathcal{A}_{ij}, i = j \\ -\mathcal{A}_{ij}, i \neq j \end{cases} \tag{1}$$

2.2 System Description

This section considers the problem of collaborative fault-tolerant tracking control of multi-agent systems with actuator fault. The discrete dynamics model of the leader node is described by:

$$\begin{aligned} x_0(k+1) &= Ax_0(k) + Br_0(k) \\ y_0(k) &= Cx_0(k) \end{aligned} \tag{2}$$

where $x_0 \in \Re^n$ is the system state of the leader, $r_0 \in \Re^l$ represents the system inputs that enable the leader to follow the desired trajectory, and $y_0 \in \Re^m$ represents the output of the leader system. $A \in \Re^{n \times n}$, $B \in \Re^{n \times l}$, $C \in \Re^{m \times n}$ represent the system matrix, input matrix, and output matrix of the appropriate dimensions of the leader respectively. The follower systems with actuator fault can be represented as:

$$\begin{aligned} x_i(k+1) &= Ax_i(k) + Bu_i(k) + Bf_i(k) \\ y_i(k) &= Cx_i(k) \end{aligned} \tag{3}$$

where $i = 1, 2, \ldots N$, $x_i \in \Re^n$, $u_i \in \Re^l$, $f_i \in \Re^l$, $y_i \in \Re^m$ indicate the system status, control input, actuator fault, and system output of the i th follower respectively.

Remark 1. In this paper, we only consider cooperative fault-tolerant control of multi-agent systems, i.e. where the system parameter matrix of the followers is the same as that of the leader.

Defining $\delta_i(k) = x_i(k) - x_0(k)$ as the tracking error between the i th follower and the leader, the tracking error system can be obtained from (2) and (3), and the tracking error system can be expressed in the following form:

$$\begin{aligned} \delta_i(k+1) &= A\delta_i(k) + Bu_i - Br_o(k) + Bf_i(k) \\ y_{\delta_i}(k) &= C\delta_i(k) \end{aligned} \tag{4}$$

The objective of this section is to solve the cooperative fault-tolerant tracking control problem of the isomorphic multi-agent systems with unknown parameters, and ensure that the state of the tracking error system is consistent and eventually bounded.

3 Design of Data-driven Distributed Intermediate Estimator

In this section, a distributed intermediate estimator based on data-driven is proposed to handle the model parameters of each subsystem that are unknown, a data-driven implementation of the error system kernel representation needs to be calculated by tracking the input-output data of the error system with no fault. Since the model parameters of the tracking error system (4) and the dynamic model parameters of the follower systems (3) are identical, the input and output data of the multi-agent systems with no fault can be used to calculate a data-driven residual generator for the tracking error system. Consider the system (3) and the parity vector, one has:

$$\alpha_s = \left[\alpha_{s,0}\ \alpha_{s,1} \cdots \alpha_{s,s}\right] \Re^{(s+1)m}, \alpha_s \begin{bmatrix} C \\ CA \\ \vdots \\ CA^s \end{bmatrix} = 0 \tag{5}$$

where $\alpha_{s,i} \in \Re^m, i = 0, 1, ..., s$.

For the following equation of Luenberger:

$$TA - A_z T = L_z C, c_z T = gC, c_z = \left[0 \cdots 0\, 1\right], g = \alpha_{s,s} \tag{6}$$

its solution can be obtained as follow: $A_z = \begin{bmatrix} 0 & 0 & \cdots & 0 \\ 1 & 0 & \cdots & 0 \\ \vdots & \ddots & \ddots & \vdots \\ 0 & \cdots & 1 & 0 \end{bmatrix} \in \Re^{s \times s}, L_z = - \begin{bmatrix} \alpha_{s,0} \\ \alpha_{s,1} \\ \vdots \\ \alpha_{s,s-1} \end{bmatrix}, T = \begin{bmatrix} \alpha_{s,1} & \alpha_{s,2} & \cdots & \alpha_{s,s-1} & \alpha_{s,s} \\ \alpha_{s,2} & \cdots & \cdots & \alpha_{s,s} & 0 \\ \vdots & \cdots & \cdots & \vdots & \vdots \\ \alpha_{s,s} & 0 & \cdots & \cdots & 0 \end{bmatrix} \begin{bmatrix} C \\ CA \\ \vdots \\ CA^{s-1} \end{bmatrix}$.

Furthermore, the residual generator for the tracking error of the i th subsystem can be expressed in the following:

$$\begin{aligned} z_i(k+1) &= A_z z_i(k) + B_z u_z(k) + L_z y_i(k) \\ r_i(k) &= g y_i(k) - c_z z_i(k) - d_z u_i(k) \end{aligned} \tag{7}$$

where $z_i \in \Re^n$, $u_i \in \Re^l$,$y_i \in \Re^m$,$r_i \in \Re^m$ indicate the system status, control input, system output, and residual signal respectively. Consider the reference input signal in the residual generator at $u_z(k) = 0$. Based on the residual generator, a new state variable is defind:

$$\varphi_i(k) = T\delta_i(k) - z_i(k) \tag{8}$$

where the matrix T satisfies the Luenberger condition, i.e.

$$TA - L_zC = A_zT, B_z = TB - L_zD \tag{9}$$

Assuming that the multi-agent systems satisfies $D = 0$, the new state variables can be expressed in the following form according to (7) and (9):

$$\begin{aligned} \varphi_i(k+1) &= A_z\varphi_i(k) + B_z(u_i + \tau_i) \\ y_{\varphi_i}(k) &= C\varphi_i(k) \end{aligned} \tag{10}$$

where $\tau_i(k) = -r_o(k) + f_i(k)$ represents the unknown input signal of the follower, it contains the reference input information of the fault information of leader and the followers. Based on (10), a distributed intermediate estimator is constructed and the intermediate variable is defined:

$$\xi_i(k) = \tau_i(k) - \omega B_z^T\varphi_i(k) \tag{11}$$

where ω is an adjustable scalar. From (10) and (11), we can obtain:

$$\xi_i(k+1) = \tau_i(k+1) - \omega B_z^T(A_z\varphi_i(k) + B_zu_i + B_z\xi_i + \omega B_zB_z^T\hat{\varphi}_i(k)) \tag{12}$$

To estimate the unknown input signal of each followers, a distributed intermediate estimator is designed for the i th node:

$$\hat{\varphi}_i(k+1) = A_z\hat{\varphi}_i(k) + B_z(u_i + \hat{\tau}_i) + L(\zeta_i(k) - \hat{\zeta}_i(k)) \tag{13}$$

$$\hat{\xi}_i(k+1) = \hat{\tau}_i(k) - \omega B_z^T(A_z\hat{\varphi}_i(k) + B_zu_i + B_z\xi_i + \omega B_zB_z^T\hat{\varphi}_i(k)) \tag{14}$$

$$\hat{\tau}_i(k) = \hat{\xi}_i(k) + \omega B_z^T\hat{\varphi}_i(k) \tag{15}$$

where $\hat{\varphi}_i(k)$, $\hat{\xi}_i(k)$ and $\hat{\tau}_i(k)$ denote the estimations of $\varphi_i(k)$, $\xi_i(k)$ and $\tau_i(k)$ respectively, and L denotes the intermediate estimator gain. In addition, $\zeta_i(k)$ and $\hat{\zeta}_i(k)$ can be expressed as follow:

$$\zeta_i(k) = \sum_{j\in \mathrm{N}_i} a_{ij}(y_i(k) - y_j(k))) + g_i(y_i(k) - y_0(k)) \tag{16}$$

$$\hat{\zeta}_i(k) = \sum_{j\in \mathrm{N}_i} a_{ij}(C\hat{\varphi}_i(k) - C\hat{\varphi}_j(k)) + g_iC\hat{\varphi}_i(k) \tag{17}$$

The state estimation error of the intermediate estimator and the estimation error of the intermediate variables are defined as follows $e_{\varphi_i}(k) = \varphi_i(k) - \hat{\varphi}_i(k)$, $e_{\xi_i}(k) = \xi_i(k) - \hat{\xi}_i(k)$, $e_{\tau_i} = \tau_i(k) - \hat{\tau}_i(k)$. Based on the intermediate variables $\xi_i(k) = \tau_i(k) - \omega B_z^Te_\varphi(k)$, we can obtain $e_{\xi_i}(k) = e_{\tau_i}(k) - \omega B_z^Te_{\varphi_i}(k)$. After defining the above variables, we can obtain the following system of intermediate estimator errors:

$$\begin{aligned} e_{\varphi_i}(k+1) &= A_z e_{\varphi_i}(k) + B_z e_{\xi_i}(k) + \omega B_z B_z^T e_{\varphi_i}(k) - L(\zeta_i(k) - \hat{\zeta}_i(k)) \\ e_{\xi_i}(k+1) &= \Delta\tau(k) - \omega B_z^T B_z e_{\xi_i}(k) - \omega B_z(A_z + \omega B_z B_z^T) e_{\varphi_i}(k) \end{aligned} \tag{18}$$

Based on the above analysis, the following synchronous tracking control protocol is designed for the i th node to meet the tracking control requirements of multiple-agent systems:

$$u_i(k) = F_z \hat{\varphi}_i(k) - \hat{\tau}_i(k) \tag{19}$$

where F_z is the observer-based state feedback control gain designed by the data-driven approach [19], and it is assumed that the calculated F_z satisfies that $A + B_z F_z$ is a Hurwitz matrix.

Substituting(19) into (10), one can obtain a closed-loop dynamic system for the i th node, whose dynamic discrete model is expressed as:

$$\varphi_i(k+1) = (A_z + B_z F)\varphi_i(k) - (B_z F + \omega B_z B_z^T) e_{\varphi_i}(k) - B_z e_{\xi_i}(k) \tag{20}$$

Further, according to (18) and (20), the global closed-loop system is expressed as:

$$\varphi(k+1) = (I_N \otimes (A_z + B_z F))\varphi(k) + (I_N \otimes S_1) e_\varphi(k) + (I_N \otimes S_2) e_\xi(k) \tag{21}$$

$$\begin{aligned} e_\varphi(k+1) &= (I_N \otimes A_z) e_\varphi(k) + (I_N \otimes B_z) e_\xi(k) \\ &\quad + (I_N \otimes \omega B_z B_z^T) e_\varphi(k) - (M \otimes LC) e_\varphi(k) \end{aligned} \tag{22}$$

$$\begin{aligned} e_\xi(k+1) &= (I_N \otimes -\omega B_z^T B_z) e_\xi(k) + (I_N \otimes -\omega B_z^T A_z) e_\varphi(k) \\ &\quad + (I_N \otimes -\omega^2 B_z^T B_z B_z^T) e_\varphi(k) + (I_N \otimes \omega B_z^T) e_\varphi(k) \\ &\quad + e_\xi(k) + \Delta f(k) \end{aligned} \tag{23}$$

where $M = L + G$, $S_1 = B_z F + \omega B_z B_z^T$, $S_2 = B_z$, $\varphi(k) = \left[\varphi_1^T(k) \ldots \varphi_N^T(k)\right]^T$, $e_\varphi(k) = \left[e_{\varphi_1}^T(k) \ldots e_{\varphi_N}^T(k)\right]^T$, $e_\xi(k) = \left[e_{\xi_1}^T(k) \ldots e_{\xi_N}^T(k)\right]$, $\Delta\tau(k) = \left[\Delta f_1^T(k) \ldots \Delta f_N^T(k)\right]^T$.

4 Convergence Analysis

Theorem 1. *For a given scalar $\omega > 0$, there exists matrices $P_1 > 0$, $P_2 > 0$, $P_3 > 0$ and a matrix H such that the following condition hold:*

$$\Phi^i = \begin{bmatrix} \Pi_{11} & \Pi_{12} & \Pi_{13} & 0 & 0 \\ * & \Pi_{22} & \Pi_{23} & A_2^T P_3 & 0 \\ * & * & \Pi_{33} & 0 & A_3^T P_2 \\ * & * & * & -I & 0 \\ * & * & * & * & -I \end{bmatrix} < 0 \tag{24}$$

where $\Pi_{11} = (A_z + B_zF)^T P_1(A_z + B_zF) - P_1$, $\Pi_{12} = (A_z + B_zF)^T P_1S_1$, $\Pi_{13} = (A_z + B_zF)^T P_1S_2$, $\Pi_{22} = S_1^T P_1S_1 + A_1^T P_2A_1 + \lambda_i^2 C^T L^T P_2LC - 2\lambda_i A_1^T P_2LC - P_2 + A_2^T P_3A_2$, $\Pi_{23} = S_1^T P_1S_2 + A_1^T P_2B_z - \lambda_i C^T L^T P_2B_z + A_2^T P_3A_3$, $\Pi_{33} = S_2^T P_1S_2 + B_z^T P_2B_z + A_3^T P_3A_3 - P_3$, $A_1 = A_z + \omega B_zB_z^T$, $A_2 = \omega B_z^T(I - A_z - \omega B_zB_z^T)$, $A_3 = (I - \omega B_zB_z^T)$, *then all states of the closed-loop system of from (21) to (23) are ultimately uniformly bounded and the observer gain can be obtained from* $L = P_2^{-1}H$, $H^T = L^T P_2$.

Proof. Given the following Lyapunov functions for the state of a closed-loop error system, the estimation error of state and the intermediate variable :

$$V(k) = \varphi^T(k)(I_N \otimes P_1)\varphi(k) + e_\varphi^T(k)(I_N \otimes P_2)e_\varphi(k) + e_\xi^T(k)(I_N \otimes P_3)e_\xi(k) \tag{25}$$

An arbitrary trajectory of estimation error system is extended, which has the difference of its Lyapunov function as:

$$\begin{aligned}
\Delta V(k+1) &= V(k+1) - V(k) \\
&= \varphi^T(k)[I_N \otimes ((A_z + B_zF)^T P_1(A_z + B_zF) - P_1)]\varphi(k) \\
&+ 2\varphi^T(k)(I_N \otimes (A_z + B_zF)^T P_1S_1)e_\varphi(k) \\
&+ 2\varphi^T(k)(I_N \otimes (A_z + B_zF)^T P_1S_2)e_\xi(k) \\
&+ e_\varphi^T(k)(I_N \otimes (S_1^T P_1S_1 + A_1^T P_2A_1 - P_2 + A_2^T P_3A_2))e_\varphi(k) \\
&+ e_\varphi^T(k)(M^T M \otimes C^T L^T P_2LC - 2M \otimes (A_1^T P_2LC))e_\varphi(k) \\
&+ e_\varphi^T(k)(I_N \otimes (S_1^T P_1S_2 + A_1^T P_2B_z + A_2^T P_3A_3))e_\xi(k) \\
&- e_\varphi^T(k)(M^T \otimes C^T L^T P_2B_z)e_\xi(k) \\
&+ e_\xi^T(k)(I_N \otimes (S_2^T P_1S_2 + B_z^T P_2B_z + A_3^T P_3A_3 - P_3))e_\xi(k) \\
&+ 2e_\varphi^T(k)(I_N \otimes A_2^T P_3)\Delta f(k) + 2e_\xi^T(k)(I_N \otimes A_3^T P_3)\Delta f(k)
\end{aligned} \tag{26}$$

Suppose $\|\Delta f(k)\| \leq \theta_f$, then the following inequality holds:

$$2e_\varphi^T(k)(I_N \otimes A_2^T P_3)\Delta f(k) \leq \frac{1}{\varepsilon}e_\varphi^T(k)(I_N \otimes A_2^T P_3P_3A_2)e_\varphi + \varepsilon\theta_f^2 \tag{27}$$

$$2e_\xi^T(k)(I_N \otimes A_3^T P_3)\Delta f(k) \leq \frac{1}{\varepsilon}e_\xi^T(k)(I_N \otimes A_3^T P_3P_3A_3)e_\xi + \varepsilon\theta_f^2 \tag{28}$$

Denote $\tilde{e}(k) = \left[\varphi^T(k)\ e_\varphi^T(k)\ e_\xi^T(k)\right]^T$, then (26) can be equated to

$$\Delta V(k+1) = \tilde{e}^T(k)\sum \tilde{e}(k) + 2\varepsilon\theta_f^2 \tag{29}$$

where

$$\sum = \begin{bmatrix} \sum_{11} & \sum_{12} & \sum_{13} \\ * & \sum_{22} & \sum_{23} \\ * & * & \sum_{33} \end{bmatrix} \tag{30}$$

in above formula, $\Pi_{11} = I_N \otimes ((A_z + B_zF)^T P_1(A_z + B_zF) - P_1)$, $\Pi_{12} = I_N \otimes (A_z + B_zF)^T P_1S_1$, $\Pi_{13} = I_N \otimes (A_z + B_zF)^T P_1S_2$, $\Pi_{22} = I_N \otimes (S_1^T P_1S_1 +$

$A_1^T P_2 A_1 - P_2 + A_2^T P_3 A_2 + A_2^T P_3 P_3 A_2) + M^T M \otimes C^T L^T P_2 LC - 2M \otimes (A_1^T P_2 LC)$, $\Pi_{23} = I_N \otimes (S_1^T P_1 S_2 + A_1^T P_2 B_z + A_2^T P_3 A_3) - M^T \otimes C^T L^T P_2 B_z$, $\Pi_{33} = I_N \otimes (S_2^T P_1 S_2 + B_z^T P_2 B_z + A_3^T P_3 A_3 - P_3 + A_3^T P_3 P_3 A_3)$.

According to the (25), we can obtain:

$$\begin{aligned} V(k) &\leq \lambda_{\max}(P_1)\|\varphi(k)\|^2 + \lambda_{\max}(P_2)\|e_\varphi(k)\|^2 + \lambda_{\max}(P_3)\|e_\xi(k)\|^2 \\ &\leq \max[\lambda_{\max}(P_1), \lambda_{\max}(P_2), \lambda_{\max}(P_3)](\|\varphi(k)\|^2 + \|e_\varphi(k)\|^2 + \|e_\xi(k)\|^2) \end{aligned} \tag{31}$$

Define $\Sigma_p = -\Sigma$, and from (29) and (31) you can obtain:

$$\Delta V(k) \leq -\alpha V(k) + \beta \tag{32}$$

where

$$\alpha = \frac{\min \Sigma_p}{\max[\lambda_{\max}(P_1), \lambda_{\max}(P_2), \lambda_{\max}(P_3)]}$$
$$\beta = 2\varepsilon\theta_f^2$$

Define the following set Z and its complement Z_s :

$$Z = \left\{ \varphi(k), e_\varphi(k), e_\xi(k) \,\middle|\, \begin{array}{l} \lambda_{\min}(P_1)\|\varphi(k)\|^2 + \lambda_{\min}(P_2)\|e_\varphi(k)\|^2 \\ +\lambda_{\min}(P_3)\|e_\xi(k)\|^2 \leq \frac{\beta}{\alpha} \end{array} \right\} \tag{33}$$

$$Z_s = \left\{ \varphi(k), e_\varphi(k), e_\xi(k) \,\middle|\, \begin{array}{l} \lambda_{\min}(P_1)\|\varphi(k)\|^2 + \lambda_{\min}(P_2)\|e_\varphi(k)\|^2 \\ +\lambda_{\min}(P_3)\|e_\xi(k)\|^2 \geq \frac{\beta}{\alpha} \end{array} \right\} \tag{34}$$

If $(\varphi(k), e_\varphi(k), e_\xi(k)) \in Z_s$, then:

$$\begin{aligned} V(k) \geq \lambda_{\min}(P_1)\|\varphi(k)\|^2 + \lambda_{\min}(P_2)\|e_\varphi(k)\|^2 + \lambda_{\min}(P_3)\|e_\xi(k)\|^2 \\ \geq \frac{\beta}{\alpha} \end{aligned} \tag{35}$$

Furthermore, from (29) and (35), one has:

$$\Delta V(k+1) < 0 \tag{36}$$

According to Lyapunov's theory, it follows that $(\varphi(k), e_\varphi(k), e_\xi(k))$ converges exponentially to Z. Since M is a symmetric real matrix, by performing a spectral decomposition of M, we get:

$$M = Q\Lambda Q^T \tag{37}$$

where Q is the matrix consisting of the vectors of the features of M, $\Lambda = diag\{\lambda_1, \ldots, \lambda_i\}$, and $\lambda_i, i = 1, 2, \ldots, N$ are the characteristic roots of the matrix M. Define a transformation matrix as follows:

$$T = \begin{bmatrix} Q^T \otimes I_n & 0 & 0 \\ 0 & Q^T \otimes I_n & 0 \\ 0 & 0 & Q^T \otimes I_m \end{bmatrix} \tag{38}$$

The transposition of the matrix Σ with the left multiplicative transformation matrix and the right multiplicative transformation matrix respectively gives:

$$\overset{o}{\sum} = \begin{bmatrix} \sum_{11} & \overset{o}{\sum}_{12} & \overset{o}{\sum}_{13} \\ * & \sum_{22} & \sum_{23} \\ * & * & \sum_{33} \end{bmatrix} \tag{39}$$

where $\Pi_{22} = I_N \otimes (S_1^T P_1 S_1 + A_1^T P_2 A_1 - P_2 + A_2^T P_3 A_2 + A_2^T P_3 P_3 A_2) + \Lambda^2 \otimes C^T L^T P_2 LC - 2\Lambda \otimes (A_1^T P_2 LC)$, $\Pi_{23} = I_N \otimes (S_1^T P_1 S_2 + A_1^T P_2 B_z + A_2^T P_3 A_3) - \Lambda \otimes C^T L^T P_2 B_z$. Using Schur complementary Lemma, for $i = 1, 2, \ldots, N$, $\sum^o < 0$, (24) is hold, when $\Sigma < 0$. All the above analysis completes the proof of Theorem 1.

5 Experimental Results

The structure of the networked multi-axis control system considered in this paper is shown in Fig. 1. There are a total of four servo motors, one of which motor is the leader, and the remaining three ones are followers that maintain synchronous motion with the leader. The system of leader is as follows:

$$x_0(k+1) = \begin{bmatrix} -0.0860 & -0.1026 \\ 0.8208 & 0.5501 \end{bmatrix} x_0(k) + \begin{bmatrix} 0.2796 \\ 0.9255 \end{bmatrix} r_0(k) \tag{40}$$

where $x_0 = \begin{bmatrix} x_{01} & x_{02} \end{bmatrix}^T$, the status x_{01} indicates the servomotor stator current and the status x_{02} indicates the servo motor rotor speed. r_0 is the reference input for the leader and the sampling time is $T_s = 0.05.[\mathrm{h}]$

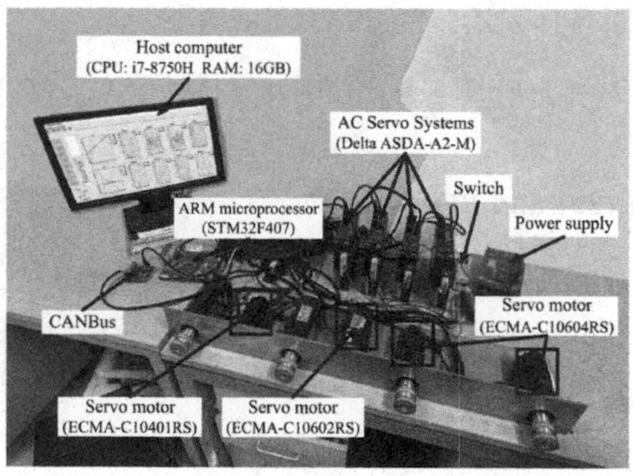

Fig. 1. Combined unknown signals and their estimations for the Followers.

Furthermore, the discrete model of the i th follower with actuator fault can be expressed as:

$$\begin{aligned} x_i(k+1) &= \begin{bmatrix} -0.0860 & -0.1026 \\ 0.8208 & 0.5501 \end{bmatrix} x_i(k) + \begin{bmatrix} 0.2796 \\ 0.9255 \end{bmatrix} u_i(k) + \begin{bmatrix} 0.2796 \\ 0.9255 \end{bmatrix} f_i(k) \\ y_i(k) &= [0\ 1] x_i(k) \end{aligned} \tag{41}$$

where x_i, u_i, f_i indicate the system status of the i th follower, the input, and the actuator fault respectively.

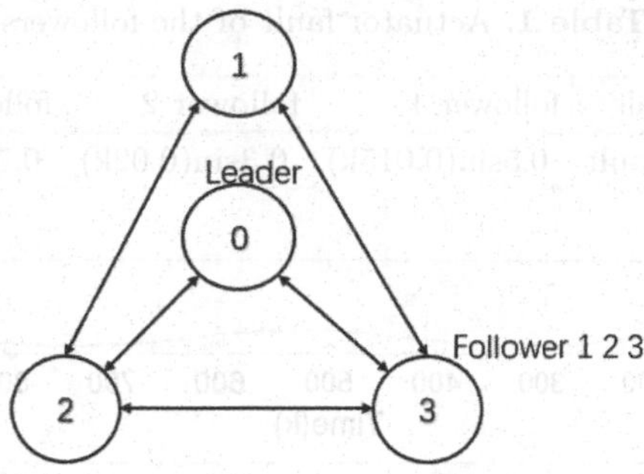

Fig. 2. Communication topology of networked multi-axis motion control system.

To design an observer-based residual generator to collect I/O process data when the system no actuator fault, the data drive implementation of the servo motor control system kernel representation is then obtained. Let $s = 2$, thus the obtained SKR data-driven implementation of servo motor system is:

$$\begin{aligned} \Psi_s^{\perp} &= \left[\Psi_{s,u}^{\perp} \; \Psi_{s,y}^{\perp}\right] \\ &= \left[-0.1397 \; -0.8466 \; -0.0380 \; 0.0061 \; -0.0570 \; 0.5089\right] \end{aligned} \tag{42}$$

where $\Psi_{s,u}^{\perp} = \left[-0.1397 \; -0.8466 \; -0.0380\right]$, $\Psi_{s,y}^{\perp} = \left[0.0061 \; -0.0570 \; 0.5089\right]$.

Further, the residual generator for the i th agent is constructed:

$$\begin{aligned} z_i(k+1) &= \begin{bmatrix} 0 & 0 \\ 1 & 0 \end{bmatrix} z_i(k) + \begin{bmatrix} 0.1397 \\ 0.8466 \end{bmatrix} u_i(k) + \begin{bmatrix} -0.0061 \\ 0.0570 \end{bmatrix} y_i(k) \\ r_i(k) &= 0.5089 y_i(k) - \begin{bmatrix} 0 & 1 \end{bmatrix} z_i(k) - (-0.0380)\mathrm{u}_i(\mathrm{k}) \end{aligned} \tag{43}$$

A new state vector $\varphi_i(k) = T\delta_i(k) - z_i(k)$ is defined for each agent and the state space of the system is $D = 0$, so that the space equation for the state vector $\varphi_i(k)$ can be obtained:

$$\begin{aligned} \varphi_i(k+1) &= \begin{bmatrix} 0 & 0 \\ 1 & 0 \end{bmatrix} \varphi_i(k) + \begin{bmatrix} 0.1397 \\ 0.8466 \end{bmatrix} (u_i(k) + \tau_i(k)) \\ y_{\varphi_i}(k) &= \begin{bmatrix} 0 & 1 \end{bmatrix} \varphi_i(k) \end{aligned} \tag{44}$$

where $u_i(k)$ is the compensating control input $u_i(k) = F_z \hat{\varphi}_i(k) - \hat{\tau}_i(k)$. $\tau_i(k)$ is the unknown input signal to the system and F_z is the observer-based state feedback controller, $\mathrm{F}_z = \left[-8.3579 \; 3.5898\right]$. To estimate the actuator fault signal, the intermediate estimator is expressed as:

$$\begin{aligned} \hat{\varphi}_i(k+1) &= \begin{bmatrix} 0 & 0 \\ 1 & 0 \end{bmatrix} \hat{\varphi}_i(k) + \begin{bmatrix} 0.1397 \\ 0.8466 \end{bmatrix} (u_i + \hat{\tau}_i) + L(\zeta_i(k) - \hat{\zeta}_i(k)) \\ \hat{y}_{\varphi_i}(k) &= \begin{bmatrix} 0 & 1 \end{bmatrix} \hat{\varphi}_i(k) \end{aligned} \tag{45}$$

[h] Servo motor experimental platform is shown in Fig. 2, the leader and the follower are connected by one-way, The followers will use the above communication

Table 1. Actuator fault of the followers

Fault signals	follower 1	follower 2	follower 3
Actuator fault	0.5sin(0.015k)	0.3sin(0.02k)	0.7sin(0.01k)

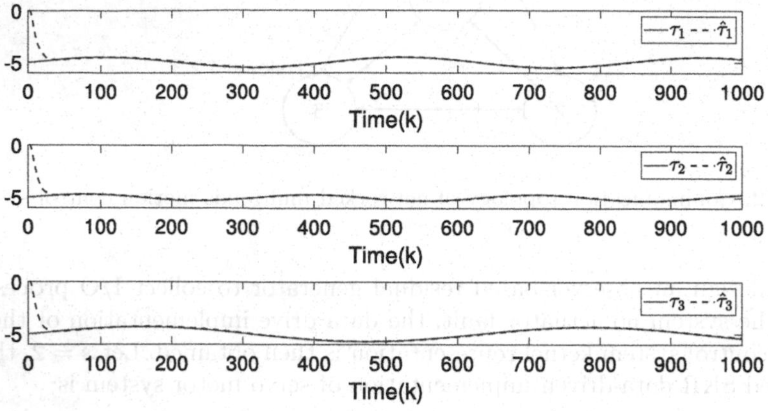

Fig. 3. Combined unknown signals and their estimations for the followers.

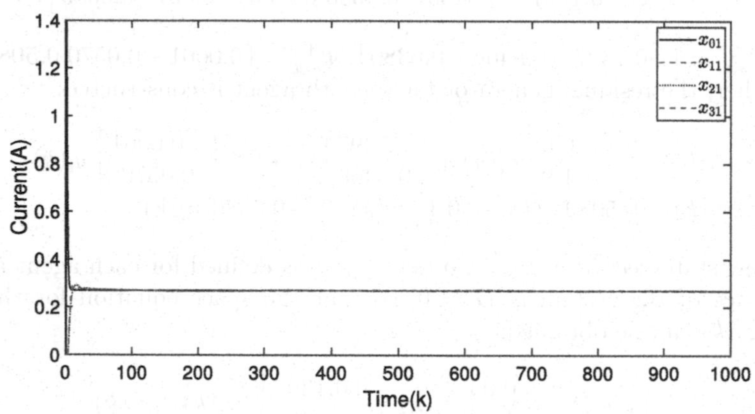

Fig. 4. Tracking curve of under fault-tolerant control.

topology to achieve cooperative tracking control. The actuator fault signals for each follower are shown in Table.1.

The experimental results of the data-driven intermediate estimator cooperative fault-tolerant control algorithm for multi-agent systems are shown in Fig. 3 and Fig. 4. Figure 3 demonstrates the algorithm's ability to accurately estimate the combined unknown input signals of followers with unknown model parameters, containing reference input information from the leader and the actuator fault information from the followers. From Fig. 4 and Fig. 5, we can see that the follower state is more than accurate enough to track the leader under

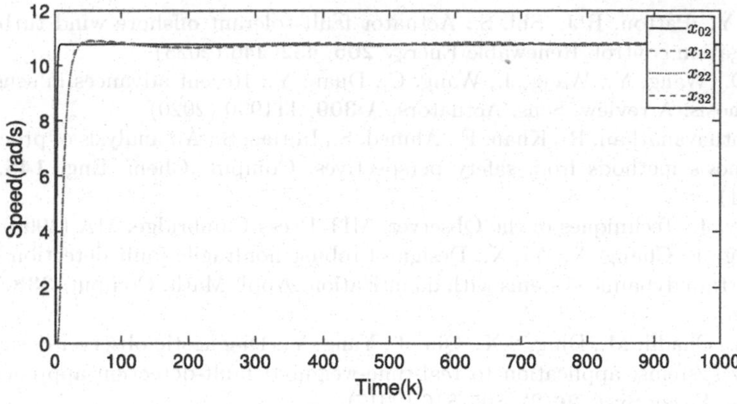

Fig. 5. Tracking curve of under x_{02} fault-tolerant control

fault-tolerant compensation control. From the above results, it can be seen that the data-driven distributed intermediate estimator proposed in this paper can achieve collaborative fault-tolerant tracking control of a multi-agent systems with unknown model parameters.

6 Conclusion

This paper presents a collaborative fault-tolerant control method based on a data-driven distributed intermediate estimator for multi-agent systems with actuator faults. Under the condition that the parameters of the system model are unknown, a data-driven residual generator can be constructed using the process input and output data of the system and the unknown input signals of each followers are estimated by the data-driven distributed intermediate estimator. Based on the estimated signals, control law is designed to compensate for the unknown inputs of the followers. Finally, the effectiveness of the algorithm is demonstrated using experiment of multi-axis motion control system. The results demonstrated that the data-driven distributed intermediate estimator has good performance of estimation and fault tolerance, and can achieve cooperative fault-tolerant tracking control of multiple-agent systems.

Acknowledgement. This work was supported in part by the Zhejiang Provincial Natural Science Foundation of China under Grant LZ21F030004, in part by the Key Research and Development Program of Zhejiang under Grant 2022C01018, and in part by the National Natural Science Foundation of China under Grant U21B2001.

References

1. van Schrick, D.: Remarks on terminology in the field of supervision, fault detection and diagnosis. IFAC Proc. Vol. **30**(18), 959–964 (1997)

2. Liu, Y., Patton, R.J., Shi, S.: Actuator fault-tolerant offshore wind turbine load mitigation control. Renewable Energy **205**, 432–446 (2023)
3. Li, D., Wang, Y., Wang, J., Wang, C., Duan, Y.: Recent advances in sensor fault diagnosis: A review. Sens. Actuators, A **309**, 111990 (2020)
4. Arunthavanathan, R., Khan, F., Ahmed, S., Imtiaz, S.: An analysis of process fault diagnosis methods from safety perspectives. Comput. Chem. Eng. **145**, 107197 (2021)
5. Crary, J.: Techniques of the Observer. MIT Press Cambridge, MA (1990)
6. Xiong, J., Chang, X., Yi, X.: Design of robust nonfragile fault detection filter for uncertain dynamic systems with quantization. Appl. Math. Comput. **338**, 774–788 (2018)
7. Li, L., Chadli, M., Ding, S.X., Qiu, J., Yang, Y.: Diagnostic observer design for t-s fuzzy systems: application to real-time-weighted fault-detection approach. IEEE Trans. Fuzzy Syst. **26**(2), 805–816 (2017)
8. Wang, M., Song, X., Song, S., Lu, J.: Diagnostic observer-based fault detection for nonlinear parabolic PDE systems via dual sampling approaches. J. Franklin Inst. **357**(12), 8203–8228 (2020)
9. Li, S., Yang, J., Chen, W.H., Chen, X.: Generalized extended state observer based control for systems with mismatched uncertainties. IEEE Trans. Industr. Electron. **59**(12), 4792–4802 (2011)
10. Zhu, J., Yang, G., Wang, H., Wang, F.: Fault estimation for a class of nonlinear systems based on intermediate estimator. IEEE Trans. Autom. Control **61**(9), 2518–2524 (2015)
11. Amin, A.A., Hasan, K.M.: A review of fault tolerant control systems: advancements and applications. Measurement **143**, 58–68 (2019)
12. Zhou, K., Ren, Z.: A new controller architecture for high performance, robust, and fault-tolerant control. IEEE Trans. Autom. Control **46**(10), 1613–1618 (2001)
13. Hua, C., Ding, S.X., Shardt, Y.A.: A new method for fault-tolerant control through q-learning. IFAC-PapersOnLine **51**(24), 38–45 (2018)
14. Zhang, K., Jiang, B., Shi, P.: Adjustable parameter-based distributed fault estimation observer design for multiagent systems with directed graphs. IEEE Trans. Cybern. **47**(2), 306–314 (2016)
15. Menon, P.P., Edwards, C.: Robust fault estimation using relative information in linear multi-agent networks. IEEE Trans. Autom. Control **59**(2), 477–482 (2013)
16. Chen, C., Lewis, F.L., Xie, S., Modares, H., Liu, Z., Zuo, S., Davoudi, A.: Resilient adaptive and h ∞ controls of multi-agent systems under sensor and actuator faults. Automatica **102**, 19–26 (2019)
17. Zhu, J., Zhang, W., Yu, L., Zhang, D.: Robust distributed tracking control for linear multi-agent systems based on distributed intermediate estimator. J. Franklin Inst. **355**(1), 31–53 (2018)
18. Zhu, J., Yang, G., Zhang, W., Yu, L.: Cooperative fault-tolerant tracking control for multiagent systems: An intermediate estimator-based approach. IEEE Trans. Cybern. **48**(10), 2972–2980 (2017)
19. Zhu, J., Xia, Z., Wang, X.: A new residual generation-based fault estimation approach for cyber-physical systems. IEEE Trans. Instrum. Meas. **72**, 1–9 (2023)

Anomaly Detection and Alarm Limit Design for In-Hole Bit Bounce Based on Interval Augmented Mahalanobis Distance

Bin Hu[1,2,3], Wenkai Hu[1,2,3(✉)], Peng Zhang[1,2,3], and Weihua Cao[1,2,3]

[1] School of Automation, China University of Geosciences, Wuhan 430074, China
{HBin,wenkaihu,zhangpengau,weihuacao}@cug.edu.cn
[2] Hubei Key Laboratory of Advanced Control and Intelligent Automation for Complex Systems, Wuhan 430074, China
[3] Engineering Research Center of Intelligent Technology for Geo-Exploration, Ministry of Education, Wuhan 430074, China

Abstract. Timely and accurate anomaly detection is of great importance for the safe operation of the drilling process. To detect bit bounce during the drilling process, this paper proposes a method based on interval augmentation Mahalanobis distance. The method first selects process variables that are closely related to bit bounce through mechanism analysis; secondly, data augmentation is performed on the selected data; then, the Mahalanobis distance statistic of normal data is calculated and its distribution threshold is designed using the Kernel Density Estimation method; finally, the Mahalanobis distance is calculated for the augmented online data and compared to the threshold to determine if a bit bounce is present.

Keywords: Geological drilling · Anomaly detection · Data augmentation · Mahalanobis distance

1 Introduction

Geological drilling is a common process used in resource and energy exploration. As an important part of the drilling system, drilling tools are responsible for the load of breaking rock throughout the drilling process [1]. Due to the abundance of underground rock types and uneven soft and hard strata, abnormal conditions of the drilling tools occur from time to time, leading to severe bit wear and even risk of breakage. It is therefore important to be able to detect the abnormal condition of the drilling tools and optimise the drilling parameters in time for safe and efficient drilling.

The existing detection methods for abnormal condition of drilling tools can be divided into model-based methods and data-driven methods [2]. The first kind of method requires the establishment of mechanical or vibration mechanism model of the drilling tools. In [3,4], different bit-rock interaction models were established to diagnose stick-slip vibration anomalies of drill strings. On this basis, the axial-torsional dynamic model of the drill strings system was established [5]. Based on this model, a

H. Zhang et al. (Eds.): NCAA 2023, CCIS 1870, pp. 545–558, 2023.
https://doi.org/10.1007/978-981-99-5847-4_39

state and output feedback control strategy was proposed to mitigate the torsional stick-slip oscillation. In addition, a distributed model of the drill strings was established in [6], and an observer-based turntable boundary controller was designed to effectively suppress the stick-slip vibration of the drill strings. Besides, it was proposed in [7–9] to establish a drilling tool vibration model with three degrees of freedom in axial, transverse, and torsional to effectively detect stick-slip and stuck drilling anomalies.

In contrast, data-driven methods detect anomalies based on process data without the need for a precise mechanistic model [10]. In [11], a fault diagnosis model based on the Kernel Principal Component Analysis (KPCA) method was proposed to realise the early detection of kick faults. Considering that the data distribution was different from the normal condition when the drill bit bounce, Kullback-Leibler Divergence (KLD) statistic of the data distribution was calculated and the alarm threshold was designed in [12]. In [13], the event diagnosis method of generalized probabilistic neural network was used to achieve the detection of abnormalities, such as stuke pipe and overpulling. In addition, there were some researches to realize the anomaly detection of drilling tools from the perspective of frequency domain. In order to detect the gyroscopic problem in the process of deep hole drilling, an improved empirical wavelet denoising algorithm combined with energy entropy was proposed in [14]. In-situ vibration signals were analyzed in the time-frequency domain in [15], and the wavelet packet energy distribution was used as an index to evaluate bit wear.

In summary, the existing abnormal condition detection methods for drilling tools are relatively extensive. However, there are still major limitations in the detection of drill bit bounce. Compared to other drilling tool anomalies, when drill bit bounce occurs, the process variable data fluctuates frequently and overlaps the normal condition in a large range, making it difficult to detect. In order to improve the detection accuracy of bit bounce, this study proposes a downhole bit bounce detection method based on Interval Augmented Mahalanobis distance (IA-MD). The contributions of this paper are twofold: 1) A method of stacking sample data to form an interval augmented vector is proposed to improve the ability to detect small changes in the data. 2) A KDE method is presented to achieve threshold design for normal data samples by calculating the MD statistic for the interval augmented data.

The rest of this article is organized as follows: The problem is described in Sect. 2. The bit bounce detection method is proposed in Sect. 3. An industrial case is presented in Sect. 4 to demonstrate the effectiveness of the proposed approach, and Sect. 5 gives the conclusions.

2 Problem Description

A schematic of a simple drilling system is shown in Fig. 1. During the drilling process, the hook bears part of the gravity of the drill string, called HooK Load (HKL), while the rest of the gravity acts on the Bottom Hole Assembly (BHA). The top drive motor rotates the drill strings, generating a certain ToRQue (TRQ), and then drives the BHA to generate a certain Rotary Per Minutes (RPM) to cut the rock. The mud pump carries the drilling mud through the drillstrings to the bottom of the hole and carries the cuttings back to the mud pit.

The bit bounce is mainly due to the uneven soft and hard formation or gravel layer encountered by the bit during the drilling process and the uneven force of the roller, resulting in the loss of contact between the bottom hole rock and the BHA. In general, the bit bounce can be reflected by the change in TRQ, HKL, and RPM. Specifically, the fluctuation range of TRQ and HKL will gradually increase, and the change in RPM will rise to a certain extent. The unusually frequent occurrence of bit bounce not only accelerates bit wear, but also causes the risk of drill fracture, which seriously affects the drilling progress and causes economic losses.

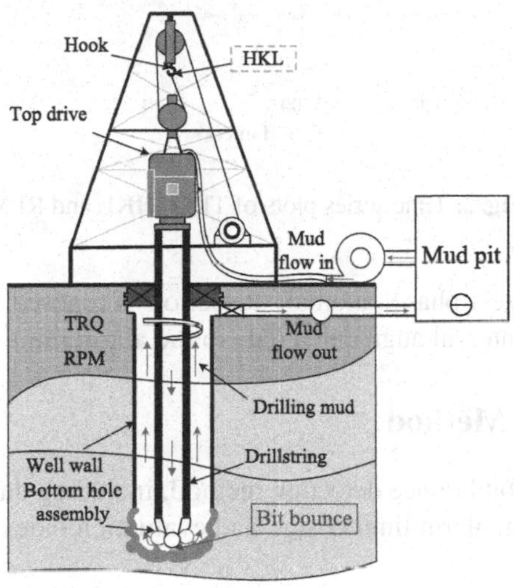

Fig. 1. Schematic diagram of drilling system

The process data extracted from a real geothermal exploration well in China is shown in Fig. 2, and the bit bounce occurs over the period from 1501 s to 3500 s. It can be clearly seen that both the TRQ and the HKL in normal and abnormal periods overlap with each other in terms of the data distribution, making it difficult to separate the abnormal data from the normal data. Although the value of RPM increases significantly at the time of 1980s, it is still difficult to distinguish the abnormal part from the normal part before this time instance.

Based on the above discussion, the purpose of this paper is to improve detection accuracy by interval augmentation of data, making full use of historical data. The proposed approach consists of the following two main steps:

1) A new vector is formed by stacking the samples with adjacent intervals, and then the MD statistic of the offline and online data is calculated by using the statistical characteristics of the normal condition.
2) The alarm limit of the MD statistic of the offline normal data is designed by the Gaussian KDE method, and the normal working zones with different intervals and times of augmentation are obtained.

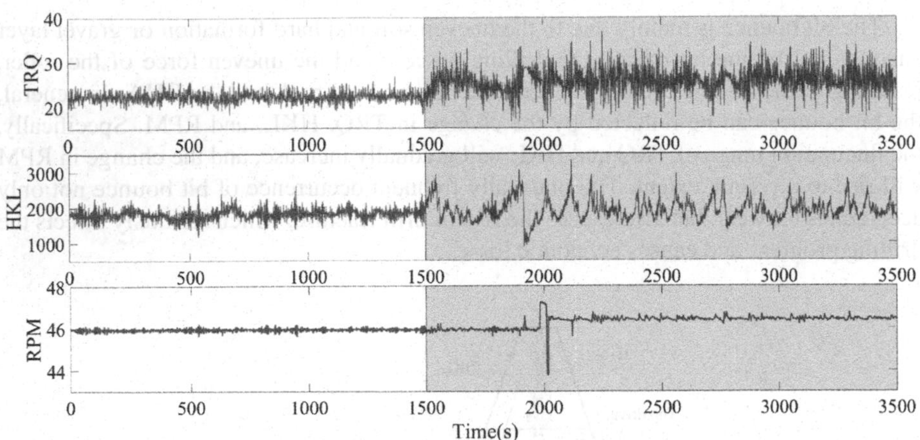

Fig. 2. Time series plots of TRQ, HKL and RPM

During the online test phase, anomaly detection is realised by comparing the MD statistic of the online interval augmented data to the alarm limit.

3 The Proposed Method

This section proposes bit bounce detection method, including data augmentation, monitoring statistics design, alarm limit design and evaluation index comprehensive evaluation method.

3.1 Data Augmentation

In this section, a data augmentation method is proposed to increase the data interval to implement sample stacking. The details are as follows:

At the time instant k, the formula for calculating a single interval augmented vector $\boldsymbol{x}_{\lambda,L}(k)$ is

$$\boldsymbol{x}_{\lambda,L}(k) = [\boldsymbol{x}(k)\cdots\boldsymbol{x}(k-\lambda(j-1))\cdots\boldsymbol{x}(k-\lambda(L-1))] \tag{1}$$

where $\boldsymbol{x}(k)$ is the vector collected by M sensors at time k, $\boldsymbol{x}(k) \in R^{1\times M}$, L-1 is the augmentation size, λ is the number of intervals. Take the training sample, $\boldsymbol{X} \in R^{N\times M}$, after the interval is augmented, it becomes $\boldsymbol{X}_{\lambda,L}$, given by

$$\boldsymbol{X}_{\lambda,L} = \begin{bmatrix} \boldsymbol{x}(1-\lambda(1-L)) & \cdots & \boldsymbol{x}(1-\lambda(j-L)) & \cdots & \boldsymbol{x}(1) \\ \vdots & \ddots & \vdots & \ddots & \vdots \\ \boldsymbol{x}(k) & \cdots & \boldsymbol{x}(k-\lambda(j-1)) & \cdots & \boldsymbol{x}(k-\lambda(L-1)) \\ \vdots & \ddots & \vdots & \ddots & \vdots \\ \boldsymbol{x}(N) & \cdots & \boldsymbol{x}(N-\lambda(j-1)) & \cdots & \boldsymbol{x}(N-\lambda(L-1)) \end{bmatrix} \tag{2}$$

where N indicates the number of samples, $\boldsymbol{X}_{\lambda,L} \in R^{[N+\lambda(1-L)]\times ML}$.

3.2 Design Abnormal Monitoring Indicator

The distribution of the selected variables during drilling is shown in Fig. 3, with the purple triangular point representing the bit bounce data; the red, green, and blue points represent the normal drilling samples; the red square point represents the centre of the mean of the normal data distribution. Compared to the green point, the purple point is obviously closer to the red point, but it is an abnormal data point. In order to overcome the influence of the overall data distribution on the classification, this section introduces the MD statistic for anomaly detection [16].

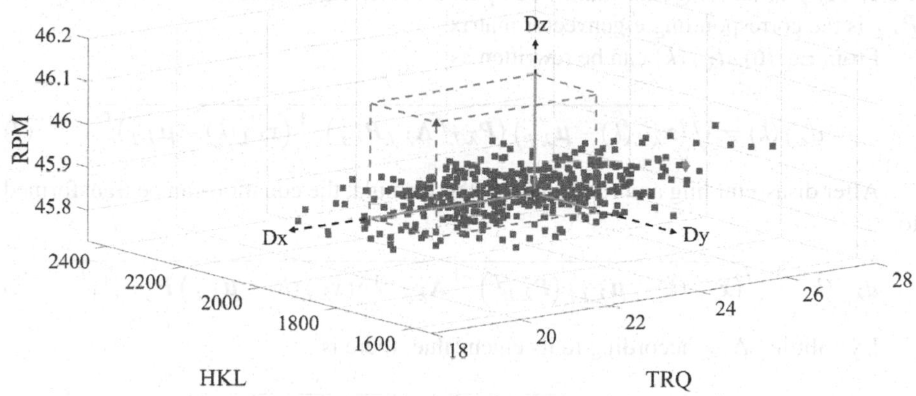

Fig. 3. Scatterplot of TRQ, HKL and RPM (Color figure online)

For the interval augmented data $\boldsymbol{x}_{\lambda,L}(k)$, the MD statistic $d_{\lambda,L}(k)$ is calculated as

$$d_{\lambda,L}(k) = \sqrt{\left(\boldsymbol{x}_{\lambda,L}(k) - \boldsymbol{\mu}_{\lambda,L}\right) \boldsymbol{\Sigma}_{\lambda,L}{}^{-1} \left(\boldsymbol{x}_{\lambda,L}(k) - \boldsymbol{\mu}_{\lambda,L}\right)^T} \tag{3}$$

where $\boldsymbol{\Sigma}_{\lambda,L}^{-1}$ represents the inverse matrix of covariance matrix $\boldsymbol{\Sigma}_{\lambda,L}$, $\boldsymbol{\mu}_{\lambda,L} = \left[\mu_1 \cdots \mu_j \cdots \mu_{ML}\right]$ indicates the mean vector of training data set $\boldsymbol{X}_{\lambda,L}$, and μ_j is the mean value of jth dimension in interval augmented training data set $\boldsymbol{X}_{\lambda,L}$. For example, given the matrix $\boldsymbol{X}_{\lambda,L}$, where each column of vectors represents a dimension, define the covariance $\mathrm{Cov}\left(\boldsymbol{a}_i, \boldsymbol{a}_j\right)$ between any two dimensions $\boldsymbol{a}_i$ and $\boldsymbol{a}_j$ as

$$\mathrm{Cov}(\boldsymbol{a}_i, \boldsymbol{a}_j) = \frac{1}{N+\lambda(1-L)-1} \sum_{k=1}^{N+\lambda(1-L)} \left(\boldsymbol{a}_i(k) - \mu_{\boldsymbol{a}_i}\right)\left(\boldsymbol{a}_j(k) - \mu_{\boldsymbol{a}_j}\right) \tag{4}$$

where $\boldsymbol{a}_i(k)$ and $\boldsymbol{a}_j(k)$ are the kth values in dimension $\boldsymbol{a}_i$ and $\boldsymbol{a}_j$ respectively [17], $\mu_{\boldsymbol{a}_i}$ and $\mu_{\boldsymbol{a}_j}$ represent the mean of dimensions $\boldsymbol{a}_i$ and $\boldsymbol{a}_j$ respectively, $N+\lambda(1-L)$ is the number of rows in $\boldsymbol{X}_{\lambda,L}$, then the covariance matrix $\boldsymbol{\Sigma}_{\lambda,L}$ is

$$\boldsymbol{\Sigma}_{\lambda,L} = \begin{bmatrix} \mathrm{Cov}(\boldsymbol{a}_1,\boldsymbol{a}_1) & \cdots & \mathrm{Cov}(\boldsymbol{a}_1,\boldsymbol{a}_i) & \cdots & \mathrm{Cov}(\boldsymbol{a}_1,\boldsymbol{a}_{ML}) \\ \vdots & \ddots & \vdots & \ddots & \vdots \\ \mathrm{Cov}(\boldsymbol{a}_i,\boldsymbol{a}_1) & \cdots & \mathrm{Cov}(\boldsymbol{a}_i,\boldsymbol{a}_i) & \cdots & \mathrm{Cov}(\boldsymbol{a}_i,\boldsymbol{a}_{ML}) \\ \vdots & \ddots & \vdots & \ddots & \vdots \\ \mathrm{Cov}(\boldsymbol{a}_{ML},\boldsymbol{a}_1) & \cdots & \mathrm{Cov}(\boldsymbol{a}_{ML},\boldsymbol{a}_i) & \cdots & \mathrm{Cov}(\boldsymbol{a}_{ML},\boldsymbol{a}_{ML}) \end{bmatrix} \tag{5}$$

where ML is the number of columns in $\boldsymbol{X}_{\lambda,L}$, the eigen decomposition of $\boldsymbol{\Sigma}_{\lambda,L}$ is given by

$$\boldsymbol{\Sigma}_{\lambda,L} = \boldsymbol{P}_{\lambda,L}{}^T \boldsymbol{\Lambda}_{\lambda,L} \boldsymbol{P}_{\lambda,L} \tag{6}$$

where $\boldsymbol{\Lambda}_{\lambda,L}$ is the diagonal matrix composed of the eigenvalues of matrix $\boldsymbol{\Sigma}_{\lambda,L}$, and $\boldsymbol{P}_{\lambda,L}$ is the corresponding eigenvector matrix.

From Eq. (6), $d_{\lambda,L}(k)$ can be rewritten as

$$d_{\lambda,L}(k) = \sqrt{\left(\boldsymbol{x}_{\lambda,L}(k) - \boldsymbol{\mu}_{\lambda,L}\right)\left(\boldsymbol{P}_{\lambda,L}{}^T \boldsymbol{\Lambda}_{\lambda,L} \boldsymbol{P}_{\lambda,L}\right)^{-1}\left(\boldsymbol{x}_{\lambda,L}(k) - \boldsymbol{\mu}_{\lambda,L}\right)^T} \tag{7}$$

After disassembling and reassembling the formula, the equation can be transformed to

$$d_{\lambda,L}(k) = \sqrt{\left(\boldsymbol{x}_{\lambda,L}(k) - \boldsymbol{\mu}_{\lambda,L}\right)\left(\boldsymbol{P}_{\lambda,L}{}^T\right)^{-1} \boldsymbol{\Lambda}_{\lambda,L}{}^{-1}\left[\left(\boldsymbol{x}_{\lambda,L}(k) - \boldsymbol{\mu}_{\lambda,L}\right)\boldsymbol{P}_{\lambda,L}{}^{-1}\right]^T} \tag{8}$$

By splitting $\boldsymbol{\Lambda}_{\lambda,L}$ according to its eigenvalue, there is

$$d_{\lambda,L}(k) = \sqrt{\sum_{i=1}^{ML} \frac{\left(\left(\boldsymbol{x}_{\lambda,L}(k) - \boldsymbol{\mu}_{\lambda,L}\right)\boldsymbol{P}^i_{\lambda,L}{}^T\right)^2}{r_i}} \tag{9}$$

where ML is the number of columns in $\boldsymbol{X}_{\lambda,L}$, r_i represents the ith eigenvalue, and $\boldsymbol{P}^i_{\lambda,L}$ represents the corresponding eigenvector of r_i.

When an anomaly occurs, changes occur in the direction of ML eigenvectors, effectively capturing the weak changes in the direction of the eigenvectors corresponding to smaller eigenvalues, and then affecting the change in the MD statistic after addition and square root.

3.3 Alarm Limit Design

The design of alarm limit directly affects the performance of anomaly monitoring. Considering that MD is the square sum of multidimensional data, the use of the Chi-square distribution curve to fit the data distribution has been proposed in [18, 19]. However, the data after interval augmented does not always conform to the Chi-square distribution, and thus fitting the sum of squares of each component with a Chi-square distribution does not work well. In order to fit the data distribution of the MD statistic after the interval augmented as far as possible, the Gaussian KDE method is used here to fit the actual data distribution under different variables. The KDE function $f^h_{\lambda,L}(x)$ is

$$f_{\lambda,L}^{h}(x)=\frac{1}{[N+\lambda(1-L)]h}\sum_{k=1-\lambda(1-L)}^{N}K\left(\frac{\breve{d_{\lambda,L}}(k)-x}{h}\right) \tag{10}$$

where x is the independent variable, representing the value of the continuous interval augmented MD, h is the bandwidth selected by KDE, k is the ordinal number of training data samples, $\breve{d_{\lambda,L}}(k)$ is the MD of the kth normal sample data after the interval is augmented. $K()$ is the Gaussian kernel function adopted, and the calculation formula is

$$K\left(\frac{\breve{d_{\lambda,L}}(k)-x}{h}\right)=\frac{1}{\sqrt{2\pi}}\exp\left[-\frac{1}{2}\left(\frac{\breve{d_{\lambda,L}}(k)-x}{h}\right)^{2}\right] \tag{11}$$

where exp is the exponential function.

After estimating the MD distribution of training samples under different λ and L, a certain confidence level δ is given, the alarm limit $A_{\lambda,L}$ is solved by the probability density integral formula, as

$$\delta=\int_{-\infty}^{A_{\lambda,L}}f_{\lambda,L}^{h}(x)d(x) \tag{12}$$

Finally, in the process of online phase, the values of $d_{\lambda,L}(k)$ and the alarm limit $A_{\lambda,L}$ are compared to determine whether an anomaly has occurred.

3.4 Procedures and Evaluation

The proposed bit bounce detection scheme is shown in Fig. 4. It is mainly divided into offline phase and online phase. The main steps of the offline phase are summarised below:

1) Step 1 - Training Set Normalization: The Z-Score method is used to normalize the training sample set $\tilde{X}$ to $\hat{X}$.

2) Step 2 - Data Interval Augmentation and MD Statistic Calculation: $\boldsymbol{X}_{\lambda,L}$ is obtained by augmenting the interval of $\hat{X}$ based on (2). $\boldsymbol{\mu}_{\lambda,L}$ is then got and $\boldsymbol{\Sigma}_{\lambda,L}$ is calculated using (6). Finally, $d_{\lambda,L}$ is computed using (9).

3) Step 3 - Design of Alarm Limit: Setting a certain confidence interval δ, the alarm limit $A_{\lambda,L}$ under different parameters λ and L is solved inversely using (12).

In the offline phase, three parameters $\boldsymbol{\mu}_{\lambda,L}$, $\boldsymbol{\Sigma}_{\lambda,L}$, and $A_{\lambda,L}$ can be obtained and used in the online phase, the main steps in the online phase are summarised below:

1) Step 1 - Online Data Normalization: The online data $\tilde{\mathbf{y}}(k)$ is normalized using the parameters of the normalization process in the offline part 1) to obtain $\hat{\mathbf{y}}(k)$.

2) Step 2 - Data Interval Augmentation and MD Statistic Calculation: The calculation method is the same as 2) in the offline phase, but the calculation of $d_{\lambda,L}(k)$ utilizes the parameters $\boldsymbol{\Sigma}_{\lambda,L}$ and $\boldsymbol{\mu}_{\lambda,L}$ from the offline phase.

3) Step 3 - Online Abnormality Detection: Comparing the magnitude of $d_{\lambda,L}(k)$ and $A_{\lambda,L}$ to determine if bit bounce has occurred:

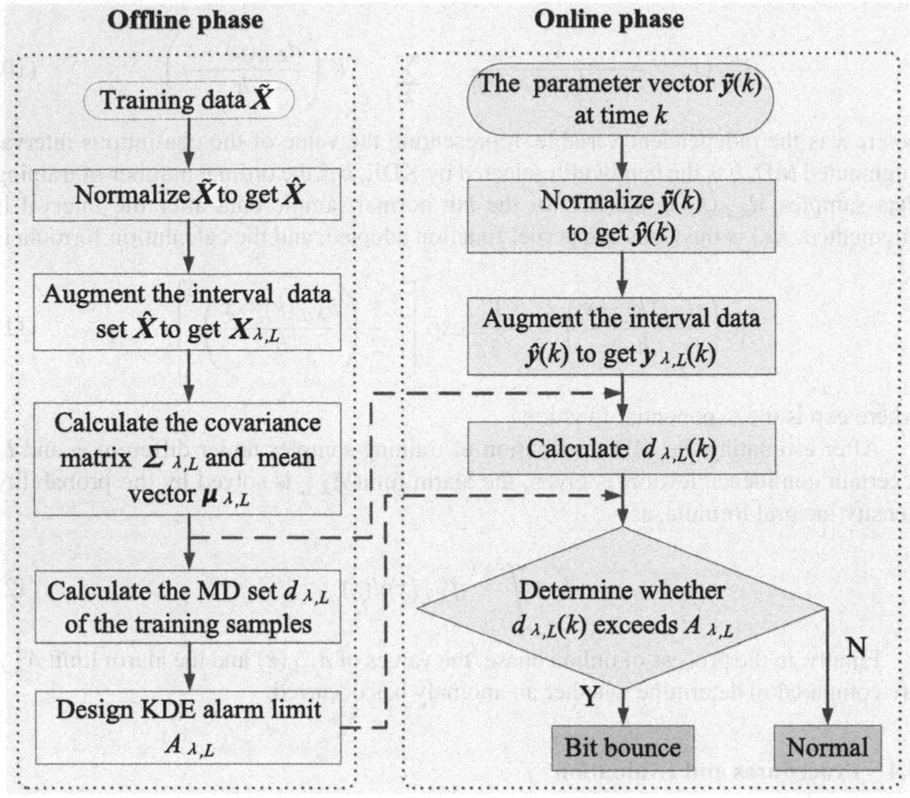

Fig. 4. Overall system flow chart

$$\begin{cases} d_{\lambda,L}(k) \geq A_{\lambda,L} & \text{Bit bounce} \\ d_{\lambda,L}(k) < A_{\lambda,L} & \text{Normal} \end{cases} \tag{13}$$

In this paper, four indexes are used to evaluate the detection effect of the bit bounce, which are as follows:

$$\widetilde{\text{Acc}} = \frac{\text{TP} + \text{TN}}{\text{TP} + \text{TN} + \text{FP} + \text{FN}} \tag{14}$$

$$\widetilde{\text{MAR}} = \frac{\text{FN}}{\text{TP} + \text{FN}} \tag{15}$$

$$\widetilde{\text{FAR}} = \frac{\text{FP}}{\text{FP} + \text{TN}} \tag{16}$$

where $\widetilde{\text{Acc}}$ is the Accuracy, $\widetilde{\text{MAR}}$ is the Missing Alarm Rate, $\widetilde{\text{FAR}}$ is the False Alarm Rate, TP, TN, FP, and FN are the number of True Positives, True Negatives, False Positives, and False Negatives respectively. In addition, the specified indicator will begin to generate an alarm at 1500 s and continue for at least 10 s at the specified time, which is the Detection Delay ($\widetilde{\text{DD}}$).

In order to comprehensively balance the above indexes and select the best λ and L, the TOPSIS method is adopted to unify the above four verification indexes into one index [20], and the calculation method is as follows. Set the counter matrix to

$$\widetilde{\text{Ind}} = \left[\widetilde{\text{Acc}}\ \widetilde{\text{MAR}}\ \widetilde{\text{FAR}}\ \widetilde{\text{DD}}\right] \in R^{P\times 4} \tag{17}$$

where P is the given tunable space, i.e. the product of λ and L. Smaller values of MAR, FAR and DD and a larger Acc imply better detection performance. In order to facilitate the uniform calculation of the following steps, all indexes are converted into indexes of the larger, better type, the conversion process is

$$\widehat{\text{MAR}} = \left[\max\left(\widetilde{\text{MAR}}_i\right) - \widetilde{\text{MAR}}_i\right] \tag{18}$$

$$\widehat{\text{FAR}} = \left[\max\left(\widetilde{\text{FAR}}_i\right) - \widetilde{\text{FAR}}_i\right] \tag{19}$$

$$\widehat{\text{DD}} = \left[\max\left(\widetilde{\text{DD}}_i\right) - \widetilde{\text{DD}}_i\right], i \in 1 \sim P \tag{20}$$

where $\widetilde{\text{MAR}}_i$, $\widetilde{\text{FAR}}_i$, and $\widetilde{\text{DD}}_i$ are the ith values of the vectors $\widetilde{\text{MAR}}$, $\widetilde{\text{FAR}}$, and $\widetilde{\text{DD}}$ respectively, max() represents the maximum value of the vector. The forward matrix is

$$\widehat{\text{Ind}} = \left[\widetilde{\text{Acc}}\ \widehat{\text{MAR}}\ \widehat{\text{FAR}}\ \widehat{\text{DD}}\right] \tag{21}$$

If the original value of the index is used directly in the analysis, the role of the index with a larger numerical scale will be strengthened in the overall evaluation and the role of the index with a smaller numerical scale will be weakened. To ensure the reliability of the results, it is necessary to standardize the index so that different metrics have the same scale. The standardized matrix Ind is

$$\text{Ind} = \left[\text{Acc MAR FAR DD}\right] \tag{22}$$

in this

$$\text{Acc} = \left[\frac{\widetilde{\text{Acc}}_i}{\sqrt{\sum_{i=1}^{P} \widetilde{\text{Acc}}_i^{\,2}}}\right] \tag{23}$$

$$\text{MAR} = \left[\frac{\widehat{\text{MAR}}_i}{\sqrt{\sum_{i=1}^{P} \widehat{\text{MAR}}_i^{\,2}}}\right] \tag{24}$$

$$\text{FAR} = \left[\frac{\widehat{\text{FAR}}_i}{\sqrt{\sum_{i=1}^{P} \widehat{\text{FAR}}_i^{\,2}}}\right] \tag{25}$$

$$\text{DD} = \left[\frac{\widehat{\text{DD}}_i}{\sqrt{\sum_{i=1}^{P} \widehat{\text{DD}}_i^{\,2}}}\right], i \in 1 \sim P \tag{26}$$

After processing the above four groups of index vectors, the matrix Ind is used to calculate the Comprehensive Index (CI).

$$\mathrm{CI}(i) = \frac{D_i^-}{D_i^+ + D_i^-} \tag{27}$$

where D_i^+ and D_i^- represent the distance between CI(i) and the locally optimal and worst solutions, respectively. CI(i) is the weighting distance of the four indexes under two adjustment parameters λ and L, and D_i^+ and D_i^- are respectively

$$D_i^+ = \sqrt{\sum_{j=1}^{4} \xi_j[\mathrm{Ind}(i,j) - \max(\mathrm{Ind}(,j))]^2} \tag{28}$$

$$D_i^- = \sqrt{\sum_{j=1}^{4} \xi_j[\mathrm{Ind}(i,j) - \min(\mathrm{Ind}(,j))]^2} \tag{29}$$

Finally, the maximum value in CI is selected as the optimal comprehensive index.

4 Industrial Case Study

This section presents a practical industrial case of collecting real geological drilling data from a real geothermal well in China, as shown in Fig. 2, to demonstrate the effectiveness of the proposed method.

As shown in Fig. 5, the KDE method is used to fit MD with the fixed parameters. The histogram shows the frequency of the IA-MD distribution, and the red dotted line represents the Gaussian KDE fitting curve. It can be seen that the KDE method has a good fitting effect. In this paper, in order to prioritise the parameters for timely anomaly detection, the weight of the DD index is set at 0.5 and the remaining weights are divided equally between the other three indexes. In this case, the λ is set to be $1 \sim 6$, and the L is set to be $2 \sim 31$, so the adjustment interval of the parameter P is set to be $1 \sim 180$. The local optimal solution of the evaluation index can be found within this interval.

The distribution of the comprehensive index under the local interval of λ and L is shown in Fig. 6. Among them, the black ellipse is the selected optimal comprehensive index. In this case, $\lambda = 1$ and $L = 23$. In order to prove the validity of the proposed bit bounce detection method, different methods are used for comparison, including PCA, KPCA [11], KLD [12], and standard MD statistics [19]. In the PCA and KPCA methods, the T^2 statistic is used to monitor the change of the uniform metrology in the principal component molecular space, and the Squared Prediction Error (SPE) statistic is used to monitor the change of the residual subspace statistics. A confidence level of $\delta = 99\%$ is set for all the above methods.

The alarm time series for different methods are shown in Fig. 7, where the series from 1 s to 1500 s is the normal drilling process, the series from 1501 s to 3500 s indicates that abnormal drilling occurs, and the red dotted lines represent the set alarm limit. The comparison of the indicators for each method is shown in the Table 1. It can

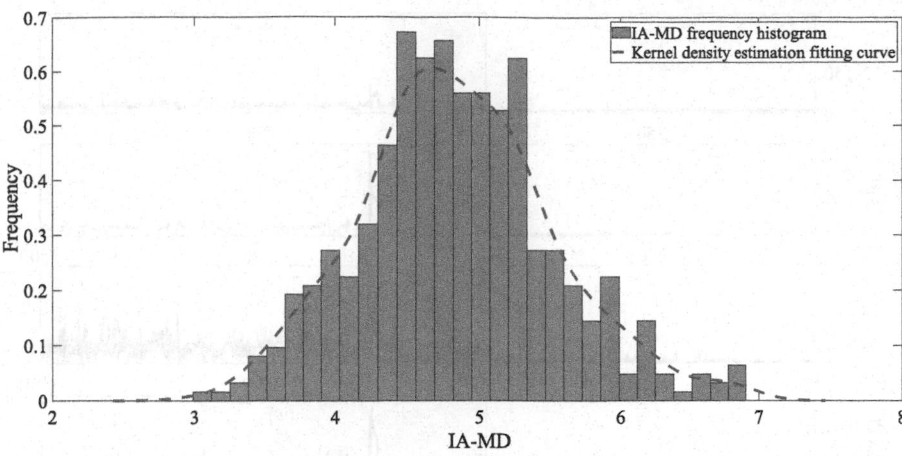

Fig. 5. Density estimation fitting effect comparison chart (Color figure online)

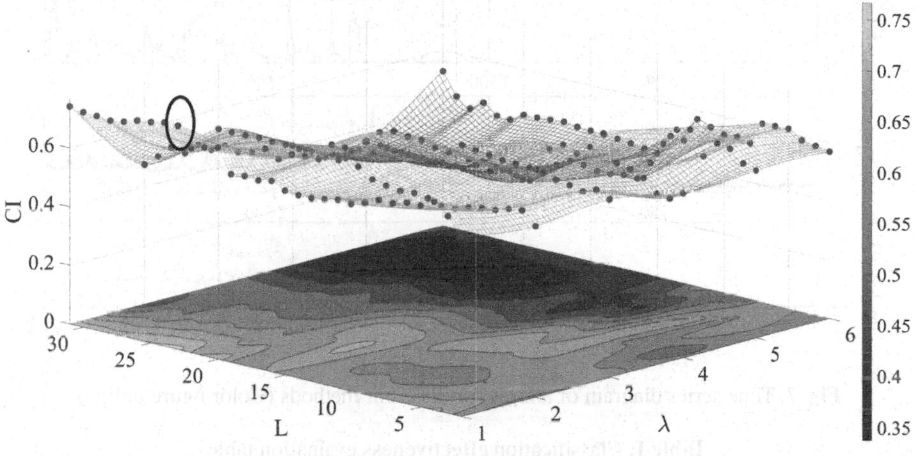

Fig. 6. Distribution of indicator CI among various λ and L

be clearly seen that the accuracy of the proposed method reached 95.9%, the missing alarm rate decreased to 0.0%, and the detection delay decreased to 0 s. Therefore, it can be seen that the proposed method based on IA-MD is effective in detecting the bit bounce.

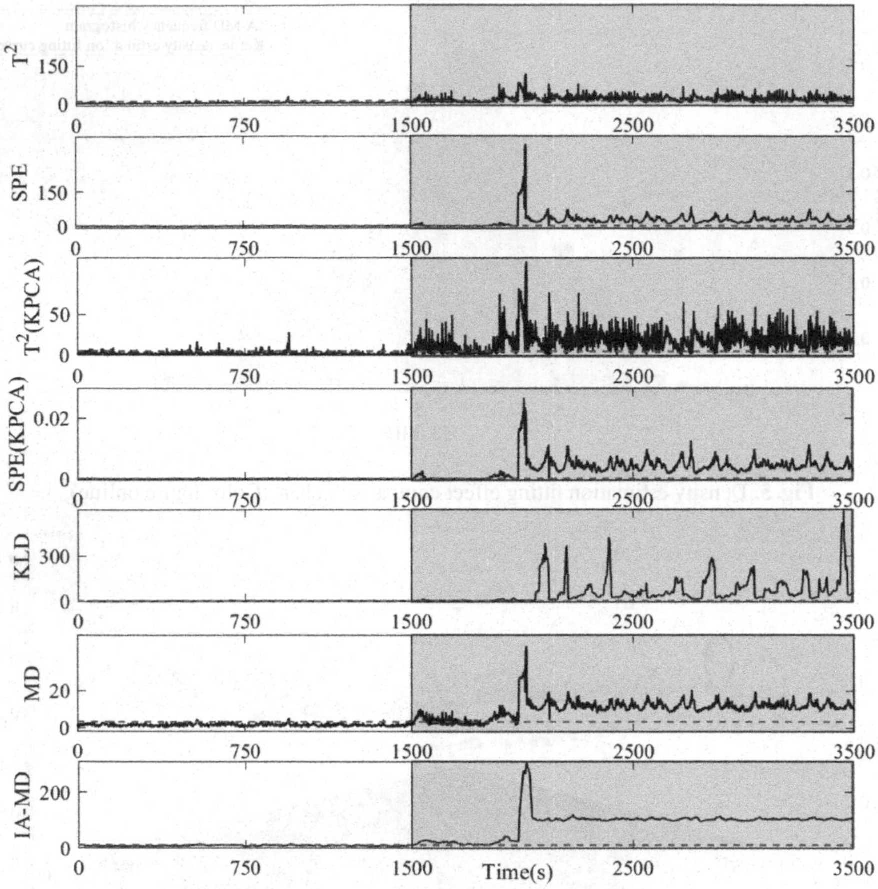

Fig. 7. Time series diagram of alarms for different methods (Color figure online)

Table 1. Classification effectiveness evaluation table

Method	Acc	FAR	MAR	DD(s)
T^2 of PCA	84.6%	2.0%	25.5%	216
SPE of PCA	87.0%	0.0%	22.7%	11
T^2 of KPCA	87.6%	9.5%	14.5%	25
SPE of KPCA	88.2%	11.4%	12.1%	9
KLD	94.7%	4.5%	5.9%	26
MD	93.5%	9.5%	2.5%	13
IA-MD	**95.9%**	**9.5%**	**0.0%**	**0**

5 Conclusion

In this paper, a method based on IA-MD was proposed to detect bit bounce. First, the interval of the selected data was augmented to improve the anomaly detection ability. Second, the MD statistic of the normal data was calculated after the interval was augmented, and the KDE method was used to design its alarm limit. The MD statistic of the online data was then calculated after the interval and compared to the threshold. Last, the optimal index was selected by parameter adjustment. The analysis results show that this method has a good detection performance for bit bounce. There are opportunities to extend this research in the future. For example, the proposed method still has a higher false alarm rate compared to other methods, so how to configure the alarm optimisation strategy is the next problem to be solved. In addition, as a promising future direction, a transfer learning method can be introduced to transfer a trained model to a different drilling process, so as to improve model versatility and ensure drilling safety.

Acknowledgements. This work was supported by the Knowledge Innovation Program of Wuhan-Shuguang Project under Grant No. 2022010801020208, the Natural Science Foundation of Hubei Province, China, under Grant 2020CFA031.

References

1. Liu, X., Long, X., Zheng, X., Meng, G., Balachandran, B.: Spatial-temporal dynamics of a drill string with complex time-delay effects: bit bounce and stick-slip oscillations. Int. J. Mech. Sci. **170**, 105338 (2020)
2. Li, Y., Cao, W., Hu, W., Wu, M.: Detection of downhole incidents for complex geological drilling processes using amplitude change detection and dynamic time warping. J. Process Control **102**, 44–53 (2021)
3. Ding, S., et al.: Axial-torsional nonlinear vibrations of bottom hole assembly in the air drilling technology. Petroleum (2023)
4. Vromen, T., Dai, C.H., van de Wouw, N., Oomen, T., Astrid, P., Nijmeijer, H.: Robust output-feedback control to eliminate stick-slip oscillations in drill-string systems. IFAC-PapersOnLine **48**(6), 266–271 (2015)
5. Besselink, B., Vromen, T., Kremers, N., Van De Wouw, N.: Analysis and control of stick-slip oscillations in drilling systems. IEEE Trans. Control Syst. Technol. **24**(5), 1582–1593 (2016)
6. Ibrahim Basturk, H.: Observer-based boundary control design for the suppression of stick-slip oscillations in drilling systems with only surface measurements. J. Dyn. Syst. Measur. Control **139**(10), 104501 (2017)
7. Liu, J., Wang, J., Guo, X., Dai, L., Zhang, C., Zhu, H.: Investigation on axial-lateral-torsion nonlinear coupling vibration model and stick-slip characteristics of drilling string in ultra-HPHT curved wells. Appl. Math. Model. **107**, 182–206 (2022)
8. De Moraes, L.P., Savi, M.A.: Drill-string vibration analysis considering an axial-torsional-lateral nonsmooth model. J. Sound Vib. **438**, 220–237 (2019)
9. Fang, P., Ding, S., Yang, K., Li, G., Xiao, D.: Dynamics characteristics of axialÍctorsionalÍ-clateral drill string system under wellbore constraints. Int. J. Non-Linear Mech. **146**, 104176 (2022)
10. Zhang, Z., Lai, X., Wu, M., Chen, L., Lu, C., Du, S.: Fault diagnosis based on feature clustering of time series data for loss and kick of drilling process. J. Process Control **102**, 24–33 (2021)

11. Peng, C., et al.: An intelligent model for early kick detection based on cost-sensitive learning. Process Saf. Environ. Prot. **169**, 398–417 (2023)
12. Li, Y., Cao, W., Hu, W., Xiong, Y., Wu, M.: Incipient fault detection for geological drilling processes using multivariate generalized gaussian distributions and kullbackÍcleibler divergence. Control. Eng. Pract. **117**, 104937 (2021)
13. Li, Y., Cao, W., Hu, W., Wu, M.: Diagnosis of downhole incidents for geological drilling processes using multi-time scale feature extraction and probabilistic neural networks. Process Saf. Environ. Prot. **137**, 106–115 (2020)
14. Si, Y., Kong, L., Chin, J.H., Guo, W., Wang, Q.: Whirling detection in deep hole drilling process based on multivariate synchrosqueezing transform of orthogonal dual-channel vibration signals. Mech. Syst. Signal Process. **167**, 108621 (2022)
15. Rafezi, H., Hassani, F.: Drill bit wear monitoring and failure prediction for mining automation. Int. J. Min. Sci. Technol. **33**, 289–296 (2023)
16. Shang, J., Chen, M., Zhang, H.: Fault detection based on augmented kernel mahalanobis distance for nonlinear dynamic processes. Comput. Chem. Eng. **109**, 311–321 (2018)
17. Li, W., Zhu, J.: CLT for spiked eigenvalues of a sample covariance matrix from high-dimensional Gaussian mean mixtures. J. Multivar. Anal. **193**, 105–127 (2023)
18. Ji, H., Huang, K., Zhou, D.: Incipient sensor fault isolation based on augmented mahalanobis distance. Control. Eng. Pract. **86**, 144–154 (2019)
19. Ji, H.: Statistics Mahalanobis distance for incipient sensor fault detection and diagnosis. Chem. Eng. Sci. **230**, 116223 (2021)
20. Chakraborty, S.: Topsis and modified topsis: a comparative analysis. Dec. Anal. J. **2**, 100021 (2022)

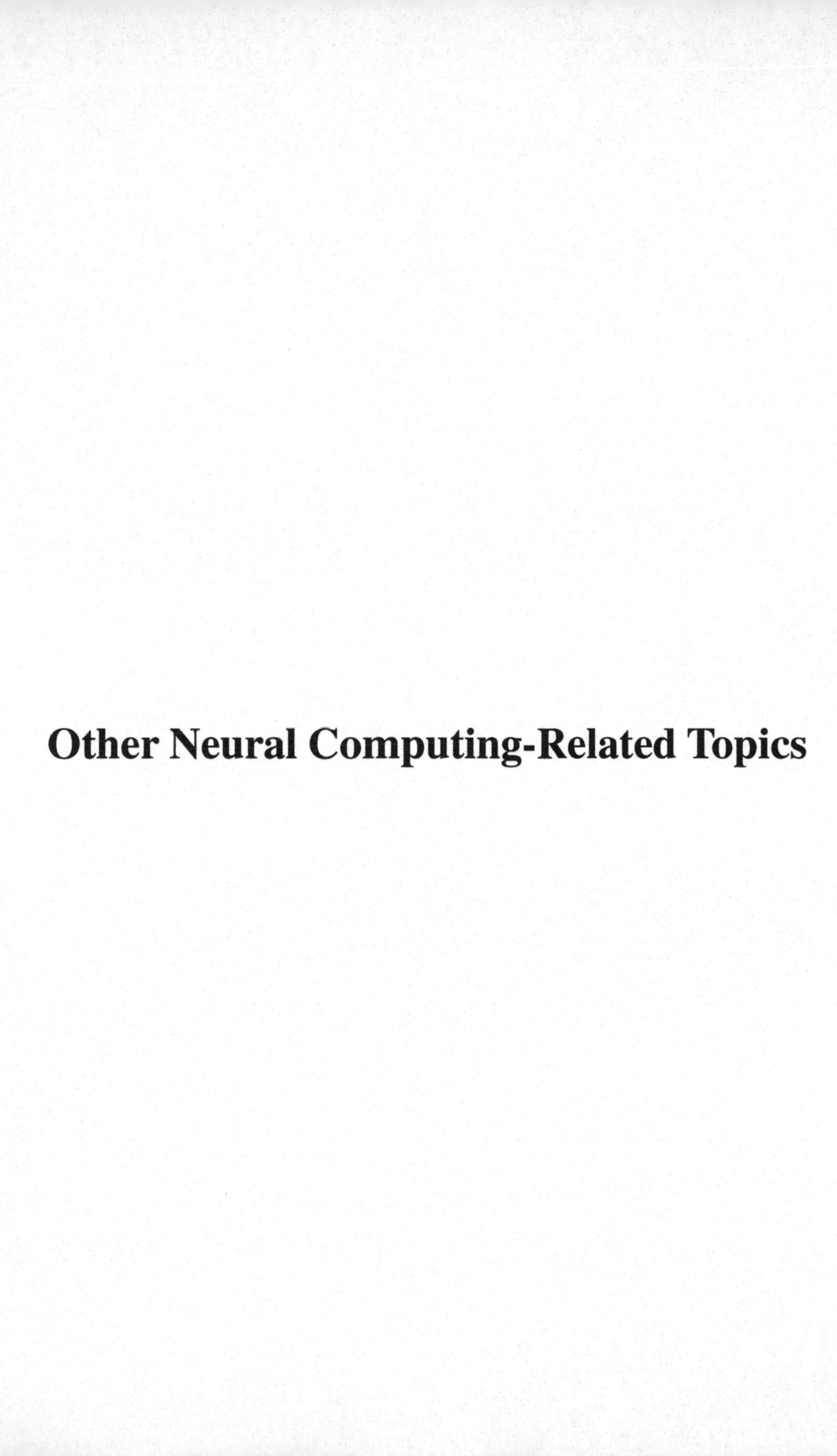

Other Neural Computing-Related Topics

A Neural Approach Towards Real-time Management for Integrated Energy System Incorporating Carbon Trading and Electrical Vehicle Scheduling

Yiying Li[1,2], Bo Wang[1,2](✉), Lijun Zhang[1,3], Lei Liu[1,2], and Huijin Fan[1,2]

[1] School of Artificial Intelligence and Automation, Huazhong University of Science and Technology, Wuhan 430074, China
{m202173329,wb8517,lijunzhang,liulei,ehjfan}@hust.edu.cn

[2] National Key Laboratory of Science and Technology on Multispectral Information Processing, Huazhong University of Science and Technology, Wuhan 430074, China

[3] Belt and Road Joint Laboratory on Measurement and Control Technology, Huazhong University of Science and Technology, Wuhan 430074, China

Abstract. This paper proposes a real-time integrated energy system (IES) management approach which aims at promoting overall energy efficiency, increasing renewable energy penetration, and smoothing load fluctuation. The electric vehicles (EVs) charging scheduling is incorporated into the IES management, where the uncertain arrivals and departures of multiple EVs are considered as a stochastic but flexible load to the IES. Furthermore, towards the carbon neutralization target, a carbon emissions trading mechanism is introduced into the IES management to incentivize the system to operate in an eco-friendlier manner. To tackle the computational complexity induced by the stochastic and intermittent nature of the renewable energy sources and EVs load, the scheduling of the IES is realized in a neural network based real-time manner, driven by a deep reinforcement learning approach that guarantees safe training and operation. The case study verifies the effectiveness of the proposed approach.

Keywords: electric vehicles · reward and punishment ladder carbon trading · flexible load · deep reinforcement learning · integrated energy system

1 Introduction

During the 75th session of the United Nations General Assembly, China put forward the national strategic goal of "carbon peak by 2030 and carbon neutrality by 2060". There are mainly two types of low-carbon approaches in the energy industry. First, build a comprehensive energy system to improve the overall efficiency of energy use. Second, introduce carbon trading mechanism to promote the transfer of carbon emission cost from enterprises with low emission reduction

H. Zhang et al. (Eds.): NCAA 2023, CCIS 1870, pp. 561–575, 2023.
https://doi.org/10.1007/978-981-99-5847-4_40

cost to enterprises with high emission reduction cost [1]. In addition, making full use of the flexible load in the system and cooperating with the energy storage system can smooth the fluctuation of renewable energy and realize peak cutting and valley filling, which is conducive to the safe and stable operation of the system. Ma et al. [2] used the modern interior point method to solve the integrated energy system(IES) model considering carbon trading mechanism, but only used the carbon trading price model with a single carbon price. Cui et al. [3] introduces the ladder carbon trading into the IES, comprehensively considers the low carbon and economy of the system, and uses CPLEX to solve the problem. Qiu et al. [4] considers both ladder carbon trading and demand response, and uses CPLEX to solve it, but only considers the demand response based on price.

To control air pollution and reduce the carbon emission of the transportation sector [5], the electric vehicle (EV) industry has developed rapidly. However, the geographical distribution of EVs coincides with the hot spots of load in the Chinese power grid [6]. Moreover, due to the large charging power of EVs, disorderly charging in local areas will cause large fluctuations of electric loads, thus increasing the system operating costs. However, orderly charging of EVs can not only avoid the above problems, but also absorb renewable energy and promote the further expansion of the scale of renewable energy [7]. Therefore, Wu and Li [8] proposed a two-layer stochastic optimization scheduling model for IES considering demand-side response and EVs synergy, which is solved using CPLEX. In [9], the charging demand and vehicle-to-grid control of EVs are considered, and a cooperative scheduling model of IES of the power grid and natural gas network is proposed, which is solved by the CPLEX solver.

Most of the current works of literature that consider carbon trading in integrated energy systems use traditional methods to solve the problem. However, due to a large number of uncertainties in the power supply and load, the traditional method will greatly increase the scene, so it takes a long time to solve the problem. However, now the power market tends to be real-time, so it needs to be solved quickly. Reinforcement learning has the advantages of rapid response and adaptability to an uncertain environment, which has been gradually popularized and applied in IES optimization scheduling. In [10], the energy supply and demand problem in the hierarchical electricity market is modeled as a discrete-finite Markov decision process, which is solved by the Q-learning greedy algorithm. In [11], because of the intermittence of renewable energy and the uncertainty of users' energy demand in IES, DDPG is used to solve the model, which does not need to predict or model the uncertainty, and can respond to the random fluctuation of source and charge dynamically.

To sum up, this paper adds EVs scheduling and carbon trading mechanism to the traditional electric-gas IES considering demand response. The EV scheduling is added to the IES, which fully considers the uncertainty of the time of EVs arrival and departure from charging pile and battery capacity when it is connected to the power grid, and not only considers the overall capacity and power of electric passenger vehicle, but also specifically considers the power distribution problem of each electric passenger vehicle. Carbon trading transforms environmental issues into economic issues and makes people voluntarily participate in emission reduction actions for their economic interests. An appropriate carbon trading mechanism, such as the carbon trading with a ladder carbon price of rewards and punishments applied, will make the more users reduce emissions, the greater the growth rate of economic benefits, and the more emissions, the greater the growth rate of economic costs, which will further increase people's enthusiasm to participate in emission reduction actions. In addition, the deep reinforcement learning method is used to solve the problem, which can not only deal with the problem that the source load is difficult to accurately predict due to a large number of uncertainties, but also does not need to decouple the coupling relationship in the model. The trained agent can quickly solve the problem only by relying on the data of the system, and realize real-time response to the load.

2 Problem Statement

2.1 System Structure

The structure of the IES is shown in Fig. 1, which is divided into four parts: source, network, load, and storage. The sources include photovoltaic generators (PV) and wind turbines (WT), which provide electricity for the system. The storage is electric energy storage (EES), which can be charged and discharged for peak regulation. The network includes the power grid and the gas network. The system can buy and sell electricity to the power grid, and the natural gas in the Gas network can be converted into electric energy supply system through Gas to Electric (G2E) Equipment. The load includes the conventional electric load and the EV load. The conventional electric load includes transferable load, interruptible load, and fixed load. The incentive-based demand response (IDR) is added to the decision-making level, and the price-based demand response (PDR) is considered according to the ladder electricity price. The EV load includes three types of EVs as shown in Fig. 1.

The EV Charging Model. The common forms of EV charging include battery replacement, fast direct current(DC) charging, and conventional alternating current(AC) charging. Electric heavy trucks and electric engineering vehicles and electric commuter buses have large battery capacities and are usually charged by fast direct current charging. Electric passenger vehicles are usually charged by conventional alternating current charging. However, to meet the requirements of fast travel under special circumstances and to better match the reinforcement learning method used, we believes that electric passenger vehicles can be charged by conventional alternating current charging and fast direct current charging. Under normal circumstances, electric passenger vehicles use conventional AC charging, but one hour before departure, it is determined whether the use of conventional AC charging can meet the travel requirements. If it can be satisfied, continue to use this mode, otherwise use fast DC charging. However,

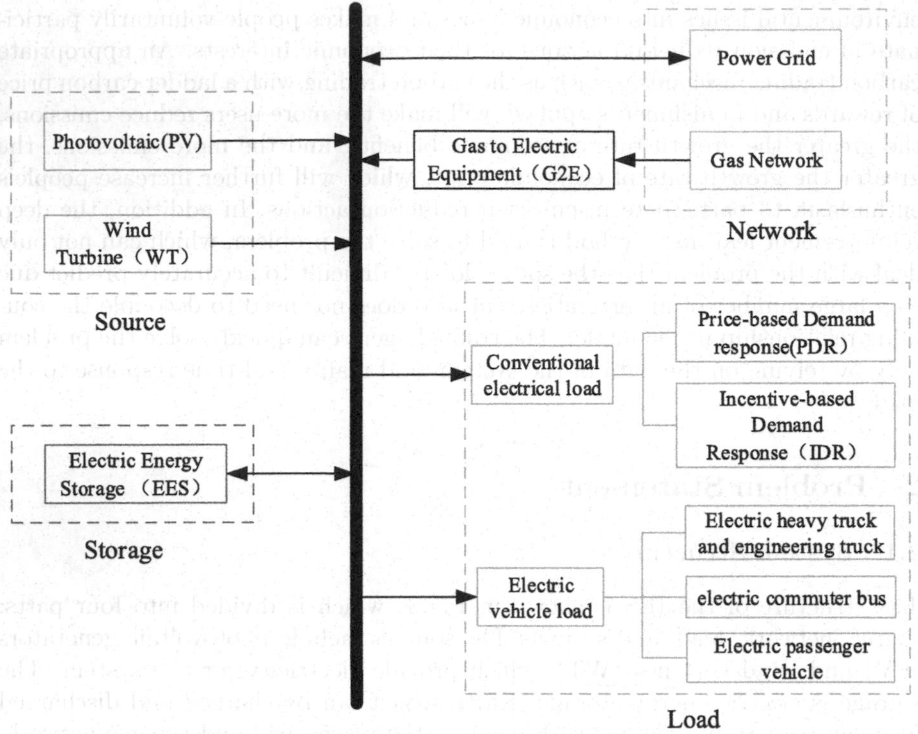

Fig. 1. The structure of the IES.

due to the large charging power of fast direct current charging, the battery of EVs will be damaged, so the owner should be compensated to some extent, and the compensation cost is shown in (1).

$$C_{EV}(t) = c_{EV}P_{EV}(t)\Delta t \tag{1}$$

where Δt is the time interval between each step, which is $1h$. $P_{EV}(t)$ is EV charging power at time t. And c_{EV} is the compensation cost coefficient of electric passenger vehicles during fast DC charging.

The relationship between the state of charge (SOC) of the EV during charging and the capacity constraint when leaving the charging pile are shown in (2) and (3) respectively.

$$SOC_{EV}(t) = SOC_{EV}(t-1) + \frac{P_{EV}(t)\eta_{EV} \cdot \Delta t}{E_{EV\max}} \tag{2}$$

$$SOC_{EV}\left(t_{EV,dept}\right) \geq SOC_{EVrequire} \tag{3}$$

Among them, $SOC_{EV}(t-1)$, $SOC_{EV}(t)$, $SOC_{EV}(t_{EV,dept})$ and $SOC_{EVrequire}$ are the SOC of EVs at the time of $t-1$, t and driving away from charging piles, and the minimum SOC required by EVs when leaving the charging pile. η_{EV} is the charging efficiency of EVs. $E_{EV\max}$ is the maximum capacity of EVs.

At the decision-making level, only the total power allocated to each class of EVs is considered. For each specific electric passenger vehicle charging power distribution problem, we adopts the method of regular charging. The randomness of electric passenger vehicles is strong, so there may be electric passenger vehicles arriving or departing at every moment, and their total number, maximum and actual capacity and maximum charging power are changing in real-time. Based on the above, the charging rules for electric passenger vehicles are formulated as follows:

1) Initialize an empty queue, and sort the electric passenger vehicles in the queue in ascending order according to their expected departure time. The earlier the vehicle is in the queue, the higher its charging priority. That is, when the power and capacity requirements are met, always choose the electric passenger vehicle that will leave first for charging.
2) At time t, for the total power provided by the system, allocate it to the electric passenger vehicles in the queue in order according to the queue's sorting, subject to the power and capacity requirements being met.
3) Add the electric passenger vehicles arriving at the charging station at time t to the queue, and the arriving electric passenger vehicle will submit an expected departure time $t_{EV,dept}$.
4) Remove the electric passenger vehicle that will depart from the charging station at $t+1$ from the queue.

Because we assumes that the arrival and departure times of all electric heavy trucks and electric engineering vehicles are fixed and the same, they are regarded as a whole when charging. And the same goes for electric commuter buses.

A Ladder Carbon Trading Model. IES and carbon trading are mutually reinforcing and promoting. The implementation and improvement of the carbon trading mechanism will further improve the economic and environmental benefits of IES. Carbon quota is the carrier of carbon trading. At present, there are mainly two ways: free allocation and paid allocation by auction. There are two main methods of free distribution: the historical method and the baseline method. We adopts the baseline method to determine the allocation of carbon allowances. We considers carbon dioxide emissions directly generated by gas-to-power equipment and indirectly generated by electricity purchased from the grid [12,13]. The actual carbon dioxide emissions and carbon quota generated by them are shown in (4)-(7).

$$E_{\mathrm{c}1} = \delta_1 P_{G2E}(t)\Delta t \tag{4}$$

$$E_{\mathrm{c}2} = \delta_1 P_E(t)\Delta t \tag{5}$$

$$E_{\mathrm{p}1} = \delta_{G2E} P_{G2E}(t)\Delta t \tag{6}$$

$$E_{\mathrm{p}2} = \delta_E P_E(t)\Delta t \tag{7}$$

E_{ci} is unit i carbon emission allocation. E_{pi} is unit i actual carbon emissions. δ_{G2E} is the actual carbon emission coefficient of G2E. δ_E is the actual carbon emission coefficient of the power grid. δ_1 is the carbon emission quota allocation coefficient of unit electricity. The weighted average of the operating margin factor (OM) and build margin factor (BM) is usually adopted as the benchmark index. OM and BM are researched and released by the National Development and Reform Commission.

The price models of carbon trading mainly include single carbon price, ladder carbon price, and market-clearing carbon price [1]. To control the total amount of carbon emissions and fully mobilize the enthusiasm of enterprises to reduce emissions, we applies a ladder carbon trading model with a reward and punishment mechanism. That is, when the actual carbon emissions are lower than the carbon emission quota allocated by the government, certain rewards will be given, while the actual carbon emissions are higher than the allocation will be punished. The details of the ladder carbon trading model based on the reward and punishment mechanism are as follows.

$$C_{co_2 i} = \begin{cases} -c_{CO_2}(1+\mu)h - c_{CO_2}(1+2\mu)\left(E_{ci} - h - E_{pi}\right) & E_{pi} \le E_{ci} - h \\ -c_{CO_2}(1+\mu)\left(E_{ci} - E_{pi}\right) & E_{ci} - h < E_{pi} \le E_{ci} \\ c_{CO_2}\left(E_{pi} - E_{ci}\right) & E_{ci} < E_{pi} \le E_{ci} + h \\ c_{CO_2}h + c_{CO_2}(1+\lambda)\left(E_{pi} - E_{ci} - h\right) & E_{ci} + h < E_{pi} \le E_{ci} + 2h \\ c_{CO_2}(2+\lambda)h + c_{CO_2}(1+2\lambda)\left(E_{pi} - E_{ci} - 2h\right) & E_{ci} + 2h < E_{pi} \le E_{ci} + 3h \\ c_{CO_2}(3+3\lambda)h + c_{CO_2}(1+3\lambda)\left(E_{pi} - E_{ci} - 3h\right) & E_{ci} + 3h < E_{pi} \end{cases} \tag{8}$$

$C_{co_2 i}$ is unit i carbon trading costs. c_{CO_2} is the price per unit of carbon traded. μ is the reward coefficient. h is the length of the carbon emission interval. λ is the penalty coefficient.

2.2 Decision Problem Formulation

State. Extracting the observation state of the system includes 11 components: $s(t) = \{P_{PV}(t), P_{WT}(t), P_L(t), SOC_{EES}(t), SOC_{EV1}(t), SOC_{EV2}(t), SOC_{EV3}(t), n_{EV3}(t), p_{bE}(t), p_{sE}(t), t\}$. Where $P_{PV}(t)$, $P_{WT}(t)$, and $P_L(t)$ are the power of PV, WT, and load at time t, respectively. $SOC_{EES}(t)$, $SOC_{EV1}(t)$, $SOC_{EV2}(t)$, and $SOC_{EV3}(t)$ are SOC of EES, all electric heavy truck and engineering truck loads, all electric commuter bus loads, and all electric passenger vehicle loads connected to the power grid at time t, respectively. $n_{EV3}(t)$ is the number of electric passenger vehicles connected to the power grid at time t. p_{bE} and p_{sE} are the prices at which electricity is bought and sold on the grid at time t. t is time series.

The EV has a high degree of uncertainty when connecting to the power grid, in terms of the uncertainty of the state of charge when the EV arrives at the charging pile and the uncertainty of the time of arrival and departure from the charging pile. As for the treatment of the uncertainty of the state of charge when EV arrives at the charging pile, we designs that the electric quantity consumed by each EV when it arrives at the charging pile is random compared with that when it left last time. As shown in (9), it follows a normal distribution with μ_{EV} as the mean value and σ_{EV} as the variance [8]. The SOC of the EV when it arrives at the charging pile is shown in (10), which is the SOC of the EV when it leaves the charging pile minus the percentage of the amount of electricity consumed during driving in the battery capacity.

$$SOC_{EV,consume} \sim N\left(\mu_{EV}, \sigma_{EV}^2\right) \tag{9}$$

$$SOC_{EV}\left(t_{EV,arr}\right) = SOC_{EV}\left(t_{EV,dept}\right) - SOC_{EV,consume} \tag{10}$$

For the treatment of the uncertainty of EV arrival time and departure time, since the daily working time of employees is determined, it is considered that the departure and arrival time of electric heavy truck, engineering vehicle load, and electric commuter bus load is also determined. The departure time and arrival time of the electric passenger vehicle at the charging pile are random, but based on nominal 9am-to-5pm commuting behavior, the arrival time of the passenger vehicle at the charging pile can be considered to follow a normal distribution $t \sim N(8.5, 1^2)$ in terms of hours [14]. And it can be considered that the time when the passenger car leaves the charging pile follows a normal distribution $t \sim N(17.5, 1^2)$.

Influenced by geographical location, wind speed, light, and other factors, the output power of wind and photovoltaic power generation has strong uncertainty. In addition, there is randomness in users' electricity consumption behavior, so there is uncertainty in electricity load. In addition, we adds the step-based electricity price to consider the demand response of users. Among them, the demand response based on price is based on the voluntary participation of users in the project, and the participation of users will change with random factors such as the incentives received by users, energy consumption habits, energy prices, communication delays, and environmental conditions. Where, the power of transferable load after price-based demand response can be obtained as follows:

$$P_{L2}(t) = P_{TL0}(t) \times \left(1 + E_{tt}\frac{\Delta p_t}{p_{t0}}\right) \tag{11}$$

where Δp_t is the change of electricity price at time t. p_{t0} is the price of electricity in a normal period. $P_{TL0}(t)$ is the operating power of transferable load before transfer at time t. $P_{L2}(t)$ is the operating power of transferable load after transfer at time t.E_{tt} is the elasticity matrix of electricity quantity and price.

Therefore, uncertainty of solar and wind power generation, load and demand response are considered in system scheduling, and randomness subject to normal distribution is added as the observed state of the environment.

Action. The decision variables were six, respectively the power of G2E equipment $P_{G2E}(t)$, the power of electric energy storage device $P_{EES}(t)$, the interruption power of interruptible load $P_{IL}(t)$, the charging power of electric heavy truck and engineering truck $P_{EV1}(t)$, the charging power of electric commuter buses $P_{EV2}(t)$ and the charging power of electric passenger car $P_{EV3}(t)$. The power of electricity bought and sold from the grid is used to achieve power balance.

Reward. The IES model considering EVs scheduling and carbon trading aims to minimize the operating costs, so the reward is designed to be a negative number of system operating costs. System operating costs include buying and selling electricity cost $C_E(t)$ from the grid, buying gas cost $C_{bG}(t)$ from the gas grid, and the operation and maintenance cost $C_{OM}(t)$ of various equipment, the loss cost $C_{EES}(t)$ of electric energy storage devices, Compensation cost $C_{IL}(t)$ after interruption of interruptible load, the cost $C_{CO_2}(t)$ of carbon trading, and the compensation cost $C_{EV}(t)$ of electric passenger vehicles during fast direct current charging.

$$r(t) = -(C_E(t)+C_{bG}(t)+C_{OM}(t)+C_{EES}(t)+C_{IL}(t)+C_{CO_2}(t)+C_{EV}(t)) \quad (12)$$

$$C_E(t) = \begin{cases} p_{bE}(t)P_E(t)\Delta t, P_E(t) \geq 0 \\ p_{sE}(t)P_E(t)\Delta t, P_E(t)<0 \end{cases} \quad (13)$$

$$C_{bG}(t) = \frac{c_G P_{G2E}(t)\Delta t}{\eta_{G2E} L} \quad (14)$$

$$C_{OM}(t) = \Delta t \sum_{m=1}^{N} [K_m |P_m(t)|] \quad (15)$$

$$C_{EES}(t) = C_1^{EES}(t) + C_2^{EES}(t) \quad (16)$$

$$C_{IL}(t) = c_{IL} P_{IL}(t)\Delta t \quad (17)$$

Among them, the electrical energy storage device loss cost, including cycle life loss cost $C_1^{EES}(t)$ and charge and discharge loss cost $C_2^{EES}(t)$, concretely explain references [15,16]. $P_E(t)$ is the power of electricity purchased and sold by the grid at time t. Buying electricity is positive while selling electricity is negative.c_G is the price of natural gas. η_{G2E} is the efficiency of G2E equipment. K_m is the operation and maintenance cost coefficient of type m equipment. c_{IL} is the unit compensation cost of the interrupted load.

3 A Neural Approach to Real-time IES Management

As shown in Fig. 2, the Actor Critic algorithm combining Policy Gradien and Function Approximation is used to solve the problem. Actor networks input state information obtained from the environment, output the probability of each action, and then choose the behavior based on the probability. Critic network evaluates the value of the policy. The probability that the Actor modifies the selection behavior according to the Critic's evaluation. The trained Actor Critic network can be solved quickly to realize real-time response.

However, during the process of using reinforcement learning for training and solving problems, due to the issue of action settings, certain actions cannot be selected in certain states. For example, the maximum charging power that can be reached differs depending on the number of electric vehicles that are connected to the power grid at different times, and considering the constraints

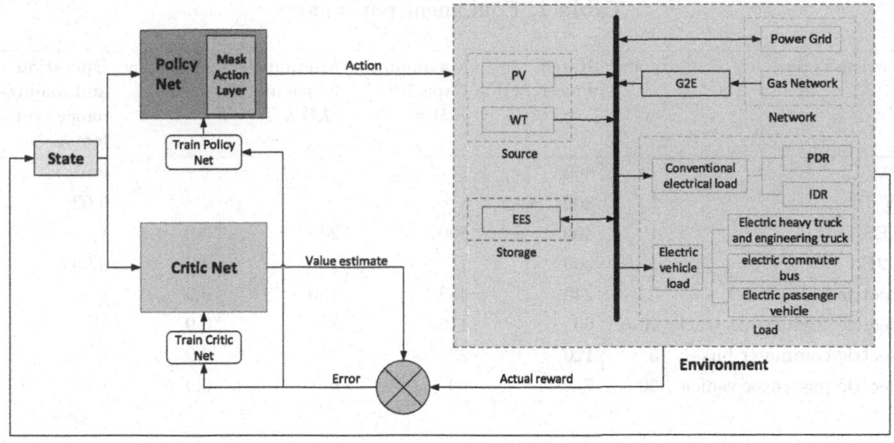

Fig. 2. The charging rules of electric passenger vehicles.

of the maximum and minimum capacity of electric vehicles and energy storage devices, at certain times, the range of charging and discharging power that can be selected for the three types of electric vehicles and energy storage devices is different. That is, when the agent chooses an action, some actions are infeasible. Therefore, to ensure the safe operation of the system, action masking [17] is used to remove infeasible actions, allowing the agent to select actions with a higher probability from the feasible action set. Here we only provide a brief overview, and for specific technical details of action masking, please refer to the literature by Zha and Wang [18].

Remark. The simulation shows that it is very difficult for the plain deep reinforcement learning approach to finish training and operating without violating constraints. The action masking is a necessary measure to guarantee safe operation. Besides, the action masking also improves the convergence speed during training. Due to the limited space, the comparisons and detailed analysis are hereby omitted.

4 Case Study

4.1 System Description

The integrated energy system of industrial and commercial parks based on HOMER PRO simulation shown in Fig. 1 is used to verify the effectiveness of the integrated energy system scheduling model considering carbon trading and flexible load. Specific equipment parameters are shown in Table 1. The time-of-use(TOU) electricity price, natural gas price, and the power of PV, WT, and electricity load on a certain day are shown in Fig. 3. The renewable energy and electric load power of a quarter used in the case study are all provided by

Table 1. Equipment parameters

Device Type	count	Rated Power $/kW$	Maximum Capacity $/kWh$	Minimum Capacity $/kWh$	Efficiency	Operation and maintenance cost factor
PV	1	400	\	\	\	0.01
WT	1	300	\	\	\	0.02
EES	1	400	800	240	0.9	\
G2E	1	500	\	\	0.7	0.064
electric heavy truck	3	240	432	130	0.9	\
electric engineering truck	10	60	110	33	0.9	\
electric commuter bus	5	120	235	71	0.9	\
electric passenger vehicle	50	7	30	9	0.9	\

HOMER PRO, and the uncertainty of renewable energy power generation, electric load itself, and transferable load when carrying out demand response based on price are fully considered in the training.

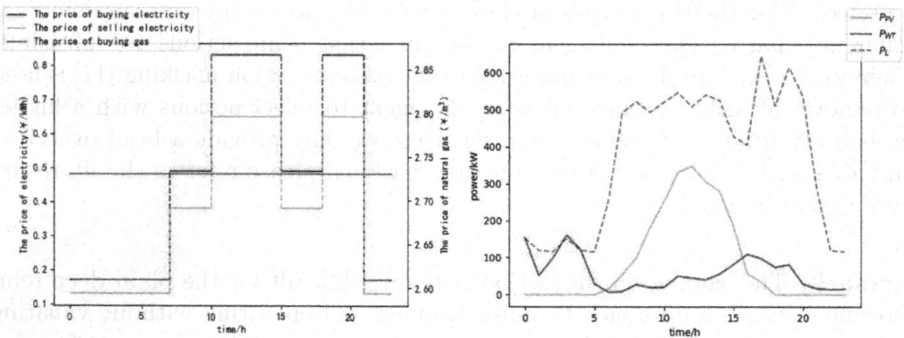

(a) Buy and sell electricity prices and natural gas prices. (b) The power of PV, WT, and electricity load on a certain day.

Fig. 3. System parameter.

4.2 Optimal Scheduling Results Considering Carbon Trading

To analyze the impact of carbon trading on system operation, three scenarios are set up for comparative analysis:

Scenario 1: Carbon trading is not considered.
Scenario 2: Consider carbon trading with a single carbon price.
Scenario 3: Consider the ladder carbon trading based on penalties and rewards.

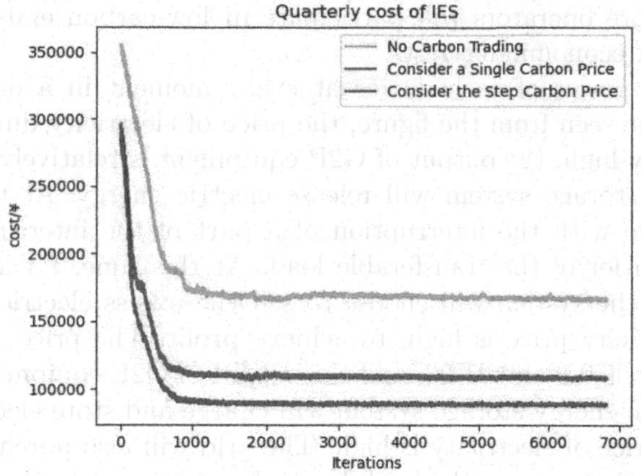

Fig. 4. Total quarterly cost curves for IES.

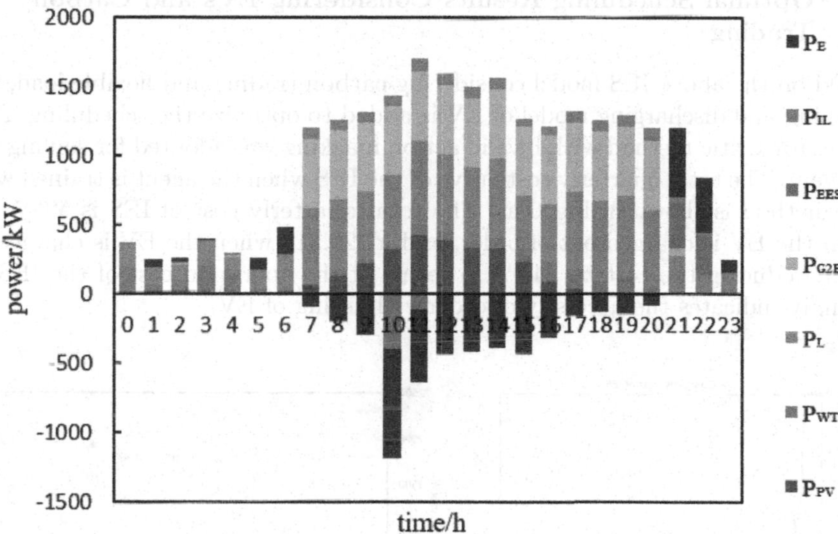

Fig. 5. The output status of each device.

The total quarterly cost curves of the IES during training for the three scenarios are shown in Fig. 4. It can be seen from the figure that the total quarterly operating cost is greatly reduced after carbon trading is added. In this way, operators will voluntarily reduce carbon emissions to reduce the operating cost of the system and maximize their interests. In addition, considering the ladder carbon trading based on penalties and rewards is less costly than carbon trading based on a single carbon price. The further reduction of system operation cost will increase the enthusiasm of operators to participate in emission reduction activities so that more operators will participate in low-carbon emission reduction actions for their economic interests.

The output power of each device at every moment in a day is shown in Fig. 5. As can be seen from the figure, the price of electricity during the period 7-20 is relatively high, the output of G2P equipment is relatively large, and the electric energy storage system will release electric energy. At the same time, it will cooperate with the interruption of a part of the interruptible load, as well as the transfer of the transferable load. At this time, PV generates more electricity, and the system will choose to sell the excess electricity to the grid when the electricity price is high, to achieve profit. The price of electricity is low during periods 0-6 and 21-23, and the output of G2E equipment is relatively low. The electric energy storage system will charge and store electric energy for use when the price of electricity is high. The grid will also purchase part of the low electricity price to meet the load demand.

4.3 Optimal Scheduling Results Considering EVs and Carbon Trading

Based on the above IES model considering carbon trading and flexible load, the charging and discharging model of EV is added to optimize the scheduling. And the Actor Critic method with invalid action masking was adopted for solving the problem. The total quarterly cost curve of the IES when the agent is trained with this method is shown in Fig. 6(a). The total quarterly cost of IES is ¥305,459 when the EV is charged out of order, and ¥229,318 when the EV is charged in order, reducing the cost by 24.9%.In terms of the operation cost of the IES, it strongly indicates the necessity of orderly charging of EV.

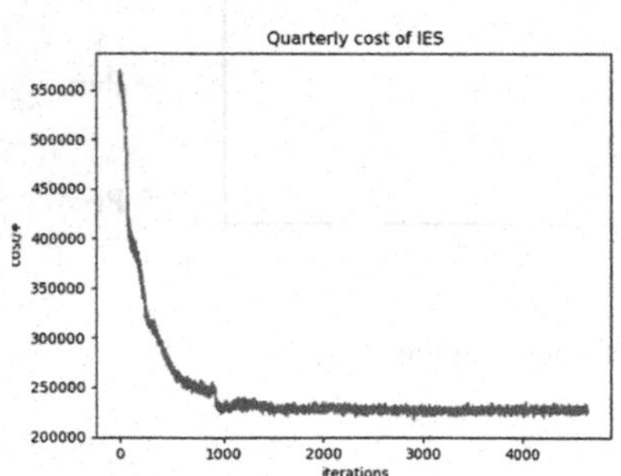

(a) Total quarterly cost curves for IES considering EV scheduling and carbon trading.

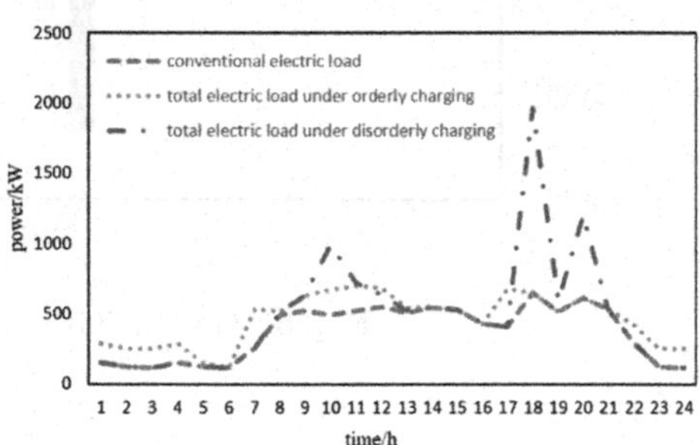

(b) Comparison curve of total load under three conditions.

Fig. 6. Output result.

Figure 6(b) shows the curves of conventional electric load, total electric load under orderly charging, and total electric load under disorderly charging. As can be seen from the figure, if EVs are allowed to use the disordered charging mode that charges immediately after connecting to the power grid, large new load peaks will appear because EVs connect to the power grid at a relatively concentrated time. The orderly charging of EVs can not only effectively avoid the emergence of new load spikes, but also have a certain smoothing effect on the conventional electric load curve.

The output power of each device at every moment in a day is shown in Fig. 7. It can be seen that when EVs are charging, they will automatically look for the time when the electricity price is lower while meeting the agreed travel requirements, to avoid new load spikes and reduce the system operation cost.

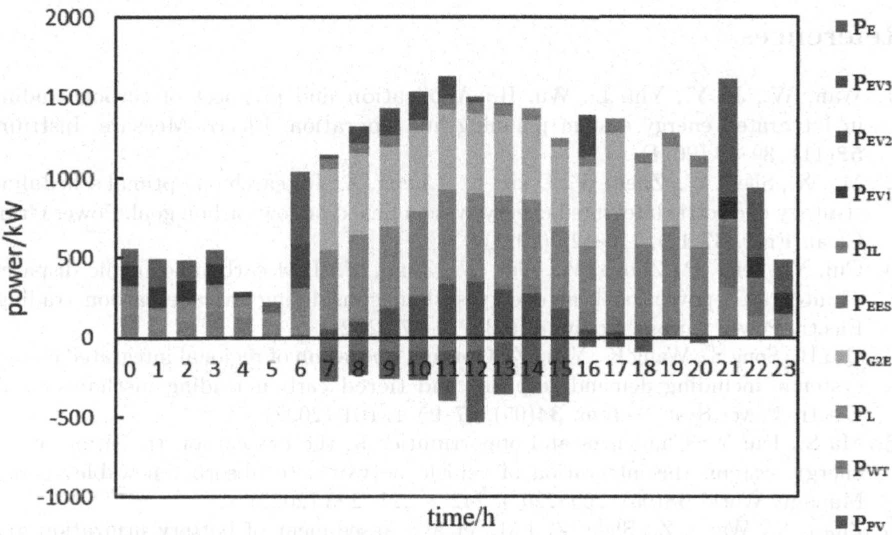

Fig. 7. The output status of EV and each device.

Conclusion

Based on the traditional IES model, this paper considers carbon trading based on punishment and reward ladder carbon price and demand response based on incentive and price. Three scenarios, carbon trading without carbon trading, carbon trading with a single carbon price, and carbon trading with a ladder carbon price based on penalty and reward, are set up for comparative analysis. The results fully demonstrate the advantages of considering carbon trading. In particular, considering ladder carbon trading based on penalty and reward can further reduce the operating costs of the system, thus making enterprises more active in participating in emission reduction actions. In addition, the orderly charging of electric vehicles can not only avoid the appearance of new peak demand but also smooth the electric load curve and reduce the system operation cost. Finally, the Actor Critic algorithm is used to solve it, which not only solves the problem of power difficult to accurately predict due to the addition of new energy and demand response based on electricity price but also can meet the fast solution requirements of the electricity market which gradually tends to real-time nowadays. In addition, when the space of invalid action is large, the method of giving a negative reward to the invalid action fails to solve the problem, and the method of masking the invalid action is used to realize the solution.

Acknowledgments. This study is supported in by the National Key R&D Program of China, Grant Number: 2022YFE0198700.

References

1. Wan, W., Ji, Y., Yin, L., Wu, H.: Application and prospect of carbon trading in integrated energy system planning and operation. Electr. Measure. Instrum. **58**(11), 39–48 (2021)
2. Ma, X., Shen, W., Zhen, W., Dong, K., Cheng, Z.: Research on optimal scheduling strategy of electric integrated energy system based on low carbon goal. Power Grids Clean Ener. **37**(12), 116–122 (2021)
3. Cui, Y., Zeng, P., Zhong, W., Cui, W., Zhao, Y.: Low carbon economic dispatch of integrated power-gas-heat energy system considering cascade carbon trading. Electr. Power Autom. Equip. **41**(03), 10–17 (2021)
4. Qiu B., Song S., Wang K., Yang Z.: Optimal operation of regional integrated energy systems, including demand response and tiered carbon trading mechanisms. J. Electr. Power Syst. Autom. **34**(05), 87–95 + 101 (2022)
5. Ma S., Fan Y.: Challenges and opportunities in the low-carbon transition of the energy system: the integration of vehicle networks to absorb renewable energy. Manage. World **38**(05), 209–220 + 242 + 221–223 (2022)
6. Zhao, Y., Wang, Z., Shen, Z.-J.M., et al.: Assessment of battery utilization and energy consumption in the large-scale development of urban electric vehicles. Proceed. Nat. Acad. Sci. **118**(17), e2017318118 (2021)
7. Zhang, X., Rao, R., Feng, Y.: Analysis of electric taxi charging behavior and cross-regional comparison of comprehensive benefits. Chin. Electr. Power **49**(2), 141–147 (2016)
8. Wu, X., Li, P.: Integrated energy system optimization scheduling considering electric vehicles and demand-side response. Sci. Technol. Eng. **22**(09), 3585–3593 (2022)
9. Wei, W., Xu, L., Xu, J., et al.: Day-ahead optimal scheduling method of park-level integrated energy system considering V2G technology. In: 2021 11th International Conference on Power and Energy Systems, pp. 827–833 (2021)
10. Wu, L., Zhang, L., Zhou, Q., Li, C.: Optimization strategy of microgrid energy scheduling based on reinforcement learning. Control. Eng. **29**(07), 1162–1172 (2022)
11. Yang, T., Zhao, L., Liu, Y., Feng, S., Pan, H.: Dynamic economic scheduling of integrated energy system based on deep reinforcement learning. Autom. Electr. Power Syst. **45**(05), 39–47 (2021)
12. Yin, P.: Calculation method of energy saving and emission reduction in CCHP system (1): calculation method of carbon emission. HVAC **46**(02), 12–17 (2016)

13. Lu, Z., Guo, K., Yan, G., He, L.: Optimal scheduling of wind power system considering demand response virtual units and carbon trading. Autom. Electr. Power Syst. **41**(15), 58–65 (2017)
14. Yang, Z.: Research on optimal scheduling of regional integrated energy system with electric vehicle. Wuhan University (2019)
15. Liu, H., Chen, X., Li, J., Xu, K.: Regional electric integrated energy system based on improved CPSO algorithm economic operation. Electr. Power Autom. Equip. **37**(06), 193–200 (2017)
16. Liu, H., Li, J., Ge, S., Zhang, P., Chen, X.: Coordinated scheduling of grid-connected integrated energy microgrid based on multi-agent game and reinforcement learning. Autom. Electr. Power Syst. **43**(01), 40–48 (2019)
17. Huang, S., Ontañón, S.: A closer look at invalid action masking in policy gradient algorithms. In: Proceeding of the International FLAIRS Conference - vol. 35, pp. 1–6 (2022)
18. Zha, Z., Wang, B., Fan, H., et al.: An improved reinforcement learning for security-constrained economic dispatch of battery energy storage in microgrids. In: Neural Computing for Advanced Applications, pp. 303–318 (2021)

Research on Chinese Diabetes Question Classification with the Integration of Different BERT Models

Zhuoyi Yu, Ye Wang, and Dajiang Lei(✉)

Chongqing Key Laboratory of Image Recognition, Chongqing University of Posts and Telecommunications, Chongqing 400065, China
2021212039@stu.cqupt.edu.cn, {wangye,leidj}@cqupt.edu.cn

Abstract. As diabetes has become one of the major public health challenges worldwide, the importance of automatic question-answering services for diabetes in daily healthcare is gradually being highlighted. To address this challenge, this paper aims to improve the accuracy of classifying Chinese diabetes questions. In order to achieve this goal, we utilized BERT pre-training models to extract text feature vectors, combined them with convolutional neural networks to extract local features, and employed pooling and fully connected layers for classification. Additionally, we also tested different BERT models and obtained the best results using voting fusion technology. The final accuracy improved by 2% compared to a single model. Experimental results demonstrate that the model can effectively classify Chinese diabetes questions, providing robust support for the implementation of automated diabetes question-answering services.

Keywords: Chinese text classification · Diabetes questions classification · BERT fusion · Convolutional neural network

1 Introduction

As one of the fundamental tasks in natural language processing, text classification aims to assign given textual data to predefined categories [1,2]. Its purpose is to automatically classify text, enabling users to quickly find the desired information. Text classification finds applications in various areas, including spam filtering, sentiment analysis, and topic classification, and is typically divided into binary or multi-classification tasks. To perform a classification task, the original text needs to be preprocessed, and features need to be extracted. This is followed by training and prediction using machine learning algorithms or building deep learning models.

As the number of diabetes patients continues to increase, providing better medical services and support for patients has become a concern for researchers. In traditional medical systems, diabetes and other non-conventional diseases typically rely on professional doctors to offer question and answer services. However,

H. Zhang et al. (Eds.): NCAA 2023, CCIS 1870, pp. 576–590, 2023.
https://doi.org/10.1007/978-981-99-5847-4_41

these services often face challenges such as high cost, inconsistent service quality, potential misdiagnosis, and slow response times. Consequently, an increasing number of researchers have begun to explore the application of natural language processing technology to automatically address these issues.

In the field of Chinese text classification, many research achievements have been made and can be divided into the following categories:

1.1 A Text Classification System Based on Machine Learning Algorithms

In the field of natural language processing, text classification is a fundamental task that aims to automatically assign text data to predefined categories. This task has wide applications in fields such as spam filtering, sentiment analysis, and topic classification. Text classification tasks are generally categorized as binary classification or multi-class classification. To perform a classification task, four steps are usually involved: text preprocessing, text representation, classifier training, and classification.

In the field of text classification, researchers have achieved significant results. For instance, F. Miao *et al.* [3] developed a Chinese news text classification system and conducted a comparative analysis of various classification algorithms, including K-nearest neighbors, Naive Bayes, and Support Vector Machine, during the training phase of the classifier. The findings demonstrated that the Chinese news text classification system based on machine learning algorithms yielded satisfactory results. Furthermore, Gonz'alez-Carvajal *et al.* [4] compared traditional machine learning methods with the use of BERT models for text classification tasks. Through a series of experiments on different scenarios, they verified the superiority of BERT over traditional TF-IDF vocabulary performance and its independence from NLP problem features.

1.2 A Text Classification System Based on Deep Learning Algorithms

Deep learning models have gained widespread usage in the domain of Chinese text classification, complementing traditional machine learning methods [5,6]. Within the realm of deep learning, appropriate neural network models are chosen to address text classification challenges. Li *et al.* [7] employed Long Short-Term Memory (LSTM) and Convolutional Neural Network (CNN) methods as deep learning techniques to classify Chinese text, and experimentally verified the model's performance. Qian Li *et al.* [8] focused on evaluating the strengths and weaknesses of various models in text classification tasks, ranging from traditional approaches to deep learning models, affirming the contributions of deep learning models in such tasks. Furthermore, deep learning algorithms for text classification models were applied in different domains: medical field by Lu, H [9], detection of aggressive language on social media platforms by P. Hajibabaee [10], and financial text analysis by Li, B [11]. These examples demonstrate the broad and effective application of deep learning algorithms in various fields for text classification tasks.

1.3 Fine Tuning of Text Classification Tasks Based on Language Pre-training Model

Language model pre-training has been proven useful for learning universal language representations. Nowadays, there are many language pre-training models available, including popular ones such as BERT [12], GPT-2 [13], RoBERTa [14], ALBERT [15], T5 [16], Bart [17], and many others. Each pre-training model has a different architecture and is applied to different domains. BERT, which stands for Bidirectional Encoder Representations from Transformers, is one of the most advanced language pre-training models that have achieved impressive results in many natural language understanding tasks. Sun *et al.* [18] conducted extensive experiments to study different fine-tuning methods of BERT for text classification tasks and provided a general solution for BERT fine-tuning. Their proposed solution achieved state-of-the-art results on eight widely studied text classification datasets. Yongjun Hu *et al.* [19] applied BERT to the text classification of psychological characteristics and found that using BERT can better capture psychological features and improve the accuracy of classification results. This will help promote the development of short text classification. F Cai *et al.* [20] used the RoBERTa model based on BERT, which has more advanced performance than BERT and showed better results in text classification tasks.

In the task of Chinese diabetes problem classification, employing natural language processing as a solution still encounters several challenges, including Chinese word segmentation, semantic expression diversity, long-tail problems, and insufficient domain knowledge. To address these challenges in the context of Chinese diabetes problems, we propose a targeted classification method that utilizes the pre-trained BERT model as a foundation. This approach effectively extracts semantic features from the input text. By freezing the parameters of the BERT model, we enhance the stability and reliability of the model.

Moreover, by incorporating Convolutional Neural Networks (CNN), we can capture local features of Chinese text, encompassing information of varying lengths. This aspect is particularly advantageous for classifying Chinese diabetes problems with diverse length distributions. Within the model, the utilization of pooling layers and ReLU activation function enhances the model's non-linear representation capability, thereby facilitating better adaptation to the classification task of Chinese text.

We optimize the model's performance and improve classification accuracy through joint training of the parameters of the BERT pre-trained model and CNN. Additionally, we employ other BERT-based models for the task and separately compare their performance. Finally, we fuse the probabilities derived from these models to obtain a more robust result.

2 Related Work

2.1 Text Classification Tasks in the Medical Field

When applying text classification to medical texts, challenges arise due to complex medical terminology and measurements, resulting in high dimensionality and data sparsity. In order to tackle these challenges, Li Qing *et al.* [21] proposed a unified neural network approach. This approach incorporates various techniques for sentence and document representation.

For sentence representation, convolutional layers are employed to extract features from sentences. Additionally, bidirectional gated recurrent units (BIGRU) are utilized to capture preceding and succeeding sentence features. An attention mechanism is employed to obtain sentence representations with important word weights.

In terms of document representation, the proposed method encodes sentences obtained from the sentence representation stage using BIGRU and decodes them through an attention mechanism, resulting in document representations with important sentence weights. Finally, a classifier is employed to categorize medical texts.

However, when encoding sentences with BIGRU, it is important to note that the output of the model can change if the order of input sentences or documents is altered. This is because BIGRU relies on recursive neural networks (RNNs), which process each element in a sequence sequentially, utilizing the output of the previous element as the input for the next element. Consequently, if the order of elements in the sequence is modified, the output of the previous element will impact the processing result of the subsequent element, ultimately affecting the final output of the model.

For instance, let's consider the context of Chinese diabetes questions. Two questions that a diabetes patient may ask are:

1. Can I eat apples?
2. What fruits can't I eat?

While these two questions may appear distinct, they essentially seek advice regarding the patient's diet. Therefore, correctly classifying them into the "healthy lifestyle" category is crucial to provide accurate answers and recommendations. To address this challenge, a Transformer-based model can be employed for sentence encoding, utilizing the encoder of BERT.

2.2 The Application of BERT Model in Classification Tasks

The BERT model has gained popularity among researchers and practitioners due to its advanced pre-training methods, which involve both masked language modeling (MLM) and next sentence prediction (NSP) tasks. With its bidirectional contextual representation and robust transfer learning capabilities, BERT is increasingly being utilized and studied in various applications. As the demand for BERT models grows, new versions and adaptations of BERT have been introduced to support different languages. For instance, there are BERT models specifically designed for languages like Arabic [22] and Marathi [23].

When fine-tuning the BERT pre-trained model directly on downstream tasks, it often yields better results compared to using recurrent neural networks (RNNs). However, the BERT model's large number of parameters makes direct fine-tuning prone to overfitting, where the model performs well on the training set but poorly on the test set. To mitigate this issue, we leverage the pre-trained BERT model to extract semantic features from the text.

Simultaneously, we employ convolutional neural networks (CNNs) to capture local features within the text. This combination allows us to benefit from both high-level abstract features provided by BERT and localized information captured by the CNN. By using a pooling layer and fully connected layer, we can perform feature mapping and classification prediction, thereby improving the model's classification accuracy and robustness.

To further enhance feature extraction, we apply multi-layer convolution operations, allowing us to compress the input sequence dimensionally. This compression facilitates better feature extraction by combining the high-level abstract features from BERT with the locally extracted features through the convolutional layer.

2.3 Different Models Based on BERT

Here are some introductions to different pre-training models based on BERT:

RoBERTa: It is a pre-training model proposed by Facebook in 2019, which optimized BERT. RoBERTa adopts a larger training set and longer training time, canceling the Next Sentence Prediction task in BERT and adding a dynamic masking strategy, thereby improving the performance and generalization ability of the model.

Briskilal J *et al.* [24] proposed an integrated model that leverages BERT and RoBERTa for the classification of periods and literals. Their model achieved an accuracy improvement of 2% compared to the baseline method. By incorporating the strengths of both BERT and RoBERTa, the integrated model demonstrated enhanced performance in accurately classifying periods and literals.

The following are RoBERTa's training objectives:

$$\mathcal{L}_{\text{RoBERTa}} = \sum_{i=1}^{N} \sum_{j=1}^{T_i} \left[-\log P\left(w_j \mid w_{<j}; \Theta\right)\right] \tag{1}$$

Among them, N represents the number of samples, and$T_i$represents the number of tokens for the i sample, $w_{<j}$ represents all tokens before the j token, $w_j$represents the j token, and $P(w_j|w_{<j})$ represents the predicted j token to be The probability of j w_j, where Θ represents the parameters of RoBERTa. The objective is to minimize the negative logarithmic likelihood of the conditional probability of all the tokens of all samples.

ALBERT: It is a pre-training model proposed by Google in 2019, which is an improvement on BERT. In ALBERT, the embedded matrix is decomposed into several small matrices through the factorization technology, so that different Transformer layers can have different embedded matrices, thus reducing the number of model parameters. The embedding matrix decomposition method of ALBERT can be expressed using a formula:

$$E = E_0 + \sum_{i=1}^{n} h_i E_i \tag{2}$$

Among them, E is the original embedding matrix, and E_0 is the global embedding matrix, E_i is the local embedding matrix, h_i is a binary indicator vector, indicating that the current token belongs to the i local embedding. Through factorization, the number of model parameters can be reduced, and the model efficiency and generalization ability can be improved.

MacBERT: MacBERT is a pre-trained language model for mixed Chinese and English text, jointly released by the Institute of Automation, Chinese Academy of Sciences and Xiaoma Zhixing. It is the first BERT-based pre-trained model for mixed Chinese and English text. Unlike BERT, MacBERT improves the model performance by optimizing the pre-training data sampling and data augmentation methods, and using a larger batch_size for pre-training. In addition, MacBERT uses a hybrid word-based and character-based tokenization method to account for the unique characteristics of the Chinese language. MacBERT outperforms BERT in multiple Chinese and English NLP tasks. Moreover, MacBERT uses a large amount of medical literature for pre-training, which may give it an advantage in the medical field compared to other models. Below is the formula for the multi-layer multi-head attention mechanism in MacBERT:

$$\text{MultiHead(Q, K, V)} = Concat(head_1, ..., head_h)W^O \tag{3}$$

Among them, Q, K, V are the input Query, Key, and Value, respectively, h is the number of attention heads, and $head_i$ represents the calculated result of the i attention head, and W^O is the weight matrix of the output layer. The calculation formula for each attention head is as follows:

$$head_i = \text{Attention}(QW_i^Q, KW_i^K, VW_i^V) \tag{4}$$

Among them, W_i^Q, W_i^K, W_i^V is the weight matrix of Query, Key, and Value, and Attention is the calculation formula for the attention mechanism.

ELECTRA: ELECTRA is a pre-training language model based on the substitution of word embeddings. Compared to other pre-training models such as BERT, ELECTRA can achieve better performance with less data and shorter training time. ELECTRA uses two neural networks: a generator, which randomly replaces some words in the input text, and a discriminator, which judges which words have been replaced.

Compared to BERT, the innovation of ELECTRA lies in its training objective. BERT is a pre-training model based on MLM, which randomly masks some words in the input text and tries to predict those masked words. ELECTRA, on the other hand, is a pre-training model based on the "discriminator objective", which randomly replaces some words in the input text and judges which words have been replaced. The goal of ELECTRA is to minimize the difference between the substituted word embeddings and the original word embeddings.

ERNIE: ERNIE is a BERT-based pre-trained language model released by Baidu. It uses richer and more diverse pre-training data, including multiple data sources such as Baidu Encyclopedia, Baidu Search logs, Baidu Zhidao, and Baidu Wenku. These data sources cover more fields and topics, making ERNIE more suitable for multi-domain and multi-task applications. In addition, unlike BERT's MLM objective, ERNIE's MLM objective is to randomly mask some words in the input and let the model predict them. In ERNIE's training, strategies such as semantic-level replacement and reversal are also added to enhance the model's understanding of language. ERNIE has shown performance comparable to or even better than BERT in multiple Chinese NLP tasks, especially in tasks that require understanding of contextual relationships. Specifically, ERNIE adds a vocabulary replacement task compared to BERT in pre-training tasks, so ERNIE's objective function has more terms than BERT, as shown below:

$$L_{ERNIE} = L_{MLM} + L_{NSP} + L_{TDMP} + L_{SDMP} \tag{5}$$

Among them, $L_{MLM}$is the loss function of MLM task, $L_{NSP}$is the loss function of NSP task, $L_{TDMP}$is the loss function of the Token Document Mutual Prediction task, $L_{SDMP}$is the loss function of the Sentence Document Mutual Prediction task. In contrast, BERT's objective function only has two terms: MLM and NSP.

After studying the advantages of different BERT models, we need to recognize that each model has its own limitations. These limitations may include poor performance on specific types of data or inefficiency in processing long sequence texts. To address these issues, I have proposed several solutions, such as integrating the results of multiple models to improve accuracy. Combining the results of multiple models can not only alleviate the limitations of individual models, but also achieve better results in tasks across different domains. Therefore, for specific tasks, it is valuable to further improve the performance of the model by combining the strengths of different models through model fusion.

3 Design Method of the Model

3.1 Brief Observation of Data

We conduct a preliminary analysis of each type of sentence in the training set to observe their approximate composition. The results are depicted in Fig. 1.

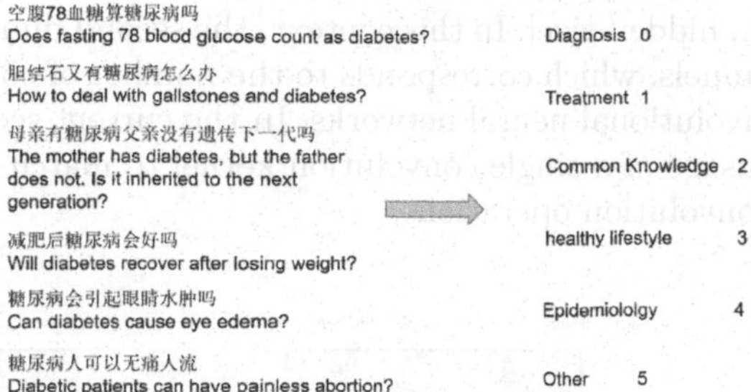

Fig. 1. The example of the diabetes data

3.2 Normalize the Input Chinese Text

The primary objective of standardizing Chinese text is to transform the input Chinese sentences into a standardized format that can be fed into the model. In the case of the Chinese diabetes problem dataset, the text is relatively clean, allowing us to skip certain text cleaning operations. As illustrated in Fig. 2, by utilizing the BERT word splitter, we can directly convert the input Chinese text into a suitable format that can be used as input for the model.

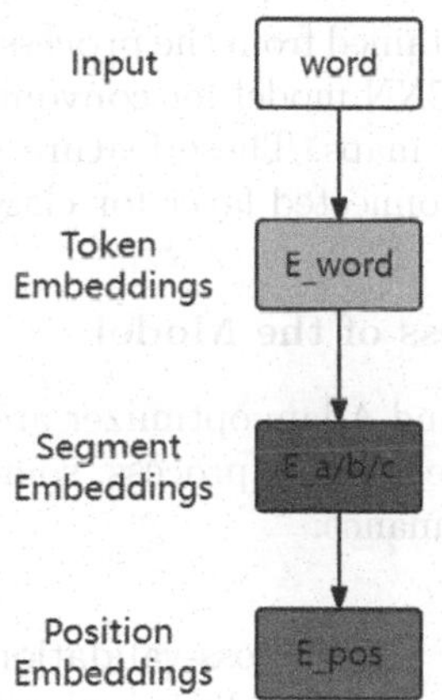

Fig. 2. BERT input representation

3.3 Using Pre-trained BERT Models as the Basis for Input

In the forward function, as depicted in Fig. 3, the transformed text data is initially fed into the BERT model. The output retrieves the hidden layer from the last layer, resulting in an encoded result of [batch_size, seq_len, hidden_size]. Subsequently, the unsqueeze function is applied to introduce an additional dimension on the second dimension, yielding a four-dimensional tensor of [batch_size,

1, sequence_length, hidden_size]. In this context, the second dimension represents the number of channels, which corresponds to the number of convolution kernels utilized in the convolutional neural networks. In the current scenario, it is set to 1, indicating the usage of a single convolution kernel to match the input format required for 2D convolution operations.

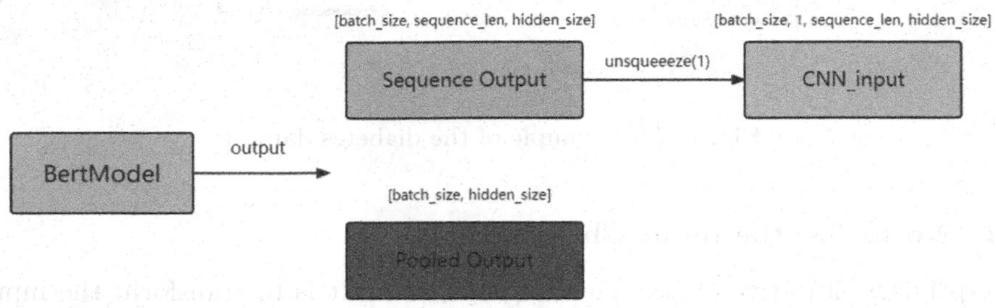

Fig. 3. BERT output to TextCNN input

3.4 Building a Convolutional Neural Network

The vector representation obtained from the processed BERT model is utilized as input and fed into the TextCNN model for convolution and pooling operations, resulting in multiple feature maps. These feature maps are then concatenated and passed through a fully connected layer for classification.

3.5 The Training Process of the Model

Cross entropy loss function and Adam optimizer are used to update the parameters of the model. During the training process, some techniques were attempted to improve the model performance:

K-Fold Cross Validation. K-fold cross-validation is a widely used method for evaluating model performance, especially when working with limited datasets. In this method, the dataset is divided into k subsets or folds. The model is then trained and evaluated k times, each time using a different fold as the validation set and the remaining k-1 folds as the training set. The evaluation results from each iteration are averaged to obtain the final performance metrics. The main advantage of k-fold cross-validation is that it provides a more reliable estimate of the model's performance by utilizing all available data for both training and validation. This approach ensures that each sample in the dataset is used for validation at least once, reducing the impact of random partitioning on the evaluation results. By averaging the performance across multiple iterations, the evaluation becomes more robust and less dependent on a single random split of the data.

Fast Gradient Method. FGM is an adversarial attack method that adds some disturbance to the input of the model to minimize interference with the predicted results of the model. FGM can enable the model to resist some malicious attacks, such as adversarial samples, and improve the robustness of the model.

Exponential Moving Average. EMA is an optimization method that can smooth model parameter updates during the training process. In the process of model optimization, EMA will weighted average the historical parameters and current parameters, making the update of model parameters more smooth and stable, and helping to avoid overfitting.

Set Parameters with Optimization Effects. Dropout refers to the random deactivation of some neurons during the training process to prevent the model from overfitting and enhance the model generalization ability. Specifically, in the training process of the neural network, some nodes are randomly selected not to participate in the forward propagation and back propagation of this round, so as to avoid overfitting of the network. **Warmup_steps** are the number of steps that control the learning rate to gradually increase to the maximum value at the beginning of training. The first few steps of the specified training steps use a smaller learning rate to help the model converge better and prevent the initial gradient change from being too large, leading to unstable training. **Weight decay** is also called L2 regularization. The sum of squares of model weights is added to the loss function of the model as the regularization term, which makes the model more inclined to select smaller weights and avoid overfitting of the model. It is equivalent to adding a penalty term for model parameters to the loss function to prevent model parameters from becoming too large.

Model Fusion. As the Fig. 4, After obtaining the training results of each model, we add up the probabilities of each model to get the final probability and assign each text to the class with the highest probability. By voting on the probability outputs of the models, the classification performance can be improved.

4 Experiment

4.1 Experiment Data

The dataset for this task includes 6 categories of Chinese diabetes-related questions, with the specific content of each category shown in the Fig. 1. The training set consists of 6,000 samples, while the validation and test sets each contain 1,000 samples. The task is to predict the classification of diabetes-related questions in the test set, and after prediction, the missing category label data in the test set needs to be filled in.

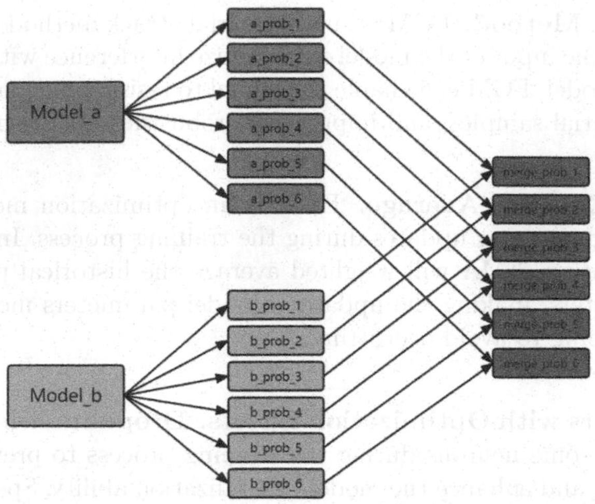

Fig. 4. BERT input representation

4.2 Evaluation Criterion

This task targets the classification labels filled in on the test set, Using accuracy as the evaluation criterion. Here is the specific defination:

$$\text{Accuracy} = \frac{\text{Predict the correct number of samples}}{\text{Total samples}} \tag{6}$$

4.3 Experimental Results and Analysis

Classification Performance of Different Pre-trained Models. When setting the hyperparameter of the model, we first observe the data and count the distribution of the text length. As shown in Table 1, the maximum length of each text is 20.

Table 1. Distribution of Text Length Ratio

Ratio	Length
mean	11.91
std	2.72
min	5.00
50%	11.00
60%	12.00
70%	13.00
80%	14.00
90%	16.00
95%	18.00
99%	19.00
max	20.00

Then, we set the learning rate to a slightly higher value to facilitate the observation of the model's convergence.

To help the model adapt quickly to the data, we will set the warmup_steps to 5% of the total number of steps. Additionally, to prevent overfitting, we will fix the dropout probability at 0.5. A higher value can enhance the model's robustness and generalization ability.

The specific hyperparameter are shown in Table 2.

Table 2. Hyperparameter Setting

Name	Value
learning_rate	5e–5
epoch	30
dropout	0.5
warmup_steps	0.05*total_steps
weight decay	5e–4
max_output_length	20

To compare the performance differences of different BERT pre-trained models, we trained multiple models with the same parameters and employed different tricks to calculate offline scores on the validation set. The original model results are shown in Table 3.

It is worth noting that except for the overall accuracy, all other evaluation indicators are calculated based on precision.

Table 3. Classification Performance of Different BERT Pretrained Models without trick(Offline Score)

Model	Acc	Diagnosis	Treatment	Common Knowledge	healthy lifestyle	Epidemiololgy	Other
RoBERTa-large	0.885	0.9500	0.9279	0.8043	0.9485	0.8718	0.6190
MacBERT-large	0.8425	0.9714	0.8772	0.7701	0.9029	0.8205	0.5
ERNIE-3.0-xbase-zh	0.75	0.7561	0.8224	0.6588	0.8400	0.7045	0.4348
ELECTRA-large	0.7475	0.8182	0.7928	0.5841	0.8673	0.7778	0.5556

Due to the superior performance of RoBERTa and MacBERT in the initial results, we focused on utilizing these two models in the subsequent experiments. The experimental results incorporating additional tricks are presented in Table 4 and Table 5.

Table 4. Classification Performance of Different BERT Pretrained Models with FGM (Offline Score)

Model	Acc	Diagnosis	Treatment	Common Knowledge	healthy lifestyle	Epidemiololgy	Other
RoBERTa-large	0.8825	0.9487	0.9279	0.7872	0.9485	0.8718	0.6500
MacBERT-large	0.8475	0.9722	0.8783	0.7791	0.9029	0.8205	0.5238

Table 5. Classification Performance of Different BERT Pretrained Models with EMA (Offline Score)

Model	Acc	Diagnosis	Treatment	Common Knowledge	healthy lifestyle	Epidemiololgy	Other
RoBERTa-large	0.88	0.9474	0.9060	0.7849	0.9579	0.8718	0.6667
MacBERT-large	0.8425	0.9714	0.8772	0.7701	0.9029	0.8205	0.5000

Based on the experimental results, it can be concluded that after using the trick, the classification performance of the model has improved in some categories, while its performance has decreased in some categories. Finally, we submitted online and the score we received increased by a point in the thousandth percentile.

The Impact of Different Techniques on Model Classification Performance. Using high-scoring models with the same parameters and different techniques, the result files were evaluated online. Due to significant differences between offline and online results, the best result offline does not always perform the best online. To address this issue, in all evaluations, the fusion of the RoBERTa-large model and the MacBERT-large model achieved the best performance when using 10-fold cross-validation. Compared to the baseline with a single model, achieving a better accuracy of 0.891 online.

5 Conclusion

Our model's main innovation lies in using different BERT models as pre-trained models and incorporating a convolutional neural network (CNN) structure on top of it. Specifically, we extract text features through multiple layers of convolution and pooling, and then connect them to fully connected layers for classification. Finally, we fuse the results of different models. Compared with traditional text classification models, our model not only uses BERT as a feature extractor, but also introduces CNN structure, which enhances the model's ability to grasp local text features, and improves the model's classification accuracy and generalization ability. Besides, model fusion can be achieved by voting on the predictions of multiple models, reducing the bias and variance of individual models, and improving the accuracy of classification. By combining the predictions of multiple models, it is possible to reduce the occurrence of misclassifications made by individual models, thereby enhancing the overall accuracy of classification.

Although we have achieved decent accuracy using the neural network model, there are still some shortcomings. For some categories with small quantities, the classification accuracy is difficult to reach the average level. At the same time, for categories with large quantities, too much training may cause overfitting, and the best results cannot be achieved. Therefore, improving the accuracy of text classification in addressing such long-tail problems is the focus of our next step.

Acknowledgements. This work was partly supported by the National Key R&D Program of China (2021YFF0704100), the National Natural Science Foundation of China (62136002, 61876027, 61936001), the Science and Technology Research Program of Chongqing Municipal Education Commission (KJQN202100627 and KJQN202100629), and the National Natural Science Foundation of Chongqing (cstc2022ycjh-bgzxm0004, cstc2019jcyj-cxttX0002), respectively.

References

1. Wang, Y., Zhou, Z., Jin, S., Liu, D., Lu, M.: Comparisons and selections of features and classifiers for short text classification. In: Iop Conference Series: Materials Science and Engineering, vol. 261, p. 012018. IOP Publishing (2017)
2. Wang, Y., Zhang, X., Mi, L., Wang, H., Choe, Y.: Attention augmentation with multi-residual in bidirectional LSTM. Neurocomputing **385**, 340–347 (2020)
3. Miao, F., Zhang, P., Jin, L., Wu, H.: Chinese news text classification based on machine learning algorithm. In: 2018 10th International Conference on Intelligent Human-Machine Systems and Cybernetics (IHMSC), vol. 2, pp. 48–51. IEEE (2018)
4. González-Carvajal, S., Garrido-Merchán, E.C.: Comparing bert against traditional machine learning text classification. arXiv preprint arXiv:2005.13012 (2020)
5. Wang, Y., Wang, H., Zhang, X., Chaspari, T., Choe, Y., Lu, M.: An attention-aware bidirectional multi-residual recurrent neural network (abmrnn): a study about better short-term text classification. In: ICASSP 2019–2019 IEEE International Conference on Acoustics, Speech and Signal Processing (ICASSP), pp. 3582–3586. IEEE (2019)
6. Wang, Y., Liao, J., Yu, H., Leng, J.: Semantic-aware conditional variational autoencoder for one-to-many dialogue generation. Neural Comput. Appl. **34**(16), 13683–13695 (2022)
7. Li, Y., Wang, X., Pengjian, X.: Chinese text classification model based on deep learning. Future Internet **10**(11), 113 (2018)
8. Li, Q., et al.: A survey on text classification: From traditional to deep learning. ACM Trans. Intell. Syst. Technol. (TIST) **13**(2), 1–41 (2022)
9. Hongxia, L., Ehwerhemuepha, L., Rakovski, C.: A comparative study on deep learning models for text classification of unstructured medical notes with various levels of class imbalance. BMC Med. Res. Methodol. **22**(1), 181 (2022)
10. Hajibabaee, P., et al.: Offensive language detection on social media based on text classification. In: 2022 IEEE 12th Annual Computing and Communication Workshop and Conference (CCWC), pp. 0092–0098. IEEE (2022)
11. Wan, C.-X., Li, B.: Financial causal sentence recognition based on bert-cnn text classification. J. Supercomput., 1–25 (2022)

12. Devlin, J., Chang, M.-W., Lee, K., Toutanova, K.: Bert: pre-training of deep bidirectional transformers for language understanding. arXiv preprint arXiv:1810.04805 (2018)
13. Radford, A., Jeffrey, W., Child, R., Luan, D., Amodei, D., Sutskever, I., et al.: Language models are unsupervised multitask learners. OpenAI blog **1**(8), 9 (2019)
14. Liu, Y., et al.: Roberta: a robustly optimized bert pretraining approach. arXiv preprint arXiv:1907.11692 (2019
15. Lan, Z., Chen, M., Goodman, S., Gimpel, K., Sharma, P., Soricut, R.: Albert: a lite bert for self-supervised learning of language representations. arXiv preprint arXiv:1909.11942 (2019)
16. Raffel, C., et al.: Exploring the limits of transfer learning with a unified text-to-text transformer. J. Mach. Learn. Res. **21**(1), 5485–5551 (2020)
17. Lewis, M., et al.: Bart: denoising sequence-to-sequence pre-training for natural language generation, translation, and comprehension. arXiv preprint arXiv:1910.13461 (2019)
18. Sun, C., Qiu, X., Xu, Y., Huang, X.: How to fine-tune BERT for text classification? In: Sun, M., Huang, X., Ji, H., Liu, Z., Liu, Y. (eds.) CCL 2019. LNCS (LNAI), vol. 11856, pp. 194–206. Springer, Cham (2019). https://doi.org/10.1007/978-3-030-32381-3_16
19. Hu, Y., Ding, J., Dou, Z., Chang, H.: Short-text classification detector: a bert-based mental approach. Computational Intelligence and Neuroscience 2022 (2022)
20. Cai, F., Ye, H.: Chinese medical text classification with roberta. In: International Symposium on Biomedical and Computational Biology, pp. 223–236. Springer, Cham (2022). https://doi.org/10.1007/978-3-031-25191-7_17
21. Qing, L., Linhong, W., Xuehai, D.: A novel neural network-based method for medical text classification. Future Internet **11**(12), 255 (2019)
22. Ali Saleh Alammary: Bert models for Arabic text classification: a systematic review. Appl. Sci. **12**(11), 5720 (2022)
23. Kulkarni, A., Mandhane, M., Likhitkar, M., Kshirsagar, G., Jagdale, J., Joshi, R.: Experimental evaluation of deep learning models for Marathi text classification. In: Gunjan, V.K., Zurada, J.M. (eds.) Proceedings of the 2nd International Conference on Recent Trends in Machine Learning, IoT, Smart Cities and Applications. LNNS, vol. 237, pp. 605–613. Springer, Singapore (2022). https://doi.org/10.1007/978-981-16-6407-6_53
24. Briskilal, J., Subalalitha, C.N.: An ensemble model for classifying idioms and literal texts using bert and roberta. Inf. Process. Manage. **59**(1), 102756 (2022)

Shared Task 1 on NCAA 2023: Chinese Diabetes Question Classification

Xiaobo Qian[1], Shunhao Li[1], Weihan Qiu[1], Tao Chen[1], Maojie Wang[2,3], and Tianyong Hao[1](✉)

[1] School of Computer Science, South China Normal University, Guangzhou, China
xiaoboqian1221@outlook.com, lishunhao99@foxmail.com, 1026100410@qq.com, 2647947112@qq.com, haoty@m.scnu.edu.cn

[2] The Second Affiliated Hospital of Guangzhou University of Chinese Medicine (Guangdong Provincial Hospital of Chinese Medicine), Guangzhou, China
maojiewang@gzucm.edu.cn

[3] State Key Laboratory of Dampness, Syndrome of Chinese Medicine, The Second Affiliated Hospital of Guangzhou University of Chinese Medicine, Guangzhou, China

Abstract. This shared task focuses on classifying Chinese diabetes questions. Medical question classification is one of important tasks in medical data processing. However, due to the unique characteristics of medical questions, such as the presence of many uncommon medical terms and ambiguous short texts, resulting in difficulty in obtaining the most ideal information for users. The goal of this shared task is to enhance the performance of classification results and promote the development of diabetes automatic question answering services. This paper presents the results of the Chinese Diabetes Question Classification Shared Task and summarizes the classification methods used by the winning teams.

Keywords: Shared Task · Question Classification · Diabetes

1 Introduction

Question classification is one of the important tasks in the field of natural language processing. The main task is to map text data to a predefined category or categories. With the rapid development of the Internet, the application of question classification is becoming increasingly widespread [1, 2]. It not only helps users extract more valuable information from text, but also helps users quickly retrieve relevant information that meets their needs [3].

According to the report of China Internet Network Information Center (CNNIC) [4], as of December 2021, the number of Internet users in China has reached 1.032 billion, and the Internet penetration rate has reached 73.0%. The Internet has become an important tool for patients to search for health information and express their needs for health information. Many online health Question and Answer communities and forums have become popular platforms for patients to post questions and share information. However, the quality of health information on the Internet varies much [5], making

H. Zhang et al. (Eds.): NCAA 2023, CCIS 1870, pp. 591–596, 2023.
https://doi.org/10.1007/978-981-99-5847-4_42

it crucial to provide reliable health information to patients. However, there are many prominent problems in the existing medical question classification, such as difficulty in identifying the semantics of professional terms and difficulty in eliminating ambiguity of polysemy words [6], which may result in users being unable to quickly find the effective information most relevant to their own condition. Therefore, satisfying the needs from patients by accurately classifying medical questions at a fine-grained level is an urgent problem to be solved in automatic question answering services.

As a typical chronic disease, diabetes has become one of the major global public health challenges. With the rapid development of the Internet, the huge group of type 2 diabetes patients and high-risk people has shown an increasing demand for specialized information on diabetes [7]. The automated diabetes Question Answering (QA) services also play a vital role in providing daily health services for patients and high-risk people. However, starting from the needs of patients and accurately classifying medical questions at a fine-grained level is still an urgent issue that needs to be addressed in automated question and answer services.

International Conference on Neural Computing for Advanced Applications (NCAA 2023) is an annual international neural computing conference, which showcases state-of-the-art R&D activities in neural computing systems and their industrial and engineering applications. The Chinese diabetes question classification shared task is one of the competition tasks of the NCAA conference. We aim to contribute to automatically and accurately classifying diabetes related questions raised by users and promote the development of diabetes automatic question answering service through this shared task.

In this shared task, participants need to predict the classification of diabetes questions in the test dataset. The task website is https://github.com/yuni-bobo/Chinese-DQC. Qian et al. [8] proposed a diabetes benchmark dataset, which is applied as the evaluation dataset in the task dataset.

Ten teams from top-level universities participated in the Chinese diabetes question classification evaluation task. Their innovative models are proposed for validation and evaluation on the test dataset. We use the ratio of the number of samples correctly classified to the total number of samples to evaluate the performance of the methods proposed by the participants. At the same time, we give three deep learning models as baselines. The scores of the top three teams in the competition are 89.9%, 89.6%, 89.1%, respectively. Most participants adopt a combination of pretrained language models and deep learning models to improve the performance of question classification.

Our datasets are described in Sect. 2, which includes the distribution of the dataset, metric evaluation and baselines. Results of baselines are provided in Sect. 3.1. The results of the methods proposed by the participating teams are provided in Sect. 3.2.

2 Datasets

2.1 Distribution of the Dataset

The Chinese diabetes questions included in the task dataset are divided into six categories, which are Diagnosis, Treatment, Common Knowledge, Healthy Lifestyle, Epidemiology, and Other. The 8000 labeled diabetes questions of the task dataset are randomly divided, of which 6000 questions are used as the training set, 1000 questions are used as the validation set, and the remaining 1000 questions are used as the test set. Participants need to predict the classification of diabetes questions in the test set. The Table 1 gives the distribution of the task dataset.

Table 1. The distribution of the shared task.

Categories	Training set	Validation set	Test set	Total
Diagnosis	527	103	87	717
Treatment	1501	260	265	2026
Common knowledge	1226	212	217	1655
Healthy lifestyle	1702	251	273	2226
Epidemiology	599	118	90	807
Other	445	56	68	569
Total	6000	1000	1000	8000

2.2 Evaluation

The Accuracy (ACC) is adopted to evaluate the prediction results of different participating teams in this competition. The computation of the Accuracy is as Eq. (1):

$$Accuracy = \frac{the\ number\ of\ samples\ correctly\ classified}{the\ total\ number\ of\ samples} \tag{1}$$

2.3 Baseline

We select three baseline models, i.e., text convolutional neural network (Text CNN), text recurrent neural network (Text RNN), and bidirectional encoder representation from transformer model (BERT) to test the models on the Chinese diabetes question dataset.

Kim [9] applied Convolutional Neural Network (CNN) to the text classification task and proposed Text CNN in 2014, which utilizes multiple kernels of different sizes to extract key information in sentences, thereby better capturing local correlations. The two key layers of CNN for text classification are convolutional layer and pooling layer. The convolutional layer can perform feature extraction, while the pooling layer can reduce dimensionality. This can reduce training parameters and extract higher-level text features. In Text CNN model, the number of filters is 256 and learning rate is set as 0.001. The filter sizes are chosen as 2, 3 and 4.

CNN focuses on local information, and therefore it is unable to obtain global dependencies between contextual vocabularies, which will limit its understanding of the semantics of entire text. Therefore, in 2016, Liu et al. [10] from Fudan University proposed Text RNN considering text as a sequence. The model focuses on the sequence relationship of each word, and stores the semantic information of each word in front of the text. In Text RNN model, the number of hidden layers is 128 and learning rate is set as 0.001.

The Google research team developed the pre-trained language model BERT (Bidirectional Encoder Representation from Transformer) in 2019 [11], which can provide context sensitive sentence representation. This model has achieved excellent performance in multiple NLP tasks, including text classification. BERT is used for pre-training large amounts of unlabeled text data to learn a language representation that can be used to fine-tune specific machine learning tasks. At the same time, this method can provide different vector representations for each identical word in different contexts, which to some extent alleviates the problem of word ambiguity. For BERT model, the number of epochs is 12 and learning rate is chosen as 5e–5.

3 The Results

Different baseline models show varying performances on the evaluation dataset, which indicates our proposed benchmarks can provide an evaluation platform for question classification algorithms. A total of ten teams from various universities participated in this competition. The scores of baseline methods and the top four teams are shown in Table 2.

Table 2. The scores of baseline methods and the top four teams.

Team	Method	ACC (%)
Baseline 1	Text CNN	86.1
Baseline 2	Text RNN	83.8
Baseline 3	BERT	87.8
Team 1 (永升队)	A fusion model of fine-tuning BERT and Text CNN	89.9
Team 2 (biubiu)	A BERT to obtain the feature vectors of sentences and conduct a majority vote on multiple machine learning models' classification results	89.6
Team 3 (雨打芭蕉队)	A sentence feature-level enhancement module and a serialized attention aggregation module to enhance sentence embedding vector	89.1
Team 4 (只会梭哈)	Different BERT models as pre-trained models with a CNN structure	89.1

The team ranked first adopts a fusion model of BERT and Text CNN. The team fine-tunes the transformer encoder layer of the last two layers of BERT, and fixes the parameters of all other layers. The team sums the output of the last two transformer encoder layers of BERT at the element level and uses it as input for subsequent Text CNN. The optimal performance of the model is 89.9% on the test dataset of the competition platform.

The team ranked second uses BERT to obtain the feature vectors of sentences. Then the team applies multiple machine learning classification models for classification and conducts a majority vote on these classification results to obtain the final result. The optimal performance of the model is 89.6% on the test dataset.

The team ranked third proposes a classification model for long-tailed distributed data, which uses a sentence feature-level enhancement module and a serialized attention aggregation module to enhance the information carried by the sentence embedding vector in expectation of better recognition of tail data. The optimal performance of the model is 89.1% on the test dataset.

Another team in third place uses different BERT models as pre-trained models with a CNN structure. They extract text features through multiple layers of convolution and pooling, and then connect them to a fully connected layer for classification. Finally, the team fuses the results from different models together. The optimal performance of the model is 89.1% on the test dataset.

4 Conclusion

This paper presents the results of the Chinese diabetes question classification shared task. The shared task is addressed by multiple teams from various universities and their innovative models are proposed for validation and evaluation on the test dataset. It helps to enhance the performance of search results and promote the development of diabetes automatic question answering services.

References

1. Hao, T., Li, X., He, Y., Wang, F.L., Qu, Y.: Recent progress in leveraging deep learning methods for question answering. In: Neural Computing and Applications, pp. 1–19 (2022)
2. Hao, T., Xie, W., Wu, Q., Weng, H., Qu, Y.: Leveraging question target word features through semantic relation expansion for answer type classification. In: Knowledge-Based Systems, pp. 43–52 (2017)
3. Liu, Y., Hao, T., Liu, H., Mu, Y., Weng, H., Wang, F.L.: OdeBERT: one-stage deep-supervised early-exiting BERT for Fast inference in user intent classification. In: ACM Transactions on Asian and Low-Resource Language Information Processing, pp. 1–18 (2023)
4. China Internet Network Information Center: China Internet Network Development State Statistic Report. https://www.cnnic.cn/n4/2022/0401/c88-1131.html. Accessed 6 Mar 2023
5. Kanthawala, S., Vermeesch, A., Given, B., Huh, J.: Answers to health questions: internet search results versus online health community responses. J. Med. Int. Res. **18**(4), e5369 (2016)
6. Luo, Y., Huang, Z., Wong, L.P., Zhan, C., Wang, F.L., Hao, T.: An early prediction and label smoothing alignment strategy for user intent classification of medical queries. In: Neural Computing for Advanced Applications: Third International Conference, pp.115–128 (2022)
7. Xie, W., Ding, R., Yan, J., Qu, Y.: A mobile-based question-answering and early warning system for assisting diabetes management. In: Wireless Communications and Mobile Computing (2018)
8. Qian, X., Xie, W., Long, S., Lan, M., Mu, Y., Hao, T.: The construction of question taxonomy and an annotated Chinese corpus for diabetes question classification. In: Proceedings of the 21st Chinese National Conference on Computational Linguistics, pp. 395–405 (2022)

9. Kim, Y.: Convolutional neural networks for sentence classification. In: Proceedings of EMNLP, pp. 1746–1751 (2014)
10. Liu, P., Qiu, X., Huang, X.: Recurrent neural network for text classification with multi-task learning. arXiv preprint arXiv:1605.05101 (2016)
11. Devlin, J., Chang, M.W., Lee, K., Toutanova, K.: BERT: pre-training of deep bidirectional transformers for language understanding. In: Proceedings of the 2019 Conference of the North American Chapter of the Association for Computational Linguistics, pp. 4171–4186 (2019)

SFDA: Chinese Diabetic Text Classification Based on Sentence Feature Level Data Augmentation

Qingyan Wang, Ye Wang, and Dajiang Lei(✉)

Chongqing Key Laboratory of Image Recognition, Chongqing University of Posts and Telecommunications, Chongqing 40065, China
S220233043@stu.cqupt.edu.cn, {wangye,leidj}@cqupt.edu.cn

Abstract. Many type 2 diabetes patients and high-risk groups has an increasing demand for specialized information on diabetes. However, the long-tail problem often generate difficulties in model training and reduced classification accuracy. In this paper, we propose enhancing semantic feature approach to solve the long-tail problem in Chinese diabetes text classification and detailed practice is as followes: we enrich the tail classes knowledge by enhancing semantic features module and then use the attention aggregation module to improve the semantic representation by fusing these semantic features. As for the enhancing semantic feature module, we proposed two strategies: using different dropouts while pre-trained language model is same and using different pre-trained language model. As for the attention aggregation module, its purpose is to better fusing the semantic features obtained previously. After processing by these two modules, we send the final feature vector into the classifier. The final accuracy of 89.1% was obtained for the classification of Chinese diabetes in the NCAA2023 assessment.

Keywords: Diabetes · Text classification · Long tail · Data augmentation

1 Introduction

Text classification is a very basic task in natural language processing, which is the process of classifying text into predefined categories according to certain criteria. Because the rise of deep learning and its excellent performance in the field of classification, classification methods have also moved from machine learning dominated such as Bayesian [1–3], decision trees [1], K-nearest neighbors [4,5], support vector machines [1,6], random forests [5], etc. are transferred to deep Learning methods [7] such as recurrent neural networks [8–10], convolutional neural networks [11–13], etc. However, there are still some problems waiting to be solved for text classification, especially in the medical field, such as the long-tail problem of Chinese diabetes classification because of the uneven distribution of category samples i.e.. Several classes have large samples while the

H. Zhang et al. (Eds.): NCAA 2023, CCIS 1870, pp. 597–611, 2023.
https://doi.org/10.1007/978-981-99-5847-4_43

remaining classes have only small samples [14,15]. However, such unbalanced training samples of classes bring difficulties for model training, because model tends to prefer the head class with many samples and does not classify well in the tail class with small samples [14,15]. In this regard, real-world applications of long-tail category imbalance make it infeasible to train deep learning using empirical risk minimization [16], making it difficult for traditional deep learning models to handle such situations, such as medical image detection [17] and text classification [18,19].

To solve the long-tail problem, many researches have been doned in recent years and they can be summarized into five categories, including class balancing method, weight-adjustment method, transfer learning method, feature representation method and module improvement method. The class balancing method is to change the number of samples from different classes, so model could learn more tail knowledge. We can oversample [20] or augment [21] minority class, undersample majority class [22] and also can combinate both oversampling and undersampling strategies [23]. Many experiments show that model will put more attention on majority class and minority class's features often not well noticed. For this problem, there are usually two solutions: weight-based adjustment [24,25]and marginal-based adjustment [24,26]. The weight-based adjust is that by adjusting the weights of samples in different class, model is more likely to emphasis the learning of majority class. The marginal-based adjustment is to find the marginal size between different classes and classifiers to adjust loss. Transfer learning method is that learning general knowledge from the majority classes and then transferring the learned knowledge to the minority classed [21,27], which can enrich minority classes knowledge to help solve long tail problem. This method can be divided into two categories according to the source of auxiliary knowledge: external transfer [28,29] and internal transfer [27,30,31]. The feature representation method aims to get better feature representation to help the following classifier to accurately identify minority classes. Specially, the feature extraction-based [32,33] approach is to improve the recognition ability of the model for minority classes by innovatively designing the feature extraction method and the feature representation-based approach [14,34] by combining, downscaling, and transforming features to obtain a mfeature representation with more semantic and generalized. The main advantage of module-based improvement approach to solve the long-tail problem is to improve the generalization and scalability of the model. Specifically, it can be divided into three categories: constructing complex classifiers [35,36], features decoupling [37,38] and ensemble learning [39,40].

According to the above five solutions for the long tail problem, we propose to use data augmentation at the sentence feature level and it can enhance the tail sample features while ensuring that the risk of overfitting can be avoided, and the technical implementation is also very simple. We proposed two ways to achieve data augmentation at the sentence feature level: the first way uses the same pre-training language model with different dropout rate to generate multiple different features embedding of the same sentence and the second way uses different pre-training language models directly. To gain feature embedding with more richer semantic information, we concate different features embedding and sent it into serialized attention module, and then to feed it into the classifier to complete the classification. Overall, the main contributions of this paper can be summarized as follows:

We innovatively propose using data augmentation at sentence feature level to solve Chinese diabetes text classification with long tail distribution.

We use serialized attention module to get sentence feature embedding with richer semantic information.

We applied the proposed method to Chinese diabetes text classification, and finally proved the effectiveness of the proposed method through experiments.

In this paper, the second section introduces the related work, the third section describes the proposed approach in detail, the fourth section introduces the analysis of the experimental results, and the fifth section summarizes the full text and making the prospect of the further work.

2 Related Work

Class rebalancing is the first way to solve the long tail problem. Liu *et al.* [22] used an undersampling approach to balance data and proposed two improved algorithms for the problem of partial samples of the head class being ignored. Estabrooks*et al.* [20] combined different oversampling methods to solve the data unbalance problem. Chawla *et al.* [23] demonstrated that the combination of undersampling for classes with many samples and oversampling for classes with small sample can result in better classifier performance. Jialun Liu *et al.* [21] argued that the distortion of the feature space due to the uneven distribution between head and tail classes impairs the model's ability to learn features, and they wanted to enhance tail classes to improve its intra-class distribution and deep representation. Jaehyung Kim *et al.* [41] enabled the classifier to assist the learning of tail classes by transferring and exploiting the head class knowledge by selecting samples from the head classes into tail class samples. Overall, although class rebalancing can solve the long tail problem to a certain extent, there are also some problems such as data loss when expanding and reducing samples, ineffective for classes with too small samples size, increasing the probability of header class misclassification and affecting the robustness of the model due to sensitivity to parameters.

Weight adjusting is the second way to solve the long tail problem and it can be subdivided into weight-based adjustment and margin-based adjustment. First, many people have explored the first method: Mateusz Buda *et al.* [42] proposed a simple and effective equalization loss to solve the long-tail problem by ignoring the gradient of rare classes, and this loss effectively protects the tail classes from being trapped in a disadvantage when the network parameters are updated. Nathalie Japkowicz *et al.* [43] studied the long-tail of face data during training and proposed a new distance loss function to better utilize the tail data during training. Seulki Park *et al.* [25] derived a new loss for balancing the training phase that mitigates the effect of samples that lead to overfitting decision boundaries. Yin Cui *et al.* [44] designed a new weighting scheme for samples that uses the effective number of samples per class to rebalance the loss to produce the current class balancing loss. Jiawei Ren*et al.* [45] suggested the use of balanced softmax to accommodate the shift in label distribution between training and testing. Second, the next study is based on marginal adjustment. The existing soft marginal loss calculation [46,47] is not applicable to the long-tail problem because it does not take into account the class imbalance. Salman Khan *et al.* [48] demonstrated that the uncertainty of class prediction is proportional to the scarcity of classes. Chengjian Feng *et al.* [26] designed a balanced loss function to increase the strength of the decision boundary for weak classes by bootstrapping the loss margin between any two classes. However, solving the long-tail problem based on adjusting the sample importance suffers from difficulties in parameter adjustment, distortion of class judgments due to data imbalance, high computational cost, and lack of in-depth understanding of a few classes and so on.

Transfer learning is the third way to solve the long tail problem and it aims to learn from classes with large samples to help the learning of classes with small classes [49,50]. According to the source of the auxiliary knowledge, we can divide it into two categories: internal knowledge migration and external knowledge migration. Internal knowledge migration such as head knowledge to tail knowledge, for example, Xinting Hu *et al.* [30] proposed a method to split the tail by dividing the overall data into a balanced number of data chunks and then applying augmented learning to learn each part. Liuyu Xiang *et al.* [27] proposed to the long tail problem Performing multiple expert learning: first train multiple expert models on a subset of samples with less balanced sample classes and then combine these expert models into one student model, which tends to show better results on unevenly distributed datasets. External knowledge migration such as the use of model pre-training [29,51] introduces external prior knowledge to assist the model in increasing the knowledge of long-tail data. Self-supervised pre-training of the model has also been shown to be a method that can improve the classification performance of long-tail problems [14,52]. Although knowledge migration-based solutions to long-tail problems have many advantages, they also have disadvantages such as migration difficulties, uncertainty of migration effects, and high training costs.

Feature representation is the fourth way to solve the long tail problem: by designing excellent feature extraction methods such as the use of deep learning models such as convolutional neural networks to extract features from images and text can improve the ability of the models to recognize a few categories. For example, Chen Huang *et al.* [32] used local feature extraction to recognize long-tailed objects in images. Bingyi Kang *et al.* [14] found that a representation model that generates a balanced feature space has better generalization ability than a representation model that generates an unbalanced feature space and can perform relatively stable on data sets with unbalanced distribution. Jiequan Cui *et al.* [34] proposed the use of parametric contrast learning to solve the long-tail recognition problem. However, Feature-based representation as a method to solve the long-tail problem also suffers from difficulties such as reliance on data quality and often complex nonlinear transformations.

Module improvement is the fifth way to solve the long tail problem and it can be subdivided into constructing complex classifiers, feature decoupling, and branch training. The first is constructing complexing classifier, Han-Jia Ye *et al.* [53] investigated that adding classrelated information when training a convolutional neural network classifier with data in a long-tail distribution can reduce model bias. Tz-Ying Wu *et al.* [35] proposed a deep realistic classifier that solves the class imbalance problem of hierarchical classification by mapping images into a class classification tree structure to solve the class imbalance problem of hierarchical classification. The second is feature decoupling, Bingyi Kang *et al.* [54] decoupled the learning process into representation learning and classification and demonstrated that data imbalance is not the main factor affecting the learning of high-quality representations and that strong long-tail recognition can be obtained by tuning the classifier with only the simplest sampled learned representations. Yifan Zhang *et al.* [39,40,55] proposed using a two-branch network to handle both representation learning and classifier learning where each branch performs its own task separately and the model uses a cumulative learning strategy i.e., learning the generic patterns first and then gradually focusing on the tail data. Hao Guo *et al.* [56] proposed to also use a two-branch network in the visual recognition task, where one branch takes uniform sampling as input and the other branch takes rebalanced sampling as input to train the model in a collaborative way with uniform and balanced samples to improve the head class and tail class performance. However, these approaches have some limitations, such as only solving the performance problems of specific modules or components, but not improving the overall performance of the system, possibly introducing new problems, increasing complexity, and not being applicable to all scenarios.

Therefore, data augmentation at the sentence feature level is very necessary, because it not only ensures a flat number of category features. At the same time, because the operation is performed at the sentence feature level, there are a large number of pre-trained language models with excellent feature extraction performance available, and the technical implementation is very simple and does not destroy the original semantics of the text, so the research based on this route is very practical.

3 Approach

The structure of our framework is shown in Fig. 1(a), which consists of a semantic feature enhancing module in Fig. 1(b), a serialized attention aggregation module in Fig. 1(c) and a separate text classifier. First, the semantic feature enhancing module generates some semantic feature embeddings for each sentence of the input and then the obtained several sentence feature embeddings are fed into the serialized attention aggregation module to obtain a representation with rich aggregated semantic information. Finally, the final feature representations are passed through a text classifier to complete text classification task.

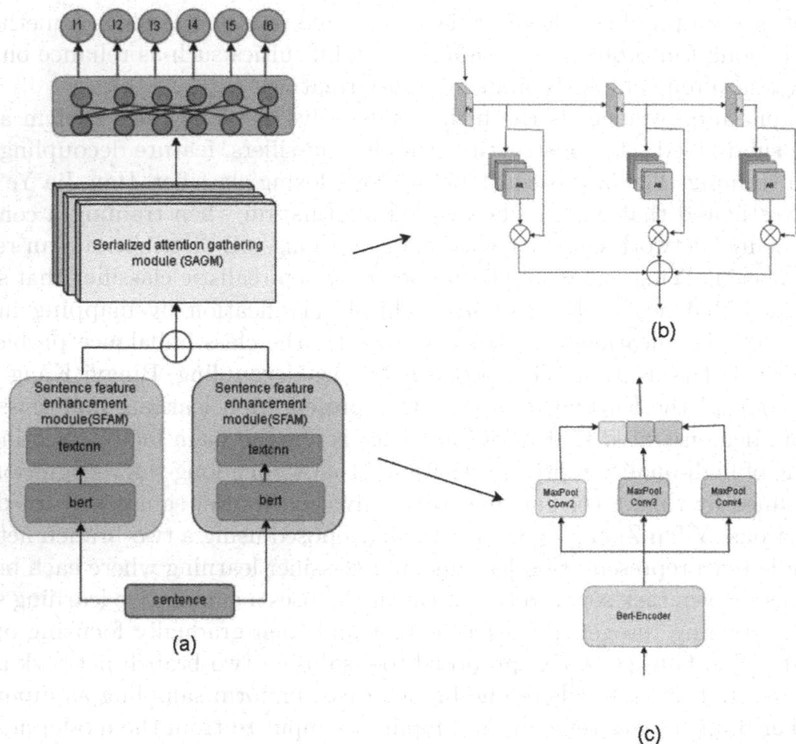

Fig. 1. The overflow of model architecture SFDA (a) and semantic feature enhancing module SFAM (b) and serialized attention aggregation module SAGM (c)

3.1 Semantic Feature Enhancing Module

The detailed structure of the semantic feature enhancing module is shown in Fig. 1(b).

When choosing a dropout-based semantic feature enhancing strategy, we follow [57] and consider dropout noise as a minimal form of data augmentation: several multibranch sentence embedding modules with identical structure and uncertainty determined only by dropout, which simultaneously accept the same sentence input $s = \{x_1, x_2, x_3......x_n\}$, producing different sentence embedding outputs as sentence feature-level data augmentation.

When choosing a semantic feature enhancing strategy based on pre-trained language models, we use different pre-trained models as different data enhancement strategies. The same sentence $s = \{x_1, x_2, x_3......x_n\}$, after different pre-training will yield different feature embedding outputs as sentence feature-level data enhancement.

The mathematical expression of these two strategies is as follows:

$$e_i = F_i(s) \tag{1}$$

Algorithm 1: SFDA model processing flow

Input: a piece of sentence $S = \{w1, w_2, w_3, ...w_n\}$,
w_i represents the i-th word in S;
Output: the sentence score about six class$\{I_1, I_2, I_3, I_4, I_5, I_6\}$
1 semantic feature enhancing module:
$E_1 \in R^{1\times d} \leftarrow$ SFAM(S) ,
where SFAM denotes the left sentence feature enhancement module
$E_2 \in R^{1\times d} \leftarrow$ SFAM(S),
where SFAM denotes the right sentence feature enhancement module
2 fusing feature embeddings:
$E_3 \in R^{1\times 2d} \leftarrow E_1 \oplus E_2$
3 serialized attention aggregation module:
$E_4 \in R^{1\times d'} \leftarrow$ SAGM(E_3),
where SAGM denotes the serialized attention gathering module
4 classifier:
$\{I_1, I_2, I_3, I_4, I_5, I_6\} \leftarrow E_4$, where I_i denotes the sentence scores of six classes.

$s = \{x_1, x_2, x_3, ...x_n\}$ denotes the input sentence, n is the sentence length. F_i is the sentence feature enhancement module for the i-th branch, either a dropout-based sentence feature-level data enhancement module or a pre-trained language model-based sentence feature-level data enhancement module, calculated as follows:

$$F_i = MaxPool_{ij}(Conv_{ij}(Encoder_i(e))) \tag{2}$$

where $Encoder_i$ denotes the pre-trained language model used in the i-th branch; $Conv_i$ denotes the convolutional layer of the i-th branch, e.g., con_{i_2} denotes the convolutional layer with convolutional kernel size 2 for the i-th data enhancement module, con_{i_3} denotes the convolutional layer with convolutional kernel size 3 for the i-th data enhancement module. $MaxPool$ denotes the maximum pooling layer, and the subscript denotes the same as the convolutional layer.

3.2 Serialized Attention Aggregation Module

The detailed structure of the serialized attention aggregation module is shown in Fig. 1(c):

The serialized attention aggregation module is designed to be used by using a multi-layer connection layer that plays constant mapping to obtain the representations of the input embedding vectors under different subspaces, and the representations under different subspaces are input to the attention aggregation layer respectively. The meaning of the attention layer in the aggregation layer is as follows: the output attention scores of multiple aggregation layers of the model and are connected with their inputs through the residuals, and finally the representations with rich aggregation semantic information are obtained, and then input to the classifier, using the weighted sum of the multilayer residuals to predict the classification. This is computed as follows:

$$last\_e = A_1(w_1(e_i) \oplus A_2(w_2(A_1(w_1(e_i))) \oplus A_3(w_3(A_2(w_2(A_1(w_1(e_i))))) \quad (3)$$

where w_i is constant mapping layer and implemented by a fully connected neural network. A_i is the attention layer, which calculates the sentence feature weights and multiplies them with the current sentence features to obtain the feature embedding vector of the sentence after focusing by the attention layer. $\oplus$represents the splicing operation, and finally the feature embedding vectors obtained in different subspaces are spliced horizontally to obtain the final feature embedding of the current sentence embedding. The attention layer is calculated as follows.

$$A_i = \sum_{i=1}^{n} a_n, e_i \quad (4)$$

where a_n computed attention weight for each word. $n$is the sentence length and a_n is computed as follows:

$$a_n = \frac{exp\{w^T tanh(Ve_j^T)\}}{\sum_{j=1}^{n} exp\{w^T tanh(Ve_j^T)\}} \quad (5)$$

w, V are the learnable parameters, $tanh$ is the activation function, and e_j is the jth word in a sentence of length n.

4 Experiment

4.1 Dataset

Chinese Diabetes Text Classification Dataset: We experimentally test the proposed method on Chinese diabetes text. This dataset contains 8000 samples, divided into six categories: diagnosis, treatment, general knowledge, healthy lifestyle, epidemiology, and others. 6000 of the samples are used for training, 1000 for validation, and 1000 for testing.

4.2 Experimental Setting

The deep learning methods involved in this experiment are all set with a fixed random number seed of 2023 to facilitate the consistency of the subsequent reproduction. Because of the design to use pre-trained language models for feature enhancement, we choose to use two pretrained language models: roberta-large and mac-bert, which generate sentences with a feature embedding dimension of 1024. Considering the length distribution of the text, the upper limit of the most frequent text interception length is set to 20. To reduce the learning process turbulence, the learning is set to 5e−5, and the sizes of the convolutional kernels used in the experiments are seted as 2,3 and 4.

4.3 Evaluation Metric

In our experiment, evaluation metrics will be used including precision, recall, f1 and accuracy. They are defined are as follows:

$$precision = \frac{TP}{TP + FP}$$

$$recall = \frac{TP}{TP + FN}$$

$$f1 = \frac{2 * precision * recall}{precision + recall}$$

$$acc = \frac{\text{The number of samples correctly predicted}}{\text{The number of total samples}} \times 100\%$$

TP is the number of samples that its prediction is true and its real label is also true in confusion matrix. FP is the number of samples that its prediction is true and its real label is false. FN is the number of samples that its prediction is false and its real label is true.

4.4 Experimental Results and Analysis

Effect Based on Different Dropout Strategy: The two-branch semantic feature enhancing module is used to enhance the input sentences at the feature level, because dropout naturally carries uncertainty when randomly deactivating neurons, so we treat it as noise for the purpose of sentence feature enhancement, and then send it to the serialized attention aggregation layer to get the final sentence embedding, and finally to the classifier. The experimental results corresponding to this strategy are shown in rows 2 to 7 of Table 1.

Effect Based on Different Pre-trained Model Strategy: Unlike the previous subsection, which uses the randomness of dropout for sentence features augmentation, then the data augmentation, this subsection uses different pre-trained language models of roberta-large and macbert-large with two-branch structure. The experiments results corresponding to this strategy are shown in rows 8 of Table 1.

The performance on the Chinese diabetes classification dataset after removing the dropout sentence-level feature enhancement module and the serialized attention aggregation module, with all experimental settings kept consistent, are shown in the last two lines in Table 1.

The pre-trained language model with macbert and dropout of 0.05, 0.1, and 0.15 respectively, is used to classify the Chinese diabetes problem. The performance of the dataset is shown in Table 2 and pre-trained language model with roberta is shown in Table 3.

Table 1. Experiment result

SFAM-bert	Variable	Acc
roberta-large	drop = 0.15	**85.75**
roberta-large	drop = 0.10	**87.00**
roberta-large	drop = 0.05	**86.25**
macbert-large	drop = 0.15	**84.50**
macbert-large	drop = 0.10	**84.50**
mabert-large	drop = 0.05	**83.75**
macbert-large+roberta-large	macbert+roberta+drop = 0.1	**86.50**
macbert-large+classifier	drop = 0.1	85.00
roberta-large+classifier	drop = 0.1	77.27

Table 2. Roberta with dropout = 0.1 in the six classes

Evaluation	Diagnosis	Treatment	Common Knowledge	Healthy Lifestyle	Epidemiology	Other
precision	94.79	92.92	76.14	93.00	87.50	52.38
recall	81.82	95.45	77.01	94.90	94.59	45.83
f1-score	87.80	94.17	76.57	93.94	90.91	48.89
acc	87.00					

Table 3. Macbert with dropout = 0.1 in the six categories

Evaluation	Diagnosis	Treatment	Common Knowledge	Healthy Lifestyle	Epidemiology	Other
precision	80.00	86.55	78.57	91.92	84.21	66.67
recall	81.82	93.64	75.86	92.86	86.49	41.67
f1-score	80.90	89.96	77.19	92.39	85.33	51.28
acc	84.50					

Analysis of Experimental Results: Since the above experimental results were conducted in the same experimental environment, it is clear from the analysis that our proposed method of feature-level data augmentation and serialized attention aggregation on the input sentences can improve the recognition of the model on the long-tail distribution data to a certain extent. Moreover, by comparing the experiments, with the addition of the sentence feature data augmentation module and the serialized attention aggregation module, the category accuracy of the framework based on the Roberta pre-trained language model improved by 2.0% over the original framework on the overall dataset. In addition, by adding the sentence feature data augmentation module and the serialized attention aggregation module, the category accuracy of the Roberta-based pre-trained language model improves by 2.0% and the macbert-based pretrained language model improves by 7.23% over the original framework on the overall dataset, which further demonstrates the effectiveness of our proposed method.

Finally, the distribution change of the inputing sentence's feature embeddings in main stage are shown in the Fig. 2: The first column black blue pictures are sentence feature embedding's(E_1) scatter picture and plot picture. The second column light blue picture are the other sentence feature embedding's(E_2) scatter picture and plot picture. The third cloumn black green pictures are the fusing sentence feature embeddings(E_3). The forth column red pictures are the sentence feature embedding experiencing SAGM(E_4). From these pictures, we can find that the fusing embedding includes more imformation than E_1 and E_2 and after passing through the attention module, the information noticed by model changes.

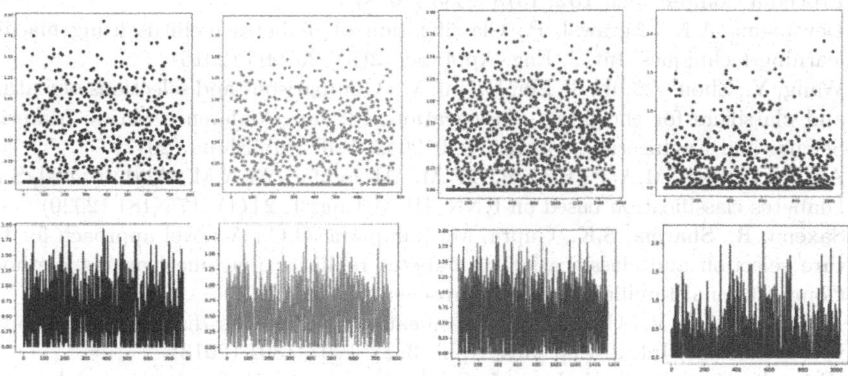

Fig. 2. Distribution of inputing sentence's feature embeddings in main stages (Color figure online)

5 Conclusion

In this paper, we propose a classification model for long-tailed distributed data, which uses a semantic feature enhancing module and a serialized attention aggregation module to enhance the information carried by the sentence embedding vector in expectation of better recognition of tail data. The sentence feature-level enhancement module achieves feature enhancement by generating multiple different sentence feature embeddings for the same sentence input, and the serialized attention aggregation module feeds the feature embedding vectors into multiple attention aggregation layers to finally obtain representations with rich aggregation semantic information for classification by subsequent classifiers. Finally, the optimal performance of the test dataset is 87% on the Chinese diabetes problem classification dataset and 89.1% on the test dataset of the competition platform.

There are still many shortcomings in the current work: such as we don't want to break the distribution of original data and want to let model to learn more knowledge about tail data, So We enhance the head data and the tail data meanwhile. Although experiment is proved is effected, we will consider methods that balance head and tail samples in feature work.

Acknowledgements. This work was partly supported by the National Key R&D Program of China (2021YFF0704100), the National Natural Science Foundation of China (62136002, 61876027, 61936001), the Science and Technology Research Program of Chongqing Municipal Education Commission (KJQN202100627 and KJQN202100629), and the National Natural Science Foundation of Chongqing (cstc2022ycjh-bgzxm0004, cstc2019jcyj-cxttX0002), respectively.

References

1. Sisodia, D., Sisodia, D.S.: Prediction of diabetes using classification algorithms. Procedia Comput. Sci. **132**, 1578–1585 (2018)
2. Dewangan, A.K., Agrawal, P.: Classification of diabetes mellitus using machine learning techniques. Int. J. Eng. Appl. Sci. **2**(5), 257905 (2015)
3. Wang, Y., Zhou, Z., Jin, S., Liu, D., Lu, M.: Comparisons and selections of features and classifiers for short text classification. In: Iop Conference Series: Materials Science and Engineering, vol. 261, p. 012018. IOP Publishing (2017)
4. Ali, A., Alrubei, M.A.T., Hassan, L.F.M., Al-Ja'afari, M.A.M., Abdulwahed, S.H.: Diabetes classification based on KNN. IIUM Eng. J. **21**(1), 175–181 (2020)
5. Saxena, R., Sharma, S.K., Gupta, M., Sampada, G.C.: A novel approach for feature selection and classification of diabetes mellitus: machine learning methods. Computational Intelligence and Neuroscience, 2022 (2022)
6. Anuja Kumari, V., Chitra, R.: Classification of diabetes disease using support vector machine. Int. J. Eng. Res. Appl. **3**(2), 1797–1801 (2013)
7. Wang, Y., Liao, J., Yu, H., Leng, J.: Semantic-aware conditional variational autoencoder for one-to-many dialogue generation. Neural Comput. Appl., 1–13 (2022). https://doi.org/10.1007/s00521-022-07182-9
8. Qiang, Y., Suresh Kumar, S.T., Brocanelli, M., Zhu, D.: Tiny RNN model with certified robustness for text classification. In: 2022 International Joint Conference on Neural Networks (IJCNN), pp. 1–8. IEEE (2022)
9. Wang, Y., Wang, H., Zhang, X., Chaspari, T., Choe, Y., Lu, M.: An attention-aware bidirectional multi-residual recurrent neural network (abmrnn): a study about better short-term text classification. In: ICASSP 2019–2019 IEEE International Conference on Acoustics, Speech and Signal Processing (ICASSP), pp. 3582–3586. IEEE (2019)
10. Wang, Y., Zhang, X., Mi, L., Wang, H., Choe, Y.: Attention augmentation with multi-residual in bidirectional LSTM. Neurocomputing **385**, 340–347 (2020)
11. Li, Q., et al.: A survey on text classification: from traditional to deep learning. ACM Trans. Intell. Syst. Technol. (TIST) **13**(2), 1–41 (2022)
12. Chen, X., Cong, P., Lv, S.: A long-text classification method of Chinese news based on bert and CNN. IEEE Access **10**, 34046–34057 (2022)
13. Liu, Z., Huang, H., Lu, C., Lyu, S.: Multichannel CNN with attention for text classification. arXiv preprint arXiv:2006.16174 (2020)
14. Kang, B., Li, Y., Xie, S., Yuan, Z., Feng, J.: Exploring balanced feature spaces for representation learning. In: International Conference on Learning Representations (2021)
15. Menon, A.K., Jayasumana, S., Rawat, A.S., Jain, H., Veit, A., Kumar, S.: Long-tail learning via logit adjustment. arXiv preprint arXiv:2007.07314 (2020)
16. Vapnik, V.: Principles of risk minimization for learning theory. Advances in neural information processing systems, 4 (1991)

17. Ju, L., et al.: Relational subsets knowledge distillation for long-tailed retinal diseases recognition. In: de Bruijne, M., Cattin, P.C., Cotin, S., Padoy, N., Speidel, S., Zheng, Y., Essert, C. (eds.) MICCAI 2021. LNCS, vol. 12908, pp. 3–12. Springer, Cham (2021). https://doi.org/10.1007/978-3-030-87237-3_1
18. Xiao, L., Zhang, X., Jing, L., Huang, C., Song, M.: Does head label help for long-tailed multi-label text classification. In: Proceedings of the AAAI Conference on Artificial Intelligence 35, pp. 14103–14111 (2021)
19. Huang, Y., Giledereli, B., Köksal, A., Özgür, A., Ozkirimli, E.: Balancing methods for multi-label text classification with long-tailed class distribution. arXiv preprint arXiv:2109.04712 (2021)
20. Estabrooks, A., Jo, T., Japkowicz, N.: A multiple resampling method for learning from imbalanced data sets. Comput. Intell. **20**(1), 18–36 (2004)
21. Liu, J., Sun, Y., Han, C., Dou, Z., Li, W.: Deep representation learning on long-tailed data: a learnable embedding augmentation perspective. In: Proceedings of the IEEE/CVF Conference on Computer Vision and Pattern Recognition, pp. 2970–2979 (2020)
22. Liu, X.-Y., Wu, J., Zhou, Z.-H.: Exploratory undersampling for class-imbalance learning. IEEE Trans. Syst. Man Cybern. Part B (Cybernetics) **39**(2), 539–550 (2008)
23. Chawla, N.V., Bowyer, K.W., Hall, L.O., Kegelmeyer, W.P.: Smote: synthetic minority over-sampling technique. J. Artif. Intell. Res. **16**, 321-357 (2002)
24. Cao, K., Wei, C., Gaidon, A., Arechiga, N., Ma, T.: Learning imbalanced datasets with label-distribution-aware margin loss. Advances in neural information processing systems, 32 (2019)
25. Park, S., Lim, J., Jeon, Y., Choi, J.Y.: Influence-balanced loss for imbalanced visual classification. In: Proceedings of the IEEE/CVF International Conference on Computer Vision, pp. 735–744 (2021)
26. Feng, C., Zhong, Y., Huang, W.: Exploring classification equilibrium in long-tailed object detection. In: Proceedings of the IEEE/CVF International Conference on Computer Vision, pp. 3417–3426 (2021)
27. Xiang, L., Ding, G., Han, J.: Learning from multiple experts: self-paced knowledge distillation for long-tailed classification. In: Vedaldi, A., Bischof, H., Brox, T., Frahm, J.-M. (eds.) ECCV 2020. LNCS, vol. 12350, pp. 247–263. Springer, Cham (2020). https://doi.org/10.1007/978-3-030-58558-7_15
28. He, K., Girshick, R., Dollár, P.: Rethinking imagenet pre-training. In Proceedings of the IEEE/CVF International Conference on Computer Vision, pp. 4918–4927 (2019)
29. Zhang, Y., Hooi, B., Dapeng, H., Liang, J., Feng, J.: Unleashing the power of contrastive self-supervised visual models via contrast-regularized fine-tuning. Adv. Neural. Inf. Process. Syst. **34**, 29848–29860 (2021)
30. Hu, X., Jiang, Y., Tang, K., Chen, J., Miao, C., Zhang, H.: Learning to segment the tail. In: Proceedings of the IEEE/CVF Conference on Computer Vision and Pattern Recognition, pp. 14045–14054 (2020)
31. Wang, J., Lukasiewicz, T., Hu, X., Cai, J., Xu, Z.: RSG: a simple but effective module for learning imbalanced datasets. In: Proceedings of the IEEE/CVF Conference on Computer Vision and Pattern Recognition, pp. 3784–3793 (2021)
32. Huang, C., Li, Y., Loy, C.C., Tang, X.: Learning deep representation for imbalanced classification. In: Proceedings of the IEEE Conference on Computer Vision and Pattern Recognition, pp. 5375–5384 (2016)

33. Zhang, X., Fang, Z., Wen, Y., Li, Z., Qiao, Y.: Range loss for deep face recognition with long-tailed training data. In: Proceedings of the IEEE International Conference on Computer Vision, pp. 5409–5418 (2017)
34. Cui, J., Zhong, Z., Liu, S., Yu, B., Jia, J.: Parametric contrastive learning. In: Proceedings of the IEEE/CVF International Conference on Computer Vision, pp. 715–724 (2021)
35. Wu, T.-Y., Morgado, P., Wang, P., Ho, C.-H., Vasconcelos, N.: Solving long-tailed recognition with deep realistic taxonomic classifier. In: Vedaldi, A., Bischof, H., Brox, T., Frahm, J.-M. (eds.) ECCV 2020. LNCS, vol. 12353, pp. 171–189. Springer, Cham (2020). https://doi.org/10.1007/978-3-030-58598-3_11
36. Liu, B., Li, H., Kang, H., Hua, G., Vasconcelos, N.: Gistnet: a geometric structure transfer network for long-tailed recognition. In: Proceedings of the IEEE/CVF International Conference on Computer Vision, pp. 8209–8218 (2021)
37. Zhong, Z., Cui, J., Liu, S., Jia, J.: Improving calibration for long-tailed recognition. In: Proceedings of the IEEE/CVF Conference on Computer Vision and Pattern Rrecognition, pp. 16489–16498 (2021)
38. Desai, A., Wu, T.-Y., Tripathi, S., Vasconcelos, N.: Learning of visual relations: the devil is in the tails. In: Proceedings of the IEEE/CVF International Conference on Computer Vision, pp. 15404–15413 (2021)
39. Zhang, Y., Hooi, B., Hong, L., Feng, J.: Self-supervised aggregation of diverse experts for test-agnostic long-tailed recognition. Adv. Neural. Inf. Process. Syst. **35**, 34077–34090 (2022)
40. Cai, J., Wang, Y., Hwang, J.-N.: Ace: ally complementary experts for solving long-tailed recognition in one-shot. In: Proceedings of the IEEE/CVF International Conference on Computer Vision, pp. 112–121 (2021)
41. Kim, J., Jeong, J., Shin, J.: M2m: imbalanced classification via major-to-minor translation. In: Proceedings of the IEEE/CVF Conference on Computer Vision and Pattern Recognition, pp. 13896–13905 (2020)
42. Buda, M., Maki, A., Mazurowski, M.A.: A systematic study of the class imbalance problem in convolutional neural networks. Neural Networks **106**, 249–259 (2018)
43. Japkowicz, N., Stephen, S.: The class imbalance problem: a systematic study. Intell. Data Anal. **6**(5), 429–449 (2002)
44. Cui, Y., Jia, M., Lin, T.-Y., Song, Y., Belongie, S.: Class-balanced loss based on effective number of samples. In: Proceedings of the IEEE/CVF Conference on Computer Vision and Pattern Recognition, pp. 9268–9277 (2019)
45. Ren, J., Cunjun, Yu., Ma, X., Zhao, H., Yi, S., et al.: Balanced meta-softmax for long-tailed visual recognition. Adv. Neural. Inf. Process. Syst. **33**, 4175–4186 (2020)
46. Wang, F., Cheng, J., Liu, W., Liu, H.: Additive margin softmax for face verification. IEEE Signal Process. Lett. **25**(7), 926–930 (2018)
47. Koltchinskii, V., Panchenko, D.: Empirical margin distributions and bounding the generalization error of combined classifiers. Ann. Stat. **30**(1), 1–50 (2002)
48. Khan, S., Hayat, M., Zamir, S.W., Shen, J., Shao, L.: Striking the right balance with uncertainty. In: Proceedings of the IEEE/CVF Conference on Computer Vision and Pattern Recognition, pp. 103–112 (2019)
49. Wang, Y.-X., Ramanan, D., Hebert, M.: Learning to model the tail. Advances in neural information processing systems, 30 (2017)
50. Tan, C., Sun, F., Kong, T., Zhang, W., Yang, C., Liu, C.: A survey on deep transfer learning. In: Kůrková, V., Manolopoulos, Y., Hammer, B., Iliadis, L., Maglogiannis, I. (eds.) ICANN 2018. LNCS, vol. 11141, pp. 270–279. Springer, Cham (2018). https://doi.org/10.1007/978-3-030-01424-7_27

51. Zoph, B., et al.: Rethinking pre-training and self-training. Advances in neural information processing systems, 33, pp. 3833–3845 (2020)
52. He, K., Fan, H., Wu, Y., Xie, S., Girshick, R.: Momentum contrast for unsupervised visual representation learning. In: Proceedings of the IEEE/CVF Conference on Computer Vision and Pattern Recognition, pp. 9729–9738 (2020)
53. Ye, H.-J., Chen, H.-Y., Zhan, D.-C., Chao, W.-L.: Identifying and compensating for feature deviation in imbalanced deep learning. arXiv preprint arXiv:2001.01385 (2020)
54. Kang, B., et al.: Decoupling representation and classifier for long-tailed recognition. arXiv preprint arXiv:1910.09217 (2019)
55. Wang, X., Lian, L., Miao, Z., Liu, Z., Yu, S.X.: Long-tailed recognition by routing diverse distribution-aware experts. arXiv preprint arXiv:2010.01809 (2020)
56. Guo, H., Wang, S.: Long-tailed multi-label visual recognition by collaborative training on uniform and re-balanced samplings. In: Proceedings of the IEEE/CVF Conference on Computer Vision and Pattern Recognition, pp. 15089–15098 (2021)
57. Gao, T., Yao, X., Chen, D.: Simcse: simple contrastive learning of sentence embeddings. arXiv preprint arXiv:2104.08821 (2021)

Author Index

H. Zhang et al. (Eds.): NCAA 2023, CCIS 1870, pp. 613–616, 2023.
https://doi.org/10.1007/978-981-99-5847-4

Printed in the United States
by Baker & Taylor Publisher Services

Printed in the United States
by Baker & Taylor Publisher Services